AF527773

Commercial Flowers

— Volume 1 —

THE EDITORS

Professor T.K. Bose obtained his M.Sc. (Ag.) and Ph.D. degree from Calcutta University, West Bengal, India. Trained at the Royal Botanic Garden, Kew, UK. Former Secretary of the Agri-Horticultural Society of India; Project Coordinator, All India Coordinated Floriculture Improvement Project (ICAR); Professor and Head, Department of Horticulture; Dean, Faculty of Agriculture; Director of Research, Bidhan Chandra Krishi Viswavidyalaya, West Bengal; Member, State Agriculture Commission, West Bengal; Member of Scientific Advisory Committees of ICAR, CSIR (Govt. of India) and Government of West Bengal. Awarded (D.Sc.) Honoris Causa from OUAT, Bhubaneshwar and UBKV, Coochbehar, India. Guided around 45 students for their Ph.D. degree, published more than 250 research papers and authored 30 books on horticultural crops.

Dr. Chenna Reddy Aswath obtained his B.Sc. (Hort), M.Sc. (Hort.) in Floriculture from University of Agricultural Sciences, Bengaluru and Ph.D. degree in Biotechnology from Guwhati University, Assam. Serving as Principal Scientist in ICAR-IIHR, Bengaluru and has made significant contributions in the field of Floriculture and Biotechnology. Recipient of "Dr BP Pal Best Scientist Award for Agriculture" from NABS, Chennai, became FELLOW of the NABS, Chennai, ISOH, New Delhi and SAB, Dharwad, CHAI, New Delhi. Received the 2012 SAB Award of Excellence in Horticultural Biotechnology. Recipient of Brain pool Post-Doctoral Fellowship from South Korea Science Federation for the outstanding research contributions in Horticultural Biotechnology. His short term fellowship in Japan under INSA has recognized him among the Orchid tissue culture group. Published more than 110 research papers besides several books. Guided more than 10 post-graduate students.

Late Dr. L.P. Yadav obtained his M.Sc. (Ag.) from Sam Higginbottom University of Agriculture, Technology and Sciences, Allahabad and Ph.D. from Bidhan Chandra Krishi Viswavidyalaya, West Bengal, India. He has published more than 100 research publications besides several books. He has guided more than 15 post-graduate students.

Prof. P. Pal obtained his M.Sc. (Ag.) and Ph.D. from Bidhan Chandra Krishi Viswavidyalaya (BCKV), West Bengal, India. He acted as Head, Department of Floriculture and Landscaping, BCKV and published more than 100 research publications besides several books. He has guided more than 25 post-graduate students.

Dr. K. Dutta obtained his M.Sc. (Hort.) from CCS Haryana Agricultural University, Hisar, Haryana and Ph.D. from Uttar Banga Krishi Viswavidyalaya, West Bengal. Acting as Assistant Secretary of The Agri-Horticultural Society of India, Kolkata, West Bengal. Recipient of Inspire Fellowship from Department of Science and Technology, New Delhi. Published several scientific papers in national journals on different aspect of horticulture particularly in tissue culture and written book chapters on nursery management and jasmine. Associated with micropropagation of different horticultural crops during last 7 years.

Dr. Rajiv Kumar obtained his B.Sc. (Ag & AH) and M.Sc. (Ag) Horticulture from C.S. Azad University of Agriculture & Technology, Kanpur, Uttar Pradesh, India and completed his Ph.D. (Horticulture) from I.A.R.I., New Delhi. Presently working as Principal Scientist at ICAR-IIHR, Bengaluru. Recipient of Young Scientist Associate Award by HTHS, U.P.; Young Scientist Award by SSDAT, U.P.; Gold Medal by HTHS, Meerut, and BR&IC Pvt. Ltd.; Fellow (Horticulture) by HAHS, Muzaffarnagar, U.P., AABS, Karnakata, and ISOH, New Delhi. Published 110 research articles, 25 popular articles, 6 edited books, 20 book chapters, 8 technical bulletins and 6 extension folders. Developed two varieties of China aster, associated in five and three varieties of Gerbera and gladiolus, respectively; registered 17 genetic stocks of China aster, Gerbera, Chrysanthemum and Gladiolus with ICAR-NBPGR, New Delhi. Guided two M.Sc. (Hort.) and two Ph.D. (Hort.) students in Floriculture.

Prof. P. Das obtained his M.Sc. (Ag.) from Calcutta University and Ph.D. from BCKV, West Bengal, India. Acted as Head, Department of Horticulture, OUAT and Director, Regional Plant Resource Centre, Bhubaneswar, Odisha. Awarded (D.Sc.) Honoris Causa from OUAT, Bhubaneshwar, Dr. Mopuri Brahman Memorial Award for Plant Diversity and Conservation by Orissa Botanical Society and recipient of Life Time Achievement Award by Indian Association of Ornamental Horticulture. Published more than 200 research publications besides several books.

Commercial Flowers

— Volume 1 —

— Senior Editor —

C. Aswath

Division of Flower and Medicinal Crops
ICAR-Indian Institute of Horticultural Research
Bengaluru 560089, Karnataka, India

— Editors —

T.K. Bose • L.P. Yadav • P. Pal

Faculty of Horticulture
Bidhan Chandra Krishi Viswavidyalaya
West Bengal 741 252, India

K. Dutta
The Agri-Horticultural Society of India
1, Alipore Road,
Kolkata 700027
West Bengal, India

Rajiv Kumar
Division of Flower and Medicinal Crops
ICAR-Indian Institute of Horticultural Research,
Bengaluru 560089,
Karnataka, India

P. Das
Regional Plant Resource Centre
Bhubaneswar,
Odisha, India

2022

Daya Publishing House®
A Division of
Astral International Pvt. Ltd.
New Delhi – 110002

Third Revised & Illustrated edition 2021

Publisher's Note:

Every possible effort has been made to ensure that the information contained in this book is accurate at the time of going to press, and the publisher and author cannot accept responsibility for any errors or omissions, however caused. No responsibility for loss or damage occasioned to any person acting, or refraining from action, as a result of the material in this publication can be accepted by the editor, the publisher or the author. The Publisher is not associated with any product or vendor mentioned in the book. The contents of this work are intended to further general scientific research, understanding and discussion only. Readers should consult with a specialist where appropriate.

Every effort has been made to trace the owners of copyright material used in this book, if any. The author and the publisher will be grateful for any omission brought to their notice for acknowledgement in the future editions of the book.

Cataloging in Publication Data--DK
Courtesy: D.K. Agencies (P) Ltd. <docinfo@dkagencies.com>

Commercial flowers / senior editor, C. Aswath ; editors, T.K. Bose [and five others]. -- Third revised & illustrated edition.
volumes cm
Contributed articles.
Includes index.
ISBN 9789390435548 (HB)

1. Floriculture. 2. Flowers--Pictorial works. I. Aswath, C. (Principal scientist at the Indian Institute of Horticulture Research), editor. II. Bose, T. K., editor.

LCC SB405.C66 2021 | DDC 635.9 23

Published by : **Daya Publishing House**®
A Division of
Astral International Pvt. Ltd.
– ISO 9001:2015 Certified Company –
4736/23, Ansari Road, Darya Ganj
New Delhi-110 002
Ph. 011-43549197, 23278134
E-mail: info@astralint.com
Website: www.astralint.com

Digitally Printed at : Replika Press Pvt Ltd

From the Desk of Prof. T.K. Bose

The undersigned had initiated the publications entitled 'Fruits: Tropical and Sub-tropical, Vegetable Crops and Commercial Flowers' in one volume each in the 1980's in collaboration with co-editors, humbly expresses from the core of his heart that the updated revised editions of the above mentioned publications with colour illustrations in thirteen volumes which was planned in June, 2020, are being printed from December this year, a rare event in Horticultural Science has been possible by the blessing of Almighty God, with the cooperation of dedicated horticulturist as editors and contributors.

T.K. Bose

Editor Emeritus

Preface to the Third Edition

The first edition of the book entitled '**Commercial Flowers**' published in 1989, reprinted in 1993 and 1998 was greatly appreciated and proved valuable scientific review of commercially important flowers to the students, teachers, researchers and extension specialists in horticulture in general and floriculture in particular, in different parts of the world. The book was revised in 2002 with the same title in two volumes due to its increasing demand, which contained updated scientific information of commercially grown flower crops and considered as valuable contribution on the subject. After about 18 years of the last revised publication in 2002, voluminous information pertaining to crop improvement, production, protection, postharvest management, biotechnology has been generated necessitating further revision of the existing chapters and inclusion of more ornamentals of commercial importance in five volumes.

First volume in this edition includes chapters on importance, uses and scope of ornamental horticulture, rose, carnation, chrysanthemum, jasmine and bird of paradise covering more than 680 pages.

Apart from own collection of photographs, editors have also taken some photographs which are in public domain from different websites as far as possible. Therefore, we acknowledge with enormous thanks to all State Agricultural Universities and Research Institutes under Indian Council of Agricultural Research, New Delhi, India; Public and Private International Agricultural Institutes across the world for using photographs from their websites to make this book a valuable source of knowledge for the readers. We think images can be powerful teaching tools, as illustrations to in-class lectures, or for studying concepts outside of the classroom. We also think that identification of flower crops, their importance and uses, species and cultivars, methods of propagation, cultivation practices including greenhouse cultivation, diseases and pests, harvesting and postharvest technology and achievement of biotechnology with the help of suitable illustrations in a consolidated manner will make a dent for the students, researchers and teachers of floriculture and ornamental horticulture, and also extension specialists.

Our thanks are due to Shri Prateek Mittal, Director, Astral International (P) Ltd., New Delhi, for his keen interest in bringing out this publication. Our special thanks go to Imprints of Kolkata for processing the manuscript.

Kolkata **Editors**

Contents

List of Contributors

Arora, J. S.

Department of Vegetable Crops and Landscaping, Punjab Agricultural University, Ludhiana 141 004, Punjab

Aswath, C.

Division of Flower and Medicinal Crops, ICAR-Indian Institute of Horticultural Research, Bengaluru 560089, Karnataka

Bennurmath, P.

Division of Flower and Medicinal Crops, ICAR-Indian Institute of Horticultural Research, Bengaluru 560089, Karnataka

Bhatt, N.R.

Department of Horticulure, Mahatma Phule Krishi Vidyapeeth, Rahuri 413 722, Maharashtra

Biswas, B.

Department of Horticulture, Bidhan Chandra Krishi Vishwavidyalaya, Mohanpur 741 252, West Bengal

Bose, T.K.

Department of Horticulture, Bidhan Chandra Krishi Vishwavidyalaya, Mohanpur 741 252, West Bengal

Chaithra, M.L.

Division of Flower and Medicinal Crops, ICAR-Indian Institute of Horticultural Research, Bengaluru 560089, Karnataka

Choudhury, T.

The Agri-Horticultural Society of India, 1, Alipore Road, Kolkata 700027, West Bengal

Dadlani, N.K.

Division of Floriculture and Landscaping, Indian Institute of Agricultural Research, Pusa, New Delhi 110012

Das, P.

Regional Plant Resource Centre, Bhubaneswar 751 015, Odisha

Dutta, K.

The Agri-Horticultural Society of India, 1, Alipore Road, Kolkata 700027, West Bengal

Gurung, A.

Division of Flower and Medicinal Crops, ICAR-Indian Institute of Horticultural Research, Bengaluru 560089, Karnataka

Jana, B. K.

Department of Horticulture, Bidhan Chandra Krishi Vishwavidyalaya, Mohanpur 741 252, West Bengal

Kher, M.A.

National Botanical Research Institute, Lucknow 226 001, Uttar Pradesh

Kumar, Rajiv

Division of Flower and Medicinal Crops, ICAR-Indian Institute of Horticultural Research, Bengaluru 560089, Karnataka

Mahapatra, A.

Regional Plant Resource Centre, Bhubaneswar 751 015, Odisha

Maity, R.G.

Department of Horticulture, Bidhan Chandra Krishi Vishwavidyalaya, Mohanpur 741 252, West Bengal

Malik, R.S.

Division of Floriculture and Landscaping, IARI, New Delhi-110012

Mitra, M.

Department of Horticulture, Bidhan Chandra Krishi Vishwavidyalaya, Mohanpur 741 252, West Bengal

Muthuswami, S.

Faculty of Horticulture, Tamil Nadu Agricultural University, Coimbatore 641 003, Tamil Nadu

Nair, S. A.

Division of Flower and Medicinal Crops, ICAR-Indian Institute of Horticultural Research, Bengaluru 560089, Karnataka

Noorjehan, J.B.

Division of Floriculture, ICAR-Indian Institute of Horticultural Research, Bengaluru 560089, Karnataka

Pal, P.

Department of Horticulture, Bidhan Chandra Krishi Vishwavidyalaya, Mohanpur 741 252, West Bengal

Palai, S.K.

Regional Plant Resource Centre, Bhubaneswar 751 015, Odisha

Parthasarathy, V.A.

ICAR-Indian Institute of Spices Research, 32/482 C Narmada Nilayam, Bharathan Bazar, Chelavoor, Calicut 673571, Kerala

Peter, G. B.

Faculty of Horticulture, Tamil Nadu Agricultural University, Coimbatore 641 003, Tamil Nadu

Ramadas, S.

Faculty of Horticulture, Tamil Nadu Agricultural University, Coimbatore 641 003, Tamil Nadu

Sangeetha Priya, S.

Division of Flower and Medicinal Crops, ICAR-Indian Institute of Horticultural Research, Bengaluru 560 089, Karnataka

Sanyal, D.

Department of Crop Improvement, Horticulture and Agricultural Botany, Institute of Agriculture, Viswa-Bharathi, Sriniketan 731 236, West Bengal, India

Soudhamani, K.

Division of Flower and Medicinal CropsICAR-Indian Institute of Horticultural Research, Bengaluru 560 089, Karnataka

Yadav, L.P.

Department of Horticulture, Bidhan Chandra Krishi Vishwavidyalaya, Mohanpur 741 252, West Bengal

1

IMPORTANCE, USES AND SCOPE OF ORNAMENTAL HORTICULTURE

T.K. Bose, C. Aswath, V.A. Parthasarathy and K. Dutta

1.0 Introduction

Ornamental plants, a symbol of love, pain, peace and tranquility as well as grace and elegance, and a feast for our eyes and soul, are integral part of our modern day living in both rural and urban areas. The importance of ornamental horticulture in social and cultural life of people, and in improving the economic and environmental condition of a country is now well recognized world over.

2.0 Importance and Uses

2.1.0 Diversity and Adaptability of Ornamental Plants

Enormous genetic diversity and large groups of flowering and foliage trees, shrubs, climbers, bulbous plants,annuals, orchids, cacti and other succulents, ferns, palms, aquatic plants, house plants, bamboos and grasses, *etc.*, make them most valuable and inseparable part in the life of both rural and urban people. Innumerable species and cultivars of ornamental plants are grown in tropical, subtropical, temperate and alpine climates. They are also adapted to grow in desert, water, swamp and snow clad mountains. Because of the diversity and wide range of adaptability of different groups of ornamental plants, production of numerous hybrid cultivators of outstanding quality and the technology of greenhouse cultivation under controlled environment have contributed immensely to the increasing use and scope of ornamental horticulture for its aesthetic value, improvement of environment and fast expanding commercial benefits world over. The importance and uses of each group of ornamental plants in our daily life are diverse and immense, which are enumerated briefly under each group (Bose *et al.*, 2015).

2.2.0 Trees

Trees are defined as perennial woody plants, more than 5m high, with a trunk and a canopy. They are the largest and most beautiful specimens in the plant kingdom. Trees evolved about 330 million years ago, much before the appearance of the first human on this planet, protect the earth and human civilization. As a nursing mother, the trees have cradled the human beings and human civilization since their infancy and are still playing the same role by providing man with food, shelter and many other things.

2.2.1 Uses

Trees are extensively used for forest plantation (afforestation), social forestry, agro-forestry, reclamation of desert land (by tree planting), protection of coastal areas and river banks, creation of tree-belt to protect habitation from natural calamities, and as a barrier against air and sound pollution. Because of their large size, elegant appearance and magnificent flowering, longevity for many years and enormous impact on environment, trees are the major components in landscaping and beautification throughout the world. They are extensively planted on i) road sides, avenues and highways, ii) arboretums, iii) botanic gardens, iv) plant resource

Trees are Beautiful and Indispensable for Life

Polyalthia longifolia 'Pendula' for screening

Neem seedlings ready for afforestation

Banyan tree, Indian Botanic Garden, Kolkata

Delonix regia in road side plantation

Cassia fistula in avenue

Agroforestry

Trees are Beautiful and Indispensable for Life

House landscaping with trees

Sal tree in forest

Trees in Japanese garden

Butea monosperma for *holy* colours

Topiary with trees

Trees reduce air pollution

Trees are Beautiful and Indispensable for Life

Colourful trees on river bank

Trees in landscaping of Victoria Memorial Garden, Kolkata

Trees are Beautiful and Indispensable for Life

Wind break reduces wind speed and drought

Pollution amelioration by plants in cities

centers, v) parks and landscaping of institutional campuses and commercial offices, viii) around apartment buildings, ix) monuments, x) in industrial complexes for beautification and improvement of environment by minimizing air and sound pollution due to tree-belts, xi) landscape development in regional and town planning (xii) around hotels and holiday resorts, along sea shores and river banks to promote eco-tourism in a better environment, (xiii) beautification of home gardens, (xiv) green verge, and (xv) also used for making bonsai, hedge, topiary and as house plants in containers (Bose *et al.*, 1998).

2.3.0 Shrubs

Shrubs are usually woody or semi-woody perennial plants with branches arising from the base of the plants, and grow up to a height a 4 m. they constitute a large group among the perennial ornamentals grown in gardens, private or public. Availability of a large number of species and cultivars, a wide range of variation in shape and size of the plants, diversified types of leaves and growth habits, remarkable ranges of size, shape and colour of flowers and the variety of fragrance have made the shrubs an important garden plant. Semi-woody dwarf or prostrate shrubs are grown in edges, as hedge or ground cover (Bose, 1965).

2.3.1 Uses

Shrubs are usually planted in shrub border (shrubbery), in groups of one species of plants of different cultivars or different species of shrubs, and also as a hedge and topiary. Many shrubs are popular pot plants throughout the world. Well-shaped tall shrubs flowering for several months are also planted as specimen plants *viz. Codiaeum variegatum, Erythrina cristagalli, Hibiscus rosasinensis, Ixora* spp., *Mussaenda erythrophylla, Portlandia grandiflora, Tecoma gaudichaudi, etc.*, in tropical gardens. Species and cultivars of *Camellia, Fuchsia, Hydrangea, Rhododendron* (azalea), *etc.*, are extensively grown as specimen plants and in shrub border in subtropical and temperate climates. Cultivars of *Bougainvillea, Brugmansia* spp., *Calliandra, Euphorbia pulcherrima* (poinsettia), *Gardenia* spp., *Hibiscus rosasinensis,* and many others are grown both in tropical and subtropical gardens in most part of the world. Shrubs are popular pot plants and extensively grown for beautification.

2.3.2 Cut Flowers and Floral Arrangements

Cut blooms of several species of shrubs such as *Camellia, Gardenia* (double), *Hydrangea, Nerium* (double), which remain fresh for a few days in vases are used in floral arrangements with other flowers. Loose flowers of *Crossandra* spp. and *Jasminum* spp. are extensively used in making floral ornaments throughout India, particularly in southern states (Bose and Bhattacharjee, 1980).

2.3.3 Essential Oil Yielding Shrubs

Essential oil yielding shrubs like *Jasminum grandiflorum, Jasminum auriculatum* also used as climbers and *Jasminum sambac* are not only popular for their scented flowers, they also yield essential oils of commercial importance (Bose *et al.*, 2002).

Shrubs are Useful Garden Plants

Ixora as hedge

Shrubbery border

Bougainvillea in pot

Hibiscus in pot

Mussaenda in residence garden

Poinsettia in border

Climbers are commonly Used on Walls, Arches, Pergolas and Trellises

Quisqualis indica **on pergola**

Pyrostegia venusta **on wall**

Beautiful blue flowering Clematis

Antigonon leptopus **on wall**

Petrea volubilis **on arch of a house**

2.4.0 Climbers

Climbers are an important group of ornamental plants. The beauty of a garden can greatly be enhanced by carefully selecting and planting them in suitable locations. Plants which have special structures to climb on supports are defined as climbers. The tendrils are formed by the modification of various plant parts such as, the terminal portion of branches as in *Antigonon leptopus* or some of the leaflets in *Pyrostegia venusia*, leaf tips in *Gloriosa* or stipules in *Smilax*. In *Clematis*, the spiral leafstalks provide necessary support for climbing. In *Bougainvillea*, thorny structures on the woody stem help the plant in securing the necessary grip.

2.4.1 Uses

Climbers are commonly used on walls, arches, pergolas, trellises, porches, arbours and pillars for beautification of houses and gardens, and in cities, their utility is increased for the purpose of screening the premises from adjacent houses and maintaining privacy. Bare walls can effectively be decorated by growing colourful climbers on them. Fences and trellises also provide scope for these plants to grow and be decorative. Large number of foliage climbers including species and cultivars of *Monstera, Philodendron, Scindapsus, Syngonium, etc.*, are popular house plants in pots; on suitable support they are also grown in semi-shady location in gardens for beautification.

2.5.0 Annuals and Biennials

Annuals are described as plants which complete their life cycle from germination, vegetative growth, flowering to seed formation in one year/season. Biennials may need more than one year to complete their life cycle.

2.5.1 Uses

2.5.1.1 Garden Beautification

Annuals and also biennials often treated as annuals are important group of ornamental plants, extensively grown in all parts of the world, outdoors and indoors, as house plants and in greenhouses, depending on the environment and purpose of cultivation. They contribute colour and beauty to gardens, and are suitable for growing in beds, herbaceous borders, shrub borders, edges, rock gardens and other locations in private gardens, parks and other public gardens, botanic gardens, and also in gardens around residential and commercial complexes around monuments and public buildings.

2.5.1.2 Annual in Pots and Hanging Baskets

Annuals are also widely grown in pots and hanging baskets and very attractive in beautifying gardens, open spaces, and are used for beautifying residential buildings, apartments, garden rooms, balcony, roof gardens, patios, living rooms, window boxes, *etc.* Pot plants of different annuals are the major exhibits in flower shows in all parts of the world. Annual climbers can be trained on arches, trellis and fences. Most of the biennial flowering plants, however, need comparatively low temperature for growth and flowering. Several annuals like *Amaranthus tricolor*, *Kochia trichophylla*, *Senecio crunetus*, *etc*., are grown for their showy foliage.

2.5.1.3 Cut Blooms and Loose Flowers

Cut flowers from large number of species and cultivars of annuals are very popular as bouquets, in vases and various floral arrangements in residential houses, offices institutions, temples, for decorating dias and surroundings, areas in public functions and meetings, in marriages and other social ceremonial occasions. Cut flowers of annuals and perennial ornamentals are the major items in local and global markets. Apart from cut flowers, loose flowers of many annuals are also extensively used in making floral ornaments for temples, decorating pandals, idols, widely used in public and private ceremonies, creating a charming atmosphere befitting the occasion. Loose flowers are universally used for worshipping almighty God. Aroma is extracted from scented flowers and natural dye from petals of annuals and perennials are now being used commercially. In addition to the use of fresh flowers in bouquets and vases, dry flowers of annuals are also important component of pot-pouri, an item in great demand in global flower trade (Bose, 2019).

2.5.1.4 Economic Benefits

Intensive cultivation of annuals for cut-flower and loose flower production outdoors, in greenhouse, also in pots is a highly remunerative practice in most parts of the world. In the developing countries, thousands of farming families are dependent on intensive flower cultivation, particularly annuals for their livelihood which also provides employment to women for flower harvesting, grading and making floral ornaments for marketing. Commercial flowers in many countries are major sources of economic development from domestic and export market. In seed industry of ornamental crops, a major turnover comes from sale of seeds of annuals, particularly in developed countries.

2.6.0 Orchids

Orchids are the most fascinating plants in the world for their unique characteristics and charms for long lasting beautiful flowers of a wide range of colours, extraordinary varieties of sizes and shape. Orchids with over 1000 described genera and 25,000 – 35,000 species available in Nature, and perhaps another 1.5 lakh man-made hybrids and cultivars constitute 7 per cent of all angiosperms and nearly 40 per cent of monocotyledons and occupy the wide range of the ecological habitats, exhibiting highly specialized morphological, structural and physiological characteristics. In particular, the most spectacular evolution is shown in the reproductive biology, *i.e.*, the flowers which vary in number, size, structure, colour and fragrance. In general, orchids are most abundant in the tropics and subtropics where prevail positive factors for their growth *i.e.*, high humidity optimum range of temperature, partial shade from thick vegetation, but they are also common in the temperate latitudes like other monocots. The tropical species are mostly epiphytes, found growing above the ground, usually perched in the trees (Bose *et al.*, 2002).

Uses of Annuals

In a home garden

In hanging baskets

In flower arrangement

Amaranthus tricolor for showy foliage

Garden beautification with annuals

Annuals in window boxes

Beautiful Orchids

Dendrobium in pot

Cymbidium in pot

Vanda in hanging pots

Orchids in flower arrangement

Orchids in greenhouse

2.6.1 Uses

2.6.1.1 Orchids in Global Trade

Orchids have occupied a significant position in world cut flower industry for its attractive, beautiful flowers of myriad colour, long shelf-life, and accounts for about 10 per cent share in global floriculture trade, highly boosting the economy of many South-East Asian countries like Malaysia, Indonesia, Thailand, Taiwan and Singapore. A number of novel and spectacular hybrids are being added every year to meet the increasing demand of changing markets. Numerous hybrids have been registered with an average of about 10,000 hybrids added every year. Netherlands is specialized in growing and breeding of *Cymbidium* and *Phalaenopsis* orchids. Considering world export value of orchid flowers, the Netherlands rank first with the export value of 111.7 million US $ (48.2 per cent), while Thailand ranks second with export value of 73.3 million US $ (31.6 per cent).

2.6.1.2 Potted Orchids

Orchids sold as pot plants are mainly the miniature hybrids which have a high demand in global floriculture trade. Cultivars, of orchids belonging to *Ascocenda, Cattleya,* dwarf cymbidiums, *Dendrobrium, Mokara, Paphiopedilum, Phalaenopsis, Vanda, etc.,* are popular pot plants through the world, grown and displayed as house plants, in greenhouses, home gardens outdoors under favourable environment, offices, hotels, airport lounges, malls, restaurants, decoration of ceremonial places and similar other locations.

2.7.0 Bulbous Plants

In horticulture, the world 'bulb' includes underground modified stems which are used for propagation, *e.g.* bulb, corm, tuber and rhizome. Tuberous roots are also covered by this term.

2.7.1 Uses

A large number of bulbous plants producing attractive flowers and foliage are extensively grown for their ornamental value. Apart from garden beautification, many bulbous plants are commercially grown for cut-flowers, pot plants and bulbs. Many of the bulbous plants, grown in tropics and subtropics, apart from their utility for beautification of gardens, produce attractive and valuable cut flowers of commercial importance and are ideal pot plants. Cut flowers of Alstroemeria, Bird of paradise, Dahlia, Etlingera, Gladiolus, Gloriosa, Heliconia, Lilium, Tuberose, Zzantedeschia, *etc.,* grown in tropics and subtropics, are very important in domestic and global flower markets; cut flowers of *Haemanthus, Hippeastrum* and *Ornithogalum* are used in floral arrangements and mostly sold in domestic markets. *Cyclamen, Haemanthus, Hippeastrum, Ornithogallum* grows and flower well in containers and is often used as house plants in partial light. Several cultivars of *Nymphaea* (water lily) in many attractive colours flower well in large containers in waterlogged soil for many months except in winter, in addition to their spectacular display of colour in tanks and lily pools. *Nelumbo* (lotus) bears spectacular flowers in water body in summer and rains. The bulbous plants being perennial, bulbs, tubers, rhizomes and corms are commercially important planting materials which have good demand throughout the world. Some of the bulbous plants have medicinal value.

2.7.1.1 Bulbous Plants in Tropics and Subtropics

A large number of bulbous ornamental plants are grown in tropical and subtropical climate. The flowering group includes *Achimenes, Acorus, Alstroemeria, Amaryllis, Canna, Crinum, Dahlia, Etlingera, Gladiolus, Gloriosa, Haemanthus, Hedychium, Heliconia, Hemerocallis, Hippeastrum, Hymenocallis, Nelumbo, Nymphaea, Ornithogalum, Pancratium, Polyanthes, Sinningia,* (gloxinia), *Zantedeschia, Zephyranthes,* while *Alocasia, Asparagus, Caladium, Calathea, Kaemferia, Sansevieria, etc.* belong to the foliage group and grown in gardens and as house plants in pots and containers. *Agapanthus, Cyclamen, Lilium* and *Ornithogalum* also grow and flower in mild climate (subtropical). Hyacinth, iris, narcissus, tulips, *etc.*, are important bulbous plants grown in temperate climate in gardens and greenhouses for cut flower and pot plant production (Bose *et al.*, 2007).

2.8.0 House Plants

House plants include annual and perennial ornamentals, both flowering and foliage plants, suitable for growing in semi-shade, bright filtered light and also in sun depending on the location and adaptability to different environments.

2.8.1 Uses

With the increasing population in cities and towns, houses with garden space are becoming limited. To add beauty to a house and to get immense pleasure and fun, a city-dweller can grow a wide range of flowering and foliage plants in every corner of his house. In recent years, there is an increasing interest in house plants which can be very effectively used in interior and outdoor decoration. Indoor plants need not necessarily be confined to the living room and garden room alone, bed room, kitchen, staircases, balcony, patios and other locations can be decorated with suitable house plants. House plants are not restricted to residential houses; there has been a tremendous scope in the use of these plants in beautifying offices, hotels, hospitals, malls, airport lounges and almost all public buildings. Natural greenery and colour by using plants help in breaking or softening the harsh lines of modern office or apartment buildings. In the modern concept and practice of indoor cultivation, the scope of growing plants has been extended to many areas in and around the house and almost all types of plants are being used depending on the environment, particularly light and temperature. Some of the hardy house plants with attractive foliage are *Calathea, Cordyline, Dieffenbachia, Dracaena, Episcia, Ficus, Fittonia, Guzmania, Heliconia, Maranta, Monstera, Neoregelia, Nidulariam, Peperomia, Philodendron* and many others (Bose *et al.*, 2004).

2.9.0 Palms

Palms are prominent for their elegance, grace and grandeur, especially remarkable where they grow naturally. There are two main classes of palm–feather or pinnate-leaved and fan or flavelliform-leaved.

Bulbous Plants are Hardy Ornamentals

Gladiolus as cut flowers

Dahlia in pots

Lilies in baskets

Alstroemeria in greenhouse

Amaryllis in pots

Zantedeschia in pot

House Plants have many Uses

Calathea cultivar

Monstera in garden

Dracaena

Dieffenbachia in garden

Fittonia

Aglaonema in pot

2.9.1 Uses

Several species of palms with elegant appearance are grown in gardens, particularly on roadsides, or in clumps, depending on the size of the garden and purpose or growing. There is scope for growing palms in parks, botanic gardens, along seashore, on rural and urban roads. Hardy palms are the major selection in landscaping in tropical regions. Species of *Bentinckia, Bismarckia, Butia, Caryota, Chrysalidocarpus, Pritchardia, Ptychosperma, Phoenix, Roystonia, Washingtonia, etc.* are grown outdoors in tropical climate in many parts of the world. Many palms are extensively grown in containers for indoor and outdoor decoration in tropical and subtropical climate, though palms as a group are tropical in origin. Palms prefer a well-drained soil rich in organic matter. Palms of decorative value suitable for growing in containers include *Aiphanes aculeate, Caryota mitis, Chamaedorea* spp., *Chamaerops* spp., *Chrysalidocarpus lutescens, Howea belmoreana, Latania* spp., *Phoenix roebelini, Rhapis* spp., *etc.* Palms can be grown in containers for several years in good condition with proper management and by repotting when they become pot bound. Apart from their ornamental importance in landscaping, indoor and outdoor decoration in pots, several species of palms like *Borassus flabellifer* (plamyra palm) *Cocos mucifera* (coconut), *Elaeis guineensis* (oil palm) and *Phoenix dactylifera* (date palm) as economic plants contribute significantly to the economy of many countries.

2.10.0 Cacti and other Succulents

The ornamental features and the unique ability of cacti and other succulents to grow under xerophytic condition make them indispensable for any outdoor or indoor landscaping. Succulent plants or 'succulents' in short are a prominent component of the vegetation in semi-arid and arid regions. Cacti are also succulents belonging to the family *Cactaceae.*

2.10.1 Uses

A well laid out rock garden at a suitable site with correct arrangement of rock and plants will be of immense pleasure and surprise for many years. A number of cacti and other succulents including *Adenium, Aloe, Agave, Beaucarnea, Cereus, Crassula, Echinocactus, Echinopsis, Ferrocactus, Howarthia, Kalanchoe, Mammillaria, Pachypodium,* and many others can be conveniently planted in rock gardens in tropics and subtropics. A bowl or tray converted into a miniature landscape garden with suitable cacti and other succulents for indoor and outdoor use is a pleasant sight. They are widely grown in pots for decoration and exhibition purpose, and many of them are ideal for growing in a rockcorner, and also in pots in a roof garden or a balcony garden receiving bright filtered light. Several succulents like *Aporocactus, Ceropegia, Epiphyllum, Hoya, Sedum, Senecio, Zygocactus, etc.*, can be grown in hanging baskets and wall baskets for an unusual display of elegance.

Many species of cacti and other succulents including *Borzicactus, Echinocereus, Echinopsis, Epiphyllum, Rebutia, Schlumbergera, Zygocactus, etc.*, among the cacti and *Adenium, Euphorbia, Kalanchoe, Stapelia, etc.*, in other succulents, produce attractive flower.

Palms are Hardy Ornamentals

Washingtonia robusta **in pot**

Palms in a residance

Red palms in landscaping

Bismarckia nobilis **as specimen plant**

Palms in a garden

Palms in seashore

Cactus and other Succulents

Cactus house

Display of succulents

Aporocactus in hanging basket

Kalanchoe in pot

Adenium in pot

Euphorbia milii in pot

2.11.0 Bamboos and Grasses

Many species of bamboos and grasses among the monocots are important ornamental plants used in gardens. Grasses are extensively used for making lawn and turf.

2.11.1 Uses

In gardens of tropics and subtropics, several species of ornamental bamboos and tall feathery grasses can be very effectively used in creating unusual diversity in the display of the landscape. Among the bamboos, bushy *Apluda aristata*, about 1 m high, is suitable as a pot plant and also clipped as hedge. The variegated *Arundo donax* 'Metallica', *Bambusa aurea* 'Variegata', the golden bamboo *Bambusa nigra* or black bamboo are the best among the tall types. *Bambusa nana*, growing up to 3 m, is used in hedge; *Bambusa aureostriata, Bambusa ruscifolia,* about 1 m high, grow excellently in semi-shade. *Bambusa pygmea* growing up to 50 cm is a showy ornamental on ground and in container as well.

Among grasses, *Panicum variegatum* with red and white variegated foliage and *Phalaris arundinacea* cv. 'Picta' are grown in sum and semi-shade. The flower sprays of *Eragostis* sp., a dwarf grass, are used in pots. *Pennisetum longistylum* is a tall grass and bears pink rush-like flower heads. *Andropogonas similis* and *Cyperus elegans* are especially suitable for small gardens.

2.11.1.1 Lawn and Turf Grasses

Lawn is one of the most important features in a garden, private or public, covering a large area. Lawn and turf are made of different species and hybrids of grasses. *Cynodon dactylon* (Bermuda grass), *Zoysia* spp., *Paspalum notatum* (bahia grass), *Axonopus* spp. (carpet grass), *Pennisetum cladatinum* (Kikuya grass) are suitable for developing lawn in warm and mild climate. Grasses most commonly used for lawn and turf in cold climate are *Lolium* spp. (perennial rye grass), *Poa partensis* (Kentuckky blue grass), *Poa trivealis* (medow grass), *Festuca rubra* (creeping red fescue), *Festuca tenuifolia, etc.*

2.12.0 Aquatic Plants

A large and diverse group of ornamental plants are grown in tanks and ponds and also in moist locations in gardens. Several species and cultivars of aquatic plants produce uncommon and spectacular flowers.

2.12.1 Uses

Water garden in a landscape, formal or informal, private or public garden, is an important feature since the conception of garden design was evolved. Water is used in various ways and forms depending on a particular style in the Italian Renaissance gardens, and became a prominent and important feature in Chinese-Japanese style Oriental gardens in a natural environment involving the ideas of meditation and relaxation. In Mughal gardens, flowering water in cascade and formal planting following symmetry for pleasure and recreation became very popular in formal style of gardening. Further, development of garden design in

Bamboos and Grasses

Bambusa nigra in front of house

Bamboo in interior decoration

Bambusa aurea in pots

Barmuda grass design

Zoysia grass in lawn

Paspalum grass in garden

Aquatic Plants and Ferns

Lotus in a tank

Victoria amazonica in a water pool

Nymphaea in a water body

Asplenium in pots

Ferns in pots

Ferns in interior decoration

different locations and climate considered different types of water body in a garden including lily pool, tank of different sizes, depending on the area of the garden, canal or streams in public gardens and resorts for providing an attractive landscape by growing water lilies (*Nymphaea*), lotus (*Nelumbo*), bearing large blooms of different colours, *Victoria amazonica* with enormous foliage and large flowers are unusual aquatic plant and water snow (*Limnantherum* sp.) like a white miniature water lily are grown in shallow tanks. Marsh loving plants like *Acanthus, Alpinia, Cyperus, Crinum, Hedychium, Typha* and many others when planted along the moist border of the tank or pond or in bog garden provide a natural landscape with diverse plant groups. Moisture loving ferns (*Asplenium, Angiopteris, Acrostichum, etc.*) are also very attractive in clumps.

2.13.0. Ferns

Most ferns are decorative and hardy ornamentals grown for their graceful appearance and beautiful foliage.

2.13.1 Uses

Since they do not need adequate light, and grow well in a wide range of environment, and many species and cultivars prefer warm humid climate and semi-shady location, ferns are popular house plants. They are grown in suitable containers in airy rooms, staircases, patios, window boxes, balcony in partial shade and other suitable locations as a specimen plant, in group of different species and also with other house plants. Species of several genera like *Asplenium, Doryopteris, Platycerium, etc.* are epiphytic. Ferns prefer porous soil rich in organic matter they are also popular greenhouse plants, propagated mainly by spores, division and from bulbils in some species.

3.0 Aesthetic Benefits of Ornamental Plants

Aesthetic value of ornamental plants is well recognized from time immemorial. Different ornamental plants carefully selected and arranged in a definite design improve the aesthetic quality of residential places, recreational areas like parks and gardens, both private and public, botanic gardens, institutions and commercial complexes, monuments and historically important locations, traffic signals, city roads, highways and railway tracks, tourists spots, river banks, *etc.* By adding a bit of natural greenery, ornamental plants will help in breaking or softening the harsh line of concrete structures like factory buildings, office buildings, hotels, hospitals, airports *etc.* To add beauty and elegance, every corner of a house or building or a garden can be decorated aesthetically by selecting suitable plants and placing them in a proper place. For this purpose trees, shrubs, climbers, palms, house plants, cacti and succulents, ferns, aquatic plants –all categories of plants play a distinct and useful role.

3.1.0 Bio-aesthetic Landscape Planning and its Implementation

Apart from architectural planning and structural development, landscape architecture involves bio-aesthetic planning and its implementation in garden and plantation development at various sites and locations mentioned before for different purposes need a thorough knowledge of the principles in designing landscape and implementations of garden features with suitable plants and appropriate materials, maintaining an overall unity in the garden. Besides development of a garden, its management and maintenance are also remarkably important to enjoy the beauty of nature thus created. The success in landscape planning, its implementation and management largely depends on the knowledge and experience of qualified horticulturists specialized in floriculture and landscaping in association with the landscape architect.

3.2.0 Interiorscaping by using House Plants

Interiorscaping with indoor or house plants makes the living or working place more enjoyable, healthy and attractive. House plants can be kept anywhere in the house – living room, bed room, study bathroom, kitchen, balcony, staircase, patio, window sill and terraces.

Ornamental plants are sometimes used as room dividers. They are quite effective in creating the feeling of a definite space. Plants suitable as room dividers are large foliage specimens such as species and cultivars of *Codiaeum, Cordyline, Dracaena, Monstera, Philodendron, Schefflera, Syngonium,* palms, *Philodendron, Monstera, Syngonium, etc.*, can be trained on canes to provide a token separation between living and dining space. The plants selected as room divider should suit the room's size and décor. The imaginative use of plants can transform a dull spot into a positive decorative place. Colour can be added by placing a few flowering plants along with the foliage plants in such places.

3.3.0 Window Garden

Plants are especially effective when used as a part of a window. They can help to break the plainness of a large expense of glass. Often, plants are used to create more interest at a window or to aid in making the outside appearance to be an extension of indoor living areas. For window boxes, species and cultivars of ornamental plants including annual flowering plants like *Antirrhinum,* Aster, Chrysanthemum, *Impatiens, Hedera, Neoregelia, Phlox, Petunia, etc.*, and many hardy foliage plants *viz., Aglaonema, Begonia, Caladium, Dracaena, Peperomia* and many others are very suitable.

3.4.0 Floral Arrangements with Cut Blooms and Loose Flowers

Besides house plants and other potted ornamentals, another interesting component in interiorscaping is floral arrangements with fresh or dry flowers and foliage and other plant parts. Loose flowers are also an integral part of decoration during festivals and social ceremonies. These are widely used for making floral carpets, floral *alpana* (design), garlands, floral ornaments like *veni, gajra, tiaras, etc.* Loose flowers are used in offering puja by a Hindu. Loose flower petals are popular floaters along with floating candles.

3.5.0 Bonsai and Topiary

Many elegant and hardy plants and their art forms like bonsai and topiary are ideally grown in home as well as in public gardens.

Interiorscaping by using House Plants

Vertical interior garden

Green wall

Air purifying plants

House plants in a restaurant

Interior decoration with plants

House plants in balcony

Value Addition in Ornamental Plants

Lucky bamboo (Dracaena)

Home made *holi* colour from flowers

Rose oil

Rose water

Rose gulkand

Pot pourri

3.6.0 Green Wall

Green wall or vertical garden or living wall is a new concept, where a wall is partially or completely covered with vegetation. Green walls, which have become very popular in recent years in urban areas, may be constructed both outdoors and indoors.

4.0 Health and Life Style Benefits of Ornamental Plants

According to The World Health Organization (WHO), health is defined as "a state of complete physical, mental and social well-being and not merely the absence of disease or infirmity". In the light of the above definition of health, the benefits of ornamental plants have been assessed in the following, paragraphs. Being around the plants generally would help people benefitted in more than one way in terms of health, leading to greater energy and better performance. Plants have been shown to reduce stresses across a broad spectrum of individuals. It has been shown that when plants are in the interior space, systolic blood pressure is reduced by one to four units. A recent study reveals that foliage plants in the office room improve the health and reduce discomfort symptoms of the office employees. Thus, by keeping ornamental plants in home and in work place one is expected to work more accurately and eminence. As flowers have a pleasing effect on mind, an elegant flower arrangement in the room can create an atmosphere which will make one to be more creative and productive. The soothing effect of natural aesthetic beauty helps people to feel relaxed and less tensed. Parks and other urban green spaces rejuvenate health by providing an area for recreational activities in the serenity of nature.

4.1.0 Effect on Healing and Recovery

Ornamental plants accelerate the healing process in ailing persons. The presence of ornamental plants or flowers in the recovery room or a well maintained garden in the hospital gives a feeling of tranquility to the patient. It is reported that the patients assigned to rooms looking out on a natural scene of trees require shorter post-operative stays and take fewer analgesics than patients facing a brick wall. Many studies have demonstrated the positive impact of plants and flowers on the mental health of a person. Many therapeutic effects of gardens and garden plants have been reported on people undergoing treatments against trauma and stress related disorders. Thus, healing gardens have been evolved to provide strength and peace to the tired body and mind.

4.2.0 Enhancement of Positive Emotions

For centuries, in cultures around the world, flowers have been used to convey emotional information among people. Flowers are used to convey sympathy, admiration, affection, romance, joy and celebration. Flowers are also used to express religious feelings and for spiritual communication. In a study it has been found that true smiles are exhibited by 100 per cent of the women who received flowers as a gift. Increase in positive moods has also been noticed.

4.3.0 Therapeutic Effects

Anecdotal evidence indicates benefits of ornamental plants for people suffering from dementia; involvement of such people in a garden can enhance their sensory pleasure, provide a sense of achievement and a sense of companionship. Hence, gardens have become an integral part of the support to people with dementia, where dementia patients would feel safe, and can relax or work and continue their normal life as much as possible. Similarly, windows view of nature in prisons is said to be associated with lower frequencies of prisoner's stress symptoms such as headache and digestive upsets in comparison to views of walls, buildings or other prisoners. *Azadirachta indica* (neem), one of the most extensively planted trees in tropics and mild subtropics, has many beneficial uses on human health and for pest control in crops, apart from its elegant appearance.

4.4.0 Horticultural Therapy

Horticultural therapy is the use of ornamental plants and garden activities to improve body, mind and spirit. Horticultural therapy provides physical, mental and social benefits to individuals working with garden plants and gardens. The Chicago Botanical Garden characterizes horticultural therapy as 'the use of professionally directed plant, gardening, and nature activities for the purpose of restoring the physical and mental health of its participants'. Horticultural therapy maintains or improves physical health by providing unlimited opportunities for exercising, increasing flexibility, improving coordination and balance, building physical strength, relieving stress, providing a non-threatening atmosphere, alleviating depression and helping people to connect with nature. Horticultural therapy or gardening proves beneficial in prisons also, since gardening has a soothing effect on criminal's mental health.

4.5.0 Use in Medicine

Ornamental plants are used extensively in medicines, both for treatment and as a preventive measure. Certain ornamental plants have been the source of modern medicines of significant economic benefits. In fact, according to World Health Organization's estimate, 80 per cent of world's population is dependent on traditional medicines involving plant materials. The utilization of flowers and ornamental plants for medicinal purpose is increasing day by day. There are hundreds of ornamental plants with medicinal value. To name a few are as: *Aloe vera* in the form of a refreshing drink has many health benefits; it controls diabetes (diabetes mellitus), cures headache, dizziness, whooping cough, hemorrhoids constipation, relieves pain from burn injury, internal injuries, reduces hair loss, *etc.* Aloe juice provides instant relief for minor burns and wounds and is used as a salve for radiation burns. *Gomphrena globosa* controls asthma, chronic bronchitis, whooping cough (pertussis). Sunflower helps in the control of high blood pressure (hypertension), headache, dizziness, stomach pain, painful menstruation (dysmenorrhea) inflammation of the breast, *etc.* Roses can control irregular menstruation, menstrual pain, bleeding during pregnancy, *etc.* *Jasminum sambac* reduces inflammation and redness of eye, fever, headache, influenza, diarrhea, shortness of breath. *Catharanthus roseus* is useful in the treatment of leukemia, bleeding due to decreased platelet counts,

Value Addition in Ornamentals

Adenium Bonsai

Bougainvillea Bonsai

Aromatherapy

Flower market

Women empowerment

World flower trade

Uses of Ornamental Plants in Gardens

In Home garden

In Parks

In Water garden

In Botanical garden

In Roof garden

In Community garden

hypertension, diabetes *etc.* The chrysanthemum tea, jasmine tea, lavender flowers are used as refreshing drinks.

4.6 Aromatherapy

Another form of medicinal therapy involving use of ornamental plants is aromatherapy. Aromatherapy is the treatment for prevention of diseases by using volatile oils including essential oils. Other stated uses include pain and anxiety reduction, enhancement of energy reduction, short-term memory, relaxation, hair loss prevention, and reduction of eczema-induced itching, removal of insomnia, *etc.* Aromatherapy does not cure conditions, but helps the body to find a natural way to cure itself, and improves the response of immune system. At the scent level, essential oils activate the emotional centres of the brain. When applied to the skin (commonly in the form of massage oils), essential oils which have anti-fungal, anti-bacterial and anti-viral effect are widely used in aromatherapy. Among the ornamental plants, rose, carnation, jasmine, tuberose and sandal are rich sources of essential oils.

5.0 Social Benefits of Ornamental Plants

The social and ecological benefits obtained by residents from community allotment gardens are summarized below:

Social Benefits

1. New horticultural skills
2. Peaceful neighbourhood
3. Nice view
4. Socializing/friendship development
5. Contact with green population
6. Exercise
7. Health
8. Relaxing hobby
9. Quality of life
10. Satisfaction from watching things to grow sense of community
11. Sense of security

5.1.0 Community Sense, Especially in Children

Children learn the value of being environment conscious and know the importance of plants in human life and their role in the conservation of environment. Researches indicate that children with greener view from home or school are better able to concentrate, inhibit impulses and delay gratification. Researches also suggest that ADD (attention deficit disorder) symptoms in children are much less mild in a greener setting. Our great poet and Nobel laureate Rabindranath Tagore set up open air class rooms for school children in close association with natural settings in Visva Bharati at Santiniketan.

6.0 Environmental Benefits of Ornamental Plants

Besides enhancing the beauty of the surroundings, ornamental plants bring about great improvement in the environment of a place.

Ecological Benefit

1. Green area for living things
2. Space between houses (providing privacy)
3. Open natural space
4. Fresh air
5. Attractive open space
6. Peace and quiet
7. Green space
8. Nice view in front and behind the house
9. Buffer zone
10. Less noise from urban road
11. More rural atmosphere

6.1.0 Oxygen Production

The fundamental benefit provided by the green plants is the production of oxygen, thus providing the atmosphere with the element that allows the animal world to breathe and survive on this planet. It has been estimated that on an average, a full grown tree can produce 117.9 kg of oxygen per year, and two mature trees can produce sufficient oxygen for a family of four. Similarly, a natural turf area measuring 15 m x 15 m releases enough oxygen to support a family of four.

6.2.0 Carbon Sinks

Green plants take in carbon dioxide and convert it to simple sugars (carbohydrates) in presence of sunlight and water. These sugars provide the plant with energy for its growth and development. When the plant or plant parts die, as a result of decomposition of plant material, the carbon returns to the soil and atmosphere. This carbon within the soil is released during agricultural tillage or soil disturbance. This cycle of removing carbon from the air and capturing it in plants and returning it to soil is known as carbon cycle. It is estimated that a tree in a forest removes 4.5 – 11 kg carbon per year. Plants' unique ability to absorb carbon dioxide and fix carbon in presence of sunlight makes it most valuable to mitigate the tremendous increase in carbon dioxide concentration in the atmosphere that has been occurring since the industrial revolution.

6.3.0 Greenhouse Gas and Global Warming

Carbon dioxide is a greenhouse gas responsible for the increase of average global temperature which has detrimental impact on climatic changes around the world. To counter this trend, planting of trees including ornamental ones in open spaces, parks, gardens, roadsides, *etc.* is suggested to build global availability of carbon sinks.

6.4.0 Energy Saving Attributes of Landscaping

Trees provide shade and reduce the amount of solar radiation hitting a building, thereby reducing energy requirement for cooling that building. It is estimated that planting one shade tree in urban setting can prevent combustion of 18 kg of carbon per year due to saving in demand for air-conditioning. In temperate climate, deciduous trees are found more beneficial because they allow radiation in the winter months to warm the building and thus assist in heating. Uses of evergreen trees in temperate climate also contribute to energy savings by protecting the building from cold winter wind as windbreaks. Reduction in wind speed can reduce infiltration of cold air into building by up to 50 per cent giving potential heating saving of 10 – 12 per cent per year. Careful planting of trees can regulate air-movement and wind in such a way that will create more comfortable and pleasant microclimates.

Trees can also alter the local climate by reducing the effect of urban 'heat islands' often found in build-up areas in metropolitan cities. The major cause of heat island phenomenon is land modification due to urban development which uses materials like concrete road, buildings, metals like steel, aluminium, *etc.* and other structures which retain heat. It has been found that heat island phenomenon causes semi-urban and sub-urban temperatures 1 – 6°C higher than nearby areas. The obvious solution for counteracting urban heat islands is to plant trees and shrubs along the city roads, traffic islands, boulevards and any open space available. The benefits from such plantings are two-fold: first, they provide cooling shade, and secondly, the trees transpire large amount of water from the leaves, thus cooling the leaves and consequently the ambient air. Experiments show that a healthy tree inside a building can have fooling effect of ten room-sized air conditioners operating twenty hours a day. In fact, tree groves have been found to be about 5°C cooler than open terrain. Similarly, grass field or lawn is 1-2°C cooler than concrete parking lot. Green roof and green wall also work to counter the urban heat island.

Energy saving can be gained by using indoor or house plants as temperature moderators. Interior plants can reduce the temperature of an office environment and maintain relative humidity in the range of 30-60 per cent which is the recommended level for human health and comfort. Correct selection and placement of plant can reduce heating and cooling cost of the office space.

6.5.0 Pollution Amelioration

Plants raised in outdoor landscaping and for indoor decoration play a very useful role as other plants including grasses are often regarded as 'lungs of the cities', because of their ability to remove contaminants from the air. Two major ways in which plants can reduce air pollution are i) collection of dusts, ashes, pollen grains and other particulate matters on their leaves, thus reducing their presence in the air breathed and, ii) absorption of gaseous pollutants through their leaves *e.g.* nitrogen dioxide, ozone, sulphur dioxide, *etc.*

The amount of air-borne pollutants removed increases with leaf surface area of a plant. Therefore, trees are better filters than shrubs and grasses. Similarly, evergreen trees serve as a much better filter than deciduous trees throughout the year. The shape of the leaf, presence of leaf-hairs, phyllotaxy, surface texture of the leaf, *etc.* are the determining factors for dust and smoke collecting efficiency of a tree. Trees having compact branching and dense foliage have more dust collecting ability. Trees like *Acacia auriculiformis, Alstonia scholaris, Azadirecta indica, Bombax ceiba, Butea monosperma. Cassia siamea, Dalbergia sisoo, Delonix regia, Derris indica, Ficus benjamina, Ficus benghalensis, Ficus religiosa, Madhuca indica, Peltophorum ferrugineum, Polyalthia longifolia, Schima wallichil, Shorea robusta, Tectona grandis, Terminalia cuneata, Thespesia populnea, etc.* have been found very efficient in this respect. Most of the grasses possess copious auricular hairs at the juncture between leaf blade and leaf sheath. Auricular hairs occupying a strategic position at the base of the lamina can absorb various chemicals from air borne particulates that are deposited on their surface. Stretch of grassland, a lawn or meadow, is a potential dust trapper, as well as air purifier, and also checks ground water contamination by reducing leaching of chemicals and fertilizers.

6.6.0 Sulphur Pollution

Sources of sulphur dioxide pollution are high sulphur containing petroleum and natural gas, thermal power generating plants, coal fertilizers, *etc.* Trees resistant to SO_2 are *Albizzia lebbeck, Ailanthus excels, Alstonia scholaris, Azadirecta indica, Ficus religiosa, Lagerstroemia flosreginae, Polyalthia longifolia, Quercus palustris, Terminalia cuneata etc.*

6.7.0 Oxides of Nitrogen

Sources of this pollution are fertilizers, fossil fuel combustions, biomass burning, *etc.* Resistant trees like *Ginkgo biloba, Pinus* sp., *Robnia pseudocacia, Quercus rubra, Sambucus nigra, Tacus baccata, Ulmus montana, etc.* are very efficient in absorbing these toxic oxides of nitrogen.

6.8.0 Indoor Air Quality Improvement

With increasing urbanization, more and more people are spending 80-90 per cent of their time indoors. Many studies indicate that the concentrations of volatile organic compounds are much higher in indoors than outdoors. Some indoor plants exhibit their efficacy in removing volatile organic compounds from the atmosphere. Ornamental plant species like *Crassula portulacea, Dieffenbachia amoena, Hydrangea morifolium, Nephrolepis exaltata, Spathiphyllum* sp., are quite efficient in removing benzene and other gaseous pollutants from the air. *Dracaena deremensis* has good capacity to remove benzene from indoor air. Another means by which indoor plants improve enclosed environment is by removing offensive odour from the air.

As with the outdoor environment, particulate matters such as dust, ash, pollen and smoke are also pollutants of indoor air. It has been found that the levels of particulate matter accumulation in a room is lower when plants are present than when they are not. NASA investigated the use of plants as air purifiers. The results of their investigation suggest that one potted plants per 100 square feet indoor space in an average home of office is sufficient to cleanse the air of pollutants.

Orchids are very effective is removing numerous pollutants during daylight hours. They are also effective at removing CO_2 and xylene at night, while at the same time releasing oxygen into the air. This is because orchids (and bromeliads also) have a unique metabolic process whereby their stomata remain open at night. This is significant because here air is continuously filtered, day and night. In addition, presence of plants in a room increases the relative humidity level, thereby increasing the comfort level in a heated room. On the other hand, indoor plants can also contribute to reduce air quality by releasing pollens and spores which may cause discomfort in the form of allergies.

6.9.0 Noise Amelioration

Shelterbelts, screens and tall hedges are effective in reducing noise pollution, especially in urban areas where noise is easily reflected from hard surfaces of pavements and buildings. Plants with thick and fleshy leaves and flexible petiole are quite efficient in absorbing high frequency sounds. A dense evergreen hedge can reduce the noise of microphone by up to 20 decibels. Plantation on both sides of roads or highways is effective in curbing noise to a significant level. In urban areas a green and thick vegetation belt and open space may bring down the noise level to tolerable level. Green roofs also reduce noise distractions from the outside world.

6.10.0 Water Management (Retention, filtration, purification, flood control) and Erosion Control

Plants play significant role in water management and soil erosion control. It is now well accepted that plants can be used to remove pollution from soil and air, and can help in flood control by retaining water. Plants stabilize soil and reduce surface run off in open space. Apart from visual appeal, lawn or grass cover is the best defense against soil erosion. Extensive root system of trees, shrubs or grasses binds the soil firmly and prevents soil erosion when there is heavy rain or floor. Similarly planting of trees and shrubs as windbreaks and shelterbelts help in reducing wind velocity and consequently soil erosion by wind. Fallen leaves, dead roots and humus bring about improvement in physical chemical and biological condition of the soil, and consequently help in gradual penetration and retention of rain water in soil. Turf provides a good growing area for many micro-organisms, which, in turn, help cleans water by digesting the pollutants in the trapped water and speeding up their normal break down.

7.0 Economic Benefits of Ornamental Plants

7.1.0 Economic Value of a Tree

The economic value of a tree on the basis of various environmental and social benefits derived during 50 years life span of the peepal tree (*Ficus religiosa*) was estimated by Prof. T. M. Das of Calcutta University in 1979 and re-evaluated in 2012.

7.2.0 Increased Property Value

Landscaping aids in appreciable increase in real and perceive value of a property. Buyers respond well to homes or office buildings with a well maintained garden, manicured lawn, or well-shaped hedge, *etc.* A tree cover between the property and its immediate neighborhood raises the value of the house. For bungalows and cottages, the4 higher the percentage of ground cover like lawn, shrub border, herbaceous border, rock gardens, *etc.*, the higher the value of the house. The green roof or green wall will raise the property value by 4-10 per cent. It is a common experience that a house or apartment building or office building with more green areas has greater demand, and therefore, is expected to earn significantly higher returns.

7.3.0 Municipal Economic Benefits

Effective landscape design can also add value to municipalities and communities. It increases community health and vitality through civic involvement in beautification projects. Beautification projects represent a sense of pride and value by residents and businesses. Plants stabilize soil and reduce run-off in open space, thus reducing maintenance cost of open space. Natural habitat, if maintained properly, can protect biodiversity.

Trees and shrubs can be used to build both physical and visual barriers, providing security and/or privacy; it also reduces crimes and accidents. A well planned city with huge green belt and meticulously maintained parks, gardens, botanical gardens, conservatories, *etc.* (as in Singapore) offer scope for eco-tourism, promoting business in hotels, transport and local handicraft and food, thus improving the financial prosperity of the area and community.

8.0 Trade of Cut Flowers, Pot Plants, Foliage and Seed

In many developing countries, in the last four or five decades, in the last four or five decades, application of science and technology has markedly improved the production, quality and availability of ornamental plants and flowers, resulting in substantial annual increase in floricultural trade.

In domestic market, apart from annuals, rose, chrysanthemum, carnation, gladiolus, orchids, anthurium, lilium, tuberose, heliconia, *etc.*, are the major cut flowers. Loose flowers of marigold, hibiscus, chrysanthemum, tuberose, jasmine, crossandra are extensively used in making floral ornaments and other purposes.

Pot plants of many flowering and foliage plants including orchids, different species of shrubs, bulbous plants, gerbera, cacti and succulents, large member of house plants are of immense commercial importance in domestic and export markets. Cut leaves of dracaena, cycas, scindapsus, palms, ferns, *etc.*, are popular items in flower arrangement.

Annuals and biennials (mainly in temperate climate) constitute a major group of ornamental plants, extensively used for garden display, as pot plants, flower arrangements and decorations throughout the world. They are generally propagated by seeds. Marked improvement in annuals has been achieved through researches on genetics and plant breeding, trials and standardization of techniques of seed production and commercialization of seed industry during the last 100 years, particularly in the developed countries. Hundreds of annuals grown in gardens provide a wide range of colour from white, red, pink, scarlet, yellow, orange to blue F_1 seeds of annuals producing plant and flower of superior quality with great demand in seed industry are of tremendous commercial importance. New cultivars and F_1 hybrids are coming in the market every year from hundreds of seed companies of repute. Nurseries are the source of planting materials of different types used in establishing gardens, private and public supply, pot plants for interior and outdoor display. Chemicals and implements are also available in the nursery.

8.1.0 World Trade of Plants and Flowers

Urbanization and modernization of the society have made flowers a part of our daily life. Thus, huge demand for flowers offers opportunity for focusing on flower trade. The total value of global cut flower market was € (euro) 31 billion in 2011. The European market (including non-EU countries) for cut flowers accounted for the majority of global market (56 per cent), or approximately € 18 billion in 2011. This regional market in EU is expected to grow further to reach € 22 billion in 2016, representing a growth of 41 per cent from 2006 to 2016. The size of European pot plant market was 12.5 billion € in 2011, and is expected to reach almost 15 billion € in 2016, representing a growth of 37 per cent during the period 2006 – 2016.

8.2.0 Flower Exporting Countries

Top ten flower exporting countries *viz.*, the Netherlands, Columbia, Italy, Belgium, Denmark, Germany, Kenya, USA, Canada and Ecuador account for over 80 per cent of the total export value of floricultural crops. Europe still maintains its position as the leading exporter in the world with a share of around 70 per cent of the total export of cut flowers and foliage. However, European exports of cut flowers are largely intra-European in nature. Re-export is quite common in Italy and the Netherlands. The market share of European exporters for bulbs, cuttings and live plants has fallen in the recent years. The major European exporters of these commodities are the Netherlands (Alsmeer flower auction centre), Belgium, Denmark, Italy, Germany, France and Spain. Again, as in case of cut flowers, pot plant export by the European countries is predominantly intra-European. European countries also export to the USA, Canada, Japan and other Asian countries.

Major flower consuming countries in the world are concentrated in Western Europe. Of the world's ten largest international markets for cut flowers, six are in the EU, namely Germany, the UK, France, Italy, the Netherlands and Spain. Other important markets are USA and Japan, accounting for around 20 per cent each. Recently, Russia and the Middle-East countries have become important markets showing rapid growth.

International trade plays a major role for floricultural products. The Netherlands, in particular, supplies almost 75 per cent of EU market and around 80-90 per cent of the UK, Germany and France's imports. In fact, the Netherlands is also the main distribution hub of floricultural products, which are imported and distributed to other EU countries. Apart from developed countries, many developing countries like Thailand, Taiwan, Columbia, Israel, Ecuador, Turkey, Kenya, Brazil, *etc.* are contributing handsomely in world trade of flowers and ornamental plants using improved production technology.

Major Floricultural Crops Exporting Countries in 2018

Sl.No.	Country	Quantity in MT	Value in US$ Mill.
1.	Netherland	25,20,354.00	8,904.00
2.	Colombia	1,78,699.00	1,381.00
3.	Germany	4,91,758.00	1,076.00
4.	Italy	3,52,418.00	941.00
5.	Ecuador	1,14,286.00	915.00
6.	Kenya	1,89,079.00	854.00
7.	Belgium	4,32,271.00	633.00
8.	Spain	1,68,688.00	512.00
9.	China P Rp	2,50,565.00	445.00
10.	U S A	53,449.00	426.00
11.	Canada	1,09,892.00	422.00
12.	Denmark	1,68,777.00	402.00
13.	Ethiopia	71,423.00	268.00
14.	Poland	1,77,984.00	257.00
15.	Other Asia, Nes	35,041.00	246.00
16.	France	2,02,165.00	196.00
17.	Thailand	49,960.00	175.00
18.	Malaysia	39,551.00	167.00
19.	Israel	20,294.00	161.00
20.	Costa Rica	47,888.00	148.00
21.	Guatemala	20,040.00	114.00
22.	Turkey	35,728.00	108.00
23.	South Africa	18,234.00	95.00
24.	Portugal	27,399.00	89.00
25.	Mexico	10,478.00	86.00
	Total	57,86,421.00	19,021.00

8.3.0 Demand in World Market

The flower and foliage market is usually segmented according to consumer use. In general, consumers purchase flowers for two main purposes: as a gift items and for their own use. The demand for cut flowers peaks around holidays or festive occasions, especially Mother's Day, Valentine's Day, Christmas Day, New Year's Day, *etc.* Moreover, certain species of flowers are associated with particular occasion; for instance, carnation for funerals in the West and tuberose in India, poppies for Remembrance Day. Consumers often buy flowers for their own use, usually with the intention of creating a pleasant atmosphere or brightening up their home, office or similar establishment. More than half of the cut flowers in Western and Southern European countries are purchased for the purpose of gift and a smaller portion for personal use. Freshly cut rose, carnation, gladiolus, alstroemeria, tulip, lilium, anthurium, chrysanthemum, heliconia, orchids, gerbera, *etc.*, are important flowers in the world market. There is a wide assortment of roses available in the EU, and they are purchased for many different occasions. Consumers' preference has gradually shifted from small-budded (*e.g.*, intermediates and sweethearts) to large ones (*e.g.*, Hybrid Tea or HT). Small budded roses are associated with generic or bulk product, mainly due to their lower price and dominant presence in the supermarket chains. Larger rose cultivars, on the other hand, are higher priced and primarily sold in more specialized channels like florists. Between 70-89 per cent of all cut flowers sold in the EU are brought in mono-bunches which contain only one variety. The rest 20-30 per cent is bought in bouquets of different varieties of flowers and foliage or in floral arrangements. The Flower Council of Holland reports that while purchasing flowers, the consumers usually look for criteria as follows: quality (36 per cent), price (13 per cent), species (13 per cent) and colour (10 per cent).

8.4.0 Uses and Markets of Dry Flowers

Along with fresh flowers there is tremendous demand for dry flowers in the domestic as well as in the international markets. In order to preserve the beauty and glory of flower and foliage or any other plant part in their original colour and shape, proper drying technology for these plant species has been developed. Presently, dry flower business ranges from making potpourri to flower bouquet and arrangement, greeting cards, garlands, page markers, table mats, and dry flower ornaments. Among these, potpourri makes maximum business in the world markets. The major flowers used for this purpose are cornflower, dahlia, globe amaranth, helichrysum, marigold, larkspur, lavender, lotus pod, rose, statice, sunflower, *etc.* With increasing demand for quality produce, advanced technology for controlled drying, colouring and processing has been developed and adopted in the recent years. Potpourri is a mixture of dried naturally fragrant plant materials, used inside the home to give the air a pleasant smell. These are esteemed as a very good gift item. A few ornamentals suitable for making potpourri are rosemary, citronella, palmarosa, juniper, rose, cypress, *etc.* The export markets for dry flowers are mainly concentrated in European Union and the USA. West Bengal (India) accounts for about 70 percent of the dried flower export from India.

8.5.0 Trade in Value Addition

Value addition can be defined as getting more returns for what is already there. Value added floriculture is the process of increasing the economic value and consumer appeal of any floricultural commodity. Value addition is especially important because of perishable nature of floricultural crops. A value added marketing approach is necessary to help spread in the market for better returns.

8.5.1 Essential Oils

Extraction of essential oils from flowers like rose, jasmine, tuberose, marigold, calendula, lavender, primrose, basil, *etc.*, are gaining momentum in floriculture business in the recent years, since it gives much higher returns that fresh flowers. Essential oils as well as related by products like absolutes or otto (*attar*) have great demand in the domestic as well as in the international markets. These are mainly used in perfumery, cosmetics, aromatherapy, fragrant commodities like soap, hair oils, candles, *etc.*, and also in pharmaceutical preparations. Hibiscus leaves and flower powder are sold as raw materials for cosmetic industries.

India is the second largest exporter of essential oils in the world after USA. Indian export of essential oil stood at US $ 298.7 million in 2007-2008. India is the largest exporter of jasmine oil in the world accounting for over 40 per cent of total export of jasmine oil. Rose (*Rosa damascenae*) is extensively cultivated in Rajasthan and Uttar Pradesh only for extraction of essential oils, rose water, attar, gulkhand, *etc.* Gulkhand or rose jelly is made keeping alternate layers of sugar and fresh rose petals, used as a sweetening agent with essence, mostly in betel leaf. Other promising floricultural crops from which essential oils are commercially extracted are kewda (*Pandanus*), maulsiri (*Mimusops*), champakka (*Michelia*) *etc.*

8.5.2 Natural Dye

Natural dyes extracted from flowers, which are used in various ways, have good demand in the market. For example, marigold pigments are widely used in the poultry industry to enhance the colour of meat and yolk of the eggs. In addition, this is also used in food and textile industry. Similarly, natural dye (pigments) has been extracted from unused flowers of wholesale market in Kolkata to prepare '*Abir*' (coloured powder) used widely during festival of colour ('*Dol*' or 'Holi') in India. These natural dyes can also be used in textile industry.

8.5.3 Value Addition

Value addition to *Dracaena* is done by twisting the otherwise straight stem or axillary shoots to a particular shape or form. Also dracaena stem cut into pieces and bundled in specific numbers are sold as 'Lucky Bamboo' for good luck.

9.0 Other Economic Benefits

Floriculture and nursery business go hand in hand. Nursery business involving seeds, plants, flowers and bulbs is a few billion dollar worth, employing thousands of people and bringing prosperity of a country. Nurseries operate selling seeds and seedlings of annuals as well as tree species, rooted cuttings of herbaceous plants like chrysanthemum, dahlia, coleus, house plants and many others, layers of shrubs and climbers and budded or grafted woody species and also bulbs, corms, suckers, *etc.*, every year. In the nursery business, potted plants occupy a very important position. Demand for potted plants is ever increasing because of wide scale use of potted flower and foliage plants in interior as well as outdoor decoration.

While flowering potted plants like poinsettia, pelargonium, geranium, chrysanthemum, marigold, dahlia, balsam, anthurium, begonia, roses and many annuals are used to add colour, foliage plants like philodendron, monstera, schefflera, palms, ornamental bamboo, ferns, *etc.* are extensively grown for indoor and outdoor display; the demand for pot plants is increasing very rapidly among apartment dwellers throughout the world for developing closeness with nature, Hanging pot plants like chlorophytum, pellionia, peperomia, sedum *etc.* are widely used in balcony gardens, in extending living rooms and porticos. As a result of

Uses of Ornamental Plants for Commercial Purpose

Rose cultivation in open

Cultivation of rose in greenhouse

Carnation cultivation in greenhouse

Chrysanthemum cultivation for cut flowers

Nursery of ornamental plants

Nursery provides planting materials

Floral Decoration for Beautification

Exhibitional arrangement

Wedding flower decoration

Welcoming flower decoration

Religious flower decoration

Indian flower rangoli

Creative flower arrangement

growing interest and demand for xeriscaping, the sale of cacti and succulents as pot plants has gained tremendous momentum.

Accessories for gardening like containers, gardening implements, garden soil, manures, fertilizers, plant protection chemicals, shading materials, *etc.*, are also available in a nursery for sale.

9.1.0 Plant Rentals Service

In recent years, the concept of plant rental service is increasing in major cities and towns in India. It involves hiring out/rental of potted plants for decoration of home or offices. Presently, most of the offices, business establishment, hotels, *etc.*, hire out plants for decorating their establishments or premises, instead of maintaining their own nursery. Rotation of plants once a week or ten days brings a new look. Thus, one can enjoy a wide variety of plants without taking the trouble of maintaining them in one's own nursery. *Aglaonema, Araucaria, Codiaeum, Cordyline, Dracaena, Ficus, Monstera, Philodendron,* palm (different species), *Syngonium, etc.*, are very popular in this service.

9.2.0 Employment Generation

The ornamental horticulture industry contributes in a huge way in employment generation, providing opportunities and entrepreneurship in semi-urban and rural areas. The creation, preservation and conservation of green habitat in regional and town planning designing and implementing garden development requires huge manpower like landscape architect, environment architect, horticulturist, gardeners, interior planners and decorators, *etc.* By providing job opportunities and business prospects, this sector can boost the local and national economy tremendously.

With the growth of aesthetic horticulture, business areas like production and sale of growing media, manures and fertilizers, plant protection chemicals, horticultural tools, and implements, pots (clay, ceramic, plastic) for plants and bonsai, flower vase and many items for floral arrangements flourish in a big way, offering further employment opportunity. Man power is also needed in the dry flower production industries, essential oil extraction and natural dye production units.

9.3.0 Ornamental Horticulture in Women Empowerment

Work gives us a sense of identity and dignity by giving us substantial income for a decent living. Floriculture offers a wide range of opportunities to women in terms of employment, income generation, empowerment and above all self-fulfillment. A large number of women in rural and urban areas are now employed in commercial flower growing, nursery production, value addition and marketing in all parts of the world.

10.0 References

Bose, T. K. (1965) *Handbook of Shrubs*, The Agri-Horticultural Society of India, Kolkata.

Bose, T. K. and Mukherjee, D. (1970) *Gardening in India*, Oxford IBH Publishing Co., Kolkata.

Bose, T. K. and Bhattacharya, S.K. (1980) *Garden Plants*, Oxford IBH Publishing Co., Kolkata.

Bose, T. K. and Bhattacharjee, S. K. (1980) *Orchids in India*, Naya Prakash, Kolkata.

Bose, T. K., Chowdhury, B. and Sharma, S.P. (1991) *Tropical Garden Plants in Colour*, Horticulture and allied publishers, Kolkata.

Bose, T. K., Maity, R.G. and Dhua, R.S. (1999) *Floriculture and Landscaping*, Naya Prakash, Kolkata.

Bose, T. K., Das, P. and Maity, G.G. (1999) *Trees of the World*, Regional Plant R esearch Centre, Bhubaneswar. India, Vol **1.**

Bose, T. K., Yadav, L.P., Pal, P., Das, P. and Parthasarathy, V.A. (2002) *Commercial Flowers*, Naya Prokash, Vols **1 and 2**.

Bose, T.K., Choudhury, B., Sharma, S.P. and Pal, P. (2004) *Garden Plants in Colour: House Plants*, Naya Udyug, Kolkata.

Bose, T. K., Maity, G.G., Sanyal, D. and Sadhu, M.K. (2007) *Dictionary of Horticulture*, Vols. **1-10**, Naya Udyug, Kolkata.

Bose, T. K., Singh, L.J., Sadhu, M.K. and Maity, T.K. (2015) *Ornamental Plants and Garden Design in Tropics and Subtropics*, Astral International, New Delhi. Vols **1 and 2**.

Bose, T. K. (2019) *Key Note Address, Bicentinary Celebration of The Agri-Horticultural Society of India*, Kolkata.

Das, T.M. (2012) *Indian Biologist*, **20**: 205.

2

ROSE

S. Sangeetha Priya, C. Aswath, N.K. Dadlani,
P. Pal and R. S. Malik

1.0 Introduction

Rose is one of the nature's beautiful creations and is universally acclaimed as the 'Queen of Flowers'. No other flower is a better symbol of love, adoration, innocence and other virtues than the rose, not only in our time, but has been for thousands of years. It is certainly the best known and most popular of all garden flowers throughout the world and has been growing on this earth for many million years before man himself appeared (Fairbrother, 1965; Gault and Synge, 1971).

Some countries have adopted rose as their national flower, *e.g.*, England, *etc.* In India, the rose was referred in old Sanskrit literature as *Tarunipushpa, Atimanjula, Simantika, etc.* and was adored not only in the royal palaces but also in*ashramas* of saints (Rode and Ogale, 1984). It has become the part and parcel of life, being connected with all phases of life right from birth to death. The cultivation of this beautiful flower in India developed with the distillation of roses as mentioned in *Ayurveda* by Charaka around 100 A.D.

The rose is an ornamental shrub with upright or climbing stems, usually prickly. The leaves are alternate, compound, oddly pinnate (*Rosa persica* has undivided leaves), withstipules adherent to the leaf stalks. Flowers are solitary or in corymbs. Calyx is five-lobed, lobes either simple or compound inserted at the top of a roundish or pear shaped fleshy tube. Petals and sepals are generally five (*R. sericea* has only four) but, many more in cultivars due to transformation of stamens. It has numerous carpels, inserted at the base of the calyx tube, free and with simple projecting style and stigma. The ripe fruit (hip or hep) consists of many one-seeded carpels in a fleshy tube and the seeds are hard. Ercisli (2007) assayed the fruits of *Rosa canina, R. dumalis* sub sp. *boissieri, R. dumalis* sub sp. *antalyensis, R. villosa, R. pulverulenta* and *R. pisiformis* for total phenolics, ascorbic acid, total soluble solids, total dry weight, total fat, fatty acids, moisture, macro and micro-elements. The highest total phenolic content was observed in *R. canina* (96 mg GAE/g DW). *Rosa dumalis* sub sp. *boisseri* had the highest total fat content (1.85 per cent) followed by *R. pulverulenta* (1.81 per cent) and *R. canina* (1.78 per cent). Nine major fatty acids were detected in rose species with α-linolenic acid as dominant in all species. Total soluble solids, total dry weight and ascorbic acid contents of rose species varied from 29.42 – 37.33 per cent, 33.85-40.35 per cent and 727-943 mg/100 g FW between *R. villosa* and *R. dumalis* sub sp. *boisseri,* respectively, whereas, moisture ranged from 59.65-66.15 per cent between *R. dumalis* sub sp. *boisseri* and *R. villosa.* The average mineral composition of rose species for N, P, K, Ca and Mg were 1.26 per cent, 513 mg/100 g DW, 639 mg/100 g DW, 196 mg/100 g DW and 114 mg/100 g DW, respectively. This showed that native rose genotypes are extremely rich sources of phenolics, carbohydrates and ascorbic acid demonstrating their potential use as a food or food additive.

2.0 Importance and Uses

The rose, because of its utility, occupies a pre-eminent place amongst the flower crops and is one of the oldest of fragrant flowers to be cultivated by man.

Its different types having beautiful flowers of exquisite shape, different sizes, bewitching colours and most delightful fragrance has made it an important flower for its varied uses. The various ways in which rose can be used for large-scale and small-scale planting is given below.

In floriculture, roses are of foremost commercial importance and cut roses have the highest demand throughout the world year round. Apart from open cultivation, roses are commercially grown under protection mainly in all the European countries, USA, Canada, Japan, Korea, Colombia, *etc.* both for domestic and export markets, the largest producer of cut roses. During the last two decades, roses are being grown in India under greenhouses to produce better quality flowers for export.

The Netherland is and will continue to be an important exporter of fresh roses. The country exported a total value of 975 million Euro cut roses in 2015 to countries like Italy, Sweden, Denmark, France, Poland, *etc.* Over the years, Royal Flora Holland played a prominent role in importing and re-exporting 40 per cent of flowers from all over the world. But in recent years, countries like Kenya and India also entered into international rose flower trade. Kenya supplies 1/3rd of the roses sold in the European Union for their extraordinary fragrance and quality and are challenging other countries like India, Columbia and Ecuador. Kenya also has excellent transport links to Europe and has specially dedicated floral cargo to transport swiftly. Ecuadorian roses are well known all over the world and contribute 74.5 per cent of all floricultural products export from Ecuador attributing to their long stem, juicy foliage and big tight bud. Top ten cut roses importing countries are The Netherlands, USA, Germany, UK, Japan, France, Belgium Switzerland, Italy and Denmark which contributed 89.74 per cent of world import in 2014-15. The market channels in France are flooded with heavy demand and around 50 per cent of rose flowers are sold in the form of bouquets mainly of red, pink and white flowers. Japanese purchase roses and prepare wine out of it which is a rising industry in the country.

Roses from India have a great demand in Europe and UK. In terms of area under cut roses in the world, India ranks first with a share of 46.54 per during 2012-2013. The total area under rose cultivation in India is 29,410 ha with the production of 301.95 thousand MT during 2015-16. India is the leading cut roses producing country and is exporting to more than 55 countries. The export of rose in terms of quantity and value was 30 MT and 141.45 lakh rupees, respectively during 2016-17 (Anonymous, 2017). However, the export of cut roses from India is less than 1 per cent of global export which might be due to import duty posed for Indian cut roses whereas, our competitor African countries had 9 per cent cost advantage over it. Major rose producing pockets in India are Medinipur, Nadia, Paraganas, Howrah (West Bengal), Bangalore, Kolar, Chikballapur, Ramanagara (Karnataka), Bharuch, Vadodara, Kheda, Ahmedabad, Valsad (Gujarat), Korba, Bilaspur, Kondagaon, Raipur, Mungeli (Chattisgarh), Nashik, Sangli, Satara, Solapur and Nagpur (Maharashtra).

Roses are also the most important perennial garden plants in almost all parts of the world. They are being grown in gardens from time immemorial. Roses in general are not only hardy plants, the variation in growth and many forms of flowers having numerous colour have made them the most suitable garden plants for different uses. They can be used as bushes, standards, climbers, hedges and edges, hangers and in rock gardens.

Roses as pot plants in suitable containers are also commercially grown and kept both indoor as well as outdoors. For beautifying the balconies or terraces which receive limited sunlight, miniatures and dwarf polyanthas should be selected. Bush roses may also be grown in pots for beautifying the compounds around the building and for display in rose shows.

Rose water is also an important commercial product obtained from rose petals. It is used as a perfume, in medicines and confectionery (Narayanswami and Biswas, 1957). It has the property of cooling the body and is often used in eye lotions and eye drops for its soothing qualities. It is also used in drinking water and sprinkled on the guests at weddings, feasts and other social functions (Pal, 1972). The roses used for preparing oil are also used for making rose water.

Rose oil is an important commercial product obtained from rose petals. Apart from sweet fragrance, it has medicinal property and is often used in Ayurveda (Rode and Ogale, 1984). Wei *et al.* (2019) revealed that rose oil might be manufactured through two pathways: direct synthesis in plastids and conversion from starch grains. Although starch grains performed as precursor andprovided energy during the synthesis process, the endoplamic reticulum primarily processed a fraction of osmiophilic oil for transport and reprocessing via vesicles of the golgi apparatus. Baydar and Baydar (2005) determined the effect of harvest date (May 24, June 1,8 and 15, 2002), fermentation duration (0, 12, 24 and 36 h at 25° C in sacks) and Tween 20 treatment (0, 1000, 2500 and 5000 ppm) on the essential oil content and composition of *Rosa damascene* petals. The highest oil content was found on May 24 harvest (0.04 per cent) and then a gradual decrease was observed upto last harvest date (0.032 per cent). The highest percentage of geraniol, nerol and phenyl alcohol were obtained from the petals harvested on May 24. However, the highest percentage of citronellol and linalool were obtained with the rose petals which were harvested on June 8 (Table 1). Moreover, highest essential oil content was found in the non-fermented petals. As fermentation duration increased, essential oil content gradually decreased. The most significant changes during the fermentation were observed in citronellol/geraniol content throughout fermentation. The highest oil content (0.045 per cent) was obtained when the distillation was treated with 2500 ppm of Tween 20. However, Tween 20 did not significantly influence the oil composition.

Rose petals are also preserved for direct consumption, by making gulkand which is prepared by pounding equal proportions of petals and white sugar. It is considered both as tonic and laxative (Pal, 1972; Rode and Ogale, 1984). *Rosa damascena, R. chinensis, R. gallica, R. pomifera* and some other scented roses, *e.g.,* Edouard are used for preparing *gulkand*.

Dried rose petals are known as pankhuri which is occasionally used for preparing sweetened cold drinks. Dried rose petals with better colour retention, flower diameter, shape and colour intactness, dry flower longevity and overall flower quality were obtained when cv. Jet Set flowers were harvested at 50 per cent open stage and dried in microwave oven embedded drying (Hemant *et al.*, 2016). A low cost high walk-in-tunnel technology has been developed for drying rose petals with better colour, fragrance and no contamination (Singh and Rohilla, 2019).

Table 1: Effect of Harvest Dates on Essential Oil Content and Composition of *Rosa damascena**

Harvest Dates	*Oil Content (Per cent)*	*Essential Oil Composition (Per cent)*						
		Citro-nellol	*Geraniol*	*Nerol*	*Linalool*	*Phenyl Ethyl Alcohol*	*Mono-terpene Alcohols*	*Citro-nellol/ Geraniol Ratio*
May 24	0.040 a	23.92 d	44.44 a	14.55 a	1.24	1.80 a	85.96 a	0.53 d
June 1	0.035 ab	25.79 c	43.66 a	11.23 b	1.15	1.34 b	83.18 b	0.58 c
June 8	0.033 ab	33.94 a	41.36 b	8.61 c	1.55	1.28 b	86.76 a	0.81 a
June 15	0.032 b	31.02 b	41.69 b	9.11 c	0.99	1.32 b	84.15 ab	0.74 b
F- values	6.08*	165.12**	10.64**	208.35**	1.84	4.73*	4.42*	141.39**
CV (per cent)	7.14	2.17	1.86	2.98	2.98	13.49	1.59	2.79

* Baydar and Baydar (2005).

It is now a well known fact that rose hips are very good sources of ascorbic acid, the life giving vitamin C (Rode and Ogale, 1984). Pal (1972) reported that every 100 g of rose hip syrup contains 150 mg of ascorbic acid compared with only 50 mg present in fresh orange juice, 20 mg in tomatoes and 5 mg in apples. The ascorbic acid content, however, varies with species and some of the promising vitamin C bearing species is *R. rugosa, R. amblyotis, R. acicularis, R. davurica, R. pendulina, R. glauca* and *R. canina* (Gribovskaya *et al.*, 1978). Apart from vitamin C, the hips of *R. roxburghii, R. acicularis, R. rugosa, R. davurica* and some other species are found to contain vitamin A1, B2, K and E (Shnyakina and Malygina, 1975). Tejaswini and Prakash (2005) stated that *R. centifolia, R. gallica, R. macrophylla, R. gigantean, R. foetida, R. moschata, R. multiflora and R. webbiana* produced hips rich in vitamin C.

In Europe, roses are also used for making pot-pourri, conserves, rose vinegar and rose petal wine. Jams, jellies and syrups have been made for centuries in Bulgaria and exported. Rose jam aids digestion and has certain curative properties. In Czechoslovakia, fruits of wild roses are used for preparing a hot drink like tea and a popular wine. In America, bushes of *R. multiflora* are utilized as shock absorber or crash barrier (Pal, 1972).

3.0 Origin and History

It is said that rose came first and man afterwards. Fossils of the rose found in Oregon and Colorado (USA) are estimated to be more than thirty millon years old (Fairbrother, 1965). According to Klougart and Fairbrother (1966), the history of man and of the rose is linked together for about 5000 years. The rose was called 'Queen of Flowers' by Sappho, a poetess more than 2500 years ago. No other flower has such a remarkable written record.

Uses of Rose

In pot

In flower arrangement

Rose bed in a garden

Rose bed in a park

Commercial cultivation of rose

Rose in greenhouse

Uses of Rose

Beautiful roses as climbers

Rose in pergola

Rose oil

Gulkand from rose

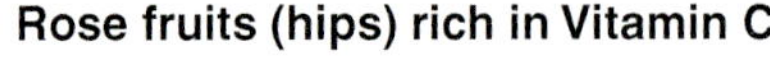

Rose fruits (hips) rich in Vitamin C

Rose pankhuri from petals

There is evidence that roses reached Rome before Christ and these were greatly admired by ancient Romans. It is said that Nero loved rose so much that he had them everywhere in his palace. Flower petals were used by the women in their baths and for the decoration of special beds, feasts and graves. Preparation of rose water and attar was also in practice even before the 17th century.

The earliest use of the rose in art is seen during the middle of the sixteenth century BC. During Minoan period (2800-1900BC) goldsmiths had used rose flowers and leaves in their designs for jewellery (Fairbrother, 1965).

About 120 known wild species, all indigenous to the temperate regions of the northern hemisphere, have been reported to be growing. Roses have not been found to be native to those countries lying on or south of Equator and have actually become known in the southern hemisphere only when introduced by man. Eleven species are reported to be growing wild in India. Three of these are endemic to India and four species have contributed to the evolution of the modern garden roses (Stewart and Brandis, 1974). The species growing wild in India are *Rosa brunonii* (Himalayan Musk rose). *R. eglanteria, R. foetida* (syn. *R. eglanteria*), *R. gigantea* (syn. *R. Odorata* var. *gigantea*) *R. involucrata* (syn. *R. clinophylla*), *R. leschenaultiana* (syn. *R. sempervirens*), *R. longicuspis, R. macrophylla, R. moschata* (Musk rose), *R. rubiginosa* (Sweet briar, Eglantine rose), *R. walpoleana, R. sericea* (Ladakh rose).

In absence of much of the reliable records during the early period of rose history, it is assumed that some of the important rose species were hybridized in nature and the present day improved forms have been evolved over centuries. Till the 19th century only four species of roses played a role in the development of varieties cultivated at that time. These are *Rosa gallica* (Red rose), *R. canina* (Dog rose), *R. moschata* (Musk rose) and *R. Phoenicia* (Phoenician rose). Theophrastus, however, made a mention of two species of roses, mainly *R. gallica and R. canina.*

Rosa gallica, the French rose came originally from Caucasus and the Middle-East and is an ancestor of Moss roses, the Centifolias (Cabbage rose) and Damasks. *Rosa centifolia* came to France in 1580 from Mediterranean. The Damask rose, *R. damascene* is a natural variation of *R. gallica* and early record was noted in 1000 BC (Bose and Mukherjee, 1977). *Rosa damascena,* in course of time, has been said to have produced a variant with a tendency to bloom in the autumn and summer and this variant came to be popularly called as the Autumn Damask (*R. damascene bifera* and *R. damascene semperflorens*).

Chinese roses were widely cultivated during the Han Dynasty (141-87 BC). In 1759, after Chinese roses (*Rosa chinensis*) were introduced into western countries, the modern roses appeared. A careful study of botanical fossils, ancient painted pottery with colors, frescos, paintings, porcelain and redwood furniture suggests that Chinese roses cultivated with characteristics of modern roses appeared about 1000 years ago (Wang, 2007).

The first Bourbon rose is believed to arise from a natural cross between the Parson's Pink China rose (*R. chinensis*) and the Autumn Damask rose (*R. damascenabifera*) while they were growing in the island of Bourbon (now Reunion) in the southern Indian Ocean during the year 1817. Parson's Pink China rose was crossed by an American grower in the early nineteenth century with White Musk rose (*R. moschata*) to obtain a pink double flowered rose of the climbling habit of Musk rose. A French nurseryman, Philippe Noisette, who stayed in the same town, *i.e.*, Charleston (South Carolina), raised several seedlings of these hybrid roses and marketed them as Noisettee roses (Gault and Synge, 1971).

In the early nineteenth century itself, one or two hybrids closely allied to *R. chinensis,* due to their recurrent flowering habit, were introduced to Europe from China. These were actually hybrids between *R. odorata* and *R. gigantea* and these gave rise to the popular Tea roses. These were of dainty form and possessed very delicate colour. These Tea roses, when crossed with the Hybrid Perpetuals, either deliberately or naturally, gave rise to the Hybrid Tea class. La France was the Hybrid Tea rose raised by M. Guillot of France and was developed out of open pollination (Ghosh, 1976; Bose and Mukjerjee, 1977; Misra, 1979). Though this first Hybrid Tea was introduced in 1867, it was only during early seventies of the nineteenth century that these came to be recognized as a promising new race of roses.

Several new cultivars developed later from these groups as bud mutations. Paul Crampel is one of the examples having brilliant orange colour which was unknown in roses before. This mutant gave rise to orange-scarlet Floribunda roses like Golden Jewel, Heat Wave and Independence. Since then, the rose has undergone many changes in its form, colour, growth and flowering habits through selection, hybridization and mutations. As a result, a very large number of cultivars have developed throughout the world. However, out of about 120 species of roses distributed in the Old as well as the New World in both the hemispheres, only eight species have played a major role in the development of the modern garden rose (Pal, 1972). An exhaustive account of about 5000 different rose cultivars has been given by McFarland (1940) in his *Modern Roses*, which is revised with additions periodically. Thus the race for the development of new cultivars will continue till the civilization exists. Dutta and Singh (1999) stated that the distribution of phenolic compounds in fresh leaves of rose provided useful data for establishing relationship between different cultivars and their probable origin. Many Gallica varieties corresponding to the first allele flow recognized through higher heterozygosity were created by productive breeders using simple cross intraspecific pollinations between Gallica cultivars. The few varieties corresponding to the second flow were created by careful breeders who practised strict interspecific pollinations between Gallicas and Chinenses roses (Gardes *et al.*, 2005).

According to some historians, the shift from Ottoman community of rose to European culture occurred during the 18th century. For example, the roses grown in Istanbul palaces lost its reputation with the proliferation of European roses. However, Alp and Koyuncu (2015) reported some 6 old Ottoman rose cultivars in the gardens and old graveyards of Lake Van region in East Anatolia.

4.0 Distribution

Ancestors of modern roses are the wild roses that grow over wide areas in the world (Ghosh, 1976). There are about 150 recognized and described species of roses which range across the temperate regions of the northern hemisphere from the Pacific coast of North America, south to Mexico and eastwards through northern Europe and North Africa, through northern Asia where there is the largest representation, through Siberia to the Himalayas, China and Japan and south to the Philippines. The most tropical species is *R. clinophylla* from Bangladesh and North India but it is rarely seen in cultivation. Roses are not found wild in the southern hemisphere and there are no native ones in Australia, New Zealand and South America, although the introduced roses grow and flower very well in some of these countries, particularly in New Zealand (Gault and Synge, 1971).

Roses are being cultivated from ancient times in France, Cyprus, Greece, India, Iran, Italy, Morocco, USA (California) and Bulgaria for the production of perfume (Rai, 1985). Modern garden roses, developed from the wild species came from France. Later, England, Germany, Belgium, Luxemberg, Holland and other countries also became important rose producing countries (Ghosh, 1976). The other rose growing countries of the world are China, Japan, Myanmar and USA, *etc.* (Pal, 1972).

In India, several species are found growing wild mostly in the Himalayan ranges. The cultivation of roses was perhaps taken up during the Mughal period. Babar introduced the Persian or Damask rose into India via Port of Bussorah in the year 1526. The Edourd rose (*R. borboniana*) was introduced in 1840 during the British rule. These two roses *viz.*, Damask and Edourd are highly scented and are being extensively cultivated in this country. Their cultivation is concentrated at Aligarh, Kannauj, Ghazipur, Ballia and Jaunpur in Uttar Pradesh, at Haldighati in Rajasthan, near Chandigarh in Punjab and in Tamil Nadu in a total area of about 6000 hectares (Srivastava and Gupta, 1984; Rai, 1985). Rajendran *et al.* (2017) documented the unexploited potential of wild or local rose genotypes found in Western Ghats of Kerala which are hardy, fragrant and resistant to major pests and diseases attacking cultivated rose varieties. This untapped reservoir could be efficiently utilized in breeding programmes by employing proper conventional and molecular approaches to obtain productive hybrids with lesser time and effort.

In India, commercial cultivation of roses for cut flower trade has increased during last four decades. The major rose producing regions are Karnataka, Maharashtra, Punjab, Uttar Pradesh, Delhi and Chandigarh, while in Gujarat, Haryana, Himachal Pradesh, Madhya Pradesh, Rajasthan, Tamil Nadu and West Bengal they are grown to a limited extent.

5.0 Classification

The genus *Rosa*, according to Rehder (1940), the American taxonomist, contains about 120 species. The genus is divided into 4 subgenera, namely, *Eurosa, Platyrhodon, Hesperhodos* and *Hulthemia*, the first one being the major one, containing species believed to be associated with the origin of modern roses. The subgenus *Eurosa* is further subdivided into 10 sections, of which 7 sections with adnate stipules are more significant as far as garden roses are concerned. Out of these seven, two sections, Indicae and Synstylae are distinguished from the other five by their exerted styles, which are free in the former and fused in the latter. These two sections are diploid (2n =14) and contain species of importance to garden rose development. The short- styled sections include Pimpinellifoliae (diploid and tetraploid) which contains *Rosa spinosissima* as also *R. foetida,* an important garden rose ancestor and the Gallicanae (tetraploid) contains the important ancestral species *R. gallica* and the ancient hybrids which originated from it, namely the Damasks, the Centifolias, the Mosses and the Albas. The largest section Cinnamomeae (48 species) contains *R. rugosa* and *R. cinnamomea,* which have played only minor role in the development of modern roses. The genus *Rosa* contains about 150 species, only 7 have contributed to the development of modern cultivars resulting in a narrow genetic background (Ueda, 1994).

There are many classes of present day garden roses, the main classes being the following:

Hybrid Tea

Originally developed from crossing between Hybrid Perpetuals (HP) and Tea roses, the present day Hybrid Teas are, by far, the most popular rose type, bearing large, highly centred flowers. 'La France' created in 1867, is believed to be the first cultivar of the type.

Floribunda

Also called Hybrid Polyanthas, the Floribundas were developed in 1924 from a cross between a Hyrbid Tea and a Polyantha. They combine the beautiful forms of the Hybrid Teas with the perpetual flowering habit of the Polyanthas. Examples-Wekplapep (2000), Jumpin Jack (1998).

Hybrid Perpetual

They are the immediate forerunners of Hybrid Tea. These are considered to be the offsprings of *Rosa chinensis, R. gallica* and *R. centifolia*. They are not perpetual flowering types, as the name suggests, but may be called remontants, being the types which flower more than once in the season.The first Hybrid Perpetual developed was perhaps Princesse Helence. Other examples of this class are Frau Karl Druschki (1901), General Jacqueminot (1853), Hagh Dickson (1905), Mrs. John Laing (1887) and Paul Neyron (1869).

Teas

Also called 'Tea scented China roses', they derive their names from their distinct aroma, believed to be felt when a chest of tea leaves is opened. They are known to have originated from *Rosa chinensis* and *R. gigantea*. These are also the free flowering types. Some notable examples are Anna Olivier (1872), Lady Hillingdon (1910) and Mme Falcot (1858).

Grandiflora

Mainly obtained from crosses between Hybrid Tea and Floribunda types, this class of roses covers large flowered and clustered cultivars of fine form or the type which produces beautifully formed Hybrid Tea-like blooms in clusters. Perhaps the first Grandiflora developed was Buccaneer (1952). Other examples are June Bride (1957), Montezuma (1955) and Queen Elizabeth (1954), Jazor (2000), Wekdykstra (2000).

Polyantha

The dwarf small flowered Polyanthas were the forerunners of the popular large flowered Floribundas of today. Their ancestry includes crosses of *R. multiflora*, *R. wichuraiana*, and the Bengal hybrid *R. indica major* (*R. chinensis*). The first cultivar to be introduced in this class was La Paquerette (1875). Other important examples are Baby Faurax (1924), Cameo, Cecile Brunner (1881), Chatillion Rose, Echo (1914) and Yvonne Rabier (1910).

Miniature

These are the popular 'Baby Roses', with small leaves and flowers. They are hardy and are multiplied by cuttings as well as propagated on rootstocks. Those raised from cuttings are ideal for growing in pots. Some important cultivars are Baby Gold Star (1940), Baby Masquerade (1956), Cinderella (1952), Peon (1936) and Pixie (1940), Meiponal (1990), Meifruije (1998).

Climber and Rambler

These old climbers, with large clusters of small, single or double flowers mainly belong to two groups, *i.e.*, the Multiflora ramblers (from *Rosa multiflora*) and the Wichuraiana ramblers (from *Rosa wichuraiana*). The latter group is more common in the present days. Some outstanding examples of Multiflora ramblers are Blush Rambler (1903), Crimson Rambler (1890), Tausendschon (1906) and Veilchenblau (1909). Among the Wichuraiana ramblers, the well known examples are Alberic Barbier (1900), American Pillar (1902), Dorothy Perkins (1901), Dr. W. van Fleet (1910) and Excelsa (1909).

China Rose

The China rose (*R. chinensis*), more than any other, is responsible for nearly all the present day popular roses and bears red to nearly white flowers in small clusters. Also called the Bengal rose or the Monthly rose, this group includes perpetually flowering types. Some of the famous hybrid Chinas include Comtesse du Cayla (1902), CramoisiSuperieur (1832) and Gruss and Teplitz (1897). The so-called Green rose (*R. chinensis viridiflora*) is included in this class.

Damask Rose

These belong mainly to the species *Rosa damascena* and are believed to have originated from the cross between *R. phoenicia* and *R. gallica*, and bear clusters of very fragrant pale pink to red, double-petalled flowers. This group includes some of the highly scented cultivars like Celsiana, Hebe's Lip (1921) and Madame Hardy (1832). A molecular study by Morteza (2014) showed that accessions of *R. damascene* from Bulgaria, Iran, and India and old European Damask rose cultivars (cvs.) have identical microsatellite profiles, suggesting a common origin. Recent studies on the molecular analysis of genetic diversity of *R. damascena* Mill. with randomly amplified polymorphic DNA (RAPD), amplified fragment length polymorphism (AFLP), and simple sequence repeat (SSR) markers did not show any polymorphism among *R.damascena* plants from various plantations in Turkey and Bulgaria. *R. damascene* can now is thought that the origin and center of diversity of Damask roses can be found in this region. In Iran, the cultivation and consumption of Damask roses has a long history. Crude distillation of roses was probably developed in Persia in the late 7th century A.D. It was found in the wild in Morocco, Andalusia, the Middle East, and the Caucasus. As Damask roses were originally introduced from the Middle East into Western Europe, it is thought that the origin and center of diversity of Damask roses can be found in this region. In Iran, the cultivation and consumption of Damask roses has a long history. Crude distillation of roses was probably developed in Persia in the late 7th century A.D.

Bourbon Rose

They have originated from a natural cross between the China rose and the Damask rose in the remote island of Reunion, in the Indian Ocean. They are so named because the island was then called the Isle de Bourbon. The Bourbons have large petals, wide and silky, and are very scented. This group includes Rose Edouard (known in India for many years), Boule de neige (1867), Bourbon Queen (1835), Gloire des Rosomanes (1825), Mme Pierre Oger (1878) and Souvenir de la Malmaison (1843).

Cabbage Rose

These belong to the species *R. centifolia* and are also called the Provence rose. They bear large, solitary, very fragrant pink flowers. Some examples are De Meaux and Chapeau de Napoleon (1827).

Moss Rose

These actually belong to the Cabbage rose class and arose as sports of the Centifolias. Their charming character is derived from little glandular growths, like modified bristles, growing closely like moss, all over the seed pod and sepals. Popular cultivars in this class include Blanche Moreau (1880), Old Pink Moss and Laneii (1845).

French Rose

Also known as the Gallica roses, having developed from *R. gallica*, these are perhaps the oldest type of cultivated roses known. They include types with varied colours (generally pink or crimson), good fragrance and are also known to have medicinal properties. Some notable examples are Belle de Crecy, Cardinal de Richelieu (1840), Gloire de France (1819), Rosa Mundi (1950) and Tuscany (the velvet rose, described as early as 1597).

Alba

Along with the Gallicas and the Damasks, the Albas constitute the truly old garden roses. These are believed to have originated from cross between *R. corymbifera* and *R. gallica*. These are generally very hardy and bear white or pinkish flowers, with delightful fragrance. Some examples are Celestial, Great Maiden's Blush, MmeLegras de Germain (1840) and Queen of Denmark.

Musk Rose

Rosa moschata is the Musk rose, deriving its name from the musky fragrance of the flowers. They make good shrub roses, as well as pillar roses. Some well known hybrid Musks are Eva (1933), Felicia (1928), Moonlight (1913), Penelope (1924), Prosperity (1919) and Vanity (1920).

Noisette Rose

These are considered to have originated from cross between *R. chinensis* and *R. moschata* or is a hybrid of the China and Musk roses. Some of the popular climbers like Lamarque (1830), Mme Alfred Carriere (1879) and Marechal Niel (1864) are included in this class.

Rugosa

Rosa rugosa contains some of the very hardy types, which do well on the seashore and are very vigorous and resistant to cold. Because of their highly thorny nature, they make good hedges besides being good shrubs. These generally bear large red or white flowers. Important examples are Blanc Double de Coubert (1892), Frau Dagmar Hartopp, Pink Grootendorst (1923), Roseraie de I'Hay (1902) and Scabrosa. The roots of this species were found to be the potential source of phytochemicals with better antioxidant activity and anti-microbial properties (Olech *et al.*, 2015).

Austrian Briar

These belong to *R. foetida* (also called *R. lutea*) and are considered to be the main source of yellows in the modern rose. The notable examples are Austrian Copper (*R. foetida bicolor*) and Persian Yellow (*R. foetidapersiana*).

6.0 Species and Cultivars

6.1.0 Species

The number of species under the genus *Rosa*, recorded in different literatures, varies from 120 to 200. Bentham and Hooker, however, estimated 30 wild species. Based on ancient descriptions available, Genders (1965) mentioned 200 species but later clarified that there were only 40 distinct species of roses. Gault and Synge (1971) reported that there are 150 recognized and described species of roses while Ghosh (1976) suggested the number of true species to be 180. According to Rehder's classification, generally accepted one, there are about 120 species of roses. The most recent estimates, however, indicate that the genus *Rosa* contains about 200 species distributed throughout the world (Srivastava and Gupta, 1984). Some important rose species are described below.

Rosa alba (probably *R. corymbifera* x *R. gallica*): White Rose of York

An erect bush with glaucous foliage, flowers white, semi-double, fragrant and usually borne in clusters. Hips bright red and oblong, Originated perhaps in Crimea in the USSR (Gault and Synge, 1971); under cultivation before 1597.

R. arvensis: Field Rose

Plants deciduous with creeping habit. Flowers white, scentless and about 5 cm in diameter.

R. banksiae: Banksian Rose

A vigorous evergreen climber reaching up to about 9 metres. Stems almost thornless with yellow-green foliage. Flowers white or pale yellow, single or double, small (about 2.5 cm diameter) but borne in large clusters. Originated in W. China (Gault and Synge, 1971) and cultivated since 1796.

R. bourboniana (R. chinensis x R. damascena semperflorens): Bourbon Rose

A vigorous shrub with erect shoots. Stem dark green with hooked prickles. Floriferous over a long season. Flowers double, deep rosy-red, scented, about 7.5 cm in diameter, and borne singly or in small clusters. Originated in the Isle of Reunion (then known as Bourbon) in 1817.

R. canina: Dog Rose

A vigorous and very prickly shrub having stout stems. Foliage toothed all round margin. Flowers single, white or pink, scented, 3.8 to 5.1 cm in diameter and borne in clusters. Hips oval, scarlet and conspicuous. Chiefly used as rootstock for budding or grafting other roses (Hartmann and Kester, 1972). Widespread throughout Britain, North Europe as well as West Asia.

R. centifolia: Rose des Peintres, Provence or Cabbage Rose

Considered to be a complex hybrid possibly containing in its parentage, the 4 species, *R. gallica, R. phoenicia, R. moschata* and *R. canina* (Wheather, 1903; Stuart, 1913; Hurst, 1929). The typical *R. centifolia* is a loose-growing shrub of medium vigour only. Flowers large, fully double with overlapping petals, deep pink with a slight purplish hue towards centre, rather globular, strongly fragrant, solitary, with long projecting sepals. Under cultivation since 1596.

R. chinensis syn. R. indica: China Rose

A large climbing evergreen shrub armed with brown, scattered and hooked prickles which are uniform and stout. Leaflets are rather small, ovate to oblong. Flowers single blush-pink, pink or crimson, rarely white in colour, nonfragrant, about 5 cm across but larger in cultivars and hybrids. Originated in China and under cultivation since 1759.

R. chinensis semperflorens: Crimson China Rose

A climbing shrub with evergreen foliage. Flowers crimson or deep pink and usually solitary. Under cultivation since 1733.

R. chinensis viridiflora, Probably R. chinensis (sport): Green Rose

Shrub of medium vigour with blue-green foliage. Flowers fully double with green petals, twisted with some purplish-brown colour towards base and centre, opening jade-green, then turning to apple-green and borne in clusters of medium size. Originated around 1855.

R. clinophylla syn. *R. involucrata*: Wild Rose of Bengal

An erect or semi-climbing compact shrub with blackish bark. Leaves composed of 5-9 leaflets. Flowers white, appear solitary or in small clusters. Fruits roundish, hairy and pale. Wild rose of the tropical parts of India, occurs especially in the eastern Himalayas in North Bengal, Sibsagar and Goalpara districts of Assam, the Garo Hills and Sylhet. In the plains of Bengal, it is found near streams and marshy places (Pal, 1972).

R. damascena (*R. gallica* x *R. phoenicia*; *R. gallica* x *R. moschata* or *R. canina*): Damask Rose

R. damascena is regarded to be of hybrid origin. Genetic analysis has shown that the summer Damask roses were derived from cross between *R. gallica* and *R. phoenicia*, a strong climber with white flowers in clusters from Turkey and Syria, while the Autumn Damask, *R. bifera* came from *R. gallica* crossed with *R. moschata* or *R. canina* (Narayanswami and Biswas, 1957).

The original *R. damascena* is an old vigorous shrub with exquisite fragrance. Branches very prickly with hooked spines as well as prickly bristles. Flowers borne in large clusters, semi-double, sweet scented, variable in colour, blush-white to deep pink, the pedicels being long and covered with small prickles. Original home unknown. Introduced to Europe from Damascus in Asia Minor during the 16th century.

R. eglanteria syn. *R. rubiginosa*: Sweet Briar

A vigorous shrub with dense prickly branches and small leaflets and scented foliage. Flowers single, in clusters, blush-pink to brighter pink, 3.8-5.1 cm across. Hips bright red, rounded with persistent sepals. Originated in North Europe including Britain, under cultivation since 1551 (Vecera, 1971).

R. foetida syn. *R. lutea*, *R. eglanteria*: Austrian Briar Rose

An erect shrub with prickly stems having delicate foliage. Flowers single, about 5.0-6.5 cm across, bright golden-yellow, with an unpleasant aroma, solitary or occasionally in clusters of 2 to 3. Native to India, Germany, Italy and South of France (Narayanswami and Biswas, 1957). It occurs in the dry inner Himalayas, in Kashmir from Kishtwar westwards in Lahoul and Ladakh at elevations of 2400 to 3300 metres (Rao, 1974). It has given rise to many modern yellow roses.

R. gallica syn. *R. rubra*: French Rose

Shrub rose with stiff erect stem. Distinguished by comparatively thornlessness but an abundance of small prickles. Foliage dark green. Flowers single, purplish-crimson, in small clusters, about 6.5 cm across. Hips subglobose, dark red, persistent. Prolific seed bearer. Many cultivars have been raised from seeds in France. Native to central and Southeast Europe, in cultivation prior to 1542. *R. gallica* var. Maxima, var. officinalis, var. Camaieux, var. Belle de Crecy, var. Georges Vibret are the surviving representatives of old gallica rose (Narayanswami and Biswas, 1957).

***R. gigantea* syn. *R. odorata* var. *gigantea*: Manipur Tea Rose**

A very vigorous climber with thick, hooked prickles on the glabrous stem. Foliage dark glossy-green, partially evergreen, comprises 5 to 7 elliptic to oblong-ovate waxy-textured leaflets. Flowers large, single, 7.5 to 12.5 cm across, white or pale yellow, fragrant, borne singly. Common in Manipur, Nagaland and Sikkim at an altitude of about 1800 metres. Also native to western China and Myanmar. An important ancestor of the Tea, Hybrid Tea and Noisette roses.

***R. hugonis*: Father Hugo Rose, Golden Rose of China**

Vigorous shrub, having delicate foliage with up to 11 leaflets. Flowers single, of medium size, saucer-shaped, 4.0-5.0 cm across, pale creamy-yellow with good boss of yellow stamens in the centre, numerous on arching branches. Subject to die back. Originated in western China.

***R. kordesii* (*R. rugosa* x *R. wichuraiana*)**

Shrub with arching branches producing semi-double, cup-shaped and deep pink flowers. Seldom grown but is of much importance as a parent.

***R. laevigata*: Cherokee Rose**

Very vigorous climber or shrub of loose growth, almost evergreen. Stems very prickly with hooked spines. Foliage very fine, glossy, bright green with red petioles, almost evergreen. Flowers single, pure white petals which do not overlap, up to 10.2 cm across. Originated in China.

***R. moschata*: Musk Rose**

A vigorous climber with reddish, spa liage dark green, with purplish-flush on first opening and up to 4 to 7 leaflets, greyish below. Flowers white, usually semi-double but sometimes single, about 5.0 cm across, with musk fragrance, in terminal clusters. Hips ovate, orange-red or dark brown. Native to Middle East and the Himalayas (Fairbrother, 1965), found at altitudes ranging from 1000 to 4700 metres extending to Europe (Pal, 1972). The persian Musk, one of the forms of R. moschata is very common in India (Steen, 1967). Introduced into England in 1599 (Wheather, 1903).

R. multiflora

A deciduous shrub with vigorous climbing branches. Foliage with up to 9 leaflets. Flowers single, white and golden centre of stamens, each about 2.5 cm across, borne in clusters, scented. Hips small, red, oval. Originated in North China, Korea and Japan. Introduced into India in 1872 or even earlier (Narayanswami and Biswas, 1957).

***R. noisettiana* (probably *R. chinensis* x *R. moschata*)**

Vigorous with climbing habit. Flowers white, pink, red or yellow, borne in clusters. Originated in about 1812.

Classification of Modern Roses

Hybrid Tea

Floribunda

Grandiflora

Hybrid Perpetual

Tea Rose

Damask Rose

Bourbon Rose

Cabbage Rose

Moss Rose

French Rose

Musk Rose

Noisette Rose

Species of Rose (*Rosa*)

Rosa alba | ***Rosa foetida*** | ***Rosa chinensis***

Rosa canina | ***Rosa centifolia*** | ***Rosa gallica***

Rosa multiflora | ***Rosa rugosa*** | ***Rosa gigantea***

Rosa moschata | ***Rosa multiflora*** | ***Rosa wichuraiana***

R. odorata (_R. chinensis_ x _R. gigantea_)

Vigorous climber with purplish red, prickly stems. Flowers slightly drooping, large, about 5-10 cm in diameter, double, pink, crimson or pale yellow in colour, scented, borne singly or 2-3 together on short stems. Used as rootstock for budding other roses (Hartmann and Kester, 1972).

R. persica syn. _berberifolia_: Rose of Persia

Dwarf shrub with silvery-grey foliage. Flowers deep golden-yellow, single, about 4.0 cm across. Hips small bristly. Difficult to grow (Narayanswami and Biswas, 1957). Native to Iran, Afghanistan and USSR.

R. rugosa: Ramanas Rose, Rugosa Rose

A vigorous erect shrub distinguished by strong, sturdy growth and bright apple-green foliage on very prickly stems. Flowers in small clusters, single, large, purplish rose to violet-carmine, about 9.0-10.0 cm across, slightly scented, variable in colour. Hips large, rounded, orange-scarlet and very decorative. Originated in China, Japan and Korea. Some of its forms are used as roostocks (Hartmann and Kester, 1972).

R. wichuraiana: Memorial Rose

A vigorous rambler producing single flower, white with yellow centre, scented, in large clusters. Hips small, ovoid, dark red. Originated in Japan, East China, Korea and Taiwan. Introduced in 1891. Valuable as a parent of garden ramblers.

6.2.0 Cultivars

Presently, there are about 20,000 cultivars of roses, differing widely in form, shape, size, colour, fragrance and flowering habit. Some beautiful flowering cultivars under different classes grown widely are mentioned below.

The cultivars are listed according to the class of roses and most of them are registered with the National Rose Registration Centres as well as the International Rose Registration Authority.

6.2.1 Exotic cultivars

Angelina-Red	Fresh moments	Pink Floyd
Albotrog White	Freeze	Pistache
African Dawn	Heide	Rose
Aque Bella	Hermosa	Red Bentley
Anouk	HeavenHippo	Red Naomi
Annakariana	Hosanna	Royal pumpkin
Aque	Hessenrose (Kantiki)	Rose cantata
Aqua Giri	Hot Merengue	Sambuca
ArmandoAspen	Lady Susanna	Safi

Aquazur	Kindness	Russian Beauty
Avaknche	Lancer	Shukrani
Audio	Leonessa	Sauvignan
Baby Love	Layla	Sanaa
Big Apple	Luciana	Soulmate
Blondine	Magadi	Sorbet Avalanche
Bizarre	Luxuria	Single Malt
Bombero	Magic avalanche	Summer Idde
Capriccio	Marathon	Summer dance
Bonbons	Marie jane	Souza
Carp Diem	Marie sun	Suraya
Cool water	Maroussia	Sweet Avalanche
Cataline	Mariyo	Taboo
Copper	Mayra's rose pink	Sweet Viviane
Cosmo	Matchpoint	Suri
Countdown	Martinique	Tibet
Dahlinova Hypontice	Mika	Test rose franko
Rose Bicolour	Mi Amor	Tacazzi
Daniel Ost	Memory Lane	Tinto
Dark vampire	Morning Dew	Tinker bell
Desert Pvcatpm	Muscat	Topsun
Ddcevita	Miss Piggy	Tribute
Downtown	Milshake	Touchee
Ebony	Natron	Trixx
Dolcetto	Myrna	Twiga
Electra	Musette	Tropea
Eros	Oh La La	Vampire
Fiha	Nordia	Vigro
Formula one	Night Queen	Video
Fortune rose	Orange Cantata	Viviane
Glow	Opium	Voodoo
	Olij rose variety test	
	Pandora	
	Orange juice	
	Outlaw	
	Pleasure	

6.2.2 Indian Varieties

Indian varieties released by public Institutes

ICAR-IIHR, Bengaluru

Dr. G. S. Randhawa, Kiran, Nishkant, Arka Kiran, Arka Nishkant, Arka Parimala, Arka Swadesh, Arka Ivory, Arka Pride, Arka Sukanya, Arka Kinnari, Arka Savi, Arka Sharmeeli, Arka Sinchana, Abhisarika, Akash Sundari.

ICAR-IARI, New Delhi

Hybrid Tea

Aravalli Princess, Aruna, Ashirwad, Anurag, Apsara, Arjun, Bhim, Chambe di Kali, Charugandha, Chitra, Chitralekha, Chitwan, Dark Boy, Delhi Apricot, Delhi Debutante, Delhi Pastel, Delhi Sunshine, Dil-Ki-Rani, Dilruba, Diva Swapna, Dr Benjamin Pal, Dr Bharat Ram, Dr Homi Bhabha, Dr R.R. Pal, Dr B.P. Pal,Dr M.S. Randhawa, Dulhan, Eastern Princess, Golconda, Gulbadan, Ganga, Golden Afternoon, Gulzar, Hans, Haseena, Homage, Indian Princess, Jawahar, Jawani, Kanakangi, Kulu Belle, Kamla Devi Chattopadhyay, Lal Makhmal, Lalima, Madhumati, Madhushala, Meghdoot, Mrs. K.B. Sharma, Madhosh, Maharani, Macia, Mechak, Mother Teresa, Mridula, Mrinalini, Nazneen, Nayika, Nandini, Nehru Centenary, Nishada, Nurjehan, PahadiDhun, Pale Hands, Pat Rani, Pink Sport of Montezuma, Pusa Christina, Poornima, Preyasi, Priyadarshini, Pusa Sonia, Pusa Sonora, Pusa Christina, Pusa Mohit, Pusa Mansij, Raat-Ki-Rani, Raja Surendra Singh of Nalagarh, Raktima, Rampa Pal, Rosy Evening, Scented Bowl, Shanti Pal, Sharmili, Sweet Innocence, Sandeepini, Rajhans, Raj Kumari, Raktagandha, Raktima, Rangshala, Ranjana, Ratnaar, Shreyasi, Sir C.V. Raman, Soma, Sugandhini, Sujata, Surabhi, Surekha, Surkhab, Uttam, Uma Rao,Vasant, White Nun.

Floribundas

Akash Nartaki, Azzez, Arunima, Banjaran, Belle of Punjab, Celestial Star, Chamba, Princess, Chingari, Chandrama, Chitchor, Deepak, Delhi Daintiness, Delhi Maid, Delhi Pink Powder puff, Delhi Princess, Delhi Rosette, Delhi Sherbet, Deepika, Deepshikha, Delhi Brightness, Dr.S.S.Bhatnagar, Fugitive, Himangini, Jantar Mantar, Kumkum, Kavita, Lahar, Loree, Manmatha, Madhura, Manasi, Mohini, Nutkhut, Navneet, Nav-Sadabahar, Neelambari, Orange Cup, Punch, Parwana, Paharan, Prema, Pusa Abhishake, Pusa Komal, Pusa Barahmasi, Pusa Manhar, Pusa Maskan, Pusa Pitambar, Pusa Virangana, Pusa Shatabdi, Pusa Urmil, Pusa Ranjana, Ragini, Rajbala, Rose Sherbet, Rupali, Sailoz Mookherjea, Sandhya Bela, Stanza, Suhashini, Suryakiran, Suryodaya, Sadabahar, Saroja, Shabnam, Shola, Shringar, Sindoor, Suchitra, Suryakiran, Temple Flame,Tarang, Usha.

Polyanthus: Swati

Minatures: Delhi Starlet

Climbers and ramblers: Climbing Sadabahar, Delhi White Pearl

HRS, TNAU, Yercaud: YCD-1 Rose, YCD-2 Rose, YCD-3 Rose

Cultivars of Hybrid Tea Rose (HT)

Rosa **HT 'Alecs Red'** ***Rosa*** **HT 'Ambassador'** ***Rosa*** **HT 'Arlene'**

Rosa **HT 'Blue Sky'** ***Rosa*** **HT 'Black Berry Nip'** ***Rosa*** **HT 'Caribbean'**

Rosa **HT 'Confidence'** ***Rosa*** **HT 'Double Delight'** ***Rosa*** **HT 'First Prize'**

Rosa **HT 'Hondo'** ***Rosa*** **HT 'Marina Marini'** ***Rosa*** **HT 'Paradise'**

Cultivars of Hybrid Tea Rose (HT)

Rosa **HT 'Sentimental'** ***Rosa*** **HT 'St. Patrick'** ***Rosa*** **HT 'Olymploid'**

Rosa **HT 'Tineke'** ***Rosa*** **HT 'Trade Wind'** ***Rosa*** **HT 'Tropical Sunset'**

Cultivars of Floribunda Rose (Flori)

Rosa **Flori. 'Angelica'** ***Rosa*** **Flori. 'Bridal Pink'** ***Rosa*** **Flori. 'Cassa Nova'**

Rosa **Flori. 'Duncans'** ***Rosa*** **Flori. 'Gold Reef'** ***Rosa*** **Flori. 'Herero'**

Cultivars of Floribunda Rose (Flori)

Rosa **Flori. 'Ivory Beauty'**

Rosa **Flori. 'Jane Burner'**

Rosa **Flori. 'Meli Melo'**

Rosa **Flori. ' Orange Sensation'**

Rosa **Flori. 'Shocking Blue'**

Rosa **Flori. 'Ruby Anniversary'**

Cultivars of Grandiflora Rose (GF)

Rosa **GF. 'Boulies Dream'**

Rosa **GF. 'Cardinal Song'**

Rosa **GF. 'Cherry Parfait'**

Rosa GF. 'Earth Song' *Rosa* GF. 'Honey Dijon' *Rosa* GF. 'Maria Shiver'

Rosa GF. 'Octoberfest' *Rosa* GF. 'Sonia Supreme' *Rosa* GF. 'Queen Elizabeth'

Cultivars of Polyantha Rose (Poly)

Rosa Poly. 'Margo Koster' *Rosa* Poly.'May Wonder' *Rosa* Poly. 'Mother's Day'

Rosa Poly. 'Yesterday' *Rosa* Poly. 'The Fairy' *Rosa* Poly. 'Sne Princess'

Cultivars of Miniature Rose (Min.)

Rosa **Min. 'Black Jade'** ***Rosa*** **Min. 'Fairy Dance'** ***Rosa*** **Min. 'Pink Carosel'**

Rosa **Min. 'Rise n Shine'** ***Rosa*** **Min. 'Tiddly Winks'** ***Rosa*** **Min. 'Zepheline'**

Cultivars of Climber Rose (Cl.)

Rosa **Cl. 'Blossom Time'** ***Rosa*** **Cl. 'Cocktail'** ***Rosa*** **Cl. 'Dynamite'**

Rosa **Cl. 'Iceberg'** ***Rosa*** **Cl. 'Yellow Blaze'** ***Rosa*** **Cl. 'Landora'**

Commercial Cultivars under Indian Polyhouse

Red: Amada, Bordeaux, Grand Amore, Master Class, Red, Ribbon, Respect, Red Corvette, Red Giant, Royal Class, Tajmahal

Yellow-Copper: Bella Donna, Chantelle, Charmant, Golden Gate, Gold Strike, Kerio, Maraia, Savitta, Sirocco, Svetlana

Light Pink: Cadillac, Cinderella, Dekora, Duet, La Belle, Sweet Avelanche, Titanic

Dark Pink: Boogy, Bugatti, Kingfisher

Orange: Kalahari

Bi Color: Ensemble, Rigoletto, Royal Circus

White: Avalanche, Sofie

7.0 Genetics and Breeding

7.1.0 Genetics

The present day garden roses are complex hybrids involving interspecific hybridization, polyploidy with high female and male sterility. Cytological abnormalities and heterozygous genotypes are perpetuated by vegetative propagation (Swarup *et al.*, 1973). The basic chromosome number of roses is 7. Classed under the name *acicularis* are plants with 14, 28, 42 and 56 chromosomes, those with the highest number (octaploid) are found in the most northern parts of Siberia and Alaska. Several important species of Asian origin are diploid, having 14 chromosomes. These include *Rosa moschata, R. gigantea, R. multiflora, R. wichuraiana* and *R. chinensis.* A number of Western species with which these Asian diploids crossed to yield several modern groups of roses are tetraploids with 28 chromosomes. These include *R. gallica, R. foetida* and their derivatives such as *R. damascene* and *R. centifolia*. The hybrids arising from the crosses between European and Asian parents were triploid having 21 chromosomes, 14 from the European and 7 from the Asian. Because of this imbalance, they were sterile but plants with 28 chromosomes arose spontaneously from them. They were fertile and gave rise to the Hybrid Perpetuals with further incorporation of some *R. damascene* genes. There are also examples where a cross between a diploid and a tetraploid does not result in triploidy and gives rise to the tetraploid state through an abnormal behaviour of the diploid parent which passes its entire complement of chromosomes to its sex cell in place of only half the number, which is the normal practice (Pal, 1972). Chromosome number of 102 rose cultivars was also determined by Shahareand Shastry (1963) who recorded that 4 cultivars were diploid (2n=14), another 4 cultivars triploid (2n=21) and the remainder tetraploid (2n=28). The first record of a heptaploid (2n=49) rose has been that of an artificial hybrid between subspecies *acicularis* in Eurasia and Sayi.

Crosses between a diploid, *Rosa rugosa* var. *typica* (E) and a tetraploid *R. spinosissima* var. *pimpinellifolia* (G) (*R. pimpinellifolia*) gave 20 triploid seedlings. Scanning electron microscopic studies showed malformed and mainly shrunken hybrid pollen grains with poorly developed exine pattern (Lata, 1983).

Debener *et al.* (2001) revealed that presence of double flowers, prickles and black spot resistance gene showed monogenic dominant inheritance whereas pink flower colour exhibited codominant and recurrent flowering as monogenic recessive inheritance. Moreover, the number of petals in double flowers inherited as quantitative trait. Hattendorf *et al.* (2004) detected and localized 3 dominant resistance genes against powdery mildew and black spot using molecular markers and some additional putative resistance genes by mapping rose resistance gene analogues (RGAs) onto rose chromosome map. Morphological characterization of rose population obtained from backcross of *Rosa* chinensis and WOB (hybrid of *Rosa wichuriana* x Old Bush) revealed that the qualitative trait, presence or absence of prickles followed monogenic inheritance whereas, the quantitative traits such as prickle density and leaflet number inherited in an quantitative manner (Shupert *et al.*, 2005). Jian *et al.* (2013) analyzed 6 species/varieties of Sect. Chinenses, 9 Chinese Old Roses and 5 modern roses through FISH using 45S rDNA. Results unraveled that species or varieties evolved from Sect. Chinenses were diploid except the triploid *R. odorata* and karyotype 2A grouped *R. chinensis* 'Tiebahong'. Three ploidy levels (2x, 3x, 4x) and 4 karyotypes (1A, 2A, 1B and 2B) were observed in Old Garden roses. All studied modern roses were tetraploid but had only 3 hybridization sites. Thus, Chinese Old Garden roses were more variable in chromosome number and structure than their progenitors. While modern cultivars were more homogenous due to repeated hybridization and inbreeding among the cultivars. Hence, it is necessary to introgress wild species and Old Garden roses into modern roses for improvement of colour, fragrance, tolerance to abiotic stress and resistance to pest and diseases.

7.1.1 Colour

Colour in roses, as in other flowers is governed by the expression of water soluble pigments called anthocyanidin. The three important anthocyanidins are pelargonidin present in orange-red to scarlet flowers, cyanidin present in crimson to bluish-red flowers and delphinidin present in blue and violet flowers. Besides these anthocyanidins, there are a number of other pigments such as flavonols and carotenoids for white, yellow or brown pigments in flowers. Variation in the colour and flavonoid pigment composition of the flowers in 4 source varieties and 38 gamma ray induced and spontaneous mutants were derived from them. Possible pathways of flavonoid pigment synthesis in roses were examined. The results suggested that the primary flavonol is quercetin and the primary anthocyanidin cyanidin, both being formed from common precursor (Zykov *et al.*, 1990). The petal flavonoids of 18 closely related cultivars of rose (*Rosa hybrida*) with almost similar flower colour were investigated. All were pigmented primarily by cyanidin 3,5-diglucoside (more than 95 per cent of total anthocyanins). Among flavonols, quercetin always exceeded 65 per cent. Using a chemometric approach with principal component and cluster analysis, it was possible to distinguish clearly each of the 18 cultivars on the basis of their flavonol glycoside patterns (revealed by high performance liquid chromatography (Biolley *et al.*, 1992). Prior to 1900, most of the garden roses were red (including pink), white and yellow in colour. The reds and pinks derived their colour from the presence of cyanidin, while the pale yellow colour was due to presence of various types of flavonols. The deep yellow colour in roses came from the crosses made in France with the bright yellow rose species *R. foetida* in 1930, the genes for the production of pelargonidin were inherited in roses. It was in early sixties that the work of Japanese scientist Arisumi (1963) showed the occurrence of delphinidin, the pigment for blue colour to be present in roses too. This has opened up the possibilities of obtaining a true blue rose. The pigments, delphinidin and myricetin which could form the basis of a true blue colour in roses are missing in rose cultivars such as Sterling Silver and Grey Pearl. It may not be possible to get true blue colour in roses till these pigments are incorporated by some means (Choudhury *et al.*, 1978). Holton and Tanaka (1994) explained the possibility of developing blue colour in rose, apparently unattainable by traditional breeding. It was suggested that the anthocyanin pathway, which produces blue flowers via delphinidin 3 glucoside in other species, was outlined. The introduction of the recently cloned flavonoid 3´, 5´-hydroxylase gene which converts dihydrokaempferol of dihydromyricetin (DHM) from petunia into pelargonidin or cyanidin producing rose cultivars should divert the anthocyanin biosynthetic pathway towards the production of delphinidin glucosides (DG) and hence the flower colourtowards blue, as the intermediate DHM can be converted to DG by rose petals. It might be possible to raise the pH of the vacuole by eliminating expression of the *Ph6* gene using antisense or co-suppression techniques, which could also increase blue colouration by affecting co-pigmented anthocyanins. Marshall *et al.* (1983) analyzed more than 1200 progeny from 47 families for anthocyanin pigments during a breeding programme for improving colour. Cyanin, peonin and pelargonin were found in 99, 52 and 31 per cent, respectively. Each pigment was highly heritable from the seed and/or pollen parent. All pigments, particularly cyanin and peonin showed quantitative inheritance. The inheritance of each of these colour pigments is reported to be mainly controlled by additive gene action (Vries *et al.*, 1980). The genetical studies on the inheritance of colour in roses have indicated that the dominant colour in rose is magenta, pink or tyrian-rose to orange-red. The pink colour is dominant over dark red, orange-yellow, white and scarlet. Deep yellow colour is recessive to light yellow and white is recessive to cream and light yellow. The greenish-white colour of cv. Message was found to be dominant over the pure white of Virgo and the crosses among whites or yellows produced hybrids having flower colours like those of the parents only. Crosses between parents having multicoloured or bicoloured flowers produced a range of variability (Lata, 1971; Swarup *et al.*, 1973).

7.1.2 Fragrance

The fragrance in rose is due to the presence of various substances. The composition of the essential oils of a cabbage rose is given below (Table 2).

Table 2: Constituents of Rose Scent*

Substance	Scent	Proportion	Also found in
Phenylethyl alcohol (soluble in water)	Rose	50 per cent	Hyacinth, Orange blossom
Citronellol (rhodinol)	Rose	30 per cent	Geranium, Eucalyptus
Geraniol	Rose	10 per cent	Geranium, Jasmine, Orange blossom, Wallflower, Green tea
Nerol	Rose	3 per cent	Cyclamen, Orange blossom, Tuberose, Wallflower
Nonylaldehyde	Rose	–	Ginger, Cinnamon, Orris root
Eugenol and Methyleu-genol	Clove	1 per cent	Cloves, Bayleaves, Hyacinth Parma, Violet, Tuberose
Citral	Lemon	-	Lemon peel, Verbena
Farnesol	Lily of the valley	-	Cyclamen, Jasmine, Lily of Valley, Mignonette
Linalool	Lily of the valley	-	Jonquil, Lavender, Orange blossom, Wallflower
Carvone	Caraway seeds	-	Caraway, Mint
Menthol	Mint	-	Mint, Pennyroyal

* Young (1971).

Though the proportions are approximate and may vary from sample to sample, the constituents give an idea about the type of fragrance present in a particular rose. Differences in fragrance chemistry in *Rosa rugosa* was investigated by Dobson *et al.* (1990). Fragrances (volatiles) from whole flowers, petals, sepals plus gynoecium, anthers and pollen were collected by heads space adsorption methods and analyzed by GC-MS. Each flower part showed a distinctive volatile profile. Petal volatiles, dominated by terpenoid and benzenoid alcohols, contributed most to the whole-flower fragrance. Sepal odours contained mainly sesquiterpenes, together with several compounds found in the petals. The major volatiles in the androecium were more diverse and overall were different from those in the pericarp. Empty anthers shared a high proportion of their volatile while with pollen. Pollen odour appeared to be self-contained, showing only minor adsorption of volatiles from sepals and petals compounds typical of the androecium were present as significant components (though in small amounts) of the whole-flower fragrance, where they may well function as signals to pollen-seeking insects. Brunke *et al.* (1992) analyzed volatile of 4 rose cultivars.The composition of the headspace of the English rose Othello (*Rosa centifolia*-based hybrid) was comparable to that of rose oil obtained from *R. damascena.* In contrast, Duchesse de Montebello (*R. alba* type) contained high concentrations of phenylethyl alcohol and phenylethyl acetate. Lichtkonigin Lucia, hybrid rose, was characterized by considerable amounts of green notes. Queen Elizabeth (Floribunda, grandiflora rose) showed a completely different pattern with 3,5-dimethoxytoluene, the theaspiranes and ionone derivatives as the main components. Moshe *et al.* (2004) analyzed two different rose cultivars, 'Fragrant Cloud' (very fragrant) and 'Golden

Gate' (non-fragant) by headspace technique during flower opening. Results showed that the volatiles monoterpene alcohols, acetates and terpene hydrocarbons were detected in 'Fragrant Cloud', while orcinol dimethyl ether was found in 'Golden Gate' which is scentless to human nose but can be perceived by honeybees. In both the cultivars, the level of volatiles increased during flower development, while the ratio of different major volatiles remained constant. Mitachi *et al.* (2005) designed an odour sensor using quartz crystal resonator (QCR) coated with organic sensory films and calcium chloride as desiccant to elimate moisture influences and use in natural environments. Aroma of roses was successfully detected of which White Christmas, Star Light and Catherine Deneuve indicated similar pattern which is distinct from the aroma pattern of Snow White, First Love and Love. It was also concluded that odour similarity has no correlation with flower colour of roses. Debener *et al.* (2010) revealed that double flowers were dependent on the direction of hybridization and showed gametophytic self incompatibility indicated by the segration of alleles resulting from the cross of diploid *R. multiflora* hybrids. Moreover, the floral scent was found to be complex due to involvement of larger number of secondary metabolites and QTLs contributing to production of 2-phenyl ethanol, gerniol, nerol, citronellol and their acetates were characterized.

Baydar *et al.* (2016) observed the variability of different floral scent molecules such as phenyl ethyl alcohol (23.26-74.54 per cent), citronellol (5.57-31.59 per cent) and geraniol (3.09-26.93 per cent) among the seed-propagated plants of *Rosa damascena.* Xiao *et al.* (2019) evaluated the impact of esters on the perception of floral aroma in rose essential oil using 10 alcohols and 9 esters. Results highlighted 5 esters: ethyl octanoate, ethyl tetradecanoate, citronellyl acetate, geranyl acetate and 2-phenethyl acetate to have prominent roles in oil perception. The 'olfactory threshold' (OT) and floral reconstitution of 5 esters as well as their mixtures were analyzed. It was obvious from the results that ethyl octanoate, ethyl tetradecanoateand citronellyl acetate added overall aroma whereas geranyl acetate masked overall aroma perception in a model floral mixture.

The inheritance of fragrance did not follow a definite pattern as its expression is governed by the ultimate effect of many genes. Therefore, considerable decrease or increase in intensity of fragrance was noticed in some of the hybrid flowers (Klemenko, 1969; Lata, 1971; Swarup *et al.*, 1973). The presence of fragrance in the blooms of hybrid progenies in the crosses between non-fragrant parents suggests the presence of complementary genes for fragrance in the parents. It has been suggested by Swarup *et al.* (1973) that in incorporating fragrance into the hybrid seedlings, both female and male parents used must be essentially fragrant.

Chen *et al.* (2015) analyzed the endogenous levels and emission of 2-phenylethanol (2PE) and 3,5-dimethoxy toluene (DMT) during flower opening along with the expression of transcript levels of key genes: PAR and OOMT2. Results revealed that emission of the two scent compounds and the relative expression of the two key genes increased during floral development at fully opened stage. Also, PAR was abundant during senescence stage. Thus, expression levels of the key genes and emission of volatiles were positively correlated and showed tissue- and development- specific patterns.

Fragrance is lost in present modern roses as breeders concentrated much on the development of varieties with novelcolour and increased vase life. Thus, Rasouli *et al.* (2018) made an attempt to investigate the physiological, phytochemical and molecular analysis of colour and scent of 4 landraces (Ghalhar, Lavasanat, Kamoor, Minab) of *Rosa damascena* during different flower development stages (A, B, C).The relative expression of genes *GGPPS* (geranyl geranyl diphosphate synthase), *DXR* and *CCD1* were characterized during flower developmental stages and was found that overall expression of *GGPPS* gene was less compared to other two genes. Further study unraveled that damask roses with light coloured flowers accumulated more carotenoid and less anthocyanin pigments. Also, the major volatile compounds citronellol, geraniol, neral, linalyl acetate and dihydro citronellol acetate increased in the early stage, hiking up during stage B and decreased thereafter. These outcomes suggested that *GGPPS* gene could be an important candidate gene for regulation of monoterpene fragrance biosynthesis pathway. Roccia *et al.* (2019) characterized a segregating population from *Rosa x hybrid* cv. H190 and *Rosa wichuriana* with contrasting scent profiles to understand rose scent inheritance. Several quantitative trait loci (QTLs) were identified among which a QTL for production of 2-phenylethanol was detected to be co-localized with a candidate gene, *RhPAAS*encoding phenyl acetaldehyde synthase in the fragrant biosynthetic pathway. On further analysis, only one allele was found highly expressed and useful for rose scent synthesis. Unfortunately, its expression was discovered to occur early during flower development, prior to the production of 2-phenyl ethanol thus leading to accumulation of glycosylated compounds in rose petals.

7.1.3 Duration of Flowering

The duration of flowering roses is determined by a gene which in its dominant form restricts blooming to once a year. The recessive form of this gene giving a far more extended period of flowering is obviously of great importance (Pal, 1972). In F_1 and F_2 crosses between 5 tetraploid (2n = 4x = 28), recurrent flowering non-Moss Hybrid Tea rose cultivars and tetraploid, non-recurrent flowering Moss rose cultivars, it was shown that recurrent flowering was controlled by a single recessive gene *r* and Moss by a single dominant gene *M*. In ancient Cabbage roses the *M* gene arose as a rare, dominant mutation. The Moss cultivars, Marie de Blois, Herman Kegal, William Lobb and Salet are triplexes for *R* and simplexes for *M*.

The inheritance of various other characters has also been worked out. According to Wylie (1954), the characters in rose dependent primarily on a single dominant gene are glossy leaves (*vs.* dull), climbing habit (*vs.* dwarf), double flowers (*vs.* single) and mildew resistance (*vs.* susceptibility), while the characters dependent on an interaction of many genes are vigour, fragrance, thorniness, strength of neck, width of leaf, length of cutting stem and shape of bud and open flowers. Dubois and Vries (1987) pollinated diploid (2n =14) polyantha-like species and cultivars with *R. chinensis* var. *minima* and 2 of its direct dwarf descendants (selections 8026 and 8027 from an F1 population). Dwarf habit was found to be controlled by a single dominant gene *D*. *R. chinensis* var. *minima* and its descendants were heterozygous for *D*, whereas the polyantha roses were homozygous for *d*. A winter hardy cultivar Simson Fraser carries genes for winter hardiness from the *R. kordesii* male parent and from both *R. laxa*and *R. spinosissima* (Ogilvie and Arnold, 1993). Investigations on the Better Times cultivar showed the bull-heading, the cause of which is unknown, is not an inherited disorder arising from the development of mutations (Kiplinger and Haramaki, 1955). Studies on phenotypic variability, heritability and genetic advance, made by Lal *et al.* (1982) indicated the existence of high heritability for several characters. High heritability was accompanied with high genetic advance for the number of branches, flower buds and petals and for average flower weight, showing an additive gene effect. High heritability and lower genetic advance showing a non-additive gene effect was obtained for flower diameter and plant spread while high heritability with moderate values of genetic advance was obtained for plant height. High degree of heritability for number of branches and thorns, branch angle and plant height was also reported by Lee and Kim (1982). The leaf length and new shoot colour, however, had a low degree of heritability.Yan and Junyu (1995) suggested the development of Reujuventaiton rose group (Rj) to introduce greater stress resistance in adverse old rose cultivars or modern roses and to develop new stress tolerant cultivars that bloom round the year.The continuous flowering (CF) behavior in modern roses is regarded to have originated from *Rosa chinensis* which is mutated with an insertion of copia-retrotransposons in the floral repressor gene and causes continous flowering by blocking the maturation of gene transcripts. But, this makes the genetic background narrower thus searching for CF roses without KSNcopia allele. Horibe *et al.* (2015) discovered wild KSN allele without such insertion in CF species (*R. rugosa*) whose seasonal expression was associated with the alteration of flowering or vegetative stages of shoot apices in *R. rugosa*. Iwata *et al.* (2012) found that continuous flowering (CF) in roses is controlled by orthologous genes of TERMINAL FLOWER 1 (*TFL1*) family. In rose, six independent pairs of CF/OF (continuous flowering/once flowering) mutants differ in the presence of a retrotransposon in the second intron of *TFL1* homologue. Because of an insertion of retrotransposon, transcription of the gene is blocked in CF roses and the absence of floral represser provokes continuous blooming. In OF-climbing mutants, the retrotransposons has recombined to give an allele bearing only the long terminal repeat element, thus restoring a functional allele. In OF roses, seasonal regulation of the *TFL1* homologue may explain the seasonal flowering, with low expression in spring to allow the first bloom.

7.1.4 Flower Development

Wild type roses have five petals (simple flowers) while, more number of petals (double flowers) are typical attributes of cultivated roses. Dubois *et al.* (2010) demonstrated that the rose ortholog of AGAMOUS (*RhAG*) is expressed differentially in double flowers compared to simple flowers. Further, *RhAG* expression is restricted towards centre in case of double flowers, wider in simple flowers and labile at border domain for more number of petals. Thus, restriction of *RhAG* expression domain was suggested as the basis of double flower selection.

7.2.0 Breeding

The creation of a new form of a plant is truly very fascinating. Creating a new rose is not very difficult and does not call for any special skill, but creating a good new rose is quite another matter. Rose breeding is now practised very extensively all over the world for creation of new forms and novelties in roses. Most of the leading nurserymen in Australia, Canada, France, Germany, the Netherlands, United Kingdom and the United States have strong research programmes for creating new forms to add to their existing collections and the quest for new cultivars continues. The work of Harkness, Wheatcroft, Meilland, Kordes, Morey, Ratsek, Flory, Yarnell and other nurserymen on rose breeding has been reviewed by Pal (1969). Wylie's Masters Memorial Lectures (1954, 1955) is a valuable document of information of history of origin and breeding of garden roses. A comprehensive review of work on rose breeding and its genetical aspects were made by Pal (1972). The origin of garden roses and trend in rose breeding were also discussed by Saakov (1965). Rose breeding including studies on genetics and cytogenetics has been reported by many workers including Pal (1969, 1972), Shahare and Shastry (1963), Banda and Banerjee (1967) and Lata (1971) in India.There are several ways in which the new roses have arisen. The two major ones being the occurrence of natural or induced mutations or sports, and the raising of roses from seeds produced as a result of artificial or natural cross pollination.

7.2.1 Hybridization

The important method for obtaining new forms of roses is the raising of plants from seeds. As the present day garden roses are very complex hybrids in origin, screening of seedlings raised from seeds obtained by open-pollination is likely to yield a wide assay of types, which are new. The inbreeding is the natural mode of reproduction in roses and a large number of wild roses have been inbred for very long time, without suffering any deterioration. In fact, it is often mentioned that due to complexity of the hybrid ancestory of various types, lack of enough inbreeding has resulted in the degeneration of many cultivars (Young, 1971).

Success of hybridization greatly relies on the quality of pollen which is genotype dependent and needs proper conservation. Giovannini *et al.* (2015) carried out an investigation to establish a preservation protocol for pollen in which anthers removed from flowers were dried on petridishes for 24 h and 24° C followed by storing in freezer and deep freezer covered in polyethylene tubes. Pollen germination was measured soon after 24 h from flower collection and several times after storage. The pollen grains were then spread on culture medium containing 0.7 per cent agar, 15 per cent sucrose, 152 mg/L calcium chloride and boric acid 40 mg/L. Results confirmed that pollen viability and pollen preservation at low temperatures varied among cultivars.

Unsuccessful pollination may occur due to several causes, the major one being the presence of sterility in the male or the female parent. Vries and Dubois (1988) investigated the effects of temperature, repeated pollination of the same flower, the time interval between emasculation and pollination, and the application of GA_4+7 on the fruit set, seed set and seed germination in the large-flowered glasshouse cv. Sonia, using Hadley or Ilona as pollinators. Optimum conditions for breeding with Sonia were (i) greenhouse temperatures between 20° and 25°C and (ii) pollinating at least twice between 1 and 4 days after emasculation. Improved fruit set was obtained by applying GA at 10, 50 or 250 ppm 7 or 14 days after emasculation. Ogilvie *et al.* (1991) studied the effect of GA_3 on a wide range of crosses between 18 garden cultivars (E parents) and 2 hardy breeding lines. Higher fruit set did not result in more seeds per pollination for many crosses and seemed to be related to the genetic background of the G parent. Both parents had a major effect on pollination success.

Polyploidy can now be induced in roses by treatment with colchicine or by including the formation of unreduced gametes, through temperature treatment or X-irradiation. There is difficulty in treating roses with colchicine, on account of the stipules which envelop the growing tips. Initially diploid or triploid ancestors in many rose groups have been raised to the tetraploid level. This has facilitated gene exchange between the different rose groups, which has led to the breakdown of the boundaries between the once distant classes of roses. Out of 200 or so species of wild roses, nearly all of which have ornamental value are hardier and also free from pests and diseases, than the cultivated forms, barely a dozen have been used in the develoment of the modern roses.

Debener *et al.* (2003) introduced a gene (*Rdr1*) conferring resistance to black spot (*Diplocarpon rosae*) from wild rose species into genetic background of cultivated roses. The diploid donor parent, *Rosa multiflora* was doubled using colchicines and crossed to cultivated tetraploid variety. The resulting hybrid genotype, 91/100-5 was crossed to Floribunda rose and resulting progeny was tested for black spot resistance. AFLP markers were then used to screen a subset of the resistant progeny for the fraction of donor genome, which segregated from 32-84 per cent in the population. The genotype with smallest number of donor-specific markers was further backcrossed to the recurrent cultivated roses and the segregation was noticed between 12.3 and 32 per cent for donor parent. Zuzek and Hokanson (2007) conducted a study on shrub rose hybridization programme with the objective of developing hardy, black spot tolerant, attractive and repeat blooming cultivars for which parental germplasm were selected from species roses, Old Garden roses, shrub roses, miniature roses and Polyantha roses. Species roses and hardy cultivars among Albas and Gallicas were selected as the source for black spot tolerance; shrub roses for their repeated blooming, cold hardiness and miniature and polyantha roses for small statue and repeated blooming. Recurrent selection was adopted and superior individuals were forwarded to subsequent hybridization. Matsumoto *et al.* (2010) observed the gene pool from *R. lavande* to *R. spp.* was high due to honeybee visit on the former species. But, the contribution of pollination between wild and cultivated types by the flies was found low due to their low pollen carrying ability.

7.2.2 Mutation

In natural mutation, a shoot with different genetic behaviour is observed on an existing plant, which when vegetatively propagated, produces a new cultivar. Budwood of 32 rose cultivars was exposed to 3-4 kradrays and dormant buds were grafted onto *Rosa indica* rootstock. Sensitivity to radiation varied among cultivars. The 3 krad treatment was most effective. Absorbedradiation doses of 50, 100 and 200 Gy were used by Arnold *et al.* (1998) to produce mutations in miniature rose (*Rosa hybrida*) cultivars Blue Blood, Mountie, Dark Red Mountie and Potluck. Depending on the absorbed dose, the mean spine number per 10 cm of stem, the mean petiole length, and the mean petal number was significantly reduced in all cultivars. The total number of mutations was highest in Potluck. From the outcome of mutation experiment carried out by Kahrizi *et al.* (2011), it was found that axillary buds of 2 rose cultivars (Apollo and Maroussia) grown on MS media suppplemented with 2 µM BAP on irradiation with gamma rays showed better survival percentage at 66 and 67 Gy whereas, 50 per cent height reduction at 70 and 68 Gy, respectively. Therefore, the most appropriate dose for irradiation could be between 60 to 70 Gy for mutation breeding programme.

7.2.3 Rose Breeding in India

The rose breeding work in India, comparatively a recent origin was taken up in right earnest in the last five decades or so. Though B.S. Bhattacharjee, a rose nurseryman from Deoghar (Bihar) is acclaimed the pioneer rose breeder in India, who evolved a fairly large number of roses, the breeding work done by Dr. B. P. Pal, evolving several outstanding cultivars like Dr. Homi Bhabha, Delhi Princess and Banjaran created the real interest in rose breeding in India. Under his guidance, a comprehensive rose breeding programme was started in 1960 at the Indian Agricultural Research Institute (IARI), leading to the evolution of over 100 rose cultivars (Malik and Dadlani, 1984). Some amateurs and private nurserymen also took up rose breeding. Attempts made by G. Kasturi Rangan, M. S. Viraraghavan, M. N. Hardikar, Mrs. P. L. Airum, Arpi Thakur, J. P. Agrawal and others led to several outstanding roses (Biswas, 1984).

Many varieties have been released by ICAR-IIHR. It is the first insitute which released Arka Swadesh for polyhouse cultivation, followed by many varieties which are on par with the imported varieties.

8.0 Growth and Flowering

There are many factors that affect the growth and flowering of roses. These factors include genetic or inherent factors, climatic or environmental factors such as light, temperature, moisture in the air and soil and aeration (Laurie *et al.*, 1958; Vries *et al.*, 1978; Hanan, 1978), managemental factors like nutrition and soil conditions. Certain operations like pruning, pinching, defoliation, *etc.* also influence the plant growth and production of flowers (Carrier, 1953; Zieslin *et al.*, 1976). Similarly, the role of certain growth regulators is well known (Ohkawa, 1979; Zieslin *et al.*, 1979; Zieslin and Khayat, 1982/83). The effects of nutrients, pruning, and soil conditions have been discussed in separate chapters and the discussion in this chapter will be restricted to other factors.

8.1.0 Genetic or Inherent Factors

Like other plants, growth and flowering behaviour of roses is governed by genetic factors. Significant different in the growth and flowering of different rose species and cultivars have been observed which are attributed to the inherent genetic variation existing amongst these species and cultivars (Zeller, 1975; Smirnov, 1975; Vries *et al.*, 1980). The range of variability in roses is very great. Cultivars are now available, suitable for almost all the purposes for which an ornamental plant is grown. There are tall cultivars growing several metres high which are suitable for planting in backdrop and as hedges. On the other hand, miniature roses with dwarf habit and producing small flowers are also available. Several cultivars are bushy in nature while many are climbers and ramblers. A wide choice of cultivars with long sturdy stems having nice flowers suitable for cut flower is now available. Similarly, there are cultivars blooming in different periods and producing varied yield and quality of flowers. Some roses also possess high quality perfume. Development of new cultivars suitable for changing demands has resulted in a very rich germplasm of roses. Thus it is possible to regulate the growth and flowering behaviour of roses by genetic manipulation.

Wang *et al.* (2015) revealed that low expression of *RTFL1C* gene induces the expression of RLFY which causes recurrent flowering in roses. It was also suggested that recurrent flowering nature is both quantitative and qualitative trait instead of only qualitative character which is ruled by recessive monogene and some multi-genes. Lian *et al.* (2016) identified and isolated a gene RoLEAFY whose deduced structure was almost similar to LFY homologues of strawberry (FaLFY-2). It was also reported that recurrent-flowering roses like *Rosa chinensis* could flower continuously without repeated accumulation of RoLFY mRNA, once the floral initaiton is complete.

8.2.0 Environmental Factors

The principal factors of the aerial environment which exert a strong effect on growth and flowering of rose are: temperature, humidity, duration and quality of light and CO_2. These factors have been discussed in details in open and greenhouse cultivation.Terfa *et al.* (2014) reported that UV radiation reduced growth of pot rose cultivars (Cygein, Snow White and Tom Tom) at 2794 and 1700 m MSL with more prominent effect in the lower altitude. While higher level of solar radiation delayed flowering by 7-10 days. Further, postharvest life and water usage were not significantly influenced by UV radiation. However, plants produced at higher altitudes showed better control on water loss and longer vase life compared to lower altitude.

8.3.0 Growth Regulators

These are organic substances other than nutrients which in small amounts promote, inhibit or otherwise modify many physiological processes. Many workers have studied the effect of various plant growth substances on growth and flowering of roses.

8.3.1 Growth Promotion

8.3.1.1 GA

Gibberellic acid has proved effective in regulating the growth and flowering of roses (Koseva and Decheva, 1976; El-Shafie *et al.*, 1980). Application of GA_3 increased the length of shoots both by extending the internode length and by increasing the number of nodes. Treatment with this growth substance also increased the number of shoots per plant, accelerated the appearance of the leaves at successive nodes, induced earlier flowering, prolonged the period of flowering and enhanced the yield of flowers (Mastalerz, 1959-60, 1965; van Onsen and Haegeman, 1962; Kljuka, 1963; Nanjan and Muthuswamy, 1975). El-Shafie (1978) reported that spraying of GA_3 four times at monthly intervals at 250 ppm on cv. Montezuma increased the number of flowers and the length, thickness and fresh weight of flower stems compared to other concentrations (50, 100, 150 and 200 ppm). He also mentioned that the total carboyhydrate content of the flowering stem and leaves increased markedly due to GA_3 treatment. El-Shafie *et al.* (1980) found the highest number of flowers in 1year-old plant (7.6-9.7) and the largest flowers in Queen Elizabeth and Baccara rose cultivars by treatment with GA_3 at 250 ppm. Gowda (1984) concluded that GA_3 spray on cv. Super Star resulted in more number of flowers with longer stems which are the important characters of a good cut flowers. On the other hand, GA_3 applied 3 times every 3 days resulted in a decreased flower yield (Wisniewska-Grzeszkiewicz and Treder, 1989). The induction of parthenocarpic development of fruits in roses was also possible by the application of GA_3 (Jackson and Prosser, 1959; Jackson and Blundell, 1964, 1966). It was also found useful for the prevention of bull-head malformation of flowers in Baccara rose (Zieslin *et al.*, 1979). Roberts *et al.* (1999) identified the gibberellins A1, A3, A5, A8, A19, A20 and A29 (GAs) in vegetative shoot tips of *Rosa canina*. Most wild roses have a short flowering season of 2-4 weeks in spring, whereas most modern cultivars flower recurrently. The cv. Felicite et Perpetue is a short-season hybrid from a cross between a wild rose and a recurrent-flowering rose, whereas its sport, Little White Pet flowers recurrently. Concentrations of GAs were measured in shoot apices of both cultivars. In March (before floral initiation in spring), the concentrations of GA_1 and GA_3 were respectively, 3 and 2-fold higher in Felicite et Perpetue than in Little White Pet. In April (after floral initiation), the concentrations of both gibberellins were substantially greater than on March 17 and 12-fold greater in Felicite et Perpetue than in Little White Pet. It was postulated that, in Felicite et Perpetue, floral initiation occurred when concentrations of GAs were low and was inhibited when concentrations of GAs were high, whereas in Little White Pet concentrations of GAs remained at permissive levels throughout the growing season. Applications of GA_1 and GA_3 to axillary shoots in March inhibited floral development in Felicite et Perpetue but not in Little White Pet. The combined concentration of exogenous and endogenous gibberellins might have been raised to inhibitory levels in the former but not in the latter cultivar. Hashemabadi (2012) obtained higher flower yield (192 flowers per sq.m), prolonged vase life (12.67 days) and longer flower stems (49.33 cm) with the application of 200 mg/L GA_3, 150 mg/L GA and 300 mg/L GA_3, respectively in the cultivar 'Poison'.

8.3.1.2 BA

Greenhouse grown rose cvs. Forever Yours and Regal Gold were either left intact or cut back to 45 cm from the ground level and 2 or 3 lower dormant buds were treated with pastes of different growth substances. It was recorded that BA + adenine (each at 0.5 per cent) in lanolin paste induced bud break and shoot development in 34-60 per cent of the treated buds on both pruned and unpruned plants (Parups, 1971). Growth promotion of lower axillary buds and the development of renewal canes for rejuvenating greenhouse rose plant of cvs. Golden Rapture, Happiness, Mary De Vor and Nordia and increase in cut flower production were achieved by treating lower buds with BA in lanolin paste introduced into tissue by scarring the bark (Ohkawa, 1979). In 2-year trials with the Miniature cultivar Meillandina were treated with BA at 50-400 ppm. A combination of pinching and treating with BA increased branching and the number of flowering shoots compared with controls. Pinching without BA treatment reduced the number of blind shoots. Increasing the BA concentration to 400 ppm also increased the number of blind shoots (Treder *et al.*, 1989). Jayroe Cournoyer and Newman (1995) applied BA (0, 75, 150 and 250 ppm) to container-grown 2-year-old Hybrid Tea rose cultivars as drenching sprays to the graft union or with floral foam applied basally or to the top of a cut cane. BA increased the number of basal shoots. Serek and Anderson (1995) treated Miniature rose cvs. Victory Parade and Elegance Parade at the marketing stage with BA and found that BA increased flowers and plant longevity.

Potted Miniature roses (4 cultivars) held after production in simulated home conditions (21°C, continuous fluorescent light) performed poorly due to premature yellowing and abscission of flower buds and leaves, and failure of flower buds to open. Spray application of 100 ppm BA reduced leaf yellowing but had no significant effect on flower opening or bud abscission.

8.3.1.3 Other Chemicals

Singh and Kumar (2003) reported that foliar application of NAA at 125 ppm on rose cv. Super Star resulted in maximum number of flower buds per shoot, flower diameter, number of flowers per bush and shoot length with improvement in plant vigour, quality and yield. Farahi and Jowkar (2016) evaluated the response of hydroponic grown 'Dulce Vita' rose to polyamines. Foliar application of putrescine (Put – 1, 2, 3 mM), spermidine (Spd – 0.5,1, 1.5 mM), spermine (Spm – 1, 2, 4 mM) and distilled water (control) was done during growth stage and various vegetative, flowering and endogenous hormone levels were analyzed. Results insisted that application of 1.5 mM Spd resulted in maximum cut flower stem length, flower bud size and increased vase life. Da *et al.* (2019) conducted an experiment to assess the accumulation of organic solutes and enzyme activity in 'Ambiance' cultivar with various physiological effect products under high temperatures like control (water), boscalid, pyraclostrobin, boscalid + pyraclostrobin, fluxapyroxad + pyraclostrobin, plant growth regulators (IBA + GA_3 + Kinetin) which were applied once in 15 days throughout the crop cycle. The treatments boscalid + pyraclostrobin, fluxapyroxad + pyraclostrobin and PGRs showed consistent responses with greater accumulation of organic solutes (carbohydrates, proteins and proline) in rose leaves. On the other hand, enzymatic activities (catalase and superoxide dismutase) were pronounced with the PGR application. Hence, the rose plants indicated the mechanism as a defense against high temperatures.

8.3.2 Growth Retardation

8.3.2.1 ABA

Stem extracts of the rose cv. Baccara also inhibited sprouting when applied to the uppermost buds of rose shoots. The level of endogenous ABA was higher in the basal than in the upper part of the rose stem (Zieslin *et al.*, 1978). While studying the involvement of cytokinin, ABA and endogenous inhibitors on sprouting of basal buds, Zieslin and Khayat (1982/83) suggested that a balance between cytokinins and the inhibitory complex is involved in the regeneration of basal bud outgrowth and plant regeneration.

8.3.2.2 CCC

Some of the growth retardants also influenced the growth and flowering of roses. Plants of cv. Gabriele Privat treated with CCC produced greater number of buds without any blind shoots compared to the untreated plants. Soil application with the chemical showed better results than spraying which also resulted in scorching of leaves. Treatment with 3 per cent CCC solution in soil of potted rose plant caused dwarfing and improved flowering (Bornemann and Muhe, 1967). Bhattacharjee and Bose (1979) reported that two spraying with CCC on pruned rose plants caused marked decrease in shoot growth, stimulated branching, induced early flowering and increased flower size. Between the two methods of application with CCC, foliar spray gave better result than soil drenching.

8.3.2.3 TIBA

Asen and Hamner (1953) reported that treatment with TIBA significantly increased the number of basal shoots per plant. At higher concentrations it caused shorter shoots and internodes but increased the number of branches and flowers in *Rosa bourboniana* (Nanjan and Muthuswamy, 1975).

8.3.2.4 Paclobutrazol

Seven cm high, 45-day-old rose seedlings cv. Paprika, in pots, was treated with paclobutrazol (as 'Bonzi') at 75 or 150 ppm. The 150 ppm paclobutrazol treatment was applied twice, the second application being two weeks after the first. Six weeks after treatment, the plants treated with 150 ppm paclobutrazol had 6 shoots with a total shoot length of 48.3 cm and 6.5 flowers, compared with 5 shoots with a total length of 72.0 cm and 4.3 flowers in the control. This treatment also resulted in plants with greener leaves and more mildew resistance (Kaminski, 1989). Jenks *et al.* (2001) reported that paclobutrazol enhanced the total wax load by 10 per cent after 11 days of application in 'Apollo Parade' which was similar to moderate drought stress causing 14 per cent increase in wax load level after 24 days of exposure to drought. Thus, it was concluded that paclobutrazol posed similar effects as that of drought conditions.

8.3.2.5 Other Chemicals

The Miniature roses (*R. hybrida* cultivars Victory Parade and Elegance Parade) at the marketing stage were treated with ethephon and silver thiosulphate (STS). Abscission and flower senescence were accelerated by spraying plants with 50 or 500 ppm ethephon, and the longevity of individual flowers and whole plants were reduced, particularly at the higher ethephon concentration. All STS concentrations (0.4, 0.8, 1.2 and 1.6 mM) increased flower and plant longevity. The higher concentration used (0.8 mM) showed the best results and it is suggested that higher concentrations may have even greater effects (Serek and Andersen, 1995). Plants of cv. Orange Sunblaze in 5-inch-pots in a greenhouse were sprayed once on 3 weeks after potting with uniconazole (Sumagic) at 10, 20 or 30 ppm a.i., each plant received 0.1, 0.2 or 0.3 mg a.i., respectively, the controls being sprayed with tap water. Sumagic-treated plants were 10-40 per cent shorter than controls. The most desirable growth retardation was achieved with 0.3 mg, these plants also having darker green foliage. Flowering was delayed by 3-4 days; the peduncles were about 70 per cent shorter than with controls and the flowers were well distributed over the plants (Pobudkiewicz and Goldsberry, 1989). Geerdink *et al.* (2020) observed maximum stem diameter and floral bud length with preharvest application of 800 mg/L Si in comparison to control. Si treatment also resulted in reduced fresh weight loss, improved relative water content and greater index of flower bud opening. Moreover, the POD activity also decreased with Si application thus favouring the quality and postharvest life of cut roses.

8.3.3 Microbial Association

El-Deeb *et al.* (2012) isolated and characterized six endophytic bacteria from different organs of rose plant (*Rosa damascene trigintipeta*) at flowering stage. These bacteria had the ability to produce hydrolytic enzymes such as cellulose, xylanase, pectinase, amylase, protease, lipase and chitinase and growth promoting enzymes for improving growth and flowering but with some variations among these isolates (Tables 3 and 4).

Table 3: Endophytic Bacterial Population Recovered at Flowering Stage of *Rosa damascene trigintipeta**

Host Plant	*TSA Medium*				*PDA Medium*			
	Root	*Stem*	*Leaves*	*Flower*	*Root*	*Stem*	*Leaves*	*Flower*
Rose	4.6 x 10^4	4.1x10^4	3.4x 10^3	1.8x 10^2	4.1 x 10^4	3.1x10^4	3.1x 10^3	1.2x 10^2

Table 4: Plant Growth Promoting Activities of the Six Endophytic Bacteria in *Rosa damascene trigintipeta**

Bacterial Strains	*Plant Organs*	*Siderophore Production*	*ACC Deaminase (μMαKB mg^{-1} h^{-1})*	*Indole Production (μg IAA/ml)*	*Phosphate Solubilisation Activity (mm)*
Planococcus sp.	Leaves	+	12.2 ± 1.2	10.2 ± 2.4	10.2 ± 2.1
Micrococcus sp.	Petals	+	14.2 ± 1.6	16.2 ± 1.8	7.6 ± 0.8
Bacillus sp.	Stem	+	18.5 ± 1.4	18.6 ± 1.6	12.2 ± 1.2
Methylococcus sp.	Stem	ND	ND	7.8 ± 0.8	6.4 ± 0.6
Acinetobacter sp.	Root	+	38.8 ± 3.6	36.6 ± 3.1	28.4 ± 3.1
Acetobacter sp.	Leaves	+	18.6 ± 2.2	12.3 + 1.5	12.2 ± 1.2

*El-deeb *et al.* (2012).

9.0 Propagation

Roses can be propagated both by seeds and by various vegetative methods like cutting, layering, budding and grafting. Specialised propagators, particularly in advanced countries also use the technique of tissue culture for producing disease-free plants in large number. The propagation methods used for greenhouse roses in Japan described by Ohkawa (1988) are: (i) grafting of scions onto dormant wild or selected *Rosa multiflora* or *R. cathayensis* rootstocks between December and February (ii) field budding on *R. multiflora* between September and October, (iii) cuttings for rockwool culture, (iv) grafted cuttings rooted under greenhouse conditions, using mist or double polyethylene tents throughout the year, (v) grafted cuttings for rooting and bud emergence outdoors, and (vi) grafting onto rootstocks using greenhouse grown scions with two 5-leaflet leaves under mist (throughout the year).

9.1.0 Seed Propagation

This method of propagation is generally adopted by breeders for developing new cultivars with desirable characters. Seedlings of rose species are also used as stock for grafting or budding. Improvement in fruit set and yield of potential seed have been possible with the use of certain growth substances. Allen (1967) reported that treatment of flowers, 10 days after pollination, with 3 drops of 250 ppm solution of GA_3 improved fruit set and the potential seed yield increased. The production of fertile achenes was also higher with this treatment. Pipino *et al.* (2013) established greater correlations between number of seeds produced per cross and the pollen diameter or percentage of normal pollen. Further, their experiments on embryo-morphogenesis unraveled that embryos of floribunda roses were fully developed at 15 and 30 days after pollination (DAP). However, ABA content in progressing embryos decreased to basal level from 30 DAP which resulted in poor seed germination.

When the fruits are fully ripe, they are harvested and thoroughly dried before the extraction of seeds. Most rose seeds when mature are in a resting condition requiring an after-ripening period before germination. Reason for delayed germination or dormancy in achenes was investigated by Bo *et al.* (1995) in cv. 'Crimson Glory'. The results indicated that dormancy is mainly associated with ABA in pericarp and testa as well as by the physical limitation posed by pericarp. It is suggested to stratify the seeds at 35° to 40°F (1.6 to 4.4°C) after harvest. Stratification for six weeks is sufficient for *Rosa multiflora, R. rugosa* and *R. hugonis* require 4 to 6 months, and *R. blanda*, 10 months (Schneider, 1960). Warm stratification followed by cold stratification has also been found beneficial. *R. canina* was found to germinate best if the seeds are held at room temperature for 2 months in moist vermiculite and then transferred to 32°F (0°C) for another 2 months (Rowley, 1956). Morey (1960) also reported that *R. canina* seeds required warm treatment at 80°F (26.6°C) for 2 months followed by 3 months at 40°F (4.4°C) for better germination. This was also confirmed by Popcov and Buc (1966) who observed that seeds of some species germinated satisfactorily by warm (20-25°C) stratification for 2-3 months and cold (5-6°C) stratification for 3-5 months. Hybrid rose seeds usually respond best to a stratification temperature of 34° to 40°F (1° to 4.4°C) for 60 to 90 days.

For raising seedlings, clean seeds are sown about 5 cm apart in pans containing porous compost with adequate organic matter or in small plots. Yambe *et al.* (1992) recorded that intact *Rosa hybrida* cv. Inspiration achenes never germinated at 25°C under continuous light whereas excised seeds germinated readily. Temperature is another very important factor influencing the germination. In a trial with different temperatures under controlled condition and also in the field, the *in vitro* seed germination percentage of *R. multiflora* was much higher under continuous white light than in complete darkness. Red (R) light was the most effective in inducing germination and far-red (FR) light was ineffective (Yambe *et al.*, 1995).

Germination percentage of *R. multiflora* achenes was greatly increased when treated with 1 per cent Driselase, a macerating enzyme, for 36 h. Seeds germinated more rapidly when the achenes were treated with the enzyme for a longer period. Treatment with Cellulase Onozuka improved seed germination at a lower concentration than did Driselase (Yambe and Takeno, 1992). Zhou and Bao (2011) reported that scarification with H_2SO_4 for 2-4 h followed by warm stratification at 20°C for 4 weeks and cold stratification at 5°C for 8 weeks enhanced the germination percentage in *R. multibracteata*, *R. hugonis* and *R. filipes* whereas, H_2SO_4 scarification for 4 h followed by cold stratification at 5°C for 12 weeks improved germination in *R. sericea*.

9.2.0 Vegetative Propagation

9.2.1 Cutting

9.2.1.1 Stem Cutting

Raising plants by stem cuttings is one of the least expensive and easiest methods of rose multiplication. Bredmose and Hansen (1996) described the growth and development of rose from single-node cuttings, from the most apical (position 1) to the most basal (position 7) with 5-leaflet leaf + its subtending internode were taken from 5-month-old plants of the cultivars Kiss, Manhattan Blue and Sonia. Only flowering shoots from which 7 single-node cutting with 5-leaflet leaves could be taken were used. Cuttings were rooted in rockwool cubes under a polyethylene tent then acclimatized and transferred to greenhouse benches. Percentage failure to root (1.9, 2.2 and 14 per cent for Kiss, Manhattan Blue and Sonia, respectively) was highest (5.8 per cent) for cuttings from position 7 and lowest (1.0 per cent) for those from positions 2-3. The percentage of blind shoots (1.38, 12.5 and 10.2 per cent for Kiss, Manhattan Blue and Sonia, respectively) was highest (16.6 per cent) for cuttings from positions 6-7 and lowest (8.1 per cent) for rose from positions 3-4.

The time from excision/planting to the time of axillary bud growth decreased as the origin of the cutting became more apical, up to position 2. Similar trends were observed of the times to visible flower bud appearance and flowering. Plant originating from cuttings taken from position 4-6 had the thickest flowering stems, the highest specific flower fresh weight (FW) had highest weight (W)/flower.

Sonia had the highest growth rate of the 3 cultivars, but Manhattan Blue had the highest specific flower FW (an average of 38.8 g/m of stem, compared with 32.0 and 29.2 g/m in Sonia and Kiss, respectively). Single-node cuttings from Samantha,

Pitica, and Landora roses rooted well. In Samantha, the percentage sprouting of the axillary bud at the 7th node was only 10 per cent, followed by 40 per cent at the 6th node, and 80 per cent for other nodes. In Pitica, sprouting of the axillary buds at the 1st to the 3rd node reached 100 per cent while sprouting at the 4th or 5th node was 40 per cent and 25 per cent, respectively, no sprouting of the axillary buds at the 6th or the 7th node occurred.

In Landora, sprouting of the axillary buds at the 1st to the 3rd node reached 90 per cent, at the 4th or 5th node it was 70 per cent and 20 per cent respectively, but buds at the 6th and the 7th nodes did not sprout. Axillary buds on Samantha and Pitica cuttings which otherwise failed to sprout could be induced by treatment with 0.75-1 per cent BA in lanolin paste, while sprouting in Landora cuttings was improved by treatment with 0.02-0.04 per cent GA_3 in lanolin paste (Chu and Lai, 1997).

Single-node cuttings with 5-leaflet leaves were excised from 9 stem positions and grown as single-stemmed rose plants. Generally, onset of axillary bud growth and initial shoot growth were promoted and flower weight and bloom quality were reduced in plants from apical bud position. Stem diameter, stem growth rate and biomass buildup were greatest from medial bud positions. Stems and internodes were longer from basal bud positions. Topophysis can determine the potential for plant development and flowering in roses and may be used to regulate plant growth. Results also indicate intrinsic mechanisms determining axillary bud growth but the physiology of topophysis is unclear. A probable role of cytokinins in the topophysic prevention of growth was observed (Bredmose *et al.*, 1999).

The morphology and physiology of cuttings of standard sizes made at different nodal levels was studied in 11 genotypes during rooting and subsequent growth. Cuttings taken towards the base of the stem generally flowered earlier and produced more flowering stems than cuttings taken closer to the stem apex. In 5 of the genotypes studied, the genotype 0-166-2 had the best percentage of shoots (54 per cent) at the end of the propagation followed by 2-121-1 (52 per cent) and Sonia (48 per cent) (Berninger *et al.*, 1990).

Costa *et al.* (2001) observed quantitative negative effect on survival and growth of cuttings after leaf removal, leaf covering or CO_2 which hampered the photosynthesis but stored photosynthates as starch with competition among sinks. The weak sink of starch accumulating organelles became the major sink mainly new roots and shoots accounting to less than 15 per cent of total dry weight of cuttings

Vegetative axillary cuttings of 15 cm long were dipped in 500 and 1000 ppm NAA and IBA powders for 20 seconds followed by planting in individual pots containing 1:1 mixture of potting soil and perlite. Among the 4 PGR treatments, IBA 500 ppm resulted in maximum survival percentage (91 per cent), strongest roots, healthier seedlings with longer sprouts of 66 cm whereas, IBA 1000 ppm produced more number of roots with maximum root length. Thus, use of 500 ppm seems to be most sustainable treatment for seedling production as well as environment conservation (Susaj *et al.*, 2012). Baksh *et al.* (2015) reported that treatment of cuttings with IBA 3000 ppm resulted in increased sprouting percentage (33.25 per cent),

Propagation of Rose

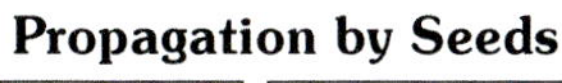

Propagation by Seeds

Hips

Rose seeds

Newly emerging seedling

Seeding after transfer in pots

Propagation by Cuttings

Cuttings of rose

Transfer of rose cutting in pot

Propagation of Rose

Propagation by Air Layering

Rose bark peeling for layering

Used sphagnum moss and clay soil for layering

Propagation by Budding

Preparation of stock for budding

T-budding

Sprouting from T-budded plants

Micropropagation of Rose

Source: Pati *et al.* (2006) *Biotechnology Advances,* 24: 94-114.

survival percentage (63.25 per cent) and earlier sprouting (14.75 days) whereas, IBA 2000 ppm produced more number of leaves (7.5), number of roots (30.25) along with maximum sprouting length of 19.16 cm). Recently, Monder and Pacholczak (2020) investigated the role of IBA, NAA and bio-stimulants based on the phenological stage of once-blooming rose shrubs and stated that cuttings taken from shoot with closed flower bud exhibited better rootingability despite low concentrations of total soluble and reducing sugars in the plant tissue while cuttings collected after 7-14 days of petal fall resulted in poor rooting ability.

Addition of VA mycorrhizal fungi (VAMF) inoculum into rooting medium during cutting propagation increased the quantity of rooting and the quality of rooted cuttings for five different cultivars of miniature roses. Four weeks after cuttings were stuck, the number of root, increased with addition of VAMF inoculum into the rooting medium (Scagel, 2001).

9.2.1.2 Root-cutting

Some rose species like *Rosa blanda, R. nitida* and *R. virginiana* can be propagated by root-cuttings (Hartmann and Kester, 1972). Kukolev (1952) observed that plants raised from root-cuttings were more resistant to frost than those from stem cuttings. The cuttings should be taken from young and healthy plants and it is important to maintain polarity when planting. The proximal end (nearest to the crown of the plant) of root piece should always be up. The cuttings should be inserted vertically with the uppermost end at level of the rooting media.

9.2.2 Layering

This type of vegetative propagation is usually limited to climbing and rambling roses. It is usually practiced during the monsoon or early spring and leads to own rooted plants. Layered plants, except those of weak growing cultivars, thrive and bloom quite well as budded or grafted plants (Gopalaswamienger, 1970). This method, as practised in rose, is of two types, ground layering and air layering. It can be used for many shrub roses as well (Hessayon, 1981).

9.2.2.1 Air-layering

It consists of removing a ring of bark, about 2.5 cm long, from around the shoot to be rooted and covering it with a rooting medium. The use of damp sphagnum moss around the ringed portion and covering with polythene film has been found beneficial. Application of some root promoting substances may further improve the root formation in air-layers. Ching *et al.* (1956) tried this method with cultivar Queen Elizabeth and observed that the most desirable roots were formed by soaking the moss in 10 ppm NAA overnight before use.

9.2.2.2 Ground Layering

It is performed by bending shoots to the ground and covering it with soil, but leaving the terminal end exposed. Cutting or notching the underside of the stem is often practiced. Root promoting substances used in cutting can also be applied at the cut point for improving root formation in layers (Pal, 1972). Rooting usually takes place in a month or so and the layered shoot is detached 15-20 days after root formation.

9.2.3 Grafting

Inarching is another method of propagating roses but has many disadvantages. This is an exacting operation, failures are common, flowering wood is sacrificed, and the cost of a grafted plant in the long run may be higher than that of budded plant. All these reasons combined have reduced the popularity of grafted rose and the number of plants that are produced by this method has dropped considerably.

The rootstocks are raised in small pots or polythene bags. The scions are selected and the stocks are brought near the selected scion shoots. The scion shoot should be of medium texture, free from pests and diseases and 1-3 eyes in length. Larger scions are not recommended. A strip of bark, about 3 cm long with a part of wood is removed from the stock as well as from the scion and the latter is attached properly on to the stock and wrapped. After union the scion shoot is cut below the graft and the rootstock above the graft union.

The method of cleft grafting has also been suggested for the multiplication of roses (McFadden, 1963-64). When fresh scions were used 16 out of 50 rose cultivars gave 80-90 per cent survival, while in others the percentage of survival ranged from 30 to 79. Baloiu (1974) recommended the method of bench grafting to save about 50 per cent of the cost of materials and land compared to the normal method. Han *et al.* (1996) propagated the rose (cvs. Mary Devor, Madelon and Lyvia) through veneer or bud-grafting methods and obtained best production of quality flowers from bud-grafted plants.

Kromwijk *et al.* (1990) reported results of stenting in which the glasshouse cultivar Sonia was side-grafted on *Rosa canina* cv. inermis rootstock. In the first trial, the cuttings were either untreated (control) or dipped in a 1:1 mixture of 0.5, 1.0 or 2.0 per cent Rhizopon AA [IBA] and captan powder, in October. Rooting was best (average 72 per cent) with 0.5 per cent Rhizopon AA. In the second trial (in weeks 48, 49, 7 and 8), half the rootstock cuttings were prepared with a cut beneath a node and half with a cut above a node, with the bud at the node being removed. The percentage rooting was higher (average 63 per cent) using a cut beneath a node. Rootstock material with one leaf, propagated in October and November gave the most satisfactory results.

The greenhouse cultivars Bingo, Mercedes, Red Success and Sonia were stented, *i.e.*, grafted on rootstocks of unrooted semi-hardwood cuttings of *Rosa canina* cv. Pava, *R. multiflora* and *R. rubiginosa*. The cuttings were inserted in 3:1 perlite and peat at 20-25°C under glass. The best results were obtained with *R. multiflora* (average 83 per cent rooted cuttings after grafting in August). Wounding of the base of *R. canina* cv. Pava cuttings improved rooting (Marczynski and Michlewicz, 1988). In a trial, cv. First Red plants were grown on their own roots or stented on 26 different rootstocks (Pol *et al.*, 1995). After 6 months, the diameter of the root collar and the number and diameter of bottom breaks were greater in plants on the rootstocks 18-31 and Natal Briar than in plants on their own roots. The number of harvested flower shoots was significantly greater (5.61 and 5.33/plant) for plants on 18-31 and

Natal Briar, respectively, compared with 3.61 for plants on their own roots. These differences in flower production continued for several years.

Rose cultivars Anna and Sari were propagated by Rodrigues *et al.* (1998) from cuttings or by grafting on *Rosa indica* [*R. chinensis*] rootstocks and grown in raised benches filled with 4:1 sand: peat mixture. Flowers were harvested for 18 months and yield of class I flowers were higher (18.9 and 14.6 flowers/plant in Anna and Sari, respectively) from grafted plants than from those raised from cuttings (12.4 and 11.6 flowers/plant, respectively). Marketable yields and flower quality were also greater from grafted plants.

9.2.4 Budding

Budding is the most popular and successful method for multiplying roses. It provides larger number of plants than cuttings, layering or grafting, as a single shoot of the desired scion furnishes a number of buds for budding. Rooted cuttings of stock or seedlings with roots are used as in grafting. Shield or T-budding is the method ordinarily used. On the selected rootstocks, the buds are inserted into a T-shaped incision and then tied with suitable wrapping material. Raffia, polythene tape, adhesive tape, binding rubber strips are common wrapping materials but strips of plastic film have been found superior to other by several workers (Kireeva, 1967; Davies *et al.*, 1980). The incision on the stock should be just enough to accommodate the bud. Plump, but not too elongated bud is selected from the middle section of the scion shoot (Swarup, 1967). Budding should preferably be done as low down as possible on the stock. When the new shoot from the grafted bud is about 10 cm long, the top portion of the stock above the union is cut off and the polythene tape is removed. It takes 3-4 weeks for the bud to unite. The side branches of the stock are removed which compete with the scion for supply of nutrients and water.

9.2.5 Rootstock

For grafting or budding, proper selection of rootstock is of immense importance. Rootstocks are known to impart marked effects on the vigour, precocity, productivity, quality of flowers, longevity of bushes, disease resistance, adaptability to soil and climatic conditions, *etc.* and therefore, it is necessary to choose the most suitable rootstocks for budding or grafting roses. Since stock-scion compatibility is not a fixed relationship for all locations and conditions, experiments on the influence of rootstock on growth, flowering, flower quality and resistance to disease and pests are considered important to standardize the rootstock in different agroecological conditions. According to Biswas (1984), the rootstock should have the following qualities:

- ☆ The rootstock should produce fibrous root system.
- ☆ It can easily be propagated from cuttings.
- ☆ It should have vigorous growth habit, be healthy and resistant to disease and frost.
- ☆ Plants should have uniform growth.
- ☆ It should have a bark, thick enough to hold the bud firmly and provide sap to the budded plant.

- It should be reasonably free from suckers.
- It should support the budded plant for a long time.
- It should withstand a wide range of soils and climatic conditions.

De Vries and Dubois (1992) attributed breaking the dormant axillary buds on the scion as the important action of rootstock and recommended clonal rootstocks over seedling types as latter one exhibits genotypical variation thus reducing the vigour of scion sometimes.

Some of the commonly used rootstocks are *Rosa borboniana* (Edouard rose), *R. canina*, *R. indica* var. odorata, *R. laxa*, *R. manetti*, *R. multiflora*, *R. rugosa*, *R. fortuneana*, *R. moschata*, *R. eglanteria*, *R. sempervirens*, *R. rubiginosa*, *R. chinensis* and *R. pollmeriana* are some of the other rootstocks tried with varied success (Rupprecht, 1974; Singh, 1974, 1980).

Rosa borboniana (Edouard Rose)

It is one of the most popular rootstock and used extensively in the northern plains of India. Plants are hardy, vigorous, propagated easily by cuttings, provide straight stems of suitable length and are found useful for budding Standards.

Rosa canina (Dog Rose)

This has been the most commonly used stock in Europe. It is usually propagated by seed, since the cuttings do not root easily. However, the seeds are difficult to germinate (Rowley, 1956). The prominent thorns make it difficult to handle. It also tends to produce sucker. Young plants on this stock grow slowly, but they are long lived. *R. canina* is adaptable to drought and alkaline soil conditions. Hazar and Baktir (2005) investigated the scion-rootstock compatibility and anatomical structure of graft union in scion varieties 'Jacaranda' and 'Kardinal' grafted on rootstock '*Rosa canina*' and observed very good compatibility with the former variety compared to the latter. However, xylem and cambium tissues were continuous in the graft union with good shoot growth in both varieties.Verma *et al.* (2015) assessed the overall diversity of 36 wild hip rose (*Rosa canina*) genotypes to detect likeness with respect to fruit quality and yield attributes. Results of multivariate analysis presented a substantial phenotypic diversity in terms of bush height (2.45-9.00 m), hip yield (0.75-7.5 kg/bush), fruit diameter (8.05-14.89 mm), fruit length (11.69-24.85 mm), fruit weight (0.46-2.20 g), fruit pulp weight (0.26-1.30 g), fruit:pulp ratio (44.05-89.37), seed number per hip (1-37) and TSS (29.8-39.6° Brix). The cluster analysis grouped genotypes into 6 different sets in which I, II and III were superior regarding fruit quality and tree characters, while I and VI were most divergent. Principal component analysis indicated that traits such as fruit diameter, fruit weight, pulp weight, number of seeds per hip imposed larger proportion of observed variability and the genotypes, 'RH-PH-1', 'RH-OAF-18', 'RH-12' and 'RH-11' were identified outstanding for yield contributing traits.

Rosa indica (var. *odorata*)

This is a large climbing shrub and used extensively as rootstock for greenhouse forcing roses. It is easily propagated from cutting and under suitable conditions produces a large symmetrical root system. It is well adapted to excessively dry or wet soil conditions and can withstand high soil pH (Grueber and Hanan, 1981). The rootstock is also quite tolerant to powdery mildew and insect pests. However, they are susceptible to *Verticillium*. In India, it has been recommended for northern plains (Yadav *et al.*, 1985a).

Rosa laxa

This is well adapted to rich heavy soils and plants budded on it have vigorous growth (Field, 1969). It can be propagated both by seed and cutting and have a very good root system. This is one of the common rootstocks used in Europe.

Rosa manetti (*R. noisettiana* var. *manetti*)

This is a very old stock commonly used in Europe for propagating dwarf roses and for planting in sandy soils. It is easily propagated from cuttings, produces plant of moderate vigour, and is resistant to some strains of verticillium wilt (Raabe and Wilhelm, 1966). This rootstock has, however, proved unsuitable for the production of cut flowers in greenhouse (Grueber and Hanan, 1981).

Rosa multiflora

A native to China and Japan, introduced into India in 1872 or even earlier in its wild form. This is widely used as rootstock for outdoor roses. Several strains are available, some giving better bud unions and bud development than others (Roberts, 1962). Cutting of this species root easily, develop a vigorous, nematode resistant root system, and do not sucker excessively (Hartmann and Kester, 1972). It is adaptable to a wide range of soil and climatic conditions. Masse and Vulinec (2010) reported that Multiflora rose reduced the species richness of North American Riparian forest-breeding song birds and was considered as invasive shrub.

Rosa rugosa

It is a vigorous erect shrub, native to Japan, North China and Korea. The form which is used as stocks produces single flowers of purplish-red colour. For bush roses, it is propagated by cuttings and for tree roses by seeds. It is widely used in Europe for standard roses (Norman, 1955). The root system is shallow and fibrous, tends to sucker badly, but the plants are long lived (Hartmann and Kester, 1972).

Ragged Robin (Gloire des Rosomanes)

This French stock is popular in California for outdoor roses. It does not sucker if the lower buds on the cuttings are removed. This stock grows vigorously during the summer, permitting budding at any time. Because of its fibrous root system, it is easy to transplant but requires good soil drainage. In some areas, however, it is difficult to propagate and is affected by leaf spot. It is resistant to nematodes and heat and dry conditions but susceptible to verticillium wilt and therefore, should not be planted on land previously planted with tomatoes.

Dr. Huey (Shafter)

This is the most common rootstock in Arizona and southern California rose districts, replacing Ragged Robin to a large extent. It is especially useful for late season budding because of its thin bark. Plants are very vigorous and well adapted to irrigated conditions and its cuttings root readily. It is very good as a stock for weak growing cultivars. This is susceptible to blackspot, mildew and *verticillium* (Hartmann and Kester, 1972).

Multiflore De La Grifferaie

This is useful as a trunk for tree roses, producing desirable straight canes. It is vigorous, extremely hardy and resistant to borers.

IXL (TausendschonVeilchenblau)

This is also used primarily as a trunk for tree roses. It is very vigorous and has no thorn. The canes tend to sunburn and are somewhat susceptible to low temperature injury.

Multiflora-type Hybrid

It is developed by crossing between *R. multiflora* (2n) and *R. hybrida* (4n). Easily propagated through cuttings. It is suitable for mild winter regions with salinity problems (Cid *et al.*, 1989).

9.2.5.1 Selection of Rootstocks

The selection of the rootstock depends on the soil and climatic conditions of the place and thus different stocks are useful under different conditions. Wright *et al.* (1957) recommended. *R. multiflora* is a vigorous stock which thrived in poor sandy soil with low pH. *R. canina*, grew well in rich heavy soils. Jicknor and Roberts (1967) reported that flowering was influenced by scion cultivars, rootstocks, location and the number of years that plants had been growing in a particular location. *R. manetti*, previously used as a rootstock in mini rose propagation, was not suitable for the production of cut flowers in greenhouse. *Rosa odorata* was considered suitable as it produced better root system andwithstood the high pH of Colorado soils. *R. indica* (vars. Major and Dr. Huey) also performed well as mini rootstocks (Grueber and Hanan, 1981). In Europe, the stocks used included the Dog Rose (*R. canina*), *R. multiflora*, *R. laxa* and *R. noisettiana* var. *manetti*. For Standards, stocks of *R. rugosa* are popular. In America, besides *R. multiflora*, which is the predominant rootstock in many parts of the country, Dr. Huey and Ragged Robin are used in some areas (Pal, 1972). According to Suskov and Mihneva (1969), in the foothill zone of Alma Ata, (earstwhile USSR) the suitable rootstocks for Hybrid Tea roses were *R. arnoldiana*, *R. corymbifera*, *R. myricantha* and *R. tomentosa* and for Miniature roses *R. canina*. In North Sweden, Lenander (1969) suggested that *R. rubrifolia* could be substituted for *R. canina* as a good rootstock. *Rosa canina* and *R. multiflora* were the two important rootstocks used in Canada (Oliver, 1954). In Czechoslovakia, wild roses, *i.e.*, *Rosa vagiana*, *R. subcanina*, *R. incana*, *R. pycracantha*, *R. vosagiaca*, *R. elliptica*, *R. tomentosa*, *R. caryophyllacea*, *R. zalana*, *R. ulozensis*, *R. glauca* and *R. pendulina*, *R. vosagica* performed well when used as rootstocks (Vetvicka, 1974). Pantopica, a rose cultivated in Nigeria, was found to be a successful rootstock for a wide range of HT cultivars (Thomas, 1980). Rootstocks of *Rosa* spp. developed at the Department of Horticulture of the Agricultural University of Wageningen were screened for their reaction to *Pratylenchus vulnus* and compared with current rootstock, proved to be a poor host for the nematode (Schneider *et al.*, 1995).

The responses of two rose (*R. chinensis* 'Major' and *R. rubiginosa*) rootstocks were investigated under salt stress. The distribution of chloride and sodium ions in all plant parts was determined. The rootstock *R. rubiginosa* showed a higher tolerance of the NaCl stress than *R. chinensis* 'Major'. The lower older leaves contained higher concentrations of Cl^- than the young upper leaves. Leaf samples had higher concentrations of Cl^- than stem samples taken from the same positions. The roots contained higher amounts of Cl^- than the stem samples. The plants accumulated higher amounts of Cl^- in comparison with Na^+. The lower leaves of *R. chinensis* 'Major' had higher amounts of Na^+ than in all other parts whereas *R. rubiginosa* had higher concentrations of Na^+ in the roots than in all other parts.

Based on the available information on the rootstock trials and the general practices followed in different parts of India, it may be said that in the northern plains of India, Rose Edouard continues to be the major rootstock for Standards (Pal, 1972). Under Delhi conditions, however, *Rosa indica* var. *odorata* has been found superior to others and has been recommended for obtaining higher yield of quality flowers (Swarup and Malik, 1974). In Bihar and in the plains and hills of West Bengal, *R. multiflora* has been found to be promising. The results of an investigation carried out in the plains of West Bengal (Bose *et al.*, 1986) showed that *R. multiflora* is the best rootstock for cvs. Double Delight, Happiness, Landora, Montezuma and Paradise. In South India (Bangalore) also, *R. multiflora* showed the best performance in Happiness, Queen Elizabeth and Super Star (Mukhopadhyay and Banker, 1982). The thornless rootstock also performed well under Bangalore conditions in Karnataka (Mukhopadhyay *et al.*, 1980). Karadi and Patil (2006) observed maximum bud take, maximum vigour and number of flowers per plant when budded on *Rosa multiflora* while early flower bud initiation with *Rosa canina*. Among various cultivars evaluated, Montezuma recorded maximum bud take whereas, Sophia Laurence exhibited maximum vigour, early bud initiation and maximum number of flowers per plant when used as rootstock.

Khosh-Khui and Zargarian (2010) evaluated the performance of three rose cultivars (Golden Gate, Hocus Pocus, Trixx) onto 4 rootstocks (*R. banksiae, R. canina, R. chinensis* 'Masquerade' and *R. multiflora*) and revealed that *R. banksiae* was incompatible to all the studied cultivars. Whereas, *R. canina* could be utilized for its highest number of flowers (6.76/plant), stem diameter (4.67 mm), flower diameter (20.21 mm), maximum flower fresh (18.15 g) and dry weight (5.26 g). Moreover, *R. chinensis* also exhibited good stem diameter, flower diameter, flower fresh and dry mass. Thus, *R. canina* and *R. chinensis* have better potential for utilization as rootstock.

9.2.5.2 Bud-take

Considerable information on rootstock-scion relationship among different rose species and cultivars are available. Budding on *Rosa canina* resulted in higher percentage bud-take than on *R. afzeliana*, but the survival was more on the latter rootstock (Veski, 1957). The Miniature roses grafted on *R. canina*, however, showed stock-scion incompatibility due to inadequate supply of organic substances by the scion to the roots. The neck of the rootstock became thick and died within 1-2 years (Suskov *et al.*, 1966). The higher percentage of bud take of Floribunda cultivars was with *R. dumetorum* but in case of HT cultivars it was with *R. canina* vars. *Pfander* and *Superb* (Meneve and Moermans, 1977). Sharma (1979) reported that out of 8 rose cultivars budded on *R. borboniana, R. multiflora* or *R. indica,* the highest take (100 per cent) was obtained with cv. Montezuma budded on *R. indica.* The bud-take on *R. multiflora* ranged from 33.3 per cent in Himangini to 75 per cent in Vienna Charm and Sadabahar and on *R. borboniana* from 50 per cent in Super Star, Himangini and Dr. Valois to 87.5 per cent in Delhi Princess and Vienna Charm. According to Mukhopadhyay *et al.* (1980), a thornless rose (obtained through mutation) was a superior rootstock in respect of bud take over *R. multiflora* and *R. indica*. In a trial with 3 *R. canina* varieties, *R. laxa* and *R. rubiginosa,* the budding success rate was highest on *R. laxa* (M.G.D., 1983). *Rosa canina* cultivars Inermis, Pfander, Pollmer, Schmidt's Ideal and Superba were compared as alternative rootstocks to *R. corymbifera* cv. Laxa (which is susceptible to nematode infestation). They were budded with scions of 14-16 cultivars. In the first year of this 2-year study, *R. corymbifera* cv. Laxa produced the highest number of salable bushes, with an average 96 per cent bud-take. In the second year, it was outperformed by *R. canina* cv. Schmidt's Ideal (Mesmaeker, 1991).

9.2.5.3 *Effect of Rootstock on Growth and Flowering*

Results of extensive investigations have been reported on the effect of different rootstocks on growth and flowering in various cultivars.

In an experiment on 10 rose rootstock species and clones, 35 Hybrid Tea cultivars and 22 Floribunda cultivars in a specialized rose nursery. Haenchen (1967) reported that rooting of stocks was influenced by the species used and by the root collar diameter at planting. The growth of the budded scion was affected by the cultivar, rootstock, the skill of the budder and temperature during the growing period following budding and during the following winter. With *R. multiflora* rootstock, there was correlation between budding and lifting of the bushes for sale and the root collar diameter of the rootstock at planting.

Growth

The influence of 6 rootstocks on the growth of cv. Motrea was determined by Kool and Pol (1992) by measuring various plant characters 4 times during early development (at the time of pinching the bottom-breaks, when laterals were harvested after 30 days, 100 days after pinching the bottom-breaks, and 258 days after planting). Clear differences were found between rootstocks in the quality of bottom-break and lateral branches 100 days after pinching out the bottom-breaks. Plants on *R. hybrida* cv. Motrea and *R. canina* cv. Inermis rootstocks were less vigorous than those on cv. Multic (a selection from *R. multiflora* cv. Cathayensis) and *R. hybrid* cv. Moonlight. The rootstocks *R. hybrida* cv. Natal Briar and *R. indica* cv. Major which by nature grows actively during the winter conferred this characteristic on the grafted plants. Plants raised from grafted cutting generally produced more basal shoots, which grew faster, than plants raised by other methods. Scions consisting of one internode with a leaf and quiescent bud were cut from 11 successive internode positions on a shoot of cv. Sonia and grafted to *Rosa indica* cv. Major and *R. canina* cv. Inermis rootstocks. Vries and Dubois (1992) found that during 12 months in greenhouse, plants performed similarly on the 2 rootstocks. Internode position had a significant effect on root collar diameter, root FW, time of emergence and number of bottom-breaks, and shoot yield. In general, the values of these traits were higher in plants which originated from internodes furthest from the shoot apex.

9.2.5.4 Flowering and Yield

Earlier flowering and more number of flowers were also recorded in plants grown on *R. multiflora* (Rowley, 1955-56; Lenander, 1956). McFadden (1956-57) studied the performance of scion of Hybrid Tea cv. Happiness on *R. fortuneana, R. multiflora, R. odorata* and var. Huey, *R. fortuneana* showed a marked superiority over the other three rootstocks in growth and flowering of scion.

In a glasshouse trials with 9 rose cultivars of 4 rootstocks, Vecera (1967a) observed that in general, the highest yields were produced by roses on the rootstock *R. cordifolia* and *R. canina* whereas, Pavuv Cerveny Sipek was also satisfactory. *R. canina* Kokulinski was suitable for the cultivar Message. In another experiment, Vecera (1967b) reported that among the rootstocks *R. rubiginosa, R. multiflora, R. coriifolia* and 6 selections of *R. canina,* the least vigorous rootstocks were Schweizer Gruss, *R. rubiginosa* and *R. multiflora*. The Kokulinski and Pav's Rote types of *R. canina* were moderately vigorous. The *R. canina* types Inermis Hort, Schmids Ideal and Broggs were susceptible to mildew. Pav's Rote and to some extent Dablice were susceptible to *Actinonema rosae*. Studies on stock-scion combinations clearly indicated that the rootstocks influenced the stem length and number of flowers. In general, the *R. canina* (Inermis and Pav's Rote) proved very useful with 8 scion cultivars tested. Most cultivars yielded heavily on *R. dumetorum* (Vecera, 1968). In an outdoor trial with 6 rose cultivars on different rootstocks, Rupprecht (1968a) obtained the best performance of Virgo on *R. canina* var. Schmids Ideal, Gloria Dei on *R. canina* var. Senff, Spek's Yellow on *R. multiflora,* New Yorker on *R. multiflora,* Rouge Meilland on *R. pollmeriana* and of Texas Centinnial on *R. multiflora.* Tsvetkov (1973) observed that cultivars Super Star, Orange Triumph, Fiesta Brava and Spartan budded on *R. canina* var. Schmids Ideal was the best scion rootstock combination. Cultivar Madelon was propagated using 27 different rootstock combinations, by various methods including cuttings and grafts (stenting and root). Plants were placed on rockwool matting early March. Kool and Pol (1991) observed that the flower production was strongly dependent on the rootstock used; no relationship was found between yield and flower quality for any of the combinations. The rootstock also had distinct effect on the winter flowering activity of the plants. In the first cropping year there were great differences in yield. Plants on the rootstock selections Helenae and Moneyway (*Rosa hybrida* cultivar), Multi 1 (*R. multiflora*), Multic (*R. multiflora* cv. *cathayensis*), propagated as grafted cuttings, gave higher yields of blooms better than average quality. Swarup and Malik (1974) concluded that *R. indica* var. *odorata* was superior to *R. borboniana* and *R. multiflora.* In field trials in 1988-91 at Pune, rose cv. Gladiator was budded onto *Rosa indica* var. *odorata* [*R. odorata*], *R. multiflora* or 2 thornless rose rootstocks (raised at IIHR and IARI, respectively). Results showed that *R. indica* var. *odorata* was the best rootstock for Gladiator roses for cut flower production on the black cotton soils of the Deccan plateau (Patil *et al.,* 1994). Scion of HT roses cv. Montezuma, Super Star and Piccadily grew more vigorously on *R. multiflora* and *R. moschata* rootstocks than on *R. borboniana*. Flower production was greatest in plants on *R. moschata* and lowest on *R. borboniana*. Piccadily produced the largest number of flowers of biggest size followed by Montezuma but it was very susceptible to die back and leaf spot diseases. Montezuma and Super Star grew best on *R. multiflora* and Piccadily on *R. moschata* (Singh, 1980). The Dutch organization STUR (Stichting Uitgangsmateriaalvoor de Kasrozenteelt) investigated the effect of several *Rosa canina* cv. Inermis rootstock clones on the yield and quality of the glasshouse cultivar Sonia. Starting initially from 573 clones, 150 were selected for a more detailed study. The flower yield of Sonia budded on a number of clonal rootstocks propagated by STUR was higher than the yield of Sonia on its own roots (*i.e.* raised from cutting) (Dood and Rademaker, 1991).

In an experiment, cultivars Mercedes and Ilseta were (1) propagated by direct rooting of cuttings, or (2) were grafted on Mercedes, Ilseta or *R. indica* cv. Major rootstocks and planted in rockwool slabs in the greenhouse. Grafting Mercedes on Ilseta rootstock, as compared with treatment 1, increased the flower bud FW in uppermost and penultimate flowers, whereas in cv. Ilseta, flower bud FW was highest when grafted on *R. indica*. Mean flower stem length and number of flowers/ stem were little affected by the choice of rootstock (Agbaria *et al.*, 1995). Seventeen clones of *R. indica*, a predominant rootstock were collected by Raviv *et al.* (1993) from various locations. Five clones were randomly chosen and grafted with scions of the commercial cultivars Mercedes, Sonia and Golden Times. The 2-year cumulative flower yields differed greatly among the combinations, suggesting that rootstock clonal selection for each cultivar should be conducted in order to ensure higher productivity. Han *et al.* (1996) grafted the rose cultivars Mary Devor, Aalsmeer Gold, Tineke and Mary Devoron *R. multiflora* which yielded the highest number of quality cut flowers.

9.2.5.5 *Effect on Diseases*

The reactions of some rootstocks to important diseases also studied by Van Der Harst (1962) who reported that *R. multiflora* was the best rootstock for plant growth and flower production but it was susceptible to *Verticillium albo-atrum* and was easily damaged by wind. On the basis of the resistance against frost an apomictic *R. canina*, No. 4 was found the most suitable rootstock (Stanko and Mihajlov, 1966).

9.3.0 Micropropagation

Propagation *in vitro* is a rapid and cheaper method producing flowers in rose (Prell and Meier-Dinkel, 1983) and extensive investigations have been carried out during the last two decades. Success has been made with regard to this technique using different plant parts such as shoot apices (Hasegawa, 1979; Khosh-Khui and Sink, 1982), leaf blades (Khosh-Khui and Sink, 1982) and axillary buds (Bressan *et al.*, 1982; Bini *et al.*, 1983; Yan *et al.*, 1996). Dormant axillary buds were found more suitable for *in vitro* culture than shoot tips and the former rooted in 2 weeks only (Prell and Meier-Dinkel, 1983). The position of the node from which axillary buds are isolated markedly affects their growth and development. Buds nearest to or farthest from the apex either failed to develop or took the longest time to develop compared with those in the middle portion of the shoot (Bressan *et al.*, 1982).

Shoot tips and lateral buds from the glasshouse rose cv. Improved Blaze proliferated multiple shoots on a basal medium (MS salts, vitamins, glycine, sucrose and agar) supplemented with 3.0 mg BA and 0.3 mg IAA/l. Roots initiated from about 50 per cent of shoots after transfer to medium containing 0.3 mg IAA and 0 or 0.3 mg BA/l. Regenerated plants were successfully transferred to soil after 2 weeks only (Hasegawa, 1979). Shoot tips and axillary buds of the pot rose cv. Rosamini were cultured on MS medium containing 30 gm sucrose/l, 8 g agar/l and supplemented with 1 mg BA/l. Multiplication rates of 6- to 7-fold were reached when shoots were subcultured every 4 weeks, and these rates decreased with prolonged maintenance of shoots without subculturing. Root induction was achieved with half-strength MS containing 20 g sucrose + 6 g agar/l and supplemented with 1 mg IBA/l. Flower bud break occurred at an early stage after transfer of rooted plants to pots, and occasionally occurred in culture vessels (Scotti-Campos and Pais, 1990). Super Star micropropagation from lateral bud explants were established on a culture medium supplemented with BA, NOXA or IAA. The highest rate of culture establishment (99.4 per cent) and growth were obtained with 2.0 mg BA and 0.1 mg NOXA acid/l. Shoots were obtained using a shoot proliferation medium supplemented with 0.1 mg NAA and 0.5 mg GA_3/l. A rooting rate of 87 per cent was obtained in a medium supplemented with 0.5 mg IBA/l (Bhat, 1992). *Rosa indica* Major, widely used in France as a rootstock for greenhouse cut rose production (difficult to propagate) has been successfully propagated by using the tissue culture technique. Nodes and shoot apices produced shoots on MS medium with 2 mg/L BA and roots on MS medium with 0.1 mg/L NAA. Knop's medium diluted by half led to slower rooting and fewer roots. Rooting of leafy shoots obtained *in vitro* was also successful in a mist chamber on a peat-vermiculite mixture (Avramis *et al.*, 1982). Stem explants (2-3 cm long) of six rose cultivars (Frisco, Europa, Kiss, Flamingo, Kardinal and Cadillac) were cultured on modified MS medium containing 20 mg BA/l, 0.004 mg NAA/l and 0.1 mg GA_3/l, although for satisfactory shoot elongation of Cadillac explants it was necessary to replace the NAA with 0.3 mg IAA/l and to omit GA_3. Culture temperature (15, 18, 21 or 25°C) significantly affected multiplication and rooting, and there were significant interactions between temperature and cultivar. Shoot formation was optimal at 21°C for all cultivars except Europa (optimal at 15 or 18°C). Rooting was optimal at 15 or 18°C (Leyhe and Horn, 1994). An efficient protocol for micropropagation of *R. damascena* was established by Anil Kumar *et al.* (2001) using single node segments from mature bushes. The effect of thidiazuron (TDZ) was compared with that of BA on *in vitro* shoot proliferation. The cultures initiated on medium supplemented with TDZ and/or cultured continuously on TDZ containing medium for 32-48 weeks exhibited considerably more shoot proliferation and growth during subsequent culture on a medium containing BA.

Ibrahim and Debergh (2001) studied adventitious bud regeneration in rose (*R. hybrida*) *in vitro* specifically to extend the protocol to different genotypes (Inka, RUI 317 and RUI 319) and to initiate production of multiple shoots per explant. The best results were obtained by using a two-stage procedure where excised leaflets were incubated on MS induction medium with 6.8 μM (TDZ) plus 0.49 μM IBA in the dark for 7 days and subsequently transferred to an MS-based regeneration medium with 2.22 µM BA plus 0.049 µM IBA exposed to a photosynthetic photon flux density of 15 µmol $m^{-2}s^{-1}$ photosynthetically active radiation.

New methods for the *in vitro* hardening of micropropagated plants were incorporated into practical experimental work using a commercially available kit (Smith *et al.*, 1992). The kit consists of cellulose plugs (sorbarods), which support and protect the roots during transfer to soil, and ventilated culture vessels to improve the resistance of the plantlets to desiccation.

The immature seeds from self-pollinated miniature rose cv. 'Shortcake' produced pale yellow and friable embryogenic calli when cultured on MS medium without any phytohormone for 3 months. The calli were then subcultured on MS medium supplemented with 1 ppm BA every 2-3 weeks for 3 months which resulted in differentiated adventitious shoots in an average of 7.7 per callus. These shoots were separated and rooted on half-strength MS medium with 0.25 ppm IBA. Although the original and regenerated plants were tetraploid, the latter showed some genetic variations in terms of flower size, colour, petal number, leaf shape and prickle number. Moreover, polymorphic DNA differences were also noted from the results of RAPD analysis (Li *et al.*, 2002). Carelli and Echeverrigaray (2002) observed a multiplication rate of 30.3 plantlets per explants taken from 2-5 year old *Rosa hybrid* cv. Baronesse after 180 days in the basal media containing sucrose 30 g/l, benzyladenine 3.0 mg/L and NAA 0.5 mg/L. This system was also found effective for six rose cultivars which showed a multiplication rate of 1.85-2.88 plantlets per explants after 60 days in culture.Wang *et al.* (2002) evalauted *in vitro* flower induction in roses cv. Orange Parade and found that the basal medium containing 400 mg/L myo-inositol and thiadiazuron (TDZ) with NAA and Zeatin (ZT) with NAA were the best choice for flower induction. Moreover, plantlets 156-561 days from original culture and 45 days from subculture were the best for flower induction (Table 5).

Table 5: Effect of different Growth Regulators on Flower Bud Induction in the Rose Cultivar Orange Parade*

Flower Induction Media RMS_2 Supplemented with				*Flower Bud Formation*
Zeatin (mg/L)	*Thiadiazuron (mg/L)*	*Benzyl Adenine (mg/L)*	*NAA (mg/L)*	
0.5	0.0	0.0	0.1	44.2 per cent
1.0	0.0	0.0	0.1	37.5 per cent
2.0	0.0	0.0	0.1	21.7 per cent
0.0	0.5	0.0	0.1	49.2 per cent
0.0	1.0	0.0	0.1	25.8 per cent
0.0	2.0	0.0	0.1	9.2 per cent
0.0	0.0	0.5	0.1	12.5 per cent
0.0	0.0	1.0	0.1	13.2 per cent
0.0	0.0	2.0	0.1	14.2 per cent
0.0	0.0	0.0	0.0	5.5 per cent

* Wang *et al.* (2002).

Singh and Dubey (2003) obtained 18 shoots per culture within 23 days of initial culture in MS medium fortified with 1 ppm BA and 16 shoots per cluster every 3 weeks upon subsequent culture for miniature rose (*Rosa chinensis* cv. Rosamini). These plantlets showed 90 and 83 per cent success on rooting followed by hardening in plastic pots containing soil and leaf mould (1:1) under greenhouse conditions, respectively.

In vitro mass multiplication for rose cv. Benjamin Paul using axillary buds resulted in 100 per cent shoot proliferation on MS medium fortified with 2 mg/L BAP + 0.1 mg/L NAA and higher rooting percentage when cultured on half-strength MS basal medium with 0.5 mg/L IBA followed by hardening in pots and transplanting at field (Maurya *et al.*, 2004). Nodal segments of miniature rose (*Rosa chinensis* var. *minima*) cvs. Candy Sunblaze and Baby Katie recorded maximum shoot multiplication rate of (6.9 shoots/explants) when inoculated on MS medium enriched with 4 mg/L BAP + 2 mg/L kinetin and 0.1 mg/L NAA following the pretreatment with 1000 mgf/l Ridomil + 200 mg/L 8-HQC for 2 h agitation and maximum rooting when cultured on half-strength MS basal medium + 1 mg/L IBA + 1 mg/L NAA. Subsequently, the regenerated plantlets were hardened in glass jar filled with peat: soilrite (1:1) with PC cap and well hardened plants were then transferred to the glasshouse (Bharadwaj *et al.*, 2006). Minas (2007) suggested a method for production of pathogen-free and genetically uniform rose rootstocks for rapid release. Plants were placed for 1 month in glasshouse under 75 per cent shade followed by removal of 5 cm long apical growing shoots in laboratory. The explants were sterilized by soaking in 20 per cent bleach solution with few drops of Tween 20 for 10 min and washed thrice with sterile distilled water. Meristem tips of 1-2 mm were excised and cultured *in vitro* onto MS medium containing MS basic salts, vitamins, 3 per cent sucrose, 4.5 ppm BAP, 9 ppb IBA, 55.7 ppm ascorbic acid solidified with phytagel 0.25 per cent. Cultures were incubated in growth room for 2 weeks in dark at constant temperature of $21 \pm 2^{\circ}$ C followed by subsequent transfer to growth room with 16 h light of 2000 lux by white and red fluorescent lamps and 8 h darkness. Proliferation occurred at a rate of 4-fold per 2 weeks and rooting was carried out in the abovesaid medium. Virus and viriod infection were checked using ELISA and RT-PCR followed by removal of infected cultures. Then, two phases of hardening was practiced, the first in which the plants were exposed to 4000 lux light *in vitro* and in the second phase, the micro-plants were placed in 5000 lux light with gradual reduction in RH from 95 to 55 per cent with the help of micro-fog system. Then, plantlets were hardened in growth chambers with 16 h day light of 5000 lux and 8 h dark at constant temperature of $21 \pm 2^{\circ}$ C.

Miniature rose plantlets at flower development stage were cultured photo-autotrophically on MS medium and subsequently exposed to water deficits (-0.23, -0.32, -0.40 or -0.67 MPa osmotic potential) by increasing mannitol concentration. *In vitro* flowering and number of flowers per plantlet declined significantly at -0.40 and -0.67 MPa water deficit levels. Growth reduction, pigment degradation and greater net photosynthetic rate were noticed at -0.67 MPa. The level of proline increased under water deficit stress. Thus, flowering stage in miniature rose plantlets susceptible to water stress suppressing floral organ development could be used as an effective index for the selection of water stress tolerant genotypes in breeding programmes (Cha-um and Kirdmanee, 2010).

Nodal segments of cv. 'Pareo' were surface sterilized with 0.1 per cent mercuric chloride for 10 mins followed by inoculation on agar-gelled MS medium enriched with 1.5 mg/L BAP. The axillary shoots regenerated were then transferred to half strength MS medium supplemented with 1 mg/L IAB for rooting. The rooted plantlets were hardened and acclimatized in the field subsequently (Mukhambetzhanov *et al.*, 2011). Explants of *Rosa hybrid* 'Al-Taif' cultured on MS medium containing 2 mg/L BAP + 1 mg/L kinetin resulted in maximum percentage of shoot initation (85 per cent) whereas, the same medium only replacing 2 mg/L BAP with 3 mg/L of BAP produced maximum number of multiple shoots (2.7) and with 1 mg/L of BAP produced maximum number of elongated shoots (66.7 per cent). Highest rooting percentages (66.7 per cent) were obtained with MS medium + 2 mg/L IBA and were subsequently transferred to pots containing sterile moss for hardening (Attia *et al.*, 2012).

Noodezh *et al.* (2012) developed an efficient and cost-effective method for rapid and high-quality shoot multiplication and *in vitro* rooting of Damask rose using nodal explants. For optimal explants establishment, shoot growth and proliferation, a modified Murashige and Skoog medium (A_{19} medium) with higher levels of nitrates, calcium and iron supplemented with 4 mg/L 6-benzyl aminopurine and 0.25 mg/L IAA was efficient. To increase shoot length, 75 days after culture initation, shoots were transferred to the same medium additionally with 0.2 mg/L GA_3. Shoots were then separated and transferred to various root induction medium (½ A_{19} medium supplemented with 0.1 mg/L IBA) for 30 days to achieve *in vitro* rooting. Shoots were cultured in this medium for 7 days in the dark, before transfer to liquid ½ A_{19} medium-A without hormone supplementation under a 16 h photoperiod. This modified protocol resulted in significant improvement in shoot regeneration and proliferation and obtained stronger shoots over a period of about 20 weeks. Alsemaan (2013) insisted the use of BA 2 mg/L + GA_3 2 mg/L for higher proliferation rate of explants and IBA 2 mg/L in ½ MS medium for high quality rooting in the micropropagation of damask rose cv. Almarah.

Kumud *et al.* (2015) observed the best multiplication of rose nodal explants in MS medium fortified with BAP 5 mg/L + KN 2.5 mg/L and effective rooting in half strength MS medium containing NAA 2 mg/L + BAP 0.5 mg/L. Further, well-rooted microshoots after subsequent dipping in bavistin 0.1 per cent, washing in running tap water followed by transferring to plastic cups with sand and gravel (1:1 ratio) showed healthy growth with 60 per cent survival in ex-vitro conditions. Ali *et al.* (2020) noted maximum number of shoots, shoot length and number of leaves with the explants cultured in MS + NAA 0.5 mg/L + BAP 2 mg/L, whereas early root initiation and maximum number of roots were recorded in ½ MS + NAA 1 mg/L + IBA 1 mg/L. Also, *in vitro* healthy and complete plantlets on subsequent transfer into pot mixture containing soil: sand: FYM (1:1:1) exhibited utmost survival percentage. Though tissue culture is believed to be advantageous for the mass production of rose plants, bacterial contamination and phenolic exudation are the main drawbacks to be addressed. Shokri *et al.* (2015) reported that surface sterilization of explants with 200 ppm nano-silver for 20 min controlled bacterial contaminations. Furthermore, addition of nano-silver 100 ppm to the medium reduced the bacterial contamination as well as phenolic exudation. However, high concentrations of nano-silver slowed down the regeneration sometimes leading to explants destruction.

Pretreatment of Raktagandha rose explants with carbendazim 0.2 per cent + dithane M-45 0.2 per cent + 8-HQC 200 mg/L for 3 h on a horizontal shaker (120 rpm) caused maximum explants survival and sucrose 30 g/l in medium improved shoot multiplication. Subsequently, MS medium fortified with 3.5 mg/L BAP + 0.2 mg/L NAA + 0.5 mg/L GA_3 resulted in better culture establishment and shoot proliferation with maximum of 4.62 shoots per explant. Further, rooting was induced on ½ MS medium + NAA 0.5 mg/L + IAA 0.5 mg/L and were hardened in glass jars filled with vermiculite + agropeat (1:2) with ½ MS medium salts followed by transfer to glasshouse (Kumari *et al.*, 2017).Hassaneinand Mahmoud (2018) reported that *in vitro* propagation could be improved by thidiazuron and silver nitrate; leaf discs could be induced with proper combination of auxin and cytokinin; GA_3 showed somatic embryogenesis without shoots in media containing thidiazuron but, 37 per cent explants regenerated shoots directly on medium containing GA_3; BA was essential for shoot elongation and IBA was better for rooting than IAA.

The tissue culture-derived plants should be true-to-its explants parent for its commercial use and this could be evaluated by genetic fidelity techniques using molecular markers. The genetic fidelity of cv. First Red, Cri Cri and Pusa Gaurav plants was assessed by using 12 RAPD and 17 ISSR markers which were multiplifed *in vitro* from the axillary bud. The markers generated a total of 119, 104 and 114 amplicons for the 3 cultivars, respectively and showed a homogenous amplification profile between explants source and TC plantlets. Thus, the result assured the clonal fidelity of TC raised plantlets and suggested axillary multiplication for safest mode without any somaclonal variation (Senapati *et al.*, 2012).

9.3.1 Acclimatization and Field Establishment

Douglas *et al.* (1989) reported that the rooting of micropropagated plants improved significantly in the compost before they were precultured for two weeks *in vitro* on sorbarods soaked in liquid medium containing IAA. Sorbarods provided the necessary protection and facility for proper handling of the microshoots and helped in the production of larger roots which facilitated 100 per cent successful establishment. The percentage survival of the micropropagated hybrid rose cvs. Landora, Queen Elizabeth, Virgo, Happiness, Seapearl varied from 92 to 98 per cent when transferred to the greenhouse at 80-85 per cent RH (Rout *et al.*, 1989).

9.3.2 Performance of Micropropagated Plants

Dubois *et al.* (1988) investigated whether plant habit might be favourably modified by *in vitro* propagation. The plant habit, growth and development of 36 dwarf cultivars propagated *in vitro* and by cuttings were compared. Plants from *in vitro* material flowered earlier, had shorter shoots with fewer and shorter internodes, and more and longer laterals than plants from cuttings.

Seven cultivars (Alexandra, Sutter's Gold, Montana, Schweizer Gruss, Madame Meilland, Duftwolke and Friesia), propagated by shoot tip culture, were planted on their own roots, at 8 sites, at altitudes of 400-1400 m (Reist and Le, 1989). They were compared with plants of the same cultivars grafted on *R. dumetorum* and *R. canina*. Vigour and flowering of the *in vitro*-propagated plants were more dependent on environmental conditions after planting than was the case for their grafted counterparts. Two soilless culture experiments were conducted on Carambole roses, one using own-rooted rose plants and the other using micropropagated plants, in a greenhouse. The adaptability of the rose to soilless culture on substrates contained in polyethylene bags and fertilized with a nutrient solution was evaluated. Pink bark, perlite, marc and rockwool were used as substrates and compared in terms of yield and quality. In both experiments the rose cultivar showed great adaptability to soilless culture. Yield was not affected by the substrate material. During the first year, the self-rooted and micropropagatedplants yielded an average of 42.1 and 39.9 stems/m^2, respectively (Carletti *et al.*, 1992).

10.0 Open Cultivation

Roses are grown in gardens in all parts of the world. In temperate countries, the most important period of flowering is summer, in tropical climate the major seasons for flowering are winter and spring while under wild subtropical environment with moderate temperature roses grow and flower almost throughout the year.In order to produce cut flower of good quality for a long duration roses are grown under protection with controlled environment.

10.1.0 Soil and Climate

10.1.1 Soil

Although any soil is good for rose cultivation provided it has proper drainage, the ideal soil should be medium loam having sufficient organic matter with a pH of 6.0 to 7.5 (Hunt, 1973; Bose and Mukherjee, 1977). Compact soils with poor structure have restricted aeration which impedes the activity of roots affecting the plant growth. The land where the external drainage is poor and water stagnates during monsoon should not be selected for rose cultivation. In very heavy soils, gravel and sand may be mixed into the bed to allow drainage. The land with high water-table is not suitable for rose beds.

10.1.2 Climate

The effects of light (photoperiod, intensity and quality), temperature (aerial and root zone), humidity, CO_2, ventilation, *etc.* and their interactions on growth, flowering, yield and flower quality have been discussed under protected cultivation. The most important point regarding environment to remember is that roses love sunshine and free ventilation (Champneys, 1956). They need bright sunshine for the whole day, if not at least for normal part of the day, the forenoon. Sunshine for six hours is ideal for better growth and flowering (Biswas, 1984). The plants should be free from shades of trees and protected from the strong winds.

Humidity also plays an important role in the incidence of pests and diseases affecting the growth and flowering. Certain diseases like mildews are associated with high relative humidity. Its importance, however, lies in the internal water conditions of the plants. Wilted plants grow very slowly and by maintaining reasonably high humidity, the internal deficiency of water will be less. Enlargement of cells is caused by water and, therefore, if the loss from the plants could be reduced, leaves and flowers would be larger and stems longer (Laurie *et al.*, 1958). Grossi *et al.* (2004) studied the influence of seasonal growing environment on growth and postharvest performance of rooted liners of pot rose. Single shoot plants were grown in controlled environment chambers simulating summer (30°C day/21°C night cycle with a 14-hour photoperiod) and winter (21°C day/16°C night cycle with a 10-hour photoperiod) greenhouse growing conditions. The plants were continuously exposed to lighting simulated at 21 ± 1° C under 15 $\mu mol\ m^{-2}\ s^{-1}$ photosynthetic photon flux from cool-white fluorescent lamps. Results revealed that quicker flowering, smaller flower diameter, more compact growth, smaller leaf area and greater flower longevity were exemplified under summer-grown plants over winter-ones.

10.2.0 Layout and Preparation of Beds

The plan of rose garden and design of the beds should be simple and formal or informal. Rose beds may be of various designs, depending upon the liking of the grower. However, rectangular beds are advantageous for maintenance. The width of the beds should be such that operations like weeding, hoeing, forking, cutting of flowers, *etc.* can be done from both the sides of the bed without stepping into the bed. The width of the bed should be 1.2-1.6 m, the length will depend on the size of the garden and plant of rose beds, preferably not exceeding 6 m each.

10.2.1 Preparation of Rose Beds

Preparation of soil is the key of success with roses. It should be rich, porous and well drained. The initial preparation of rose beds should preferably be started during summer season so that soil gets exposure to sun and air and during the monsoon it gets a chance to settle down before planting. The area selected for rose growing should be cleaned of weeds and according to the plan, the borders of the proposed beds should be marked. Grasses along with their roots, within the marked border, should be completely weeded out by deep digging. With the help of digging spade, the soil up to a depth of about 30 cm, should be dug out and heaped on the ground adjoining the beds. The soil should be pulverized; gravels, stones, brick pieces and other foreign materials removed and exposed to sun for at least a week.

If the soil is light, sandy and stony, the next 30 cm of the soil in the trench should be dug as deep as possible, pulverized and levelled in the trench itself. After that about half portion of the dug out soil, heaped on the ground surface, should be returned to the trench, worked with digging fork and levelled. This layer needs manuring with organic manure at the rate of 50-60 tonnes of farmyard manure or compost per hectare. The remaining portion of the dug out soil should be returned to the trench, worked thoroughly with digging fork and levelled. Superphosphate at the rate of 30 kg P_2O_5 per hectare should be incorporated and beds irrigated thoroughly to allow the soil to settle before planting.

On the other hand, if the soil is heavy, it would be necessary to dig out the second 30 cm layer of the soil and heap on the ground surface. The next layer of the soil should be deeply dug, pulverized and levelled. The second layer of soil, after it is pulverized and sun-dried, should be returned to the trench and manured with organic manures at the rate of 50 tonnes per hectare. The remaining operations will be similar as in the case of sandy soil.

When the rose beds are prepared after monsoon, at least two heavy soaking should be given for the soil to settle, otherwise the rose plants after planting and irrigation will sink too much, creating a lot of problems for management. About 10 days before planting roses, the soil should be again manured with farmyard manure or compost at the rate of 50 tonnes per hectare and the beds should be irrigated (Biswas, 1984). Chloropyriphos should be used to prevent attack of termites and other soil pest.

10.3.0 Procuring Rose Plants

Rose growers, who can raise their own nurseries, may plant suitable rootstocks and bud the cultivars of their choice. The specification also indicates the minimum dimensions for rootstock, root system and shoots. Root shall arise within 6.5 cm of the base of the union and the root system should include a minimum of three major roots. If the plants are lifted with earthball, the diameter of the earthball should be at least 15 cm. The plants should bear at least two shoots arising from the union or one shoot which produces branches not more than 6.5 cm above the union. The sum of the diameters of the shoots or shoots arising directly from the union shall exceed 3 cm. These shoots should be hard and ripe and shall not yield to normal pressure of forefinger and thumb at 7.5 cm above the budded portion. In Standard rose plants, the stem should have a minimum diameter of 2.5 cm below the lower union. The earthball should measure at least 20 cm in diameter and 25 cm in height. The root system should not be exposed in any case. In case of double budded rose, the unions shall be a close together as possible on opposite sides and shall not be more than 7.5 cm apart. At each of the two unions, the sum of the diameter of the shoots arising from the union should not be less than four-fifths of the diameter of the stem, measured 2.5 cm below the lower union. Like bush roses, the shoots of Standard roses should also be ripe and hard and should not yield to normal pressure of forefinger and thumb at 7.5 cm above the union. In case of climbers, ramblers and pillar roses, the plants should have at least two shoots of not less than 50 cm in length, measured from the union.

10.4.0 Selection of Cultivars

The success of rose cultivation depends on the proper choice of the classes and cultivars. There are several thousand cultivars of roses and several hundred new ones are being added every year. Therefore, selection of suitable ones is very important. The first step in the selection will be the kind of roses which will suit the location and fulfil the object. Apart from this, form, size, shape, floriferousness, colour and fragrance of the blooms should also be taken into consideration. Resistance to insect-pests and diseases is another important factor to be kept in mind while making the selection of cultivars. Polara *et al.* (2004) reported that rose cultivars like Peace, Eiffel Tower, Gladiator of Hydrid Tea group and Samba, Dickson's Flame and Zorina of Floribunda group gave best performance with respect to maximum yield, stem length, flower diameter and flower weight, respectively under south Saurashtra region of Gujarat. A study conducted by Priya *et al.* (2019) revealed that 'Taj Mahal' was superior with respect to plant height (52.83 cm) and number of petals (79.54) whereas, more number of flowers was recorded with 'Revival' (12.45) which was statistically similar to that of Arka Ivory (11.38), Arka Swadesh (10.10) and Peach Avalanche (9.87). Vase life was found highest with Revival (4.83 days), Taj Mahal (4.75 days) and Noblesse (4.75 days). Further, the genetic estimates of heritability and genetic advances were greater for the traits: fresh weight of flower, number of petals per flower and vase life, indicating their additive gene action. Thus, varieties such as Taj Mahal and Noblesse which were outstanding in terms of these attributes could be used in breeding programmes and also for cultivation.

Suganthi *et al.* (2019) evaluated *Rosa hybrid* L. varieties for growth, yield and quality. Among the varieties highest stem length was recorded in Top Secret followed by Rock Star, similarly maximum vase life was recorded in Top Secret and Rock star. The remaining parameters like plant height, number of leaves, number of shoots, bud length, flower diameter, number of flowers per plant, number of petals per flower were also higher in the variety Top Secret closely followed by Rock Star. The study revealed that the Top Secret and Rock Star was the best performing varieties under poly house in shevaroy hills (Tables 6 and 7).

Table 6: Vegetative Parameters of Dutch Rose Varieties*

Varieties	*Plant Height (cm)*	*No. of Leaves*	*No. of Shoots*	*Thorn Density*	*Days to Flower Bud Initiation*
TopSecret	121.03	99.45	6.95	2.03	28.81
Brilliant	99.61	89.30	4.10	3.70	32.41
Solair	94.20	85.90	3.70	2.03	35.40
Avalanche	106.03	96.80	6.10	4.40	32.60
RockStar	118.30	99.03	6.93	3.03	30.41
Peach Avalanche	106.90	90.60	5.60	4.60	36.71
Candy avalanche	91.96	80.30	4.80	7.30	40.60
Espana	99.00	80.10	5.40	5.90	33.85
Jumilia	96.90	96.90	4.10	11.40	31.62
Hot shot	98.30	91.30	4.09	3.61	42.46
BonHeiur	88.50	85.60	3.10	3.09	37.89
MEAN	101.88	90.48	4.99	4.64	34.80
SE	6.17	4.58	0.35	0.46	2.33
CD	12.63	9.38	0.72	0.94	4.76

Table 7: Flower Yield and Quality Parameters of Dutch Rose Varieties*

Varieties	*Bud Length (cm)*	*Diameter of Opened Flower (cm)*	*Neck Length (cm)*	*Stalk Length (cm)*	*No. of Petals/ Flowers*	*No. of Flowers/ Plant*	*Vaselife (Days)*
TopSecret	5.64	12.42	9.12	70.63	53.60	26.16	17.67
Brilliant	5.10	8.06	7.48	56.30	51.30	21.92	15.33
Solair	5.20	8.52	8.13	50.86	34.30	21.32	15.00
Avalanche	5.13	10.80	7.94	58.23	42.30	23.82	14.05
RockStar	6.01	11.80	9.02	68.15	54.60	25.06	16.33
Peach Avalanche	5.51	11.40	8.70	61.34	52.00	23.97	13.01
Candy avalanche	5.40	10.62	7.60	49.63	37.30	20.56	14.67
Espana	5.79	10.41	8.83	56.40	42.00	23.55	15.61
Jumilia	4.99	11.01	8.00	54.40	36.90	15.07	12.90
Hot shot	5.21	9.31	7.90	54.13	28.80	19.05	15.00
BonHeiur	4.90	9.05	8.76	48.63	36.60	18.22	13.21
MEAN	5.35	10.31	8.32	57.15	42.70	21.70	14.80
SE	0.27	0.44	0.39	2.17	0.99	1.51	0.65
CD	0.56	0.91	0.80	4.45	2.03	3.10	1.32

*Suganthi *et al.* (2019).

10.5.0 Planting

This operation should receive very careful attention and to do this job well, there are few important preliminaries which are given below:

- ✰ To cut away all immature, dead, inward growing or diseased shoots and unduly long, dead snags. Also, as far as possible, each shoot to the next outward growing bud should be cut.
- ✰ To remove all suckers growing below the point of union.
- ✰ To reduce the possibility of loss of moisture, some leaves including dried and yellow ones should be removed.
- ✰ If the roses are delivered with shrivelled bark, it is suggested to immerse the plants in water for 24 hours to plump up.
- ✰ To immerse each plant in a suspension of 1 g blitox in 1 litre of water to lessen the risk of attack of fungi.

Planting should be done in thoroughly prepared beds. At the planting time, the soil should neither be too wet nor too dry. At each marked spot, a hole just large enough to accommodate the earthball around the roots of the plant or the roots themselves in case of bare-rooted plants should be dug and the soil taken out, heaped on the surface near the hole. A hole, measuring 20 to 30 cm in diameter and 30 cm deep, has been found to be sufficient in most of the cases (Biswas, 1984). The earthball should be scrapped lightly and gradually lowered in the hole and allowed to rest on its bottom. The gap in the hole is to be filled with the soil which was dug out and pressed properly to anchor the plant firmly. Then the soil around the stem should be rammed firm by treading over with feet. This will press out air pockets in the soil which will help the roots coming in contact with the soil particle and intake of water and nutrients through the rootlets. The bed after planting should be thoroughly irrigated.

If the rose plants, received for planting, are with bare roots, the roots along with stem up to the bud union should be dipped in water for about six hours before planting. This will restore the loss of moisture which might have taken place during the transit period. The method of planting is rather similar to that of plants with earthball except that when hole is prepared for receiving the plants, a mound of soil is made at centre of the bottom of the hole. The plant is placed on this mound and roots are spread out evenly on all sides. The hole is then filled in with soil and pressed as mentioned earlier.

While planting the Standard roses, it is necessary to stake the plants by fixing at points at the top, middle and bottom of the main shoot. The position of the stake preferably should be on the rear side of the normal viewing side and the knots should be tied loosely on the stem but firmly on the stake. A bamboo stake painted green will present a good look. For climbing roses, the trellis should be made before planting. If it is to be planted near a wall, the plants should always be selected with side-growing roots because the soil at the foot of a wall is generally dry. Planting is to be done about 30 to 40 cm away from the wall, with the tips of the roots outwards where there is moisture, and trained back to the wall.

10.5.1 Depth of Planting

While planting, the position of the bud union should be looked into. Own-rooted plants are set at the same depth as grown in the nursery. Budded plants generally are planted so that the bud union is slightly above the ground level in the bed (Laurie *et al.*, 1958). In temperate countries where winters are severe, it is usually recommended that the bud union should be placed slightly below the soil. But in India, better results are obtained if the planting is done with the bud union 2.0 to 3.0 cm above the soil level (Pal, 1978).

10.5.2 Season of Planting

Planting time depends mainly on the climatic conditions of the region. In countries with a severe winter, planting may be done either in autumn or in the spring when plants are dormant and easy to handle (Pal, 1978). Three year's trial on time of planting carried out by Lundstad (1956c), however, showed casualty due to frost in autumn-planted roses. There was no casualty among the spring-planted roses. Autumn planting, on the other hand, resulted in earlier flowering. In general, roses can be planted any time except in very hot summer and during heavy rains when the soil is very wet. But winter months are the best and safest. The suitable time for planting roses in the most plains of India is during September-October and in the hills it is during October-November or February-March in plains (Malik and Dadlani, 1984). Under the climatic conditions of northern India, October is the best month for planting. However, planting can be extended upto January. But with late planting, flowering is delayed and the quality of bloom also becomes poor. In the plains of eastern India, rose can be planted from October to February (Bose and Mukherjee, 1977). According to Biswas (1984), the period of September to December will be quite suitable for rose planting in the plains of West Bengal, Bihar and Orissa. In UK, the season for planting roses is from late October to March (Hessayon, 1981).

10.5.3 Planting Distance

Plant density plays an important role in the production of quality blooms per unit area. Different spacings are adopted for different types of roses. It also differs from place to place. Pal (1972) suggested a spacing of 60-75 cm between plants and rows, for HT and Floribunda, mainly for garden display, depending upon the vigour of the cultivars. The more vigorously growing cultivars like Golden Giant, Super Star, South Seas, First Prize and Queen Elizabeth require a greater planting distance than others. The dwarf Polyanthas are planted 45 cm apart, Miniatures 30 cm apart and climbing roses, 3 m or more apart (Malik and Dadlani, 1984). Harkness (1979) suggested planting of HT and Floribunda at 55 cm apart.

From a trial with rose cv. Montezuma spaced at 20 x 30 cm, 30 x 30 cm and 30 x 40 cm, the highest yield was obtained with a spacing of 20 x 30 cm (Table 8). The production of flowers with longer stems was more with this spacing as compared to others (Yadav *et al.*, 1985a). According to Swarup (1967) closer spacing tends to make the plants grow erect and produce long stalks. Higher yields due to increased plant population per unit area were also recorded by several workers in different cultivars of roses (Pessala, 1977; Bose, 1984; Han *et al.*, 1999). Kengond *et al.* (2015) revealed that planting at a distance of 0.6 x 0.6 m increased the flower yield per ha compared to 0.9 x 0.9 m, but reduced the plant height and plant spread in Gladiator and Sophia.

Table 8: Effect of Spacing on Growth and Flowering of Rose Cultivar Montezuma*

Spacing	*Number of Flowers per Bush/Year*					*Flower Diameter*
	Plant height	*First Grade Flower*	*Second Grade Flower*	*Third Grade Flower*	*Total (cm)*	
20 x 30 cm	105.0	4.1	7.7	15.0	26.8	8.5
30 x 30 cm	101.6	5.0	9.0	19.3	33.3	8.8
30 x 40 cm	95.3	5.7	12.0	20.5	38.2	9.0
C.D. at 5 per cent	1.93	0.50	0.86	0.77	1.00	0.21

* Yadav *et al.* (1985a).

10.6.0 Irrigation

Adequate soil moisture at all stages of vegetative growth and flowering is essential for the rose plants. Excess of water, on the other hand, is harmful to rose plants which cannot tolerate waterlogging. The total volume of water through irrigation should be sufficient to meet the water loss through the processes like transpiration, evaporation from the soil and deep percolation. According to Allen

(1956), an average sized Hybrid Tea rose bush might lose through transpiration about 30 gallons of water during the growing period. A well established Rambler like Dorothy Perkins may transpire about 100 gallons of water.

The frequency of irrigation depends on stage of growth, soil texture, climate and whether the plants are grown in the field or pot. In lighter soil more frequent irrigation is required than that in heavy soil. After manuring each time, watering is necessary. The plants should not be irrigated before and after pruning. Biswas (1984) suggested to water the rose beds once a week or 10 days in winter and twice a week during the summer season. During the rainy season watering depends upon rainfall. If the rainfall is regular and adequate, there is no need for irrigation. However, one should be careful about drought and sporadic light rainfall. If such days prolong, irrigation should be given. In most of the hilly regions of India, it rains heavily from June to September. But in April and May watering is necessary. In India, usually the surface or channel system for ground culture and manual watering-can system or soaking with hose pipe are practiced. From practical point of view, the water should be allowed to flow into rose beds and care should be taken that water spreads evenly all over the bed.

The effects of no irrigation for 0, 4 or 8 weeks on yield and quality of cut flowers of roses (cultivars Mary Devor and Aalsmeer Gold) harvested in autumn and winter, were investigated in Korea by Han *et al.* (1997). Total yields of cut flowers were 27 flowers/plant in continuous irrigation (control), and 30.4 and 30.9 flowers/plant in the treatments not irrigated for 4 and 8 weeks respectively. Chen *et al.* (2015) developed an improved drip irrigation method for Chinese rose (*Rosa chinensis*) which utilized a non-saline root environment during early stages of salt-leaching and a gravel sand layer below the saline soil but above water table. This method controlled the soil matric potential (SMP) threshold of -5 kPa in the first year, -10 kPa in second year and -20 kPa in third year which could be used to reclaim coastal urban saline land for vegetation rehabilitation. Al-Yasi *et al.* (2020) observed that irrigating plants to 25 per cent field capacity (FC) caused water stress and significantly decreased fresh and dry weight of *Rosa damascena* as well as influenced the photosynthetic parameters. Severe water stress (25 per cent FC) resulted in more negative osmotic potential at turgor pressure and turgor loss point than mild stress (50 per cent FC) and control. Moreover, the leaf bulk modulus of elasticitiyexpanded from 2.5 M Pa under well-watered conditions to 2.82 and 3.5 M Pa under mild and severe stress, respectively. Also, *R. damascena* experienced osmotic adjustment due to the active accumulation of soluble carhobydrates and proline with tissue dehydration under severe stress. Thus, two important mechanisms of drought tolerance were observed: osmotic and elastic adjustment but failed to offer resistance to water stress beyond 25 per cent FC. Li *et al.* (2016) determined the effects of salinity on rose (*Rosa chinensis*) growth and ion uptake using five salinity levels of 0.8, 3.1, 4.7, 6.3 and 7.8 dS/m of drip irrigation after planting on the soil reclaimed with soil tillage, drip irrigation and gravel-sand layer. Tensiometers were buried 20 cm deep to control SMP (-5 kPa in first year and -10 kPa in second year). It was clear from the results that the increasing salinity of irrigation water had adverse effects and salt stress had the greatest impact on relative leaf water deficit value.

When irrigated with saline water, most roots penetrated beyond 16-19 cm depth into the high-salinity subsoil, which was disadvantageous to the absorption of water and nutrients. Moreover, rose plants stored more absorbed Na and Cl ions in roots and stems; Ca, Mg and K in leaves. Increased Na concentration and decreasing K/Na ratio had adverse impact on dry matter production. Therefore, soluble potash should be applied for saline water irrigation to increase the selective absorption ratio of K to better counteract the effect of high Na concentrations in such soils (Table 9).

Table 9: Chinese Rose Growth Characteristics for different Irrigation Water Quality (EC)*

Treatments	*Survival Rate (Per cent)*	*Plant Height (cm)*	*Stem Diameter (cm)*	*Growth Ratio*	*Total Length of Branches (cm)*	*Length of Branch (cm)*
0.8 dS/m	96.67a	106.60 ab	12.57 ab	1.19 a	563.81 ab	40.73 b
3.1 dS/m	68.89b	117.03 a	13.43 a	1.17 a	602.11 a	50.47 a
4.7 dS/m	40.00c	101.05 ab	12.23 ab	1.21 a	468.09 b	37.09 bc
6.3 dS/m	10.00d	96.40 ab	11.83 b	1.23 a	441.65 b	34.35 c
7.8 dS/m	8.89d	92.90 b	11.46 b	1.23 a	436.00 b	34.25 c

* Li *et al.* (2016).

10.7.0 Manuring and Fertilization

Rose is a nutrient loving plant and about 16 elements have been known to play important role for its proper growth and development. In addition to the major nutrients like nitrogen, phosphorus, potassium, calcium, magnesium and sulphur, roses require micronutrients like iron, manganese, copper, boron, molybdenum, zinc, *etc.* The yield of rose flowers can be markedly increased through supply of adequate level of nutrients (Zieslin and Halevy, 1972; Zieslin *et al.*, 1973). Inadequate plant nutrition causes serious disorders in rose and may eventually lead to decline of the plants.

The chemical composition of plant may be an index of their nutritional status and requirement. Carbon, hydrogen and oxygen constitute the bulk (about 95 per cent) of the rose plant. The remaining 5 per cent consists of mineral matter and nitrogen (Biswas, 1984). Benedetto *et al.* (1995) analyzed leaf samples of *Rosa hybrid* cultivars for N, P, K, Ca, Mg and micronutrients. P was not detected at limiting levels in any samples, probably owing to the annual incorporation of large amounts of organic matter. K was found to be deficient in all cases. N levels were high at the start of the cultivation cycle but decreased appreciably in a short time owing to high plant growth rates. Foliar N, P and K concentrations tended to follow distinct patterns for most cultivars and cultural regimes, but Ca, Mg and most of the microelements were more variable. If the soil fertility is to be maintained, the quantities of the nutrient elements removed by the plants from the soil need be replenished through the appli-cation of manures and fertilizers. Since the nutrient availability and requirement depend on many factors like the types and cultivars, soil type, soil reaction, soil ferti-lity, extent of leaching of nutrients and soil erosion, and on differentclimatic factors, proper maintenance of fertility of soil is somewhat a complex problem.

Judicious fertilizer schedules of different locations can thus be worked out only through location based nutritional trials coupled with continuous determination of soil and plant nutrient status. Nagaraju *et al.* (2003) observed significant increase in the number and diameter of first (L_1), second (L_2) order laterals and plant spread with the independent application of 30 g N, 24 g K and 12 g multiplex per plant in field grown rose cultivar Gladiator. However, the number and diameter of bottom breaks increased with the treatment of 30 g N in addition to number of laterals and plant spread. Furthermore, plants receiving 30 g N + 24 g K improved number and fresh weight of flowers, stem and bud length, bud diameter and production of good quality flowers whereas multiplex exhibited improvement only with respect to flower number and bud diameter. Da *et al.* (2014) stated that the application of sewage sludge obtained from pond stabilization enhanced the root and aerial biomass, leaflets area and nitrogen content in leaves of rose cv. 'Carola' and no statistical difference was observed between application of mineral fertilizer and that of sewage sludge 36 Mg/ha

Trivedi *et al.* (2016) proclaimed that 6 per cent Panchgavya + 1:5 times dilution of vermiwash + 30 per cent Jivamrita exhibited early bud initiation, maximum number of flowers/plant, maximum flower bud length, maximum number of petals/flower, bigger flowers, extended flowering duration with increased stem length and higher yield in cv. 'Grand Gala'. Butcaru *et al.* (2019) insisted the use of ameliorative plant species in organic rose cultivation. Before planting roses, plant species like *Sinapis alba* L., *Tagetes patula* L. and *Phacelia tanacetifolia* L. were planted and incorporated into the soil when they reached flowering and seed formation. After rose planting, the ameliorative plant species were planted between the rose rows. These practice improved the microbiological activity, soil organic matter content, potassium, optimized pH value and stimulated the vegetative growth and flowering of plants in terms of shoot length and number of flowering sprouts.

10.7.1 Role of Nutrients

10.7.1.1 Nitrogen

Nitrogen plays a key role on growth and development of rose plant (Bik, 1970). Due to nitrogen deficiency, on the other hand, the young leaves become small and develop pale green colour. Sometimes, there are reddish spots on the leaves. The stems are thin and reduced in number. The buds develop poorly and flowers are small and light coloured (Laurie *et al.*, 1958; Pal, 1972; Johansson, 1978). If nitrogen dressing is excessive, the nutrient balance is disturbed and non-protein nitrogen accumulates, thereby decreasing the yield (Dougall, 1959). Normally, severe chlorosis of leaves and defoliation are typical of the situation. Unlike deficiency which can be corrected by addition of the nutrient in some form, toxicity is difficult to correct. The best and simplest way is to leach out the excess nutrients through heavy irrigation and proper drainage. The maximum flower production and stem length of roses under glass occurred when the nitrogen level of the soil was maintained between 25 and 100 ppm all the year round (Seeley, 1948). According to Young

(1971) about 7 g of inorganic nitrogen is required by a rose plant during a season. However, this will vary depending upon the size and vigour of the plant, fertility status of the soil. Nitrogen from nitrates is more readily available to the plants and continued application of ammonium sulphate may increase the subsoil acidity. It is advisable to apply soluble forms of nitrogen in split doses, particularly in the high rainfall areas, and it is commonly recommended that half of the required nitrogen should be applied in the form of organic matter. Foliar spray of urea of low biuret content may be beneficial to a limited extent.

10.7.1.2 Phosphorus

Phosphorus is also one of the important nutrients necessary for the proper growth and flowering of rose. It is an active ingredient of cell nucleus and encourages root growth and stimulates flowering (Allen, 1959). Vigorous growth and the production of sturdy and healthy foliage in rose plants can be obtained by application of this element as superphosphate (Ledlie, 1923). Soukup and Stankova-Opocenska (1974) reported that a 50 per cent rise in the phosphorus rate reduced black spot (*Actinonema diplocarponrosae*) incidence by 33 per cent and omission of phosphorus increased it by 33 per cent. Experiments conducted by Zal'tsfas and Chemarin (1974) indicated that when phosphorus was increased from 60 to 180 kg/ha, the yield of flowers increased by 11 per cent and oil yield by 12 per cent. A remarkable increase in flower number, fresh weight, flower stem length and number of strong flower shoots per plant was also recorded by Bakley (1974). In case of phosphorus deficiency, rose leaves turn dark green and may drop before turning yellow. Stems are weak and bud development is slow due to ill developed root systems (Heatly, 1959; Pal, 1972). Application of soluble forms of phosphorus has been found beneficial and superphosphate (single) should be the first choice for phosphatic fertilizer, otherwise calcium phosphate can be used.

10.7.1.3 Potassium

Potassium is an important and indispensable element for the growth and development of rose. It ensures hardiness of plants, helps the wood to ripen, increases the quality of blooms and resistance to cold and diseases, particularly powdery mildew (West, 1966; Bik, 1972). Reduction in NH4 toxicity symptoms in Forever Yours roses was also recorded due to the supply of an increased amount of potassium (Woodson and Boodley, 1982a). The deficiency of potassium causes marginal browning of lower leaves resulting in marginal leaf scorch. Stems are weak and the growth of the plants is stunted. Buds fail to open and the flower colour is pale (Pal, 1972; Woodson and Boodley, 1982b).

Seeley (1950) was of the view that when rose plants were grown with insufficient supply of potassium, they produced short flowering shoots and small-sized flowers than normal, followed by chlorosis resembling that of a mild deficiency of iron. Post and Fischer (1951) reported that a potassium level of 100 lb per acre was insufficient to obtain higher yield of flowers. However, there was no advantage in maintaining a level higher than 300 lb per acre. Excess potassium is injurious and should be avoided. Since in Indian soils potassium content ranges from medium to high, potassium should be used judiciously, otherwise it may even induce magnesium, manganese and zinc disorders.

10.7.1.4 Calcium

Calcium is an integral part of the cell wall materials. It stimulates root growth and helps in translocation of carbohydrate. The deficiency of this element results in hard and stubby growth and is generally associated with a low soil pH. The root system becomes short and somewhat thickened (Laurie *et al.*, 1958). The terminal buds die and the plant becomes defoliated. The margins of leaflet die, rest of the leaf turns yellowish-green and dark brown patches appear between the mid-rib and margin. Ultimately, leaves drop off before they are dead. The flowers are deformed, reduced in number and have brown spots near margin of the petals (Johansson, 1978; Biswas, 1984). The deficiency can be corrected by the application of domestic lime or gypsum. Post and Fischer (1951) conducted experiments on calcium nutrition of glasshouse roses and recorded reduction in flowering by the application of higher amount (7500 lb/acre) of this nutrient. Teradi *et al.* (1997) studied the effect of Ca concentration in nutrient solution on growth and nutrient uptake of roses cv. Sonia. Ca uptake increased from 0.12 to 0.20 g per 100 g FW when Ca concentration increased from 20 to 160 ppm. The amounts of N, P and K uptake per 100 g FW averaged for all Ca treatments were 0.47, 0.1 and 0.5 g, respectively. Mg uptake decreased from 0.06 to 0.03 g per 100 g FW as Ca concentration increased from 20 to 160 ppm. Increased calcium concentration in the nutrient solution during production increased the concentration of calcium in the buds and flowers in roses with low levels of ammonium (<27 per cent). The number of wilted flowers until day 21 of the postharvest period was reduced by increasing calcium concentration. By adding more lime to the peat at low ammonium levels the number of wilted flowers and abscised yellow leaves decreased (Nielsen and Starkey, 1998). At 5 per cent ammonium, which was optimal for flower and leaf longevity in roses, the concentration of calcium in the plants was almost constant for all the ranges of lime. Postharvest life of potted *Rosa hybrida* (cv. Purple Parade) plant was studied by Nielsen and Starkey (1999). At the end of production, the buds and flowers were analysed for Ca and total N and 3 weeks later, postharvest life was determined by flower wilting and leaf abscission. With increasing Ca concentration in the nutrient solution the percentage of wilted flowers decreased, and at low concentrations of NH_4 (< 3.5 mM) the Ca concentration in buds and flowers increased.

Etemadi *et al.* (2012) observed improved root development, increased root length, diameter, number and root area, increased Ca content in leaves and petals with increasing Ca (2.5-7.5 mM) and B (25-75 μM) concentration in the nutrient solution when rose cultivar 'Easy Lover' was grown in soilless culture. However, high Ca concentration showed antagonistic effect against B content in leaf and petal. Mitali *et al.* (2020) reported that application of $Ca(NO_3)_2$ at 10.0 g/l registered siginificantly higher flower yield (993.56 g/bush) compared to controlled treatment, but remained statistically at par with 7.5 and 12.5 g/l in *Rosa damascena*.

10.7.1.5 Magnesium

Magnesium aids in the dark green colour of leaves and promotes uptake and trans-location of phosphrous and other elements. Insufficient magnesium causes chlorosis between veins of older leaves. The stems are reduced in number and flowers are smaller in size and poor in colour (Pal, 1972). Johansson (1978) studied the effect of nutrient levels on growth, flowering and leaf nutrient content of glasshouse roses and noted that low magnesium reduced root weight and flower yield. The foliar nutrient levels depended on leaf position on the flower stem, and the magnesium level increased consistently from basal leaves upwards.

10.7.1.6 Sulphur

No clear cut and distinct symptoms could be identified for sulphur deficiency and the requirement appears to be met indirectly through the N, P and K fertilizers carrying S. According to Biswas (1984), however, sulphur deficiency appears as yellowing of foliage.

10.7.1.7 Manganese

Manganese acts as a catalyst in the enzymic activity in plant metabolism. As a micronutrient it regulates the growth and development and is not replaced by any other element in its function. Its deficiency causes deformities in tissue formation and development of chlorosis between the veins. The veins, however, remain green (Pal, 1972). Bik (1961) reported that chlorosis of roses could be corrected by spraying of 0.5 per cent $MnSO_4$ once every 4-5 weeks.

10.7.1.8 Iron

Iron, which is essential for chlorophyll formation, is also very useful element for maintaining normal growth of rose plants. Lack of iron results in chlorosis between the veins of the terminal leaves. The younger leaves become more chlorotic and appear transparent (Laurie *et al.*, 1958; Pal, 1972). Bik (1961) reported that lime-induced chlorotic roses could be rectified by applying 5 g chelated Fe per plant.

10.7.1.9 Boron

It promotes uptake and utilization of calcium by plants. Insufficient boron affects the meristematic tissues of the rose plants. The terminal buds die, resulting in excessive branching. The leaves are crowded together at the tips of stem giving a tufted appea-rance. The flowers are bull-headed or malformed and the extreme deficiency causes the death of the stem tip (Biswas, 1984). Since the range of deficiency and toxicity is very narrow in case of boron, only soil and plant analysis can be used as a correct guide for boron application. Deficiencies can be corrected by foliar application of borax or boric acid. An excess of boron may occur if boric acid or other sources of this nutirent are used indiscriminately. To correct the toxicity, the soil should be acidified to make the boron soluble and then leached thoroughly (Laurie *et al.*, 1958). According to Knoblauch (1971), the leaf of rose should contain 20-25 ppm of B for maximum production.

10.7.1.10 Copper

The response of rose to copper has been noted in its effects on increasing vigour, yield and quality of blooms. McFadden and Poole (1968) stated that in addition to the macronutrients, foliar spray of 53 per cent basic $CuSO_4$ at 1.4 lb or 635 g/100 gallons of water resulted in marked improvement in growth and quality and regulated the size of the scion. Baccara roses have been found to be more sensitive to copper (Penningsfeld, 1968). Copper-deficient plants are dwarf and have a bushy habit. The leaves are narrow and twisted and leaf tips become white. The deficiencies may be corrected by foliar sprays with a mixture of copper sulphate (0.3 to 0.5 per cent), hydrated lime and water.

10.7.1.11 Molybdenum

Baccara roses have also been reported to be sensitive to molybdenum (Penningsfeld, 1966). Since Mo is intimately involved in N metabolism of plants, its deficiency resembles N deficiency. Plants have restricted growth, their leaves become pale and eventually wither. The deficiency can be corrected through foliar application of sodium or ammonium molybdate.

Lack of proper aeration, waterlogging, excess nitrogen, deficiency of so many elements and several virus infections may also produce symptoms similar to those described above, thus complicating the diagnosis. For example, the chlorosis may also be induced by lime and due to soil alkalinity. Alkalinity of soil lowers the availability of some nutrient elements, leading to deficiency. It is, therefore, necessary to study the symptoms carefully before coming to any conclusion.

Franco-Hermida *et al.* (2020) effectively validated the DRIS and CND norms previously generated for greenhouse rose crops (Franco-Hermida *et al.*, 2013). Compared to conventional FRN and SNR methodologies, the integrative nutrient diagnostic capacity of DRIS and CND significantly enhance the ability to identify those elements with the most probability of crop response. Results from this study support the notion that the intrinsic consideration of relationships between all elements in DRIS and CND significantly enhance the identification of their interactions and imbalances (*e.g.* antagonisms and synergies), particularly when yields or plant growth parameters are used as the response criterion. There are proposals that nutrient interactions are among the most important factors affecting crop yields (Rietra *et al.*, 2017), before, if ever, nutrient disorders can be expressedvisually during a growth crop cycle, a phenomenon often referred to ashidden hunger As an example, the strong Fe:Mn interaction observed for roses in this study has been incorrectly diagnosed and addressed in the Bogota Plateau as a simple incipient or severe Fe deficiency when using conventional SNR and FNR methods (according to anecdotal referencesfrom growers and horticultural consultants). Application of integrative DRIS and CND methods has underlined the significance of this Fe:Mninteraction, modulated by changes in the supply and balance of micronutrients (Fe, Mn, Zn, Cu), the inherently low Mn levels in the soilsof the region, and a major role of the dominant rose rootstock ('Natal Briar') in use. The present study supports the contention that the use ofDRIS and CND methods, to guide the introduction of sensible (gradualand iterative) changes to nutrient supplies in fertilization programs offer a higher probability of enhancing fertilizer use efficiency.

10.7.2 Inorganic Fertilizers

For sustained production and maintenance of proper plant health, it is essential to apply fertilizers to roses. Ammonium sulphate, a nitrogenous fertilizer, was very popular few years back, but now it is being gradually replaced by urea. The most popular phosphatic fertilizer is single superphosphate and both potassium sulphate and muriate of potash have been widely used as potassic fertilizer. The most common practice is to apply these fertilizers mixed to the soil in certain proportions, known as fertilizer mixtures. Micronutrients are usually supplied through foliar sprays.

10.7.2.1 Fertilizer Rates

The most important point to remember in the fertilizer practice is the need of using balanced fertilizers. During the last several decades recommendations of inorganic fertilizers have been suggested for roses from different parts of the world. Laurie and Ries (1950) recommended a complete fertilizer for roses to be applied 2-3 times during the season. A good formula was 10: 6: 4 or 6: 8: 6 or 1: 3: 1 of N, P and K at the rate of 3 lb per 100 sq.ft.

Bakley (1974) reported that flower yields and quality were highest with a NPK mixture of 90: 180: 30 g per plant. Johansson (1979) obtained the highest return from cut flowers of greenhouse rose treated with 60 g N as ammonium sulphate and 39 g K as potassium sulphate per sq.m. annually. Phosphorus was found not effective in this respect. In the alluvial soil zone of West Bengal, a fertilizer dose of 600 kg N along with 200 kg each of P_2O_5 and K_2O per hectare has been recommended for cultivar Montezuma (Table 10).

Table 10: Effect of Nitrogen and Phosphorus on Flower Production in Cultivar Montezuma*

Nitrogen Levels (kg/ha)	*Flower Number/Bush Phosphorus Levels (kg/ha)*			
	0	*50*	*100*	*200*
0	10.2	14.3	10.8	13.5
200	15.5	21.6	25.2	29.0
400	22.7	27.4	30.2	30.4
600	25.3	31.4	34.1	36.4
C.D. at 5 per cent	6.43			

200 kg K_2O was the common dose.

*Yadav *et al.* (1985b).

Damke *et al.* (1995) found that chlorophyll concentration increased with leaf age and was highest during the second flush of flowering. The highest flower yield (53.75 flowers/plant) was recorded with 75 g N, 125 g P and 100 g K/1.44 m^2.

10.7.2.2 *Time and Method of Application*

It is advisable to apply the fertilizers in three split doses and applications should be made when there is sufficient moisture in the soil. Normally, the first application of the mixture is given about 15 days after pruning when the new growth has started (Raghav, 1969; Pal, 1972; Biswas, 1984). After the first flush is over, the second dose of fertilizer mixture is applied. When the second flush of blooms is over and before the spring blooming, application of third dose of fertilizer should be done.

The method of application varies according to plant density and size of the plants. If the plant density is low, the requisite quantity of fertilizer mixture can be applied in a continuous band on the soil around the plant, 20-25 cm away from the stem. If the plant population is fairly large, the fertilizer dose should be fixed according to smaller units and placed in continuous band in between the rows of plants, also between the plants and edges of the borders. The applied fertilizers should be worked lightly into the surface soil but the digging should not be deep.

10.7.3 Liquid Fertilizers

Application of fertilizers to the soil in solution has also been practised by some rosa-rians for hardening the limp stem and getting good blooms for exhibition purposes (Park, 1956; Champneys, 1956). The advantage of this method is that the nutrients in solution reach the plant roots immediately for utilization. The disadvantages are the extra cost of application, the risk of toxicity and fixation of phosphates to soil, rendering it unavailable to plants. As a regular feeding of rose plants, this practice seems unnecessary. However, if immediate effect is the object, this may be tried. If the rose plants have been supplied with adequate manures and fertilizers, there may not be any need for this practice. The liquid fertilizer is put in moist soil around the plant and the dose be repeated, if necessary.

10.7.3.1 *Foliar Feeding*

Foliar feeding or applying nutrients to the foliage through sprays has become popular in recent years as an additional method of improving rose plants. This is an excellent supplement to root-feeding because the leaves can directly absorb the nutrients. This is an effective method and helps in producing healthy and abundant foliage and flowers of enormous size and colour. The effect is the same as that of soil application and sometimes better under certain adverse soil conditions (Raghav, 1969; Biswas, 1984). It also minimizes the amounts of fertilizer mixture used for obtaining result comparable with those obtained from soil application. However, it should not be used as a substitute for good cultivation but to augment it is an endeavour to attain perfection (Gault and Synge, 1971). The concentration of solution should not be very high otherwise there is risk of damage to the leaves. A simple nutrient solution or a combination of nutrients may be applied as foliar spray.

A comparison of 4 NPK treatments with water spray showed that all the foliar treatments increased the flowering shoot length, number of shoots per plant and flower yield of Edouard rose. The treatment supplying 1.72 g N, 0.88 g P_2O_5 and 2.01 g K_2O per plant showed the highest flowering shoot length, while the treatment supplying 3.08 g N, 1.85 g P_2O_5 and 1.75 g K_2O produced the largest number of shoots/plant. The highest flower yield was obtained from the treatment supplying 1.93 g N, 1.34 g P_2O_5 and 1.5 g K_2O (Nanjan and Muthuswami, 1974). In a trial with 1-year-old Edouard roses, bimonthly foliar applications of N, P_2O_5 and K_2O at 2.5, 1.5, 1.5 g per plant, respectively gave the highest flower yield of about 4.5 kg per plant. The essential oil content was highest (0.575 per cent) in plants receiving foliar NPK at 1.9, 3.74, 3.35 g per plant, which had a total flower yield of about 4.3 kg per plant (Akbar and Rao, 1982).

In soils, which are deficient in micronutrients, it will be worthwhile to supplement the macronutrient spray along with a spray of micronutrients. The symptoms, however, should be properly identified before undertaking the spray. Foliar application of a solution prepared from the mixture containing 15 g manganese sulphate, 20 g magnesium sulphate, 10 g chelated iron and 5 g borax at the rate of 2 g per litre has been found very effective for obtaining brighter colour in flowers and foliage (Pal, 1972).

10.7.4 Organic Manures

Organic manures made from the plant, animal and other sources, are very essential for successful growing of rose plants in soil. Under tropical conditions, it is necessary to add these manures in order to supply plant nutrients and to help to maintain a good soil structure. Roses like farmyard manure, this is an indisputable fact. Oil-cakes are good sources of organic nitrogen and are excellent for this crop. Poultry manure, compost and activated sludge are also used. Bonemeal is very useful in rose culture (Harkness, 1979). Owen (1976) suggested regular application of humus forming materials such as manure, peat, compost or leaf mould for successfully growing roses.

10.7.4.1 Rate of Application

Information regarding the manurial requirement of roses based on well planned and systematic trials are meagre. Personal experience has played a significant role in this matter. The rate of application of organic manures varies depending upon their source and nature. Sandy soils need heavier and more frequent application of organic manure than heavier soils. Well decomposed farmyard manure has been the choice of the majority of the rosarians. Application of 8-10 kg of this manure per plant (Raghav, 1969; Swarup, 1980) or 45 tonnes per hectare (Thomas, 1969) has been found useful for the growth and flowering of roses.

10.7.4.2 Time and Method of Application

Apart from using bulky organic matter at the time of preparation of rose beds and planting, concentrated organic manures like oil-cake, cowdung slurry from biogas plant, poultry manure, bonemeal, *etc.* are used after pruning or between first and second flush of blooming either around each plant or between two rows, thoroughly mixed with soil followed by irrigation.

10.7.4.3 Liquid Manures

Some rose growers are of the opinion that liquid manuring improves the quality of bloom. Manures are decomposed in water and applied in dilute solution so that plant nutrients from complex organic forms are converted to simple soluble forms. The soil should be moist at the time of liquid manuring. The application may be repeated at an interval of 5 to 7 days till the flower buds start opening. The liquid manure should not be applied when the soil is dry and in hot season. The organic manure during decomposition emits foul smell and the rose growers are reluctant to handle and use it. In view of the recent advances in the fertilizer usage technology, the method appears to be out of date.

10.7.4.4 Combination of Organic and Inorganic Manures

When roses are grown in soil either for garden display or for cut flowers, both organic and inorganic fertilizations are commonly practised to maintain good soil condition and nutrient availability. White and Richter (1974) reported that sand in mixture with sphagnum peat, fertilized with 57 to 71 per cent of N as nitrate gave the best results.Application of farmyard manure followed by a fertilizer mixture comprising potassium nitrate 2 parts, ammonium sulphate 1 part, superphosphate 5 parts and potassium sulphate 2 parts by weight, after pruning, was also found beneficial (Fairbrother, 1965).

10.7.5 Leaf Nutrient Analyses

Puccini (1958) studied the nutrition of hybrid teas and found that the N: P: K ratio in the leaves was approximately 1: 0.25: 0.60. Chan (1961) reported that for good growth of roses the leaf should contain 2.8 to 3 per cent N, 0.28 to 0.30 per cent P and 2.40 to 2.60 per cent K. According to Bik (1972), the optimum leaf N and K contents appeared to be 3.8 and 2.1 per cent, respectively. He reported that in the pot-grown roses, the optimum leaf N and K contents were 3.2 to 3.5 and 2.05 to 2.30 per cent, respectively.

Young *et al.* (1973) found that with the increase in N application, leaf N content increased. Boodley (1975) concluded that the optimum yield was obtained when recently matured leaves contained 3.5 to 4.5 per cent N and 3.5 to 6.0 per cent K. Cabrera (2000) observed a quadratic relationship between rose leaf N concentration and flower yields and quality. Leaf N concentrations increased asymptotically with increased N applications, season of application had an effect on overall leaf N concentrations. Higher values were observed in the winter and lower values in the summer. Best yields were possible during spring and summer with leaf N concentrations below the recommended critical level of 3 per cent (as low as 2.4-2.5 per cent). This result suggested that, as currently utilized leaf N concentration *per se* is not a dependable indicator of rose productivity.

10.8.0 Mulching

Mulches are used for a number of purposes on rose beds or in greenhouses. They conserve soil moisture, supply humus-making material and suppress weeds resulting in improved growth and flowering (Carpenter and Watson, 1955; Cikalov, 1962; Atanasov and Cingova, 1967). Mulches also keep the soil somewhat cooler in summer. Roots will permeate a mulch, and it adds to the area where roots can develop. The survival of spring-planted roses was found to be increased by mulching as compared to unmulched plants (Miller and Waggoner, 1962). Rose *et al.* (1982), however, reported that plants of the rose cvs. Peace and Garden Party covered with mulch (perlite + paper) in October and November for 2 weeks were less cold-hardy than unmulched plants. Field injury of mulched stems was observed in December and January. Plants covered with mid-December were less cold-hardy than unprotected plants but exhibited less field injury in January than October and November mulched plants. Among the substances used for mulching, well rotten garden compost, farmyard manure, peat, straw, sawdust, ground or whole corncobs and other similar materials make satisfactory mulches (Widmoyer and Sherwood, 1954; Aleksandrova *et al.*, 1977). Chadwick (1950) studied the effect of organic mulches on soils and noted that organic matter content of a soil mulched with about 6.5 cm of ground corncobs was more than double that of unmulched soil after a three-year period. The percentage of stable soil aggregates was also higher with mulching. According to Watson (1954) application of ground corncobs resulted in rapid reduction of the nitrate concentration in the soil. The mulches caused reduced oxygen diffusion but better aggregation of soil particles, better retention and lower and more uniform soil temperature during the summer months. Reduction in nitrate and oxygen supply was only a temporary effect. Growth of the mulched plants was reduced during the first few months but thereafter it was much better than under controls. Quantity and quality of flowers were also greatly increased by the use of corncobs.

In a 6-year trial with essential oil yielding rose, straw mulch applied in the rows in a layer 12-15 cm deep at the rate of 5-6 tonnes per hectare, restricted weed growth, improved the soil moisture regime, reduced soil compactness, protected the root systems of the bushes, promoted shoot growth and increased flowering by 71 per cent and labour requirements were reduced from 28-30 man days to 7-9 man days per hectare. Cost of flower production was reduced by 25 per cent (Cikalov, 1962). Mulching along plant rows with 80-100 cm wide polythene film was also found to cause remarkable improvement in growth and flowering of roses (Paskalev and Tsachev, 1982; Paskalev, 1983). Black polythene sheet (0.18 mm) mulching also increased 29 or 56 per cent flower yield of Anna and Sari after a harvest period of 27 months, respectively (Rodrigues *et al.*, 1999).

10.9.0 Weed Control

These chemicals are applied either as a pre-plant or pre-emergence of weeds after the roses are planted or post-emergence to established weeds.Effective control of broad-leaved weeds in an essential oil-bearing rose plantation was obtained by spraying with the Na salt of 2,4-D 25-30 days before flowering at the rate of 2 kg per 600 l per hectare followed by cultivation of the soil to a depth of 6-10 cm. Soil moisture, shoot growth and the average weight and yield of flowers were increased by the treatment without any reduction in the yield of essential oils (Salimov and Mincov, 1963). The effects of atrazine (2.0-3.0 kg/ha), simazine (2.0-3.0 kg/ha), diuron (2.0-2.5 kg/ha), oxyfluorfen (0.5-1.0 kg/ha) and glyphosate (0.5-1.0 kg/ha) on the control of weeds in roses cv. Happiness were evaluated by Rajamani *et al.* (1992). The most problematic weeds were *Bidens pilosa*, *Oxalis latifolia*, *Ephorbiahirta*, *Ageratum conyzoides*, *Apetonia cordifolia*, *Anaphalis subdecurens* and *Cynodon dactylon*. Glyphosate at 0.5 kg and oxyfluorfen at 1.0 kg resulted in the greatest control of monocotyledonous and dicotyledonous weeds, respectively, reducing weed densities to 16.83 and 23.58 weeds/m^2, respectively. No phytotoxicity was observed in roses. Selection of cheap, easily available and effective chemicals, doses, time, and method of application and phytotoxicity are very important factors for successful control of weeds by herbicides.

10.10.0 Plant Management Practices

Pruning is an important operation regulating the growth and flowering of roses. Following are some other plant management practices which influence the vegetative growth and production of flowers.

10.10.1 Pinching

Removal of a part of terminal growing portion of stem is called pinching. This operation was found to reduce the plant height but promote axillary branching. In an experiment on 55 dwarf roses, Dubois and Vries (1994) found that the number of days to anthesis, plant height and width and the numbers of flower buds and bottom-breaks were significantly greater for pinched plants than for non-pinched plants, and growth rate (mm/day) was lower. Without pinching, the quality of all cultivars was poor. Maturity of flower bud was also prevented by this practice (Laurie *et al.*, 1958).

The usefulness of chemical pinching for substituting manual pruning was investi- gated by Grzesik and Rudnicki (1989). Chemicals like BA, promlin (BA + GA_3), ethephon, alar (daminozide), PP-333 (Paclobutrazol or dikegulac) were employed on rose cultivars Sonia and Marcedes. Ethephon at 1500-2000 ppm and PP-333 at 250-600 ppm promoted branching on budded roses in nurseries and eliminate the need for hand pinching.

10.10.2 Pruning

Pruning refers to the removal of certain portion of a plant. It is an important operation for maintenance of floriferousness and flower quality along with vigour of rose plants. It should not be confused with training which mainly concerns with giving a definite shape to the plants. The training determines the form while pruning effects the functioning of the plant. There is a marked difference in growth habit of roses and each type has its own particular method and season of pruning. The practice of rose pruning consists of two operations, thinning out and shortening of stems. Thinning out comprises removal of old weak, dry, twiggy and diseased stems and branches from the point of start while shortening of the remaining shoots aims at cutting down of last year's growth to a desirable height. Correct pruning is an essential factor for success in rose growing and therefore, it should be done with precision and care (Gault and Synge, 1971).

10.10.2.1 Objects of Pruning

The main object of pruning rose plants is to remove the unproductive growth, ensure production of large number of strong and healthy shoots, which will bear flowers and improve the quality of blooms. Pruning serves the purpose of forcing the strongest shoot bud to break in growth. It keeps the rose bush in proper shape and size. Another important purpose served by judicious pruning is to allow light and air to reach the centre of the plant. Annual removal of the wood on which flowers were produced will encourage growth of new healthy shoots which bear more flowers than old branches. According to Lindemann (1972), the rose blooms harvested after pruning the bushes have longer stem than those cut from unpruned bushes. Pruning facilitates various cultural operations like hoeing, weeding, *etc.* It also helps in rejuvenation of the old plants.

10.10.2.2 Pruning Time

The purpose of pruning will not serve, if it is done at the wrong time. Sufficient time must be allowed for the new shoots to mature and flower. Late pruning delays flowering as well as reduces the production considerably (Kohl and Post, 1952; Srivastava, 1973; Depauw, 1975; Gowda *et al.*, 1984). Diseased and dry wood must be cut from the base of the affected branches whenever detected. The branches which grow irregularly and affect the shape of a neat bushy plant should also be removed without waiting for proper season. The best time of pruning is the period when the activity of the rose plant is least and the plant is at dormant to near dormant stage. Pruning time will depend entirely on the climatic conditions of the region. In temperate climate, it is normally done in spring though some growers do it at the end of autumn. Early spring pruning has been recommended for autumn and winter planted roses and for established roses in UK (Hessayon, 1981). Harkness (1979) suggested late autumn to early spring for pruning roses. In Israel, Zieslin *et al.* (1975) recorded the largest number of flowers by pruning in autumn and without removing flower. Pruning in spring and autumn and removal of flower bud did not show any appreciable variation in flower yield (Table 11).

Table 11: Effect of Summer Treatments on the Number of Main Branches and on the Winter Yield of Flowers*

Sl. No.	*Treatments*	*No. of Plants*	*No. of Stems*		*Flower Yield*	
		Per sq m	*Per sq m*	*Per sq m*	*Per Plant*	*Per Stem*
1.	Autumn pruning – no flower removal	13.1	2.4a+	137.0a	10.4a	4.3
2.	Autumn pruning – with flower bud removal	13.6	2.4a	113.2b	8.3b	3.4
3.	Spring + Autumn pruning – no flower removal	13.5	1.9b	127.7ab	9.4ab	5.0
4.	Spring + Autumn pruning – with flower bud removal	13.1	2.0b	128.9ab	9.9ab	4.8

+ Means not followed by the same letter within each column are significantly different at $P < 0.05$.

* Zieslin *et al.* (1975)

Over a large area in India, and especially in the Indo-Gangetic plains, pruning is done only once. The most usual time for pruning is during October-November after the rains are well over and the cold season is approaching. Under the soil and climatic conditions of Delhi, the highest yield of cut flowers in cultivars like

Cultivation of Rose

Cultivation in greenhouse

Fertilizer application

Cultivation in open field

Watering

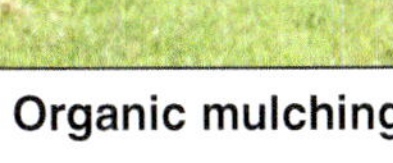

Organic mulching

Hydroponics system in rose

Super Star, Happiness and Queen Elizabeth was obtained when the pruning was done on or about 8 October. Staggering of pruning, *i.e.*, at weekly intervals from 23rd September to 16th October, seemed to provide a regular supply of flowers throughout winter from December to March in a commercial garden (Malik and Dadlani, 1984; Malik *et al.*, 1984). In case of *Rosa damascena*, the rose grown for perfume, pruning during the end of December was found to be useful. Experiments conducted at National Botanical Research Institute at Lucknow showed that plants pruned during December to 10th January yielded the highest number of flowers (Srivastava and Gupta, 1984). In the plains of West Bengal and the adjoining areas, pruning is best done in late October or early November (Bose and Mukherjee, 1977). In some regions, pruning is done twice a year. In Bangalore, roses are pruned at the end of November and again at the end of June and hence there are two distinct flowering seasons (Mukhopadhyay and Banker, 1983). In a 2-year trial, 1-year old plants of Edouard rose (*Rosa borboniana*) were pruned in the second fortnight of September, October, November, December or January. Pruning in November of both years gave the highest number of flowers, namely, 216 and 223/plant. The lowest number of flowers (138 and 123/plant in 1st and 2nd year, respectively) was obtained in plants pruned in September (Irulappan *et al.*, 1983). In Bombay area, pruning is practised in the first week of November for winter season blooms and in the first week of June for monsoon flush.

10.10.2.3 How to Prune

Every rose stem has eyes (buds) alternating on opposite sides, usually outward and inward. The basic rule in pruning is always to make the cut at about half a centimetre above a vigorous bud that points in the direction one desires the new shoot to grow. Since a bush rose is to be kept open in the centre, the cut is made at an outward growing bud whereas in case of climbing rose, the pruning is done at a bud pointing more or less upwards (Barton, 1975).

10.10.2.4 Where to Prune

Every rose stem has eyes (buds) alternating on opposite sides, usually outward and inward. The basic rule in pruning is always to make the cut at about half a centimetre above a vigorous bud that points in the direction one desires the new shoot to grow. Since a bush rose is to be kept open in the centre, the cut is made at an outward growing bud whereas in case of climbing rose, the pruning is done at a bud pointing more or less upwards (Barton, 1975). Whichever bud is selected, the cut should be slightly slanting. A horizontal cut retains moisture and, therefore, is liable to cause fungal growth. While making the cut, care should be taken not to make it too high above the eye as there may be chance of die back of the shoot.

If the cut is on the other hand, very near the eye, the bud may die for want of sap flow. It is absolutely necessary to cut sharp and clean. Broken tissues, bruises or hanging shreds of bark are an invitation to pests and diseases. Hassanein (2010) evaluated different pruning types (light, medium, heavy and control without pruning) at four different times during autumn at three weeks interval. It was found that light pruning at the beginning of autumn or three weeks later were the best pruning times allowing highest flowering rate and quality compared to other tested times. Comparison of no pruning and light pruning is given in Table 12.

Table 12: Improvement in Rose Flowering during Winter under the Pruning Conditions Compared to Control*

Pruning Type and Timing	*Control without Pruning (A)*		*Light Pruning (B)*		*Increment (B-A)*	
	First	*Second*	*First*	*Second*	*First*	*Second*
Number of flowers per plant	57.00	47.00	88.00	79.30	31.00	32.30
Flower stem length (cm)	19.00	18.00	51.90	50.70	32.90	32.70
Flower weight (g)	6.10	4.40	14.60	16.70	8.50	12.30
Petals weight per flower (g)	4.00	3.60	10.90	11.50	6.90	7.90
Essential oil (ml/100 g)	0.22	0.20	0.28	0.31	0.06	0.11
Flower weight per plant (g)	347.70	206.80	1284.80	1324.30	937.10	1117.50
Petal weight per plant (g)	228.00	169.20	959.20	912.00	731.20	742.80

* Hassanein (2010).

Downer *et al.* (2015) studied four pruning treatments based on pruning height which were no pruning, pruning to a height of 0.91 m with 5 primocanes, pruning to 0.46 m with 4 canes and pruning to 0.15 m with 4 canes. Results revealed that plant quality and stem length were moderately influenced by pruning whereas flower production was greatly affected. Maximum number of flowers was obtained from unpruned plants followed by pruning to 0.91 m and 0.46 m. While significantly fewer flowers were recorded with severe pruning to 0.15 m.

All wounds are dressed with cane sealer to protect against attack of fungus and caneboring insects. The cut surface may be painted with fungicide. The other alternative is to spray the whole rose plants, immediately after pruning, with insecticide like rogor or malathion. Besides insecticides, fungicide like benlate or bavistin is also recommended as a spray after pruning (Barton, 1975).

Some rose growers follow a practice known as wintering. In this practice, the base of the rose plants is exposed to sun and air by removing the soil to a depth of about 10-15 cm and a diameter of 20-30 cm around the plants. After a few days, the dug out soil mixed with well decomposed farmyard manure is put back, firmed by tramping and irrigated. This practice does not offer any significant advantage in the semi-arid zone of North India. In the humid regions of the country, however, vigorously growing old established plants thrive and produce good blooms due to the treatment of wintering.

Plant Management Practices

Disbudding in rose

Pinching in rose

Pruning

Rose plant after pruning

Zn deficiency

Boron deficiency

10.10.2.5 Types of Pruning

The intensity of pruning markedly influences the growth and flowering of roses (Rupprecht, 1968b; Borrelli, 1978; Irulappan *et al.*, 1983). Depending upon the extent and level of shortening, there are three types of pruning: light, moderate and hard. In light pruning, the healthy shoots, left after thinning of diseased and unwanted portion, are cut either at the second or third eye (bud) immediately below the flower-bearing footstalk. Moderate pruning is done by cutting back the ripe main and lateral shoots of the previous year's growth at an outward-growing eye, at about half the length of the growth. Hard pruning consists of keeping only three or four shoots of the last year and heading back at about three or four eyes from the base.

Sun *et al.* (1998) conducted an experiment on the effects of pruning severity (1.15-5.43 cm of stem retained above bud) and position on sprouting and flowering of rose buds. The longer the section of stem left above a bud, the longer the bud took to sprout. Pruning severity had little influence on bud development, flowering and cut flower quality. In an investigation, Elenes-Fonseca and Hanan (1987) found that the renewal canes arising from the bottom 6 inches of the plants pinched back at the basal shoot break, when the bud was visible, to the second 5-leaflet leaf from the top of the cane resulted in the highest total flower yield and the best quality flowers, especially when plants were grown in the single-layer plastic greenhouse (with or without screens). The cropping cycle was also more regular under the single-layer greenhouse, particularly when no screen was provided.

The methods of pruning generally adopted for different types of roses are discussed below.

10.10.2.6 Pruning of One-year-old Rose Plants

Newly planted roses, generally utilize most of the time in the first year in establishing themselves in the new habitat and adapting to the new environment. Rose plants should never be pruned until they are properly established. The real pruning starts in the second year of the growth of plants. In the first year of planting, the pruning is done to give a definite shape to the plants and the operation will vary according to the class of the rose. For the plants in Hybrid Tea and Floribunda groups, first of all, weak, dead and crowded branches should be removed from the base. The remaining healthy shoots should be cut back to less than half the length of the shoot, the cut being always above an outward pointing eye. Nitransky (1974), however, observed that severely pruned transplants (cut back to 2 buds) produced better and earlier cropping bushes than lightly pruned (cut back to 1/2 shoot length) ones. Climbing and rambling roses do not require any pruning except the removal of weak growth and unhealthy, dead and interlaced twigs. The Polyanthas are pruned lightly whereas the Miniatures are generally not pruned (Hunt, 1973).

10.10.2.7 Pruning of Established Rose Plants

The pruning of established plants is done to remove the weak, dead and damaged shoots and to obtain the larger number of quality blooms by regulating the shape and size of the plants. The method adopted varies in the various type of roses. The method of pruning of a particular type under specific situation is described below:

Hybrid Tea

The established Hybrid Teas, whatever type of pruning is adopted, should have all dead, weak, damaged and crossing shoots removed from their point of origin. It is useful to retain only four or five healthy basal shoots and remove the remaining ones from their bases. It has been observed that hard pruning, *i.e.*, cutting above three or four eyes from the base reduced the number of flowers per bush but increased the flowers size and length of stem (El-Gamassy *et al.*, 1960; Moe, 1972; Bose and Mukherjee, 1977; Borrelli, 1978). Another group of rose workers recommends light pruning which causes earlier flowering and more number of flowers (Rupprecht, 1968b; Gault and Synge, 1971). To obtain higher yield of good quality flowers with long stem, however, hard pruning, *i.e.*, cutting at 10 cm height from the bud joint, was reported to be the best for cut flower cultivars like Super Star and Happiness (Malik and Dadlani, 1984). Gowda and Jayanthi (1985) recorded the maximum flower production by light pruning in cultivars Christian Dior, Oklahoma, Pink Parfait and Montezuma. They also noted variation in response to flower yield in different cultivars and found that pruning in the rainy season produced larger number of flowers than summer pruning. The practice should be to keep the previous year's growth which had produced flowers and to prune them, cutting off the old stems which had become almsot unproductive. Usually, the cultivars having vigorous and sturdy growth need light pruning whereas those growing slender and comparatively less vigorous are cut severely at lower length. However, certain cultivars like Crimson Glory and Peace resent even moderate pruning. Similarly, cultivars producing yellow and white flowers are pruned lightly (Pal, 1978).

Floribunda

Pruning of Floribunda rose will be different from that of Hybrid Teas. The main object of pruning roses of this class is the production of abundant flowers with mass effect in the bed. This trend can be encouraged by removal of older growths at every opportunity and the young growths be shortened by about a third or half of their height. It is necessary to remove the mass of twiggy growth at the end of the main stems (Pal, 1978). In order to maintain the symmetry of the plants, extra growths are trimmed. Thus, light or moderate pruning is the general recommendation to obtain higher yield of flowers (Nanjan *et al.*, 1971; Hunt, 1973; Irulappan *et al.*, 1983). In the first season, however, pruning may be severe to encourage strong healthy growth from the base and to regulate the shape of the plant. While pruning, it is important to know the growth and flowering habit of a particular cultivar.

Polyantha

The primary objects of pruning roses of this group are to obtain large number of flowers and to give a good shape to the plant. Therefore, every little pruning should be practised. Thinning of dead, weak and overcrowded shoots and tipping of the end of branches that have already flowered are all that is really necessary (Pal, 1972). In the first year, they should not be pruned so hard as Hybrid Tea or Floribunda roses (Hunt, 1973).

Miniature

Miniature roses are generally not pruned except the removal of dead, diseased and overcrowding branches. Some cultivars are apt to produce one or two rather strong growths which upset the symmetry of the plant. These should be removed entirely or cut back severely to maintain the low bushy compact habit which is the most desirable character of this class. Some of the older stems may also be pruned to make a well balanced bush (Hunt, 1973).

Climber and Rambler

The Hybrid Tea climbers hardly need any pruning because it is desirable to promote the growth of the lateral and side branches to bear flowers. Therefore, only some branches from the base should be cut to give room for the new growths. Some of the cultivars grow very strongly and if pruned, produce more strong growths rather than a good crop of flowers. The best flowers are produced on the short branches that emerge from older canes. So, they require only removal of tips. One or two main stems may be cut back severely after 3-4 years to encourage more lateral and sublateral branches. Pillar roses and the large-flowered climbers also should not be pruned in their first year, and afterwards pruning merely consists of removal of unwanted wood in controlling size and shape. Many Ramblers and some shrubs of the Hybrid Musk group only bloom on the new wood. The older stems which have completed flowering once may, therefore, be removed (Gregory, 1984).

Standard

Pruning of Standards which are generally Hybrid Teas or Floribundas should follow the same pattern as advocated for bushes. However, they do not need drastic pruning. But sometimes, when there are two unions, the shoots from one are weaker than the other growths and these should be encouraged to develop by somewhat harder pruning. Growth which comes in the way of its main stem should be removed to give symmetry to the plant (Hunt, 1973).

10.10.2.8 Removal of Young Vegetative Shoots

This practice, also known as deshooting, is generally followed in Hybrid Tea roses. Young vegetative shoots developing from the axils of leaves of basal and lateral shoots are removed to allow only one terminal shoot. In several rose cultivars such side shoots develop from the eyes down the stem below the terminal flower bud. If these shoots are not pinched off at the initial stage, plants become much branched and produce a number of small flowers. Deshooting in cvs. Sonia and Belinda were found to increase the flower production by 50 and 75 per cent, respectively. In case of cv. Baccara, however, it did not affect the number of flowers. The formation ofstructural shoot was also enhanced by this practice (Zieslin *et al.*, 1976). A survey was conducted over 12 months to determine productivity and compare the relationships between the different shoot types of three rose cutivars 1-year-old Sonia, 8-year-old Gabrielle and 7-year-old Kardinal under glasshouse conditions. The plants were trained into vase shapes. The ratio of basal shoots: branches: shoots (flowering and non-flowering shoots) over one season was 1:2:7 (Sonia), 1:3:8 (Gabrielle) and 1:3:11 (Kardinal). Of the shoots produced, 44 per cent

(Sonia), 29 per cent (Gabrielle) and 19 per cent (Kardinal) were marketable stems. A high number of unmarketable stems occurred in summer and the marketing requirement to harvest the entire crop over within 2-3 days. Although harvested stem production was positively correlated with the number of branches, increasing branch numbers do not necessarily increase the number of harvested stems in direct proportion (Mosher and Turner, 2000).

10.10.3 Disbudding

Removal of undesirable buds is known as disbudding. Keeping only the central bud and removal of others cause development of a quality bloom. Removal of flower buds of structural shoots (bottom breaks) increased the number of flowers in Baccara rose (Zieslin *et al.*, 1976).

10.10.4 Defoliation

Several attempts were also made to study the effect of leaf removal on subsequent growth and flowering of roses. Although defoliated plants produced about twice as many shoots as undefoliated, many of them were blind and the total number of flowers was less. Complete defoliation of mature and young leaves caused atrophy of almost all flower buds. Removal of only mature leaves caused about 50 per cent blindness. Removal of only young leaves, however, did not cause blindness (Zieslin and Halevy, 1976).

10.10.5 Removal of Faded Flowers

If the spent blooms are not removed in time, there is a chance of developing fruits bearing seeds. Once the hips are formed and reach the advanced stage of develop-ment, growth and flowering are severely reduced during the season. Cutting of faded flowers forced strong laterals which produced good quality flower (Biswas, 1984).

11.0 Greenhouse Cultivation

Rose is a commercially important cut flower, has demand throughout the year. In field cultivation, plant faced great climatic hazards during its growth and development stage. Good quality blemish free and better postharvest life of cut-bloom of roses may be obtained through greenhouse cultivation. In this culture, several external factors like temperature, light intensity and duration, relative humidity and CO_2 concentrations influence the growth and yield of roses. Rodriguez *et al.* (2017) designed a wireless sensor network in greenhouse rose and implemented for allowing agricultural environment data collection such as temperature, humidity and light. Each sensor node can transmit monitoring data to the cloud. Data mining techniques were used with the purpose of identifying behavioural patterns given the environment conditions captured by the sensor network.

11.1.0 Location and Orientation

The first consideration in establishing a greenhouse for rose cultivation is that of location. A good site can make a difference in the functional and environmental operations of greenhouse. The factors are (i) level and well-drained site, (ii) a climate favourable to the crop intended, (iii) reasonable poximity to utilities and transport routes and (iv) sufficient supply of good quality of water.

Orientation of the greenhouse is another important factor. An east-west oriented free standing greenhouse maintains better winter light level as compared to a north-south oriented greenhouse. Gutter connected greenhouses should be oriented north-south to avoid continuous shading of certain portions of the greenhouse due to structural features.

11.2.0 Structure

A greenhouse structure has three distinct segments *i.e.*, frame, glazing material and monitoring equipments. The greenhouse frame is the most important component of the system. It provides support for glazing materials and a place for fitting environmental control equipments. The frames are designed to withstand wind and snow load with minimum obstruction to sunlight. Now-a-days, aluminium sections are the most preferred for greenhouse frames.

The design of the greenhouse should be cost effective with better cooling and heating efficiency. In China, Wang *et al.* (1999) developed a freely ventilated solar greenhouse with double thermal insulation with more space available, higher vertical height one meter from front slope and better thermal conservation structure.

Flexible films of plastic including myler, vinyl and polyethylene have been used for greenhouse covering. Ultra violet (UV) stabilized polyethylene film of 150 to 200 micro thickness is recommended for greenhouse glazing which has a life expectancy of four years. From a field study conducted to study the effect of different cladding materials (UV-A 205/N, UV-A milky white and IR-504), it was found that maximum amount of PAR was recorded with UV-A milky white material whereas, more number of long grade flowers and maximum dry matter accumulation of stem and flower heads were noticed with UV-A 205/N. Moreover, the temperature and humidity as influenced by covering materials had greater implication on stem length of cut rose flowers (Oloo-Abucheli *et al.*, 2015).

11.3.0 Benches and Beds

Roses are grown in either ground beds or raised benches. Benches are 1.0-1.30 m wide and 15-30 cm deep. These are oriented along the length of the greenhouse with 45 cm passage between. This arrangement of beds allows 67 per cent utilization of floor space for growing the crops.

Pot plants can be grown on benches or directly on the floor. Raised benches should be constructed with open bottoms using aluminium net (2.5 cm x 5.0 cm mesh) which will provide better drainage and aeration. Step-benches are 1.6-2.0 m wide and 75 cm high from the floor. In order to maximum space efficiency for greenhouses pot plant can be grown in various movable bench systems. Such systems increase the production space to about 90 per cent of the floor space. In this arrangement only one passage is allowed which can be shifted to any position.

Victoria *et al.* (2007) described mobile rose cultivation by using movable benches which facilitates only the movement of plants that are under same developmental stage thus reducing the movement in greenhouse by two-thirds. This technique also encompasses the provision of specific climatic conditions for each stage group hence making maintenance and care easier.

11.4.0 Heating

In temperate zones, greenhouse cultivation of roses require heating systems to maintain the ideal temperature (15.6°C) inside. Heat is supplied to a greenhouse at the same rate with which it is lost to maintain desired temperature. A central heating system is more efficient than localized unit heaters in large greenhouse ranges. Pipe coil heat distribution systems are used with high-pressure hot-water boiler. To reduce the expense of conventional pipes and installation cost, fin pipes are used. One linear meter of fin pipe is equivalent to four or more linear meter of conventional pipe. The heat released from fin pipe is much more intense than from conventional pipe. Localized heating system is popular in small greenhouse ranges due to less initial investment.

Natural gas is a more desirable fuel because the initial installation is cheaper, burns clean, which reduce the labour for adjusting and cleaning the boiler. The operation of boiler or heater is controlled by a thermostat. When the temperature drops below the desired level, the heating system is activated by the thermostat.

11.5.0 Cooling

In warm region, summer cooling system is essential for greenhouse cultivation of roses. But in temperate climate, cooling is often required during winter when days are bright. Temperatures inside the greenhouse are 10 to 20°C higher than those outside inspite of open ventilators. An evaporative cooling system, known as fan and pad cooling is used for this purpose. It consists of pads on one wall, through which water is circulated and exhaust fans on the opposite wall. Air enters through the pads is 10-25°C cooler than the outside temperature if the humidity is low and then drawn across the greenhouse to the exhaust fans.

The distance between pad and fans is an important consideration. A distance of 30 to 60 m is best. The pad is extended the entire length of the wall in which it is mounted. The necessary height of the pad is determined by dividing the total area of the pads by the length of the pads. Pad should be located at and slightly above the plant height. During the winter, external air is introduced indirectly and mixed with the warm air within the greenhouse prior to making contact with the plants in order to prevent cold spot at the plant level.

11.6.0 Light

Light is one of the most important factors influencing the growth and flowering of roses (Byrne *et al.*, 1978; Vries *et al.*, 1982; Khayat and Zieslin, 1982). This may vary in intensity, duration and quality and all these various aspects of light have marked influence on growth and development in one way or other (Carpenter and

Anderson, 1972; Asaoka and Heins, 1982; Mor and Halevy, 1984). Under high light intensity and the resultant rise in temperature in summer, growers apply a shading material on the roof of the glasshouse. A reduction of light intensity from 10000-12000 foot-candles to about 6000 to 8000 is satisfacotry. Too much shade markedly reduced growth (Laurie *et al.*, 1958).

11.6.1 Supplemental Lighting

Rose plants grow well under full summer light intensities. Light is often a limiting factor to photosynthesis and growth during the winter at northern latitudes. Light intensities below the optimum level affect the growth and increase the production of blind shoots in rose. This situation can be rectified by using supplemental lighting. There are many lamps that can be used in the greenhouse. Basically fall into three groups – incandescent, fluorescent and high intensity discharge (high-pressure mercury, metal halide; low-pressure sodium and high-pressure sodium).

In greenhouse, total plant weight of cultivar Beter Time was greatly reduced by 65 per cent decrease in light intensity, but very little growth reduction occurred with 42 per cent decrease in light intensity. Total leaf structure was greatly modified by the degree of light intensity. Plants grown under shade produced thinner leaves with low chlorophyll content than plants grown in full sunlight (Chandler and Watson, 1954). Supplementary high-intensity lighting, especially during periods of low solar radiation, increased the number of flowers (Zieslin and Mor, 1990). Supplementary PPFD (100 $\mu mol\ m^{-2}s^{-1}$) from HPS or MH lamps enhanced vegetative and reproductive growth of rose cultivars After Glow, Obsession and Royalty (Menard and Dansereau, 1995). In general, yield of 6 greenhouse rose cultivars increased as the PPFD increased up to 370 $\mu mol\ m^{-2}s^{-1}$, which was given for 20 h/day from September to June. During summer period yield increased by 40-50 per cent with 210 $\mu mol\ m^{-2}s^{-1}$ supplementary lighting, when the natural light inside the greenhouse decreased below 300 $\mu mol\ m^{-2}s^{-1}$, the percentage of class I roses with long stem (> 40 cm), dry weight of shoot or weight per cm of stem increased as the PPF increased during winter (Gislerod and Mortensen, 1997).

Supplementary lighting (210 $\mu mol\ m^{-2}s^{-1}$) for 20 h/day (2.00-22.00 h) during the first 2 flushes increased yield per m^2 up to 5 flushes by 44 per cent in Baronesse and by 47 per cent in Kiss (Mortensen and Gislerod, 1994). In a trial, Bredmose (1998, 2000) grew single stem rose plant in rockwool cubes on ebb and flood benches. When daily quantum integral was increased from 17.8 to 21.0 mol $m^{-2}s^{-1}$, fresh biomass, stem diameter and specific fresh mass increased, but the number of nodes, number of 5 leafter leaves, plastochron value and stem length at anthesis decreased. Increasing plant population from 100 to 178 plants/m^2 increased stem length and reduced fresh biomass efficiency and specific fresh mass. High quantum integral slightly reduced flower diameter but quantum integral or plant density did not affect flower quality or vase life. It is suggested that these effects are related to assimilate supply and translocation, light perception of the roses (Tables 13 and 14).

Table 13: Stem Length and Total Node Number at Visible Flower Bud, and Stem Length, Internode Length, Five-leaflet-leaf Number, and Flower Height at Anthesis of Single-stemmed Cut Roses (Lambada, Manhattan Blue, Red Velvet and Sonia) grown at Either Low (17.8) or High (21.0 mol $m^{-2}s^{-1}$) Daily PPF Integral and Low (100) or High (178 plants/m^2) Plant Population Density*

Treatment	*At Visible Flower Bud*		*At Anthesis*			
	Stem Length (cm)	*Total Nodes (No.)*	*Stem Length (cm)*	*Internode Length (cm)*	*Five-Leaflet Leaves (No.)*	*Flower Height (cm)*
PPF integral (mol $m^{-2}s^{-1}$)						
17.8	26.8 az	13.1 b	45.1 b	3.44 a	8.28 b	3.57 a
21.0	25.4 a	12.8 a	42.6 a	3.32 a	7.82 a	3.52 a
Plant population density (plants/m^2)						
100	25.1 a	12.9 a	43.0 a	3.34 a	7.94 a	3.55 a
178	27.3 b	13.1 a	44.8 a	3.44 a	8.17 a	3.53 a

z Numbers within the same column and section followed by the same letter are not significantly different at P = 0.05.

* Bredmose (1998).

Table 14: Stem Diameter and Fresh Mass Accumulated at Anthesis, and Growth Rate, Fresh Biomass Efficiency, and Specific Fresh Mass of Single-Stemmed Cut Roses (Lambada, Manhattan Blue, Red Velvet and Sonia) grown at Either Low (17.8) or High (21.0 mol $m^{-2}d^{-1}$) Daily PPF Integral and Low (100) or High (178 plants/m^2) Plant Population Density*

Treatment	*Stem Diameter at Anthesis (mm)*	*Stem Growth Rate (mm d^{-1})*	*Fresh Mass Accumulation at Anthesis (g/bloom)*	*Fresh Biomass Efficiency (g d^{-1})*	*Specific Fresh Mass (gm^{-2})*
PPF integral (mol $m^{-2}d^{-1}$)					
17.8	4.46 az	9.25 a	19.2 a	0.392 a	41.7 a
21.0	4.58 b	9.57 a	19.0 a	0.427 b	43.7 b
Plant population density (plants/m^2)					
100	4.53 a	9.45 a	19.5 a	0.426 b	44.3 b
178	4.50 a	9.38 a	18.7 a	0.390 a	40.8 a

z Numbers within the same column and section followed by the same letter are not significantly different at P = 0.05.

*Bredmose (1998)

The effect of 2 supplementary lighting regimes (30 and 60 µmol $m^{-2}s^{-1}$) and 2 cultural techniques (bending or decapitation) on the development of new stems, dry matter distribution and carbohydrate content of *Rosa hybrida* cv. Frisco plants was evaluated by Labeke *et al.* (2000). Bending of the primary shoot resulted in significantly longer stems and increased flower stem weight and diameter with no effect of the supplementary lighting on flower stem quality during harvest. For decapitated plants, however, a supplementary lighting of 60 µmol $m^{-2}s^{-1}$ significantly reduced the percentage of flower abortion and increased the flower stem development rate and weight in comparison with 30 µmol $m^{-2}s^{-1}$.

Abidi *et al.* (2013) studied the impact of blue light on three rose cultivars *Rosa hybrid* 'Radrazz', 'Meiratcan' and *Rosa chinensis* 'Old Bush' grown from single nodal cuttings under white or blue light at an equal photosynthetic efficiency of 110 µmol m^{-2} s^{-1} till the end of flowering of second order axes. Results indicated that vegetative, floral development, length and diameter of first and second axes were normal and similar in both the light treatments for 'Radrazz' and 'Old Bush'. Whereas organogensis and growth of 'Meiratcan' was slightly influenced by blue light with minimum axis length and diameter of first axes and reduced number of second order axes. These results explain that rose plants posses strong adaptive properties to their light environments, but the effects of blue light were cultivar dependent.

11.6.2 Use of Filter

Filters are used to regulate the quality (wave length) of light inside the greenhouse. The quality of light has an influence on plants other than photosynthesis. Mor and Halevy (1984) concluded that the effect of light on bud sprouting is related mainly to red: far-red ratios, while the effect on flower bud development is related to mainly photon flux density. The miniature cultivar meirutral, grown under reduced FR/R filter, produced short plant with higher leaf chlorophyll content. The number of flower buds/plant and number of buds showing colour were not affected by the light quality treatments (McMahon and Kelly, 1990). In 2 experiments, the growth of the miniature pot-plant of cultivar Meirutral under different spectral filters was evaluated by McMahon and Kelly (1990) with the aim of manipulating light quality to control growth of these greenhouse-raised plants that currently require pinching and/or growth retardant treatment to maintain the desired compact form. Two filters which altered far-red (730 nm)/red (660 nm) light were developed. A third filter which did not alter light quality served as control. Plants were significantly shorter and had a higher leaf chlorophyll content when grown under the reduced FR/R filter. The number of flower buds/plant and number of buds showing colour by the end of the experiments (15 June-10 July and 10 August-7 September) were not affected by the light quality treatment.

Plants of *Rosa hybrida* cv. Victory Parade were grown under shade for approximately 46 per cent of the day with 400 w/m^2 set point produced more number of flower buds in summer (Andersson, 1994). Colour of rose flowers is also influenced by the intensity of light. In trials with Floribunda Masquerade, coloured cellophane was placed over the flower buds before they opened and was removed 2 weeks later. No red colour development in the petals indicated that high light intensity is necessary for anthocyanin formation in this cultivar.

11.6.3 Duration of Light

The duration of light has also pronounced effect on growth of vegetative as well as reproductive structures. The influence of duration of light is most marked and spectacular in inducing or suppressing flowering. Polyantha rose cultivar Charles Bonnet produced the largest number of flowers under 16 h light. At a photoperiod of 12 h, the length, weight and flowering of the shoots of cv. Mercedes markedly increased with irradiance. At the highest photosynthetically active radiant (PAR) level (270 µmol $m^{-2}s^{-1}$) flowering occurred in 89 and 33 per cent of the uppermost and second shoots respectively. At an irradiance level of 90 µmol $m^{-2}s^{-1}$ these values were 6 and 0 per cent respectively (Maas and Bakx, 1995; 1997). Uninterrupted light induced the earliest bud formation but resulted in fewer flowers and leaves than 16 hours photo-period (Gorjackina, 1967). *Rosa hybrid* cultivars, 'Frisco' and 'Madelon' plants were grown during winter in natural light (NL) and two supplementary lighting regimes: NL + 6 h and NL + 12 h per day (PAR 90-100 µmol m^{-2} s^{-1}).Results indicated that supplementary lighting had significant effect on number of flowers with the harvested stems being longer and more with 12 h additional light compared to NL + 6 h light and NL in 'Frisco'. Whereas the cultivar 'Madelon' exhibited lower demand to supplementary light with the similar productivity in 6 and 12 h additional light (Wisniewska and Treder, 2003).

11.7.0 Temperature

Temperature is another important factor regulating the growth and flowering of rose plants. Though its influence on the many processes occurring in the plant, temperature markedly affects the quantity and quality of the plant and flower (Semeniuk, 1964; Hanan, 1978; Ginzburg *et al.*, 1983).

11.7.1 Effect of Temperature on Growth

Vries and Smeets (1979) studied the effect of different temperatures on growth of Hybrid Tea rose seedlings and recorded that increase in temperature resulted in reduction in the juvenile period, length of leafy shoots and the fresh weight of shoot but that of the root. Leaf number was not affected. The cultivar Talisman and its sports were grown at 56°F (13.3°C) and the stems became heavier and stiffer (Laurie *et al.*, 1958). In greenhouse studies with Cara Mia rose, Pasian and Lieth (1989) measured the leaf net photosynthesis rate at 4 leaf temperatures, with the 4 leaf ages (10, 20, 30 and 40 days old) over a range of irradiance levels and recorded maximum net photosynthetic rates at 30 and 37°C. Pasian and Lieth (1996) also developed a model for predicting the stages of development of Cara Mia flowering shoots, but required adjustments for successful use with Royalty and Sonia.

11.7.2 Effect of Temperature on Flowering

Most rose cultivars are grown at a night temperature of 60°F (15.5°C) because at that temperature the quality and quantity of roses produced are considered satisfactory (Kiplinger and Nelson, 1957; Laurie *et al.*, 1958). When cultivar Talisman and its sports were grown at 56°F (13.3°C) produced flowers with better colour, compared with plants at 60°F (15.5°C). The cultivar Happiness produced more flowers at 65°F (18.4°C) than at 60°F (15.5°C), and quality was not reduced at warmer temperature (Laurie *et al.*, 1958). The highest yield of flowers in cv. Sonia was recorded with night temperature of 21°C but the quality was poor, while the longest stems were obtained at 15°C (Geers, 1978). Low temperature also brings about the development of pigment in the leaves, stems or flowers. The formation of red pigment in cv. Crimson Glory was studied in plants growing in a phytotron at temperatures ranging from 10° to 30°C. At 10° and 20°C the petals were dark red and resembled velvet, the upper epidermis being much thicker than that of petals grown at 30°C. At a day and/or night temperature of 30°C the upper epidermis was flattened and the red colour of petals was higher at 13°C than at 23°C (Shisa and Takano, 1964). The desirable deep golden colour in Golden Rupture roses was achieved by maintaining the greenhouse night temperature at 65°F (18.3°C) or at 60°F (15.5°C) rising to 65°F (18.3°C), when the buds were formed. Lower temperature increased the percentage of pale-off coloured flowers (Kiplinger and Nelson, 1957). A drop in night temperature from 62° to 40°F (16.7-4.4°C) resulted a reduction in stem length with an increase in the number of misshapen flowers in Forever Yours roses (Hanan, 1978).

The stem and flower buds in some roses do not elongate or expand until there has been an exposure to cool temperature. Winter chilling was found to be necessary for flower bud formation in *R. damascena* (Staikov, 1955). Evaporative cooling, however, reduced the formation of flowers in cvs. Baccara, Marimba, Yellow Belinda and Magic Moment. This effect was ascribed to a reduction in shoot growth rate, resulting in fewer flowering flushes than in uncooled control plants during the same period. However, flower quality was significantly improved by cooling, both the weight of the entire flower and the stem length being increased. Flower bud weight was less affected and increased only in autumn and spring. Cooling raised the anthocyanin content of flower petals, especially in summer (Plaut *et al.*, 1979). At higher temperatures, production of flowers was greater but the quality was not good (Mathis, 1972; Berg and Valentin, 1975; Geers, 1978). Vries and Smeets (1979) recorded that increase in temperature reduced the number of days to first flowering and the length of flower stalk.

Trials with cvs. Belinda, Mercedes and Sonia grown under glass showed that after a resting period at 2°C the plants gave higher flower yields than they did after a period at 8-12°C, ventilation above 4°C showed better yield (Jensen, 1979a, 1979b, 1980a). Glasshosue grown rose cultivars Samantha and Cara Mia were subjected to minimum night temperatures of 32°F (0°C) for varied periods in September/ October, December/January or February/March, the control being maintained at 62°F (16.7°C) at night. For both cultivars the yield and flower stem quality for October/June following the September/October treatment were considerably lower for the cold treated plants than the controls. The number of bottom breaks was, however, increased by cold treatment in December/January or February/March, producing 5.2-6.5 bottom breaks per 12 plants compared with 0-1.3 for controls and cold treatments in October/December (Schrock and Hanan, 1979).

Effects of temperature on postharvest changes in flowers of cultivars Meitakilor and Meired grown in greenhouse at 21°C/13°C or 22.5°C/17°C day/ night temperature were studied by Gudin (1992). Flower stems were cut at 2 well-defined morphological stages (tight bud and cracked bud stages). For both stages and cultivars, flower stems with longer buds (1.2 to 1.4 times) and more petals were harvested 6-9 days earlier from the warmer greenhouse than from the cooler greenhouse. Flower stems harvested at tight bud stage from later ones, 100 per cent and 87 per cent of cv. Meitakilor flowers reached the half open and fully open stages respectively compared with 63 per cent and 60 per cent, respectively for Meired. This difference between cultivars almost disappeared at the cracked bud stage harvest.

In cultivars Mercedes and Baccara, Hendriks *et al.* (1987) found that reversing the temperature from 23°C/14°C to 14°C/23°C (day/night) for a period of 14 days led to a large increase in the number of flowering shoots, provided that this treatment was commenced in the 2nd or 3rd week after pruning.

11.7.3 Root Zone Temperature

Pink Radiance rose was subjected to various temperature treatments and temperature below 8°F (13.3°C) was found to be lethal to root growth. Root injury was apparent to plants subjected to a temperature of 16°F (8.88°C) (Carrier, 1951). Low temperature tolerance of plant parts influenced by their diameter and position in the plant. On budded plants the union was more susceptible to low temperatures than any other above ground parts (Carrier, 1952).

Compared with control, root zone warming in winter resulted in a yield increase by 7.5-47 per cent in cultivars Golden Emblem, Dolores and Tineke. Accumulated yields after 42 days in Dolores and Tineke by 47 and 19 per cent respectively, following warming. Root growth (1.94 mg/100 cm^2) of Golden Emblem was higher than the control (0.14 mg/100 cm^2) (Chung *et al.*, 1996, 1997). Soil surface covers increased plant growth only when radiation was high (Chandler and Watson, 1954).

The effects of interaction of factors on growth and flowering of cultivar Jaguar were studied by Mortensen and Gislerod (1996) in greenhouse with supplementary lighting at PPD of 175 µmol $m^{-2}s^{-1}$ for 20 h/day, root temperature (RT) 12, 15, 18 or 21°C, air temperature (AT) 18 and 22°C and CO_2 concentration of 360 and 850 µmol. As the RT reduced, there was a significant decrease in the number of flowering shoots, shoot fresh weight and stem length, irrespective of AT and CO_2 concentration. Vase life of flower increased by a decrease in RT or an increase in AT, but was not affected by CO_2 concentration.

11.8.0 Carbon Dioxide (CO_2)

Carbon dioxide content of the air is a limiting factor in photosynthesis. Several attempts have also been made to add CO_2 to the atmosphere. Rose cv. Better Times was grown in an atmosphere containing 1200 to 2000 ppm CO_2 produced larger and grew more flowers than the control plants (Lindstrom, 1965). Supplementary CO_2 (3000 ppm) applied to Baccara roses grown under reduced light, increased the flower yield by reducing the incidence of flower abortion (blind shoots) in the upper buds of the stems (Zieslin *et al.*, 1972). Carbon dioxide-enriched plants of rose cv. Sonia also yielded on an average 25 per cent more marketable blooms over the period from December to March than non-enriched plants (Hand and Cockshull, 1975). Jensen (1982) studied the effect of CO_2 (1000-1200 v.p.m.) on Belinda, Mercedes and

Sonia roses on *Rosa canina* rootstocks forced on 15th January, 7th February or 1st March and reported that CO_2 always increased the yield (up to 31 per cent) and the greatest response was with cv. Mercedes. The highest percentage increase in yield was with crops forced on 1st March but the highest actual yield was from plants forced on 7th February. Carbon dioxide also increased flower stem length and the proportion of marketable high quality blooms. In the growth chamber, raising the CO_2 concentration to 950 µl/l enhanced the number of flowering shoots by 19 and 50 per cent, increased the dry weight of flowering shoots by 27 and 24 per cent, and increased shoot length by 3.7 and 4.6 cm in cvs. Garnette and Mercedes, respectively. Raising the CO_2 concentration also increased the ratio of dry matter of flowering: non-flowering shoots from 3.8 to 7.0 in Garnette and from 0.5 to 0.9 in Mercedes (Mortensen and Moe, 1983). Zeroni and Gale (1989) found that salinity tolerance increased at high CO_2. Some combinations of CO_2 and salt gave higher yields than those obtained with CO_2 alone, due to a decrease in blindness. This suggested that salt, in the presence of high CO_2, changed the distribution of assimilates. Pandey *et al.* (2009) stated that the indigenous rose cultivar 'Arjun' responded positively with respect to growth and photosynthetic rate to CO_2 enrichment (370, 1000 and 1500 µmol mol^{-1}) under optimum growing conditions in phytotron. Exposure of plants to 1000 µmol mol^{-1} CO_2 recorded maximum plant height (95.0 cm), photosynthetic rates (11.58 µmol CO_2 m^{-2} s^{-1}, maximum specific leaf weight and root dry weight. Whereas, maximum number of leaves, leaf area, chlorophyll content and root volume (28.2 cm^3/plant) were observed with 1500 µmol mol^{-1}.

11.9.0 Humidity

Darlington *et al.* (1992) reported the influence of humidity on the production of rose (cut flowers) in greenhouses either controlled at 77 ± 5 per cent RH by high pressure fogging systems or at ambient humidities (30-80 per cent RH). During the summer months, there was a significant increase in production due to an increase both in the numbers and size (length and other morphological characteristics) of the roses harvested. This increase in production was not associated with an extension of the growth cycle. However, during the winter months the use of humidity control did not lead to the same general increase in production seen in summer. Furthermore, the winter response to humidity control wasdependent on the lighting regime. Humidity control with 18 h of light increased the number of flowers harvested by 7 per cent whereas with 24 h of light the number decreased by 4.1 per cent. Cultivars Royalty and Red Success grown in climatically-controlled greenhouses at 4 levels of RH (5, 10, 15 or 20 mb VPD) did not show significant difference in flower yield and average stem length. With both cultivars the leaf area increased with increasing RH (Harkess and Hanan, 1988). Bouranis *et al.* (2019) observed the effect of air humidity on plant performance, transpiration, stomatal closing and anatomy of *Rosa hybrid* 'Pasadena'. As RH rose from 40 to 60 per cent as well as from 60 to 90 per cent, plants exhibited larger leaf area and thinner leaves. With rising RH, plant transpiration declined at growth environment. Larger stomata were observed at 90 per cent RH as compared to 40 or 60 per cent. Stomata characteristics were considerably influenced at 90 per cent humidity with reduced gas oscillations within photoperiod, attenuated opening response following dark/

light transition and reduced closing response upon darkening. The plants exposed to 90 per cent RH exemplified lesser ability to manage water loss upon water deficit compared to 60 per cent, whereas plants under 40 per cent RH showed ready response to water stress. It was concluded that low RH exposure to rose plants resulted in thicker leaves with very responsive stomata.

11.9.0 Interaction of Factors

11.9.1 Temperature and Light

Kyalo *et al.* (1996) assessed the effect of high temperature long-day (HTLD) *vs.* low temperature short-day (LTSD) on Miniature cultivars Meirutral and Meijikata and recorded more number of flowering shoots under HTLD (30°C day/21°C night plus 725 µmol $m^{-2}s^{-1}$ photosynthetic photon flux (PPF) for 14 h/day) environment than those grown under LTSD (21° day/16°C night plus 725 µmol $m^{-2}s^{-1}$ PPF for 10 h/day) environment (Table 15). A central composite rotatable design was used by Hopper and Hammer (1991) to estimate the relationship between PPF and day (DT) and night (NT) temperatures on growth and development of cv. Royalty in controlled environment. Low DT and NT delayed both visible bud formation and development to flowering. Increased PPF improved overall flower stem quality by increasing stem diameter and fresh and dry weight of all plants at flowering, as well as enhanced the flower bud formation and flowering.

Table 15: Effect of Growing Environment on Plant Growth and Flowering Characteristics of 'Meirutral' and 'Meijikatar' Roses***

Growing Environment	*Breaks (Number)*	*Flowering Shoots (Number)*	*Blind Shoots (Number)*	*Flowers/Plant (Number)*	*Flower Diameter (mm)*
HTLDz	16.1 ay	13.2 a	2.9 b	14.3 a	29 b
LTSDz	15.5 a	8.8 b	6.7 a	10.8 b	43 a
Significance					
Environment	NS	**	*	*	**
Cultivar (cv)	NS	**	*	NS	NS
cv environment	NS	NS	NS	NS	NS

z HTLD: High-temperature long day; LTSD: Low-temperature short day.

y Means in columns significant by F test, P (0.05).

NS,*,** = Nonsignificant or significant at P (0.05 or 0.01), respectively.

***Kyalo *et al.* (1996).

11.9.2 Light and CO_2

Supplementary lighting and CO_2 enrichment increased the total yields of marketable flowers by 34 and 26 per cent, respectively, over yields of control plants (natural lighting and CO_2 concentration). The combined effects of the treatments increased yields by 69 per cent compared with the control. Improved flower quality and shortened production cycles were also observed (Balazsovits *et al.*, 1989).

Jiao and Grodzinski (1998) studied the photosynthesis and concurrent export rates of expanded leaves on the flowering shoot of cv. Samantha were measured at 3 stages of shoot and flower bud development. At 35 and 90 pa CO_2 photosynthesis and concurrent export rates of the upper expanded leaves were greater at stage 3 (when petal colour of the flower bud was visible) than at the 2 earlier stages of shoot and flower development. The optimum for leaf photosynthesis and concurrent export at ambient CO_2 and saturating irradiance was 25°C. Export was more sensitive to increased temperature than was carbon fixation. For example, at 40°C photosynthesis was 40 per cent lower while the export rate during photosynthesis was reduced by 80 per cent. Increasing the photon fluence flux rate from 200 to 1000 µmol $m^{-2}s^{-1}$ PAR increased the photosynthetic rate and the concurrent export rate at 35 and 90 pa CO_2, but the increase in export was proportionally greater than that of photosynthesis.

Naing *et al.* (2016) observed that the combination of supplementary lighting using high-pressure sodium (HSP) with CO_2 improved flower stem yield, flower stem diameter, fresh weight and number of petals per flower in rose cultivars 'Loving Heart' and 'Top Grace' compared to supplementary lighting. Moreover, stomatal density and chlorophyll fluorescence were affected by either supplementary lighting alone or in combination with CO_2 (Table 16).

Table 16: Flower Quality of Rose Cultivars under Supplementary Lighting and CO_2 Conditions*

Treatments	*Length of Flower Stem (cm)*	*Fresh Weight (g)*	*Stem Diameter (mm)*	*Flower Diameter (mm)*	*No. of Petals*
		'Loving Heart'			
Natural Light (control)	>70	29.4 c	6.0 b	10.2 c	54.2 c
HSP supplementary lighting	>70	31.4 b	6.3 ab	11.5 b	56.8 b
HSP supplementary lighting + CO_2	>70	34.2 a	6.7 a	13.4 a	59.4 a
		'Top Grace'			
Natural Light (control)	>70	29.0 c	5.9 b	10.8 b	41.8 c
HSP supplementary lighting	>70	31.1 b	6.2 b	11.5 a	44.4 b
HSP supplementary lighting + CO_2	>70	35.0 a	6.7 a	12.9 a	48.2 a

* Naing *et al.* (2016).

11.9.3 Light and Humidity

The influence of relative humidity (75, 83 and 91 per cent corresponding to vapour pressure deficits of 550, 390 and 200 Pa, respectively) and lighting periods (18 and 24 h per day) was studied in 14 rose cultivars by Mortensen and Gislerod (1999). Results revealed that rasing the RH from 83 to 91 per cent decreased shoot fresh weight by 11 per cent, whereas had no effect on number of flowering shoots.

Increase in the lighting period enhanced number of flowering shoots and shoot fresh weight by 12 and 5 per cent, respectively. Moreover, vase life was found positively correlated with increasing RH and negatively correlated with increasing lighting period. Thus, it was concluded that high RH and continuous lighting should be avoided in the production of greenhouse roses.

11.9.4 Light and EC of Growing Medium

The yield and quality of greenhouse rose cultivars *viz.* Kardinal, Frisco, Jaguar, Kiss and Madelon increased as the PPFD level increased from 130 to 250 or 370 µmol $m^{-2}s^{-1}$. The response of PPFD varied between cultivars and as an average the yield increased by 18, 41 and 53 per cent at 190, 250 and 370 µmol $m^{-2}s^{-1}$ PPFD respectively. Raising the electrical conductivity in the soil-based growing medium from 2-3 to 4-6 mS/cm increased the mean yield by 9 per cent (Gislerod *et al.*, 1993).

11.9.5 Light, Temperature and CO_2

Jiao *et al.* (1991b) developed a daily growth model for rose cv. Samantha based on non-destructive measurements of whole plant net CO2 exchange rate (NCER) under various environmental conditions. The relative contribution of irradiance (100-1200 µmol $m^{-2}s^{-1}$), day length (8-16 h), CO_2 concentration (350-150 µl/l), day temperature (15-30°C) and night temperature (15-25°C) to daily plant growth was 64, 31, 4, 0.3 and 0.7 per cent, respectively.

The role of temperature on net CO_2 exchange of source and sink tissues of flowering shoots and whole plants was examined using single-stemmed cv. Samantha roses (Jiao *et al.*, 1991a). The optimum temperature range for whole-plant C gain at saturating irradiance at ambient CO_2 level was between 20° and 25°C. Dark respiration increased more markedly than photosynthesis with temperatures between 15 and 35°C. At 25°C loss due to respiration from the flower bud at the colour bud stage account for 45 per cent of the C loss of the flowering shoot. At low irradiation levels (200 µmol $m^{-2}s^{-1}$) whole plant net photosynthesis was greater at 16° than at 22°C because of greater reduction in respiration. Whole plant net photosynthesis of plants grown and measured at enriched CO_2 concentrations (1000 ± 100 ml CO_2/l) was greater (30-40 per cent) than of those grown and measured at ambient CO2 concentrations (150 ± 50 ml CO_2/l) at temperatures ranging from 15 to 35°C.

11.9.6 Light, Temperature and Humidity

Pettersen *et al.* (2007) investigated the effect of air humidity and temperature on growth and flowering of two pot rose cultivars under continuous lighting. Results indicated that air humidity had no effect on number of flowers and days to flowering under continuous lighting whereas increasing light period from 18 to 24 h/day enhanced the number of flowers by 34 per cent and hastened flowering by 12 per cent. Moreover, the increasing temperature promoted earlier flowering and also increased the mean growth rate. The plants grown under 80 per cent RH failed to close stomata under post-production conditions and darkness. Thus, high constant RH under continuous lighting caused malfunctioning of stomata which could be subjugated by daily variation in vapour pressure deficit.

11.10.0 Ventilation

An exchange of air in greenhouse is desirable for normal growth and development of roses. Air circulation ensures an adequate supply of carbon dioxide and oxygen for physiological processes occurring in the plant and also reduces the relative humidity which may cause serious diseases affecting the growth and flowering (Laurie *et al.*, 1958).

11.11.0 Shading

Shading will help to control the temperature of the greenhouse if the ventilation is insufficient. It protects the plants from too much direct sunlight, reducing the rise of leaf scorch and prevent the fading of flower colour.

The amount of shade required in a greenhouse depends on the season. Shading washes are often the most effective method of reducing heat from sun, while allowing adequate light to penetrate for good growth of plant. The influence of whitening a greenhouse roof on microclimate and canopy behaviour of rose cv. First Red was studied during summer in a greenhouse located in the coastal area of eastern Greece (Baille *et al.*, 2001). Whitening reduced the average glasshouse transmission coefficient for solar radiation from 0.62 to 0.31. As a consequence, air temperature, vapour pressure deficit and canopy to air temperature difference experienced drastic changes while transpiration rate was not strongly affected. Globally, glass whitening is inexpensive, presents positive effects on both microclimate and crop behaviour and can be consi- dered as an efficient means for alleviating the large heat load during summer in warm countries.

11.12.0 Selection of Cultivars

Selection of species and cultivars is of paramount importance for successful greenhouse rose cultivation. Although many new Dutch glasshouse cultivars are released each year, a survey made by Oprel (1988) showed that only about 30 per cent of total of around 200 cultivars raised in the last 25 years command 1 per cent or more of the total market sales. The most successful cultivars were those with both a high yield and prolonged period of cropping.

Flower production and quality as influenced by different rose varieties under polyhouse conditions was investigated by Mantur *et al.* (2005). Significantly higher number of flowers was recorded with Sweetness (114.50/m^2) followed by Passion (105.98/m^2) and Polo (103.47/m^2). Regarding flower stem length, Polo (64.18 cm) was significantly superior followed by First Red (59.63 cm) and Sky Line (55.95 cm), whereas the flower diameter was maximum with First Red (2.62 cm) followed by Polo (2.52 cm) and Passion (2.45 cm). Kumbhar *et al.* (2015) studied the vase life of rose cultivars grown under polyhouse condition and found that maximum vase life was observed with Mericlear (10.66 days) along with maximum diameter (0.51 cm) and stalk length (61.23 cm). Whereas, minimum vase life (6.00 days) in Iceberg and Snow white which was at par with rest of the genotype except Mericlear, Spinx and Naranga (Table 17).

Table 17: Vase Life of Flowers as Affected different Cultivars of Rose grown Polyhouse Conditions*

Colour of the Genotypes	*Genotypes*	*Vase Life of Flower (Days)*
Red	Passion	7.33
	Grand Gala	7.66
	First Red	6.66
	Mean	7.21
Yellow	Gold Strike	6.33
	Spinx	9.00
	Sunking	7.33
	Mean	7.55
Orange	Naranga	8.66
	Mericlear	10.66
	Sakira	6.66
	Mean	8.66
White	Bianca	6.66
	Iceberg	6.00
	Snow White	6.00
	Mean	6.22
S. E. ±		0.57
C.D. (5 per cent)		1.68

* Kumbhar *et al.* (2015).

The Hybrid Tea cultivars producing beautifully shaped blooms of lasting quality, borne singly on long straight stems are suitable for use as cut flowers. The cultivars which open slowly and the blooms last for a long period when placed in water are suitable for export as cut flowers. But the cultivars preference in export market keeps on changing and at present Amadeu, Amazone, Astra, Assorted, Bianca, Black Magic, Cappuccino, Capri, Charon, Confetti, Dallas, Dream, Escada, Escimo, Femma, First Lady, First Red, Frisco, Golden Eye, Golden Gate, Grand Gala, Grand Prix, Hollywood, Kiss, Lambada, Maasai, Miracle, Movie Star, Nicole, Noblesse, Osiana, Pailine, Pappillon, Ravel, Red Belvet, Red Devil, Renate, Rising Sun, Sacha, Skyline, Starlite, Steffi, Sundance, Sweetnesse, Taxas, Timeless, Tineke, Toplesse, Vanilla, Versillia, Vision, Vivaldi, White Noblesse are popular.

11.13.0 Substrate

Other than soil, organic or inorganic substrates have been used for rose cultivation in greenhouse conditions. Rockwool and soil cultures of cut roses were compared by Chung *et al.* (1998). First crop yields were similar on both the substrates. In the second crop, yields on rockwool were significantly higher than yields on soil. The stem length of cut flowers of the first crop of cv. Carl Red were 61 and 65.9 cm on rockwool and soil, respectively. In the second crop, the length decreased to 57 cm on soil, whereas on rockwool the length was 61 cm. Generally, stem length of cut flower of all cultivars decreased in continuous culture on soil. However, Nagesh *et al.* (2004) found that the pretreatment of soil with dazomet followed by neem cake amendment and introduction of antagonistic fungus such as *Pochonia chlamydosporia* aided in soil health restoration and also exhibited antagonistic effect for long-term benefit to achieve good quality flowers.

Syros *et al.* (2001) compared the performance of rose on coco-soil and pumice with the recirculation of nutrient solution. The amount of fertigation per plant was 800 and 1000 ml for coco-soil and pumice, respectively. Moreover, maximum number of flowers were harvested from plants grown on coco-soil whereas, the stem length and weight were equivalent in rose flowers produced on both susbstrates.

Fascella and Zizzo (2005) revealed that perlite:coir dust media produced more number of flowers (17.7 stems/plant) with longest length of 65 cm. The water consumption also varied among substrates with 0.62 l/plant/day in perlite:coir dust and 0.78 l/plant/day in perlite, respectively. The best results in the mixture was attributed to the higher water holding capacity and cation exchange capacity of coir dust proving its alternative potential to peat in hydroponic culture.Waikar *et al.* (2012) proclaimed that soil drenching with 40 per cent more K_2O over regular farmer practice improved the flower stalk length and girth of flower stalk in addition to enhanced vegetative growth, yield and vase life.Kalbhor*et al.* (2015) illustrated that application of 18 g single super phosphate in two equal splits (basal and 50 DAP) significantly increased the number of leaves (32.5), branches (3.75), plant height (49.7 cm), number of flowers (3.75) per plant, fresh weight (24.98 g) and dry weight (5.5 g) of flower with minimum days taken for flowering from bud initiation (8.56) over monoammonium phosphate (control).

Jaffrin and Champeroux (1994) used sand substrate for growing rose cv. Vivaldi under greenhouse conditions. Rodrigues *et al.* (1999) grew rose cultivars Anna and Sari with success in raised branches filled with sand: peat (4:1 v/v). Bisht *et al.* (2013) noticed maximum number of flowers per m^2 (79.05) with soil:cocopeat (2:1) and increased flower bud length (5.34 cm) and number of petals (30.50) with soil:vermicompost (2:1) along with common fertigation of 150 ppm NPK. Rose cultivars Sonia and Laser were grown by Blom (1999) for two years in a recirculating system in either coir or granulated rockwool substrate. In the first year, coir produced about 15.6 per cent more marketable (> 30 cm length) stems as well as 18.0 per cent more fresh weight than granulated rockwool. There were no significant differences between the substrates during the second year. Cut flower numbers and weights tended to be the highest with plants grown on coconut fibre mats (mean of 322 and 261 stems/m^2) over the 3 years for the two cultivars Cream Dream and Dream, respectively (Blindeman, 1998). Raviv *et al.* (1999) grew rose cv. Mercedes on yellow tuff (YT) (volcanic ash), pumice from Italy (PI) and Greece (PG). Irrigation was controlled using electronic tensiometers. Yields of cut flowers were significantly higher in YT and than in PI and even lower in PG (Table 18).

Table 18: Effect of Various Volcanic Growth Media on Yield Parameters of Cut 'Mercedes' Roses*

Medium	Main Season			Summer Season		
	FPW	Avg wt. (g)	Marketable Flowers (Per cent)	FPW	Avg wt. (g)	Marketable Flowers (Per cent)
YT	2.51 a	17.9 a	55 a	4.27 a	14.1 a	50 a
PI	2.15 b	18.3 a	55 a	2.92 b	14.8 a	46 a
PG	1.97 c	18.2 a	55 a	2.68 b	14.4 a	45 a

YT: Yellow tuff; PI: Pumice from Italy; PG: Pumice from Greece; FPW: Average number of flower per plant per week. Within columns, figures followed by the same letter do not differ significantly at $P > 0.05$.

* Raviv *et al.* (1999).

It is suggested that the limiting factor was dynamic water availability, which was affected mainly by the unsaturated hydraulic conductivity. Rose cultivar Dallas, was grown in a greenhouse in perlite (4-6 mm in diameter) or lapilli (from local volcanic deposits, 6-8 mm in diameter) in linear plastic container (12 plants/ linear m, equivalent to 6 plants/m^2). Cut flower yield was highest in lapilli than in perlite (128 and 118 cut flowers/m, respectively). More than 85 per cent of the yield from all except the first harvest was in the Extra or First quality grades (Pascale and Paradiso, 1999).

11.13.1 Soilless Culture

Rooted rose cv. Sonia cuttings were grown in solution culture at various N concentrations and the optimal range of N concentration for normal growth, and the relationship between growth and N uptake were investigated by Terada *et al.* (1997b). The N concentration of the culture solution was adjusted weekly to 50, 100 or 200 ppm. Plant fresh weight (FW) increased with decreasing N concentration in the culture solution. The yield (number and weight) and quality of cut flowers were higher at a low N concentration than at a high N concentration. N uptake per 100 g of FW growth decreased with decreasing solution of N concentration. It was concluded that for normal growth of rose plants, the N concentration in the nutrient solution should be maintained relatively low. The weight ratio of cut flowers to whole plant in solution containing 50 ppm N, where plant growth and cut flower yield were highest, was 47 per cent. Therefore, it was calculated that to yield 1 kg cut flowers, the plant must increase 2.1 kg in FW and absorb 15 g N.

An experimental hydroponic nutrient film technique (NFT) system was used by Brun and Morisot (1996) to measure water and nitrate uptake rate of cultivars Sonia and Vivaldi on a short term basis. Water absorption closely followed the climatic demand with a large peak during the day, whereas nitrate absorption seemed to start slightly before. In order to determine the relationship between plant growth and nutient uptake, the effect of K and Ca concentration in nutrient solution on growth and nutrient uptake of roses (cv. Sonia), grown in hydroponics was studied by Teradi *et al.* (1997). K and Ca content of nutrient solutions did not influence growth. Uptake increased with increasing concentration in the nutrient solution. K uptake per 100 g fresh weight (FW) increased from 0.32 to 0.47 g when K concentration increased from 25 to 200 ppm. The average amount of N uptake per 100 g FW was 0.66 g. P and Mg uptake per 100 g FW averaged for all K treatments were 0.09 and 0.08 g, respectively. Ca uptake decreased from 0.16 to 0.11 g per 100 g FW when K concentration increased from 25 to 200 ppm. Excess potassium is injurious and should be avoided. Kajihara *et al.* (2005) worked to study the effect of recycled nutrient solution in a closed system of rockwool cultivation of cut roses by diluting the drainage solution with water, both with and without addition of NH_4-N, Fe and Mn nutrient components following the slow sand and active charcoal filtration. Results indicated that the yield of cut flower (35 per plant) obtained from plants provided with recycled solution was similar to that of conventional method. Under this system, the nitrogen use and disposal quantity were reduced to 151.8 and 110.5 kg, which were 290.3 and 135.9 kg, respectively in conventional method. However, deficiency symptoms were not noticed in rose plants and no further supplementation of nutrient components were required in closed system when the drainage solution was diluted by 2.5 times after slow filtration.

Fascella *et al.* (2007) evaluated the vegetative and productive behavior of red rose cultivars on hydroponics in unheated greenhouse with polyethylene cover. Plants were grown in polypropylene containers filled with volcanic lapillus and coir dust (1:1) with a final density of 6.7 plants per m^2. Results revealed that 'Dallas' recorded maximum stem length (84.1 cm), stem diameter (0.78 cm) and highest percentage of flowers with length > 80 cm (46.6 per cent) whereas, 'Maira' showed early flowering (108 DAP) and 'Red France' produced stems with manimum number of petals per bud (47.0).

Khoshgoftarmanesh *et al.* (2008) studied the effect of micronutrients namely, Fe, Mn, Zn and Cu on quality of 3 cut rose cultivars, 'Modern Girl', 'Orange Juice' and 'Aqua Fresh'. The different treatments employed were control, 150 µ mol/l Fe, 150 µ mol/l Fe + 10 µ mol/l Mn, 150 µ mol/l Fe + 10 µ mol/l Mn + 12 µ mol/l Zn and 150 µ mol/l Fe + 10 µ mol/l Mn + 12 µ mol/l Zn + 5 µ mol/l Cu. All the micronutrient treatments significantly improved the stem length in all cultivars except 'Aqua Fresh'. Treatments containing Zn extended the vase life as well as did not show any mildew symptom. Thus, results indicated the significance of Zn in augmenting rose flower quality.

Os *et al.* (2012) suggested a better technique to degrade plant protection chemicals and to discharge the nutrient solution in hydroponics without affecting the plant growth. The method envisaged that addition of hydrogen peroxide (0-25 mg/L) to the nutrient solution followed by exposure to UV-C light (0-900 mJ/cm^{-3}) could degrade large organic molecules. Moroever, it was obvious that this water treatment method eliminated pathogens and decreased growth inhibition without changing the composition of nutrient solution except for iron.

11.14.0 Planting System and Plant Density

Rose plants were grown in two planting systems and evaluated for stem yield and quality for cut flower production. In the trellised system, rose cultivars Gabrielle and Kardinal were planted in a commercial glasshouse in 3 m sections of bed. Within-row spacing was varied to give 6-16 plants/m^2. In trellised system after a 3 monthesta-blishment phase, the basal shoots were bent to an angle of 30° above horizontal. Compared with conventional vase-shaped rose plants at the same density, trellised roses produced 24 per cent more basal shoots, 46 per cent more flowering shoots (Gabrielle) and approximately 46 per cent less blind shoots per plant over 5 months. The maximum number of basal shoots was formed at plant densities of 4.20 and 3.32 plants/m^2 for Gabrielle and Kardinal, respectively. Within-row spacing did not affect the number of flowering shoots per basal shoot. The single shoot planting system contained a mixed population of single-stemmed cvs. Gabrielle and Gerdo. Plants were grown in a field over a range of within-row spacing to give 20-105 plants/m^2. Over 3 harvest, increasing the number of plants by 10 plants/m^2 reduced the proportion of flowering shoots by 4.4 per cent.

Expressed on a unit area basis, a 5-fold increase in plants/m^2 produced a 3-fold increase in stem production (Mosher and Turner, 1999). Increasing plant density from 4 to 8 plants/m^2 increased the proportion of stem length and yield by 31 per cent in Baronesse and 39 per cent in Kiss (Mortensen and Gislerod, 1994).

11.15.0 Irrigation and Fertigation

11.15.1 Irrigation

Langhans (1957) reported that plants grown under mist lost almost no water by transpiration, although their stomata remained open the whole day. Similarly, the water loss from plants in the non-cooled greenhouse was 15-40 per cent higher than in the air-cooled greenhouse. Production of roses from July to September showed a reduction of 10 per cent with cooling but there was a marked improvement in flower and foliage colour and size (Durkin and Kiplinger, 1960). Waikar *et al.* (2012) observed water requirement of cv. Passion to be 495.10, 596.38 and 696.70 ml/pot/day during February, March and April, respectively under greenhouse condition. The effects of growing season and irrigation regimes on growth, water use (soil evaporation and transpiration) and yield of roses were investigated by Terada *et al.* (1997c). Rooted cuttings of rose cultivar Carl Red were grown in a greenhouse and watered at soil moisture tensions of 3, 10, 32 and 50 kPa (pF 1.5, 2.0, 2.5 and 2.7, respectively) during summer, autumn-winter and winter-spring season. In all seasons, the best plant growth was obtained using the 3-kPa treatment, which involved using the largest amount of water. Experiments carried out in a commercial greenhouse planted with roses, cv. Royalty, in Watsonville, California, showed that both timing and volume of irrigation could be accurately determined by measuring evaporation of water from a 22-inch diameter open pan at canopy height. Monitoring of evaporation enabled water use to be optimized and soil moisture tension to be kept below 10 centibars which was essential for maximum flower production (Tjosvold and Schulbach, 1991). A novel tensiometer, sensitive in the millibar range, was developed by Raviv *et al.* (1993) when used in conjunction with an adjustable contactor; it enabled the study of plant response to small changes in soil water potential. When tension controlled was compared with timer controlled-irrigation, net photosynthesis rates (Pn) were consistently higher during the warmer hours of the day when irrigation intervals were determined according to a preset water potential, provided that its value was low enough. Rose cv. Madelon development under 4 irrigation regimes was investigated with particular emphasis on the effect of applied water stress on the quantity and quality of roses produced. Soil moisture deficit (SMD) and leaf stomatal resistance (LSR) were highly correlated under the 4 irrigation and the 3 shading levels. However, under mild to moderate water stress (SMD up to-100 kPa and corresponding LSR up to 45 s/cm) flower stem length was reduced. Stem length of the flowering shoots was increased with increasing shading during spring and summer (Chimonidou-Pavlidou, 1998).

Different systems of irrigation are prevalent in various parts of the world. Constant-level sub-irrigation gave a higher production of total saleable blossoms for a 1-year period and longer average stem length than other methods, *viz.*, injection-sub-irrigation, the Ohio State Special Nozzle system and surface watering (Wright and Volz, 1950). Sprinkling is not advised as it brings the feeding roots to the surface and causes burning (Laurie and Ries, 1950). In greenhouse, plants grown under mist grew faster and produced more flowers with longer stem than non-misted plants (Langhans, 1957).

High concentration of salts in the water is harmful to the rose plants. Ishida *et al.* (1978, 1979) studied the salt tolerance of roses irrigated with diluted sea water containing concentrations of chlorides from 0 to 3000 ppm. Salt injury was observed above 500 ppm and the degree of injury increased with the rise in the concentration of chloride. Interveinal chlorosis, marginal burn, tip burn and dieback first appeared on the lower leaves, and spread upwards, and were generally more severe on the grafted plants. The number and height of cut flowers decreased with increasing sea water concentration.

To test effect of EC on rose, an experiment was conducted by Kreij *et al.* (1989) with cv. Sonia grown in a rockwool in 6 treatments involving EC values ranging from 1.0 to 5.0 dS/m in the drip-irrigation water and from 1.1 to 7.1 dS/m in the drainage water; the leaching fraction was 30-42 per cent. Stem thickness, length and firmness, and vase life were negatively affected by increasing EC. Optimum flower production was obtained at an EC in the irrigation water of 1.4 dS/m (2.4 dS/m in the drainage water). Drainage water recycling had no negative effect on cut flower yield or quality. Frequent monitoring revealed no proliferation of pathogenic microorganisms (Raviv *et al.*, 1998). Almeida *et al.* (2015) evaluated the effect of water deficit through different soil water tensions (15, 30, 45, 60, 90 and 120 kPa) on quality of 'Carolla' rose flower stems. It was clear from the results that bud diameter decreased with increasing tensions with maximum (36.44 mm) and minimum (34.96 mm) under 15 and 120 kPa, respectively. Postharvest life remained unaffected by the water deficit with the average being 11 days. The average fresh weight loss and water absorption were 5 g and 74 ml in the 11 days period. Thus, to obtain good quality flower stems, soil water tension of 15 and 30 kPa would be beneficial.

Nirit *et al.* (2006) studied the effect of treated sewage water for irrigation on roses cultivated in perlite and coir. Results showed that application of treated water exhibited similar plant appearance, growth, size of flowering stems and postharvest performance as that of usual irrigation. Morever, contents of macroelements remained unaffected in leaves, whereas Mn, Cu, B and Cl content increased at the rate of 47 per cent and 73 per cent Cl contents in perlite and coir pith grown plants, respectively.

11.15.2 Fertigation and Salinity

The root media pH level should be between 6.0 to 7.5 (Hunt, 1973; Bose and Mukherjee, 1977). Nutrient availability is controlled by the soil pH level. Most greenhouse media are acidic due to the use of acidic amendments (Koths, 1973). Levi-Minzi *et al.* (2000) recorded high salinity values in greenhouse soils and suggested that excessive use of fertilizers should be avoided and salinity should be monitored during crop growth.

Nitrogen is not retained long enough in the root media, need periodical application except where slow-release fertilizer is used. The effect of varying levels (50, 100 and 150 ppm) and source of N fertigation (ammonium nitrate, aqueous ammonia, nitric acid and urea) on the flowering of cut rose cv. First Red was investigated under protected conditions (Ashok *et al.*, 1999). Ammonium nitrate at 150 ppm gave the highest values for bud circumference (6.09 cm), flower diameter (7.33 cm), petal length (4.01 cm), petal breadth (3.84 cm), and flower yield (153/m2) compared with the other treatments. Gurav *et al.* (2005) recorded maximum number of flowers per plant (29.42, 27.46 and 33.46) and maximum stalk length (72.86, 62.53 and 75.60 cm) with high levels of nitrogen and phosphorous *i.e.* 400 ppm N, 200 ppm P and interaction, 400:200:200 ppm NPK/plant/week, respectively. Moreover, more number of flowering shoots per plant (4.40) and larger flower buds (2.40 cm) were obtained with the combination 400:200:200 ppm NPK/plant/week. Rajesh *et al.* (2020) observed significantly maximum plant height, mean number of branches per plant, number of leaves per plant and intermodal length when rose plants were treated with 120 per cent of recommended dose of fertilizers as fertigation in combination with 300 ppm GA_3 (Table 19).

Table 19: Number of Leaves per Plant and Intermodal Length as Influenced by Fertigation and Growth Regulators*

Particulars	*Mean Number of Leaves*		*Mean Internodal Length (cm)*	
G x F	*Season 1*	*Season 2*	*Season 1*	*Season 2*
G_1F_0	164.75	148.11	3.97	3.81
G_1F_1	190.12	185.42	4.41	4.30
G_1F_2	219.76	231.09	5.25	4.93
G_2F_0	177.16	161.80	4.08	3.90
G_2F_1	212.91	202.80	4.72	4.55
G_2F_2	225.08	242.24	5.51	5.07
G_3F_0	147.81	149.75	3.42	3.28

Particulars	Mean Number of Leaves		Mean Internodal Length (cm)	
G x F	Season 1	Season 2	Season 1	Season 2
G_3F_1	185.88	180.05	3.76	3.80
G_3F_2	196.44	217.78	4.28	4.16
G_4F_0	163.66	155.20	3.07	3.11
G_4F_1	182.94	181.47	3.46	3.48
G_4F_2	195.03	224.01	3.81	3.83
F-test	NS	*	*	*
S.E.m ±	8.93	2.86	0.03	0.06
C.D. (5 per cent)	-	9.91	0.12	0.22

G_1: 200 ppm GA_3; G_2: 300 ppm GA_3; G_3: 0.2 per cent humic acid; G_4: 0.4 per cent humic acid; F_0: 100 per cent RDF soil application; F_1: 80 per cent fertigation; F_2: 120 per cent fertigation.

* Rajesh *et al.* (2020).

Royalty roses were grown under different leaching rates and applied N concentrations for 1 year (Cabrera *et al.*, 1993). During this period, there were 8 synchronized flowering flushes at approximately 6-week intervals. Analysis of leachates collected from the microlysimeters indicated leaching losses of 21, 40 and 49 per cent of the applied N from plants irrigated at a leaching fraction of 25 per cent with nutrient solutions containing 77, 154 and 231 ppm N, respectively. Average leachate N concentrations were 79, 257 and 441 ppm. There were no significant differences among treatments in number of flowers or harvested DM/plant, except for the 77 ppm N treatment. Only minor difference in leaf N content (3.2-3.4 per cent) was noted.

To develop a new method of fertigation in cut flower production of roses, the relationships between plant growth and nutrient and water uptake, and the ratio of cut flower growth to total plant growth on a FW basis, were investigated in two-year-old rose plant cultivars Sonia and Carl Red, grafted on *Rosa multiflora*, grown hydroponically for about a year (Terada *et al.*, 1997a). The growth rate increased immediately after harvesting of the cut flowers and pruning. On a long term basis, however, the growth rate was slower in summer than during the other seasons. Growth of Sonia plants was higher than that of Carl Red plants. After flower production had become stable, the ratio of cut flowers to plant growth on a FW basis was always about 40 per cent in both cultivars. The amounts of nutrients absorbed per 100 g plant FW in both cultivars were about 1.0 g N, 0.10 g P, 0.40 g K, 0.17 g Ca and 0.04 g Mg. The amounts of nutrients absorbed for 100 g cut flowers were 2.5 times greater than those on a per plant FW basis. Sonia plants absorbed 2-3 litres of water in winter, 4-5 litres in spring and autumn, and 8-9 litres in summer per 100 g FW. Carl Red plants absorbed more water than Sonia plants. The concentrations of nutrients absorbed by the plants (nutrient absorption/water absorption) were lower in summer than in winter, and in Carl Red were approximately 80 per cent of those of Sonia. The amount of N absorbed by Sonia was about 230 ppm in winter, 145 ppm in spring and autumn, and 110 ppm in summer. The values for P, K, Ca and Mg uptake by plants were, on range 14, 60, 24 and 6 per cent of that for N, respectively.

Vogli and Sigrimis (1999) examined the plant quality response to solar radiation plus supplementary lighting, temperature, and the potassium to nitrogen ratio, during the hydroponic cultivation of roses cv. First Red and Grand Prix in cocosoil and pumice substrates, in a greenhouse. The hydroponic system was monitored and controlled using a modern control and management system for greenhouses (MACQU). Results showed that Grand Prix absorbs fewer nutrients than First Red under the same climate. Quality parameters were related to temperature and solar radiation, with the flower size of Grand Prix being affected more than that of First Red, and *vice versa* for stem length. Simple equipment for managing the nutrition of rose plants in a closed soilless culture system was evaluated by Farina *et al.* (1998). Electrical conductivity (EC) of the nutrient solution was kept at 2100 mS/cm in winter and at 1400 mS/cm in summer. Daily water consumption averaged 0.28 l/plant and 90-95 per cent of the nutrients supplied were utilized. Total cut flower yields for the 15 month harvesting period averaged 26.5 flower/plant. This system can be used to cut costs and reduce pollution. A model was designed by Brun *et al.* (1993) which proved satisfactory for managing the daily EC of the nutrient solution supplied in soilless culture so as to maintain it at or near a fixed level in the leachate and thus to synchronize the plant demand for and supply of nutrients. The cultivar Sonia was grown on rockwool slabs, with a comparison of 3 EC levels in the leachate (1400, 1800 or 3800 dS cm^{-1} at 25°C). Watering frequency was adjusted to the crop transpiration rate, so that each day the amount of leachate produced was about 30 per cent of the amount of water supplied. Fertigation was set according to the system of Brun and Tramier scheduled by means of a commercial computer linked to a solar energy sensor. Management of mineral elements in roses grown in a closed rockwool system was reported by Bloemhard and Moolenbroek (1995) in cultivars Jacaranda and Madelon. The crop absorbed 772 mm of water/year. The total supply of nutrients in Kg/ha annually was 906 N, 263 P, 1117 K, 204 Mg, 668 Ca and 332 S, and the corresponding drainage waste was 265 N, 53 P, 429 K, 88 Mg, 299 Ca and 148 S.

Irrigation with nutrient solution was controlled by a 24-h time switch, float-less relay switch, the sensor and a time limit switch. With equal volumes of rockwool, only the float-less relay switch and sensor were used. Growth and flower yields of roses grown in the rockwool did not differ with different settings to maintain water levels at -4 or -8 cm (Watanabe and Yokoi, 1991).

In comparison with other plants and crops, roses have been generally characterized for having a low to moderate tolerance to salinity. However, to the significant modulating effect of rootstock selection on the productivity and quality of roses exposed to salinity. A combined analysis of the results from all these studies confirm that on a productivity basis (biomass and flower yield), roses could withstand overall maximum salinities (osmotic effects) of applied fertigation solutions of 3.0 ± 0.5 dSm^{-1} (equivalent to ECe _4.5 ± 0.7 dSm^{-1}), which indeed fall within the sensitive division in the normative classification of crop salinity tolerance (Maas, 1990).On the other hand, and considering theaesthetic responses (*i.e.*, salt burn damage) of on-plant and harvested foliage (flower shoots), greenhouse rose tolerance to applied Na^+ and Cl^- concentrations (ion-specific effects) could range up to 10 ± 2 mM, which surpasses the older recommended threshold of 2 to 4 mM. This difference is primarily attributed to the use of well-drained soilless media and intensive irrigation practices in current production systems for greenhouse roses. It should be clarified that continuous exposure to these overall EC and Na^+ and Cl^- thresholds for periods exceeding more than four to six flowering cycles could eventually result in worse rose growth, yield, and quality responses, eventually leading to plant death.

Raul *et al.* (2009) showed 'Manetti' as the most salt-tolerant rootstock for greenhouse rose production, particularly under challenging soil-based growing systems. Its relatively recent and almost complete displacement by 'Natal Briar' in soilless and hydroponic-based rose production systems. Lastly, the results from these experiments highlight the critical importance of carefully considering the chemical composition of the irrigation solutions in experimental salinity studies aimed at generating ecologically and practically relevant results and recommendation. The $Na^+/(Na^+ + Ca^{2+})$ ratios of the salinized solutions in Expt. 1, at 0.81 and 0.91 for 15 and 30 mM NaCl, respectively, would be representative of those in irrigation waters in coastal areas, whereas the 0.27 to 0.44 ratios in the solutions from Expt. 2 would be more representative of typical inland irrigation waters. The differential supply of Na^+, Cl^-, and Ca^{2+} indeed led to a differential accumulation in plant tissues in these studies and interacted with the modulating effect of the rootstocks.

11.16.0 Training

The arching technique was developed by Japanese rose producers in the late 1980's. The technique has been also developed in Holland, Israel, USA, Korea and Taiwan. About 214 ha, 35 per cent of the total area (609 ha), of cut rose production in Japan uses rockwool as a substrate, and 60 per cent of them employ the arching technique (Ohkawa and Suematsu, 1999). The basal shoots that appear in the early stage of growth are bent in an arch, and the subsequent axillary basal shoots (thick shoots growing from the basal part) that appear later are harvested as cut flowers from the basal branching part. Since these shoots are cut, instead of being pinched, according to conventional methods, the rose stems are thick, long, and of superior grade. Although fewer roses can be cut from a plant using this method, as flowers are harvested before the plant is sufficiently established, a certain number of flowers cut per unit area can be assured simply by increasing the planting density. This growing technique makes it easier to harvest compared to older conventional methods. The effects of 3 training methods on glasshouse cut flower production of roses grown in rockwool culture were studied by Han *et al.* (1999) investigated the effects of training method (no arching, low arching, high arching or a combination of low and high arching) and obtained 228-230 cut flowers/3.3m^2 in high arching and non-arching treatments and was reduced by 21 per cent in the low arching treatment. The yield of high grade flowers with stem lengths of > 60 cm was about 62 flowers/3.3m^2 in the non-arching treatment and was greatly increased (by 121-123 per cent) in low arching and high arching treatments. Mosher and Turner (2000) compared the productivity of three rose cultivars of different ages: 1 year-old 'Sonia', 8 year-old 'Gabrielle' and 7 year-old 'Kardinal' which were trained to 'vase shaped' architecture. The ratio of basal shoots: branches: shoots (flowering and non-flowering) over one season were 1:2:7 (Sonia), 1:3:8 (Gabrielle) and 1:3:11 (Kardinal). Of the shoots produced, the marketable stems accounted for 44, 29 and 19 per cent for three cultivars, respectively. Alos, maximum number of unmarketable stems or blind shoots occurred in summer which was associated with high temperatures. Although, harvested stem portion was positively related with number of branches, increasing branch numbers did not correlate with number of harvested stems. Mantur *et al.* (2005) recorded significantly maximum number of flowers per plant (16.22 and 16.09), stem length (53.70 and 54.64 cm) and flower diameter (2.43 and 2.52 cm) with single bending and double bending, respectively compared to pruning. Vasudevan and Kannan (2014) narrated tht bending at shoot junction bud (B_1) followed by application with BA 200 ppm significantly enhanced the plant height (165.33 cm), plant spread (45.19 sq. cm), hastened bud appearance (26.40 days), harvesting (48.80 days), increased stem length (86.79 cm), flower bud circumference (13.13 cm), flower stem girth (8.7 mm), produced more number of flowers per plant (26.47) and prolonged vase life (12.37 days) of var. Tajmahal under greenhouse.

11.17.0 Automation and Models

Based on the survey, 4 mechanization models for rose cultivation in Korea for 0.3 and 0.5 ha pipe greenhouses with soil culture, and 0.5 and 1.0 ha pipe or steel greenhouses with soilless culture, Kang *et al.* (1998) summarized the requirements. The models suggested that 0.3 ha house needed one boiler unit of 500000 kcal/ ha, the 0.5 ha needed 2 boiler units, and the 1.0 ha, 4 units. The number of rotary cultivators required was 1 in each greenhouse for group use, along with 1 insulated truck for soil culture. The 0.5 ha pipe house needed 1 trickle irrigation system, 1 automatic pest control machine, a steam sulphur fumigator, a liquid CO_2 generator, a central environmental controller and a binding machine. The 0.5 ha steel house with soilless culture required a nutrient solution supply system, 1 automatic pest control machine, 1 stearasulphur fumigator, 1 liquid CO_2 generator, 1 complex environmental controller and 1 binding machine. Bredmose and Hansen (1997) explained that cut rose production can be automated if plants are gown single-stemmed for robotic harvesting and handling. The physiological basis for single-stemmed cut rose production in glasshouses and the synchronizationof plant growth and flowering were studied. Topophysis, the effect on growth and differentiation of position of axillary buds along the shoot, was studied in 3 cultivars Korokis Kiss, Tanettahn Manhattan Blue and Sweet Promise. In roses topophysis is intrinsic to the bud, determines growth potential, and is a decisive tool for the synchronization of development and flowering, which is essential to automation. The development and verification of ROSESIM, a computer simulation model of the growth of rose cv. Royalty plants based on experimentally observed growth responses from pinch unit flowering under 15 combinations of constant photosynthetic photon flux (PPF), day temperature (DT) and night temperature (NT) was described by Hopper *et al.* (1994). ROSESIM incorporates regression models of 4 flower development characteristics (day from pinch to visible bud, first colour, sepal reflex and flowering) that are full quadratic polynomials in PPF, DT and NT. ROSESIM also incorporates mathematical models of 9 plant growth characteristics (stem length and the following fresh and dry weights; stem, leaf, flower and total) based on data recorded every 10 days and at flowering.

Hopper (1996) described ROSESIM, a model designed to predict daily rose growth increments in response to set ranges of photosynthetic photon flux and day and night temperatures, developed in SLAM II and FORTRAN to run on a mainframe computer. As use of models on pCs has become more popular for instruction and simulation, ROSESIM was translated into the American Standard Code for Information Interchange (ASCII) to run in BASIC in the Microsoft Disk Operating System (MS-DOS).

In order to help growers and advisors to define schedules and sets of growing conditions, a predictive model, PP Rose (potential production of roses), was developed by Morisot (1996). It is an empirical model based on observed response functions. Production is assumed to be linearly related to the energy intercepted by the rose canopy. Temperature effects, supplementary lighting, CO_2 enrichment, cultivars and locations were taken into account. Harvesting dates and yields in relation to given quality grades were predicted. Lieth (1996) described model for optimizing cut rose production, which runs on a personal computer and is based on the effects of light, temperature and leaf age on photosynthesis in cv. Cara Mia. Photosynthesis rate for each leaf is used to compute the amount of carbohydrate produced, and then the amounts of accumulated carbohydrate used in the leaves, stem and flower are calculated. Thus the weights of these plant parts can be determined. The model can simulate the effects of factors such as pest damage on shoot growth.

Victoria *et al.* (2007) recommended movable benches, Ebb and flood irrigation and synchronized Y-bush training along with camera vision technique for facilitating robotic rose harvesting in greenhouses. However, synchronized Y-bush limited the production due to two stems per plant, but greater plant densities could improve the production with reduction in quality after a few production flushes.

12.0 Diseases and Pests

12.1.0 Diseases

Fungal Diseases

Like any other crops, roses are also prone to many diseases, some of which have been found to be very destructive. The most important diseases and the measures for their control described below.

12.1.1 Die-back

Die-back is one of the major diseases of rose, particularly in India. As the name indicates, the disease causes death of the plant from top to downwards. Generally observed after pruning, one finds blackening of twigs below the pruned surface, which at times extends further down, killing the whole plant. The older plants are more prone to the attack of the disease as compared to younger ones. The stems and roots of the dead plants show browning of internal tissues. *Diplodia rosarum* is the causal organism. Some other fungi such as *Colletotricum* sp. have been found associated with this malady. The pathogen gain entry into the host tissue through the minute injuries caused by the digger wasp (Pal, 1972; Dharam Vir and Sharma, 1985).

Die-back type of disorder may occur through improper use of fertilizers, improper watering, borer damage, poor drainage, lack of sunlight, continuous outbreak of black-spot or mite attack and importantly scale infestation. Die-back may not become apparent directly with these troubles but as after-effect of these maladies. In fact, anything that slows down the general growth processes can be a contributory factor to die-back. Therefore, proper cultural practice is the prevention for die-back in roses (Ramakrishnan, 1984).

Attempts have also been made to identify rose cultivars resistant to die-back. Of 56 cultivars assessed for resistance to die-back caused by *D. rosarum*, 11 were found highly resistant including White Christmas, Royal Ascot, Blue Moon and Crimson Glory (Kore *et al.*, 1977). The common control measure is to cut away the affected parts and burn it. The secateur should be disinfected with spirit and the cut ends immediately coated with a paste containing 4 parts of copper carbonate, 4 parts of red lead and 5 parts of linseed oil. The diseased twigs should always be pruned upto 4-5 cm below the visible blackened portion of the tissue. The pruning should preferably be done obliquely, which will prevent the accumulation of water drop (Dharam Vir and Sharma, 1985). Some systemic fungicides like carbendazim (1 g/l) may also be used as soil drench.

12.1.2 Powdery Mildew

This major disease of rose, causing much damage to roses all over the world, is caused by the fungus *Sphaerotheca pannosa* var. *rosae*. Almost all the aerial parts of the plants are generally affected and show white powdery growth of the fungus particularly on the lower side of the leaves. Affected flower buds fail to open and are damaged. The fungus appears whenever the days are warm and nights are cool. Highest spore release is reported to occur between mid-day and 4 pm (Cobb *et al.*, 1978). Gupta *et al.* (2004) observed significantly positive correlation between disease severity and average relative humidity (0.873) and negative correlation with temperature (-0.737) and light (-0.623). Also, coefficient of determination between disease severity and environmental factors explained 80.34 per cent change in the disease and Triadimefon reduced the severity to 6.25 per cent compared to control (64.37 per cent).

In earlier years, this disease was used to be controlled by applying sulphur dust but most of the roses are susceptible to sulphur injury, especially when it is done during hot days (Hahn, 1959; Cobb *et al.*, 1978). It is also not effective once the disease has become established on the plant and therefore, is very effective as a preventive method. Triforine, bupirimate, triadimefon and sulphur, sprayed fortnightly, had only a preventive effect. Of 8 cultivars, Pascall was the least susceptible to infection of powdery mildew (Gamboa *et al.*, 1986). Sulphur needs to be applied at about 10-day intervals (Anon., 1975). Benlate (0.05 per cent), nimrod (0.05 per cent), bavistin (0.5 per cent) and karathane (1000 ppm) provide fairly effective control of the disease, if sprays are given when the first initial symptoms appear. Applications need to be made every 10 to 14 days throughout the autumn to keep the disease under control. Several other fungicides like zineb, ziram and captan are also reported to be effective in controlling the disease (Marziano *et al.*, 1973). Woidyea (1997) found that covering both sides of leaves with fungicides (saprol, sportak alpha 380 EC, topas MZ) significantly increased their efficacy. Spraying of plants with a fungicide suspension or solution at a temperature of 20° and 30°C was more effective than at 10°C. Penconazole (as topaz), diclobutrazol (as vigil), flutriafol (as impact), fenpropimorph (as corbel) and tridemorph (as calixin) were tested on greenhouse-grown roses naturally infected by *S. pannosa* var. *rosae* in Bulgaria. All fungicides tested except flutriafol control disease (Nikolov, 1992). The best control of the disease was obtained with the highest concentration (1 ml/l) of hexaconazole applied at 8-day intervals (Perez-Olvera and Lozoya-Saldana, 1994). *Sporothrix flocculosa,* reported to be antagonistic against several members of the Erysiphales, was used under strict commercial conditions to control rose powdery mildew caused by *S. pannosa* var. *rosea* (Belanger *et al.*, 1994). The efficacy of *Tillettiopsis pallescens*, a naturally occurring ballistospore-forming yeast isolated from mildew-infected leaves, was evaluated as a biological control agent against rose powdery mildew (Ng *et al.*, 1997). Sporulation was significantly reduced (by 97-98 per cent) on plants treated with 3 applications of *T. pallescens* spore suspension, compared with a 47-57 per cent reduction on plants treated with 3 applications of distilled water. Sanitation is important to reduce the source of infection. All the infected leaves and branches should be cut away and burnt. The disease spreads quickly as spores are carried by wind.

Kakade *et al.* (2006) inferred that spraying Difenoconazole 0.05 per cent at 8 days interval reduced the disease incidence by 89.17 per cent and increased the number of 'A' grade flowers (0.79 lakh/m^2) with maximum net returns (Rs. 86,290/500 m^2) and BC ratio (1.45) followed by dinocap 0.05 per cent and tridemorph 0.05 per cent.Seddigh (2018) stated that application of aerated compost tea at 1:8 dilution twice a week aided in controlling rose powdery mildew. Elsharkawy *et al.* (2015) stated that polymer form of sodium silicate aqueous solution was effective against powdery mildew and root rot diseases in miniature rose than its monomer form and potassium silicate.

Efforts are also being made to develop or identify cultivars resistant to the attack of the causal organism. A hybrid Rugosa rose 'Martin Frobisher', selected from seedlings obtained by open pollination of hybrid Schneezwerg and the cultivar Care-free Beauty, has a high field tolerance to powdery mildew (Svejda, 1969; Buck, 1979). Susceptibility of 146 rose cultivars and species was evaluated in field trials conducted during 1993-95. Wojdyla and Wisniewska-Grzeszkiewicz (1999) found that the cultivars/species like Sky Violet (hybrid tea), Montana, Pussta (floribunda), Degenhardt (miniature), Friihlingsduft, Handel, *Rosa rugosa* Hansa, Heidekonigin, Lichkonigin Lucia, *R. rugosa* Pink Grootendorst, *R. nitida* and *R. glauca* (park and climbing) were immune to *S. pannosa* var. *rosae*. Szekely*et al.* (1981) observed a correlation between degree of infection and certain morphological and anatomical features of the cultivars. Among 170 cultivars, about 30 cultivars with red, pink and brick-red flowers were resistant, the remainder being moderately resistant or highly susceptible, while in those with yellow, mauve and white flowers the proportion of susceptible and highly susceptible cultivars was 60-75 per cent. Stomata size was inversely proportional to cultivar resistance, 21-23 per cent in resistant and 2-29 per cent in susceptible cultivars. The thickness of the epidermis was directly proportional to varietal resistance (28-30 per cent and 22-24 per cent for resistant and susceptible cultivars, respectively). Wen *et al.* (2006) categorized *R. sterilis, R. kweichonensis, R. laevigata, R. lucidissima, R. chinensis* and *R. roxburghii* 'Guinong No.6' as powdery mildew resistant sources and *R. multiflora var. multiflora, R. multiflora var. cathayensis, R. roxburghi* 'Guinong No. 5' and 'Guinong No. 2' as susceptible cultivars by evaluating their disease duration, infection date, peak disease index (PDI) and area under disease progressive curve (AUDPC).

12.1.3 Black Spot

This is a fungal disease caused by *Diplocarpon rosae*. In India, it is reported from Bangalore, Bombay, Delhi, Jammu, West Bengal and other areas and also causes serious problem in dull cold climate of the temperate region. The disease is characterized by dark brown, circular spots with fringe borders, present on both sides of leaflets. The spots multiply, expand and join together. At a later stage, the leaves become yellow and fall. If left uncontrolled the whole stem is defoliated and the plant becomes weak. Leaf buds are not properly developed and the flower buds are also affected.Yamaguchi and Boontiang (2010) reported the dynamic change of photosynthetic rate in rose after artificial inoculation of black spot disease. Significant breakdown of photosynthetic activity occurred 4-7 days after inoculation in cv. 'Tineke', whereas the epidemic change began 14 days after inoculation. Thus, this disorder could be helpful for rapid selection of black spot resistant lines during breeding programme.

Since the black spot fungus is located under the protecting cuticle, no fungicide can kill it without destroying the leaf (Pirone *et al.*, 1960). Hence, cure of the this disease is difficult and prevention seems to be the only hope. The infected leaves, as soon as they are observed should be clipped off, taken out of field and burnt. Preventive sprays of fungicides like zineb reduce the damage to a large extent (Glenn *et al.*, 1972; Nichols and Nelson, 1974). Benlate which controls powdery mildew attack can also be used very effectively to control this disease (Gojniczek and Laszcz, 1974; Shah *et al.*, 1981; Dharam Vir and Sharma, 1985). Of a larger number of fungicides evaluated, daconil 2787 (chlorothalonil) gave the most consistent control of *D. rosae*. Chlorothalonil application at 1 lb of 75 per cent W.P. formulation per 100 gal water starting at disease appearance and repeating at 7 or 14-day intervals throughout the season is recommended (Morrison, 1982). Bowen and Roark (2001) reported that decreased black spot disease symptoms and defoliation of rose plants early in the 1998 and 1999 seasons following winter applications of cyproconazole and triforine compared with myclobutanil or non-treated plants. Winter treatments with these fungicides did not consistently provided season long reduction of black spot. However, season long plant vigour and flower production on plants were improved at one site following winter applications of any of the above systemic fungicides compared with non-treated plants. In tests with 10 fungicides used to control black spot in China, the best were propiconazole at 0.19 g a.i./l, followed by prochloraz (0.25 g) and chlorothalonil (0.8 g), giving 91.2 per cent, 86.7 per cent and

> 70 per cent control respectively. Applications of chlorothalonil combined with a nutrient mixture at 16.7 g/l gave better control than the fungicide alone, and also promoted plant growth and improved flower quality (Jiang *et al.*, 1996). Different cultivars of rose vary in their susceptibility to the black spot and the disease can cause much less damage in covered plants than in exposed plants (Lundstad, 1957; Nazarenko and Zhalnina, 1975). Hybrid Rugosa rose 'Martin Frobisher' has been reported to be immune to black spot (Svejda, 1969). The cultivars Belaya (*Rosa alba*), John Cabot and Carefree Beauty are also found to be resistant to black spot disease (Nazarenko and Zhalnina, 1975; Svejda, 1978; Buck, 1979). Carbendazim was the most effective fungicide to control black spot with diseaseindices of 15.84 per cent and 24.47 per cent for the 2 experiments compared with 32.81 per cent and 42.19 per cent for the untreated controls (Maljaja, 1997). Nicetic *et al.* (1999) observed low incidence of black spot on cvs. Bridal Pink, Sonia, Friendship and Tournamut using 1.0 per cent C24 petroleum spray oil in combination with $NaHCO_3$. It provides sustainable and cost effective control. Rol *et al.* (2000) reported the efficacy of antitranspirants (Nu Film 17, Vapor Gard, Wilt Pruf, Stressguard and Transfilm) in combination with the fungicide chlorothalonil. Defoliation was reduced in all the fungicide treatments with and without anti-transpirants over non-treated control plants. Moreover, the combinations minimized the season-long fungicide use by 40-69 per cent without any decrease in disease control. However, plant vigour was found lower in anti-transpirant treatments compared to chlorothalonil alone. Gachomo *et al.* (2009) revealed that strobilurins inhibited the conidial germination completely when applied protectively 1 week before inoculation whereas, triazoles slowed down the germination rate when applied protectively and suppressed the disease symptom when applied after fungal establishement but before expression of symptoms. Moreover, post-infection application of triazoles caused breakdown of subcuticular, intercellular mycelia and hyphae in epidermal cells while strobilurins restricted to the subcuticular mycelia thus stating that triazoles were more effective and systemic than strobilurins.

12.1.4 Leaf Spot (*Alternaria alternata*)

Leaf spot caused by *Alternaria alternata* has been reported from some parts of India where it causes heavy losses during the rainy season. Spraying at weekly intervals with captan or zineb (2000 ppm) has been recommended for its control (Dharam Vir and Sharma, 1985). *A. circinans* also causes leaf spots in rose (Pirone *et al.*, 1960). Khatun *et al.* (2009) studied the role of phenols and their oxidizing enzymes in defense against leaf spot of *Rosa centifolia* and found an increased peroxidase and polyphenol oxidase activity in the infected leaves which mediate the oxidation of phenols rendering toxic effect against pathogen (Table 20).

12.1.5 Stem Blight

This disease is also caused by the fungus associated with die back. Several tan-coloured spots appear on the stems, particularly in older plants which merge together and spread to the whole bark. Older plants are more susceptible than younger ones. Regular sprays of captan (2000 ppm) and such other fungicides control the infection.

Table 20: Phenols and their Oxidizing Enzymes in Healthy and Alternaria Infected Leaves of *Rosa centifolia* at different Periods of Infection*

Days after Infection	*Peroxidase Activity (g/min)*		*Polyphenol Oxidase (g/min)*		*Total Phenol (at 650 nm)*	
	Healthy	*Infected*	*Healthy*	*Infected*	*Healthy*	*Infected*
0 days	0.040 ± 0.000	0.042 ± 0.002	0.036 ± 0.002	0.038 ± 0.002	0.522 ± 0.009	0.550 ± 0.016
7 days	0.062 ± 0.002	0.222 ± 0.007	0.048 ± 0.002	0.248 ± 0.014	0.536 ± 0.007	0.850 ± 0.000
14 days	0.052 ± 0.002	0.282 ± 0.002	0.054 ± 0.002	0.328 ± 0.012	0.542 ± 0.004	0.984 ± 0.002
21 days	0.046 ± 0.002	0.330 ± 0.012	0.056 ± 0.002	0.460 ± 0.018	0.532 ± 0.010	1.152 ± 0.142

* Khatun *et al.* (2009).

12.1.6 Rust

Rose rust is considered as a serious disease of rose and is most common in warm and humid areas. It is caused by the fungus *Phragmidium*, several species of which have been reported to be causing the disease, the most common being *P. mucronatum*. Among the other species of rust fungi found on cultivated roses are *P. americanum*, *P. rosae-pimpenellifoliae* and *P. speciosum*. Wild roses are subjected to *P. montivagum*, *P. fusiforme*, *P. rosae-arkansanae*, *P. rosae-califlornicae* and *P. rosicola*. These rusts are rarely serious. Plakhova (1975) reported that rust on essential oil yielding and other roses in Crimea is caused by *P. disciflorum* and *P. tuberculatum*. The disease is characterized by the presence of reddish-orange pustules on leaflets and sometimes on petioles. The colour of these pustules changes to black when teleutospores are formed. In severe cases of infection, defoliation occurs. The production of flowers from diseased plants is much less.

Weekly sprays of cycloheximide, RP-26019 (iprodione) and triforine gave adequate control of *P. mucronatum*. Oxycarboxin gave excellent control with only 3 bi-monthly sprays (Wallace and Yule, 1977). Spraying with zineb (2000 ppm) at fortnightly intervals was recommended for the control of rust (Dharam Vir and Sharma, 1985).

12.1.7 Botrytis Blight

Infection starts from the sepals as black-brown specks that cover the flower in due course. This disease is caused by *Botrytis cinerea*. The buds turn brown and decay. Sometimes partially opened buds are infected and individual petals turn brown and shrivel. In cool moist weather the flower is covered with greenish-grey or darkish, mouldy growth. Infection persists on old blooms, winter-killed canes also harbour infection. *B. cinerea* causes latent infections of roses, which can develop into aggressive rot (*Botrytis* blight) at pre- and postharvest stages. Botrytis blight is the cause of major flower losses. *Trichoderma harzianum*, polyoxine, fenpiclonil, tebuconazole with or without dichlofluanid, and fenbuconazole sprayed once a week, were evaluated for control of *B. cinerea* on roses. Fenpiclonil, tebu-conazole, dichlofluanid alone or alternated with *T. harzianum*, and *T. harzianum* alone also reduced development of grey mould on flowers that were harvested soon after the treatment and incubated under Botrytis-conducive conditions (Elad *et al.*, 1993). The effectiveness of fungistatic atmospheres for postharvest control of *B. cinerea* infections on cut rose flowers (*Rosa hybrida*) was investigated by Hammer *et al.* (1990). Storing cut Sonia, Royalty and Gold Rush roses at 2.5°C with 10 per cent CO_2 for 5 days followed by 2 days of cold storage in air, reduced the number of *B. cinerea* lesions that developed on inoculated and non-inoculated flower petals by 77 per cent and 82 per cent, respectively, compared with cold storage for 7 days in air. Thermal dip treatment for flower heads was effective against grey mould of roses. Dipping of flowers in tap-water at 50°C for 20-40s was more effective than at higher or lower temperature and than for longer treatments (Elad and Volpin, 1991). There is a direct relationship between postharvest senescence processing in Mercedes rose flowers and sensitivity to botrytis blight disease. It is suggested that shortening the handling time after harvest or delaying the senescence process will help to decrease damage caused by *B. cinerea* (Shaul *et al.*, 1993). Pyrrolnitrin (an antibiotic isolated from *Pseudomonas cepacia*) was tested for postharvest control of *B. cinerea* infections on cut Sonia and Royalty rose flowers. Dip treatment with 100 mg/L reduced lesion development by c. 90 per cent compared with inoculated control flowers and prevented post-storage flower rot. This degree of disease control was comparable to that achieved with 1800 mg vinclozolin/l (the max. label, rate) (Hammer *et al.*, 1993). Yohalem (2000) suggested that *B. cinerea* that several filamentous fungi can be applied to suppress sporulation of the pathogen during the establishment of pot roses. *Ulocladium atrum* and *Clonostachys rosea* inhibited grey mould better than the fungicide iprodione. Commercial *T. harzianum* preparations failed to inhibit disease under the conditions of the experiments. The development of botrytis blight was significantly suppressed in detached petals of cv. Mercedes and Sonata treated for 24 h with 20 mg/L solution of GA3. To obtain a similar level of suppression in intact flower buds to that obtained in detached petals, a postharvest spray of GA3, with a concentration of 346 mg/L (1 mM) was required (Shaul *et al.*, 1995, 1996). The development of the disease was promoted by application of abscisic acid (ABA). The effects of environmental conditions during crop production on the susceptibility of cut roses (cv. Moonlight) to postharvest infection by *B. cinerea* were investigated. For flowers harvested from a commercial greenhouse, susceptibility was linearly correlated with mean air velocity during the 5-week period before each harvest. Susceptibility was also correlated with mean leaf-to-air temperature gradient (Hammer and Evensen, 1996).A single application of *Ulocladium atrum* significantly reduced gray mould incidence and sporulation, but increasing number of applications increased its effectiveness (Yohalem and Kristensen, 2004).

12.1.8 Root Fungus

The roots are attacked by a fungus identified as *Trichoderma viride* and get decayed. The leaves turn yellow and the growth of the plant is affected, and the plant ultimately dies. As a control measure, the roots may be exposed, some quantity of dry wood ashes incorporated and covered back with soil. In case of excessive evaporation of moisture, the plant should be irrigated (Biswas, 1984).

12.1.9 Crown Gall Disease

Crown gall disease is caused by a soil borne bacterium, *Agrobacterium tumefaciens*. This disease development occurs when the ideal environment, the virulent pathogen and the susceptible plant host interact at the same time. Its pathogenesis is unique and includes the transfer of part of tumour inducing (Ti) plasmid in to plant. As a result, the plant produces increased amount of hormones leading to uncontrolled tissue proliferation and synthesis of unusual compounds like opines, derivatives of sugars and special amino acids. Gitari *et al.* (2018) investigated the biological and chemical approach in the crown gall disease management of rose cultivars under greenhouse conditions. Results showed that minimum incidence (6 per cent) of crown gall were observed when rose varieties were inoculated with *Agrobacterium radiobacter* compared to control (54 per cent) (Table 21). However, no significant difference was observed among the different chemical treatments (copper oxychloride and corn oil).

Table 21: Effectiveness of Antagonist (*Agrobacterium radiobacter*) on Tumour Formation*

Treatment	*Number of Wounds with Galls*	*Gall Size (diameter in cm)*
Agrobacterium radiobacter	0.1875 ± 0.1001	0.06875 ± 0.0435
A.tumefaciens + A. radicans	0.0625 ± 0.0625	0.00625 ± 0.0063
A.tumefaciens	0.4375 ± 0.1281	0.36875 ± 0.1306
Control	0.5625 ± 0.1281	0.79375 ± 0.2242

*Gitari *et al.* (2018).

Viral Diseases

12.1.10 Rose Wilt

Rose wilt is caused due to infection with a virus *Marmorflaccum fasciens*. The first symptoms appear in the form of recurving of leaflets at the tip of young shoots which are brittle. Defoliation may happen, with leaves turning yellow and falling off. The stem also gets infected and the whole plant dies. In epidemic form, one infected plant could ruin the whole garden within a couple of years. Control of aphids, which transmit the disease, helps to control the infection. The affected plants should be removed and burnt (Harkness, 1979).

12.1.11 Rose Mosaic Virus

Rose mosaic occurs mostly on greenhouse roses. Chlorotic areas along with midribs of the leaflets and localized distortion are the principal symptoms. Sometimes one finds ring and watermark pattern. Another strain of this virus causes brighter and light yellow patterns. Diseased plants should be destroyed and never used for propagation (Pirone *et al.*, 1960).

Diseases of Rose

Die-back

Stem blight

Black spot in rose

Powdery mildew

Alternaria leaf spots

Mosaic virus in rose

Pests of Rose

Red spider mite
White fly infestation

Aphids infestation
Mealy bugs on rose

Thrips attack
Rose beetle
Root knot nematode

12.1.12 Rose Rosette Virus

Bello *et al.* (2015) examined the cause and vector for Rose rosette disease (RRD) which was first indicated in 1942 in USA. The suspected and viruliferous eriophyid mite, *Phyllocoptes frutiphilius* were collected from RRD-infected roses and introduced into RRD-free rose plants. Two months after infestation, the latter roses developed RRD symptoms proving the above-said mite to be the vector. Further, the total nucleic acids from infected plants were subjected to deep sequencing which revealed that none of viruses other than RRV was detected. Thus, rose rosette virus and eriophyid mite were detected as the causal organism and vector, respectively.

12.1.13 Rose Spring Dwarf Disease

Rose spring dwarf disease caused by Rose Spring Dwarf associated Virus (RSDaV) which could infect both monocots and dicots belongs to the family Luteoviridae and has 5 ORFs in different coding regions for RNA-dependent RNA polymerase (ORF 1, 2), coat protein (ORF 3), movement protein (ORF 4) and readthrough protein (ORF-5). *Metapolophium dirhodum* (rose-grass aphid) and *Rhodobium porosum* (yellow rose aphid) were reported as vectors for RSDaV (Salem *et al.*, 2007, 2008),

12.1.14 Other Diseases

Several other diseases affect rose cultivation. They are generally of minor importance. Various diseases reported include rose anthracnose (*Elsinoe rosarum*), black mould (*Chalaropsis thielavioides*), leaf spots (*Cercospora puderi, C. rosicola, Mycosphaerella rosicola, Septoria rosae, Curvularia brachyspora*) and downy mildew (*Peronospora sparsa*) among the fungal diseases; bacterial leaf spot (*Pseudomonas syringae*) and crown gall (*Agrobacterium tumefaciens*) among bacterial diseases; and Rose Streak Virus, Strawberry Latent Ring Spot Virus, Rose Leaf Curl, Rose Spring Dwarf and Rose rosette among the important viral diseases (McClellan, 1949; Boelema, 1973; Khristova, 1974; Salac and Palmer, 1975; Paterson *et al.*, 1975; Hoof and Caron, 1975; Secor and Nyland, 1978; Semina *et al.*, 1982). Besides, the fungus *Verticillium alboartrum* has also been isolated from several cultivars of rose (Dinock, 1951; Nelson and Wilhelm, 1958).

He *et al.* (2016) reported the infection of *Rosa chinensis* with Apple Stem Grooving Virus (ASGV) for the first time. This virus has extended its host range and could be a potential threat to rose production in China. Madhupriya (2017) observed phytoplasma-suspected symptoms like leaf chlorosis, phyllody, virescence and little leaf in 35 rose cultivars and 13 plants budded with the above infected plants were symptomatic when detected using nested PCR. The resultant amplicons on further BLASTn analysis and phylogenetic analysis showed that sequences fitted to '*Candidatus phytoplasma asteris*' subgroup 16Srl-B and 'Ca. P. *australasia*' subgroup 16SrII-D.

12.2.0 Pests

Roses attract a large number of pests which cause considerable damage. At almost every stage, be in the nursery or in beds, rose plants are prone to attack by various pests. Smitha *et al.* (2017) carried out an investigation to document the insect pest complex of rose at Regional Agricultural Research Station (RARS), Wayanad, The major insect pests identified were *Macrosiphum rosae, Thrips tabaci, Anomala orientalis, Aonidiella aurantii, Achaea janata, Orygia posticus, Megachile anthracina, Odontotermes obesus, etc.* Bout *et al.* (2010) recommended a sampling unit for pest monitoring which consisted of a single stem with flower and corresponding basal foliage. Generally a 90-unit regular grid (1 U/6.4 m^2) was used with weekly assessments in a spatial distribution approach. Various grid sizes were modeled and evaluated to determine the number of sampling units. A grid size of 1 U/21 m^2) was found to be suitable in improving the efficiency of crop monitoring and in taking more rational decisions.Some of thecommonly reported insects and other pests attacking roses are described below.

12.2.1 White Ants

White ants also known as termites, cause damage to rose plants even before they are established. Small colonies are formed inside the soil and the insects multiply very rapidly during the rainy season. They destroy the underground parts including the roots of the young as well as old plants. Sometimes, it is found that apparently healthy looking rose plants appear to wither. The best measure to avoid damage is to apply 5 chlorpyriphos dust in the pits before planting roses. Drenching of rose beds with chlorpyriphos 4 ml/l also controls the damage in established gardens. Termiseal is another insecticide used for controlling the termites (Pal, 1972; Biswas, 1984).

12.2.2 Aphids

Aphids or plant lice attack most garden plants, and roses are no exception. Four types of aphids, *i.e.*, greenpeach, melon, potato and rose infest (Pirone *et al.*, 1960), the common species being *Macrosiphum rosae, Wahlgreniella nervata*, a North American aphid, has been found to infest roses in Britain (Prior, 1975). These small green insects appear early in the spring and feed on the tender shoots, buds and flowers. They damage the plants by sucking plant sap. Flowers when attacked, become malformed. The aphids can be easily controlled by spraying malathion (0.1 per cent) or metasystox (0.1 to 0.2 per cent) or rogor (0.1 to 0.2 per cent). Blumel and Hausdorf (1996) conducted a trial to control the aphid species *M. euphorbiae* on greenhouse rose cultivars Frisco, Kardinal and Vivaldi in Austria with parasitoid *Aphelinus abdominalis*. Flint and Dreistadt (2005) observed 66-88 per cent aphid control with a single release of 100 lady bird beetles (*Hippodamia convergens*) in potted roses, whereas reduction in aphid density was not observed even with release of 175 or 250 beetles per 0.5-1 m tall shrub in the landscape. However, release of 1400-1750 beetles per shrub reduced aphid density in the landscape. Caged shrubs controlled aphids with 10-20 beetles per flower bud, while the uncaged rose shoots required > 100 beetles per bud. Thus, predative convergent lady beetles could be effectively used for aphid management on rose.

12.2.3 Thrips

Thrips are very minute, long and slender insects which inhabit the undersurface of tender leaves and suck sap from leaves and flower buds. Mottled and deformed young leaves with brown or silvery patches or burnt margins and deformed flowers are the main signs of the attack. Bergh and Blanc (1997) observed the performance of western flower thrips on cultivars of miniature rose, *Rosa chinensis* 'Minima' and revealed that larval development was slower in petals of red and orange flowers, whereas faster in white and yellow flowers. Similar trend was observed with respect to adult oviposition. Moreover, leaf tissues were preferred more for oviposition than petals.In all, six different thrips, namely Florida flower, flower, greenhouse, onion, tobacco and western flower attack roses, the most common being *Thrips maginis*. Jayalaxmi *et al.* (2016) stated that thrips infestation was more in polyhouse than open conditions. Further more, the thrips density was positively correlated with temperature and sunshine hours and negatively correlated with relative humidity, rainfall and wind velocity. Rani *et al.* (2003) studied the response of various rose varieties to thrips and found that Versilia was relatively resistant; First Red and Grand Gala were susceptible and rest of the varieties was moderately resistant. In addition, young leaves and red flowers were more preferred by the thrips compared to mature leaves, other floral parts and orange or yellow flowers.

Higher fertilizer rates resulting in accelerated floral growth exhibited more damage by thrips with greater density in buds followed by flowers and leaves. The thrips infestation was found maximum in larger flowers compared to smaller flowers however the population density being similar (Mannion *et al.*, 2013).The chemicals used for aphid control are also effective in controlling thrips (Pirone *et al.*, 1960).

Seven days after application, populations of *R. syriacus* treated with 0.15 or 0.18 g a.i./plant and those of *A. aurantii* and *T. neocaledonicus* treated with 0.18 g a.i./plant were significantly reduced. Leaf damage by *S. dorsalis* was reduced 7 days after treatment with 0.12, 0.15 and 0.18 g a.i./plant, while flower damage was reduced with 0.09 and 0.18 g a.i./plant after 14 days (Nair *et al.*, 1990). In field studies conducted by Nair *et al.* (1991), 0.05 per cent monocrotophos, quinalphos or phosphamidon effectively suppressed leaf damage to roses caused by *S. dorsalis*. Dimethoate and phosphamidon at 0.05 per cent, and monocrotophos and endosulfan at 0.1 per cent controlled flower bud damage. Treatment of roses in commercial greenhouses with *Beauveria bassiana* WP resulted in a significant decrease in numbers of *Frankliniello occidentalis* (Murphy *et al.*, 1998).

Phosphamidon 85 EC (0.02 per cent), oxydemeton-methyl 25 EC (0.025 per cent), monocrotophos 36 EC (0.05 per cent), neem seed kernel extract (NSKE, 5 per cent) and 1500 ppm azadirachtin (achook 1 per cent) significantly reduced *R. cruentatus* population on roses, but monocrotophos was the most effective, reducing the population by 95.6 per cent after 72 h (Dadmal *et al.*, 1999). Park *et al.* (2002) investigated the occurrence of thrips in experimental and commercial greenhouse. In the experimental greenhouse, thrips were first detected on yellow sticky trap during early April and were severe during June each year. In commercial greenhouses where insecticides were applied > 2.2 times per month showed relatively low levels of population which was 40 per cent less than experimental greenhouse. Furthermore, yellow followed by white rose cultivars were more preferred. Among various stages, Stage F_2 when sepals uncover the bud partly and corolla slightly protrudes was identified for first habitation. The number of adults increased from this stage to senescence stage (F_4).

Broughton *et al.* (2015) reported that use of traps baited with thrip lures, ThripLineams (male sex aggregation hormone) and Lurem-TR (kairomone) trapped 1.2-4 times more Western flower thrips (*Frankliniella occidentalis*) compared to unbaited traps. Also, female thrips were caught more in both baited and unbaited trap than males. Thus, the use of pheromones together with sticky traps for mass control of thrips was suggested for glasshouse grown rose plants.

Pizzol *et al.* (2010) demonstrated that estimation of thrips population is more accurate using Yellow Sticky Trap (YST) in greenhouse rose crops as YST takes 2 times lesser time compared to other monitoring methods like knock down techniques (i.e tapping flower heads, FT) and actual counts of entire plants.

A polyhouse experiment was conducted by Jagadish and Purnima (2011) to find out the efficacy of botanicals and entomopathogens against *Scirtothrips dorsalis* Hood on different stages of rose *viz.*, bud, half opened and full opened flower. Resultsrevealed that among different stages of rose, half opened flower was found superior to control *S. dorsalis.*Among different botanicals NSKE (2 per cent) was recorded 74.37 per cent mortality to thrips. Among different entomopathogens, *Heterorhabditis indica* (2000IJs/ml) was found next best to botanicals by recording 72.08 per cent mortality of thrips and chlothianidin 50 WDG (0.06 g/l) was recorded 98.58 per cent mortality of thrips (Table 22).

12.2.4 Jassids (leaf-hoppers)

These are very small, light grey or pale green insects which suck the cell sap from the leaves, causing yellowing or whitening of the attacked surface, giving the plants a sickly appearance. Jassids can easily be controlled by spraying with 0.03 per cent parathion (Pal, 1972), 0.1 per cent metacid or 8 per cent thimet as soil application (Prasad and Verma, 1980). The use of 0.05 per cent monocrotophos is also effective in controlling jassids.

12.2.5 Chafer Beetles

These insects feed upon the growing points, making irregular holes and punches on the leaves. The grubs feed on the roots. Flowering is affected if not controlled in time. The soil should be treated with 3 per cent heptachlor dust for controlling grubs, and spraying the plants 0.2 per cent metacid will control the beetles. Certain indigenous plant materials were also tested for their antifeeding and insecticidal properties against the beetles, and treatments with petroleum ether extract of dried rhizomes of Acorus calamus, Centretherum anthelminticum, dried and immature seeds of Linum usitatissimum having the concentrations of 0.1 to 0.5 per cent, 0.2 per cent and 2 per cent, respectively, were found very useful (Verma and Singh, 1980).

12.2.6 Red Scales

A number of scale insects infest roses in the greenhouse and in the garden. They can easily be detected by the reddish-brown encrustations on the lower portion of old stems and at times on the younger shoots also. Several pox-mark

Table 22:Effect of different Treatments against *S. dorsal* is on Rose Bud under Polyhouse Conditions*

Treatment No.	*Treatment*	*Dose*	*Number of Thrips/Bud*				*Per cent Reduction*		
			Pre Count		*Post Count*				
			I-DBS	*I-DAS*	*III-DAS*	*V-DAS*	*I-DAS*	*III-DAS*	*V-DAS*
T1	NSKE	2 per cent	25.00 (5.04)	20.00 (04.58) b	20.00 (4.58)b	15.00 (4.00)c	20.0	20.00	40.00
T2	Aqueous leaf extract of Pongamia glabra	2 per cent	26.60 (5.23)	24.20 (05.02)c	21.00 (4.69)b	20.00 (4.58)d	9.02	21.00	24.00
T3	Metarhizium anisopliae	4 gL^{-1}	26.00 (5.13)	25.00 (05.10) c	24.20 (5.02)c	22.60 (4.86)e	3.80	8.46	13.07
T4	Heterorhabditis indica	2000 $IJsml^{-1}$	26.40 (5.21)	22.60 (04.86) c	19.00 (4.47)b	13.00 (3.74)b	14.4	28.03	50.00
T5	Clothianidin 50 WDG	0.06 gL^{-1}	25.60 (5.14)	15.00 (4.00)a	9.20 (3.19)a	1.80 (1.66)a	41.4	64.00	92.96
T6	Water spray		27 (5.23)	24.60 (5.06)c	26.40 (5.27)c	26.80 (5.27)f	8.80	2.22	0.74
T7	Control		27.60 (5.32)	28.00 (5.38)d	26.00 (5.19)c	29.00 (5.48)f	-1.40	5.70	-5.07

DBS: Day before spray DAS: Days afterspray figures in parentheses are "X+1 transformation value. Values in the column followed by common letters are non significant at P=0.05 as per DMRT.

* Jagadish and Poornima (2011).

scales appear on shoots and bark gets covered with layers of overlapping scales. Each scale represents an individual insect which sucks out the plant sap. The tiny insects occur in thousands and the affected branches become dry. They appear more severely before and after monsoon and are often choosy in their preference to certain cultivars of roses. Some preferred cultivars are Versailles, Incense, Kordes Perfecta, First Prize, Camelot among Hybrid Teas and Zorina, Zambra, Ice White, Red Pinocchio in Floribundas. No rose plant is immune to it (Ramakrishnan, 1984). The menace can be removed by rubbing the scurfy encrustations with a used tooth brush or a cotton swab dipped in methylated spirit. Spraying the affected plants with insecticides like malathion, rogor will control the insect (Pal, 1972; Pirone *et al.*, 1960). Application of granular insecticides like carbofuron at 1.5 kg a.i./ha and disulfoton and phorate at 3.0 kg a.i./ha also gave good control of rose scales (Srinivasan *et al.*, 1974). Use of permethrin contained in bio flydown is also very effective (Hessayon, 1981).

12.2.7 Red Spider Mite

These are minute, polyphagous pests found in large colonies on the underside of leaves covered with fine silky webs. Protonymph is the most active feeding stage. As a result of their feeding, white specks appear on the leaves and these specks coalesce and appear as white patches. Ultimately, affected leaves become mottled, turn yellow and fall. Cultivars like Summer Snow, First Prize, Oregold and Maria Callas are more susceptible than certain others (Ramakrishnan, 1984). Kumar *et al.* (2009) reported Spinx and temptation as moderately susceptible; Aqua, Passion, Milwa, Noblesse, Konfetti, Gold Strikes, Grand Gala and Biyanca as susceptible and First Red as highly susceptible cultivars to two-spotted spider mite.Some Polyantha and Floribunda roses completely lose their leaves prematurely due to the attack of this pest. The damage is more pronounced during warm and dry season. Mites infestation has significantly positive correlation with maximum temperature (0.684) and morning relative humidity (0.21) whereas, negatively correlated with minimum temperature (-0.49) and evening relative humidity (-0.783) as stated by Deshmukh *et al.*, 2017.Frequent spray with sulphur helps to control this mite. Several mite-killers such as kelthane, morestan, dimite, tedion V-18 have been reported to be very effective in controlling rose mites (Pirone *et al.*, 1960; Ramakrishnan, 1984). Eight ml kelthane is thoroughly mixed with 10 litres of water with 10-15 drops of any sticker and application of this chemical every five or six days may be sufficient to control the mites. Spraying the affected plants with 0.05 per cent parathion (Pal, 1972) or 0.1 per cent metasystox (Misra, 1981) is also suggested. Field and Hoy (1984) tried the biological control method and reported that strains of Typhlodromus occidentalis could be integrated into pest control programme without being seriously affected by chemicals used against other pests or against Tetranychus urticae itself during the establishment of predator. In a protected one-year-round crop of roses, T. urticae was successfully suppressed by the predator Phyoseiulusper similis applied at 26/m^2, the results were better than for the use of caricides (Bluel, 1990). P. persimilis successfully controlled T. urticae on rose hedges in an integrated mite control programme in 2 unheated plastic-covered tunnels at Cleveland in southern Queensland (Gough, 1991). Abamectin applications confined to the upper portion of the canopy (from where the marketed portion of the crop is taken) provided chemical control of the mite in the upper canopy. The lower canopy supported a rapidly decreasing population of T. urticae and a slightly increasing population of P. persimilis. These results show the potential for integration of biological control of T. urticae with chemical control of T.urticae and other pests of greenhouse roses (Sanderson and Zhang, 1995). Samples of spider mite population collected from ten rose houses were tested for resistance to parathion, tetradifon and dicofol. Despite the fact that for a period of 7 years tetradifon and dicofol had not been used, resistance to these acaricides was still present in most population. Reversion to susceptibility in practice is apparently very slow and excludes the possibility of applying a rotational spray system (Overmeer *et al.*, 1975). Thuringiensin was more effective than cyhexatinin controlling red spider mite but there were no significant differences between azocyclotin, dienochlor and thuringiensin (Ochoa *et al.*, 1990). Diafenthiuron was highly efficient at 40 and 30 g a.i/100 l in controlling spider mite. Mortalities obtained were comparable with those obtained with abamectin at 20 ml a.i./100 l of water (Aguiar *et al.*, 1993). Jasmine *et al.* (2008) reported that abamectin 1.9 EC at 9.6 and 12.0 g a.i ha-1was effective in controlling two-spotted spider mite increasing yield as wells as safe to predatory mites under polyhouse conditions. Clofentezine at 25 g and abamectin at 3.6 g a.i./100 l of water were the most effective treatments against the mite (Aguiar *et al.*, 1995). A concentration of 0.20 to 0.25 per cent Cinnamite controlled two-spotted adult spider mites (T. urticae) and eggs by 80 per cent and 100 per cent respectively (Parrella *et al.*, 1999). Spraying rose plants with bifenazate 0.24 g/l along with dicofol 18 EC 2.5 ml/l effectively controlled mite with the minimum survival of 3.56 per cent over two season. Moreover, bifenazate was illustrated to be safe for indigenous predatory mite (Amblyseius sp.) and non-phytotoxic effect on rose (Rani and Sridhar, 2005). Clofentezine 50 SC @ 300 g a.i./ha which was tested at par with abamectin (8.55 g a.i./ha) and superior over dicofol (231.25 g a.i./ha) reduced mite infestation and was found to be relatively safe to natural enemies as well as non-phytotoxic to rose under greenhouse (Sridhar *et al.*, 2011). Patke *et al.* (2015) revealed that propargite 0.1 per cent recorded 82.37 per cent reduction of nymphal population with the production of 29 flowers/plot followed by acetamiprid 0.005 per cent (79.39 per cent reduction and 34 flowers/plot) and dimethoate 0.03 per cent (77.63 per cent and 31 flowers/plot). Moreover, the biopesticide Verticillium lacanii, bioagent Chrysoperla carnea and botanical Nimbicidin1500 ppm reduced nymphal population by 58.60, 51.48 and 63.69 per cent, respectively. However, dimethoate 0.03 per cent recorded highest ICBR of 141.21 indicating its economic viability. Singh *et al.* (2017) evaluated the efficacy of different acaricides against two-spotted spider mite and found that the order of effectiveness based on per cent mortality were fenzaquin (0.0025 per cent) >propargite (0.0057 per cent) > Horticultural Mineral Oil - HMO (2.5 per cent) > dimethoate (0.03 per cent) > HMO (2 per cent) > dicofol (0.05 per cent) > imidacloprid (0.015 per cent) > HMO (1.5 per cent) > HMO (1 per cent) > HMO (0.5 per cent), respectively (Table 23). Nyalala and Grout (2015) reported that the species Gynandropsis gynandra and Cleome hassleriana could be effectively used for controlling two-spotted spider mite by producing acetonitrile following foliar disruption in temperate and tropical regions, respectively. Moroover, this species could be exploited as potential gene donot for genetic manipulation as well.Nyalala and Grout (2007) found that density of 15 Cleome gynandra plants in a 1.8 m^2 bed was effective when planted around bed perimeter or within rows of roses.

Table 23: Efficacy of different Pesticides on per cent Mortality of Two-spotted Spider Mites under Polyhouse Condition*

Treatments	*Concentration (per cent)*	*Percent Mortality of Mites at Indicated Days*				
		1	*3*	*7*	*10*	*Mean*
Dicofol	0.05	34.89	39.44	41.59	45.98	40.48
Imidacloprid	0.015	28.67	33.60	42.63	52.79	39.42
Dimethoate	0.03	31.99	47.06	51.75	55.86	46.67
Propargite	0.0057	35.11	49.40	56.64	59.94	50.27
Fenzaquin	0.0025	35.55	50.10	57.84	63.04	51.63
HMO	0.5	20.16	26.69	35.52	38.71	30.27
HMO	1	21.33	28.56	37.16	40.10	31.79
HMO	1.5	24.61	33.38	39.13	40.78	34.47
HMO	2	32.29	45.77	43.86	52.25	43.56
HMO	2.5	32.38	48.18	54.69	57.65	48.23

* Singh *et al.* (2017).

12.2.8 Digger Wasps

Digger wasps (*Crabro* sp.) damage the plants after pruning. They dig a hole into the pith of the stem through the cut end and make a nest there. Because of this burrowing, a fungus, causing die back disease, often gains entry into the stem tissue and consequently, the pruned stem shows sign of drying up from the tip downwards. The fungus can be controlled by painting the pruning cuts with a fungicidal paint containing 4 parts each of copper carbonate and red lead along with 5 parts of linseed oil (Pal, 1972). Few drops of rogor or dimecron in the hole and plugging with a small piece of cotton is an excellent way to kill the insect. The portion of the stem which is almost dry should be cut away and painted with the above fungicide at the cut surface (Biswas, 1984).

12.2.9 Mealy Bugs

Mealy bugs are generally observed on the stalks of buds and flowers. They suck plant sap due to which the buds do not open and the flowers wither away. The control measure is the same as that of red scales. Fumigation with methylbromide (32 mg/L) at atmospheric pressure and at temperatures of 24-25°C for 1.5 hours was also found suitable for treating rose against this pest (Junaid and Nasir, 1956).

12.2.10 Other Minor Insect Pests

Besides the above mentioned major pests, there may be occasional infestation with caterpillars, rose midge (*Dasyneura rhodophaga*), rose stem girdler (*Agrilus aurichalceus*), Japanese beetle (*Popillia japonica*), mossy rose-gall (*Rhodites rosae*), rose budworm (*Pyrrhia umbra*), sawflies (*Arge berberidis*), *etc.*, causing damage to the plants at different stages of the growth (Pirone *et al.*, 1960; Stenseth, 1983; Harranger, 1984). The routine spray with rogoror malathion will control most of these pests. Shorthouse (2001) observed 4 species of gall-inducing wasps of genus *Diplolepis* endemic to wild roses in Canadian gardens. Among that 2 species, *D. spinosa* and *D. radiucm* became pests whereas, *D. rosaefolii* and *D. polita* were less common and not considered as pests.

12.3.0 Nematodes

12.3.1 Root-knot Nematode (*Meloidogyne hapla, M. incognita*)

These are most common plant parasitic nematodes in India and their infestation is easily recognized by the characteristic root galls they cause. The attack of the nematode results in symptoms of malnutrition, decreased shoot growth and leaf chlorosis (Pirone *et al.*, 1960). Flowering is adversely affected. Certain preventive measures such as keeping the beds fallow during summer months, deep ploughing and application of poultry manures reduce the nematode population.

Soil fumigation with DD (a mixture of 2,3-dichloropropane and 1,3-dichloropropane), EDB (ethylene dibromide) or DBCP (1,2-dibromo-3-chloropropane) is very effective in controlling the nematodes. In order to ensure a healthy growth of the rose plants, Pal (1972) suggested to fumigate the rose beds 4-6 weeks before planting with DD at the rate of 50 litres per hectare. Nemagon, however, can be used even in the planted beds through irrigation water at the rate of 10-12 litres per hectare. Hayashi (1976) reported that soil drenching with DBCP in established roses controlled *M. hapla* well and increased the flower yield. The application method recommended was incorporation of DBCP at low concentration in the watering system, 3 times a week. Bareroot dip in the chemical can also eliminate root-knot nematodes from roses. Dormant budded rose rootstocks from a nursery infested with *M. hapla* were washed and dipped for 30 min in 0.1 per cent prophos or nemacur [ethyl 4(methylthio)-m-tolylisopropylphosphoramidate] at 16°C. *M. hapla* was eliminated by either treatment without phytotoxicity (Dale, 1973).

12.3.2 Root-lesion Nematode

These are elongate, more or less cylindrical pests belonging to the genus *Pratylenchus*. The common species to which roses are susceptible are *P. vulnus* and *P. penetrans*. They cause root decay and affected plants become stunted and chlorotic having leaves with low amount of iron, copper and potassium (Jakobsen, 1975).

Beuzenberg (1972) studied the efficacy of three nematides namely nemafos, nemagon and temik (aldicarb) in Garnette roses on *Rosa canina* rootstocks and found that soil disinfection with nemagon or temik before planting gave satisfactory control of nematodes. Nemagon, however, checked rose growth, if used before the plants were 2 years old, but it killed nematodes both in the soil and the roots and had a long residual action. It was, therefore, recommended for older plants. Temik could be used without damage even on young plants, nematode control was effective and plant growth was satisfactory, but it remained active for a shorter period than nemagon.

In a trial, vydate (oxamyl) at 2.5-10 ppm a.i. mixed with the potting soil or at 1.25 + 500 ppm applied as a combined soil mix and leaf spray treatment gave virtually complete nematode control, similar to that obtained with the standard temik applied as a 5 ppm soil mix or combined treatments of 2.5 + 1000 ppm. Foliar sprays alone of 1000-4000 ppm vydate or 2000 ppm temik were relatively ineffective. The top growth (fresh weight) of the plants was markedly increased by the 25 ppm vydate soil admixture, compared with that of controls (Bunt, 1972). Treatment with fenamiphos at 100 and 200 ppm for 10 minutes and at 50 ppm for 30 minutes, achieved 100 per cent control of *Pratylenchus vulnus* (Veld *et al.*, 1988).

Some of the other nematode species like *Longidorus* spp., *Pratylenchus* spp. and *Trichodorus christiei* have also been reported to attack roses and cause much damage to the plants (Pirone*et al.*, 1960; Gabriel, 1979). The unproductive flowers in *Rosa borboniana* and *R. chinensis* were attributed to the attack of the ectoparasites *Hemicycliphore labiata* and *Xiphinema basiri* (Muthukrishnan *et al.*, 1975; Rajendran and Muthukrishnan, 1980).

13.0 Harvesting and Yield

13.1.0 Harvesting

The stage at which flowers should be cut, either for decoration or for despatch, is the tight-bud stage when the buds show full colour but the petals have not yet started unfolding (Swarup, 1972; Kaicker, 1984). If harvested at this stage, they last longer in vases or during transportation, retain colour and freshness. The optimum stage may vary slightly depending on cultivar and one has to be experienced to judge the right stage for cutting. A flower bud of a red cultivar when cut at a little earlier stage may fail to open later. In this respect the pink and yellow cultivars are generally better than the reds (Pal, 1972; Nalwadi, 1981). Most pink and red cultivars should be allowed to develop to a stage where one or two of the outer petals begin to unfurl. Cultivar (Dia, Dayroo, MarydeVor and Red Sandra) and seasonal differences in the quality of cut rose flowers were investigated by Ahn and Park (1996b). The vase life of flowers cut in winter was twice as long as that of flowers cut in summer. Winter-harvested flowers took longer to open and exhibited less bent neck than summer-harvested flowers. Of all the cultivars screened, Red Sandra had the longest vase life. Loose flowers, used for making garlands, preparing perfumes and various other products and for worshipping are harvested only when they are fully open and collected in large open baskets. The optimum stage to cut cv. Mercedes rose blooms was when the sepals moved away from the buds and the tips of the petals separated (Carow and Sommer, 1962).

The flowers should be cut in the early morning before sunrise or late in the afternoon when the sun is about to set so as to avoid damage of buds due to high temperature during the day. Late harvest results in short vase life of cut flowers (Nalwadi, 1981) and low oil content in loose flowers used for preparing perfumes (Srivastava and Gupta, 1984). For cut flowers, the stem length should be little more than what is required and bear fairly good number of leaves. The cut should always be above a healthy outward pointing bud with a clean and sharp secateur. The general rule observed by the most rose growers in cutting the stem is to allow two five leaflet leaves to remain below the cut. Ahn and Um (1991a) observed that the optimum leaf number on the cut stem was 1 (flowers did not open when all the leaves were removed). Vase life was longest when the stems were air-cut and the ends burned, followed by air cutting, cutting under water, air cutting + crushing and air cutting + boiling. Tsujita (1982) reported that cutting just above the knuckle (the point at which the shoot originates) significantly reduced flower yield per plant compared with cutting above the first five leaflet leaf.

The effect of planting density and harvesting method on flower production and flower stem quality of *Rosa hybrida* cv. First Red was studied for a 3-year cropping period (Dambre *et al.*, 2000). Two planting densities (6 and 8-plants/m^2) were combined with three types of harvesting methods in a field experiment. The primary shoot was bent to allow the outgrowth of basal and lateral buds. Harvesting of the newly formed shoots was as follows: (i) all shoots were harvested as low as possible to the point of outgrowth; (ii) the first basal shoots were harvested on the first, most basal leaf, the following emerging shoots were harvested as low as possible to the point of outgrowth; (iii) the first basal shoots were harvested on the second most basal 5-leaf, the following emerging shoots were harvested on the first most basal 5-leaf. Harvesting technique had a pronounced effect on formation of new flower shoots. Harvesting method 3 resulted in the highest number of flower stems and biomass production [kg fresh highest number of flower stems and biomass production (kg fresh weight/m^2)]for the three successive years while the lowest productivity was noted for harvesting method (i). Flower stem quality, however, was highest in method (i). Planting density affected productivity but not flower stem quality, the highest yield was noted for a planting density of 8 plants/m^2. Planting density did not interact with harvesting method. The results indicated that the grower can use harvesting techniques as a tool for increasing productivity (lower flower stem quality) or flower stem quality (lower productivity). Bastos *et al.* (2017) reported that the density of 3.3 plants/m^2 increased the flower yield without affecting flower quality of cv. Avalanche and Carola. Moreover, Carola produced longer flower stems than that of Avalanche.

Flowers of cultivars 'Duett', 'Golden Gate' and 'Cream Prophyta' harvested in the afternoon recorded low respiration rate (89 ml CO_2/kg/h) and extended life (7.83 days) while, flowers harvested in the morning showed greater respiration rate (96 ml CO_2/kg/h) and shorter life of 5.5 days (Tshwenyane and Bishop, 2010). Sharma (2018) revealed that harvesting damask roses petal whorl opened stage recorded higher cis-rose oxide, citronellol + nerol, geraniol and lower nonadecane content than harvesting at petal whorl loosened or fulled opened stage. On further distillation, the maximum essential oil content (0.067 per cent), citronellol + nerol and geraniol content were recorded by extending the distillation time upto 5 h.

13.2.0 Yield

The yield of cut flower depends on a number of factors such as cultivar, plant density per unit area, flower quality, duration of flowering, pruning, fertilization and other cultural practices adopted from time to time. In Europe, cultivars of Hybrid Tea and floribunda roses, growing in glasshouses having perfectly controlled environments, as quoted by Gill (1984), are known to yield 144 cut flowers per square metre area per year, whereas in India plants grown in the open field conditions yield about 13.5 long stemmed cut blooms from the same area. Yield comparison of 20 outdoor rose cultivars, made by Maatsch (1967) recorded a yield ranging from 1340 flowers per hundred plants in cv. Isabell de Ortiz to 3134 flowers per hundred plants in Bel Ange. In a trial in Sweden with 15 glasshouse rose cultivars, the highest yield (16 flowers/plant/year) were obtained from GeheimratDuisberg, but only 3 per cent was of grade 1 quality. Carina averaged 12 flowers per plant/year and 48 per cent of those was grade 1. In a second trial of 13 Garnette and similar cultivars, the highest yielding cultivar was Zorina which produced 34 flowers/plant/year, about 40 per cent of these was grade 1. Other high yielding cultivars were Starlet, Golden Garnette, Garnette, Carol, Gold Strike and Pink Garnette, which produced 24-28 flowers per plant. Sonora and Coral Princess gave low yields (15 flowers/plant/year) but 60 per cent was grade 1 (Lenander, 1968). Plant density has much influence on the yield of quality blooms per unit area. Gill (1984) obtained 120000 cut flower for exportable quality from an area of one hectare by following close planting and providing plastic cover over the plants during November-February. In an essential oil bearing rose, Dernovic and Zutucnyi (1956) recorded a yield of 6977 kg petal per hectare when planting was done at 2.5 ± 0.5 m apart. One of the best methods for maximization in yield and quality of flower is through balanced application of fertilizers. A yield of 35.4 flowers per bush was obtained in cv. Montezuma by the application of 600 kg nitrogen along with 200 kg each of phosphorus and potassium (Yadav *et al.*, 1985b). Higher yield as a result of fertilization of plants was also reported by Bik, 1972; Johansson, 1978; Malik, 1980. Investigations with Miniature roses, carried out in Sicily, revealed that cv. Carol produced 1816 flowers/square metre in a period of 6 years (Roxas, 1982). Mitali *et al.* (2020) observed significantly higher flower yield in *Rosa damascene* during first cropping season (1044.55 g/bush) than second cropping season (699.89 g/bush).

14.0 Postharvest Management

Immediately after cutting, the stem should be dipped in clean water upto the neck or base of the flower buds. An ordinary plastic bucket is convenient for the purpose. The delay in keeping the flowers in a bucket of water will cause air entry and result in vascular blockage. The flower stalk should be kept separately to avoid bruising and in shade until the cutting is completed. Soon after cutting, the stems should be re-cut in water, about 2 cm above the previous cut end. If they are not required for immediate use the cut flowers along with stems dipped in water in the bucket should be stored at a cool air temperature of 4.4-7.2°C for about 6 to 12 hours to harden the buds and thus enhances their keeping quality. During winter (December-January) a shorter treatment for 6 to 8 hours is sufficient while a longer duration of 8 to 10 or 12 hours is required in March-April. The pulsed and precooled roses have longer vase life, and bent neck is prevented (Halevy *et al.*, 1978). The flowers which are uniform in colour, stem length and development should be grouped together at the time of cutting and kept in separate containers. For the facility of handling, the foliage and also the thorns may be removed from the base of the stem upto about 20 cm at the time of cutting the flowers. If cut-blooms are to be despatched to distant places it is better to send flowers by direct air flights than other means of transport.

14.1.0 Grading and Packaging

The cut-blooms are graded according to the length of stem and then packed in corrugated cardboard boxes. The size of box varies with the quality and quantity of roses to be packed. A box of 100 cm long, 32.5 cm wide, 6.5 cm high can accommodate 80 roses having 65-70 cm long stem. The inside area of box is lined with thin polythene and very fine moist tissue paper shavings are spread at each end of the box to provide a cushion to the blooms. The blooms are generally packed in bundles of 20 each and the bundles are tied with string or rubber band but not too tight. The upper half of each bundle having flower buds is wrapped in a corrugated paper which is fixed with an adhesive tape. The lebelling of cultivar is made on this paper with a band of moistened tissue paper. The lower half of the bundle should preferably be wrapped in tissue paper. It is also useful to put a wad of moistened tissue paper at the bottom of each stem. Two bundles each having 20 flowers are placed opposite to another along the length of the box in such a way that their flower buds face the sides of the box and their stem ends towards the centre of the box. The stem ends of the two bundles on either side are secured firmly with a wooden stick fixed along the width of the box. This wooden stick is placed over a strip of foam rubber to avoid damage of stems. The inside of the box is then finally covered with a sheet of tissue paper before puttings on the cover of the box. Along the outer edges of the box, adhesive tapes can be fixed to close it. Labels of address are provided at the top of each box. Packing materials and transportation time influence the quality of cut flowers of rose cv. Mary de Vor (Ahn, 1997). Cut flowers were packed in corrugated cardboard boxes (CCB) alone or in combination with a polyethylene film (PEF) lining and/or wet newspaper (WNP). Cut flowers experienced simulated transport for 12, 24 or 48 h at 25°C. Vase life was most prolonged following packing in CCB + PEF + WNP; flower opening was promoted and the incidence of bent neck decreased in this treatment. The second best treatment was packing in CCB + PEF. In general, keeping quality and vase life increased as transportation time decreased. The greatest variation in vase life with transportation time was observed in flowers packed in CCB. Singh and Mirza (2004) evaluated the postharvest life and quality of cut rose cv. Super Star influenced by different packaging materials. Results affirmed that the pulsed (DMSO 2 per cent for 15 mins) and precooled (4°C for 6 h) flowers wrapped with butter paper for 12 h in CFB box exemplified vase life of 10.17 days whereas,butter paper wrapping for 20 h resulted in vase life of 10 days. Flowers wrapped in brown paper and CFB sheet showed a vase life of 8.50 and 8.17 days at 12 h, respectively. However, the flowers which were not packed after pulsing and precooling exhibited maximum increase in vase life upto 11 days together with maximum gain in fresh weight (0.88 g), flower diameter (8.86 cm) and water uptake (21.60 ml). While the control flowers showed a poor vase life of 6.33 days.

14.2.0 Pulsing

Postharvest pulsing for 20 h with solutions containing 20 mg GA_3/l extended the vase life and promoted bud opening of unstored and stored flower of the rose cv. Mercedes, while continuous treatment with GA_3 was detrimental. Holding solution did not improve vase life. Vase life of Christian Dior roses was longest (11.7 days compared with 6.25 in controls) with 2.5 ppm kinetin + 3 per cent D-fructose (Bhattacharjee, 1998). Pulsing flowers of cv. Avalanche and Fiesta with 200 mg/L of nano-silver extended the vase life upto 18 days by suppressing stomatal opening, reducing transpiration loss from leaves and inhibiting bacterial growth (Rafi and Ramezanian, 2013). Kumar and Ponnuswami (2014) reported that pulsing with 200 ppm 8-HQC + 10 per cent sucrose showed better performance forrelative water content (82.32 per cent), water uptake (6 g/stalk) and vase life (4.1 days) with suppressed peroxidase activity, minimum loss in physiological weight, transpiration and membrane integrity.

14.3.0 Storage

Low temperature storage is absolutely necessary for roses (Staby *et al.*, 1978). Lenander (1960) reported that cut-roses could satisfactorily be cool stored at least for 10 days and still last for about 5 days at room temperature. The results were best at 1°C and 5°C. Roses were stored successfully at 31°F (0.9°C) and storage in a moisture proof package was superior to that with cut-stems in water (Fischer, 1953). In another experiment, Amariutei and Panait (1981) observed that cold storage for 14 or 21 days (1°C) shortened the vase life of Baccara rose by 2 days compared to control (kept in water at room temperature) but in Sonia even after 21 days of storage, the flowers lasted for 4 days. The poor performance of Baccara rose was atrributed to a low level of total and reducing sugars in its petal.Storing cut rose flowers cv. First Red in cold chamber at 15 ± 2° C and 90-95 per cent RH in holding solution containing sucrose + HQC for a period 4 days reduced angle of neck bending, retained flower pigments, lowered respiration rate and reduced ethylene evolution resulting in overall better quality of cut flowers (Pal *et al.*, 2003). Tshwenyane *et al.* (2012) reported that storage of greenhouse roses with CO_2 significantly reduced grey mould rot compared to storage under normal air. However, 20 per cent CO_2 showed foliage damage in addition to reduction of rot, whereas 10 per cent CO_2 controlled *Botrytis* rot without affecting foliage quality. Demircioglu *et al.* (2018) recorded a prolonged vase life upto 21 days in cv. 'First Red' on treatment with 1-MCP which inhibited decay followed by storage at 4° C and 70 per cent RH in 1 per cent sucrose solution.

14.4.0 Senescence of Flower and Vase Life

Senescence of cut-roses in water was characterized by a decrease in concentration of anthocyanin, protein N, asparagine and tannic acid and an increase in most aminoacids, glutamine, malic acid and free ammonia in the petal tissue (Weinstein, 1957). Senescence of cut-rose flowers cv. Mercedes at 22°C occurred earlier in flowers previously held at 2°C for 10 or 17 days than in freshly cut flowers. The advance in senescence was observed as an earlier increase in both ethylene production and membrane permeability. The rise in ethylene production preceded the rise in the level of ion leakage in the petals and this in turn preceded visible sysmptoms of petal death (Faragher and Mayak, 1984).

The moisture content, fresh weight and soluble protein mass fraction of the petals increased during early stage of vase holding, and decreased rapidly 4 days later. The membrane permeability and peroxidase activity increased rapidly while the activities of catalase and superoxide dismutase increased for the first two days during vase holding and then decreased rapidly. It was concluded that water loss and rising of lipid peroxidation level were considered to be the main physiological causes of senescence of cut flowers of *R. hybrida* cv. Samantha (Xue and Lin, 1999).

Doorn and Vojinovic (1996) studied the effect of light intensity and light quality on petal abscission. The time to petal abscission was not affected by adding sucrose, plus an antimicrobial compound, to the vase water, at any light intensity. Far-red light also had no effect, compared with red or white light. Garello *et al.* (1995) found that vase life was longer when plants had been grown under high pressure sodium (HPS) lamp than under metal halide (MH) lamps. A change in light source during plant culture did not lead to a change in postharvest leaf ABA content but significantly affected petal ABA content and its pattern of change during vase life (overall, petal ABA content was greater in flowers which had been grown under MH lamp). The higher the ABA contents at harvest, the shorter the vase life.

Cytokinin activity was detected in extracts from petals of cut glasshouse roses, cv. Sonia, by submitting the HPLC eluates corresponding to the standards *Amaranthus tricolor* bioassay. When cytokinin was applied (at 10-5, 10-6 or 10-7 M) in holding solutions, delayed flower senescence by 34-56 per cent and prolonged overall longevity compared with controls held in distilled water (Lukaszewska *et al.*, 1994). Foliar spraying with cytokinin delayed wilting and senescence, increased the vase life and resulted in higher FW/DW ratio and lower pH, osmolarity and conductivity after 10 days compared with control roses (Barthe *et al.*, 1991). Rio *et al.* (1989) studied the effect of pretreatment and storage condition of cut flowers of cv. Sonia sprayed with or held in solutions containing kinetin (25 or 50 ppm), prior to keeping in simulated home conditions.

The ethylene production of cut rose cultivars Samantha, Angetique, Gabriella and Athena was low at the bud stage, increased gradually as the outer petals opened, and reached a peak at full opening, before decreasing. This pattern was similar to that of ethylene climacteric cut flowers and fruits. The respiratory rate of all cultivars was like a typical respiratory climacteric type, being low at the bud stage and increasing markedly to a peak before full opening, before decreasing. The respiratory climacteric peak was reached 1-2 days earlier than the ethylene climacteric peak in Samantha and Athena (Gao *et al.*, 1997).

In cut roses, cultivars Dame de Coeur and Lady X, the soluble protein concentration in petals increased during the early stages of vase life then decreased. The concentrations of total and basic in amino acids increased throughout vase life. Addition of a preservative containing 2 per cent sucrose, 250 ppm 8-HQ (8-hydroxyquinoline), 500 ppm citric acid and 25 ppm silver nitrate reduced the level of increase in amino acid concentrations. During petal senescence, the free ammonia concentration and pH value of petals increased while anthocyanin concentration decreased (Gao and Wu, 1990).

Ca-induced delay in petal senescence involves the protection of membrane proteins and phospholipids from degradation, thus preserving the integrity of the membranes, reducing ethylene production, and hence maintaining solute transport and tissue vitality (Torre *et al.*, 1999). The $CaCl_2$ increased the relative water content and the activities of superoxide dismulase, catalase and peroxidase decreased the rate of protein decomposition and malondialdehyde content and maintained membrane stability in petals during the vase life period (Hong and Zhao, 1999).

Kazuo *et al.* (2005) studied the postharvest characteristics of *Rosa hybrid* L. cv. 'Delilah' and 'Sonia' whose vase life were 10.6 and 5.6 days, respectively. Both the cultivars showed a rapid decline in hydraulic conductance of stem segments after harvest which was subdued with HQS treatment. Petals of 'Sonia' kept in water did not reflect completely but showed blueing effect whereas the response was*vice versa* with the case of 'Delilah'. Even though starch concentrations were higher in 'Sonia', the vase life was longer with 'Delilah' which might be attributed to its higher soluble carbohydrate concentrations. Laird *et al.* (2005) evaluated the vase life of cut rose cultivars 'Golden Gate' and 'Snowy Jewel' which were regarded as long-lived and short-lived by industry. *Pseudomonas fluorescens* was used to contaminate vase water, secateurs, vase surface and results indicated that contamination of secateurs is least detrimental whereas contamination onto dried vase surface was most detrimental in case of 'Golden Gate', whereas, the vase life of 'Snowy Jewel' was reduced in a uniform manner for all treatment with contaminated secateurs being exception.

Nabigol *et al.* (2010) stated that higher soluble carbohydrate concentration in petals and low ethylene production of the flowers could extend the vase life. For example, the flowers of cultivar 'Black Magic' lasted for 5.6 days while that of 'Maroussia' remained fresh for 14.3 days.

14.4.1 Factors Influencing Vase Life

14.4.1.1 Water Stress

The termination of vase life of cut roses is also characterized by wilting, even they are constantly held in water. The physiological processes like water uptake and transport, water loss and the capacity of flower tissues to retain its water are inter-related and caused to this phenomenon. Measurements of the transpiration rates of cut roses of 40 cultivars showed wide differences among them. It was considered that leaf diffusive conductance could be used as an indicator of water uptake in cut roses (Jodo *et al.*, 1989). Microscopic examination of the xylem of cut stems of the cv. Samantha held in a vase showed physical blockage appearing within 200 mm of the cut end during the first 2-4 days after harvest. At distances further from the cut end, little or no evidence of physical blockage could be found. It is concluded by Dixon and Peterson (1989) that initial water deficits are caused by some degree of physical blockage at the cut end; thereafter, the continued decline in stem conductance is caused by the onset and extent of xylem cavitation induced by water stress. Stem plugging in cut roses cv. Sonia was due to bacterial growth (*Pseudomonas* or *Alcaligenes faecalis*) (Witte and Doorn, 1988). In cut roses (cv. Norena), *P. fluorescens*, *Klebsiella oxytoca* and *Aeroma hydrophila* were responsible for the vascular blockage in the xylem vessels of stem end in distilled water. These bacteria were not observed in stem of freshly harvested flower (Kim *et al.*, 1997). Doorn and Vojinovic (1996) placed cut stem rose cultivars Frisco, Sonia and Cara Mia in water within minutes of cutting and kept in glasshouse along-side non-cut-flower (still on plant). Flowers on plants showed no water stress, while cut flowers showed symptoms of water stress depending on the weather and cultivar and petal abscission occurred in only a few flowers.

De and Bhattacharjee (1999) also noted that total water uptake and loss in weight at senescence were positively correlated with vase life of cut roses. After a dry period of 1 h water uptake by Sonia roses was fastest with cold (0°C) water. It was slowest with water at 20°C. The effect of water at 40°C was variable; sometimes as fast as with cold water, sometimes slower than with water at 20°C (Slootweg, 1995).

Spinarova and Hendriks (2007) developed a standard testing procedure for characterization of water stress tolerance of rose cultivars using ultrasonic acoustic emission technique. The method consisted of 3 h dehydration followed by 3 h rehydration repeated twice to determine the water status of cut roses. The flowers would be tested for their dehydration resistance, rehydration performance and cavitation control during the experiment. Rose flowers with dehydration resistance expressed as relative low weight loss during the first 3 h of dehydration followed by complete regain to original weight during subsequent rehydration were classified as 'water stress tolerant' cultivars. Moreover, good cavitation control was determined based on the strong signal occurrence during dehydration and relatively few hits during rehydration phases. Low recovery cultivars exhibited low dehydration resistance, no/partial refilling and cavitation occurrence during rehydrations.

Fanourakis *et al.* (2012) observed extended vase life (10-21 days) with the genotypes with high stomatal responsiveness whereas, shorter vase life of 8 days was recorded with genotypes having low stomatal responsiveness and 2 out of 3 such genotypes failed to open flowers. Thus, it was insisted to consider the stomatal hydrosensitivity during breeding for longer vase life. Almeida *et al.* (2015) evaluated the response of cut rose (cv. Carolla) quality on exposure to water stress of soil tensions (15, 30, 45, 60, 90 and 120 kPa). Results indicated that bud diameter showed an inverse relationship with increasing soil tensions whereas, other parameters like vase life, fresh weight loss and water absorption remained unaffected. Moreover, irrigating the field with soil water tension of 15 and 30 kPa was recommended for getting better quality flower stems.

14.4.1.2 Light

Cut roses cv. Sonia grown with supplementary lighting (20 h SON/T; 24.5 $\mu mol\ m^{-2}s^{-1}$) (SL-flowers) rehydrated more slowly after a dry period of 2 hours than roses grown under natural light (NL-flowers). This difference in rehydration rate remained when transpiration was blocked (Slootweg, 1997). Vase life of rose cultivars, Cardinal and Madelon increased when the supplementary irradiance level increased from 130 to 370 $\mu mol\ m^{-2}s^{-1}$. Signifiant cultivar differences in vase life and days to leaf drop and bent beck were observed. Madelon had the longest vase life (Fjeld *et al.*, 1994). Kuiper *et al.* (1997) found that cut rose cv. Madelonbuds do not open satisfactorily when kept under low light (LL) conditions, in contrast to cv. Sonia buds, adding sucrose to the keeping solution or storage in high light (HL) helped Madelon buds to open but it was thought that 2 different mechanisms were involved.

14.4.1.3 Temperature

Thermal dip treatment for flower heads was effective against grey mould of roses. Dipping of flowers in tap-water at 50°C for 20-40 s was more effective than at higher or lower temperature and than for longer treatments (Elad and Volpin, 1991). Roses grown under 20° C lasted significantly longer than at 16° C with the 10 per cent higher sucrose levels and 50 per cent lower starch levels in the latter in relation to the former temeperature condition (Marissen, 2001).

14.4.1.4 Relative Humidity

In an experiment, Mortensen and Fjeld (1998) observed that increasing the air RH reduced the vase life of the roses from 8-13 to 2-5 days, and increasing the lighting period from 16 to 24 h/day at 65 per cent. RH also reduced the vase life from 13 to 8 days. Blindeman (2000) recorded best keeping quality (vase life of 16-21 days) of roses cv. Souvenir grown at 64 per cent RH, compared to (vase life of 8 days) grown at 91 per cent RH. Fanourakis *et al.* (2012) reported that flowers of 'Pink Prophyta' grown at high RH (95 per cent) failed to open throughout the vase life which was reduced to about 80 per cent. This reduced flower opening and shorter vase life might be due to higher transpiration rate attributed to high RH. Moreover, the leaves partly closed their stomata upon reducing water potential or on exposure to ABA at high RH whereas, remained open in plants grown at moderate RH (60 per cent). Fanourakis *et al.* (2012) evaluated the stomatal response of to leaflet desiccation among 60 tetraploid rose genotypes grown at 85 per cent RH. Results indicated that genotypes with low stomatal responsiveness had shorter vase life of 8 days with reduced flower opening due to high postharvest water loss rates whereas, genotypes with high stomatal responsiveness extended the vase life upto 10-21 days with maximum opening and minimum water loss. Thus, the authors suggested to select genotypes with high stomatal responsiveness for cultivation under high RH for prolonged vase life and normal flower opening.

14.4.1.5 Chemicals

Sugar

YooYongkweon *et al.* (1999) found that addition of 5 mg chitosan/l and 3 per cent sucrose to the holding solution prolonged the vase life of roses, cv. Cardinal, for 3 days as compared to the control, and improved the quality of cut roses in terms of flower diameter and fresh weight. Petals were brighter when 3-5 per cent sucrose was added in holding solution (Cho and Lee, 1979; Ahn and Um, 1991b). Barman and Rajni (2006) evaluated the vase life of cut rose cv. First Red harvested at three different stages with respect to preservative chemicals. Results revealed that 8-HQC 100 ppm + $AgNO_3$ 100 ppm + sucrose 1 per cent, $AgNO_3$ 100 ppm + sucrose 2 per cent and 8-HQC 100 ppm + sucrose 2 per cent extended the vase life upto 12 days when harvested at half-open stage. While 8-HQC 100 ppm + Ag NO_3

100 ppm + sucrose 1 per cent, sucrose 2 per cent + 8-HQC 100 ppm and KCl 200 ppm + $Al_2(SO_4)_3$ 200 ppm + boric acid 200 ppm enhanced the vase life for 12 days.

Calcium

Calcium was applied either alone or together with 2 per cent sucrose (S) and 8-hydroxy-quinoline citrate (HQC) to cut flowers of greenhouse roses, extended longevity and promoted bud opening (Michalczuk *et al.*, 1989). The longevity of roses held continuously in the preservative solution [8-HQC (200 ppm) + 2 per cent sucrose + calcium nitrate 0.1-1.0 per cent] was 11 days compared with 5 days for roses held in water (control) (Goszczynska *et al.*, 1989). Cut Samantha roses were placed in deionized water or a 200 mM $Ca(NO3)_2$ pulsing solution for 72 h. Flowers were then held in preservative solutions containing 0 or 4 mg F (floride)/l. Fresh weight gain, solution uptake, degree of flower opening, and flower longevity were reduced in the presence of F in the holding solution (Pearson-Mims and Lohr, 1990). Torre *et al.* (1999) found that $CaCl_2$ treated cut roses cvs. Mercedes and Baroness remained turgid and continued their initial postharvest growth for longer period of time.

Pre- and postharvest treatment of cut roses cv. Ariana grown in plastic house were (i) 0.017 per cent CaO as calcium chelate, applied 3, 2 or 1 week before harvest, (ii) 8.6 ppm NAA, applied one week before harvest to the upper third of plants, (iii) Ca + NAA, and (iv) control. After harvest, flowers were pulsed with the commercial cut flower preservatives Chrysal professional (0.2 per cent) or Florissima (1 per cent) after hydration at 4°C in $Ca(ClO)_2$. Cut flowers were then stored for 17 or 24 h at 2°C or underwent transport simulation for 24 h at 18°C. The longest vase life was obtained with the treatment Ca + NAA + Chrysal and Ca + NAA + Florissima (11.4 and 9.6 days, respectively), compared to 7.3 days with control (Bolivar *et al.*, 1999). Fertilization of cut roses cv. 'Alexander' with 10 mM $CaCl_2$ at 40, 30 and 25 days before harvest improved the vase life and exhibited good quality even after 10 days of harvesting whereas, control flowers started senescing from 5th day of harvest (Mehran *et al.*, 2008). Pre-harvest spray of $CaSO_4$ at 200 mg/L + 0.01 per cent Tween promoted bud opening, delayed senescence, maintained turgidity, suppressed ethylene production and ion leakage in addition to increase in Ca content in stems, leaves and petals of cv. 'Cool Water' and 'Pretty Blinda' flowers (Nabigol, 2012) (Table 24). Asfanani *et al.* (2008) revealed that application of 10 mM $CaCl_2$ to rose cut flowers cv. Alexander at 40, 30 and 25 days before harvest improved vase life compared to untreated controls. Moreover, the cut roses treated with $CaCl_2$

Table 24: Longevity and Flower Diameter (after 6 days) of Cut Flowers Treated Pre-harvest with Water (control) and $CaSO_4$*

Cultivar	*Treatment*	*Longevity (day)*	*Flower Diameter (mm)*
'Cool Water'	Calcium	10.6 ± 0.3 a	82 ± 2 a
	Control	7.6 ± 0.6 b	72 ± 3 b
'PrettyBlinda'	Calcium	10.3 ± 0.3 a	87 ± 3 a
	Control	8.6 ± 0.3 b	79 ± 2 b

* Nabigol (2012).

had good quality even 10 days after harvesting while the quality of control flowers decreased on 5th day after harvested (Table 25).

Table 25: Effect of $CaCl_2$ on Quality Factors of Cut Rose cv. Alexander*

Parameters	*0 mM $CaCl_2$*	*10 mM $CaCl_2$*	*20 mM $CaCl_2$*
Flower diameter (mm)	57.68 a	43.08 b	45.89 b
Stem diameter (mm)	6.037 a	5.880 a	5.800 a
Flower length (cm)	57.09 a	52.82 b	52.14 b
Fresh weight (g)	35.66 a	29.56 b	28.88 b
Dry weight (g)	5.063 a	6.101 a	6.039 a
Leaf chlorophyll (SPAD number)	48.29 a	45.53 b	47.99 a
Flower diameter (15 days preharvest)	1.863 b	2.062 a	2.100 a
Flower length (15 days preharvest)	2.780 b	2.873 ab	3.058 a
Flower wilting (4th day of vase life)	1.887 a	1.979 a	1.962 a
Flower wilting (8th of vase life)	1.500 ab	1.726 a	1.397 b
Flower browning (4th day of vase life)	2.220 a	2.266 a	2.243 a
Flower browning (8th day of vase life)	1.833 b	2.271 a	1.866 b

Means in the same row followed by same letter are not significantly different ($P<0.05$).

* Asfanani *et al.* (2008).

Aluminium

Vase life of cut rose cv. Mary De Vor was extended from 6 to 9 days when 3-5 per cent sucrose and 300 ppm aluminium sulphate were used. There was a positive correlation between fresh weight increase and longevity (Cho and Lee, 1979). Ahn and Um (1991b) found that the combination of 5 per cent sucrose with 300 ppm aluminium sulphate gave a vase life of 5 days compared with 3 days for control (tap water). Pretreatment with $Al_2(SO4)_3$ and 5 per cent sucrose was effective at maintaining the quality and vase life of 'transported' and control cut flowers, among 'transported' flowers, it was more effective at 25°C than at 5°C (Ahn, 1996, 1997).

Hasek (1981) reported that after grading, the flowers are returned to the refrigerator, where the stems are immersed in a preservative solution. The solution usually contains 1 to 3 per cent sugar and 100 to 200 ppm 8-quinoline citrate plus aluminium sulphate, citric acid or silver nitrate. Flowers should be held in the preservative solutions at least for 3 to 4 hours, preferably longer. Flowers that have been stored dry for any length of time should have the stems re-cut and should be placed in preservative for 4 to 6 hours before shipping. In preservative, flowers with 4 foliage leaves had a vase life of 16 days compared with 12 days for flower in defoliated stems. Vase life was longest in flowers held in 4 per cent sucrose + 0.1 per cent aluminium and potassium sulphate + 0.02 per cent potassium chloride, being 14.2, 12.7 and 11.7 days in cvs. Mercedes, Zorina and Marimba, respectively, compared with 10.2, 7.2 and 8.4 days, respectively, in the control (Gherghi *et al.*, 1983). Gupta *et al.* (2007) reported that cut rose cultivars, Janter Manter, Imperator, Lahar, Pandit Nehru, Sugandha, Arjun, Kentucky Durbey and Fetura exhibited better postharvest quality when placed in holding solutions, Aluminium sulphate (300 ppm) and sucrose (3 per cent) + Aluminium sulphate (300 ppm). Paramagoudar *et al.* (2014) revealed that keeping cut stems of Dutch rose cv. Naranga in Aluminium sulphate 300 ppm exhibited longer vase life (12.89 days), maximum solution uptake (8.83 g/flower) and greater flower fresh weight (8.64 g/flower) at 12th day.

Silver

Ferreira and Swardt (1980) in an experiment on cv. Sonia obtained the best result with silver nitrate (2.5 mg/dm^3) + 8-hydroxyquinoline sulphate (130 mg/dm^3) + citric acid (200 mg/dm^3) + sucrose (30 g/dm^3). The solution extended the vase life from 7 days in deionized water to 17 days, associated with enhanced respiration rate of the petals. They (1981) also observed that in deionized water bent neck occurred in Sonia, Carina, Lara and Tommy Bright but in preservative it was absent. There was no correlation between the respiration rate and vase life of flowers. Rose flower cv. Sonia with stems bearing 0-4 leaves were placed in deionized water or in a flower preservative containing silver nitrate + 8-hydroxyquinoline sulphate + citric acid + sucrose. Silver nitrate and RNA-Ag + tris (a ribonucleic acid-silver complex and trishydroxymethylaminomethane) increased the vase life by 2.7 days and prevented bent neck of cut rose cv. Asami Red compared with the control, whereas STS did not have a significant effect on longevity (Ohkawa *et al.*, 1999) (Table 26).

Table 26: Water Uptake, Vase Life, and Bent Neck of Cut 'Asami Red' Rose Flowers after Pretreating with Silver Thiosulfate (STS), Silver Nitrate or RNA-Ag + tris*

Treatment	*Water Uptakez (ml/stem)*	*Vase Life (d)*	*Bent Neck (Per cent)*
Water control	2.56	9.3	100
STS	2.02	8.3	100
Silver nitrate	2.13	12.0	0
RNA-Ag + tris	2.56	12.0	0

z Measured immediately after the pretreatment of 4 h.

y Standard error (n = 10).

* Ohkawa *et al.* (1999).

Son *et al.* (2003) from their experiment on effect of different pulsing chemicals on vase life of cut rose cv. Red Sandra reported that silver ion absorption treated with $AgNO_3$ was restricted within the basal stem while found in all parts when treated with STS. Moreover, the vase life of flowers treated with STS was prolonged by 1 day compared to $AgNO_3$ or distilled water when 15 cm portion was removed from the basal stem. Thus, it was highlighted that treatment of STS is more effective than that of silver nitrate for cv. Red Sandra.

NaOCl

Treatment with 25 per cent carbonated salt + 10 ppm NaOCl developed less bent neck in cut roses and produced less ethylene than other treatments. This was also the most effective treatment for prolonging vase life and maintaining flower quality (Ahn, 1996).

Cobalt

Mahr and Hanan (1980) reported that cobalt nitrate at 270 or 360 ppm and 8-(HQC) increased the vase life of cv. Samantha from 7.2 days in deionized water to 11.5, 11.7 and 13.2 days, respectively. In another experiment with cvs. Samantha, Cara Mia, Red Success, Royalty, Visa and Volare, 8-HQC at 200 ppm + Co at 360 ppm were more effective than either preservative alone. Cut rose of cv. Samantha had a longer vase life when opened in solution containing cobaltous ion (Co^{2+}) and it was related to increased water uptake into cut flower, delay in loss of fresh weight and prevention of occurrence of bent neck (Venkatarayappa *et al.*, 1980). Nagarajaiah *et al.* (1989) reported that 0.5 mM cobalt sulphate + 4 per cent sucrose in combination prolonged the vase life of cut flower of cv. Queen Elizabeth than individually and control.

Germicide and Antioxidant

Flower stems of *R. hybrida* cv. Serena were harvested at the bud stage and pulsed for 6 h with combinations of 200 ppm 8-HQS, sucrose at 40 or 80 g/l and deionized water, and stored for 3 or 6 days at 2°C in darkness. Flower stems were then transferred to deionized water at room temperature. The longest vase life (13.3 days) and greatest flower diameter (18 mm) were obtained with HQS + 80 g sucrose/l + 3 days of dark storage at 2°C. Similar results were obtained for control flowers (kept in deionized water at room temperature) (Accati *et al.*, 1992). Doorn *et al.* (1991) found that inclusion of sucrose and an antimicrobial compound in the solution resulted in growth and opening as in unsevered flowers. Cut rose cv. Mercedes flowers, held in a preservative solution containing germicides and sucrose showed an improvement in flower quality and an increase in petal water content. The treatment also resulted in a slowing down of the age-related decrease in membrane proteins and in membrane lipid fluidity (Goszczynska *et al.*, 1990). Three water soluble antioxidants morin, kaempferol and glutathione (GSH) were applied as pulsing solutions for 20 h at 22°C to cut flowers of rose (cultivars Frisco, Jaguar and Mercedes). All 3 antioxidants improved the appearance and opening of the petals and extended vase life in cut rose flowers. The response, however, was cultivar dependent (Meir *et al.*, 1994).

14.4.1.6 Other Chemicals

The effects of ethionine, methionine, aminooxyacetic acid (AOA) and calcium nitrate on the polyamine content, FW, flower diameter, water balance and longevity of cut flowers of rose cv. Red Sandra were investigated by Son *et al.* (1995) The longevities of cut roses in distilled water (control), basal solution (BS) (2 per cent sucrose + 200 ppm 8-HQS), BS + 0.1 mM ethionine, BS + 0.1 mM methionine, BS + 0.1 mM methionine + $Ca(NO3)_2$, and BS + 1 mM AOA were 8.8, 10.6, 18.3, 11.1, 13.7 and 10.7 days, respectively. The effect of ethionine in preservative solution (BS: 2 per cent sucrose + 200 ppm 8-HQS), Sonk 1 solution (BS + 0.1 mM ethionine), Chrysal or DW (distilled water) in the physiological changes of petals during vase life of cut rose cv. Red Sandra flowers was investigated. Vase life of flowers maintained in DW, Chrysal, BS and Sonk 1 solutions were 5.3, 6.1, 11.6 and 15.4 days, respectively. Sonk 1 solution was very effective in maintaining fresh weight, solution uptake, water potential and real colour compared with the other treatments. It was suggested that ethionine indirectly influenced physiological damages of petal cells and had a synergistic effect with sucrose (Son KiCheol *et al.*, 1997). Cut rose (cv. Sonia) flowers, with or without leaves, were held in distilled water (controls) or in an aqueous solution of 500 mg 2-hydroxy-3-ionene chloride polymer (HICP)/l, a quaternary ammonium compound, for 4 h and then transferred to tap water. The pulse treatment with HICP extended the vase life of cut flowers with leaves from 4.6 days (control) to 9.5 days, but had no effect on the vase life of cut flowers without leaves (approx. 10 days) (Ueyama and Ichimura, 1998). When zeatin and zeatin riboside were applied as holding solutions to cut roses, longevity of 6 cultivars was increased by an average of 20 per cent (Lukaszewska *et al.*, 1995).

The effects of methyl glucoside *myo*-inositol on the opening of cut roses (cv. Sonia with stem lengths of 12, 20 or 40 cm) were investigated by Ichimura *et al.* (1999). In stem and petals, the concentrations of methyl glucoside and *myo*-inositol decreased before flower opening. The addition of glucose and *myo*-inositol (4 per cent) to the vase water promoted flower opening, which was accompanied by increased concentrations of *myo*-inositol and xylose in petals.

Flowers treated with 50 µM sodium nitroprusside as a source of nitric oxide exhibited maximum soluble protein content, greater relative fresh weight and prolonged vase life upto 13.3 days in cut rose cv. 'Utopia' (Seyf *et al.*, 2012). GABA (γ-aminobutyric acid) was found to have dose-dependent effect on postharvest indices of cut rose 'Red Naomi'. One per cent GABA extended the vase life as well as posed significant effect on relative solution uptake, chlorophyll content, relative fresh weight and ion leakage compared to control. Hence, it can be used as a safe chemical in the postharvest chain to prolong vase life (Mashhoud *et al.*, 2016).

14.4.1.7 Chemicals and Temperature

In a trial with 3 rose cvs. Tanbedee Blinda, Merko Mercedes and Sweet Promise Sonia, the use of Krislite VB during cold storage (3°C) of cut flowers increased subsequent flower life at room temperature by 1/2 to 1 and 1/2 days. Krislite VB added to the water at room temperature (21°C, 60 per cent RH) increased the vase life of flowers by 6 to 10 days (Bredmose, 1979). Results of a systematic investigation on storage and vase-life of roses was reported by Ruting (1991). Freshly cut roses (cultivars Madelon, Sonia, Cocktail, Idole, Jacaranda and Frisco) were obtained at the prescribed harvesting stage. The flowers were divided into bunches of 5 wrapped in paper and placed in solutions of Agral LN (2 mg/L; a wetting agent), aluminium sulphate (0.8 g/l; a bacteriostatic solution) or a combination of the 2, for 24 h at 2°C. Control roses were placed in water. Transport was simulated by storing the flowers in cardboard boxes for 24 h at 20°C. After transport simulation, the stems were recut, placed in vases and maintained at 20°C, 60 per cent RH and illuminated with light at 1000 lux for 12 h/day. In a second experiment, cut stems of cultivars Cocktail, Madelon and Gerdo were placed in vases (5 stems/vase) containing cut flower food solution (Chrysal) or water and vase life was measured. Vase life of cultivars Madelon, Sonia, Cocktail and Idole was increased by treatment with Agral LN or aluminium sulphate and further increased when treated with a combination of the 2. Vase life of Cocktail flowers was almost doubled by the combined treatment (from 4.9 days in water to 8.3 days) and for the other 3 cultivars, vase life increased by 32-54 per cent. The treatment had no significant effect on vase life of cultivars Jacaranda and Frisco.

14.4.1.8 Fungicides

Kim *et al.* (1997) added myclobutanil (6-8 mg/L), a triazole fungicide to the holding water which had been inoculated with the bacteria the percentage of bent neck was significantly reduced to low level (< 5 per cent). The effects of fungicides on bent-neck of cut *Rosa hybrida* cultivars Mary de Vor and Norena were investigated at 3 bud development stages. These were: (i) bud tightly closed, (ii) calyx separated from petals and petals beginning to open, and (iii) calyx fully open and 2-3 petals open. The best treatment for inhibition of bent-neck in both cultivars was 120 mg myclobutanil/l for 4 h at stage 2 of bud development (Kim *et al.*, 1998). Put and Jansen (1989) reported a marked decrease in the vase life of cut roses when water-washed viable or heat-inactivated cells of *Bacillus subtilis* (2 strains), *Enterobacter agglomerans*, *Pseudomonas fluorescens* or *P. putida* were added to the vase water. An 18-h pulse in a solution 3 per cent sucrose and a bactericide (50 mg/L sodium dichloroisocyanate) before or after storage, increased vase life of cv. Gabriella by one day (Mor *et al.*, 1989). Methyl jasmonate (MJ), a natural growth regulator postulated to induce plant defence responses, was tested by Meir *et al.* (1998) for postharvest control of *Botrytis cinerea* in 6 cut flower rose cultivars (Mercedes, Europa, Lambada, Frisco, Sacha and Eskimo). Systemic protection against *B. cinerea* was evident in flowers of all cultivars pulsed with 200 mM MJ following either natural or artificial infection. Cold storage of cut Sonia, Royalty and Gold Rush roses at reduced humidities (50-80 per cent RH) significantly decreased the severity of *B. cinerea* infections that developed from naturally occurring or experimentally applied inocula, compared with storage at saturated humidity (Hammer and Marois, 1989).

14.4.2 Effect of Growth Substances

14.4.2.1 Gibberellic Acid (GA_3)

The fresh and dry weights and water content of petals of GA_3 treated flowers during the vase-life period were higher than in flowers treated with deionized water or preservative solution (Goszczynska *et al.*, 1990). The initial postharvest increase in FW and DW of rose cv. Mercedes petals was promoted by placing in GA_3 solution (20 mg/L) for 24 hours. The GA_3 enhanced increase in DW of petals kept in both light and dark conditions (Sabehat and Zieslin, 1995).

14.4.2.2 Benzyladenine (BA)

Rio *et al.* (1989) studied the effects of pretreatment and storage condition of cut flowers of cv. Sonia sprayed with or held in solutions containing BA (25 or 50 ppm), prior to keeping in simulated home conditions. The vase life ranged from 5.6 to 7.4

days (with 6.2 days in the control). Holding in a BA solution did not improve vase-life and the best results were obtained by spraying especially with BA at 25 ppm.

14.4.2.3 Ethylene

Changes in the concentrations of endogenous polyamines and ethylene were determined in two diverse species of rose *viz., Rosa damascene* and *R.* bourboniana during postharvest periods. At full bloom stage, the concentration of free putrescine was significantly higher than rest of the polyamines *i.e.* sperimine and spermidine in both species. The concentrations of all polyamines decreased subsequently upto 48 h after full bloom. Similar trend was observed with conjugated form, but in bound fraction, the concentration of spermine was higher than rest of the polyamines during full bloom. In both the species, ethylene showed higher levels during full bloom with maximum in *R. damascene* which increased during postharvest periods (Sood and Nagar, 2008).

Reid *et al.* (1989) found that the presence of very low concentrations of ethylene had marked effects on the opening of cut flowers (from glasshouse crops). Depending on cultivar, the rate of opening was unaffected (*e.g.* Gold Rush), accelerated (*e.g.* Sterling Silver), or inhibited (*e.g.* Lovely Girl). The K*m* for the inhibition of opening of Lovely Girl by ethylene was 4 p.p.b. Flowers of some cultivars (*e.g.* Royalty) had an abnormal shape when opened in the presence of ethylene. The effects of exogenous ethylene could be overcome by pretreatment of the flowers with 0.5 µmol silver thiosulphate per stem. Cut roses (cv. Sonia) were treated with ethylene (12 µl/l for 24 h at 20°C in darkness) and subsequently kept at 20°C, 60 per cent RH under continuous fluorescent light in order to accelerate senescence or stored at 2°C, 85-95 per cent RH under continuous fluorescent light to retard senescence. Untreated roses kept at 27°C were used as a control. The total amino acid content of ethylene-treated and control rose corollas decreased from a reference value (100 per cent) on day 0 to approximately 60 per cent on day 3 and then remained constant until the end of vase life.

The total amino acid content of corollas of cold-stored roses increased from 100 per cent on day 0 to 135 per cent on day 9 and then decreased to 82 per cent on day 12. There was no correlation between senescence of cut roses and accumulation of total free amino acids in the corollas. Of the individual free amino acids, only leucine content was significantly affected by ethylene treatment (Gorin *et al.*, 1989). Ethylene treatment (10-12 µl/dm^3 for 24 h at 20°C in darkness) when the petals and sepals were tightly folded affected neither subsequent opening nor vase life of cv. Sonia and therefore, would not be greatly affected by it during handling and marketing. It was suggested that cut Sonia roses harvested at an early stage were insensitive to ethylene (Lukaszewska and Gorin, 1988, 1989). Exogenous applicaion of ethylene (10 µl/l for 24 h at 20°C) to cut roses harvested simultaneously at 4 stages of opening, accelerated petal shedding. Pulsing with ethephon (5 µl/l/flower) or placing the flowers in a solution of 50 or 100 mg/L was more harmful than direct application of ethylene, inhibiting bud opening and accelerating leaf and corolla abscission (Lukaszewska *et al.*, 1990). Silver thiosulphate (STS) increased vase life of fresh and cold-stored roses, whether or not they were treated with ethylene. The effect of STS was increased as ethylene treatment was delayed. During the storage of rose cv. Visa flowers at 4°C (Serrano *et al.*, 1992) the rate of ethylene production was maintained at very low levels and was unaffected by the length of the cold storage period. Respiration, in contrast, decreased during storage. The level of endogenous ACC increased throughout the storage period, most of it in its conjugate form, whereas the activity of ethylene-forming enzyme (EFE) decreased. During flower senescence at 20°C, ethylene emission was also very low and was not affected by the previous storage period of 4°C, however, the rate of respiration became lower as storage time increased, and at no time were respiratory or ethylene crises observed associated with senescence. ACC levels increased with senescence, both in the free and conjugate forms, and were higher in the roses which had previously been in cold storage. On the other hand, EFE activity became lower as the storage period increased subsequently decreasing further with senescence. Jambor-Benczur *et al.* (2010) reported that the flowers treated witn 1-MCP for 18 hours at 4° C resulted in larger flower diameter and prolonged vase life (11.2 days) in cv. 'Bordeaux' compared to treatment for 6 hours at 17° C (8.25 days).

Muller *et al.* (2001) assessed the ethylene production in *Rosa hybrid* and found the upregulation of ACC oxidase and ACC synthase on exposure to ethylene. Ethylene exposure posed a significant increase in ACC content indicating ACC synthase activity. In the cultivar 'Bronze', ethylene production increased with flower senescence even in absence of exogenous ethylene. Treatment of 1-MCP inhibited the production of ethylene in response to external ethylene, while less effective in senescing flowers. Moreover, the cultivar 'Vanilla' was found producing lower amounts of ethylene endogenously or through external stimuli.

Huang *et al.* (2017) reported that cut rose flowers subjected to preharvest 1-MCP significantly extended storage life (26 days), while those subjected to PE 50 (MAP packaging), 1-MCP + PE50 and the untreated control had storage lives of 22, 19 and 16 days, respectively. Moreover, the preharvest 1-MCP spray prolonged vase life of cold-stored flowers for 0 or 6 days, by 27.08 per cent and 42.55 per cent. Preharvest 1-MCP spray treatment significantly suppressed respiration intensity, ethylene production, soluble protein degradation, ACC production and ACC oxidase activity. The 1-MCP + PE50 compromised all the effects of 1-MCP except the effect of soluble protein content. Thus, it was concluded that preharvest 1-MCP spray treatment had positive effects on vase life and storage longevity of cut rose and its combination with PE50 package diminishes these effects (Table 27).

Mueller *et al.* (2001) explained the differences in postharvest characteristics of miniature potted roses (*Rosa hybrida*) due to variation in endogenous ethylene and/or sensitivity to exogenous ethylene. Comparing cultivars with different postharvest characteristics, dramatic differences in flower longevity of the cultivars Bronze and Vanilla were found, even though both cultivars exhibited a climacteric rise in ethylene production in senescing flowers. Four rose ethylene receptor gene fragments, termed *RHETR1-4*, were isolated and characterized. *RhETR2* (88 per cent amino acid identity to *Arabidopsis thaliana* ETR1) was barely regulated during flower development, but the transcript level differed between the two rose cultivars. *RhETR1* (72 per cent amino acid identity to *A. thaliana* ERS1) and *RhETR3* (67 per cent amino acid identity to *A. thaliana* ETR2) exhibited differential expression. These results indicate that variation in flower life and ethylene sensitivity among rose cultivars- in ethylene free environment and in response to exogenous ethylene-may be due to differences in receptor levels.

Table 27: Effect of Preharvest 1-MCP (1), Postharvest PE50 Package (P) and 1-MCP + PE50 (1+P) on Flower Maximum Diameter and Vase Life of Cut Rose after 0, 6, 12, 18, 24 and 30 Days of Cold Storage*

Parameter	*Treatment*	*Storage (days)*					
		0	*6*	*12*	*18*	*24*	*30*
Maximum diameter (mm)	Control	49.35 ± 8.84 a	55.76 ± 10.22 a	57.74 ± 9.69 b	49.48 ± 4.77 b	39.04 ± 9.71 a	-
	1	48.59 ± 12.43 a	54.10 ± 6.68	54.73 ± 6.68 ab	48.26 ± 5.00 ab	40.35 ± 7.71 a	31.86 ± 4.19 a
	P	49.35 ± 8.84 a	54.37 ± 9.70 a	55.19 ± 8.05 ab	48.41 ± 3.87 ab	39.96 ± 5.91 a	32.13 ± 8.35 a
	1+P	48.59 ± 12.43 a	48.85 ± 6.88 a	49.54 ± 3.83 a	40.74 ± 4.39 a	38.72 ± 10.67 a	26.47 ± 5.12 a
Vase life (days)	Control	4.8 ± 0.63 a	4.7 ± 0.67 a	4.8 ± 0.67 b	4.6 ± 0.52 b	3.6 ± 0.70 a	-
	1	6.1 ± 0.32 b	6.7 ± 0.67 c	5.1 ± 0.74 bc	4.9 ± 0.74 bc	4.4 ± 0.70 bc	2.1 ± 0.32 a
	P	4.8 ± 0.63 a	6.0 ± 0.73 bc	5.1 ± 1.02 bc	5.1 ± 0.80 bc	4.7 ± 0.57 bc	2.2 ± 0.43 a
	1+P	6.1 ± 0.32 b	4.9 ± 0.57 a	3.7 ± 0.88 a	3.1 ± 0.74 a	4.0 ± 1.15 ab	1.8 ± 0.63 a

* Huang *et al.* (2017).

Exogenous ethylene inhibited floral opening in cut rose cv. Kardinal and slightly promoted opening in cv. Samantha. The ethylene production rate was much higher in 'Kardinal' thanin'Samantha' following the ethylene treatment whereas the expression of ETR fragments remained unaffected in the former and up-regulated in latter. However, no unequivocal effect was observed with the expression of CTR and EIN3 genes which are involved in ethylene signal transduction pathways. Thus, Tan *et al.* (2006) reported floral opening trait to be associated with the differential ethylene effect on expression of ethylene receptor genes.

Ahmadi *et al.* (2009) evaluated 231 F_1 genotypes of 'Lavender' and 'Vanilla' reciprocal crossing for ethylene sensitivity. Initially, low and high sensitive genotypes were sorted based on the ethylene induced organ abscission. Expression of ethylene receptor genes (*RhETR1/3*) and signal transduction genes (*RhCTR1/2*) were found not correlated with ethylene sensitivity. While, the relative expression of laccase (*RhLAC*) was found correlated with lignin deposition and abscission. This gene expression was found more in low sensitive genotypes than that of high sensitive ones. Thus, enhanced expression of laccase gene by ethylene treatment might be correlated with abscission process.

Hajizadeh *et al.* (2011) observed the relative expression of *RhCTR1 and RhCTR2* genes involved in ethylene signal transduction pathway in cv. 'Black Magic' and 'Maroussia'. *RhCTR1* expression amplified significantly during flower opening in both cultivars with more pronounced effect in 'Maroussia' whereas, no such expression was observed with *RhCTR2* thus concluding that vase life was not associated with *RhCTR2* which behaved like constituitive gene. Al-Salem and Serek (2017) compared the expression of ethylene biosynthetic genes and ethylene signal transduction genes in low ethylene sensitive 'Vanilla', high ethylene sensitive 'Lavender' and their F_1 progenies. Results revealed that *RhETR1, RhETR2, RhETR3 and RhEIN3* were expressed more in most of the tissues of 'Vanilla' than in 'Lavender' whereas, the genes *RhETR4, RhACS3, RhACS4 and RhACS5* were not expressed in the tissues.

14.4.2.4 Abscisic Acid (ABA)

Barthe *et al.* (1991) in an experiment on vase life of roses found that foliar spraying with ABA delayed wilting and senescence, increased the vase life and resulted in higher FW/DW ratio and lower pH, osmolarity and conductivity after 10 days compared with control roses.

14.4.2.5 Salicylic Acid

Cut flowers of rose cultivar 'Black Magic' sprayed with 50 200 µM salicylic acid in controlled greenhouses 2 weeks before harvest followed by placing in vase solution with 50-400 µM SA showed significant increase in cumulative uptake, relative fresh weight, catalase activity, prolonged vase life and improved reactive oxygen species scavenging capacity by facilitating better water balance in the stems (Alaey *et al.*, 2011). Kazemi *et al.* (2018) reported that preharvest application of salicylic acid at 1.5 mM concentration extended the vase life of cut roses by improving its membrane stability and decreasing lipid peroxidation activity. Also, flowers exposed continuously to 1.5 mM and pulsed with 15 mM SA reduced the weight loss, enhanced antioxidant (CAT and POD) activity and decreased lipid peroxidation.

Mazaro *et al.* (2018) evaluated the response of cut rose cv. Vega on treatment with different concentrations of salicylic acid (SA): 0, 0.5, 1.0, 1.5 and 2.0 mM when used as vase solution. Results revealed that SA treatment maintained the postharvest quality, reduced fresh weigh loss, stem curvature, improved turgescence and minimized darkening of petals. Moreover, SA application increased total protein content and phenylalanine ammonia-lyase (PAL) which characterized the activation of plant defense mechanisms to the senescence process. Cocetta *et al.* (2019) examined the effect of salicylic acid (SA) and its functional analogue benzothiadiozole (BTH) on postharvest quality and life of cut roses (*Rosa hybrid*). Flowers and leaves of cut roses were treated with SA and BTH for 24 h and analyzed for vase life and pheolic index. Results revealed that SA treatment extended the vase life of flowers without affecting phenolic content whereas, BTH significantly enhanced the phenolic content with negative effect on stem quality due to the impairment of water uptake.

14.2.4.6 Methyl Jasmonate

Pulsing cut flowers of rose cultivars 'Frisco','Jazz', 'Mercedes' and 'Red Charm' with 350 µM methyl jasmonate and spaying with 500 µM methyl jasmonate suppressed gray mould (*Botrytis cinerea*) without causing any adverse changes in ethylene production. Moreover, this treatment improved the petarcolour by inhibiting colour fading in yellow, orange and pink cultivars (Meir *et al.*, 2005). De Pietro *et al.* (2012) stated that spraying methyl jasmonate 500 µM on 'Vega' rose flowers resulted in improved flower quality, reduced fresh weight loss, lowered respiratory rate and extended shelf life with better retention of colour, relative water content and reducing sugars.

Almeida *et al.* (2017) investigated the efficiency of methyl jasmonate to prolong vase life of 'Avalanche' cut rose flowers. The treatments studied were 125, 150, 500 and 1000 µM of methyl jasmonate along with a control. The flower buds were sprayed with 4 ml of the above solution according to the treatments and stored in cold chamner (1° C) for 2 and 6 days. Results revealed that better quality, maximum fresh weight and water absorption were obtained in flowers stored for 2 days. However, application of methyl jasmonate caused less turgidity, greater petal darkening and had no effect of maintenance of cut rose quality.

14.5.0 Pot Roses

Exogenous application of ABA (abscisic acid) to intact Miniature potted rose plants (cultivars Bronze and Vanilla) increased leaf drop and accelerated flower senescence in both cultivars, while bud drop was only induced in Bronze. Application of ABA to detached rose flowers accelerated their senescence. ABA treatment increased ethylene production in Bronze flowers, while no ethylene production was measured in flowers of Vanilla, or in the leaves of both cultivars. Pretreatment with 1-MCP (1-methylcyclopropene) delayed ABA promoted flower senescence in Bronze, ABA is at least partly mediated by ethylene (Muller *et al.*, 1999). In an interior environment, the diploy life of Miniature potted rose (*R. hybrida* cv. Bronze) was 7 days, while that of cv. Vanilla was 23 days. Where the plants were subjected to stimulated transpor-tation stress, postharvest life of both cultivars was reduced (to 6 days for Bronze) and to 17 days for Vanilla.

Six cultivars of potted rose plants were evaluated by Clark *et al.* (1991) for shipping stress-induced leaf chlorosis during holding at 8°, 16° or 28°C for 2, 4 or 6 days. Cv. Meijikatar showed more leaf chlorosis than the similar cv. Meirutral at the higher simulated shipping temperatures and longer durations. Plants of Meijikatar were treated before simulated shipping with BA, thidiazuron or Promalin [GA + BA] at 0, 25, 50 or 100 mg/L each, then placed in paper sleeves and stored in the dark in fibreboard boxes at 16°C for 5 days. Plant quality 5 days after removal from storage was better with BA at 50 or 100 mg/L than at 0 mg/L. All cytokinin-treated plants showed less leaf chlorosis than controls. Pot grown Miniature roses (*R. hybrid* cvs. Bronze and Vanilla) were subjected to simulated transportation stress, resulted in modest increase in ethylene production in Bronze. Pretreatment with 1-MCP before transport simula-tion improved display life of potted roses (Muller *et al.*, 2000). Experiments were conducted by Cushman *et al.* (1998) to determine

Table 28: Floral Development during Storage of Pot Roses as Affected by Flower Stage at the Beginning of Storage (Beginning flower stage or BFS), Temperature (Temp), Cultivar (Cv), and Storage Duration (Time)*

Beginning Flower Cultivar	*Temp. (°C)*	*Floral Development during storage (stage) stage of*		*Summer experiment*	
		Storage Duration (d)	*Experiment*		*Winter*
	1	*2*	*4*	*6*	
Beginning flower stage					
1	-	0.6 b	1.1 a	1.3 ab	0.7 a
2	-	1.0 a	1.4 a	1.4 a	0.5 ab
3	-	0.8 ab	1.0 a	1.0 b	0.3b
Cultivar					
Meijikatar	4	0.3	0.8	0.3	-
	16	1.0	1.1	1.5	-
	28	2.1	2.1	2.5	-
Meirutral	4	0.0	0.0	0.1	-
	16	0.3	0.7	0.9	-
	28	1.1	2.1	2.2	-
	4	-	-	-	0.1
	16	-	-	-	0.4
	28	-	-	-	1.0
Significance			F values		
Temp	60.90**				30.39**
TempL	-				58.52**
TempQ	-				2.28**
BFS	11.43**				22.63**
Other main effects were not interest due to significant interactions					
Time BFS	3.05*				1.78**
TimeL was significant at P < 0.01 for BFS 1 and 2, but was NS for BFS 3					
Time Temp Cv	5.56*				3.17**
The TimeQ TempL effect was significant at P < 0.05 for both cultivars per interaction effects were NS					

Stage 1 = tight bud, calyx not reflexing; stage 2 = showing colour, calyx reflexing, no petals reflexed; stage 3 = full colour, petals beginning to reflex, traditional bud stage; stage 4 = several petals reflexed, traditional inhibition stage; stage 5 = fully open; stage 6 = postharvest life over, petal wilting or abscission, can separation within columns and factors by Duncan's multiple range test at P = 0.05.

***,** Non significant or significant at P = 0.05 or 0.01, respectively.

L = Linear or Q = quadratic

* Cushman *et al.* (1998).

the effects of simulated shipping (storage) temperature and duration on flower longevity and leaf abscission of pot rose cultivars Meijikatar and Meirutral. Three flower stages (1= tight bud, calyx not refle-xing; 2 = showing colour, calyx reflexing, no petals reflexed, 3 = full colour, petals beginning to reflex, traditional bud stage) were selected immediately prior to storing plants at 4, 16 or 28°C for 2, 4 or 6 days. Results that plants stored at 4°C had the longest post-storage floral longevity, the best flower quality, and the least leaf absci-ssion, regardless of cultivar, storage duration, or flower stage at the beginning of storage (Table 28).

A correlation between dry matter distribution and post-production quality in 3 cultivars (Charming Parade, Royal Parade and Dreaming Parade) of potted miniature roses was evaluated under 3 different environmental conditions (Hoyer and Salhoj, 2001). Post-production quality was evaluated after simulated transport (16°C 90 per cent RH, in the dark for 3-4 days) and then under indoor conditions (20°C, 60 per cent RH, in a 12-h photoperiod). Significant correlations were observed between the number of days to 50 per cent wilting and the dry weight of different plant parts of the pot roses at the marketing stage. A few correlations were found for the other 2 cultivars. A negative correlation was observed between the dry weight of single flowers and the number of days to 50 per cent wilting. The results indicated that post-production quality of Dreaming Parade and Charming Parade was correlated to the distribution of dry matter at the marketing stage.

The miniature rose 'Vanilla Kordana' is a useful genetic resource for enhancing the vase life and ethylene resistance. Muller *et al.* (2001) investigated the significance of ethylene production and sensitivity as determinants of vase life of two mutants (Goldy, Safari) and an offspring from 'Vanilla' x 'Amber' (Champagner). Results showed that two mutants had unequivocally higher ethylene production than 'Vanilla' suggesting the cause for its shorter flower life. The ethylene production was lower in 'Champagner' than its parents and the mutants. Moreover, autocatalytic ethylene production was observed in 'Amber' and 'Champagner' after exposure to exogenous ethylene indicating that both initial ethylene production and autocatalytic stimulation might be the factor determining vase life.

15.0 Biotechnology

A review on biotechnology of rose by Das and Rout (2000) has clearly indicated that extensive investigations have been carried out on the tissue culture of roses and there is adequate scope to intensify work on molecular markers and genetic transformation.

15.1.0 Tissue Culture

15.1.1 Anther Culture

Tabaeezadeh and Khosh-Khui (1981) first reported induction of callus from the anthers of two tetraploid *Rosa* species (4n = 28). They cultured anthers at various stages of development in different media, with varying concentrations of auxins and kinetin, under different light conditions. The MS medium with 2.0 g/l IAA and 0.4 mg/L kinetin was the best for anther culture of *R. damascena* while the medium with 7.5 mg/L IAA and 0.8 mg/L kinetin was found to be optimal for *R. hybrida* cultivars. Reports on the regeneration of plants from anther-derived callus are still to come.

15.1.2 Protoplast Culture

Mathews *et al.* (1991) reported high yields of protoplasts from embryogenic suspension cultures of *R. persica* x *R. xanthina* and *R. wichuraiana*. About 30 per cent of embryos developed into plantlets on transfer to basal Schenk and Hildebrandt's medium. Mathews *et al.* (1994) also reported successful isolation and culture of protoplast from transformed cell lines of *R. persica* x *R. xanthina* and cv. All Gold. They successfully regenerated shoots from the protoplast derived calli and the regenerants were analyzed using different molecular techniques.

15.1.3 Protoplast Fusion

Callus-derived protoplasts from *R. hybrida* cv. Frensham were fused using polyethylene glycol to protoplasts from cherry (*Prunus avium* x *pseudocerasus* cv. Colt) or blackberry (*Rubus laciniatus* cv. Thornless Oregon). *Rosa persica* x *R. xanthina* was self-fused and tetraploid plants were obtained. These tetraploids exhibited unusual morphology, contained a chromosome complement of 28 which is double the diploid number for this hybrid and had significantly increased guard cell lengths and chloroplast numbers. Putative somatic hybrids, involving either Frensham and blackberry or Frensham and cherry, were also regenerated. The morphological differences, root-tip squash preparation and RAPD analysis revealed genetic similarities to Frensham (Mottley *et al.*, 1996).

Schum and Hofmann (2001) analyzed the feasibility of using isolated protoplasts through somatic hybridization. They reported that combination of 1.5 per cent cellulysin, 0.5 per cent driselase and 0.5 per cent macerase could be used to isolate the protoplast from *in vitro* shoot cultures, callus, embryogenic as well as non-embryogenic cell suspension cultures. Protoplasts isolated from non-embryogenic cultures regenerated non-morphogenic callus whereas, embryogenic cultures ('Heckenzauber' and 'Pariser Charme') resulted in embryogenic callus which was later developed into plantlets. Schum *et al.* (2002) generated callus from protoplasts of non-embryogenic sources (*Rosa canina, R. caudate, R. corymbifera* 'Laxa', *R. multiflora, R. roxburghii, R. spinosissima* and *R. wichuriana*) and embryogenic sources (*Rosa x hybrid* 'Elina' and Pariser Charme'). Specific genotypes carrying black spot resistant genes were identified and their protoplasts were fused. Protoplasts were pretreated with rhodamine (0.1 mmol) or iodoacetate (0.5-1 mmol) for 15 minutes along with X-rays 300 Gy to suppress cell divisions of non-fused protoplasts and homologous fusion products. Ultimately, shoots were regenerated from fused somatic hybrid of 'Pariser Charme' and *Rosa wichuriana* and the traits were authenticated with flow cytometry and AFLP-analysis.

15.1.4 Somatic Embryogenesis

Somatic embryogenesis in rose has been achieved by various researchers using leaf, internode, filament of stamen, root and zygotic embryo (Rout *et al.*, 1991; van-der-Salm *et al.*, 1996; Marchant *et al.*, 1996). Van-der-Salm *et al.* (1996) reported somatic embryogenesis and shoot regeneration from an excised adventitious root-derived callus of *R. hybrida* cv. Moneyway on hormone-free SH medium supplemented with 500 mg/L casein hydrolysate, 87 mM sucrose, 1000 mg/L inositol, 600 mg/1 L-proline, 5 mg/L nicotinic acid, 0.5 mg/L pyridoxine-HCl, 5 mg/L thiamine-HCl. Induction of secondary somatic embryos and their maintenance for long periods (16 months) was achieved by subculturing the embryogenic callus mass at 4-week intervals on MS medium supplemented with 2.2 µM BA, 0.05 µM NAA and 0.05 µM GA_3 (Rout *et al.*, 1991).Castillon and Kamo (2002) studied the effect of various carbohydrates and osmotically active compounds on somatic embryo maturation and conversion. Results indicated that 27 per cent conversion was achieved for the Floribunda 'Trumpeter' embryos following their maturation on 3 per cent fructose; 36 and 61 per cent conversion were attained with 'Dr. Huey' (Multiflora) embryos in maturation medium containing 3 per cent sucrose with either 2.5 or 5 per cent mannitol, respectively. The Hybrid Tea 'Tineke' embryos showed 12 per cent conversion on 3 per cent sucrose or glucose or fructose. Burrell *et al.* (2006) assessed the ability of 15 genetically diverse roses to endure somatic embryogenesis and found that the cultivar 'Tournament of Roses' produced 20-30 per cent embryogenic callus capable of producing somatic embryos whereas, 'Baby Love', 'Ingrid Bergman', 'Perfume Delight', 'Prominent', 'Sunflare' and '90-102' produced crystalline callus which failed to produce somatic embryos. Surprisingly, the F_1 progeny of Tournament of Roses' x 'Baby love'produced somatic embryos proving its heritability.

15.1.4.1 Embryo Development and Germination of Somatic Embryos

Embryo development and germination of somatic embryos are prerequisites for plantlet production. Omission or lowering the concentration of 2,4-D from the culture medium helped embryo development and germination as reported in many cultivars of rose (Noriega and Sondahl, 1991). Mathews *et al.* (1991) reported that about 30 per cent of somatic embryos of hybrid roses germinated into plantlets after transfer to hormone-free SH medium.

15.1.4.2 Somaclonal Variation

Larkin and Scowcroft (1981) extensively reviewed the phenotypic variations found among plants regenerated after a passage through tissue and cell culture. Hasegawa (1980) reported one 'abnormal looking' plant amongst 600 tissue-culture-propagated plants of cv. Improve Blaze. Martin *et al.* (1981) found no variation among 2125 rose plants raised in the field for three years. The lack of somaclonal variability suggests that rose is relatively stable when propagated via axillary buds.

Lloyd *et al.* (1988) reported that callus-derived shoots of *R. persica* x *R. xanthina* exhibited a considerable degree of variation in leaf shape. Arene *et al.* (1993) observed somaclonal variation in plants regenerated from callus of *R. hybrida* cv. Meirutral. They also observed that among 152 recorded variants, 83 differed significantly from the original cultivar by the number of petals, 16 by the dwarf growth habit, 4 by the round shape petals, 12 by the colour, 33 by both growth habit and number of petals and 4 by both the petal number and colour.

15.2.0 Biochemical and Molecular Markers

15.2.1 Genetic Linkage and Identification of Genotypes

15.2.1.1 Isozymes

In the investigations on the use of isozymes and molecular markers of rose, Kim and Byrne (1996) examined the hybrid origin of 23 rose (including 3 *Rosa* spp. thought to be interspecific hybrids) accessions with three isozymes, acid phosphatase, malate dehydrogenase, and phosphoglucose isomerase. All three isozymes were useful for interspecific hybrid verification. This procedure was effective if the putative parents were known and differed in isozyme phenotype.

15.2.2 Molecular Markers

Moreno *et al.* (1996) analyzed twenty-two wild species of rose from the Royal Botanical Garden, Madrid, using isoenzymes and RAPD markers. Each 10-base arbitrary primer produced a specific DNA banding pattern that grouped together plants belonging to the same species and botanical sections, as predicted from their genetic background. Although variation was observed, the isoenzyme technique was not powerful enough to distinguish between the different sections.

Ben Meir *et al.* (1997) assessed the applicability of DNA fingerprinting (DFP) using mini- and micro-satellite sequences to the identification of genotypes and to the establishment of genetic distances between rose (*R. hybrida*) plants. The probability of 2 offsprings from the crossing of similar genotypes having identical DNA fingerprints was 2×10^{-8}. A comparison of genetic relationships within and between phenotypic categories, based on known genetic history, with genetic relationships derived from DFPs revealed a perfect match for flower types. Werlemark *et al.* (1999) studied both morphological and molecular markers in a pair of reciprocal crosses between *R. dumalis* and *R. rubiginosa* (2n = 35). A canonical discriminant analysis based on seven morphological characters showed only a minor overlapping between the two progeny groups. In addition, the *R. dumalis* x *R. rubiginosa* offspring was more heterogeneous than the offspring from the reciprocal cross in each of the characters analyzed. Eleven RAPD markers specific for the *R. dumalis* parent and 10 RAPD markers specific for the *R. rubiginosa* parent were scored in the offspring. Each of the offspring exhibited either all or all-but-one of the seed parent markers. The average number of pollen donor markers found in the offspring was 3.2 (*R. dumalis* x *R. rubiginosa*) and 2.7 (*R. rubiginosa* x *R. dumalis*). About half of the pollen donor markers were never transmitted to the progeny.

Some 119 accessions and 213 markers of 36 rose species that include eight sections of the subgenus *Eurosa* and one species each from the subgenera *Hesperhodos* and *Platyrhodon* were used by Jan *et al.* (1999) to calculate a similarity matrix, which was clustered with the unweighted pair group method using arithmetic means (UPGMA). The RAPD markers distinguished between all the rose accessions and species grouped into their respective sections.

The *matK* DNA was sequenced in each of the 15 Japanese *Rosa* taxa, 17 other Asian taxa, and two American taxa to analyse their phylogenetic relationship

(Wu *et al.*, 2000). All the 14 tested taxa in the section *Synstylae*, associated with the section *Indicae*, formed one clade. It was confirmed that sections *Synstylae* and *Indicae*, with the common styles exerting from the hypanthium, are closely related to each other. *R. brunonii* and *R. moschata* in the section *Synstylae* formed an inner clade which is consistent with their common morphological traits. Two clones of *R. acicularis* formed one single clade which was separated from the remaining species by a distinctive species in the section *Rosa*. This finding suggests that *R. acicularis* might be a distinctive species in the section *Rosa*. The remaining species of the section *Rosa* formed one clade which was further subdivided into two inner clades: *R. rugosa* and *R. manetti* [*R. noisettiana*] formed one clade, while *R. acicularis* var. *nipponensis* and *R. willmottiae* formed another clade. Genetic variability among 100 old cultivated rose varieties from 13 horticultural groups was estimated by arbitrary primed (AP) PCR (Martin *et al.*, 2001). Using five long (20-mer) PCR primers, 58 polymorphic DNA fragments were produced, of which 55 was highly discriminant, allowing differentiation of the quasi-totality of the 100 cultivars. Adendrogram was constructed displaying the related genetic similarities between cultivars estimated from the presence or absence of PCR fragments. It showed that relationships between the Chinese and European founder roses, hybrid groups of the first (Bourbons, Noisettes, Portlands) and second (Hybrid Perpetuals and Teas) generations, and the most modern Hybrid Teas, produced during the history of domestication.

Parental linkage map of a segregating population of diploid rose hybrids derived from two half-sib parents, P119 and p117 comprising a total of 365 uni-parental AFLP and SSR markers resulted in 157 P119 markers (85 per cent) on 8 linkage groups and 133 P117 markers (78 per cent) on 7 linkage groups spanning 464 cM and 491 cM with average interval of 2.9 cM and 3.7 cM between markers, respectively. Moreover, three QTLs for leaf area and 2 QTLs for chlorophyll content were identified with the phenotypic variance of 50.8 and 25.8 per cent, respectively (Yan *et al.*, 2003).

Leus *et al.* (2004) obtained clear molecular polymorphism among 88 rose species and cultivars using AFLP. Other molecular techniques based on the sequences of evolutive markers like nrITS, rbcL and matK revealed more information about phylogenetic relationships in which rbcL were less polymorphic compared to other two regions for phylogenetic work in roses focused in *Rosa multiflora, R. wichuriana* and *R. chinensis*. Kumar *et al.* (2005) assessed the polymorphism and genetic relatedness among 21 Indian-bred rose cultivars with the help of 40 random decamer RAPD primers. Results revealed that out of 40 random primers, 18 primers amplified the scorable banding patterns from which 113 polymorphic bands were obtained from a total of 131 bands. The percentage polymorphism ranged from 66-100 per cent and size of the bands varied between 0.40 and 2.4 kb. Moreover, four clusters were distinguished from 21 Indian bred rose based on the Jaccard's similarity coefficient.

Twenty one fragrant rose cultivars and 2 non-fragrant cultivars were profiled using RAPD primers. According to dendrogram formed, cultivar Nurjehan was genetically most diverse whereas, the most similar cultivars were Sir C. V. Raman and Arunima. Cluster A had a minimum of 3 cultivrs, while cluster D contained maximum of 9 cultivars. Cluster B and C possessed 5 and 6 cultivars, respectively. Thus, RAPD profile distinguished and established the inter-relationship which would be highly helpful for parental selection in breeding programmes (Prasad *et al.*, 2006). Mohapatra and Rout (2006) found that A and N series of RAPD decamer primes exhibited good amplification of DNA fragments when used for screening and optimization of RAPD analysis across 34 rose cultivars. The primers OPN-07 and OPN-15 produced maximum number of fragments in *Rosa hybrid* cv. Anuraag. On further selection of 10 primers for genetic analysis, a total of 162 distinct bands ranging from 100-3400 bp were amplified out of which primer OPN-15 amplified 21 fragments.

Sasikumar *et al.* (2007) analysed genetic diversity among 25 rose cultivars and 5 species by employing 35 random decamer RAPD primers in which 28 generated polymorphic bands. Out of these 28 primers, 25 amplified all rose cultivars with the fragment size ranging between 0.25 and 3 kbp. From a total of 226 bands, 209 polymorphic and 17 monomorphic bands were obtained with the average band per primer being 8.1. The highest polymorphic bands were generated by OBA11 and OPA10 primers. Results of UPGMA cluster analysis grouped 25 cultivars into 3 clusters and 5 species into a separate cluster in which 'Abhisarika' was found to exist farther.

Fourty four EST-SSRs and 20 genomic-SSRs were developed from database of rose bud EST containing 2556 unigenes and a genomic library, respectively. These SSRs were employed to expand genetic maps of interspecific F_1 progeny and parents using pseudo-testcross mapping strategy. Results revealed that maternal map consisted of 105 markers covery 432 cM and paternal map involved 136 markers covering 438 cM with a total of 7 linkage groups. Moreover, loci controlling flowering traits were localized as major gene or QTL for number of petals and a QTL for blooming date which could be used for establishment of consensus linkage map that combines traits and markers in different rose genetic maps (Oyant *et al.*, 2008).

Nine ISSR primers were selected to fingerprint 33 cultivars which included unrelated cultivars, sports and offspring obtained by sexual propagation. A total of 149 fragments showed polymorphism out of 159 fragments generated from 9 primers with polymorphic percentage being 93.7 per cent. Low dissimilarity was found among the cultivars according to Jaccard's dissimilarity coefficient whereas; four groups were obtained from UPGMA and PCA analysis based on their horticultural classification and pedigree. Therefore, ISSR could be a useful tool for cultivar identification as well as assessment of genetic relationships due to low cost and high reliability (Crespel *et al.*, 2009). Clear resolution of ISSR markers were obtained in the optimal conditions of 2 mM $MgCl_2$, 1 U Taq DNA polymerase, 30 ng template DNA and 1 µM primer concentration with reaction programme of initial denaturation at 94° C for 5 mins followed by 35 cylces of denaturation of 94 ° C for 30s, annealing for 45 s at primer annealing temperature, extension for 2 min at 74° C with a final 10 min extension at 72°C. These conditions resulted in 66 bands in which 50 were polymorphic in nature (Jabbarzadeh *et al.*, 2010).

Samiei *et al.* (2010) evaluated the genetic variation among accessions from *R. persica, R. foetida, R. pimpinellifolia, R. hemisphaerica, R. canina, R. iberica, R. damascene, R. beggeriana* and *R. orientalis* using 10 SSR markers. SSR markers amplified alleles in all species irrespective of different sections within the genus. Results of UPGMA analysis sorted five main groups with no variation between *R. canina* and *R. iberica*. Moreover, accessions of Sect. Pimpinellifoliae were grouped with plants from Sect. Rosa and Cinnamomeae in two distinct groups. Whereas *R. persica* was group separately from other species with few alleles common with other taxa suggesting its potential in phylogenetic analysis.

Azeem *et al.* (2012) developed RAPD markers to assess genetic diversity among 4 species and 30 rose accessions using 25 RAPD primers. A sum of 110 polymorphic bands out of 146 bands with an average of 7 bands per primer was obtained in which maximum allele frequency (10) was found with primer GLD-20 and miminum(2) with GLC-02. Maximum polymorphism (100 per cent) was found with GLA-03, GLA-05, GLA-07, GLA-10, GLC-02, GLC-06, GLC-08, GLC-10, OPG-11 and OPE-19 whereas lesser polymorphism (42.85 per cent) with the primer GLB-11.

Wang *et al.* (2012) employed matK sequence to analyse genetic relationship among 37 oil-bearing roses grown in China and reported the two clades, A and B among various rose types. Clade A was subdivided into A1 inlcuding *R. serata x R. rugosa* and *R. davurica* and A2 containing most rugosa cultivars which were consistent with classification of Sect. Cinnamomeae. Similarly, Clade B was categorized into B1 (R. 'Tujue', R. 'Hongshuangxi', R. 'Baojialiya Hong', R. 'Hezeyang', R. 'Crimson Glory' and R. 'Baojialiya Bai'), B2 (some Gallicanae cultivars introduced from abroad) and some fragrant wild species including *R. odorata* var. *gigantea, R. multiflora var. cathayensis, R. rubus*and *R. rugosa* assuming the introgression of wild species in the evolution of oil-bearing cultivars of B1 and B2.

Ghose *et al.* (2013) characterized 30 wild rose ecotypes by using SSR markers, SNP of chloroplast *trna*L intron and the nuclear gene *GAPDH*, ploidy determination and metabolite profiling for major phenolics and fatty acids. Results showed a total of 244 SSR alleles with an average of 12.2 alleles per marker and average PIC content of 0.70. Further, cluster analysis grouped the ecotypes into 3 major clusters among which cluster 3 exhibited greater diversity. Also, the ecotypes possessed higher metabolite diversity within each cluster.

Qiu *et al.* (2013) investigated genetic diversity and genetic relationships among 39 wild species, 21 old garden roses and 29 modern cultivars using internal transcribed spacers (ITS) of the nuclear rDNA and the chloroplast gene matK sequence. Results of dendrogram analysis revealed that the 39 wild species studied were almost concurrent with that of botanical sections; most of the wild genotypes were separated from rose cultivars and the relationships among cultivated roses did not correlate with horticultural groups as inferred by ITS and matK sequences.

Vukosavljev *et al.* (2013) evaluated the genetic diversity of 138 rose cultivars using 24 SSR markers which obtained 518 different alleles indicating the subgroup of garden roses. Genetic differentiation was found to be 4 times more in garden roses compared to cut roses. Moreover, Svejda's cultivars were noted similar to European roses with clear inheritance of winter hardiness from wild species. Whereas roses of Harkness and Olesen shared similar gene pool. Among various linkage groups, group 5 possessed QTLs for winter hardiness, group 6 for greater diversity and group 2 was most differentiated among garden rose types. Zhang *et al.* (2013) used an SNP-STR marker, Rc06 in the flanking sequence from 20 diploid and 35 polyploid accessions of various *Rosa* species in which most accessions of diploid species had only one allele whereas polyploidy had 2-5 different alleles. A total of 18 haplotypes are formed only from 12 SNPs in the flanking sequence. Results of dendrogram analysis sorted haplotypes into 5 groups of which diploid species in Section of genus Rosa were grouped as a haplotype whereas, polyploidy species contained different haplotype groups with few identical SNP haplotypes shared commonly between diploid and polyploidy species. Three different polymorphic repeat regions in STR were found with 8 additional SNPs. Thus, the hypotheses on diploid Rosa species that contributed to polyploid taxa was supported.

Otagaki *et al.* (2015) isolated three *RoFT* alleles and correlated the *RoFT* genotypes and flowering time phenotypes in the mapping population. Results of which showed that *RoFT f* and*RoFT g* alleles contribute to early flowering whereas, *RoFT e* allele contributes for late flowering. Two novel CAPS markers were developed based on the genome sequences of aforesaid alleles and analysed allele specific expression using RT-CAPS assay which suggested that these *RoFT* alleles are regulated differentially at the transcription level. It was clear from the results that *RoFT*was continuously expressed after floral bud formation and its transcript reached its peak after expression of *RoAP1b* floral meristem identity gene. Therefore, it was concluded by that authors that *RoFT* is essential during floral transition, flower development and flowering in roses.

Rai *et al.* (2015) studied the genetic diversity amoung 23 rose genotypes both by using morphological and RAPD markers. Results of morphological characterization showed highest diversity among *R. damascene* var. Rani Sahiba and *R. moschata* whereas, molecular characterization done with 50 RAPD primers revealed higher similarities between *R. moschata* and *R. tomentosa* and 168 polymorphic bands from 26 primers.Both the marker types were concurrent in which *R. damascena* var. Rani Sahiba, Himroz showed distinction from Jwala. Moreover, the members of Sect. Indicae, *R. indica* var. *odorata*, *R. indica major*, *R. chinensisviridiflora* and *R. bourboniana* were clustered together by molecular characterization whereas, Kakinada rose was clustered with Himroz both by morphological and molecular characterization.

Sapna *et al.* (2015) investigated the genetic variability among 32 rose cultivars by using 20 RAPD and 10 ISSR markers from which a total of 306 bands with a range of 6-18 bands per marker were detected. Moreover, ISSR markers were considered more useful in terms of average polymorphic information content (PIC), resolving power of primer (RP) and marker index (MI) over RAPD markers. Genetic distance coefficients ranged from 0.44 to 0.91 and the UPGMA cluster analysis differentiated all the genotypes even the Suryodaya and Suryakiran which are almost morphologically identical. Outcomes from diversity analysis indicated that cultivated rose cultivars exhibit more genetic variability but are less polymorphic with respect to morphological and physiological attributes. Thus, the molecular fingerprinting would be highly functional for identification and conservation of varieties.

Vogt *et al.* (2015) assessed the epigenetic patterns in genetically imbalanced polyploidy Sect.Caninae using cDNA-AFLP and methylation-sensitive amplified polymorphism (MSAP). Results revealed maternal bias of DNA methylation and expression patterns causing apomictic mode of inheritance which was assumed from the genetic composition of pentaploid and hexaploid hybrids. However, more number of paternally expressed fragments were found in sepals than leaves insisting the rapid subfunctionalization of parental genetic information in the hybrids. Thus, paternal genomes were not silenced but transcribed and differentially expressed in various plant organs.

Jian *et al.* (2018) reported that endangered status of *Rosa praelucens* (decaploid alpine rose) might be due to habitat fragmentation, habitat loss and inherent reproductive difficulties than low genetic diversity. Among the seven population assessed for genetic variation, the populations possessing higher cpDNA genetic diversity, representing more AFLP genotypes and surrounding cpDNA haplotypes were given importance for conservation by providing plant microclimate.

Alsalem and Serek (2019) evaluated tetraploid miniature potted cultivars, 'Vanilla' (slightly sensitive), 'Lavender' (very sensitive) and 'Vanilla x Lavender' F_1 progeny plants (4 with low sensitivity and 4 with high sensitivity) to develop markers for distinguishing their ethylene sensitivity at an early stage. The ethylene receptor gene *RhETR3* and the transcription factor *RhEIN3* were used for development of CAPS (Cleaved Amplified Polymorphic Sequence) markers. Amplified fragments of *RhETR3* and *RhEIN3* cDNA were cloned, sequenced and digested with restriction enzymes specific for the recognition sites containing SNPs which were identified by comparing 'Vanilla' and 'Lavender' sequences. Although the SNPs located in restriction site resulted in different fragment sizes, the outcome were not correlated with their ethylene sensitivity. Since the plants are tetraploid, characterization of more alleles for the identified genes and use of more SNPs may be expected to find a useful marker system.

Aparna *et al.* (2019) assessed the genetic diversity among 25 fragrant rose cultivars of Indian and exotic origin which were characterized using 31 SSR markers to aid cultivar identification, conservation and frangrant rose breeding. A total of 96 alleles were detected among the test population with an average of 3.9 alleles per loci. From the results of diversity analysis, it is clear that the cultivars of Indian origin exhibited higher diversity compared to that of American and European origin. Effective number of alleles (Ne) and expected heterozygosity (He) values were more in Indian cultivars (1.93, 0.49) over American (1.45, 0.45) and European cultivars (1.29, 0.29). Matching coefficient values ranged from 0.55-0.86 implying moderate variability among the fragrant rose cultivars. Unweighted Pair Group Method using Arithmetic Mean (UPGMA) dendrogram grouped the 25 genotypes into 7 clusters. Rose Sherbet, Century Two Seedling and Brandy showed distant relationship and were sorted in cluster V, VI and VII, respectively. Highest similarity of 86 per cent was noted between Jawahar and Double Delight, whereas lowest similarity of 55 per cent with Century Two Seedling and Brandy.

Veluru *et al.* (2019) analysed diversity among 148 rose genotypes which included wild species and cultivated varieties of Indian and exotic origin using SSR primers. A total of 88 alleles were scored with 30 polymorphic loci which resulted in 2.9 ± 1 alleles per locus. Polymorphic information content (PIC) values ranged from 0.08 to 0.80 with the mean being 0.5±0.2. The neighbor-joining tree developed based on Nei's genetic distance values sorted the population into 3 major clusters, cluster I and II with modern rose cultivars of Indian origin and cluster III with exotic cultivars, wild species and some Indian cultivars. From the results of STRUCTURE analysis, four different sub-populations with estimates between 0.01 and 0.15 indicating low to moderate level of divergence were found in which some individual genotypes exhibited genomic admixture. Population differentiation among the genotypes was associated with their geographical location and lineages. In addition, results of AMOVA (Analysis of Molecular Variance) established 83.12 per cent variance within sub-groups, 10.42 per cent among populations and 6.46 per cent among individuals within groups.

Nguyen *et al.* (2020) performed genome-wide association study across 95 rose genotypes *in vitro* and *in vivo* to evaluate the rooting behavior with the help of genetic information from 68 K Axiom WagRhSNP chip. For *in vitro* parameters, 49 SNPs were significantly related to root length whereas for *in vivo* parameters, 98, 218 and 4 SNPs were obtained for root number, root length and root biomass, respectively. Few SNPs were present in genes with homology related genes such WUSCHEL-related homeobox 8-like and ETHYLENE INSENSITIVE 3-like for *invitro* root length; auxin response factor 19, protein AUXIN RESPONSE 4 and MYC2 transcription factor (AtMYC2) for *in vivo* root number and root length. On further mapping of SNPs to recently developed genome sequence, additional homologous genes like SCARECROW (SCR) on chromosome 3 and WUSCHEL-related homeobox genes on chromosome 1,4 and 6 were identified. These markers could help to verify and annotate the genes in other populations.

15.2.2.1 Detection of Pathogen

Ballard *et al.* (1995) compared RFLP and RAPD patterns and characterized 20 of the 22 cultivars. In addition, a genetic mapping programme was initiated to identify molecular markers associated with blackspot (*Diplocarpon rosae*) resistance in roses using genotypes of different ploidy level and differing in resistance/susceptibility. Preliminary RAPD analysis of the parental rose plants of the crosses showed a high degree of genetic polymorphism between all the resistant amphidiploids and susceptible tetraploid parent cultivars. They suggested that progeny produced from a cross between any 2 resistant and susceptible parents would be suitable for genetic map construction. Smulders *et al.* (2015) developed the WagRhSNP Axiom array in rose to develop genetic maps in tetraploid garden roses and cut roses which could be employed for inheritance studies as well. Jason *et al.* (2020) assessed the novelty of Rdr3 which remained unmapped among black spot resistant genes Rdr1, Rdr2, Rdr3 and Rdr4. A mapping population was developed by hybridizing Rdr3 consisting cultivar 'George Vancouver' and susceptible cultivar 'Modern Blush'. The mapping population was genotyped with WagRhSNP 68 K Axion array and mapped with the help of 'PolymapR' package. It was obvious from the results that Rdr3 was mapped to chromosome 6 homologs thus proving it distinct from Rdr1 and Rdr2 which are found on chromosome 1 and Rdr4 found on Chromosome 5. This mapping data was used along with *Rosa chinensis* genome assembly to generate novel tightly linked SSRs for marker-assisted breeding which obtained 3 markers to forecast the presence of Rdr3 gene. In addition, 12 cultivars possessed resistance genes other than Rdr3. Thus, the identified diagnositc markers could be used for developing black spot resistance varieties.

15.3.0 Genetic Transformation

While there is an abundance of literature available on *in vitro* regeneration of rose, information of transformation studies is scanty. Firoozabady *et al.* (1991) published a short report on successful *Agrobacterium*-mediated transformation of *R. hybrida* cv. Royalty. Transgenic rose plants were developed using an embryogenic callus line (Noriega and Sondahl, 1991). Friable embryogenic callus was cocultivated with *A. tumefaciens* strain LBA4404 carrying the binary vector (pnos NPTII/p35SLUC) or *A. rhizogenes* strain 15834 carrying vector (pnos NPTII/ p35SGUS), both of which contained a *nptII* and a *gusA* gene, followed by kanamycin selection. The transformation efficiencies were very high; about 40-60 independent kanamycin-resistant calli were produced for each gram of rose callus inoculated with *Agrobacterium*. Almost 100 per cent of the calli were found to be transformed based on GUS or LUS assays and PCR. From the transformed calli, transgenic somatic embryos were continuously produced. Mathews *et al.* (1991) also successfully reported *Agrobacterium*-mediated transformation of *R. persica* x *R. xanthina* using protoplasts of embryogenic cell lines on cocultivation with *A. tumefaciens* LBA-4404, followed by hygromycin selection.

Further, Firoozabady *et al.* (1994) reported regeneration of transgenic rose cv. Royalty from embryogenic tissues. Fifty to sixty transferred embryogenic callus lines were reproducibly obtained from each gram of friable embryogenic calli inoculated with *Agrobacterium*. Transformed embryogenic calli were transferred to maturation medium to form somatic embryos, which subsequently produced flowering plants. The transgenic nature of the plants was confirmed by enzyme assays and PCR. More than 100 transgenic plants were established in soil and flowered in the greenhouse.

Matsuda *et al.* (1996) bombarded somatic embryos obtained from leaf calli of rose (*R. hybrida*) with tungsten particles coated with plasmid DNA carrying the selec-table marker gene *nptII* and the *gus* reporter gene. The most efficient expression was achieved when tungsten particles coated with 8 g of DNA were bombarded. Rose embryos to which the plasmid was successfully transferred were selected in the presence of kanamycin. Marchant *et al.* (1998a) described a reproducible method for biolistic gene delivery into embryogenic callus of roses, using *R. hybrida* cv. Glad Tidings. DNA delivery was optimized by bombardment with the β-glucuronidase (*gus-uidA*) gene. All transgenic plants were regenerated successfully and were morphologically true-to-type.

The exact role and mode of action of the *rol* genes from *A. rhizogenes*, though not very clear, indicated marked implication on phenotype of the plant.

Van-der-Salm *et al.* (1998) introduced the *rolA*, *B* and *C* genes from *A. rhizogenes* in the cultivar Moneyway, for improving the root characteristics of the rootstock. Four independent *rolA* + *B* + *C* transformants were produced. One *rolC* expressing transformant, showing good growth and clearly decreased apical dominance, was selected for a grafting experiment with the cut rose cultivar Madelon as a scion. Grafting on this *rolA* + *B* + *C* transformed rootstock resulted in a stimulation of both root development of the rootstock and axillary bud release of the untransformed scion. Since basal shoots form flowers, it is suggested that this altered plant architecture should lead to a more efficient flower production.

Pale and friable calli obtained from immature seeds of self-pollinated miniature rose 'Shortcake' exhibited 50-60 per cent transient GUS expression in Agrobacterium-mediated transformation. Among 65 putative transgenic shoots grown *in vitro*, seven plants established and flowered in greenhouse with unstable flowers on primary shoots and flowers similar to non-transgenic regenerated plants on 2nd and later emerged shoots. In addition, the integration of T-DNA fragment in transgenic plants was confirmed by PCR-Southern blot analysis with GUS probe and with hygromycin phosphotransferase RNA probe (Asano and Tanimoto, 2003). Nakamura *et al.* (2011) developed a transgenic blue rose by expressing genes of viola F3'5'H and torenia anthocyanin 5-AT followed by environmental risk assessment for its further release. Results from biological assay indicated that absence of allelopathic compound production and gene flow towards wild roses and cultivated modern roses which might be due to that fact that the transgenes were present only on L_1 layer cells and not in L_2 cells which are responsible for reproductive cell generation thus supporting the commercial release of transgenic rose.

15.3.2 Resistance to Disease

Resistance to the fungal pathogen (*Diplocarpon rosae*) causing blackspot was introduced into rose cultivar Gland Tidings by Marchant *et al.* (1998b) using genetic transformation with a basic chitinase gene. Somatic embryogenic callus was initiated from leaf petioles of *in vitro* shoots. Transformation was performed using plasmid Z56 containing the –90 to –800 enhancer region of the CaMV 35S promoter upstream of the -160 to + 114 promoter and coding region of the rice chitinase gene together with *nos* terminator. Leaf disc from the transformed plants were inoculated with conidial suspension of *D. rosae*. Lesions on 65 per cent of the transformed discs had diameters significantly smaller than those of control lesions.

15.4.0 Other Molecular Studies

15.4.1. Cold Stress

Reis *et al.* (2020) assessed the genome-wide expression profiles in *Rosa hybrid* exposed to different low temperature treatments. Transcriptome profiles are determined in floral buds at 0, 2 and 12 h of low temperature (4° C). It is noticed that 134 and 169 transccripts were up- and down- regulated, respectively on low temperature exposure. Out of that, a total of 8 upregulated genes including 2 each of cytochrome p450 proteins, ankyrin repeat family proteins, metal ion binding proteins and zinc finger protein related transcription factors together with a downregulated

Disorders of Rose

Bull head

Bent neck

Harvesting and Grading of Rose

Harvesting

Tight bed stage

Pre cooling

Grading of rose

Packaging of Rose

In polythene packs

Packaging in bags

Packaging in Cardboad boxes

Marketing of Rose

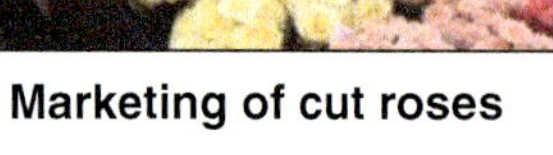

Marketing of cut roses

Loose rose for sale

Flower arrangement with rose

Exhibitions of cut rose

dynamin-like protein were detected. These genes which are involved in cold stress response were validated using qRT-PCR. In addition, the expression patterns of AP2/ERF gene family of transcription factors were analysed in floral buds and leaves which showed that the genes were rapidly induced in leaves than in floral buds. Likewise, differential expression of several such genes was detected earlier in vegetative rather than in reproductive tissues. These results emphasize the vital function of different low temperature response genes in facilitating cold acclimation, thereby rendering roses the ability to tolerate low temperatures without negatively influencing flower bud development and consequent flowering whilst the vegetative part experience premature adaptation to low temperatures.

15.4.2. Powdery Mildew

Qiu *et al.* (2015) notified that phylogenetically closely related *Rosa hybrid* Mildew Resistance Locus O (*RhMLO*) genes exhibited alike or overlapping tissue specificity and corresponding response to external stimuli. The *RhMLO1* and *RhMLO2* transcriptional levels were upregulated by additional 2-fold on inoculation with powdery mildew fungus, whereas *RhMLO3* and *RhMLO4* remained unaffected at early phases. Thus, it was concluded that *RhMLO1 and RhMLO2* might play vital role in rose-mildew interaction which could be beneficial to study the disease susceptibility mechanisms in rose.

Xiang *et al.* (2019) sequenced and assembled the transcriptome of susceptible *Rosa gigantea*and the resistant *Rosa longicuspis* using de novo assembly and scrutinized the mechanisms lying behind Rosa-powdery mildew pathogen interactions. The results discovered that active response to the pathogen occurred later in *R. gigantea* than in *R. longicuspis*. Also, the regulatory network of susceptible species was more complex and diverse compared to the resistant one. Moreover, two MLO members RgMLO6 and RlMLO7 were identified which possibly possessed vital role in rose-pathogen interaction. These outcomes could also pave route for studying the functional evolution of disease resistance genes as well developing new varieties resistant to powdery mildew pathogen.

15.4.3 Genome Sequencing

Like Arabdiopsis, rose is considered as model plant species especiallly to study the ornamental traits like recurrent blooming, flower morphogenesis, floral scent emission because of it small genome size (560 Mb). Various groups comprised in rose genetics and genomics congregated to obtain 'Rose Genome Sequence Initiative'. Foucher *et al.* (2015) attempted to acquire high quality rose genome sequence of diploid *Rosa chinensis* 'Old Bush' with the help of haploid production for overcoming the high level of heterozygosity. A genetic map was developed from a cross between 'Old Bush' and *Rosa wichuriana*. The genotype of 'Old Bush' was sequenced using NGS approach. The obtained data would be assembled and arranged with the aid of high density map and ESTs from various tissues of the cultivar would be developed to improve ESTs and to facilitate genome annotation. Thus, the genome sequence will be a grand move to detect the molecular basis of ornamental traits and to assess the genetic diversity and genome evolution in the genus *Rosa*.

Nakamura *et al.* (2018) determined the draft genome sequence of wild rose *Rosa multiflora* Thunb. using Illumina MiSeq and HiSeq platforms. Total scaffold length was 739 Mbp including 83,189 scaffolds which was near 711 Mbp determined by k-mer analysis. N50 length of the scaffolds was 90,830 bp and that of longest was 1.133 Mbp. Moreover, the average GC content of scaffolds was 38.9 per cent. Further, 67,380 candidates showing sequence homology to known genes and domains were isolated which consisted of complete and partial gene structures. The large number of genes identified for this diploid plant reflects heterogeneity of the genome originating from self-incompatibility in *R. multiflora*. Additionally, 91.9 and 98.0 per cent core eukaryotic genes were completely and partially conserved in the scaffolds according to CEGMA analysis. The results of this study would serve as valuable resource for breeding and phylogenetic studies of cultivated roses.

15.5.0 *In vitro* Conservation

Lynch *et al.* (1996) were successful in the cryopreservation of shoot-tips using encapsulation/dehydration procedure. The shoot-tips recovered from liquid nitrogen only grew after a 24 h pretreatment in medium containing 0.5 M sucrose followed by 2 h of drying with silica gel and rapid freezing. Jayasree and Devi (1997) reported production of synthetic seeds and plant regeneration in *R. hybrida* cv. King's Ransom. They could store the synthetic seeds in sterile dry petri dishes sealed with parafilm for a period up to 40 days at 4°C in the dark. The germination ability was higher on MS medium supplemented with 1.0 mg/L BA + 0.1 mg/L NAA.

16.0 References

Abidi, F., Douillet, O., Guillemain, G., Sintes, G., Huche-Thelier, L., Girault, T., Leduc, N., Laffaire, M., Ahmed, H. B. and Smiti, S. (2013) *Acta Hortic.*, **990**: 306-309.

Accati, E., Barni, E., Lavagno, M., Montoneri, E. and Savarino, P. (1992) *Adv. Hort. Sci.*, **6**: 155-159.

Agbaria, H., Heuer, B. and Zieslin, N. (1995) *J. Hort. Sci.*, **70**: 651-656.

Aguiar, E.L., Carvalho, G.A., Menezes, E.B. and Machado, C.A. (1993) *Anais da Sociedade Entomo-logica do Brasil*, **22**: 577-582.

Aguiar, E.L., Carvalho, G.A.De. and Menezes, E.B. (1995) *Anais da SociedadeEntomologica do Brasil*, **24**: 557-562.

Ahmadi, N., Mibus, H. and Serek, M. (2009) *Postharvest Biol. Tec.*, **52**(3): 260-266.

Ahn, G. (1997) *J. Korean Soc. Hort. Sci.*, **38**: 597-602.

Ahn, G. (1997) *J. Korean Soc. Hort. Sci.*, **38**: 803-807.

Ahn, G.Y. (1996) *J. Korean Soc. Hort. Sci.* **37**: 719-725.

Ahn, G.Y. and Park, J.C. (1996b) *J. Korean Soc. Hort. Sci.*, **37**: 598-602.

Ahn, K.Y. and Um, S.K. (1991a) *J. Korean Soc. Hort. Sci.*, **32**: 490-496.

Ahn, K.Y. and Um, S.K. (1991b) *J. Korean Soc. Hort. Sci.*, **32**: 497-505.

Akbar, S.M. and Rao, V.N.M. (1982) *South Indian Hort.*, **30**: 166-167.

Alaey, M., Babalar, M., Naderi, R. and Kafi, M. (2011) *Postharvest Biol. Tec.*, **61**(1): 91-94.

Aleksandrova, N.M., Kuznetsova, G.E. and Sergeev, B.F. (1977) *OzelenenieGorodovKrainSevera*, Apatity, USSR, pp. 61067.

Ali, M. A. and Mangrio, G. S. (2020) *Trakia J. Sci.*, **18**(1): 106-110.

Allen, E.F. (1959) *The Rose Annual*, The Royal National Rose Society, UK.

Allen, E.F. (1967) *The Rose Annual*, The Royal National Rose Society, UK.

Allen, R.C. (1956) *Rose for Every Garden*, M. Barrows and Co. Inc., New York.

Almeida, E. F. A., Manfredini, G. M., Paiva, P. D. O., Reis, M. V. D. and Maia, M. O. (2017) *Ornam.Hortic.*, **23**(2): 207-211.

Almeida, E. F. A., Oliveira, E. C., Carvalho, J. A., Mimura, S. N., Fuzzato, D. O., Lessa, M. A., Carvalho, L. M. and Figueiredo, J. R. M. (2015) *Acta Hortic.*, 1060.

Alp, S. and Koyuncu, M. (2015) *Acta Hortic.*, **1064**: 406-409.

Al-Salem, M. and Serek, M. (2017) *Sci.Hortic.*, **216**: 22-28.

Alsalem, M. and Serek, M. (2019) *Acta Hortic.*, **1256**: 222-228.

Alsemaan, T. (2013) *Int. J. Agric. Res.*, **8**(4): 172-177.

Al-Yasi, H., Attia, H., Alamer, K., Hassan, F., Ali, E., Elshazly, S., Siddique, K. H. M. and Hassin, K. (2020) *Plant Physiol. Biochem.*, **150.**

Amariutei, A. and Panait, E. (1981) *LucrariStiintifice, Institutul de CeretarisiProiectariPentru-ValorificareasiIndustrializareaLegumelorsiFructelor,* 12, pp. 61-70.

Andersson, N.E. (1994) *Acta Agric. Scandinavica Section B, Soil and Plant Sci.*, **44**: 244-247.

Anil Kumar, SnilSood, Palni, U.T., Gupta, A.K. and Palni, L.M.S. (2001) *J. Hort. Sci. and Biotech.*, **76**: 30-34.

Anonymous (1975) *American Rose Annual*, American Rose Society, **60**: 31-33.

Anonymous (2017) *Horticultural Statistics at a Glance 2017*, Government of India, Ministry of Agriculture and Farmers Welfare, Department of Agriculture, Cooperation and Farmers Welfare, pp. 514.

Aparna, V., Bhat, K. V., Janakiram, T., Prasad, K. V., Raju, D. V. S., Sapna, P., Namitha, Singh, K. P. (2019) *Indian J. Agric. Sci.*, **89**(11).

Arene, L., Pellegrino, C. and Gudin, S. (1993) *Euphytica*, **71**: 83-92.

Arisumi, K. (1963) *Sci. Bull. Fac. Agric. Kyushu Univ.*, **30**: 131-149.

Arnold, N.P., Barthakur, N.N. and Tanguay, M. (1998) *Hort Science*, **33**: 127-129.

Asano, G. and Tanimoto, S. (2003) *Plant Biotechnol.*, **20**(4): 291-296.

Asaoka, M. and Heins, R.D. (1982) *J. Amer. Soc. Hort. Sci.*, **107**: 548-552.

Asen, S. and Hamner, C.L. (1953) *Bot. Gaz.*, **115**: 86-89.

Asfanani, M., Davarnejad, G. H. and Tehranifar, A. (2008) *Acta Hortic.*, **804.**

Ashok, A.D., Arun, D.S. and Rengasamy, P. (1999) *South Indian Hort.*, **47**: 115-118.

Atanasov, Z. and Cingova, B. (1967) *Rasten. Nauki*, **4**: 81-87.

Attia, A. O., Dessoky, E. S. and El-tarras, A. E. (2012) *African J. Biotechnol.*, **11**(48): 10888-10893.

Avramis, T., Hugard, J. and Jonard, R. (1982) *ComptesRendus des Seances de l'Academie des Sciences, III Sciences de La Vie*, **294**: 679-682.

Azeem, S., Khan, A. I., Awan, F. S., Riaz, A. and Bahadur, S. (2012) *African J. Biotechnol.*, **11**(47): 10650-10654.

Baille, A., Kittas, C. and Katsoulas, N. (2001) *Agric. For.Meteorol.*, **107**: 293-306.

Bakley, S.A. (1974) *Agric. Res. Rev. Cairo*, **52**: 95-99.

Baksh, M. M. A. M. A., Javaid, S. A. M. A. and Rehman, A. (2015) *J. Agric. Res.*, **53**(2).

Balazsovits, J., Dansereau, B., Gagnon, S. and Beeson, R.C.Jr. (1989) *Can. J. Plant Sci.*, **69**: 979-989.

Ballard, R., Rajapakse, S., Abbott, A. and Bryne, D.H. (1995) *Acta Hortic.*, **424**: 265-268.

Baloiu, I. (1974) *Productia Veg. Hort.*, **23**: 30-34.

Banda, G.K. and Banerjee, S.K. (1967) *J. Palyn.*, **4**: 36-41.

Barman, D. and Rajni, K. (2006) *J. Ornam.Hortic.*, **9**(4): 262-265.

Barthe, P., Vaillant, V. and Gudin, S. (1991) *Acta Hortic.*, **298**: 61-68.

Barton, A.J. (1975) *American Rose Annual*, American Rose Society, **60**: 89-94.

Bastos, F. E. A., Lerin, S., Allebrandt, R., Kretzschmar, A. A. and Rufato, L. (2017) *Ornam.Hortic.*, **23**(1): 58-62.

Baydar, H. and Baydar, N. G. (2005) *Ind. Crops Prod.*, **21**: 251-255.

Baydar, H., Erbas, S. and Kazaz, S. (2016) *Turkish J. Agric. Fores. Sci.*, **40**(4): 560-569.

Belanger, R.R., Labbe, C. and Jarvis, W.R. (1994) *Plant Dis.*, **78**: 420-424.

Bello, P. L. D., Ho, T. and Tzanetakis, I. E. (2015) *Acta Hortic.*, 1064.

Benedetto, A.Di, Befumo, M., Rossi, G. and Boschi, C. (1995) *Revista de la Facultad de Agronomia*, **14**: 229-234.

Ben-Meir, H., Scovel, G., Ovadis, M. and Vainstein, A. (1997) *Acta Hortic.*, **447**: 599-601.

Berg, G.A.V.D. and Valentin, J.C.M. (1975) *Vakbladvoor de Bloemisterij*, **30**: 10-11.

Bergh, C. J. and Blanc, J. R. L. (1997) *J. Econ.Entomol.*, **90**(2): 679-688.

Berninger, E., Barrade, R. and Pudelska, K. (1990) *Agronomie*, **10**: 787-796.

Beuzenberg, M.P. (1972) *Michigan Florist*, No. 500, p. 3.

Bharadwaj, R., Singh, S. K., Pal, S. and Kumar, S. (2006) *J. Ornam.Hortic.*, **9**(4): 238-242.

Bhat, M.S. (1992) *Indian Hort.*, **37**: 17-19.

Bhattacharjee, S.K. (1998) *Ann. Plant Physiol.*, **12**: 161-163.

Bhattacharjee, S.K. and Bose, T.K. (1979) *Haryana J. Hort. Sci.*, **8**: 209-215.

Bik, R.A. (1961) *Schweiz. Gartnerztg.*, **64**: 113-117.

Bik, R.A. (1970) *Vakbladvoor de Bloemisterij*, **25**: 478-479.

Bik, R.A. (1972) *Vakbladvoor de Bloemisterij*, **27**: 8-9.

Bini, G., Leva, A.R.C. and Nicese, F.P. (1983) *Riv. Ortoflorofruttic. Ital.*, **67**: 1-13.

Biolley, J.P., Jay, M. and Barbe, J.P. (1992) *Biochem. System. Ecol.*, **20**: 697-705.

Bisht, D. Singh, C.P. Kumar, S. and Singh, N. (2013) *Progress.Hortic.*, **45**(2).

Biswas, T.D. (1984) *Rose Growing, Principles and Practices*, Associated Publishing Company, New Delhi.

Blindeman, L. (1998) *Verbondsnieuws*, **42**: 26-27.

Blindeman, L. (2000) *Verbondsnieuws*, **44**: 18.

Bloemhard, C.M.J. and Moolenbrock, J. Van (1995) *Acta Hortic.*, **401**: 481-490.

Blom, TJ. (1999) *Acta Hortic.*, **481**: 503-507.

Bluel, S. (1990) *Pflanzenschutzberichte*, **51**: 25-35.

Blumel, S. and Hausdorf, H. (1996) *Anzeiger fur Schadlingskunde, Pflanzenschulz, Umweltschulz*, **69**: 64-69.

Bo, J., Huiru, D. and Xiaohan, Y. (1995) *Acta Hortic.*, **404**.

Boelema, B.H. (1973) *Phytophylactia*, **5**: 7-12.

Bolivar, P., Fischer, G., Florez, V.J. and Mora, A. (1999) *Acta Hortic.*, **482**: 83-87.

Boodley, J.W. (1975) *HortScience*, **10**: 41-42.

Bornemann, E. and Muhe, R. (1967) *DtscheGartenb.*, **14**: 328-329.

Borrelli, A. (1978) *Ann. Fac. Sci. Agrar. Napoli*, **12**: 172-181.

Bose, T. K. (1984) *Indian Agric.*, **28**: 53-72

Bose, T.K. and Mukherjee, D. (1977) *Gardening in India*, Oxford and IBH Publishing Co., Calcutta, Bombay, New Delhi.

Bose, T.K., Yadav, L.P. and Maity, R.G. (1986) *Proc. 22nd International Hort. Congr.*, Davis, California, Abstract No. 1148.

Bouranis, D., Ottosen, O. C., Giday, H., Fanourakis, D., Korner, O. and Hyldgaard, B. (2019) *Eur. J. Hortic. Sci.*, **84**(4): 245-252.

Bout, A., Boll, R., Mailleret, L. and Poncet, C. (2010) *J. Econ.Entomol.*, **103**(6): 2242-2248.

Bowen, K.L. and Roark, R.S. (2001) *Plant Dis.*, **85**: 393-398.

Bredmose, N. (1979) *Gartner Tidende*, **45**: 701.

Bredmose, N. (2000) *Acta Hortic.*, **515**: 105-109.

Bredmose, N. and Hansen, J. (1996) *Ann. Bot.*, **78**: 215-222.

Bredmose, N. and Hansen, J. (1997) *Acta Hortic.*, **440**: 99-104.

Bredmose, N., Hansen, J. and Nielsen, J. (1999) *Int. J. Plant Sci.*, **160**: 819-825.

Bredmose, N.B. (1998) *J. Amer. Soc. Hort. Sci.*, **123**: 569-576.

Bressan, P.H., Y.J., Hyndman, S.E., Hasegawa, P.M. and Bressan, R.A. (1982) *J. Amer. Soc. Hort. Sci.*, **107**: 979-990.

Broughton, S., Cousins, D. A. and Rahman, T. (2015) *Crop Prot.*, **67**: 130-135.

Brun, R. and Morisot, A. (1996) *Acta Hortic.*, **417**: 55-63.

Brun, R., Paris, B. and Hammelin, I. (1993) *Adv. Hort. Sci.*, **7**: 145-148.

Brunke, E.J., Hammerschmidt, F.J. and Schmaus, G. (1992) *FlavourFragr. J.*, **7**: 195-198.

Buck, G.J. (1979) *HortScience*, **14**: 196-197.

Bunt, J.A. (1972) *Mededelingen van de FaculteitLandbouwwetenschappen, Rijksuniversiteit*, Gent, **37**: 798-807.

Burrell, A. M., Lineberger, R. D., Rathore, K. S. and Byrne, D. H. (2006) *HortScience*, **41**(5): 1165-1168.

Butcaru, A. C., Stancia, F., Matei, G. M. and Matei, S. (2019) *Acta Hortic.*, 1242.

Byrne, T.G., Doss, R.P. and Tse, A.T.Y. (1978) *J. Amer. Soc. Hort. Sci.*, **103**: 500-502.

Cabrera, R.I. (2000) *Acta Hortic.*, **511**: 133-140.

Cabrera, R.I., Evans, R.Y. and Paul, J.L. (1993) *Sci. Hort.*, **53**: 333-345.

Carelli, B. P. and Echeverrigaray, S. (2002) *Sci. Hort.*, **92**(1): 69-74.

Carletti, M.G., Leoni, S., Pisanu, B. and Grudina, R. (1992) *ColtureProtette.*, **21**: 101-106.

Carow, B. and Sommer, G. (1962) *DeutscherGartenbau*, **33**: 1177-1178.

Carpenter, W.J. and Anderson, G.A. (1972) *J. Amer. Soc. Hort. Sci.*, **97**: 331-334.

Carpenter, W.J. and Waston, D.P. (1955) *American Rose Annual*, American Rose Society, **40**: 45-54.

Carrier, L.E. (1951) *Proc. Amer. Soc. Hort. Sci.*, **57**: 381-386.

Carrier, L.E. (1952) *Proc. Amer. Soc. Hort. Sci.*, **59**: 501-508.

Carrier, L.E. (1953) *Proc. Amer. Soc. Hort. Sci.*, **61**: 573-580.

Chadwick, L.C. (1950) *Proc. 13th International Hort. Congr.*, London.

Champneys, H.P. (1956) *Pearson's Encyclopaedia of Roses, C. Arthur Pearson Ltd.*, London.

Chan, A.P. (1961) *Flor. Exch.*, **136**: 56-58.

Chandler, E.L. and Watson, D.P. (1954) *Proc. Amer. Soc. Hort. Sci.*, **64**: 441-447.

Cha-um, S. and Kirdmanee, C. (2010) *J. Hortic. Sci. Biotechnol.*, **85**(6): 533-538.

Chen, X., Baldermann, S., Cao, S., Lu, Y., Liu, C., Hirata, H. and Watanabe, N. (2015) *Plant Physiol.*, **87**: 109-114.

Chen, X., Kang, Y., Wan, S., Chu, L. and Li, X. (2015) *Agric. Waste Mngg.*, **158**: 99-111.

Chimonidou-Pavlidou, D. (1998) *Acta Hortic.*, **358**: 95-102.

Ching, F., Hammer, C.L. and Widmoyer, F.B. (1956) *Quart. Bull. Mich. Agric. Exp. Stat.*, **39**: 3-9.

Cho, H.K. and Lee, J.M. (1979) *J. Korean Soc. Hort. Sci.*, **20**: 106-110.

Choudhury, B., Malik, R.S. and Singh, A.P. (1978) *First All India Rose Convention*, Commemoration volume, pp. 22-25.

Chu, C. and Lai, C. (1997) *J. Agri. and Forestry*, **46**: 19-29.

Chung, S., Kim, W., Park, M. and Park, Y. (1997) *J. Korean Soc. Hort. Sci.*, **38**: 766-770.

Chung, S.K., Kim, W.H. and Park, Y.J. (1996) *Proc. 9th Inter. Cong. Soilless Culture, St. Helier, Jersey, Channel Islands*, April 12-19, pp. 123-127.

Chung, S.K., Park, Y.J., Kim, W.H., Oh, Y.N., Lee, E.K. and Kwack, B.H. (1998) *J. Korean Soc. Hort. Sci.*, **39**: 203-206.

Cid, M.C., Caballero, M. and Reimann-Philipp, R. (1989) *Acta Hortic.*, **246**: 345-351.

Cikalov, P.M. (1962) *Vestn. Sel'-hoz. Nauki*, **7**: 127-131.

Clark, D.G., Kelly, J.W. and Pemberton, H.B. (1991) *HortScience*, **26**: 1195-1197.

Cobb, G.S., Hanan, J.J. and Baker, R. (1978) *HortScience*, **13**: 464-466.

Cocetta, G., Trivellini, A. and Ferrante, A. (2019) *Acta Hortic.*, 1257.

Costa, J. M., Challa, H., Meeteren, U. V. and Pol, P. A. (2001) *Acta Hortic.*, **547.**

Crespel, L., Pernet, A., Bris, M. L., Gudin, S. and Oyant, L. H. S. (2009) *Plant Breed.*, **128**(5): 501-506.

Cushman, L.C., Pemberton, H.B., Miller, J.C. Jr. and Kelly, J.W. (1998) *HortScience*, **33**: 736-740.

Da, P. F. J. L., Giuliano, G., Bruno, C. F. and Adriano, T. L. (2014) *RevistaBrasiliera de Engenharia Agricola e Ambiental*, **18**(1): 85-89.

Da, S. M. N., Correia, S. M., de Melo, B. M. S., Ferreira, S. J. H., Coelho, R. H. L., Pereira, R. A. R., Da, S. M. A. and Zunete, B. M. (2019) *Ciencia Rural*, **49**(12).

Dadmal, S.M., Ghawade, S.M., Taral, B.W., Mahajan, R.K. (1999) *Agri. Sci. Digest (Karnal)*, **19**: 172-174.

Dale, P.S. (1973) *New Zealand J. Exp. Agric.*, **1**: 121-122.

Dambre, P., Blindeman, l. and Labeke, M.C. van (2000) *Acta Hortic.*, **513**: 129-135.

Damke, M.M., Deshmukh, P.S., Chatterjee, S.R. and Bhattacharjee, S.K. (1995) *Gujrat Agril. Univ. Res. J.*, **20**: 112-118.

Darlington, A.B., Dixon, M.A. and Tsujita, M.J. (1992) *Sci. Hort.*, **49**: 291-303.

Das, P and Rout, G. (2000) In: *Biotechnology of Horticultural Crops*, vol. 3 (Eds. V.A. Parthasarathy, T. K. Bose, P. Das) pp. 1-28.

Davies, F.T.Jr., Fann, Y.S. and Lazarje, J.E. (1980) *HortScience*, **15**: 817-818.

De Pietro, J., Mattiuz, B. H., Rodrigues, T. J. D. and Mattiuz, C. F. M. (2012) *Acta Hortic.*, **934.**

De Vries, D. P. and Dubois, L. A. M. (1992) *Acta Hortic.*, 320.

De, L.C. and Bhattacharjee, S.K. (1999) *Orissa J. Hort.*, **27**: 8-10.

Debener, T., Bretzke, M., Dreier, K., Linde, M., Kaufmann, H., Berger, R. G. and Krings, U. (2010) *Acta Hortic.*, **870.**

Debener, T., Malek, B. V., Mattiesch, L. and Kaufmann, H. (2001) *Acta Hortic.*, **547**.

Debener, T., Malek, B. V., Schreiber, M. and Drewes-Alvarez, R. (2003) *Eur. J. Hortic. Sci.*, **68**(6).

Demircioglu, H., Dundar, O. and Ozkaya, O. (2018) *Acta Hortic.*, **1194.**

Depauw, B. (1975) *Bull. Colo. Flower Grs' Ass.*, No. 300, pp. 1-3.

Deshmukh, A. P., Kharbade, S. B. and Kulkarni, K. V. (2017) *J. Entomol. Res.*, **41**(3).

Dhara Vir and Sharma, R.K. (1985) *Indian Hort.*, **29**: 14-15.

Dinock, A.W. (1951) *Phytopathology*, **41**: 781-784.

Dixon, M.A. and Peterson, C.A. (1989) *Sci. Hort.*, **38**: 277-285.

Dobson, H.E.A., Bergstrom, G. and Groth, I. (1990) *Israel J. Bot.*, **39**: 143-156.

Dood, J.De and Rademaker, J. (1991) *VakbladVoor de Bloemisterij*, **46**: 52-55.

Doorn, W.G. Van and Vojinovic, A. (1996) *Ann. Bot.*, **78**: 619-623.

Doorn, W.G. Van, Groenewegen, G., Pol, P.A. Van De and Berkholst, C.E.M. (1991) *Postharvest Biol. Technol.*, **1**: 47-57.

Dougall, B.M. (1959) *The Roe Annual*, The Royal National Rose Society, UK.

Douglas, G. C., Rutledge, C. B. and Casey, A. D. (1989) *Plant Cell Tissue Organ Cult.*, **19**: 55-64.

Downer, A. J., Howell, A. D. and Karlik, J. F. (2015) *Acta Hortic.*, **1064.**

Dubois, A., Raymond, O., Maene, M., Baudino, S., Langlade, N. B., Boltz, V., Vergne, P. and Bendahmane, M. (2010) *PLoS One.*

Dubois, L.A.M. and Vries, D.P.De (1987) *Euphytica*, **36**: 535-539.

Dubois, L.A.M. and Vries, D.P.De (1994) *Gartenbauwissenschaft*, **59**: 18-20.

Dubois, L.A.M., Roggemans, J., Soyeurt, G. and Vries, D.P.De (1988) *Sci. Hort.*, **35**: 293-299.

Durkin, D. and Kiplinger, D.C. (1960) *Flor. Exch.*, **134**: 17, 51.

Dutta, S.K. and Singh, M.S. (1999) *Scientific Hort.*, **6**: 157-164.

Elad, Y. and Volpin, H. (1991) *Plant Pathol.*, **40**: 278-286.

Elad, Y., Kirshner, B. and Gotlib, Y. (1993) *Crop Prot.*, **12**: 69-73.

El-Deeb, B., Bazaid, S., Gherbawy, Y. and Elhariry, H. (2012) *J. Plant Interact.*, **7**(3): 248-253.

Elenes-Fonseca, C. and Hanan, J.J. (1987) *Res. Bull., Colorado Greenhouse Growers' Asso.*, **450**: 1-4.

El-Gamassy, A., El-Hakim, S. and El-Shafie, S. (1960) *Ann. Agric. Sci.*, Cairo, **5**: 159-170.

El-Shafie, S.A. (1978) *Arch. Gartenb.*, **26**: 287-296.

El-Shafie, S.A., El-Kholy, S.A. and Afify, M.M. (1980) *Monufeia J. Agric. Res.*, **3**: 291-310.

Elsharkawy, M. M., Hase, T., Shimizu, M. and Hyakumachi, M. (2015) *African J. Biotechnol.*, **14**(42): 2917-2927.

Ercisli, S. (2007) *Food Chem.*, **104**: 1379-1384.

Etemadi, N., Shams, M. and Baninasab, B. (2012) *Acta Hortic.*, **937.**

Fairbrother, F. (1965) *Roses*, Penguin, Great Britain.

Fanourakis, D., Carvalho, D. R. A., Gitonga, V. W., Heusden, A. W. V., Almeida, D. P. F., Heuvelink, E.and Carvalho, S. M. P. (2012) *Acta Hortic.*, **937.**

Fanourakis, D., Carvalho, S. M. P., Almeida, D. P. F., Kooten, O. V., Doorn, W. G. V. and Heuvelink, E. (2012) *Postharvest Biol. and Tec.*, **64**(1): 64-73.

Faragher, T.D. and Mayak, S. (1984) *J. Expt. Bot.*, **35**: 965-974.

Farahi, M. H. and Jowkar, M. M. (2016) *Acta Hortic.*, 1131.

Farina, E., Rodrigues, E.J.R., Paterniani, T. and Palagi, M. (1998) *ColtureProtett.*, **27**: 93-100.

Fascella, G. and Zizzo, G. V. (2005) *Acta Hortic.*, **697.**

Fascella, G., Zizzo, G. V. and Agnello, S. (2007) *Acta Hortic.*, **751.**

Ferreira, D.I. and Swardt, G.H.De (1980) *Agroplantae*, **12**: 53-59.

Ferreira, D.I. and Swardt, G.H.De (1981) *Agroplantae*, **13**: 73-81.

Field, R.P. and Hoy, M.A. (1984) *Calif. Agric.*, **38**: 29-32.

Field, Z. (1969) *Book of Roses*, Paul Hamlyn, London.

Firoozabady, E., Lemieux, C.S., Moy, Y.S., Moll. B., Nicholas, J.A. and Robinson, K.E.P. (1991) *In vitro*, **27**: 96A.

Firoozabady, E., Moy, Y., Couson, N. and Robinson, K. (1994) *Biotechnol.*, **12**: 609-613.

Fischer, C. (1953) *Proc. Amer. Soc. Hort. Sci.*, **61**: 585-589.

Fjeld, T., Gislerod, H.R., Revhaug, V. and Mortensen, L.M. (1994) *Sci. Hort.*, **57**: 157-164.

Flint, M. L. and Dreistadt, S. H. (2005) *Biol. Control*, **34**(1): 38-46.

Foucher, F., Oyant, L. H., Hamama, L., Sakr, S., Nybom, H., Baudino, S., Caissard, J. P., Byrne, D. M., Sumlder, J. M. S., Desnoye, B., Debener, T., Bruneau, A., Riek, J. D., Matsumoto, S., Torres, A., Millan, T., Amaya, I., Yamada, K., Wincker, P., Zamir, D., Gouzy, J., Sargent, D., Bendahmane, M., Raymond, O., Vergne, P., Dubois, A. and Just, J. (2015) *Acta Hortic.*, 1064.

Franco-Hermida, J.J., Quintero, M.F., Cabrera, R.I. and Guzman, J.M. (2013) *PLoS One*, **12**: e0178500.

Franco-Hermida, J.J., Quintero, M.F., Guzmán, G. M. and Cabrera, R (2020) *Sci. Hort.*, **264**: 109094.

Gabriel, G. (1979) *Gb + GW*, **79**: 14.

Gachomo, E. W., Dehne, H. W. and Steiner, U. (2009) *Ann. Appl. Biol.*, **154**(2): 259-267.

Gamboa, B.S., Larroque, O.R. and Ronco, B.L. (1986) *Revista de la Facultad de Agronomia Universidad Nacional de La Plata*, **61/62**: 175-179.

Gao, J., Zhang, X., Huang, M., Ye, X. and Sun, Z. (1997) *Acta Hortic. Sinica*, **24**: 274-278.

Gao, Y. and Wu, S.J. (1990) *Hort. Sinica*, **17**: 71-75.

Gardes, L., Hizmann, P. and Joyaux, F. (2005) *Eur. J. Hortic. Sci.*, **70**(4).

Garello, G., Menard, C., Dansereau, B. and Page-degivry, M.T.Le (1995) *Plant Growth Regul.*, **16**: 135-139.

Gault, S.M. and Synge, P.M. (1971) *The Dictionary of Roses in Colour*, Rainbird Publishing Group Ltd., London.

Geerdink, G. M., Orsi, B., Tezotto-Uliana, J. V., Pessoa, C. O., Sasaki, F. F. C. and Kluge, R. A. (2020) *J. Plant Nutr.*, **43**(10).

Geers, F.A.M. (1978) *Vakbladvoor de Bloemisterij*, **33**: 22-23.

Genders, R. (1965) *The Rose: A Complete Handbook*, The Bobbs-Merril Co. Inc., New York.

Gherghi, A., Amariutei, A. and Baloiu, I. (1983) *LucrariStiintificeInstitutul de CercitarisiProiectariPentruValorificareasiIndustrializareaLegumelor. Si Fructelor*, **14**: 125-129.

Ghose, K., Mccallum, J., Fillmore, S., Kirby, C., Sanderson, K., Joly, S., Bruneau, A. and Fofana, B. (2013) *Sci. Hort.*, **160**: 251-263.

Ghosh, S.N. (1976) *Rose World*, **1**: 27-29.

Gill, A.P.S. (1984) *Indian Hort.*, **29**: 37-42.

Ginzburg, C., Gilad, S. and Ackerman, A. (1983) *Hassadeh*, **63**: 768-769.

Giovannini, A., Macovei, A., Dona, M., Valassi, A., Caser, M., Mansuino, A., Ghione, G. G., Carbonera, D., Scariot, V. and Balestrazzi, A. (2015) *Acta Hortic.*, **1064.**

Gislerod, H.R. and Mortensen, L.M. (1997) *Acta Hortic.*, **418**: 25-31.

Gislerod, H.R., Fjeld, T. and Mortensen, L.M. (1993) *Acta Hortic.*, **342**: 51-60.

Gitari, J. M., Maingi, J. M. and Onyango, B. O. (2018) *J. Pharm. Biol. Sci.*, **13**(6): 25-33.

Glenn, W.K.Hr., Baxter, L.W. and Witcher, W. (1972) *Phytopathology*, **62**: 803.

Gojniczek, B. and Laszez, E. (1974) *PraceInstytutuSadownictwa W. Skierniewicach*, No. 4, pp. 8-18.

Gopalaswamiengar (1970) *Complete Gardening in Indiai*, 3rd edition, Kalyani Printers, Bangalore.

Gorin, N., Dreise, G.r., Lukaszewska, A. and Perezzuniga, F.J. (1989) *Acta Hortic.*, **251**: 381-388.

Gorjackina, N.A. (1967) *Bjull. glav. bot. Sada*, No. 64, pp. 32-34.

Goszczynska, D., Itzhaki, H., Borochov, a. and Halevy, A.H. (1990) *Sci. Hort.*, **43**: 313-320.

Goszczynska, D.M., Michalczuk, B.R. and Rudnicki, R.M. (1989) *Acta Hortic.*, **261**: 281-286.

Goszczynska, D.M., Ziezlin, N., Mor, Y. and Halevy, A.E. (1990) *Plant Growth Regul.*, **9**: 293-303.

Gough, N. (1991) *Exp. Appl.Acarol.*, **12**: 83-101.

Gowda, J.V.N. (1984) *The Indian Rose Annual*, **3**: 54-55.

Gowda, J.V.N. and Jayanthi, R. (1985) *The Indian Rose Annual*, **4**: 185-187.

Gowda, J.V.N., Gopal, A. and Sulladmath, U.V. (1984) *The Indian Rose Annual*, **3**: 8-61.

Gregory, T. (1984) *The Pocket Encyclopaedia of Modern Roses*, Blandford Press, Poole, Dorset, UK.

Gribovskaya, I.F., Ugulaya, N.A. and Karyakin, A.V. (1978) *Agrokhimiya*, No. 4, pp. 108-111.

Grossi, J. A. S., Pemberton, H. B. and Lang, H. J. (2004) *HortScience*, **39**(1): 138-141.

Grueber, K.L. and Hanan, J.J. (1981) *Res. Bull. Colorado Greenhouse Growers' Association*, No. 371, p. 1.

Grzesik, M. and Rudnicki, R.M. (1989) *Acta Hortic.*, **251**: 411-414.

Gudin, S. (1992) *Postharvest Biol. Tec.*, **2**: 155-161.

Gupta, S.K. Gupta, Y.C and Thakur, R. (2004) *J. Ornam. Hort.*, 7(3-4): 366-369.

Gupta, V. N., Chakrabarty, D. and Datta, S. K. (2007) *J. Ornam. Hort.*, **10**(2): 122-124.

Gurav, S.B., Singh, Katwate B.R. Sable, S.M.Kakade, D.S. and Dhane, A.V. (2005) *J. Ornam. Hort.*, 8(3): 228-229

Haenchen, E. (1967) *Grower*, **71**: 655-666.

Hahn, F. (1959) *Gartenwelt*, **59**: 298.

Hajizadeh, H., Razavi, K., Mostofi, Y., Cacco, G., Mousavi, A., Zamani, Z. and Stevanato, P. (2011) *Iran. J.Biotechnol.*, 9(3): 230-233.

Halevy, A.H., Byrne, T., Kofranek, A., Farnham, D., Thomson, J. and Hardenburg, R. (1978) *J. Amer. Soc. Hort. Sci.*, **103**: 151.

Hammer, P.E. and Evensen, K.B. (1996) *J. Amer. Soc. Hort. Sci.*, **121**: 314-320.

Hammer, P.E. and Marois, J.J. (1989) *J. Amer. Soc. Hort. Sci.*, **114**: 100-106.

Hammer, P.E., Evensen, K.B. and Janisiewicz, W.J. (1993) *Plant Dis.*, **77**: 283-286.

Hammer, P.E., Yang, S.F., Reid, M.s. and Marois, J.J. (1990) *J. Amer. Soc. Hort. Sci.*, **115**: 102-107.

Han, Y., Sim, Y., Song, I., Yoon, J. and Choi, B. (1996) *J. Korean Soc. Hort. Sci.*, **37**: 99-150.

Han, Y., Sim, Y., Woo, J., Choi, K. and Choi, B. (1999) *J. Korean Soc. Hort. Sci.*, **40**: 727-730.

Han, Y., Woo, J., Sim, Y., Choi, K. and Choi, B. (1997) *J.Korean Soc. Hort. Sci.*, **38**: 527-532.

Hanan, J.J. (1978) *Bull. Colo. Flower Grs'. Ass.*, No. 337, pp. 1-4.

Hanan, J.J. (1979) *J. Amer. Soc. Hort. Sci.*, **104**: 37-40.

Hand, D.W. and Cockshull, K.E. (1975) *ADAS Quarterly Review*, No. 7, pp. 26-32.

Harkess, R.L. and Hanan, J.J. (1988) *Res. Bull.*, Colorado Greenhouse Growers' Assoc., **461**: 1-3.

Harkness, J.L. (1979) *The World's Favourite Roses and How to Grow Them*, McGraw Hill Book Company, New York, San Francisco.

Harranger, J. (1984) *Phytoma*, No. 355, pp. 36-37.

Hartmann, H.T. and Kester, D.E. (1972) *Plant Propagation, Principles and Practices*, 3rd edition, Prentice-Hall of India Pvt. Ltd., New Delhi.

Hasegawa, P.M. (1979) *HortScience*, **14**: 610-612.

Hasegawa, P.M. (1980) *J. Amer. Soc. Hort. Sci.*, **105**: 216-220.

Hasek, R.F. (1981) *Rose*, In: *Introduction of Floriculture* (Ed. R.A. Larsen), Academic Press.

Hashemabadi, D. (2012) *Acta Hortic.*, **937.**

Hassanein, A. M. A. (2010) *World J. Agric. Sci.*, **6**(3): 260-267.

Hassanein, A. M. A. and Mahmoud, I. M. A. (2018) *Hortic. Sci.*, **45**(2): 83-91.

Hattendorf, A., Linde, M., Mattiesch, L., Debener, T. and Kaufmann, H. (2004) *Acta Hortic.*, **651.**

Hayashi, I. (1976) *Bull. Kanagawa Hort. Expt. Stat.*, No. 23, pp. 54-63.

Hazar, D. and Baktir, I. (2005) *Acta Hortic.*, **690.**

He, Y., Chen, Y. S., Wang, Z. H., Wang, L. P., Wang, G. P., Hong, N. and Xu, W. X. (2016) *Plant Dis.*, **100**(6): 1252.

Heatly, S. (1959) *The Rose Annual*, The Royal National Rose Society, UK.

Hemant, U., Singh, A., Ahlawat, T. and Palagani, N. (2016) *J. Ornam. Hort.*, **19**(1-2).

Hendriks, L., Scharpf, H.C. and Hackbarth, H.J. (1987) *Deutscher Gartenbau*, **41**: 272-276.

Hessayon, D.G. (1981) *The Rose Expert*, pbi Publication, England.

Holton, T.A. and Tanaka, Y. (1994) *Trends Biotechnol.*, **12**: 40-42

Hong, F. and Zhao, H. (1999) *Acta Hortic. Sinica*, **26**: 62-64.

Hoof, H.A. van and Caron, J.E.A. (1975) *Mededelingen van de Faculteit Land bowwetenschappen, Rijksuniversiteit Gent.*, **40**: 759-763.

Hopper, D.A. (1996) *Hort. Technol.*, **6**: 76-79.

Hopper, D.A. and Hammer, P.A. (1991) *J. Amer. Soc. Hort. sci.*, **116**: 609-617.

Hopper, D.A., Hammer, P.a. and Wilson, J.r. (1994) *J. Amer. Soc. Hort. Sci.*, **119**: 903-914.

Horibe, T., Yamada, K., Otagaki, S., Matsumoto, S. and Kawamura, K. (2015) *Acta Hortic.*, **1064.**

Hoyer, L. and Salhoj, K. (2001) *Acta Hortic.*, **543**: 119-124.

Huang, S., Gong, B., Wei, F. and Ma, H. (2017) *Hortic. Environ. Biotechnol.*, **58**(2): 144-151.

Hunt, P. (1973) *150 of The World's Most Beautiful Roses*, Marshall Cavendish Ltd., London.

Hurst, M. (1929) *The Rose Annual*, The Royal National Rose Society, UK.

Ibrahim, R. and Debergh, P.C. (2001) *Sci. Hort.*, **88**: 41-57.

Ichimura, K., Mukasa, Y., Fujiwara, T., Kohata, K., Goto, R. and Suto, K. (1999) *Ann. Bot.*, **83**: 551-557.

Irulappan, I., Muthuswami, S. and Pappiah, C.M. (1983) *South Indian Hort.*, **31**: 305-306.

Ishida, A., Masui, M., Nukaya, A. and Ogura, T. (1979) *J. Japanese Soc. Hort. Sci.*, **47**: 517-523.

Ishida, A., Shigeoka, H., Nukaya, A. and Masui, M. (1978) *Bull. Fac. Agric.*, Shizuoka Univ., No. 28, pp. 1-6.

Iwata, H., Gaston, A., Remay, A., Thouroude, T., Jeauffre, J., Kawamura, K., Oyant, L. H., Araki, T., Denoyes, B. and Foucher, F. (2012) *Plant J.*, **69**: 116-125.

Jabbarzadeh, Z., Khosh-Khui, M., Salehi, H. and Saberivand, A. (2010) *Afr. J. Biotechnol.*, **9**(37): 6091-6095.

Jackson, G.A.D. and Blundell, J.B. (1964) *Nature*, **202**: 1027-1028.

Jackson, G.A.D. and Blundell, J.B. (1966) *Nature*, **212**: 1470-1471.

Jackson, G.A.D. and Prosser, M.V. (1959) *Naturwissenschaften*, **46**: 407-408.

Jaffrin, A. and Champeroux, A. (1994) *Acta Hortic.*, **361**: 258-266.

Jagdish, E.J and A. P. Purnima (2011)*J.Biopest.*, **4**(1): 81 – 85.

Jakobsen, J. (1975) *TidsskrPlanteavl*, **79**: 489-494.

Jambor-Benczur, E., Terek, O., Mathe, A. and Boronkay, G. (2010) *Acta Hortic.*, **877.**

Jan, C.H., Byrne, D.H., Manhart, J. and Wilson, H. (1999) *HortScience*, **34**: 341-345.

Jasmine, R.S., Kuttalam, S. and Stanley, J. (2008) *Ann. Plant Prot. Sci.*, **16**(1)

Jason, Z. D., David, Z. C., Matthew, H., James, B. M., Stan, H. C. and Nahla, B. V. (2020) *Theor. Appl. Genet.*, **133**(6).

Jayalaxmi, N.H., Chakravarthy, A. K. Kumar, N.G. Ashok Kumar, C.T.Thyagaraj, N.E. Jayanthi, R. and Surendra, H.S. (2016) *Entomon.*, 41(3): 215-226.

Jayasree, N. and Devi, B.P. (1997) *Indian J. Exp. Biol.*, **35**: 310-312.

Jayroe-Cournoyer, L. and Newman, S.E. (1995) *J. Enbviron. Hort.*, **13**: 47-50.

Jenks, M. A., Andersen, L., Teusink, R. S. and Williams, M. H. (2001) *Physiol. Plant.*, **112**(1): 62-70.

Jensen, H.E.K. (1979a) *Tidsskr. Planteavl*, **83**: 432-440.

Jensen, H.E.K. (1979b) *Gartner Tidende*, **50**: 772-773.

Jensen, H.E.K. (1980a) *Tidsskr. Planteavl*, **84**: 229-236.

Jensen, H.E.K. (1982) *GB + GW*, **82**: 99-101.

Jian, H. Y., Min, T., Ting, Z., Li, S. B., Zhang, H. and Tang, K. X. (2013) *Acta Hortic.*, **977.**

Jian, H., Li, S., Guo, J., Li, S., Wang, Q., Yan, H., Qiu, X., Zhang, Y., Cai, Z., Volis, S. and Tang, K. (2018) *Conserv. Genet.*, **19**(4): 761-776.

Jiang, Z., Qi, P.K., Li, X.F., Li, M.Z. and Qiu, H.S. (1996) *Acta PhytophylacicaSin.*, **23**: 115-120.

Jiao, J. and Grodzinski, B. (1998) *J. Amer. Soc. Hort. Sci.*, **123**: 1081-1088.

Jiao, J., Tsujita, M.J. and Grodzinski, B. (1991a) *Can. J. Plant Sci.*, **71**: 235-243.

Jiao, J., Tsujita, M.J. and Grodzinski, B. (1991b) *Can. J. Plant Sci.*, **71**: 253-261.

Jicknor, R.L. and Roberts, A.N. (1967) *Comb. Proc. Int. Plant. Prop. Soc.*, **17**: 397-403.

Jodo, S., Kato, M., Fujioka, M. and Fukuyama, T. (1989) *Memoirs College Agri. Ehime Uni.*, **33**: 181-188.

Johansson, J. (1978) *Acta Agric. Scand.*, **28**: 363-386.

Johansson, J. (1979) *Acta Agric. Scand.*, **29**: 191-208.

Junaid, A.H.M. and Nasir, M.M. (1956) *Agric. Pakist.*, **6**: 80-103.

Kahrizi, Z. A., Kermani, M. J., Amiri, M. E. and Vedadi, S. (2011) *Acta Hortic.*, **923.**

Kaicker, U.S. (1984) *Indian Hort.*, **29**: 47-41.

Kajihara, S., Nobuyasu, H., Fujiwara, T., Kunita, H. and Koguchi, T. (2005) *Hortic. Res.: Japan*, **4**(2): 181-186.

Kakade, D.S. Gurav, S.B. Singh, B.R. and Nimbalker, C.A. (2006) *J. Ornam. Hort.*, 9(4): 293-295.

Kalbhor, H.B. Patil, D.E. and Patil, A.V. (2015) *Asian J. Soil Sci.*, 10(2): 283-287.

Kaminski, W. (1989) *Acta Hortic.*, **251**: 407-410.

Kang, G., Park, W., Yan, J. and Kang, C. (1998) *Farm Mngg. Agri. Engg.*, **40**: 193-199.

Karadi, R. V. and Patil, V. S. (2006) *J. Ornam. Hort.*, **9**(2): 126-129.

Kazemi, M., Abdossi, Jari, S. K. and Moghadam, A. R. L. (2018) *J. Hortic. Sci. Biotechnol.*, **93**(1): 81-90.

Kazuo, I., Masayuki, K., Ryo, N., Yoshihiko, K. and Kunio, Y. (2005) *J. Hortic. Sci. Biotechnol.*, **80**(3): 280-286.

Kengond, S. R., Bintory, M. A., Ibaad, M. H., Humajahan, S. V. and Bawoor, S. (2015) *Trends Biosci.*, **8**(11): 3000-3003.

Khatun, S., Bandyopadhyay, P. K. and Chatterjee, N. C. (2009) *Asian J. Exp. Sci.*, **23**(1): 249-252.

Khayat, E. and Zieslin, N. (1982) *J. Expt. Bot.*, **33**: 1286-1291.

Khoshgoftarmanesh, A. H., Khademi, H., Hosseini, F. and Aghajani, R. (2008) *J. Plant Nutr.*, **31**(9): 1543-1554.

Khosh-Khui, M. and Sink, K.C. (1982) *Sci. Hort.*, **17**: 371-376.

Khosh-Khui, M. and Zargarian, M. (2010) *Acta Hortic.*, **870.**

Khristova, D. (1974) *VirusniBolesti po Gasteniyata*, Sofia, Bulgaria, pp. 77-87.

Kim, K., Kim, W. and Kim, S. (1997) *J. Korean Soc. Hort. Sci.*, **38**: 592-596.

Kim, K., Kim, W. and Kim, S. (1998) *J. Korean Soc. Hort. Sci.*, **39**: 98-102.

Kim, Y.J. and Byrne, D.H. (1996) *HortScience*, **31**: 1207-1209.

Kiplinger, D.C. and Haramaki, C. (1955) *Ohio Fm. And Home Res.*, **40**: 55.

Kiplinger, D.C. and Nelson, K.S. (1957) *Ohio Fm. And Home Res.*, **42**: 70.

Kireeva, M. F. (1967) *I. V. Micurina*, **12**: 110-13.

Klemenko, V. (1969) *The Rose Annual, The Royal National Rose Society*, U.K.

Kljuka, V. I. (1963) *Fiziol. Rast.*, **10**: 595-597.

Klougart, A. and Fairbrother, F. (1966) *The Culture History of the Roses*, In: Methuen Hand Book of Roses, Methuen and Co.

Knoblauch, F. (1971) *Tidsskr. Planteavl*, **75**: 293-300.

Kohl, H. C. and Post, K. (1952) *Proc. Amer. Soc. Hort. Sci.*, **59**: 527-30.

Kool, M.T.N. and Pol, P.A. Van De (1991) *Vakb. Voor de Bloemisterij*, **46**: 62-64.

Kool, M.T.N. and Pol, P.A. Van de (1992) *Gartenbauwissenschaft*, **57**: 120-125.

Kore, S. S., Nirmal, D. D. and Ansarwadckar, K. W. (1977) *South Indian Hort.*, **25**: 167.

Koseva, D. and Decheva, R. (1976) *Fiziol. Rast.*, **23**: 537-42.

Koths, J.S. (1973) *Connecticut Greenhouse Newsletter*, No. 54, p. 11.

Kreij, C.De, Berg, T.J.M. Van Den (1989) *Proc. Xith International Plant Nutrition Colloquium, Wageningen, Netherlands*, 30 July – 4 August, pp. 519-523.

Kromwijk, A., Mourik, N.Van, Beelen, C. and Rademaker, J. (1990) *VakbladVoor de Bloemisterij*, **45**: 40-41.

Kuiper, D., Reenen, H.s.Van and Ribot, S.A. (1997) *Postharvest Biol. Tec.*, **9**: 75-86.

Kukolev, P. A. (1952) *Agrobiologija*, No. 4, pp. 133-4.

Kumar, S., Prasad, K. V., Kumar, S., Chauhan, V. S. and Choudhary, M. L. (2005) *Indian J. Hortic.*, **62**(4).

Kumar, S. M. and Ponnuswami, V. (2014) *Green Farming*, **5**(5): 854-857.

Kumar, S. S., Shelke, S. S. and Bhojne, I. (2009) *J. Entomol. Res.*, **33**(3).

Kumari, S., Singh, K. P., Singh, S. K., Kumar, S. and Sarkhel, S. (2017) *Indian J. Hortic.*, **74**(2).

Kumbhar, A. V., Yamgar, S. V. and Pawar, A. U. (2015) *J. Agric. Veter. Sci.*, **8**(1): 39-40.

Kumud, S., Hem, P. and Vijay, R. (2015) *J. Environ. Resear. and Devel.*, **10**(1): 40-46.

Kyalo, T.M., Pemberton, H.B. and Zajicek, J.M. (1996) *HortScience*, **31**: 120-122.

Labeke, M.C. van, Dambre, P. and Bodson, M. (2000) *Acta Hortic.*, **515**: 245-255.

Laird, G., John, P., Pearson, S. (2005) *Acta Hortic.*, **687**: 602-610.

Lal, S. D. Seth, J. N., Yadav, J. P. and Danu, N. S. (1982) *Progress. Hort.*, **14**: 234-6.

Langhans, R. W. (1957) *Diss. Abstr.*, **17**: 10.

Larkin, P.J. and Scowcroft, W.R. (1981) *Theor. Appl. Genet.*, **60**: 197-214.

Lata, P. (1971) *Curr. Sci.*, **40**: 4-6.

Lata, P. (1983) *Environ. Exp. Bot.*, **23**: 85-92.

Laurie, A. and Ries, V. H. (1950) *Floriculture Fundamentals and Practices*, McGraw-Hill Book Company, Inc., New York.

Laurie, A., Kiplinger, D. C. and Nelson, K. S. (1958) *Commercial Flower Forcing, 6th edition*, McGraw-Hill Book Co., Inc., New York.

Ledlie, R. (1923) A Handbook on Rose Culture in India, Thacker Spink and Co., Calcutta.

Lee, J. S. and Kim, Y. R. (1982) *J. Korean Soc. Hort. Sci.*, **23**: 141-62.

Lenander, S. E. (1956) *Medd. Tradgarsfors. Malmo*, **62**: 32.

Lenander, S. E. (1960) *SartrycksserTradgardsfors. Alnarp*, **117**: 13.

Lenander, S. E. (1968) *Gartneryket*, **58**: 59-66.

Lenander, S. E. (1969) *Viola*, **75**: 2-7.

Leus, Jeanneteau, F., Huylendroeck, J. V., Bockstaele, E. V. and Riek, J. D. (2004) *Acta Hortic.*, **651.**

Levi-Minzi, R., Cardelli, R., Saviozzi, A., Riffaldi, R. and Bassi, S. (2000) *ColtureProtette*, **29**: 79-83.

Leyhe, V. and Horn, W. (1994) *Gartenbauwissenschaft*, **59**: 85-88.

Li, X., Kim, S. J., An, S. H. and Kim, S. Y. (2002) *Plant Biotechnol.*, **19**(4): 271-275.

Li, X., Wan, S., Kang, Y., Chen, X. and Chu, L. (2016) *Agric. Water Mngg.*, **163**: 180-189.

Lian, L., Wang, F., Zhang, Y., Fang, R. and Liu, Q. (2016) *Eur. J. Hortic. Sci.*, **81**(2): 122-132.

Lieth, H. (1996) *Grower Talks*, **60**: 42-50.

Lindemann, A. (1972) *Gartenwelt*, **72**: 409-11.

Lindstrom, R. S. (1965) *Proc. Amer. Soc. Hort. Sci.*, **87**: 521-4.

Lloyd, D., Roberts, A.V. and Short, K.C. (1988) *Euphytica*, **37**: 31-36.

Lukaszewska, A.J., Barthe, P. and Page-devigry, M.T.Le (1995) *Acta Hortic.*, **405**: 422-426.

Lukaszewska, A.J., Bianco, J., Barthe, P. and Page-degivry, M.T.Le. (1994) *Plant Growth Regul.*, **14**: 119-126.

Lukaszewska, A.J., Tonecki, J., Woltering, E.J. and Gorin, N. (1990) *Gartenbauwissenschaft*, **55**: 118-121.

Lukaszewska, J. and Gorin, N. (1988) *Bull. Polish Acad. Sci. Biolo. Sci.*, **36**: 113-119.

Lukszewska, A. and Gorin, N. (1989) *Acta Hortic.*, **261**: 197-200.

Lundstad, A. (1956) *Gartneryrket*, **46**: 877-84.

Lundstad, A. (1957) *Gartneryrket*, **47**: 366-7.

Lynch, P.T., Harris, W.C. and Chareir-Hollis, J.M. (1996) *Plant Growth Regul.*, **20**: 43-45.

M.G.D. (1983) *Verbond. Voor de Belgis. Siert.*, **27**: 907, 909.

Maas, E.V. (1990) *Crop salt tolerance*, In: Tanji, K.K. (ed.). Agricultural salinity assessment and management. Amer. Soc. Of Civil Engineers, ASCE Manuals and reports on engineering practice No. 71. New York, NY, pp. 262–304.

Maas, F.M. and Bakx, R.J. (1995) *J. Amer. Soc. Hort. Sci.*, **120**: 571-576.

Maatsch, R. (1967) *Gartenwelt*, **67**: 248-9.

Madhupriya (2017) *J. Hortic. Sci. Biotechnol.*, **92**(4): 424-4431.

Mahr, J. and Hann, J. J. (1980) *Res. bull. Colorado Greenhouse Growers' Association*, No. 365, pp. 1-3.

Malik, R. S. (1980) *Nat. Sem. Prodn. Tech. comm. Flower Crops*, Coimbatore.

Malik, R. S. and Dadlani, N. K. (1984) *Indian Hort.*, **29**: 27-30.

Malik, R. S., Singh, B. and Dadlani, N. K. (1984) *Indian Hort.*, **29**: 9-13, 15, 22.

Maljaja, *E.G.E.P.* (1997) *J. Myco. Plant Path.*, **27**: 93-94.

Mannion, C. M., Derksen, A. I., Seal, D. R., Osborne, L. S. and Martin, C. G. (2013) *Florida Entomologist*, **96**(2): 403-411.

Mantur, S.M., Bagali, A.N. and Patil, S.R. (2005) *J. Farm Sci.*, **18**(2).

Marchant, R., Davey, M.R., Lucas, J.A. and Power, J.B. (1996) *Plant Sci.*, **120**: 95-105.

Marchant, R., Davey, M.R., Lucas, J.A., Lamb, C.J., Dixon, R.A. and Power, J.B. (1998b) *Breed.*, **4**: 187-194.

Marchant, R., Power, J.B., Lucs, J.A. and Davy, M.R. (1998a) *Ann. Bot.*, **81**: 109-114.

Marczynski, S. and Michlewicz, A. (1988) *Acta Hortic.*, **226**: 675-677.

Marissen, N. (2001) *Acta Hortic.*, **543.**

Marshall, H. H., Campbell, C. G. and Collicutt, L. M. (1983) *Euphytica*, **32**: 205-216.

Martin, C., Carre, M. and Vernoy, R. (1981) *C.R. Acad. Sci. Paris III*, **293**: 175-177.

Martin, M., Piola, F., Chessel, D., Jay, M. and Meizmann, P. (2001) *Theor. Appl. Genet.*, **102**: 398-404.

Marziano, F., Calarese, S. and Stefanis, D. (1973) *AnnalidellaFacolta di ScienczeAgrariedella University degliStudi di Napoli, Portici*, **7**: 42-56.

Mashhoud, M. M., Aelaei, M. and Mortazavi, S. N. (2016) *Acta Hortic.*, 1131.

Masse, R. J. and Vulinec, K. (2010) *Northeastern Naturalist*, **17**(4): 647-658.

Mastalerz, J. W. (1959-60) *Science for the Farmer*, **7**: 7.

Mastalerz, J. W. (1965) *Proc. Amer. Soc. Hort. Sci.*, **87**: 825-830.

Mathews, D., Mottley, J., Horan, I. and Roberts, A.V. (1991) *Plant Cell Tissue Organ Cult.*, **24**: 173-180.

Mathews, D., Mottley, J., Yokoya, K. and Roberts, A.V. (1994) In: *Biotechnology in Agriculture and Forestry*, Vol. 29 (Ed. Y.P.S. Bajaj), Springer-Verlag, Heidelberg, pp. 146-160.

Mathis, V. (1972) *Bull. Colo. Flower Grs'. Ass.*, No. 270, pp. 1-3.

Matsuda, Y., Yoyoda, H., Nakahara, T., Ogata, Y. and Ouchi, S. (1996) *Bull. Inst. Comprehensive Agril. Sci. Kinki University*, **4:** 23-28.

Matsumoto, S., Yamada, K., Shiratake, K., Koketsu, T., Negishi, H., Taneda, A., Fukui, H. and Ueda, Y. (2010) *Acta Hortic.*, **870.**

Maurya, R. P., Godara, N. R. and Yadav, R. C. (2004) *J. Ornam. Hort.*, **7**(2): 169-173.

Mazaro, S. M., Bertoldo, E., Dalacosta, N. L., Borsatti, F. C., da Cruz, M. P., Junior, A. W. and Haas, J. (2018) *J. Agric. Sci.*, **10**(8).

McClellan, W. D. (1949) *Plant Dis. Reptr.*, **33**: 137-138.

McFadden, S. E. (1956-57) *Proc. Fla. St. Hort. Soc.*, **69:** 368-370.

McFadden, S. E. (1963-64) *Proc. Fla. St. Hort. Soc.*, **76**: 412-416.

McFadden, S. E. and Poole, R. T. (1968) *Proc. Fla. St. Hort. Soc.*, **80**: 480-486.

McFarland, J.H. (1940) *Modern Roses II*, Macmillan Co.

McMahon, M.J. and Kelly, J.W. (1990) *J. Environ. Hort.*, **8**: 209-211.

Mehran, A., Hossein, D. G. and Tehranifar, A. (2008) *Acta Hortic.*, **804**: 503-509.

Meir, S., Droby, S., Davidson, H., Alsevia, S., Cohen, L., Horev, B. and Phlosoph-Hadas, S. (1998) *Postharvest Biol. Tec.*, **13**: 235-243.

Meir, S., Droby, S., Kochnek, B., Salim, S. and Philosoph-Hadas, S. (2005) *Acta Hortic*, **669**.

Meir, S., Reuveni, Y., Rosenberger, I., Davidson, H. and Philosoph-Hadas, S. (1994) *Acta Hortic.*, **368**: 355-364.

Menard, C. and Dansereau, B. (1995) *Sci. Hort.*, **64**: 117-132.

Meneve, I. and Moermans, R. (1977) *Rev. del' Agric.*, **32**: 1173-1182.

Mesmaeker, G.De (1991) *Combined Proc. Inter. Plant Propa. Soc.*, **40**: 251-254.

Michalczuk, B., Goszczynsky, D.M., Rudnicki, R.M. and Halevy, A.H. (1989) *Israel J. Bot.*, **38**: 209-215.

Miller, P. M. and Waggoner, P. E. (1962) *Amer. Nurserymen*, **115**: 150.

Minas, G. J. (2007) *Acta Hortic.*, **755.**

Misra, R. L. (1979) *The New Zealand Rose Annual*, pp. 58-66.

Misra, R. L. (1981) *The Lal-Baugh*, **26**: 40-48.

Mitachi, S., Sasaki, K., Kondoh, M. and Sugimoto, I. (2005) *Acta Hortic.*, **679.**

Mitali, M. and Kumar, P. P. (2020) *Acta Physiol. Plant.*, **42**(2).

Moe, R. (1972) *J. Amer. Soc. Hort. Sci.*, **97**: 796-800.

Mohapatra, A. and Rout, G. R. (2006) *Biol. Plant.*, **50**(2): 295-299.

Monder, M. J. and Pacholczak, A. (2020) *Biol. Agric. Hortic.*, **36**(1).

Mor, Y. and Halevy, A. H. (1984) *Physiol. Plant.*, **61**: 119-124.

Mor, Y., Johnson, F. and Faragher, J.D. (1989) *Acta Hortic.*, **261**: 271-279.

Moreno, M.T., Torres, A.M., Millan, T., armada, J. and Cubero, J.I. (1996) *Acta Hortic.*, **424**: 293-295.

Morey, D. (1960) *Proc. 10thAnnu. Mtg. Plant Prop.Soc.*, : 267-273.

Morisot, A. (1996) *Acta Hortic.*, **417**: 127-138.

Morrison, L. S. (1982) *Res. Rep. Agric Exp. Stn.*, Oklahoma State Univ., No. 826, pp. 20.

Mortensen, L. M. and Moe, R. (1983) *MeldingerfraNorgeslandbrukshogskole*, **62**: 11.

Mortensen, L.M. and Fjeld, T. (1998) *Sci. Hort.*, **73**: 229-237.

Mortensen, L.M. and Gislerod, H.r. (1994) *Gartenbauwissenschaft*, **59**: 275-279.

Mortensen, L.M. and Gislerod, H.R. (1996) *Gartenbauwissenschaft*, **61**: 211-214.

Mortensen, L.M. and Gislerod, H.R. (1999) *Sci. Hort.*, **82**: 289-298.

Morteza, K. (2014) *Int. J. Hortic. Sci.Tec.*, **1**(1): 20.

Moshe, S., Sharoni, S., Olga, L., Einat, B. Dalit, K., Zach, A., Dani, Z., Alexander, V., David, W., Uzi, R. and Efraim, L. (2004) *Israel J. Plant Sci.*, **52**(3): 245-255.

Mosher, J. M. and Turner, D. W. (2000) *Sci. Hort.*, **83**(3-4): 311-324.

Mosher, J.M. and Turner, D.W. (2000) *Sci. Hort.*, **83**: 311-324.

Mosher, J.M., Turner, D.W. (1999) *J. Hor. Sci. Biotechnol.*, **74**: 721-728.

Mottley, J., Yokoya, K., Mathews, D., Squirrell, J., Wentworth, J.E. and Morisot, A. (1996) *Acta Hortic.*, **424**: 393-397.

Mueller, R., Andersen, A.S., Serek, M. and Stummann, B.M. (2001) *Acta Hortic.*, **543**: 153-157.

Mukhambetzhanov, S., Nam, S., Mursalieva, V. and Rakhimbaev, I. (2011) *Afr. J.Biotechnol.*, **10**(46): 9344-9346.

Mukhopadhyay, A. and Banker, G. J. (1982) *South Indian Hort.*, **30**: 61-63.

Mukhopadhyay, A. and Banker, G. J. (1983) *Indian Hort.*, **82**: 8-9.

Mukhopadhyay, A. and Banker, G. J. and Shukla, K. S. (1980) *Indian Hort.*, **25**: 2-3.

Muller, R., Andersen, A.S., Stummann, B.M. and Serek, M. (1999) *Dordrecht, Netherlands, Kluwer Academic Publishers*, pp. 443-444.

Muller, R., Sisler, E.C. and Serek, M. (2000) *Sci. Hort.*, **83**: 51-59.

Muller, R., Stummann, B. M. and Andersen, A. S. (2001) *Sci. Hort.*, **91**(3-4): 325-337.

Muller, R., Stummann, B. M., Sisler, E. C. and Serek, M. (2001) *Eur. J. Hortic. Sci.*, **66**(1): 34-38.

Murphy, B.C., Morisawa, T.A., Newman, J.P., Tjosvold, S.A. and Parrella, M.P. (1998) *California Agric.*, **52**: 32-36.

Muthukrishnan, T. S., Chandrasekaran, J., Lakshmanan, P. L. and Chinnarajan, A. M. (1975) *South Indian Hort.*, **25**: 75-76.

Nabigol, A. (2012) *Afr. J. Biotechnol.*, **11**(5): 1078-1083.

Nabigol, A., Naderi, R. and Mostofi, Y. (2010) *Acta Hortic.*, **858.**

Nagarajaiah, C., Reddy, T.V. and Seenappa, K. (1989) *Progress. Hort.*, **21**: 83-88.

Nagaraju, C. G., Reddy, T. V. and Madaiah, D. (2003) *Journal of Ornamental Horticulture*, **6**(4): 287-293.

Nagesh, M. and Akiram, T.J. (2004) *J. Ornam.Hortic.*, **7**(2): 147-152.

Naing, A. H., Jeon, S. M., Park, J. S. and Kim, C. K. (2016) *Can. J. Plant Sci.*, **96**: 503-510.

Nair, R.V., Mathew, T.B. and Visalakshi, A. (1990) *Entomon.*, **15**: 273-275.

Nair, V.V., Reghunath, P. and Visalakshi, A. (1991) *Entomon.*, **16**: 327-329.

Nakamura, N., Fukuchi-Mizutani, M., Katsumoto, Y., Togami, J., Senior, M., Matsuda, Y., Furuichi, K., Yoshimoto, M., Matsunaga, A., Ishiguro, K., Aida, M., Tasaka, M., Fukui, H., Tsuda, S., Chandler, S. and Tanaka, Y. (2011) *Plant Biotechnol.*, **28**(2): 251-261.

Nakamura, N., Hirakawa, S., Sato, S., Otagaki, S., Matsumoto, S., Tabata, S. and Tanaka, Y. (2018) *DNA Res.*, **25**(2): 113-121.

Nalwadi, U. G. (1981) *The Lal-Baugh*, **26**: 58-62.

Nanjan, K. and Muthuswami, S. (1974) *South Indian Hort.*, **22**: 73-76.

Nanjan, K. and Muthuswami, S. (1975) *South Indian Hort.*, **23**: 94-99.

Nanjan, K., Hiriyan, M. A. and Ravindran, R. (1971) *Madras Agric. J.*, **58**: 526-528.

Narayanswami, V. and Biswas, K. (1957) *Rose Growing Centres and Rose Industry in India*, C.S.I.R., New Delhi.

Nazarenko, L. G. and Zhalnina, L. S. (1975) *Trudy po PrikladnoiBotanike, GenetikeSelektsii*, **54**: 226-231.

Nelson, P.E. and Wilhelm, S. (1958) *Phytopathology*, **48**: 613-616.

Ng, K.K., MacDonald, L. and Punja, Z.K. (1997) *HortScience*, **32**: 262-266.

Nguyen, T. H. N., Tanzer, S., Rudeck, J., Winkelmann, T. and Debener, T. (2020) *Sci. Hort.*, **266**.

Nicetic, O., Watson. D. M. and Beattie, G. A. C. (1999) *Plant Prot. Quart.*, **14**: 131-135.

Nichols, L.P. and Nelson, P.E. (1974) *Bull. Pennsylvania Flower Growers*, No. 273, p. 5.

Nielsen, B. and Starkey, K.R. (1998) *Acta Hortic.*, **456**: 215-221.

Nielsen, B. and Starkey, K.R. (1999) *Postharvest Biol. Tec.*, **16**: 157-167.

Nikolov, A. (1992) *Rasteniev'dniNauki*, **29**: 3-6.

Nirit, B., Asher, B. T., Haya, F., Pini, S., Ilona, R., Amram, C. and Marina, I. (2006) *Sci.Hortic.*, **108**(2): 185-193.

Nitransky, S. (1974) *TechnickychInformacij*, **2**: 98-102.

Noodezh, H. M., Moieni, A. and Baghizadeh, A. (2012) *In Vitro Cell. Dev. Bio. – Plant*, **48**: 530-538.

Noriega, C. and Sondahl, M.R. (1991) *Biotechnol.*, **9**: 991-993.

Norman, A. (1955) *Successful Rose Growing*, W.H. and L.C. Ltd., London.

Nyalala, S. and Grout, B. (2007) *Sci. Hortic.*, **114**(3): 194-198.

Nyalala, S. and Grout, B. W. W. (2015) *Acta Hortic.*, **1064.**

Ochoa, R., Aguilar, H. and Merind, F.L. (1990) *AgronomiaCostarricense*, **14**: 103-108.

Ogilvie, I.S. and Arnold, N.P. (1993) *HortScience*, **28**: 680.

Ogilvie, I.S., Arnold, N.P. and Cloutier, D.C. (1991) *HortScience*, **26**: 1344-1345.

Ohkawa, K. (1979) *HortScience*, **14**: 612-613.

Ohkawa, K. (1988) *Acta Hortic.*, **226**: 567-572.

Ohkawa, K. and Suematsu, M. (1999) *Acta Hortic.*, **482**: 47-51.

Ohkawa, K., Kasahara, Y. and Suh, J.N. (1999) *HortScience*, **34**: 112-113.

Olech, M., Nowak, R., Nowacka, N., Pecio, L., Oleszek, W., Los, R., Malm, A. and Rzymowska, J. (2015) *Ind. Crops Prod.*, **69**: 129-136.

Oliver, R.W. (1954) *Garden Rose Growing*, Canada Dept. Agric., p. 23.

Oloo-Abucheli, G., Aguyoh, J. N. and Gaoquiong, L. (2015) *Acta Hortic.*, **1077.**

Oprel, L. (1988) *Prophyta*, **42**: 59-60.

Os, V., Maas, A. A. D., Meijer, R. J. M., Khodabaks, M. R., Blok, C. and Enthoven, N. L. M. (2012) *Acta Hortic.*, **927**.

Otagaki, S., Ogawa, Y., Oyant, L. H., Foucher, F., Kawamura, K., Horibe, T. and Matsumoto, S. (2015) *Plant Biol.*, **17**(4): 808-815.

Overmeer, W.P.J., Zon, A.Q. Van and Helle, W. (1975) *Entomologia Experimentalis et Applicata*, **18**: 68-74.

Owen, F.E. (1976) *The Rose Annual*, The Royal National Rose Society, UK.

Oyant, L. H., Crespel, L., Rajapakse, S., Zhang, L. and Foucher, F. (2008) *Tree Genet. Genom.*, **4**(1): 11-23.

Pal, B.P. (1969) *Indian J. Hort.*, **26**: 35-41.

Pal, B.P. (1972) *The Rose in India*, Indian Council of Agricultural Research, New Delhi.

Pal, B.P. (1978) *All About Roses*, Vikas Publishing House Pvt. Ltd., New Delhi.

Pal, R. K., Lahiji, M. F. and Roy, S. K. (2003) *J.Ornam.Hortic.*, **6**(2): 130-133.

Pandey, R., Chacko, P. M., Prasad, K. V., Pal, M. and Choudhary, M. L. (2009) *J. Hortic. Sci.Biotechnol.*, **84**(1): 35-40.

Paramagoudar, P., Patil, S. and Chougala, V. (2014) *Trends Biosci.*, **7**(14).

Park, B. (1956) *Guide to Roses*, Collins, London.

Park, J., Kim, S., Kim, D. and Cho, K. (2002) *J. Asia-Pacific Entomol.*, **5**(1): 97-102.

Parrella, M.P., Damm-Kattari, D. Van, Damm-Kattari, G. Van and Murphy, B. (1999) *Grower Talks*, **63**: 62-90.

Parups, E.V. (1971) *HortScience*, **6**: 456-457.

Pascale, S. De and Paradiso, R. (1999) *ColtureProtette*, **28**: 85-91.

Pasian, C.C. and Lieth, J.H. (1989) *J. Amer. Soc. Hort. Sci.*, **114**: 581-586.

Pasian, C.C. and Lieth, J.H. (1996) *Sci. Hort.*, **66**: 117-124.

Paskalev, G. (1983) *Rasten. Nauki*, **20**: 67-71.

Paskalev, G. and Tsachev, S. (1982) *Rasten. Nauki*, **19**: 36-39.

Paterson, D.R., Lyle, E.W. and Earhart, D.R. (1975) *HortScience*, **10**: 381.

Patil, M.T., Singh, B.R. and Bluijbal, B.G. (1994) *J. Mahatma Agril. Univ.*, **19**: 291-292.

Patke, Y. N. Gawande, M. B. and Budhvat, K. P. (2015) *Trends Biosci.*, **8**(18): 4855-4860.

Pearson-Mims, C.H. and Lohr, V.I. (1990) *HortScience*, **25**: 1270-1271.

Penningsfeld, F. (1966) *Proc. 17th International Hort. Cong. Md.*, **1**: 491.

Penningsfeld, F. (1968) *Wiss. Z. Humboldt Univ.*, Berlin, **17**: 261-265.

Perez-Olvera, M.A. and Lozoya Saldana, H. (1994) *RevistaChapingo, Serie Horticultura*, **1**: 71-74.

Pessala, T. (1977) *Annales AgriculturaeFenniae*, **6**: 72-79.

Pettersen, R. I., Moe, R. and Gislerod, H. R. (2007) *Sci. Hort.*, **114**(3): 207-213.

Pipino, L., Scariot, V., Labeke, M. C. V. and Leus, L. (2013) *Acta Hortic.*, 990.

Pirone, P.P., Bernard, O.D. and Rickett, H.W. (1960) *Diseases and Pests of Ornamental Plants*, Constable and Co. Ltd., London.

Pizzol, J., Nammour, D., Hervouet, P., Bout, A., Desneux, N. and Mailleret, L. (2010) *J. Pest Sci.*, **83**: 191-196.

Plakhova, T.M. (1975) *ByulletenVsesoyuznogoOrdenaLenina Institute Rastenievodstavimeni*, N.I. Vavilova, No. **47**: 86-88.

Plaut, Z., Zieslin, N., Grawa, A. and Gazit, M. (1979) *Sci. Hort.*, **11**: 183-190.

Pobudkiewicz, A. and Goldsberry, K.L. (1989) *Res. Bull. Colorado Greenhouse Grower's Assoc.*, **471**: 1-2.

Pol, P.A. Van De and Pierik, R.L.M. (1995) *RevistaChapingo. Serie Horticultura*, **1**: 15-22.

Polara, N. D., Viradia, R. R. and Khimani, R. M. (2004) *J. Ornam.Hortic.*, **7**(3-4): 102-105.

Popcov, A.V. and Buc, T.C. (1966) *Bjull. Glav. Bot. Sada*, No. 62, pp. 30-34.

Post, K. and Fischer, C.W.Jr. (1951) *Proc. Amer. Soc. Hort. Sci.*, **57**: 561-568.

Prasad, A. and Verma, G.S. (1980) *Nat. Sem. Prodn. Tech. Comm. Flower Crops*, Coimbatore, pp. 105-106.

Prasad, K. V., Kumar, S. and Choudhary, M. L. (2006) *Indian J. Hortic.*, **63**(3).

Prell, W. and Meir-Dinkel, A. (1983) *DtscheBaumschule*, **35**: 202.

Prior, R.N.B. (1975) *Plant Pathol.*, **24**: 123-124.

Priya, P., Sankar, M., Sreelatha, U., Minimol, J. S. and Anupama, T. V. (2019) *J. Trop. Agric.*, **57**(2).

Puccini, G. (1958) *Ann. Sper. Agar.*, **12**: 1655-1676.

Put, H.M.C. and Jansen, L. (1989) *Sci. Hort.*, **39**: 167-179.

Qiu, X., Jian, H., Wang, Q., Tang, K. and Bao, M. (2015) *J. Amer. Soc. Hortic. Sci.*, **140**(4): 333-338.

Qiu, X., Zhang, H., Jian, H., Wang, Q., Zhou, N., Yan, H., Zhang, T. and Tang, K. (2013) *HortScience*, **48**(12): 1445-1451.

Raabe, R.D. and Wilhelm, S. (1966) *Calif. Agric.*, **20**: 5-6.

Rafi, Z. N. and Ramezanian, A. (2013) *South Afr. J. Bot.*, **86**.

Raghav, S.P.S. (1969) *Indian Hort.*, **14**: 23-30.

Rai, B. (1985) *Indian Hort.*, **29**: 13, 24.

Rai, H., Raju, D. V. S., Kumar, A. M. B., Janakiram, T., Namita, Krishnan, S. G. and Rana, J. C. (2015) *Indian J. Agric. Sci.*, **85**(2).

Rajamani, K., Thamburaj, S., Thangaraj, T. and Murugesan, S. (1992) *S. Indian Hort.*, **40**: 121-122.

Rajendran, G. and Muthukrishnan, T.S. (1980) *Nat. Sem. Prodn. Tech. Comm. Flower Crops*, Coimbatore, pp. 111-112.

Rajendran, P., Ajithkumar, K. and Smitha, R. (2017) *Acta Hortic.*, 1165.

Rajesh, A. M., Jayaprasad, K. V., Yathindra, H. A. and Harshavardhan, M. (2020) *J. Pharmacog.Phytochem.*, **9**(2): 142.146.

Ramakrishnan, T.A. (1984) *The Indian Rose Annual*, **3**: 64-70.

Rani B. J. and Sridhar, V. (2003) *J. Ornam.Hortic.*, 6(3): 165-171.

Rani, B. J. and Sridhar, V. (2005) *J. Ornam.Hortic.*, **8**(1): 55-61.

Rao, G.R. (1974) *The Rose Annual*, The Rose Society, Ranchi (India), pp. 67-72.

Rasouli, O., Ahmadi, N., Monfared, S. R. and Sefidkon, F. (2018) *Sci. Hort.*, **231**: 144-150.

Raul, C, Alma R. and John J. S. (2009) *Hort Science*, **44**(7): 2000–2008.

Raviv, M., Krasnovsky, A., Medina, S. and Reuveni, R. (1998) *J. Hort. Sci. Bitechnol.*, **73**: 485-491.

Raviv, M., Medina, S., Shamir, Y. and Ben-Ner, Z. (1993) *Acta Hortic.*, **342**: 111-119.

Raviv, M., Medina, S., Shamir, Y., Gil'ab, S., Durdevani, O., Shor, Y. and Schayer, R. (1993) *Sci. Hort.*, **53**: 141-148.

Raviv, M., Wallach, R., Silber, A., Medina, S. and Krasnovsky, A. (1999) *J. Amer. Soc. Hort. Sci.*, **124**: 205-209.

Rehder, A. (1940) *Manual of Cutivated Trees and Shrubs Hardy in North America*, The Macmillan Co., New York.

Reid, M.S., Evans, R.Y., Dodge, L.L. and Mor, Y. (1989) *J. Amer. Soci. Hort. Sci.*, **114**: 436-440.

Reis, M., Rouhana, L. V., Sadeque, A., Koga, L., Clough, S. J., Calla, B., Paiva, P. and Korban, S. S. (2020) *Plant Physiol.Biochem.*, **146**.

Reist, A. and Le, L.C. (1989) *Revue Suisse de Viticulture, d'Arboriculture et d'Hort.*, **21**: 19-23.

Rietra, R.P.J.J., Heinen, M., Dimkpa, C.O. and Bindraban, P.S. (2017) *Plant Anal.*, **48**: 1895–1920.

Rio, M.A.Del, Navarro, P. and Mateos, M. (1989) *Acta Hortic.*, **246**: 319-325.

Roberts, A.N. (1962) *Proc. Amer. Soc. Hort. Sci.*, **80**: 605-614.

Roberts, A.V., Blake, P.S., Lewis, R., Taylor, J.M. and Dunstan, D.I. (1999) *J. Plant Growth Regul.*, **18**: 113-119.

Roccia, A., Oyant, L. H., Cavel, E., Caissard, J., Machenaud, J., Thouroude, T., Jeauffre, J., Bony, A., Dubois, A., Vergne, P., Szecsi, J., Foucher, F., Bendahmane, M. and Baudino, S. (2019) *Plant Physiol.*, **179**(3): 1064-1079.

Rode, V.A. and Ogale, V.K. (1984) *The Indian Rose Annual*, **3**: 89-99.

Rodrigues, E.J.R., Farina, E., Paterniani, T. and Palagi, M. (1999) *ColtureProtette.*, **28**: 85.88.

Rodrigues, E.J.r., Paterniani, T., Palagi, M. and Farina, E. (1998) *ColtureProtette.*, **27**: 79-82.

Rodriguez, S., Gualotuna, T. and Grilo, C. (2017) *Procedia Computer Science*, **121**: 306-313.

Rol, Roark, S., Bowen, K. L. and Behe, B. K. (2000) *J. Environ.Hortic.*, **18**(3).

Rose, A.E., Pellett, N.E. and Aleong, J. (1982) *J. Amer. Soc. Hort. Sci.*, **107**: 1173-1177.

Rout, G. R., Debata, B. K. and Das, P. (1989) *Curr. Sci.*, **58**(15): 876-878.

Rout, G.R., Debata, B.K. and Das, P. (1991) *Plant Cell Tissue Organ Cult.*, **27**: 65-69.

Rowley, G.D. (1955-56) *A. R. John Innes Hort. Inst.*, pp. 12-13.

Rowley, G.D. (1956) *American Rose Annual*, American Rose Society, **41**: 70-73.

Roxas, U.A. (1982) *ColtureProtette*, **11**: 31-40.

Rupprecht, H. (1968a) *DtscheGartenb.*, **15**: 219-221.

Rupprecht, H. (1968b) *DtscheGartenb.*, **15**: 241-245.

Rupprecht, H. (1974) *Acta Pruhon.*, No. 32, pp. 8-26.

Ruting, A. (1991) *Acta Hortic.*, **298**: 69-74.

Saakov, S.G. (1965) *The Origin of Garden Roses and the Trend of their Breeding*, Ist Annual Kamarov Lecture, USSR Acad. Sci.

Sabehat, A. and Zieslin, N. (1995) *J. Plant Physiol.*, **145**: 296-298.

Salac, S.S. and Palmer, L.T. (1975) *Nebraska Univ. Inst. Agric. Nat. Res.*, No. SB 536, p. 10.

Salem, N., Golino, D. A., Falk, B. W. and Rowhani, A. (2008) *Plant Dis.*, **92**(4): 508-512.

Salem, N., Rowhani, A., Moyne, A. L., Sim, S. T. and Golino, D. A. (2007) *Acta Hortic.*, **751.**

Salimov, V.N. and Minkov, B.P. (1963) *Zasc. Rast. Vred. Boler*, **8**: 27.

Samiei, L., Naderi, R., Khalighi, A., Shahnejat-bushehri, A. A., Mozaffarian, V., Esselink, G. D., Osaloo, S. K. and Smulders, M. J. M. (2010) *J. Hortic. Sci. Biotechnol.*, **85**(3): 231-237.

Sanderson, J.P. and Zhang, Z.Q. (1995) *J. Econom.Ento.*, **88**: 343-351.

Sapna, P., Singh, K. P., Humira, S., Deshmukh, R. K., Namita, Prasad, K. V. and Sharma, T. R. (2015) *Indian J.Biotechnol.*, **14**(4): 518-524.

Sasikumar, K., Singh, S. K., Mohapatra, T. and Prasad, K. V. (2007) *Indian J.Hortic.*, **64**(4).

Scagel, C.F. (2001) *Environ. Hort.*, **19**: 15-20.

Schneider, G. (1960) *Proc. Pl. Prop. Soc.*, **10**: 282-285.

Schneider, J.H.M., Jacob, J.J.S. and Pol, P.A. Van De (1995) *Sci. Hort.*, **63**: 37-45.

Schrock, D. and Hanan, J.J. (1979) *Bull. Colo. Flower Grs'. Ass.*, No. 352, pp. 1-2.

Schum, A. and Hofmann, K. (2001) *Acta Hortic.*, 547.

Schum, A., Hofmann, K. and Felten, R. (2002) *Acta Hortic.*, 572.

Scotti-Campos, P. and Pais, M.S.S. (1990) *Sci. Hort.*, **43**: 321-330.

Secor, G.A. and Nyland, G. (1978) *Phytopathology*, **68**: 1005-1110.

Seddigh, S. (2018) *Int. J. Pest Mngg.*, **64**(2): 178-184.

Seeley, J.G. (1948) *Proc. Amer. Soc. Hort. Sci.*, **51**: 613-617.

Seeley, J.G. (1950) *Proc. Amer. Soc. Hort. Sci.*, **56**: 460-470.

Semeniuk, P. (1964) *Agric. Res.*, Washington, **12**: 15.

Semina, S.N., Timoshenko, N.M. and Klimenko, Z.K. (1982) In: *VrediteliiBolezniPlodovykhiDekorativnykhKul'turKryma*, Yalta, USSR, pp. 110-116.

Senapati, S. K., Aparajita, S. and Rout, G. R. (2012) *Afr. J. Biotechnol.*, **11**(100): 16532-16538.

Serek, M. and Andersen, A.S. (1995) *Acta Hortic.*, **404**: 48-57.

Serrano, M., Martinez, G., Pretel, M.T., Riquelme, F. and Romojaro, F. (1992) *Sci. Hort.*, **51**: 129-137.

Seyf, M., Khalighi, A., Mostofi, Y. and Naderi, R. (2012) *J. Agric. Sci.*, **4**(12): 174.

Shah, A., Srivastava, K.K. and Roy, A.J. (1981) *Progress. Hort.*, **13**: 133-136.

Shahare, M.L. and Shastry, S.V.S. (1963) *Chromosoma* (*Berl.*), **13**: 702-724.

Sharma, S. (2018) *J. Essent. Oil Bear. Plants*, **21**(1): 92-102.

Sharma, S.S. (1979) *Haryana J. Hor. Sci.*, **8**: 220-222.

Shaul, O., Elad, Y. and Zieslin, N. (1993) *Hassadeh.*, **74**: 71-73.

Shaul, O., Elad, Y. and Zieslin, N. (1995) *Postharvest Biol. Tec.*, **6**: 321-330.

Shaul, O., Elad, Y. and Zieslin, N. (1996) *Postharvest Biol. Tec.*, **7**: 145-150.

Shisa, M. and Takano (1964) *J. Japanese Soc. Hort. Sci.*, **33**: 140-146.

Shnyakina, G.P. and Malygina, E.P. (1975) *Rastitel. Resursy*, **11**: 390-394.

Shokri, S., Babaei, A., Ahmadian, M., Arab, M. M. and Hessami, S. (2015) *Acta Hortic.*, 1083.

Shorthouse, J. D. (2001) *Acta Hortic.*, **547.**

Shupert, D., Byrne, D. H. and Pemberton, H. B. (2005) *HortScience*, **40**(4): 1102.

Singh, A. K. and Dubey, A. K. (2003) *J. Ornam. Hort.*, **6**(3): 234-238.

Singh, A. K. and Kumar, S. (2003) *J. Ornam. Hort.*, **6**(3): 248-251.

Singh, I. and Rohilla, P. P. (2019) *Indian Hortic.*, **64**(3).

Singh, P. V. and Mirza, A. A. (2004) *J. Ornam. Hort.*, **7**(1): 58-63.

Singh, S., Rana, R. S., Sharma, K. C., Sharma, A. and Kumar, A. (2017) *J. Entomol. Zool. Studies*, **5**(6): 104-107.

Singh, S.N. (1974) *Hort. Adv.*, **9**: 39-42.

Singh, S.N. (1980) *Progress. Hort.*, **12**: 5-13.

Slootweg, G. (1995) *Acta Hortic.*, **405**: 67-74.

Slootweg, G. (1997) *Acta Hortic.*, **418**: 107-110.

Smirnov, A.E. (1975) *ByulletenVsesoyuznogoOrdenaLeninaInstitutaRastenievodstvaImeni N.I. Vavilova*, No. 47, pp. 84-86.

Smith, E.F., Mottley, J. and Roberts, A.V. (1992) *Biotech. Edn.*, **2**: 56-58.

Smitha, R., Rajendran, P., Sandhya, P. T., Aparna, V. S. and Rajees, P. C. (2017) *Acta Hortic.*, **1165**: 306-312.

Smulders, M. J. M., Voorrips, R. E., Essellink, G. D., Leonardo, S. T. M., Westende, W. P. C., Vukosavljev, M., Koning-Boucoiran, C. F. S., Weg, W. E., Arens, P., Schulz, D., Debener, T., Bellon, L., Mittmann, M., Pirani, A., Webster, T., Brew, F., Cox, P. and Maliepaard, C. (2015) *Acta Hortic.*, **1064**: 508-517.

Son Kicheol, ByounHyejin and Kim Mikyoung (1997) *J. Korean Soc. Hort. Sci.*, **38**: 309-314.

Son, K. C., Byoun, H. J. and Yoo, M. H. (2003) *Acta Hortic.*, **624**: 407-412.

Sood, S. and Nagar, P. K. (2008) *Acta Physiol. Plant.*, **30**: 243-248.

Soukup, J. and Stankova-Opocenska, E. (1974) *Acta Pruhon.*, No. 32, pp. 8-92.

Spinarova, S. and Hendriks, L. (2007) *Acta Hortic.*, **751**: 601-608.

Sridhar, V. and Rani, B.J. (2011) *Pest Mngg. Hortic. Ecosystems*, 17(2): 127-131.

Srinivasan, P.M., Govindarajan, R. and Paul, A.V.N. (1974) *South Indian Hort.*, **22**: 124.

Srivastava, H.P. (1973) *Indian J. Hort.*, **17**: 40-45.

Srivastava, H.P. and Gupta, M.N. (1984) *The Indian Rose Annual*, **3**: 44-51.

Staby, G.L., Robertson, J.L., Kiplinger, D.C. and Conover, C.A. (1978) *Chain of Life, Ohio State Florist's Assoc. Ohio State Univ.* p. 71

Staikov, V.M. (1955) *C. R. Acad. Bulg. Sci.*, **8**: 49-52.

Stanko, I.I. and Mihajlov, N.L. (1966) *Bjull. Glav. Bot. Sada*, No. 63, pp. 19-21.

Steen, N. (1967) *The Charm of Old Roses*, Herbert Jenkins Ltd., London.

Stenseth, C. (1983) *Gartneryrket*, **73**: 86-91.

Stewart, L. and Brandis, D. (1974) *The Forest Flora of North, West and Central India*, London.

Stuart, C.M.V. (1913) *Gardens of the Great Mughals.*

Suganthi, P., Pugalendhi, L., Subramanian, S., Anand, M. and Kalarani, K. (2019) *Int. J. Chem. Stud.*, **7**(2): 1477-1479.

Sun, Z., Li, S., Li, C., Lin, M. and Chen, L. (1998) *Adv. Hort.*, **2**: 711-715.

Susaj, E., Susaj, L. and Kallco, I. (2012) *Res. J. Agric. Sci.*, **44**(3): 121-127.

Suskov, K.L. and Mihneva, T.N. (1969) *Fmg. S. Afr.*, **44**: 120-125.

Suskov, K.L., Besscetnova, M.V. and Mihneva, T.N. (1966) *Trudy Bot. Sadov. Akad. NaukKaz. SSR*, **9**: 52-58.

Svejda, F. (1969) *Can. J. Plant Sci.*, **40**: 100.

Svejda, F. (1978) *Can. J. Plant Sci.*, **58**: 907-908.

Swarup, V. (1967) *Garden Flowers*, National Book Trust of India, New Delhi.

Swarup, V. (1972) *Delhi Garden Mag.*, Flower Show Number, pp. 38-41.

Swarup, V. (1980) *Ornamentals*, In: *Handbook of Agricuture*, Indian Council of Agricultural Research, New Delhi.

Swarup, V. and Malik, R.S. (1974) *Indian J. Hort.*, **31**: 268-273.

Swarup, V., Kaicker, U.S., Malik, R.S. and Singh, A.P. (1973) *Indian J. Genet.*, **33**: 43-51.

Syros, T., Economou, A., Galagtis, A., Tsicritsis, G. and Ralli, P. (2001) *Acta Hortic.*, **548.**

Szekely, I., Wagner, S. and Dragan, M. (1981) In: *Principles of Plant Resistance to Disease and Pestsi.* Bucharest, Romania, pp. 41-45.

Tabaeezadeh, Z. and Khosh-Khui, M. (1981) *Sci. Hort.*, **15**: 61-66.

Tan, H., Liu, X., Ma, N., Xue, J., Lu, W., Bai, J. and Gao, J. (2006) *Postharvest Biol. Tec.*, **40**(2): 97-105.

Tejaswini, M. and Prakash, S. (2005) *Acta Hortic.*, **690.**

Terada, M., Kageyama, Y. and Konishi, K. (1997a) *J. Japanese Soc. Hort. Sci.*, **66**: 149-155.

Terada, M., Kageyama, Y. and Konishi, K. (1997b) *J. Japanese Soc. Hort. Sci.*, **66**: 379-383.

Terada, M., Tamura, A., Kageyama, Y. and Konishi, K. (1997c) *J. Japanese Soc. Hort. Sci.*, **66**: 569-574.

Teradi, M., Goto, T., Kageyama, Y. and Konishi, K. (1997) *Acta Hortic.*, **440**: 366-370.

Terfa, M. T., Roro, A. G., Olsen, J. E. and Torre, S. (2014) *Sci. Hort.*, **178**: 184-191.

Thomas, A.S. (1969) *Better Roses: Rose Growing for Everyone*, Angus and Robertson Let., London.

Thomas, C.A. (1980) *J. Hort. Sci.*, **55**: 15-17.

Tjosvold, S.A. and Schulbach, K.F. (1991) *California Agri.*, **45**: 31-32.

Torre, S., Borochov, A. and Halevy, A.H. (1999) *Physio-plantarum*, **107**: 214-219.

Treder, J., Wisniewska-Grzeszkiewicz, H., Podwyszynzka, M. and Hempel, M. (1989) *Proc. InstytutuSadownictwaiKwiaciarstwa W skierniewicachSeria B, RoslinyOzdobne*, **14**: 71-76.

Trivedi, H., Kumar, P., Kapoor, A. and Parween, S. (2016) *Indian J. Agric. Sci.*, **86**(8).

Tshwenyane, S. O. and Bishop, C. F. H. (2010) *Acta Hortic.*, **877.**

Tshwenyane, S. O., Cullum, F. J., Bishop, C. F. H. and Gash, A. F. J. (2012) *Acta Hortic.*, **934.**

Tsujita, M.J. (1982) *HortScience*, **17**: 44-45.

Tsvetkov, R. (1973) *Rasten. Nauki*, **10**: 37-47.

Ueda, Y. (1994) *Tech. Bull. Fac. Hort. Chiba University*, No. 48, pp. 241-328.

Ueyama, S. and Ichimura, K. (1998) *Postharvest Biol. Tec.*, **14**: 65-70.

Van Der Harst, J. (1962) *Proc. 16th International Hort. Cong.*, Brussels, 1, pp. 372-373.

Van Onsen, J and Haegeman, J. (1962) *Proc. 16th Int. Hort. Congri.*, Brussels, 1, p. 395.

van-der-Salm, T.P.M., Bouwer, R., Van-Di, V.J., Keizer, L.C.P., Ten-Cate, C.H.H., Van-der-Plas, L.H.W. and Dons, J.J.M. (1998) *J. Expt. Bot.*, **49**: 322.

van-der-Salm, T.P.M., Van-der-Toorn, C.J.G., Ten-Cate, C.H.H., Van-der-Krieken, W.M. and Dons, H.J.M. (1996) *Plant Growth Regul.*, **9**: 123-131.

Vasudevan, V. and Kannan, M. (2014) *Trends Biosci.*, **7**(13).

Vecera, L. (1967a) *Ved. Pr. Vyzk. Ust. Okrasn. Zahrad. V. Prahonicich*, **4**: 67-77.

Vecera, L. (1967b) *Gartenwelt*, **67**: 437-438.

Vecera, L. (1968) *Wiss. Z. Humboldt Univ.*, Berlin, **17**: 289-292.

Vecera, L. (1971) *A Concise Guide in Colour-Roses*, Hamlyn, Middx., England.

Veld, E.J.M.H. In't, Egberink, M.B., Pol, P.A. Van de and Jacob, J.J.S. (1988) *39th Inter. Symp. on Crop Protection*, Belgium, **53**: 893-901.

Veluru, A., Bhat, K. V., Janakiram, T., Prasad, K. V., Raju, D. V. S., Bharadwaj, C., Gayacharan, C., Singh, K. P., Namita, Panwar, S. (2019) *Indian J. Genet. Plant Breed.*, **79**(3): 101-112.

Venkatarayappa, T., Tsujita, K.J. and Murr, D.P. (1980) *J. Amer. Soc. Hort. Sci.*, **105**: 148-151.

Verma, G.S. and Singh, P.D. (1980) *Nat. Sem. Prodn. Tech. Comm. Flower Crops*, Coimbatore, pp. 101-103.

Verma, M. K., Lal, S., Ahmed, N., Kumar, D., Singh D. B. and Sagoo, P. A. (2015) *Indian J. Hortic.*, **72**(2): 250-256.

Veski, V. (1957) *Sad iOgorod*, No. 8, p. 69.

Vetvicka, V. (1974) *Acta Pruhon.*, No. 32, pp. 75-82.

Victoria, N. G., Eveleens, B. and Telgen, H. J. V. (2007) *Acta Hortic.*, **751**: 602-608.

Victoria, N. G., Eveleens, B., Telgen, H. J. V. and Weel, P. V. (2007) *Acta Hortic.*, **751**: 616-626.

Vogli, V. and Sigrimis, N. (1999) *Proc. IFAC Workshop, Athens, Greece*, June 14-17, 1998. Oxford, U.K., pp. 225-229.

Vogt, J. C., Herklotz, V. and Ritz, C. M. (2015) *Int. J. Plant Sci.*, **176**(5): 433-445.

Vries, D.P. De and Smeets, L. (1979) *Sci. Hort.*, **11**: 261-268.

Vries, D.P.De and Dubois, L.A.M. (1988) *Acta Hortic.*, **226**: 223-230.

Vries, D.P.De and Dubois, L.A.M. (1992) *Gartenbauwissenschaft*, **57**: 64-

Vries, D.P.De, Dubois, L.A.M. and Smeets, L. (1978) *Netherlands J. Agric. Sci.*, **26**: 399-404.

Vries, D.P.De, Smeets, L. and Dubois, L.A.M. (1980) *Sci. Hort.*, **13**: 61-66.

Vries, D.P.De, Smeets, L. and Dubois, L.A.M. (1982) *Sci. Hort.*, **17**: 377-382.

Vukosavljev, M., Zhang, J., Esselink, G. D., Westende, W. P. C. V., Cox, P., Visser, R. G. F., Arens, P. and Smulder, M. J. M. (2013) *Sci. Hort.*, **162**: 320-332.

Waikar, A. U. and Jadhav, S. B. (2012) *Int. J. Agric.Engg.*, **5**(1): 12.

Wallace, D.B. and Yule, J.A. (1977) *Proc. Amer. Phytopath. Soc.*, **4**: 200.

Wang, F., Lian, L. J. and Liu, Q. L. (2015) *Acta Hortic.*, **1064**: 204-206.

Wang, G. (2007) *Acta Hortic.*, **751**: 507-512.

Wang, G. Y., Yuan, M. F. and Hong, Y. (2002) *In Vitro Cell. Develop. Biol. – Plant*, **38**: 513-518.

Wang, G., Wang, Y. and Li, X. (1999) *J. Jiangsu Fores. Sci. Tec.*, **26**: 1-6.

Wang, H., Yao, L., Cai, R., Pan, J. and Chen, X. (2012) *Sci. Hort.*, **137**: 121-124.

Watanabe, H. and Yokoi, K. (1991) *Bull. Nara Agril. Expen. Station*, **22**: 73-80.

Watson, D.P. (1954) *Amer. Nurserymen*, **100**: 36-37.

Wei, T., Zhu, M., Zhou, J., Chen, L., Sheng, L. and Fen, L. (2019) *J. Essen. Oil Bear. Plants*, **22**(2): 50-56.

Weinstein, L.H. (1957) *Contr. Boyce Thompson Inst.*, **19**: 33-48.

Wen, X., Xu, Q., Cao, Q. and Deng, X. (2006) *N. Z. J. Crop Hortic. Sci.*, **34**(2): 118-122.

Werlemark, G., Uggla, M. and Nybom, H. (1999) *Theor. Appl. Genet.*, **98**: 557-563.

West, G. (1966) *The Rose Annual*, The Royal National Rose Society, UK.

Wheather, J. (1903) *Beautiful Roses for Garden and Greenhouse.*

White, J.M. and Richter, D. (1974) *J. Amer. Soc. Hort. Sci.*, **98**: 655-660.

Widmoyer, F.B. and Sherwood, C.H. (1954) *Proc. Amer. Soc. Hort. Sci.*, **63**: 462-464.

Wisniewska-Grzeszkiewicz, H. and Treder, J. (1989) *Acta Hortic.*, **251**: 389-392.

Wisniewska-Grzeszkiewicz, H. and Treder, J. (2003) *Acta Hortic.*, **614**.

Witte, Y.De and Doorn, W.G.Van (1988) *Sci. Hort.*, **35**: 285-291.

Woidyla, A.T. (1997) *Proc. 49th Inter. Symposium Crop Protectio, Genet., Belgium*, May 7, Part IV, pp. 1135-1139.

Wojdyla, A.T.andWisniewska-Grzeszkiewicz, H. (1999) *ZeszytyNaukoweInstytutu Sadownictwai Kwiaciarstwa w Skierniewicach*, **6**: 177-187.

Woodson, W.R. and Boodley, J.W. (1982a) *J. Amer. Soc. Hort. Sci.*, **107**: 275-278.

Woodson, W.R. and Boodley, J.W. (1982b) *HortScience*, **17**: 46-47.

Wright, J.A. and Volz, E.C. (1950) *Proc. Amer. Soc. Hort. Sci.*, **55**: 486-488.

Wright, R.C.M., Brett, W.S., Anstiss, L.A., Pearson, C.E. and Champneys, H.P. (1957) *Roses*, Ward Lock and Ltd., London.

Wu, S., Ueda, Y., He, H., Nishihara, S. and Matsumoto, S. (2000) *Breed. Sci.*, **50**: 275-281.

Wylie, A.P. (1954) *J. Roy. Hort. Soc.*, **79**: 551-572.

Wylie, A.P. (1955) *J. Roy. Hort. Soci.*, **80**: 77-87.

Xiang, G., Zhang, J., Jian, J., Yan, H., Wang, Q., Zhou, N., Li, S., Tang, K. and Qiu, K. (2019) *Scientia Horticulturae*, **257**: 110-118.

Xiao, Z., Luo, J., Niu, Y., Wang, P. and Sun, X. (2019) *Food Res. Int.*, **116**: 211-222.

Xue, Quihua and Lin, Ru (1999) *J. Fujian Agric. Univ.*, **28**: 304-308.

Yadav, L.P., Bose, T.K. and Dhua, R.S. (1985a) *The Indian Rose Annual*, **4**: 166-171.

Yadav, L.P., Bose, T.K. and Mukhopadhyay, T.P. (1985b) *The Indian Rose Annual*, **4**: 172-176.

Yamaguchi, S. and Boontiang, K. (2010) *Acta Hortic.*, **870**: 503-510.

Yambe, Y. and Takeno, K. (1992) *HortScience*, **27**: 1018-1020.

Yambe, Y., Hori, Y. and Takeno, K. (1992) *J. Japanese Soc. Hort. Sci.*, **62**: 383-387.

Yambe, Y., Takeno, K. and Saito, T. (1995) *J. Amer. Soc. Hort. Sci.*, **120**: 953-955.

Yan, M. and Junyyu, C. (1995) *Acta Hortic.*, 404.

Yan, M., Byren, D.H. and Jing, C. (1996) *Biol. Plant.*, **32**: 103-108.

Yan, Z., Dolstra, O., Dennebom, C., Stam, P. and Visser, P. B. (2003) *Acta Hortic.*, **612.**

Yohalem, D. S. and Kristensen, K. (2004) *Biol. Cont.*, **29**(2): 256-259.

Yohalem, D.S. (2000) In: *17th Danish Plant Protection Conference, Horticulture*, No. 12, pp. 97-102.

YooYong Kweon, Park, Hyunjin, Kang, SangWook and Kim, HyunKyung (1999) *Korean J. Hort. Sci. Tec.*, **17**: 481-484.

Young, N. (1971) *The Complete Rosarian*, Hodder and Stoughton, London, Sydney, Toronto.

Young, T.W., Snyder, G.H., Martin, F.G. and Hayslip, N.C. (1973) *J. Amer. Soc. Hort. Sci.*, **98**: 109-112.

Zal'tsfas, A.A. and Chemarin, N.G. (1974) *ReferativnyiZhurnal*, **12**: 994.

Zeller, O. (1975) *Gartenbauwissenschaft*, **40**: 276-284.

Zeroni, M. and Gale, J. (1989) *J. Hort. Sci.*, **64**: 503-511.

Zhang, J., Esselink, G. D., Che, D., Fougere-danzen, M., Arens, P. and Smulders, M. J. M. (2013) *J. Hortic. Sci. Biotechnol.*, **88**(1): 85-92.

Zhou, Z. and Bao, W. (2011) *Sci.Hortic.*, **129**(4): 818-824.

Zieslin, N. and Halevy, A.H. (1976) *Sci. Hort.*, **4**: 73-78.

Zieslin, N. and Khayat, E. (1982-83) *Plant Growth Regul.*, **1**: 279-288.

Zieslin, N. and Mor, Y. (1990) *Sci. Hortic.*, **43**: 1-14.

Zieslin, N., Halevy, A.H. and Biran, I. (1973) *J. Amer. Soc. Hort. Sci.*, **98**: 321-324.

Zieslin, N., Halevy, A.H. and Enoch, Z. (1972) *Hort. Res.*, **12**: 97-100.

Zieslin, N., Hurwitz, A. and Halevy, A.H. (1975) *J. Hort. Sci.*, **50**: 39-47.

Zieslin, N., Madori, G. and Halevy, A.H. (1979) *J. Expt. Bot.*, **30**: 1325.

Zieslin, N., Mor, Y., Bachiach, A. and Haaze, H. and Kofranek, A.R. (1976) *Sci. Hort.*, **4**: 63-72.

Zieslin, N., Spiegelstein, H. and Halevy, A.H. (1978) *Bot. Gaz.*, **139**: 64-68.

Zuzek, K. and Hokanson, S. C. (2007) *Acta Hortic.*, **751**: 108-115.

Zykov, K.I., Rat'Kin, A.V. and Khimenko, Z.K. (1990) *IsvestiyaAkademiiNauk SSR SeriyaBiologicheskaya*, No.4, p. 524.

3

CARNATION

C. Aswath, M.L. Chaithra and
V.A. Parthasarathy

1.0 Introduction

Carnation (*Dianthus caryophyllus* L.) is a herbaceous perennial flowering plant, belonging to the family Caryophyllaceae. Carnation is preferred to other flowers due to its excellent keeping quality, wide range of forms, ability to withstand long distance transportation and remarkable ability to rehydrate after continuous shipping.

2.0 Importance and Uses

Carnation in the modern times has become one of the most important cut flower crop. Some of its varieties are used for bedding, pots, rock gardens, window boxes and edging too. It gives a unique softness in the rock gardens. Not only standard carnations, even spray type carnations are popular in flower arrangements for decorating homes.

In India, the major production centers are located around Bangalore, Pune, Delhi, Trivandrum and a few cities of Andhra Pradesh. Many private organizations and foreign collaboration companies are being set up in clusters around Bangalore, Pune, Hyderabad and Delhi.

Besides aesthetic value, carnation flowers are considered to be cardio tonic, diaphoretic, and alexiteric. The whole plant is used as vermifuge in China while in France it is used for perfume extraction (Poucher, 1950).

Ceron Madrigal *et al.* (1995) reported the possible use of by-products (leaves, stems, buds and low-quality flowers) of carnation production in Spain as animal feed. Whole fresh plant contained 13.9 to 16.1 per cent crude proteins (CP) and 28.9 to 32.8 per cent crudefibre. Variation in composition of different by-products was not great. When by-products were prepared as silage or hay, CP decreased from 14 to 9.57 and crude fibreincreased from 32.53 to 36.31.

3.0 Origin, History and Distribution

Carnation has been under cultivation by man for over 2000 years. Theophrastus, who lived about 300 B.C. gave the name *Dianthus* (Greek *dios*, divine; *anthos*, flower) to the group. The common name carnation (Latin, *carnatio*, from caro; *carnis*, flesh) is believed from a Greek word coronation because these flowers were used in decorating the crowns of Greek athletes.

Carnation is indigenous to the Mediterranean area. Some people believed that it was cultivated by the Muslims of Africa and introduced to Europe by Tunis in the 13th century (Coats, 1956). According to Montagau Allwood, a pioneer in carnation breeding, the present day perpetual flowering carnation was derived from cross between *D. caryophyllus* (female parent) and *D. chinensis* (male parent). He had further suggested that cross between *D. chinensis* (seed parent) and *D. caryophyllus* gave rise to annual marguerite carnation. Malmaison types were developed in 1857 from seedlings of perpetual carnation, which bore flowers markedly larger and more handsome than previous cultivars.

Major work on carnation improvement was carried in France and USA.The cultivar 'William Sim' produced in 1938 by William Sim of USA is the greatest contribution to the present carnation industry. From the red flowering 'William Sim' there have been mutations to all the possible colours and several variegated forms.

3.1.0 Distribution

The natural climate for carnation occurs near 30° North or South latitude. Southern California, the Mediterranean area, Australia, Valparaiso, Chile and the Republic of South Africa are the places where large quantities of carnation are produced, in addition to Mexico, Central America and parts of Kenya.

Carnations are also cultivated on a large scale in France, Holland, Italy, Israel and Germany for their flowers. The clove-scented marguerite carnations are quite popular in India. However, the cultivation of Sim carnations which are of great commercial importance has been taken up widely around Pune (Maharashtra), Bangalore (Karnataka), Himachal Pradesh, Punjab and Jammu and Kashmir.

4.0 Species and Cultivars

4.1.0 Species

There are about 250 species of *Dianthus* of which only a few are cultivated, the most common being *D. caryophyllus, D. barbatus, D. chinensis.* The species *barbatus* commonly known as Sweet William grows readily from seeds. The stem in this species is glabrous; four-angled and branched at the top. The leaves are opposite, broad, and flat. The flowers are borne in dense cyme. The flowers are solitary, the petals are broad with filled margins and the calyx is cylindrical with bracts at the base. The present day florist's carnations are the results of genetic improvement with their stamens transformed into petaloids. The pistil in florist's carnation are unchanged. The growth of pistil is hormonally controlled at the later stages of flower development and is considered to be responsible for wilting and death of flower. The petals are toothed and bearded. The flower stalk is short and forms rounded top.

The *D. chinensis* is commonly known as Indian Pink, Chinensis or Japanese pink. The plants are glabrous more or less creeping at base. The stems are pubescent and more or less grooved. The leaves are flat and broad. The individual petals are hairy towards base. The pinks are excellent for beds, borders, edging, rock gardens, pots and cut flowers. The two most commonly grown varieties are *heddewigii,* the Japanese Pink (*D. chinensis* var. *heddewigii*) and *lacinatus*, the fringed pink (*D. chinensis* var. *lacinatus*).

Dianthus aydogduii from Salt Lake, Central Anatolia, Turkey, is formally described as a new species. *D. aydogduii* is very closely allied to *D. cyri* and *D. tripunctatus* but clear differences exist between the species. *D. aydogduii* grows up to 15 cm tall with a single stem whereas the other 2 species grow 12-50 cm with many branches. *D. aydogduii* also differs in having bracteoles shorter than the calyx tube which is 8-10 mm long with teeth apex acumunate, *D. tripunctatus* has a calyx margin up to 2 cm wide but in *D. cyri* and *D. aydogduii* it is 0.5 mm wide (Menemen and Hamzaoglu, 2000).

Uses of Carnation

Carnation in pot

As cut flowers

In garden beds

In border

In window boxes

In flower arrangement

Species of Genus *Dianthus*

Dianthus alpinus

Dianthus barbatus

Dianthus caryophyllus **(Carnation)**

Dianthus grandiflorus

Dianthus chinensis

Dianthus superbus

Dianthus sylvestris

Dianthus hybridus

Some other species of *Dianthus* are *D. alpinus, D. arenarius, D. attenua, D. caesius, D. callizonus, D. capitatus, D. carthusianorum, D. cruentus, D. deltoids, D. diadematus, D. fimbriatus, D. giganteus, D. glacialis, D. grandiflorus, D. heddewigii, D. hybridus, D. knappii, D. laciniatus, D. latifolius, D. monspessulanus, D. nobilis, D. pancicii, D. petraeus, D. plumarius, D. squarrosus, D. superbus, D. sylvestris, D. versicolor, D. viscidus and D. winteri.*

4.2.0 Types of Carnation

There are several types of carnation, which are described below:

4.2.1 Chabaud or Marguerite

This type of carnation was developed by crossing *D. chinensis* x *D. caryophyllus*. These are annuals with single or double flowers reproducing from seed. The flowers are large and have fringed petals. They do not last very long in vase.

4.2.2 Border and Picotee

The border carnation is the easiest to grow and has symmetrical flowers. The flower colour varies from single to blended colour with irregular markings. The petals are broad and smooth-edged. The flowers are generally frilled with open centers. They produce single stem during the first year and become bushy in the subsequent years. They are classified according to the colours as Bizarres, Flakes, Selfs, Fancies and Picotees.

4.2.3 Malmaison

Malmaison has stiff and massive habit with broad leaves and large fully developed flowers mainly pink with well filled centers. The buds are rounded and the calyx is short compared to perpetuals. Flowers have rich fragrance. The pollen production in this cultivar is very scanty and not viable.

In Japan, annual carnation cultivars with double and attractive flowers, dwarf in height and suitable for cultivation in pots are available.

4.2.4 Perpetuals

Perpetuals are hybrids involving many *Dianthus* species. The plants are not hardy and flower all round the year. The flower stalks in this group are longer which make them suitable for cut flowers. They produce better quality flowers and withstand long transportation.

They are grouped into standard and spray, based on flower size. The standard types produce large blooms and longer flower stalks. They suffer from diseases when grown under warm humid climates. The miniature or spray types, on the other hand produce many flowers of smaller size and are better adopted to warm climate than standards. These are preferred to standard types in European markets.

4.2.5 New Types

Umiel *et al.* (1987) observed new phenotypes in segregating populations manifesting 3 groups of characteristics. The first group included plants with new patterns of colours which resemble the wing colours of exotic butterflies. These colour combinations were based on variations in shape of the basal macule and of the central spot on the petals, and their combinations of various colours and shades. The second group had different numbers of flower buds per stem, before disbudding, ranging from 1 to 30 or more. The third group showed variations in the arrangement of flower buds on the stem and included standard and spray types. Some genotypes had flowers arranged on very short secondary stems, giving a gladiolus type inflorescence.

4.3.0 Cultivars

There are lot of cultivarers released from private companies.The cataloges of the varieties are available on the websites of the companies.

Two varieties have been released from ICAR-IIHR Bengaluru.

Variety	*Year of Release*	*Breeding Method*	*Pedigree/ Parentage*	*Important Traits*
Arka Flame (Standard)	2008	Mutation breeding (*in vitro*) mutagenesis)	*In vitro* mutation selection from IIHRS-1	Red with smooth petal edges with long flower stalks. Yield: 300 to 325 flowers/ m^2/yearImportant character: Tolerant to *Fusarium* wilt and nematode
Arka Tejas (Spray)	2010	Hybridization and selection	*D. caryophyllus*x *D. chinensis.*	Red coloured single whorl flowers with white anthersYield: 20-25 flowers/ stalkUse: For pot cultureImportant character: Dark red colour profuse bloom and mass effect. Released in 2010

5.0 Genetics and Breeding

5.1.0 Genetics

Basic chromosome number in *Dianthus* is 15 and most of the species are diploid (2n = 30). However, *D. chinensis* is tetraploid (2n = 60). *D. gratianopolitanus* is found in the forms of tetraploid and hexaploid. Most of the F1 hybrids are triploid and sterile due to production of unreduced gametes in *D. caryophyllus*. The behaviour of fertile tetraploid was as stable as amphidiploid. Backcross derivatives of three generations looked like autotetraploid. They were self and infertile and set seed when pollinated by either diploid or tetraploid (Mehlquist, 1945). Triploids in carnations are produced with difficulty from reciprocal diploid-tetraploid crosses, but carnation tolerates a high degree of aneuploid. Sparnaaij (1979) observed that polyploid carnations had larger flower, slower growth, shorter stem and lower yield than original diploid.

5.1.1 Inheritance of Characters

5.1.1.1 Single vs Double

Carnation is grouped into three categories namely, single which is commonly five petals, rarely four or six, semi-doubles are commercial carnation with 30-60 petals and double or superdouble or bullhead with 100-350 petals. Singleness in

Cultivars of Carnation

Antigua | Arevalo | Bizat

Farida | Crimpson Tempo | Francino

Grand Slam | Malaga | Nambo

Pink Candy | Pink Sim | White Sim

Spray Cultivars of Carnation

Albatros

Cato

Cerise

Milkyway

Osiris

Peach

carnation is a monogenic character and is incompletely recessive to doubleness, while the heterozygotes yield semi-double flowers. The dominant gene for doubleness was designated as *Da* by Imai (1938). Detailed work of Dumalo (1969) threw further light on the inheritance of doubleness in carnation. There are two types of single- lowered forms, the basic single (*dd*) and variable single (*Ee*). Upon selfing, the basic single gives 5-7 petaled progenies, while *Ee* single type segregates into 3 singles (*EE* or *Ee*) to 1 semi-double (*ee*). The gene *e* represents a recessive doubleness. Similarly, the semi-doubles of lower petalled represent two different types depending on the genotype. The recessive genotype*ee* produces 8-15 petals and *FF* and *Ff* genotypes produce 16-45 petals. The *F* gene appears to be incompletely dominant to *f*. The semi-doubles of higher petaled known as doubles are due to *D* gene. The *Dd* genotype on selfing gives rise to progenies with 5 to 100 petals in the proportion of 3 semi-doubles (*Dd* or *DD*) to single (*dd*). The *Dd* individuals usually have 6-50 petals. Phenotypically, a true double (*DD*) may not be distinguishable from a commercial double of heterozygous genotypes *Dd, EeFF* or *eeff*. Only by selfing or appropriate crosses, true double can be identified.

5.1.1.2 Flower Colour

Carnation is divided into three main groups based on flower colour. The first Acyanic group containing white and yellows. The second group is cyanic containing red, salmon, lavender, pink and crimson and the third transition group containing salmon yellow, orange and maroon. Six independent factors are responsible for expression of self colour. They are:*Y* controls the production of yellow anthoxanthin; *I* controls the production ivory anthoxanthin. It is epistatic to *Y*. *A* is the basic anthocyanin factor. It is effective in the presence of *Y* and *I*. *S* controls the production of scarlet-red anthocyanin, its recessive allele *s* dilutes red or salmon. *R* controls the production of crimson-red anthocyanin, its recessive allele *r* permits the production of scarlet-red anthocyanin modifies the coloursof the red series to the corresponding colours of magenta series.

Gimelli *et al.* (1982) from a study with nine selfed cultivars of the Mediterranean carnations found the presence of a gene homozygous recessive for white colour, epistatic on other genes for colour. Variegation and light colours (both recessives) are also controlled by recessive genes. Gimelli *et al.* (1984) further reported heterozygosity for at least one character (colour, dwarfness or flower structure) in 41 of 43 selfed cultivars. The genetic control of colour appeared to differ from the American Sim ecotype in that the colour range was wider in the Mediterranean carnation, and in the occurrence of epistasis for variegation and various other gene interactions.

5.1.1.3 Abnormal Types

Mehlquist (1941) reported genetics for albino and yellow seedlings *vs* green, lutescent seedlings *vs* normal, club-neck *vs* normal, virescent*vs* normal, lazy *vs* normal, dwarf *vs* normal, thin *vs* normal.

5.1.1.4 Albino and Yellow Seedlings vs Green

Albino and yellow seedlings have occurred in same proportion in several cultures. The proportions, approximately 9 green: 3 yellow: 4 albino, indicated that albino and yellow differ from normal green in a single recessive factor, the double recessive also being albino. The factor albino is designated as *alb* and yellow as *yel*. Both the albino and yellow seedlings die in the cotyledon stage.

5.1.1.5 Lutescent Seedling vs Normal

Lutescent seedlings are characterized by greenish yellow colour and weak growth. Most of them die in the cotyledon stage but some grow up to 7 or 8 pairs of leaves. None reaches maturity. This character is monogenic in nature and lutescent differs from normal in a single recessive factor. The segregation for this type is 3 normal to 1 lutescent. This gene has been designated as *lut*.

5.1.1.6 Club-neck vs Normal

Club-neck, upper portion of the root enlarges or swells. This inherited tendency might be due to some physiological disturbances resulting in the swelling and eventual death of the tissue in hypocotyl region. This club-neck is due to single recessive factor, 3 normal: 1 club-neck. The gene has been designated as *cb*.

5.1.1.7 Virescent vs Normal

Virescent is characterized by distinct deficiency of chlorophyll in longitudinal stripes on the leaves and stems. This also is due to single recessive factor (3 normal: 1 virescent) and has been designated as *vir*. Virescent plants may grow and flower.

5.1.1.8 Lazy Virescent vs Normal

In this case chlorosis occurs not in stripes but towards the edges of leaves. The stem is nearly white and internodes are greatly elongated resulting in a weak plant. This character is also monogenic in nature and differs from a normal in a single recessive factor (3 normal: 1 lazy virescent). The gene has been designated as *laz*.

5.1.1.9 Dwarf vs Normal

Dwarfness is characterized by short internodes and finally in a very short plant. It is due to single recessive factor (3 normal: 1 dwarf) and the gene has been designated as *dw*.

5.1.1.10 Thin vs Normal

Thinness in carnation means the total absence of branches at the base. This abnor-mality is also approximately in the ratio of 3 normal: 1 thin differs from normal in only one factor.

5.2.0 Breeding

The objective of breeding is to produce varieties with pleasing colours, symmetrical in form, resembling a half sphere with just enough petals to fill the bloom nicely without crowding. A good variety should have flower diameter of 8-10 cm and 30-45 cm long flower stalk. The plant should have a free-growing habit, throwing blooming shoots freely after the shoot is topped or a bloom is cut. The other main aims of the carnation breeding are to evolve cultivars that do not produce shoot at upper nodes. This can save labour in disbudding operation. The other objective in standard carnation is to develop cultivars, which is more tolerant or resistant to calyx splitting. In such situation the ideal cultivars should have broader calyx and moderate number of petals (Besemer, 1980). Rymon *et al.* (1987) carried out a joint project between market researchers and flower breeders, to optimize breeding towards an 'ideal carnation', or a mix of future varieties. They found that out of the 12 characteristics ranked, colour, number of petals, number of flowers per stem and stem structure were the most important. In the past more attention was given to evolve cultivars with better vegetative and flower characteristics and very little importance for disease resistance.

Introduction, selection, hybridization and mutation breeding are the different methods used for carnation improvement. Recently, the biotechnological tools like, *in vitro* breeding, somatic embryogenesis and transformation of genes have been used for the improvement of carnation.

5.2.1 Floral Biology

Before breeding any flower the understanding of its flower structure, floral biology and seed set is important. Carnation flowers are pedicellate, complete, hermaphrodite, actinomorphic, pentamerous, hypogynous and borne in terminal or solitary regions of the growing stem. Petals and stamens vary from 5-50 and 10-30, respectively. The pistil consists of 2-5 carpels. The gamosepalous calyx forms a tube-like structure enclosing the corolla and essential organs before anthesis. Anthesis commences at 06.00 h with the highest number of flowers (31.98 per cent) opening between 09.00 and 10.00 h. Anther dehiscence commences immediately after anthesis with a peak (63.33) between 09.00 and 10.00 h. Pollen ability, as ascertained by staining test, ranges from 89.18 per cent for fresh pollen to 31.94 per cent after 10 days of storage under laboratory conditions (20-25°C). Stigmas become receptive 48 h after anthesis and remain viable up to 224 h with a peak at 72 h after anthesis. The highest number of seeds/capsule (86) was obtained with hand cross-pollination compared with hand self-pollination (42). An average of 39.16 days was required for capsule maturity (Shafi Bhat *et al.*, 1991).

Sparnaaij and Beeger (1973) found carnation plants on which 3-4 flowers were allowed to develop on each stem had a higher percentage of fruit set and 33-50 per cent more seeds/fruit after hand pollination than normally disbudded plants, resulting in a 7-fold increase in seed production/plant. In another trial on autumn seed production hand pollination from October until December, at temperaturebetween 18 and 20°C and supplementary lighting, gave good results and enabled the breeding cycle to be reduced from 3 to 2 years. Lundqvist (1995) raised the plants of *D. arenarius* in the greenhouse from seed collected in the wild or from botanical gardens. High seed set under both self- and cross-pollination regimes indicates absence of self-incompatibility genes. The loss of self-incompatibility was thought to be due to natural selection pressures.

Rupali and Ranjan (2008) evaluated of sixteen genotypes of carnation indicated that plant height of full flowering stage and stem length was found to be maximum in var. Master. Maximum number of leaf pair per stem was found in var. Ferato. For flowering characters, var. Piax Dover took minimum days taken to bud opening, while maximum flower diameter, flower duration and minimum per cent of incidence of insect pest damage was observed in var. Tabour and minimum per cent of incidence of disease was found for var. Ferado. The coefficient of variation, both at genotypic and phenotypic levels, was maximum for per cent of incidence of disease, incidence of insect pest damage, flower yield/m^2 and leaf pair/stem. Higher values for genetic advance as per cent mean were recorded for incidence of disease (53.85 per cent), flower yield/m^2 (39.20 per cent), leaf pair/stem (35.16 per cent). Pratibha *et al.* (2017) recorded the widest range of variation in plant height (68.15 to 99.85 cm), followed by stem length (46.38 to 67.07 cm), number of days to bud formation (131.92 to 148.00), number of days to first flowering (156.54 to 174.20), flower size (5.81 to 7.78 cm), duration of flowering (11.84 to 15.71 days) and vase life (8.84 to 11.79 days). Whereas, narrowest range was observed for number of flower stems per plant (4.76 to 7.72). The magnitude of phenotypic coefficient of variation (PCV) was higher than genotypic coefficient of variation (GCV) for all the characters, indicating high degree of environment influence. A moderate value of PCV along with GCV was observed for plant height, number of flowers per plant and stem length. High values for heritability was recorded for plant height followed by stem length and flower size. High heritability associated with high values of genetic advance was observed in plant height and stem length. The selection on the basis of plant height, stem length, flower yield and vase life will be more effective for further breeding programme. Pratibha *et al.* (2017) studied the genetic diversity analysis and grouped 50 genotypes into 5 clusters revealing the presence of diversity for different growth and flowering traits. Among which, maximum number of 16 genotypes was accommodated in cluster III, followed by cluster V accommodating 15 genotypes, cluster IV having 10 genotypes, cluster II having 8 genotypes, whereas, cluster I consists of only one genotype. The genotypes falling in the same cluster are more closely related than those belonging to another cluster. The divergence at inter and intra cluster level revealed the highest inter cluster distance (8.426) between cluster-I and cluster-IV, whereas, the lowest (2.944) was found between cluster-II and cluster-IV. The intra cluster distance was maximum in cluster-III (3.036) followed by cluster-V (2.910) and minimum in cluster-I (0.001) followed by cluster-II (2.442). The diversity observed within and between the clusters can be utilized in breeding of carnation varieties/hybrids for desirable traits. Thus, genetic divergence can also be used as an indirect parameter in selecting parents to produce heterotic high yielding progenies.

Narayan *et al.* (2015) found the phenotypic coefficient of variation was found higher than genotypic coefficient of variation for all the traits under study. High phenotypic and genotypic coefficients of variation were observed for total flowers/hill and number of flowers/stem. Significant variations among the genotypes were observed for most of the traits studied. Highest number of flowers/hill (104.50), maximum flower diameter (62.75 mm) and minimum days for bud to flower opening (8.00 days) were recorded in cv. Master. In general, PCV was found higher than the corresponding GCV and ranged from 9.27 (in stem girth) to 89.35 (in total flowers/hill) and 9.16 (in stem girth) to 89.22 (in total flowers/hill), respectively. Total flowers/hill, stem girth, number of leaves and flower opening duration showed high heritability (99.72 per cent, 99.13 per cent, 98.80 per cent and 94.48 per cent, respectively) coupled with high genetic advance as per cent of mean indicating presence of additive gene action showing scope for improving these characters through selection.

5.2.2 Use of Wild Species

Atanasova (1998) made crosses between cultivated spray carnation (*D. caryophyllus*) and some wild *Dianthus* spp. Crosses between *D. silvestris, D. knappii, D. sequierii, D. carthusianorum* and *D. caryophyllus* were successful in producing seeds and hybrid plants. Crosses involving *D. chinensis, D. plumarius and D. gratiano-politanus* were successful when the spray carnation cultivars were used as the maternal plants. Crosses between cultivated carnation and *D. giganteus, D. barbatus* and *D. deltoides* did not produce seeds.

Bacterial wilt (*Pseudomonas caryophylli*) is one of the most important and damaging diseases of carnations in Japan and almost all carnation cultivars are susceptible to bacterial wilt. Onozaki*et al.* (1999) screened 70 wild *Dianthus* accessions for resistance to bacterial wilt by using the cut-root soaking method with an inoculum concentration of 10^7cfu (colony-forming units) per ml and found two highly resistant wild species, *D. capitatus* ssp. *andrzejowskianus* and *D. nenteri*. These 2 species showed no disease symptoms throughout the experiment.

5.2.3 Resistance Breeding

Diseases, which cause severe damage to carnation, are fusarium wilt (*Fusarium dianthi*), leaf blight (*Alternaria dianthi*), powdery mildew (*Oidium dianthi*), rust (*Uromyces dianthi*), stem rot (*Rhizoctonia solani*) and bacterial wilt (*Pseudomonas caryophyllus*).

Sparnaaij and Demminck (1976) selected resistant clones from seed grown cultivars which served as resistant parents. In three generations adequate level of resistance had reached and selection for quality and productivity was made.

5.2.3.1 Fusarium

Arus *et al.* (1992) compared individual plants from different progenies of crosses between resistant and susceptible genotypes to race 2 with a set of controls of known level of resistance. The distribution pattern of susceptible and resistant phenotypes among the progenies tested suggested that the genetic basis of resistance is complex. Ben Yephet *et al.* (1993) evaluated response of carnation cultivars in an artificially infested field for 3 years. Disease incidence was highly correlated with disease seve-rity, suggesting that disease incidence may be used to estimate the impact on the host. Pawliczuk and Orlikowski (1994) found that hybrids differed in their resistance and the greatest number of highly resistant clones (no more than 10 per cent of plants infected) was derived from the cross Pallas x Sacha. Resistant cultivars gave clones with higher resistance when they were used as maternal parents than as pollinators.

Maria *et al.* (2007) inoculated eight isolates of *F. oxysporum* fsp. *dianthi* (Fod) from a collection of isolates obtained from diseased plants surveyed in the main growing area of Spain. Susceptible reactions were shown in most cultivars inoculated with six isolates characterized as race 2 when tested on differentials. However, cultivars Elsy and Westcristal were fully resistant to the six isolates, whereas 'Scarlet King' showed a variable response, suggesting genetic diversity within race 2 of Fod. In contrast, eight cultivars inoculated with three isolates of races 1 or 8 were usually resistant except for cultivars Nelson and Solar, which were fully susceptible, and 'Elsy', which showed resistance to only one of the isolates. Partial polygenic inheritance of resistance to race 2 determines the complexity in the host reactions.

Soto *et al.* (2012) results found that the individual F_1 population for *Fusarium* wilt were distributed between two extreme ranges, resistant and susceptible; this shows that there is segregation for the trait resistant to *F. oxysporum* f.sp *dianthi*. They did not observe clearly differentiated classes, *i.e.* with complete absence or presence of the disease, indicating a possible control of the resistance in the evaluated carnation material, governed by more than one gene and with a possible additive genetic action. Sunita (2015) revealed that neem cake and pine needles out of nine organic amendments gave 77.49 and 72.49 per cent *Fusarium* disease control with minimum disease incidence. Combined application of two antagonists *Trichoderma viride* and *T. harzianum* after 60 days of soil solarization practices registered minimum incidence (16.25 per cent) of the disease followed by *T. viride* and *T. harzianum*. The growth characteristics stem length, number of flowers and flower size per plant also increased significantly in these treatments.

5.2.3.2 Rust

Semina and Shestachenko (1981) studied the resistance of 23 carnation species, 22 cultivars and 42 hybrids to *Uromyces dianthi*. In the Shabo group 10 cultivars were found to be unaffected by the rust; in the Sim group there were no unaffected cultivars but Rosa and Arthur Sim were highly resistant. All the hybrid forms were susceptible.

5.2.3.3 Blight

Madhu Meeta *et al.* (1996) assessed carnation cultivar for their resistance to blight caused by *Alternaria dianthi*. Cultivars Yellow Dusty, Harvest Moon, Leena and Shocking Pink were highly susceptible and cultivar Scania was the least susceptible.

5.2.3.4 Bacterial Wilt

Onozaki *et al.* (1998) carried out interspecific hybridization between *D. caryophyllus* and *D. capitatus*, a wild species highly resistant to bacterial wilt. Interspecific F1 seedlings were resistant to bacterial wilt. New bacterial wilt resistant line 91B04-2 was selected out from the cross of spray-type carnation cv. Super Gold x *D. capitatus*. This line was highly resistant to bacterial wilt. Onozaki *et al.* (1999) screened 70 wild accessions for bacterial wilt (*Pseudomonas caryophylli*) by using the cut root soaking method with an inoculation concentration of 10 cfu/ml. Two highly resistant wild species *D. capitatus* ssp. *andrzejowskianus* and *D. henteri* were found. Retesting by using 4 isolates showed *D. capitatus* ssp. *andrzejowskianus* to be the most resistant.

Cultivars of Carnation

***Source*: www.florenceflora.com**

Yagi *et al.* (2012) breed new cultivar 'Karen Rouge' having resistance derived from *Dianthus capitatus* ssp. andrzejowskianus, a wild species highly resistant to bacterial wilt. It is a deep yellowish-red standard-type cultivar selected from a cross between carnation cultivar 'Miracle Rouge' and resistant line 4AZ31-5. 'Karen Rouge' has a high degree of resistance to bacterial wilt, and its productivity and other traits desirable for cut flowers are similar to those of modern commercial cultivars.

5.2.4 Flower Colour

Alpi and Buiatti (1972) carried out genetic and biochemical analyses on segregating progenies derived from selfing carnation cvs. 5774 and 4800. They observed divergent selection with increase in vase life of 1.28 days for 5774 and increase in petal number of 26.55 for 5774 and 10.69 for 4800. Pigment analysis showed the presence of pelargonidin-3-monoglucoside with a varying number (0-2) of phenolic radicals. Genetic analysis of flower colour revealed 2 heterozygous genes in 5774. Gatt *et al.* (1998) carried out interspecific hybridization with the aim of transferring yellow flower colour from yellow carnation (2n = 2x = 30) and *D.knappii* (2n = 2x = 30) to a white flowering cultivar and garden pink *D. plumarus* (2n = 6x = 90). All the progeny from the crosses with carnation were pink but those from crosses with *D. knappii* were pale cream yellow with some variation in intensity between plants. Differences were observed in chromosome number between hybrids, which were either tetraploid, or pentaploid. The F1 hybrids with *D. knappii* were yellow because they contained the same pigments as *D. knappi* but the hybrids with the carnation were pink due to their ability to convert chalcones through dihydroflavones and then to anthocyanins.

The biosynthetic pathway to anthocyanins and flavonols in carnation is briefly reviewed by Forkmann *et al.* (1995). In carnation flower specific cDNA library was constructed and heterologous probes were used to isolate and characterize clones for the enzymes chalcone synthase [naringenin-chalcone synthase], chalcone isomerase, flavanone 3-hydroxylase, dihydroflavonol 4-reductase (DFR) and anthocyanidin synthase. The lack of DFR activity in the stable acyanic mutants Virginie and Pallas was caused by a transcriptional block of the DFR gene. This was also the case in the white areas of the unstable mutant Vienna whereas a functional gene expression occurred in the red stripes.

5.2.5 Dwarfing

Sparnaaij and Demmink (1985) used principal wild species like *D. carthusianorum, D. cruentis* and *D. giganteus* and found that all have a markedly compact growth habit; the last species showed most promise in producing more bushy hybrids. Other crosses involve *D. knappii, D. superbus, D. sylvestris, D. chinensis, D. barbatus* and *D. trifasculatus* were also of interesting. Sparnaaij *et al.* (1990c) obtained two dwarf regenerants by ovule culture of 2 genotypes of tetraploid carnation. Parents and regenerants were selfed and intercrossed in all combinations. Analysis of progenies showed that premature generative development in the regenerants, as expressed in dwarfing, excessive branching and premature flowering, was genetic in nature with inheritance being either additive and polygenic or monogenic with an intermediate heterozygote. The degree and frequency of dwarfing depended on explant, genotype and/or environment. Wen GuoBao *et al.* (1995) tried embryo rescue between *D. caryophyllus* x *D. chinensis* and *D. caryophyllus* x *D. barbatus*. Backcrossing for 3-4 generations resulted in the production of 30 new varieties and 130 hybrids with potential for use as parental materials in breeding programmes. One male sterile line was observed in the F1 generation of the cross *D. caryophyllus* x *D. chinensis* and this line was propagated for use as a pot plant.Codianki, a repeat flowering, compact carnation (*Dianthus hybrid*) cultivar was derived from the cross X93.1.1 x Alpine White (Anon., 2000).

5.2.6 Biomass Production

Improvement of dry matter production in the vegetative phase is a breeding objective in many crops, especially where whole plants or parts of plants in the vegetative phase constitute the commercial product. Dry matter production is a complex process and plant dry weight, as measured at a given date, is a complex character, of which initial weight of planting material and mean relative growth rate are the major components. Sparnaaij *et al.* (1996) demonstrated that a method of component analysis can be used to facilitate breeding for a high rate of vegetative growth. The procedure was illustrated by data from a study of variation in forcing ability among 13 carnation cultivars. It involves a detailed analysis of mean relative growth rate over the period of vegetative growth, including components derived from observations on net rate of photosynthesis and dark respiration. It is shown how component analysis can be used: (i) to quantify the relative contribution of initial plant weight and of mean relative growth rate and its components to the genetic variation in plant dry weight, thus allowing a judicious choice of selection criteria for breeding parents; and (ii) to predict mean plant dry weight of F1 hybrids obtained by crossing these parent genotypes, thus facilitating the choice of crosses to be made.

5.2.7 Ethylene Production and Vase Life

Cross-breeding technique to develop carnation lines with long vase life and either low ethylene production or low ethylene sensitivity was reported by Onozaki *et al.* (2001). Two cycles of selection and crossing to improve vase life led to a 3.6-day increase in mean vase life. All 39 selected lines had significantly longer vase life than the control cultivar White Sim. In particular, second generation lines 63-3, 63-12, 66-15 and 63-41 had a mean vase life of more than 15 days without chemical treatment. Measurements of ethylene production indicated that flowers of all second generation selected lines had a greatly reduced capacity to produce ethylene.

Burchi *et al.* (1999) studied the postharvest flower life in a cross between carnation (*Dianthus caryophyllus*) cultivar Roland, with long-lasting flowers that do not release ethylene during senescence, and cultivar Milady, with short-lasting flowers that produce ethylene during senescence. Flower life of the F_1 hybrid and F_1 Milady backcross were evaluated during different seasons. In addition, flowers were evaluat-ed both intact on the plant and cut in a vase. An interesting relationship between flower colour and flower life was observed; the F_1 plants that had a similar flower colour to the longest-lasting parent (violet) showed a longer postharvest life than plants with a similar flower colour to the other parent (pink). The F_1 progenies with a flower colour darker than violet (purple) had the longest postharvest life, whereas F_1 progenies with a flower colour lighter than pink (pale pink) had the shortest life. A loss in the linkage between the genes controlling the two traits was observed in the backcross offspring.

Takashi *et al.* (2006) developed many carnation lines with extremely long vase life by using the conventional cross-breeding technique. Selected lines from second-, third- and fourth-generation have vase lives that are 2.7 to 4.1 times longer than the control cultivar, 'White Sim'. The mean vase life of line 108-44 was 23.6 days in 2003 and 19.1 days in 2004 (414 per cent and 341 per cent of 'White Sim' flower longevity, respectively). All selected lines showed low ethylene production in whole flowers during senescence. In addition, the petals and the gynoecia of three selected lines 908-46, 702-21, and 006-13, produced only trace amounts of ethylene during senescence, indicating that the ethylene biosynthesis pathway in these lines was almost completely blocked during senescence. Although ethylene sensitivity of the selected lines was generally high on day 0, immediately after harvesting, ethylene sensitivity after anthesis rapidly decreased with age in the three selected lines. Takashi (2008) developed two new cultivars, 'Miracle Rouge' and 'Miracle Symphony', with a genetically determined long vase life. They had vase lives of 17.7 to 20.7 days (3.2 to 3.6 times the vase length of the 'White Sim' cultivar) under standard conditions (23°C, 12-h photoperiod, 70 per cent RH). The ethylene biosynthesis pathway in these cultivars was almost completely blocked during natural senescence, which was responsible for the long vase life. Changes in ethylene sensitivity with flower senescence were evaluated using a time-lapse video recording system, which provides a simple and accurate way of evaluating ethylene sensitivity. The video system revealed that the ethylene sensitivity of various cultivars including 'Miracle Rouge' and 'Miracle Symphony' after anthesis decreased with increasing age.

Juan *et al.* (2017) worked out the strategies to prevent postharvest losses include the use of genotypes that have a longer life. The objective of this studywas to develop a screening test for the estimation of cut flower postharvest life and the response to exogenous ethyleneof different carnation cultivars at an early stage of plant growth. Ethylene sensitivity and production in different cutflower cultivars was evaluated, and a similar response in the vegetative stage was studied. Also, the possible relationship between the morphological parameters of cuttings and flower postharvest life was studied. Ethylene production of cuttings may beauseful tool for estimating ethylene production of cut flowers, there is a strong relationship between cut flower vaselife and the root length of cuttings as well as cut flower ethylene sensitivity and the number of internodes the cuttings have. Applications of exogenous ethylene to cutting cultivars have an effect on the growth parameters of the cuttings, but the response to ethylene in cut flowers does not behave in the same way in the vegetative stage (Juan *et al.*, 2017).

Maike *et al.* (2018) found indirect selection is a useful tool for the selection of complex traits like vase life (VL) of cut carnations. Phenotypic and genetic correlations were calculated for traits such as VL, flower diameter (FD) of standard carnation, stem length (SL) and count of flowers (CF) in mini carnations, phenotyped in three different experiments. The experiments are special in that some traits are measured in a second phase. Data were analyzed by bivariate mixed models considering in which phase the trait was measured. Some relevant genetic correlations were found in VL and FD and the traits SL and CF, but their results indicate that indirect selection would show no benefit compared to direct selection. However, the combination of direct and indirect selection could improve breeding of long-lasting cut carnations

5.2.8 Embryo Rescue

Demmink *et al.* (1987a) collected ovules prior to anthesis from the flower buds of *D. trifasciculatus* (tetraploid), *Dianthi* types (tetraploid interspecific hybrids) and of an F_1 cross between the two were cultured. A few of the ovules, however, developed into plantlets. Some diploid plants were regenerated from unpollinated ovules of a tetraploid F_1 plant and it is suggested that they originated via gynogenesis. They produced no pollen. Attempts to utilize these plants as female parents in crosses succeeded but hybrids with diploid glasshouse carnations were not obtained.

5.2.9 Mutation Breeding

5.2.9.1 Natural Mutation

Mutation breeding is one of the most reliable technique in improving crop plants. Mutagens are used to bring the variability in floricultural crop, like *Dianthus*. In breeding programme aimed at improving flowering and flower architecture, two mutants heterozygous for a mutation in a gene termed *evergreen* (*e*) were identified by Scovel *et al.* (2000). In these mutants, spike-like clusters of bracteoles subtend each flower. Genetic analysis of the mutants confirmed the semidominant nature of this nuclear mutation and that the two original mutants were allelic at the *evergreen* locus. In homozygous mutant plants, a more severe phenotype was observed. Flower formation was completely blocked and spike-like clusters of bracteoles did not sub-tend any flowers. Morphological characterization of mutant plants revealed that vegetative growth and inflorescence structure are not affected by the mutant allele. In plants heterozygous for the evergreenmutation, fertility, petal and pistil length, calyx diameter, and stamen number were not affected. However, flowers from these heterozygous plants had a reduced number of petals, suggesting an intriguing link between *evergreen* and the *double flower* (*d*) gene that determines petal number in carnation.

Among chemical mutagens, the efficiency and effectiveness are necessary to study the effective dose that can bring the broad spectrum of variability. Rajib and Jagatpati (2011) studied three different concentrations (0.1, 0.4 and 0.7 per cent) of colchicine (Col), ethyl methane sulphonate (EMS) and sodium azide (SA) were used to treated *Dianthus* seeds to assess seed germination behaviour, lethality, pollen sterility and mutagenic effectiveness. It was noted that increase in the dose of EMS and SA, germination percentage and survivability were decreased; whereas colchicine doses were proportional to increase germination percentage at seedling stage, but they were not survived till maturity. Higher lethality over control (32.89) was shown by 0.7 per cent EMS. Pollen sterility also increased with increasing mutagenic doses. The maximum pollen sterility (61.1 per cent) was observed under 0.7 per cent colchicine. So, the effect of chemical mutagenesis on seedling and pollen sterility with EMS (especially 0.7 per cent) treatment is much more beneficial as compared to colchicine and SA. 0.4 per cent colchicine is effective for other agronomical characters. The highest mutagenic frequency (13.953) was observed at 0.4 per cent Col and lowest one (4.464) at 0.1 per cent Col. The mutagenic effectiveness was maximum (86.42 per cent) at 0.1 per cent EMS and minimum (13.824 per cent) in 0.7 per cent Col. The highest mutagenic efficiency (6.977) was recorded in 0.4 per cent Col and lowest (0.995) in 0.7 per cent SA on the basis of survivability (Rajib and Jagatpati, 2011).

5.2.9.2 Irradiation

Irradiation has been used as a breeding and diagnostic tool in carnation, especially for determining whether a mutant is periclinal chimera or homogeneous. Mehlquist (1957) irradiated four Sim cultivars with dosages 40KR and found a number of reversions due to the destruction of the epidermal layer of the growing point and regeneration from underlying unmutated tissue, whereas Sagawa and Mehlquist (1957) found extremely high number of changes (47 to 91 per cent) when treated with 2.5 KR to 5 KR X-rays. It was observed that reversion was due to morphological alteration of the shoot apex and not due to mutation as the mutant condition is con-fined to outer layer of cells and do not penetrate deep enough to be included in the gamete forming tissue. Harmful effect on flower yield and quality by X-rays at 2.5 or 5 KR have been observed by Sparnaaji and Demmink (1970). The overall effect on the yield had disappeared after vegetative multiplication in the 3rd layer. But quality continued to be adversely affected. Samata *et al.* (1979) found that gamma irradiation on 30 carnation cultivars led to 3.3-30 per cent colour changes in various directions, frequently from recessive to dominant colours. In cv. Clear Pink the antho-cyanidin was changed from cyanidin to pelargonidin and this was the only instance reported to be caused by gene mutation. Much of the colour variability might have been caused by the uncovering or rearrangement of chimeral layers rather than by mutation.

Johnson (1980) reported that gamma irradiation (6800 rads in 4 min) on rooted cuttings of carnation produced non-chimeral shoots regenerated from each genotypic layer of 5 periclinal chimera. Ferrero *et al.* (1987) found that application of 40 and 60 Gy produced the most effective results, although the 60 Gy dosage was often lethal. Comparison of the chromatograms confirmed morphological observations regarding the identity of rival cultivars of the same colouring, and reversion to the cv. Londorga chemotype.

Kanyarat Chaisut (1995) irradiated five carnation cultivars William Sim, Harvest Moon, Cocomo, Jacky and Jacqueline with 3000, 5000 and 7000 rads of gamma rays. Whole plants and cut branches were exposed to the radiation. The change in flower colour was evident in Jacky and Jacqueline cultivars. Treated Jacqueline produced flowers of different colors; solid red like William Sim's, yellow with red streaks like Cocomo's and Panache flower. When the treated plants were exposed to the radiation, Jacqueline branches which gave original flower colour, produced orange, red and yellow flowers. It was proposed that in cv. Jacqueline, genotypes controlling flower colors are existing in shoot tip meristematic layers. That is, genotype for yellow colour is in *tunica-1* and genotype for red color is in *tunica-2* and *corpus.*

In vitro propagation techniques are of significance for the production of solid, non-chimeric mutants in mutation breeding programmes of vegetatively propagated crops. Irradiated explants of carnation were used for the production of (adventitious) shoots and plantlets to determine the number and frequency of solid mutants and chimeras respectively. It was demonstrated that high numbers of solid, non-chimeric mutants can be obtained *in vitro* and that the percentage of chimeras is comparable to the low figures reported after use of *in vivo* adventitious bud techniques. Consequently, the micropropagation techniques seem very promising.

Oradee and Chaichoompon (1985) exposed shoot tips of White Sim of about 1.5- 2.0 cm long gamma irradiation which were cultured in MS medium. With the increasing dose, the survival as well as growth rate were decreased. Shoots died 12 weeks after gamma ray treatment at the dose of 5 KR. The plants irradiated with 1 and 3 KR differed from the control by having the pink streak on white petals of the flowers. Buiatti *et al.* (1986) analyzed genetic (interclonal) variability for flower colour and three quantitative characters (number of flowers, flowering date and plant height) in clones of the Mediterranean carnation cv. Corrida obtained from cuttings, meristems (control and X-irradiated) and *in vitro* petal culture. Use of X-rays and petal culture increased heritability estimates for plant height and flowering date but not for number of flowers. Flower colour variations, probably resulting from deletions, were present in all plants except those derived from cuttings, their frequency increasing with X-ray dosage.

5.2.9.3 Induction of Mutation *In vitro*

Induced mutations have never been predictable or repeatable. Nodal segment was the most convenient and susceptible material for induction of mutations (Roest, 1977; Roest and Bokelman, 1981). X-ray was used to induce mutation. More than 10 plant-lets were obtained from single nodal explants 4 months after inoculation on MS medium containing 4.4 µM BA and 0.57µM IAA. Cassells *et al.* (1993) reported that X-irradiation of nodes of cv. Mystere derived from *in vitro* cultures resulted in the production of useful flower color variants at a rate of approximately 2 per cent. The resultant mutants were vigorous and most were stable although pleiotropy in respect to Alter-naria disease was observed. Simard *et al.* (1992) treated the *in vitro* petal callus of cv. Niky with gamma rays. Explants of 50 Gy administered on the fourth day of culture produced variants of horticultural interest in Niky (having white flowers with pink dots, streaks and border, assumed to have arisen by mutation of a recessive gene encoding anthocyanins). This period corresponded to differentiation of cells that subsequently developed into buds.

5.2.9.4 Chemicals

Chemical mutagens have also been found promising in inducing variation in carnation. Vosamae and Kask (1972) observed changes in flower colour when treated with 2.5 mM nitrosomethylurea in 2nd or 3rd generation. Numerous mutagenic effects occurred in the shoots arising from treated axils with methyl ethane sulphonate (MES) (1-2.5 per cent). Sarafi and Mahdavian (1978) treated seeds of a diploid carnation with 2 or 4 g colchicine/l for 7, 14 or 24 h. They obtained tetraploid (4n = 60) seedlings following 7 h of treatment with 4 g/l colchicine. Tetraploids later flowered and had a greater flower diameter, more branching, longer pistils and larger petals and leaves than diploid controls. No differences in number of stomata or plant height were observed. Hentrich and Glawe (1982) applied ethyl methane sulphonate (EMS) at 1.0 or 2.5 per cent to the leaf axils of about 250 plants of 5 carnation cultivars differing in colour. Numerous mutagenic effects occurred in the shoots arising from treated axils, particularly if these were already formed before treatment. Fifteen new colour variants were selected from the 3303 axillary shoots propagated.

5.2.9.5 Resistance to Diseases

Carnation is also subjected to many bacterial and fungal diseases (Forsberg, 1963; Holley and Beker, 1963; Pirone, 1970). Three fungi (*Phialophora cinerescens, Fusarium oxysporum* f.sp. *dianthi*, and *Rhizoctonia solani*) cause important vascular diseases. Demmink *et al.* (1982) raised two cultivars, namely Revada and Novada which showed resistance to wilt caused by *Phialophora* sp. and *Fusarium* sp. Garibaldi (1983) screened 112 cultivars of carnation and found that the cultivar Duca was resistant to eight pathotypes of *F. oxysporum* f. sp. *dianthi*. Guda and D'Aquila (1980) grafted susceptible cultivars on rootstocks resistant to *F. oxysporum* f. sp. *dianthi* and the resulting plants were resistant. Arai and Takeuchi (1993) studied the influence of culture filtrates of *F. oxysporum* f.sp. *dianthi* on growth and viability of carnation callus. Culture filtrates of virulant race 1 did not affect calluses of both susceptible (Nora) and resistant (Pallas) cultivars to fusarium wilt. But, culture filtrates of virulent race 2 decreased viability and suppressed growth of callus of susceptible cultivars. In contrast, viability of Pallas callus was not greatly inhibited by culture filtrates. They recommended adoption of this procedure as a method of selection of new wilt resistant carnation cultivars *in vitro*.

The phenol compositions of two cultivars of carnation (*Dianthus caryophyllus*) namely "Gloriana" and"Roland", which are partially and highly resistant, respectively, to *Fusarium oxysporum* f.sp. *dianthi* havebeen investigated by Paolo *et al.* (2003) with the aim of determining if endogenous phenols could have an anti-fungal effect againstthe pathogen. Analyses were performed on healthy and *F. oxysporum*-inoculated *in vitro* tissues, and on *in vivo* plants. Two benzoic acid derivatives, protocatechuic acid (3,4-dihydroxybenzoic acid) and vanillic acid (4-hydroxy-3-methoxybenzoic acid), were found within healthy and inoculated tissues of both cultivars,together with the ûavonol glycoside peltatoside (3-[6-O-(-L-arabinopyranosyl)—D-glucopyranosyl] quercetin).These molecules proved to be only slightly inhibitory towards the pathogen. 2,6-Dimethoxybenzoicacid was detected in small amounts only in the inoculated cultivar "Gloriana", while the highly resistantcultivar "Roland" showed the presence of the avone datiscetin (3,5,7,2-tetrahydroxyavone). The lattercompound exhibited an appreciable fungitoxic activity towards *F. oxysporum* f.sp. *dianthi*.

6.0 Growth and Flowering

Apart from genetical factors growing environment plays an important role on growth and development of carnation.

6.1.0 Environmental Factors

6.1.1 Temperature

Temperature is another major factor that influences the growth and flowering, colour, grade and the quality of cut carnation. Day temperature of 18.3°C was found to be optimum for getting maximum number of fancy grade carnation (Hanan, 1959). Higher day temperature of 23.9°C with cool night 11.1°C causes most rapid flower bud development. Calyx splitting of flowers could be reduced by controlling temperature of the houses being heated to 15.6°C on cool days and cooled to 18.3°C on sunny days (Holley, 1960). Hopkins and Holley (1962) observed that the proportion of flowers showing petal burns was higher in plants grown at 10-15.6°C than those grown at 15-21°C. The optimum range of night temperature during winter is 10-11°C while the same during spring and summer is 12.7°C and 13- 15.4°C respectively (Laurie *et al.*, 1968). Kho and Baer (1973) studied several carnation selections and found the optimum temperature for pollen production to be 23°C. An increase in the growing temperature aids early establishment of the crop, but same temperature for more than 3 months after this period may cause thin leaves and weak stems, leaves senesced early and the plants were noticeably shorter and bore more leaves than plants grown under cool conditions. Warm-grown plants, especially cv. Harvest Moon, were more prone to red spider mite attack during the summer, whereas in winter cool-grown plants were more prone to *Botrytis* infection. The growing temperature alone had a slight effect on the rate of bud initiation. A relatively low temperature (cool regime) delayed flower development and delayed and increased the flowering flush (main cropping period). Flower diameter was increased, but in some cultivars a greater proportion of calyx spiltting occurred. Stem quality in winter was improved, but the total yield for an 18-months crop was likely to be slightly reduced (Bunt, 1974).

Ito *et al.* (1996) investigated the interaction between carnation growth and air and soil temperatures, the dry weight and nutrient absorption of Scania, Kibouno and Hikari. Plants were influenced more strongly by air temperature than soil temperature, although a favourable effect of cooler soil was noticed when the air temperature increased.

Jensen (1997) concluded that short day + 16°C without high temperature delayed flower formation and promoted lateral shoot growth. Koganezawa *et al.* (1998) reported a new disease of carnation, called carnation stunting and proliferation syndrome, when plants were grown under high temperature and excess moisture, almost all the plants, including the virus-free ones, displayed typical symptoms. However, all the plants grown in a greenhouse, in the shade, at a constant temperature of 22°C and with optimum moisture remained healthy, irrespective of virus infection. These results indicate that carnation stunting and proliferation syndrome is not caused by transmissible agents like a viroid, but by a combination of high temperature and high soil moisture.

Masaru *et al.* (2020) suggested that an end of day-cooling treatment at 21°C for 4 h from sunset was as effective as an overnight cooling treatment to improve the quality of cut flowers by promoting stem hardness and flowering. This method is a cost-effective and efficient thermal management strategy to produce high-quality carnation cutflowers in hot conditions.

6.1.2 Light

6.1.2.1 Photoperiod

Genetically carnation is a long day plant, but the present day Florist's carnations are considered to be relatively insensitive to photoperiod. Generally, long photoperiods promote flowering while short days tend to delay it. It is suggested that to develop carnation cultivars suitable for year-round cultivation, the interspecific hybrids, especially *Dianthi types* and *D. chinensis* hybrids should be selected, since they are promising for flowering under low (winter) light conditions. Sparnaaij *et al.* (1987) studied the Vari *et al.*, differences in response to long days (24 h/day), given for 2 or 3 weeks in December and/or January. Average flowering date of the first flush was used as the main parameter. LD did not induce flowering in all shoots. LD effect was expressed as the percentage of shoots induced to early flowering. Reaction of individual shoots appeared to depend on: (i) position of the shoot on the main stem, (ii) number of leaf pairs on the shoot at the start of the LD treatment, and (iii) speed of shoot development (in terms of leaf pairs unfolded per unit of time). Sparnaaij *et al.* (1990a) conducted a trial on 20 genotypes (cultivars and backcrosses of interspecific hybrids), representing wide variation in flowering time and response to long days, when planted in September in a phytotron under a winter light intensity of 15 W/m^2 for 8 h per day. Long day (LD) response was most pronounced in the earliest flowering Dianthi selections and least in the traditional commercial cultivars. Proportion of the shoots actually responding to LD varied from 28 per cent in the traditional cultivars to 54 per cent in the *Dianthi* x *D. allwoodii* cv. Doris group of genotypes.

Supplementary Light

Namikawa (1980) found that supplementary incandescent lighting (increasing the natural day length to 16 h from November to January), increased the number of flowers in January and in March but reduced yields in May, and had no effect on total yields. This treatment tended to reduce flower stem length. Kadman Zahavi and Ephrat (1982) found that light from Gro-lux fluorescent lamps, as a 4 h night break, was more effective than incandescent light in promoting spray carnation flowering under natural day light conditions. When the illuminations were applied for 4 h in the middle of the night, the effectiveness of a certain amount of radiant energy from incandescent light was found to be the same whether applied as intermittent or as continuous illumination. Andersen and Aabrandt (1989) grew potted plants of hybrid carnation under glass with or without lighting from 03.00 to 21.00 h daily beginning on 1 December. Lighting markedly increased the number of main and side shoots and the number of flower buds. The economic treatment was around 25 W/m^2. Sakai *et al.* (1993) recorded that the optimum night break for a crop planted in June was 3 h/night for 3 weeks from mid-August. These plants flowered more rapidly than those under natural day-length or with a 2 h night interruption. The night break was less effective if started later in the season, while if it was started earlier, the bud number/stem increased but the stems became weaker. Carnation planted between September and February did not require the night break treatment. Koyama and Uda (1994) reported that Nora developed normal petalcolour at light intensities of 0.2 or 1 klx whereas Coral required 2 and 3 klx to force flowers with normal petalcolour. Kumar *et al.* (1999) exposed carnation to artificial illumination at 4-5th, 6-7th and 8-9th leaf pair stage for 10, 15 and 20 days to determine the growth stage. When exposed at 6-7th leaf pair stage for duration of 15 days produced higher number of quality flowers/plant and gave an early crop.

6.1.2.2 Light Intensity

Bunt (1974) found that solar radiation had the greatest effect on growth, in winter, with an average 30 cal/cm^2/day of radiation the growth of rooted cuttings was at the rate of 0.01 g increase in dry weight/cutting/day. In summer the average radiation increased to 300 cal/cm^2/day and there was a corresponding 10-fold rise in the rate of dry weight production to 0.1 g. Byrne (1982) reported that increasing light intensity (from dusk to dawn) from 0 to 10-15 foot candle when shoots had 4-6 leaf pairs decreased the number of days to flowering in Sim carnations from 123 to 93 and the number of leaf pairs below the flower buds from 15.7 to 13.0. A 4-hour night light break (10-foot candle) for 6 weeks decreased the number of days to bloom, from 147 with no night light to 123 and the number of leaf pairs below the flower bud from 15.4 to13.0. Powell and Bunt (1983) observed that reduction in yield of flower/plant at high density was great due to a reduction in light intensity.

Bylov and Smirnova (1984) reported that photo-induction (4-hr light at night) for a month followed by supplementary light (to provide 18-hr photoperiod) for 6 months produced more and larger flowers. Kageyama *et al.* (1985) stated that plants when pinched and 3 shoots/plant were allowed to grow under different sunlight intensities; the development of lateral shoots at the middle and lower nodes of the main shoots was markedly reduced under 60 and 20 per cent sunlight.

Mulching with reflective film improved the light environment amongst the plants (73 per cent of incident radiation was reflected back); over 85 per cent of the reflected light was in the range of photosynthetically active radiation. Mulching promoted flower bud development; 100 per cent flower bud development was observed 16 weeks after planting, compared with 50 per cent bud development in control. Mulched plants also produced more number of cut flowers of better quality (lower percentage of calyx splitting) and with long stems (46-65 cm) than unmulched plants (Huh and Ki, 1995).

6.1.3 Shade

Kohl and Rundle (1974) found plants receiving full light formed far more disbud (shoots that would produce secondary flowering stems) shoots than those grown under cheesecloth shading which reduced the light by 40 per cent. In full light the number of vegetative shoots below the 6 top nodes was also greater. The response to light intensity could thus be taken into account when indexing seedlings for this characteristic, which is valuable in breeding single-flowered carnations that need no disbudding. Bunt (1974) recorded seasonal changes in natural light intensity and found that it did not affect the rate of flower development, plants grown in full light and those under netting (giving a 50 per cent light reduction) flowered at the same time. Dalbir Singh *et al.* (1994) planted rooted cuttings and provided 0, 25 or 50 per cent shading using rambo nets until five months. The results revealed that shading treatments significantly increased plant spread, height, number of shoots and flower stem length (Tables 1, 2 and 3).

Table 1: Effect of Summer Shading on Plant Growth of Standard Carnation cv. Espana*

Shading Treatment (Per cent)	*Plant Spread (cm)*	*Plant Height (cm)*	*No. of Shoots per Plant*	*Dry Weight per Plant (g)*
0	15.67	49.42	3.42	22.22
25	24.42	54.47	4.17	27.45
50	17.35	49.97	4.42	17.90
C.D. at 5 per cent	1.42	2.25	0.49	1.46

* Dalbir Singh *et al.* (1994).

Table 2: Effect of Summer Shading on Yield and Quality of Standard Carnation cv. Espana Produced in the First Flush after Planting*

Shading Treatment (Per cent)	*No. of Days taken to Flowering after Picking*	*Flower Yield (Per plant)*	*Stem Length (cm)*	*Weight of Cut Flower (g)*	*Flower Diameter (cm)*	*Vase Life (Days)*
0	93.75	5.45	45.95	18.85	6.47	7.22
25	82.00	4.35	52.90	13.67	5.35	8.25
50	84.75	2.75	50.92	13.42	5.22	5.52
C.D. at 5 per cent	3.77	0.52	3.26	0.91	0.22	0.30

* Dalbir Singh *et al.* (1994).

Wang *et al.* (2020) observed better root formation from cuttings grown under the supplementary lights after 10 days. After 25 days of propagation, Metal Halide lamps significantly increased the root length, root number, root fresh and dry weights, and shoot biomass, shoot length, and shoot fresh and dry weights. The best root ball formation and the highest root activities were also found in cuttings propagated with supplementary MH light. Supplementary light increased the plant temperature, quantum yield, stomatal conductance, and the contents of chlorophyll, soluble proteins, and carbohydrates.

Table 3: Effect of Summer Shading on Yield and Quality of Flowers of Standard Carnation cv. Espana, Second Flush of Flowering*

Shading Treatment (per cent)	*No. of Days taken to Flowering after First Flush*	*No. of Flowers Harvested (per plant)*	*Final Stem Length (cm)*	*Final Weight of Cut Flowers (g)*	*Final Flower Diameter (cm)*	*Vase Life (Days)*
0	142.00	12.87	58.78	28.82	7.45	10.01
25	140.00	13.27	58.80	28.35	7.42	9.12
50	143.00	13.02	59.05	28.02	7.40	9.10
C.D. at 5 per cent	N.S.	N.S.	N.S.	N.S.	N.S.	N.S.

* Dalbir Singh *et al.* (1994).

6.1.4 CO_2

CO_2 enrichment in greenhouse improved the flower quality (Cox, 1967). The response to CO_2 enrichment greatly depends on existing temperature and light condition. The best quality flowers were produced when CO_2 concentration in the greenhouse was maintained 500 ppm along with a day temperature of 14-15.4°C (Holley *et al.*, 1964). It is a common practice to inject CO_2 during day light hours, 3-4 times higher concentration than normal atmospheric levels of CO_2 are frequently used. Additional CO_2 under suitable conditions increased flower production from 10-30 per cent (Besemer, 1980).

Nemeth (1982) observed that the CO_2 concentration of 0.1 per cent increased the temperature at night or in sunny weather by 2°C but had no effect on temperature on cloudy days. Carnations planted during August yielded 11 per cent more cuttings by 10 months than the controls. The cuttings were also 20 per cent heavier and CO_2 nutrition was economic.

6.2.0 Growth Regulators

Growth regulators have proved effective in increasing the yield mainly by increasing the side shoots by overcoming apical dominance. There are reports about the enhancement of quality of flowers especially by increasing colour brightness. Accati *et al.* (1979) found that BA (50 ppm) applied as a foliar spray once or twice during rooting of cuttings stimulated lateral bud growth resulting in the same flower yields as hand-pinched plants. In another trial the effect of BA at 50 ppm was compared with that of zeatin, zeatin riboside and kinetin (all at 10 and 20 ppm). Zeatin at 10 and 20 ppm and zeatin riboside at 10 ppm also gave a significant increase in the number of flowers, but only when applied after transplanting rooted cuttings. Flower quality was not affected by any of the treatments. Wickel and Boerner (1979) reported that spraying of morphactin solutions at 1.1 and 2.2 ppm of active ingredients promoted growth of lateral shoots and resulted in higher yields as compared to control plants. When 6-10 shoots/plant were obtained, every additional shoot caused an increase in yields of 6.3 flowers per 100 plants. Mynett

(1979) studied the effect of growth regulators BA, GA and NAA on flower bud development. The BA unaffected or slightly promoted the flower bud development both under short and long days. GA retarded the bud development more under short days than under long days. NAA retarded carnation bud development and caused abnormalities and deformation of flowers. Atherton and Harris (1980) observed an increase in incidence of secondary growing centers within the flowers, when plants were chilled at 5°C for 1-3 weeks during flower development. Application of GA, to shoots in the early stages of flower formation also caused an increase in the number of secondary growing centres formed. Foliar spray of 100 ppm ethychlozate at 10 ml/plant just before flowering increased stem strength and improved flower quality (Sakai *et al.*, 1993).

Amar (2019) found cultivar King Lion had significantly highest plant height (20.66 cm), while cultivar Eskimo had the highest flower stem length (53.59 cm) and stem diameter (5.63 cm) followed by Delson (51.66 cm). The lower flower stems lengths of cultivars: White Liberty (42.48 cm) and King Lion (43.87 cm) were at par. The effect of mineral nutrition and hormone on plant height, flowerstem length and flower head diameter was non-significant, however, IAA (25 ppm) foliar spray was found superior over N 40 g/m^2 and IAA 50 ppm applications.Thus, Eskimo is recommended as the excellent cultivar for its bigger flower stem, straightness and attractive flower, while, Delson and White Liberty as promising for higher flower yield. Similarly, soil application of nitrogen @ 40 g/m^2 and foliar spray of auxin @ 25 ppm are recommended for higher flower yield.

6.2.1 Growth Retardants

Pot grown carnation cv. Hanacy sprayed with chloromequat (CCC) at the concentration ranging from 10-100 ppm enhanced floral initiation and increased the yield of flower considerably (El-Fouly *et al.*, 1977). Muench and Fritzche (1974) recorded that application of CCC at 0.25 per cent and daminozide (SADH) at 0.4 per cent caused earlier flowering and improved flower quality, reduced the number of flowers showing calyx splitting, whereas SADH at 0.25 per cent delayed flowering. Dubey (1972) noted that pot-grown carnations sprayed with MH at 500-1000 ppm increased the number of flowers but flowering was delayed by 8-28 days depending on the variety. Scorching of the leaf margins and curling of leaves were caused by concentration between 1500-2000 ppm.

Pobudkiewicz and Goldsberry (1989a) pinched back the plant leaving 3-4 pairs of leaves and sprayed 10, 21, 35 or 63 days later with one of 3 growth retardants, *viz.* sumagic [uniconazole] at 2.5, 5.0, 7.5, 10, 15 or 20 ppm, paclobutrazol at 3 or 10 ppm or CCC at 2500 or 3500 ppm. All treatments significantly reduced plant height, but the optimum height of 15-18 cm was achieved with sumagic (uniconazole) at 15 ppm when lateral shoots were 2-4 inches long. Pobudkiewicz and Goldsberry (1989b) pinched back after 11 days of planting and sprayed after 20,40 and 60 days with either unicona-zole or paclobutrazol, each at 5, 10 or 15 ppm. The lateral shoot length, flower diameter, number of days to flower and plant FW were recorded 16 weeks after planting. Responses to treatment did not differ between the 2 growth retardants. The most desirable plant height (15-18 cm) was achieved when either compound was applied at 15 ppm on 4 November, 6 weeks after pinching, when the vegetative laterals were 7.5-10 cm long.

6.3.0 Physiology of Growth and Flowering

Leaf area is one of the most important factors which affect the photosynthesis of the plant. Stereology is a set of simple methods for the quantification of 3-dimensional structures. The use of such methods to estimate volume and leaf surface areas of carnation plants was compared with values obtained using other analytical techniques by Bierman *et al.* (1997). Stereological methods could not be used to estimate volume, but gave good estimates of leaf area. Such values could be used in risk assessment studies involving exposure to pesticides.

6.3.1 Photosynthesis

Martinez *et al.* (1993) observed decreased carbon assimilation rates when growing media was flooded with water. The decrease in photosynthesis was greater in plants in which half the root system was flooded than in those in which the full root system was flooded. This decline was not linked to a decrease in stomatal conductance indicating that it was due to a decline in photosynthetic capacity. Flooding slightly lowered shoot water potential; this decrease was higher in treatments affecting full root systems. This suggests that water potential is not the factor causing the observed decline in photosynthesis. Keun *et al.* (2019) found at 200 mM NaCl, stressed leaves appeared to have smaller cell size and larger intercellular spaces. Stomata subsided at lower and upper epiderma, while stomatal size and density decreased at higher NaCl. Epicuticular wax layer was particularly thicker at 200 mM NaCl and some stomata were completely covered with the epicuticular wax. Overall, the reduced photosynthetic capacity under salinity stress could be attributed to a decreased stomatal conductance.

6.3.2 Flower Development

The mobilization of polysaccharides from leaves to developing flower buds and mature flowers occurs after initiation of flower primordia (Fawzi and Fouly, 1979). This process functioned by integrative activities of amylase and invertase, ensuring abundant flow of the mobilized carbohydrates from the source to the sink. Successive stages of floral development exhibited different sink strengths as indicated by amylase and invertase activities and carbohydrate contents in stem apices, which increased with progression of floral development.

Mor *et al.* (1980) observed that ovary growth in carnation flowers was inhibited by pre-treatments with silver thiosulfate which retarded petal senescence. Although the start of ovary growth preceded visible wilting of untreated flowers, the ovary does not appear to control carnation senescence. Its removal did not alter the time of senescence, and isolated petals wilted at the same time as those on intact flowers. The ovary normally competes weakly for metabolites with the petals; growth of the ovary occurred only when the petals were removed or started to senesce, if the sink strength of the ovary was increased by injecting growth regulators or if sucrose was supplied exogenously. It seems unlikely that the growth of the ovary is dependent on reallocation of carbohydrate materials from the petals.

Propagation of Carnation

Seeds of carnation

Seedlings

Planting of Cuttings

Mass propagation by cutting

Micropropagation of carnation

Source: Gutiérrez-Miceli *et al.* (2010) *In Vitro Cellular and Developmental Biology Plant*, 46: 57-63.

6.3.3 Flowering

Nimura *et al.* (1996) investigated the possibility of predicting flowering in the short-term culture (single-pinched culture) of carnation cv. Francesco. The developmental rate (DVR), which was based on environmental factors, was computed and flowering was predicted by a developmental index (DVI), which was an accumulation of DVR. The DVI value was set to 0.0 at the date of pinching, and 1.0 at the date of flowering. Air temperature and photoperiod were analysed as environmental factors, and the meteorological prediction model was also calculated. They found that it was possible to apply the prediction model to plants grown in environments at 5 or 10°C. The DVR function was calculated to predict the flowering date using data from 9 crops based on different planting date and pinching date.

They concluded that it is possible to advance or delay the flowering time to the target time by temperature control based on growth diagnosis.

7.0 Propagation

Carnation may be propagated by both sexual and asexual methods. Perpetual flowering carnations are multiplied vegetatively by means of stem cuttings while seed propagation is normally practiced in raising plants of marguerite carnations and for the purpose of hybridization. Large scale multiplication through micropropagation for producing disease-free plants is also standardized and followed.

7.1.0 Seed

Carnations, especially the Marguerite or Chabaud types, are easily raised from seeds. In the plains of India, seeds should be sown in July or August under protection from rains and it may extend up to October. Seedlings raised from seeds sown earlier than this are weak. In Bangalore, it is possible to raise seedlings during May and June or even earlier, but the best sowing time is between August and October. In the hills, sowing should be done during August to October and March to April depending upon temperature. The seeds are sown thinly in seed pans or seed beds. Watering should be done judiciously to avoid damping off. The seeds germinate in about 5-8 days. At four-leaf stage, the seedlings are pricked or transplanted singly in 7.5-cm nursery pots using compost of equal parts of leaf-mould and good garden soil till final potting or transplanting. The seedlings are hardened by gradually exposing them to sunlight. When these grow to a height of 15 cm, they are transferred to 25-cm pots or in beds. Laurie *et al.* (1968) suggested to sow the seeds as soon as they ripe. Sowing at a temperature of 20°C gave the maximum germination. However, Mastalerz (1978) found the optium temperature for germination of seeds at 21°C.

7.2.0 Vegetative Propagation

Vegetative propagation by using soft terminal cuttings, is the commonest method of multiplication used by commercial growers throughout the world.

7.2.1 Treatment of Stock Plant

In *Dianthus allwoodii* and *D. alpinus*, Wurr *et al.* (2000) reported that keeping stock plants at low temperatures, before cuttings were taken advanced flowering.

The earliest flowering in Doris was achieved by storing stock plants at 3.9°C and then rooting cuttings at 14.9°C (rather than at 10.0°C), suggesting that it was possible to satisfy their flower induction requirement entirely during cold storage. In Pike's Pink earliest flowering was achieved by storing stock plants at 3.9°C and then rooting cuttings at 10.0°C, suggesting that the cold requirement for flower induction was not entirely satisfied by cold storing stock plants but in addition needed some time at low temperature during rooting.

7.2.2 Cutting

Terminal cuttings measuring about 10 cm, with four leaves are taken from the side shoots which have not yet developed any flower buds. The lower cut is given below a node and leaves at the base are removed before inserting the cuttings in the rooting media. The leaves should not come in touch with the media. The cuttings may be raised in a mist-chamber, a small plastic house or in the open if humidity is sufficient. In about a month's time, the cuttings develop sufficient roots, they are transferred singly to 7.5 cm pots containing a potting mixture of equal parts of soil and leaf-mould and half part sand.

7.2.2.1 Selection of Cutting

It is necessary to segregate the stock plants and maintain them under careful obser-vations. Undesirable plants should be rouged and cuttings are to be made from top quality stock. Cuttings taken from plants fertilized with adequate nitrogen were found to root better than those obtained from low nitrogen stock (Odom, 1953). Khalighy (1976), however, observed high rooting activity in stem extracts of non-fertilized plants and those which received low doses of N and P. The rooting in cuttings were suppressed when K was applied or by treatment with higher rate of NPK to mother plants. Polyanista (1972), reported that cuttings taken from plants which had not been allowed to flower rooted better as compared to those from plants weakened by flowering. Terminal cuttings measuring about 10-15 cm long with 4-5 pairs of leaves are commonly selected. According to Nakasone and Kamemoto (1957), cuttings with leaves rooted better than defoliated buds. Plants from long cuttings produced more and earlier flowers than those from small cuttings with no difference in quality.

7.2.2.2 Storage of Cutting

Gill and Grewal (1987) planted stock plants in October in the open in Punjab and found that cuttings started yielding in January and by mid-April each plant produced a total of 24-25 cuttings. The cuttings could be stored successfully for 140 days at 2° to 4°C without any appreciable adverse effect on the rooting potential. Prolonged storage of cuttings increased losses from <10 per cent after 12 weeks to 25 per cent after 20 weeks. Suh Jeungkeun *et al.* (1997) studied initiation of root primordia and physiochemical changes after low temperature storage (0-6 weeks at 5°C) in cuttings of carnation. Sucrose content markedly increased after 2 weeks, but starch and total carbohydrates decreased with storage. Activity of auxin-like substan-ces in stem tissue increased after 2 and 4 weeks of storage. Root primordia were initiated and differentiated from pericycle parenchyma cells after low temperature storage and NAA (1000 mg/l) treatment. Skipina (1983) establishedthat cuttings for rooting should have not less than a 10 per cent carbohydrate content on dry weight basis, leaves with high water-retaining capacity and a morphologically sound apical meristem. Goto *et al.* (1998) stored plants for 0-30 days at 1 or 10°C in the dark. Plants stored at 10°C showed increased node number and stem length with increased storage duration. At 30 days after transplanting, lateral shoot number of plants stored for 20 and 30 days at 1°C decreased. According to Garrido *et al.* (1998), the duration of storage and auxin treatment (immersion of the basal 3 cm in 15 mM IBA + 9.88 mM NAA) modified rooting process and root growth. Long storage periods (8 or 12 weeks) delayed and reduced the rates of root formation. It was suggested that some changes occur in the endogenous auxin concentration and auxin sensitivity during cold storage, and that this could be responsible for modifying the distinct phases of the rooting process. It was also explained that the requirement for auxin by the cuttings differed in each phase, depending on the cultivar.

7.2.2.3 Rooting Media

Several workers tried damp sphagnum moss (Sweet, 1952), sand (Juga, 1963), perlite (Matzke, 1965), vermiculite (Tinga, 1965) *etc.*, for rooting of carnation. Ozawa and Kikugawa (1953) observed better rooting in sand and vermiculite mixture than either alone or various sand-soil mixtures. They found that 19-35 per cent soil moisture, 60 per cent porosity and 30-40 per cent air content were the best physical condition of the media used for rooting. Higher percentage of rooting was also recorded in cuttings planted in peat-perlite mixture than perlite alone (Cheever, 1969b). Bailey (1971) recommended perlite while Besemer (1980) suggested a rooting medium containing 1 part of peat moss and 2 parts perlite plus sufficient calcium carbonate to bring the pH near 7.0. According to Hartley *et al.* (1978), carnation cuttings rooted best in a mixture of 40 per cent perlite and 60 per cent sphagnum peat moss + $CaCO_3$, followed by those in 100 per cent perlite, perlite + peat alone, and perlite + peat + $CaSO_4$. Lal and Danu (1985) recorded best rooting (50-76.6 per cent) and plant survival (43.3-63.3 per cent) with sand alone and this was attributed to its high pH (8.02) and low water holding capacity. According to Mirzaev (1988), the most suitable growth substrate among sand, perlite, peat was peat + perlite (1:1 or 1:2), followed by peat + perlite + sand (1:1:1). Hanic (1990) found that higher percentage of vermiculite had a better effect on rooting quality. Vermiculite can substitute perlite in rooting of carnation cuttings.

7.2.2.4 Propagation Time

The propagating season extends from early November until the end of March. It is possible to set the time of propagation quite accurately if it is known when the carnations are going to be planted. Rooting of cuttings varied according to the planting time, but in general, the percentage of root formation was found over 90 during cooler months under intermittent mist (Nakasone and Kamemoto, 1957). The most suitable time for collecting cuttings of Red Sim carnation was reported to be early summer or autumn. A reduction in rooting capacity at the end of August was associated with partial lignification of parenchyma (Polyanitsa, 1971).

According to Bailey (1971), January, February and March were most suitable months for propagation of carnation through cuttings. Bagaturiya and Tavberidze (1976) reported that cuttings set in summer rooted more rapidly than in the other seasons, but produced a smaller and weaker and less floriferous plants. The production of cuttings in carnation cvs. Scania, Elliott White, Crowley Sim and White No.1 was the highest in April and the lowest in January (Kiyota and Hanan, 1981).

7.2.2.5 Sterilization of Cutting

The rooting medium should be sterilized for each successive group of cuttings. Treatment of cuttings with several fungicides also improved rooting. Disinfection of soil 3 days before planting with 50 g of PCNB (20 per cent a.i.)/m^2 increased the rooting of cuttings from 64.7 to 87 per cent and dipping the base of the cuttings in 65 per cent zineb also increased rooting to 76.5 per cent, and the latter chemical proved very effective against the infection caused by *Fusarium* (Andreucci, 1964). Mitrofanob *et al.* (1982) reported that treatment with fundazol (benomyl) + cuprosan at 0.2 per cent proved more effective to control rust (*Uromyces dianthi*) on the rooted perpetual carnation cuttings than Bordeaux mixture at 1 per cent. Vigodsky-Haas *et al.* (1982) also reported that incidence of *Rhizoctonia* in carnations was reduced by drenching rooted cuttings with fungicide before planting in infected soil.

Cecilia *et al.* (2020) found microwave and ultraviolet treatments did not sufficiently sterilize leaf tissue, however, exposure to CG for 2 h resulted in no growth of either fungi or bacteria. Exposure times of carnation leaves to CG were also evaluated for spore production, spore size and morphological characteristics of five *Fusarium* spp. Only carnation leaves exposed to CG for 45, 60 or 90 min were completely free of microorganism contamination. This study identified leaves sterilized by CG as a reliable substitute for gamma irradiation sterilization. The method described here is suitable for most laboratories and will provide a means for *Fusarium* identification when gamma irradiated leaves are not available.

7.2.2.6 Treatment with Growth Regulators

Farina (1984) observed that cuttings sprayed with 50 ppm BA at the stage of callus formation during mist propagation produced more flowers throughout the season than unsprayed controls. Montero *et al.* (1985) dipped the basal ends of cuttings (cv. Scania) for 10 minutes at 20°C in solutions of H_3BO_3 (0.05-1000 mg/l), IAA, NAA or IBA (10, 25 or 50 mg/l), Exuberone (based on IBA) or in water. They were then rooted in perlite. The best results in respect of rooting and subsequent plant development (number of internodes and buds) were produced by IBA and Exuberone. Suh Jeungkeun (1997) observed that root ball diameter and root length were low in cuttings from stock plants which were grown under 8 hour photoperiod. These were promoted by hot water treatment at 40°C at 5 or 10 minutes. Dipping the cuttings in NAA (1000 mg/l) for 10 seconds increased the root diameter, length, rooting zone, shoot length and percentage of rooting.

Some workers used bacterial cultures instead of growth hormones. Jacob and Hamdan (1992) found that the vigour and quality of carnation cuttings was promoted by 14- to 16-h exposure to granular suspensions of *Bacillus subtilis* applied to the growth substrate, compared with cuttings treated or not with NAA. It was concluded that the bacterium was an effective substitute for synthetic rooting hormone. Obieglo (1992) found that the combined application of an antagonistic strain of *Bacillus subtilis* and 0.025 per cent potassium permanganate to the perlite substrate before planting gave the best control of *Fusarium oxysporum* f. sp. *dianthi* during rooting of carnation cuttings. This procedure is recommended to replace the standard captan dip.

7.2.2.7 Effect of Environment

Light, temperature, carbon dioxide, *etc.*, play important roles in rooting of carnation cuttings. Full sunlight with proper misting was considered ideal for root formation in cuttings.

Champeroux (1983) found that climatic variations had little effect on cutting weight but did affect the rate of production by the mother plants, and the heterogeneity of the cuttings for particular morphological criteria could be controlled by mother plant management. On this basis vegetative cuttings weighing 7-10 g or more were suitable as they ensured a high yield of standard planting material with a high productivity potential.

7.2.3 Layering

Stout lateral shoots not having any flower buds are selected for layering. Layering is done in pots or directly in the ground. The lower leaves on the selected branches are removed. A sharp cut is given with a knife about half the thickness of the branch, between two nodes starting just above the lower node and ending a little below the upper node to form a tongue. The tongue is kept open with the help of a wedge and inserted in the soil and held in position by a small hair-pin or layering pin made out of split bamboo. The layers generally root earlier than the stem cuttings. After rooting, these are separated from the mother plants. Layering although not very common, can also be used to propagate carnation. For this purpose ground layering was found to be most suitable method.

7.2.4 Grafting

Guda and D'Aquila (1980) recorded about 100 per cent success by grafting several cultivars on wilt resistant rootstocks. Garibaldi and Garibaldi (1982) observed that oblique cutting and use of clips was the best. Depending on the graft combination the percen-tage take varied from 65 to 97 and depended on differences in diameter of the two stems and their woodiness rather than on the affinity between the two components. Union was completed in 15 days. According to Kiyota and Hanan (1982), grafted plants produced as many flowers as self-rooted plants. Plants grafted on the wilt-resistant cultivars Gus Royalette and Scania did not succumb when planted in soil inoculated with *Fusarium oxysporum* f. sp. *dianthi*, whereas plants grafted on susceptible or partially susceptible cultivars reacted like self-rooted plants of the same cultivars.

7.2.5 Micropropagation

7.2.5.1 Type of Explants

Almost all parts of the plant were used as explants in carnation except the root. In carnation, as in many other plants, the concentration of viruses in the shoot tips region decreased towards meristematic dome

Naglaa *et al.* (2013) screened ten cultivars using leaf base explants and four media, for their capability of callus induction, embryogenic callus and plant regeneration. MS medium supplemented with 0.5 mg/l of NAA + 0.5 mg/l of BAP and medium supplemented with 0.01 mg/l of NAA + 1.0 mg/l of BAP gave the highest average of callus induction. Results also showed that the cultivars Lia and White Liberty produced the highest mean values of shoot formation (15.5 and 13.8 per cent, respectively) when calli differentiated on MS medium supplemented with 0.2 mg/l NAA and 1.0 mg/l BAP. Suman *et al.* (2013) found highest shoot induction when the concentration of $CuSO_4$ in the medium was 1 µM (10 times higher than the MS level) and 0.5 µM (5 times higher than the MS level) for axillary shoots. In proliferation medium highest number of axillary shoots 27 formed at 1.0 µM CuSO4.A significant improvement in shoots induction and proliferation was observed, when the concentration of B, Zn, and Mn was higher in the medium.

7.2.5.2 Culture Conditions

In most of the experiments the pH of the media was adjusted to range from 5.5 (Davis *et al.*, 1977) to 6.5 (Phillips and Matthews, 1964) and temperature ranged from 20°C (Hauzinska, 1975) to 26°C (Spinsky *et al.*, 1974; Jelaska and Sutina, 1977; Malczewska *et al.*, 1979). Few published reports on the effects of light were available. Gimelli *et al.* (1983, 1984) did not find any difference between long day (16-h/8-h) and short day (8-h/16-h) conditions as far as shoot regeneration from petals was concerned. The different light intensities had great effect on callus proliferation and shoot bud regeneration. The light intensities ranged from total darkness (Kakehi, 1970) to 1600 lux (Spinsky *et al.*, 1974) or 6000 lux (Malczewska *et al.*, 1979) or continuous light (Phillips, 1968). Kim *et al.* (1988) obtained normal plantlets on MS solid medium of pH 5.5 containing 20-30 g sucrose/l at a light intensity of 1700 lux for 16-h/day. Zia *et al.* (2020) analyzed the effect of chemically synthesized silver nanoparticles (AgNPs) on *in vitro* regeneration of carnation cultivars cvs. Noblessa, Antigua and Mariposa. Number of shoots/explant of cvs. Noblesse and Antigua significantly increased at 6 mg/L AgNPs (average 7.33 shoots per explant) when supplemented in MS media. While cv. Mariposa showed highest regeneration rate at 8 mg/L (average 10 shoots per explant). High concentration of AgNPs (12 mg/L) in the medium enhanced rooting response, number of roots/plant, and root length as compared with control. The fresh and dry weight of regenerated plants significantly ($P< 0.05$) increased at 6 mg/L. DPPH based free radical scavenging activity, total antioxidant activity and reducing power potential of regenerated plants varied depending upon concentration of AgNPs in the media. To find non-enzymatic antioxidants to combat oxidative damage, total phenolics and flavonoids were also determined in the regenerated plants. The study concludes that m*et al*lic nanoparticles has significant effect on *in vitro* growth of carnation cultivars however concentration dependent. Furthermore, nanoparticles can be effectively used for increased *in vitro* shoot multiplication and regeneration of floriculture plants.

7.2.5.3 Meristem Culture and Shoot Multiplication

Genotypes and growth regulators influence the morphogenetic process in shoot apex culture (Kallak *et al.*, 1966) (Table 8). Shoot tips were widely used for micropropagation as well as for obtaining virus free plants in various plant species because of their highly regenerative nature (Van Os, 1961; Murashige, 1974). Hackett and Anderson (1967) demonstrated propagation of carnation cultivars by multiple shoot formation via shoot tip culture. They placed 0.5 mm long shoot apices on a medium containing 5 times White's inorganic salts supplemented with 11 µM NAA and obtained the callus tissues that eventually developed dark green meristematic regions. Montes *et al.* (1997) recommended the use of Brassinosteroids (BB-6) in MS medium, which showed better result than IAA. Usually, hormonal combinations with high concentrations of cytokinin (2 to 10 µM) and low concentrations of auxin (0.1 to 0.5 µM) were effective for multiple shoot formation. Bull and Garton (1985) observed regeneration of only roots from internodal segments in 5 cultivars of carnation in the medium containing IAA (10M). Slight addition of Kin and NAA had positive effect on rooting of shoots obtained from shoot tips (Hempel, 1977). Early root initiation was however, recorded with 5.0 µM of NAA. On the contrary, Pennazio (1975) reported that addition of 0.1 ppm Kin to a liquid medium inhibited root formation and 1 ppm Kin suppressed it completely.

Sharma *et al.* (2010) found that plants grown in 6 per cent Panchagavya + 6 per cent Manchurian mushroom tea recorded maximum plant height (71.00cm), stem length (64.77 cm), number of flowers (6.60), flower size (4.99 cm), vase life (15.30 days) and 'A' grade flowers based on stem strength (100 per cent) and length (100 per cent). Minimum days taken to flower (173.47 days) were recorded with 4 per cent Panchagavya + 6 per cent Manchurian mushroom tea. Minimum duration of flowering (23.57 days) was recorded with 6 per cent Panchagavya + 6 per cent Manchurian mushroom tea. Minimum calyx splitting (0.00 per cent) was observed with 6 per cent Manchurian mushroom tea. Panchagavya (4 per cent) resulted in minimum (0.00 per cent) incidence of disease.

Kaviarasu and Selvarajv (2015) obtained best shoot formation response in Domingo after 7 to 8 days of inoculation from shoot tip on MS medium supplemented with (4.0 mg/l BAP and Kin) in nodal explant 9 to 11 days for shoot formation and combination of (BAP 1.0 mg/l+ Kin 0.5 mg/l) least response of shoot formation in

Bizet (11 days) and Domingo (19 days). Shoot tip showed more pronounced effect for shoot formation than nodal cuttings. Welldeveloped shoots were shifted for their multiplication. Maximum number of multiple shoots was obtained on MS medium containing (1.0 mg/l BAP and Kin) and (BAP 1.0 mg/l + Kin 0.5 mg/l) in Domingo. These multiple shoots increased in their number when were given subsequent incubation period. Addition of Kinetin to BAP shows fair shoot multiplication response. Shoots after attaining the size of 5.0 to 7.0 cm were shifted for rooting. Best rooting response was obtained on Domingo MS Basal + NAA 1.5

Cultural Operations in Carnation

Pinching

Flowering after pinching

Disbudding

Shoot after disbudding

Nylon netting

Wire netting for support

mg/l), Bizet (MS Basal + IBA 1.0 mg/l) and combination of (NAA 1.0 mg/l+ IBA 1.0 mg/l) Baltico.

Kaviarasu and Selvarajv (2015) obtained best shoot formation response in Domingo after 7 to 8 days of inoculation from shoot tip on MS medium supplemented with (4.0 mg/l BAP and Kin) in nodal explant 9 to 11 days for shoot formation and combination of (BAP 1.0 mg/l+ Kin 0.5 mg/l) least response of shoot formation in Bizet (11 days) and Domingo (19 days). Shoot tip showed more pronounced effect for shoot formation than nodal cuttings. Welldeveloped shoots were shifted for their multiplication. Maximum number of multiple shoots was obtained on MS medium containing (1.0 mg/l BAP and Kin) and (BAP 1.0 mg/l + Kin 0.5 mg/l) in Domingo. These multiple shoots increased in their number when were given subsequent incubation period. Addition of Kinetin to BAP shows fair shoot multiplication response. Shoots after attaining the size of 5.0 to 7.0 cm were shifted for rooting. Best rooting response was obtained on Domingo (MS Basal + NAA 1.5 mg/l), Bizet (MS Basal + IBA 1.0 mg/l) and combination of NAA 1.0 mg/l + IBA 1.0 mg/l (Baltico).

Kumar *et al.* (2016) achieved the highest number of direct adventitious shoots from axillary buds with 9.08 µM TDZ and 5.37 µM NAA. The maximum callus induction was achieved from leaf with 2.06 µM 2, 4-D and 2.68 µM NAA. The calli derived from MS medium supplemented with 5.37 µM NAA and 2.27 µM TDZ were found to have maximum shoot regeneration potential. The highest number of calli producing shoots and average number of shoots per callus was recorded on MS medium supplemented with 9.12 µM zeatin and 5.07 µM IAA. The calli derived from leaf explants cultured on medium containing 2.06 µM 2, 4-D and 2.68 µM NAA were highly friable, had poor regeneration potential and were selected for cell suspension studies. Plantlets were obtained from *in vitro* derived shoots/ somatic embryos with 60-80 per cent survival after 30 day of transfer to pots. The plants regenerated from axillary buds, callus and somatic embryos were compared with mother plant to assess genetic fidelity using RAPD and 2.94 per cent, 26.47 per cent and 20.58 per cent variation was observed, respectively.

7.2.5.4 Virus-free Plantlets

Commercial growers propagated carnation vegetatively by stem cuttings. The plants were mostly infected with one or more viruses which included carnation mottle (CaMV), ring spot (CRSP), vein mottle (CVMV), latent (CaLV), *etc.*hed ring (CaERV), streak (CaSV), Italian ring spot (CIRV), necrotic fleck and Arabic mosaic (Holley and Beker, 1963; Takeda, 1974). Growers were most concerned with the first threeviruses, because of the wide distribution and serious damages they cause to plant growth and flower quality. Because of the damage caused by virus infections, commercial producers of carnation were compelled to start from virus-free plants and meristem culture was regarded as the most efficient method to acheive the objective (Quak, 1977). Moshari and Danesh (1982) obtained carnation mottle virus-free plants when 0.6-1 mm long meristem tips were cultured *in vitro*. Pop *et al.* (1994) developed a system to obtain, maintain and propagate virus-free planting material for carnation by heat treatment, meristem culture and selection until the last step of propagation. Zhola and Ushkaura (1983) suggested t the use of ELISA test for detection of carnation mottle virus in meristems, and to obtain healthy plants through *in vitro* culture. Based on ELISA test Ko *et al.* (1997) found that the plants propagated by seed or somatic embryos were free from carnation mottle virus.

7.2.5.5 Vitrification in Carnation

A problem encountered during the *in vitro* culture of carnation was the formation of abnormal leaves and stem with thickened and translucent structures. Earle and Langhans (1975) reported that subculturing of the inoculum was important for shoot multiplication; the rate of shoot multiplication was maximum within 3-6 weeks of culture in the liquid medium when the cultures were continuously and slowly agitated at 1 rpm. An increase of about 60-fold in shoot multiplication was obtained at every 6 week intervals with a higher cytokinin/auxin ratio (9.3 µM KN and 0.11 µM NAA). Good quality shoots were obtained consistently when subculturing was done at 20-day intervals without much callusing at the basal end of the shoots. A nutrient mist bioreactor system was significantly better for tissue culture of carnation than an agar medium and similar to that in liquid medium (Wao and Park, 1993).

Manoj *et al.* (2012) studied the effect of ABA and GA_3 on hyperhydricity and shootprolifeartion on three varieties of carnation (White Sim, Exquisite, Scania) on MS medium. The culture system consisted of a full strength Murashige and Skoog medium with an optimum concentration of NAA (0.15 ppm) and Kinetin (1 ppm) for all three varieties. Different concentrations of ABA and GA_3 were tested for their effect on hyperhydricity and proliferationon each variety for 21 days. NAA (0.4 ppm) with GA_3 (2.0 ppm) and NAA (0.5 ppm) plus GA_3 (2.5 ppm) recorded 0 per cent hyperhydricity in all the threevarieties but also reduced the shoot proliferation. All the non-hyperhydrified shoots tested for *in vitro* and *ex vitro* rooting revealed that *ex vitro* rooting is better.

Luc *et al.* (2019) found shoot length, root length, and number of nodes of plantlets cultured in the TIS were highest when the immersion interval was 8 h. compared with that of plantlets cultured in the conventional solid medium culture, the fresh weight of plantlets cultured in the TIS was at least 3 times greater. The greatest total chlorophyll content, stomata with normal shapes was observed for plantlets grown in the TIS with an 8-h immersion interval. The lowest H_2O_2 level was recorded in plantlets cultured with the 8-h immersion interval. In the second study, growth traits such as the shoot length, root length, and stem diameter, as well as the number of shoots and roots tended to increase with immersion durations, and reached their peaks when the immersion duration was 90 s. Excessive water accumulation in tissues and a higher incidence of hyperhydricity were observed in plantlets where the immersion duration was 120 and 150 s.

Sato *et al.* (1993) recovered vitrified shoots of carnation by culturing them in a medium containing 3.0g/l Bacto-peptone. They fractionated the Bacto-peptone solution and the efficiency of each fraction for the recovery of vitrification was examined. They found that only basic, non-high molecular fractions whose molecular weight was less than 10,000 were effective. The recovered shoots rooted and acclimatized successfully.

Jayashree *et al.* (2004) subcultured *in vitro* on MS medium with BAP and kinetin at different concentrations. The highest number of multiple shoots and fresh weight were observed on 20 μM BAP/1. But the shoots showed vitrification. The shoots proliferated on kinetin containing media were healthy and normal. But on medium without cytokinins, single shoots of maximum length were produced. Also when BAP and kinetin were combined with NAA at different concentrations higher number of multiple shoots were observed. NAA at higher concentration showed vitrification and also reduced the length of the shoots. Rooting of microshoot was induced on half strength MS medium with different levels of auxins and rooting was observed even in auxin free medium. Maximum number of roots with better shoot and root characters were observed on medium with 5μM NAA/1. NAA at higher concentration proved to be detrimental to root growth by suppressing its growth. Soilrite (TC) medium gave the highest survival percentage at 15 and 30 days after transfer to hardening media and plantlets so hardened were also more vigorous.

Mahmoud (2011) reported that hyperhydricity could be controlled, in part, by ventilation, and was completely eliminated by increasing the phosphate concentration of the MS growth medium to 6.45 mM. Moreover, explant growth and morphogenesis were also promoted by increasing the level of phosphate in the MS medium to 6.45 mM. In contrast to MS medium with the standard 1.25 mM phosphate, which produced 7.9 – 10.0 shoots explant-1, MS medium with an additional 5.20 mM phosphate produced 20.8 – 34.7 shoots explant-1, depending on providing ventilation and on the sucrose concentration. The results suggested that container ventilation and MS medium enriched with 6.45 mM phosphate plus 3 per cent (w/v) sucrose significantly improved the efficiency of mass micropropagation of *D. caryophyllus* cv. White Sim.

Kazaz *et al.* (2010) obtained highest yields per plant in both cultivars ('Falcon'; 4.87 stems, 'Silkroad'; 4.68 stems) from 0-0.5 mm sizes of clinoptilolite + peat mixture at the ratio of 3c:1p, v/v. The lowest yield was in 0-0.5 mm sizes of pure clinoptilolite medium for both cultivars. Cultivar 'Silkroad' provided better yield and quality than that of 'Falcon'. Mixtures of clinoptilolite and peat can be successfully used to grow carnations. However, 0-0.5 mm size of clinoptilolite alone may not be suggested for carnation production in Isparta Province.

Thu *et al.* (2020) attempted to induce normal and vitrified leaves from three different carnation cultivars using three different media compositions and evaluated shoot induction in different PGR combinations. The shoot tips cultured on Murashige and Skoog (MS) basal media (Type 1 media) produced normal leaves, while those cultured-on media supplemented with plant growth regulators and/or vitamin (Type 2 media and Type 3 media) produced vitrified leaves for all cultivars. Culture of normal leaf segments on MS medium containing differerent combinations of IBA and TDZ concentrations induced callus in all treatments; however, the callus was unable to induce shoots and finally became necrotic. In contrast, no callus induction was observed in the control (hormone-free treatment). induced from the

Cultivation of Carnation

Planting in greenhouse

Growth in polyhouse

Development of carnation in polyhouse

Flowering in greenhouse

Cultivation in open

Drip irrigation

leaves of Type 3 media transformed into normal shoots and survived well under greenhouse conditions.

When vitrified leaf segments underwent the same treatments, shoots were induced from the vitrified leaves (derived from Type 2 media) but were unhealthy and gradually died, whereas those induced from Type 3 media were vitrified and healthy. The optimal combination for the best shoot regeneration and number of shoots per explants varied depending on the genotypes used. The vitrified shoots

7.2.5.6 Induction of Rooting

In spite of the common rooting hormones Kim *et al.* (1995) supplemented the medium with aminobenzotriazole (ABT) and obtained greater frequency of rooting and plantlet survival as compared with NAA or IBA. Regeneration of root was achieved on media containing high concentration of auxin (IAA and NAA) and various concen-trations of cytokinins (BA or Kin) (Engvild, 1972; Debergh, 1972, 1973; Malczewska *et al.*, 1979; Mii and Cheng, 1982) or in which auxin concentration was reduced after the callus had been cultured on media with auxin alone (Earle and Langhans, 1975).

7.2.5.7 Acclimatization and Field Establishment

There were several studies on acclimatization and field establishment of *in vitro* raised plants.

Marzieh *et al.* (2017) compared temporary Immersion Bioreactor (TIB) and solid culture. By applying different levels of BAP, at the concentration of 3 mg L^{-1}, we observed 14.3 new shoots in TIB, but the number of new shoots on solid medium reached to 5.7 at the same treatment. Our results also showed that with 3 mg L^{-1} BAP in TIB, the initial fresh weight of plant material increased from 10 g to 450 g after 15 days. It is concluded that TIB showed more than 10 times shoot production of carnation. Shoot elongation and rooting induction was successfully stimulated in TIB by applying 1 mg L^{-1} IBA. Rooting of proliferated plantlets from TIB and solid culture were successfully happened, and led to highest number of roots (4.6 cm) and highest length of roots (6.87 cm) in TIB. More than 90 per cent of plantlet was acclimatized to *ex vitro*.

8.0 Greenhouse Cultivation

Annual carnations are often grown in the open in an optimum range of temperature (10-25°C) with low or no rainfall during the period of growth and flowering, in a rich and porous growing media for garden decoration, cut flower and as pot plants. For commercial cultivation of perennial types, control of environment and protection from rains are considered essential for higher yield of good quality flower for a longer period.

8.1.0 Soil and Growing Media

Growing media is the material which is organic or inorganic provides anchorage to the plantsby holding the root system. It provides the essential plant nutrients required for themetabolism, growth and development of the plants. Growing media are an integral part ofmost horticultural production systems. There is a wide range of media available, this leafletconsiders the purpose of growing media and the qualities that growers should look for whenselecting media for different purposes. Growing media are the substrates in which a plant will grow. They provide anchorage for the plant's roots, air spaces to allow respiration, and retainsufficient available water to enable plant growth. When selecting media, the grower needs tofind the optimum balance between their requirements and those of the plants to be grown.

Criteria for Selecting Media

- Serves as reservoir of plant nutrients.
- Sufficiently firm enough to encourage or support the plants.
- Provides aeration for exchange of gases.
- Should not shrink or expand easily.
- Should have good drainage, porosity, aeration, *etc.*
- Should be easily available and economic.
- Should be sterilized easily.
- Free from pathogens, pests and weed seeds *etc.*

Desirable Properties of Growing Media

1. Should possess high organic matter content that will not diminish quickly. The organicmatter content should be more than 0.5 per cent.
2. The carbon and nitrogen (C: N) ratio should be narrow (10:1 to 20:1) to facilitate high rateof decomposition.
3. Optimum pH for soil based and soilless substrate is 6.2 to 6.8 and 5.4- 6.0.
4. EC levels of 0.4 to 1.4 ds/m is optimum.
4. Sufficiently porous to permit good aeration, mobility of water, root penetration as well asshould retains enough water.
5. Should possess high cation exchange capacity for nutrient reserve (6-15 me/100cc).
6. Sufficient amendments and nutrients should be added to root substrate for better results
7. Medium should be free from weeds, nematodes, diseases and insects
8. Should be readily available or easily formulated within the acceptable cost

Types of Growing Media

1. Soil based media
2. Soil less media

Soil

Soil is the basic material/ingredient of the media. It forms themajor portion in the combination of different media. Soil ischeaply available, economicand easy to handle.Most soils on the average are composed of 46-49 per cent mineralparticles (often called separates), 1-6 per cent organic matter and50 per cent air and water. The mineral particles of soil are sand, siltand clay.

Soil Less Growing Media

Soil less growing media is of two types:

1. Organic growing media (derived from living things *i.e.*plants).
2. Inorganic growing media (mined or man-made).

Organic growing media- Organic growing mediumamendments usually are derived from plants or plant productsthat occur naturally (peat moss from peat bogs), or are the byproducts of processing plants or mills (sawdust, cedar chips, bark, bagasse, rice hulls) or waste disposal plants (compost, processed sewage sludge, biosolids). The main purpose ofusing organic amendments are to loosen the soil and createlarge pores to increase. These media provides aeration, drainage, usable water holding capacity, nutrient holding capacity and decrease growing medium weight (Sapna and Poonam, 2020).

1. Peat

Peat is obtained from remains of aquatic, marsh, bog, swampvegetation found under water. It is formed when partiallydecomposed plants accumulate under water in areas with lowtemperatures and low oxygen and nutrient levels. Theimportant property of peat are retaining moisture in soil when it's dry and preventing the excess of water from killing rootswhen it's wet. Peat has long been a favoured growing mediumor constituent in mixed composts. It has excellent air andwater holding qualities, is relatively uniform, attractive, safeto handle and is generally free of weeds and pathogens.Because it has been used for many years, the properties andperformance of peat are well known and growers feelcomfortable and confident when using it.However, peat does have some disadvantages, particularlythat it can shrink significantly if left to dry out too much, andwill then be difficult to re-wet. Perhaps of more concern thanpotential production problems are the growing concerns aboutthe environmental impacts of the use of peat. Peat takes manyyears to form and its use is far outstripping its regeneration.Most of the peat used in Wales is imported from Ireland, butincreasing amounts are being sourced from Finland and theBaltic States. Peat based composts usually contain lime,nutrients, a wetting agent, sand, perlite, vermiculite and grit.The ratios of each component vary by purpose and brand.

Different Types of Peat

A. Peat Moss

Peat moss is most commonly used coarsegrade peat in horticulture and it is derived fromsphagnum or other mosses. Its colour varies from lighttan to dark brown.Peat moss has highest moisture holding capacity (about 60 per cent) among various types and it decomposes slowly and contains 6-14 per cent nitrogen. Peat moss is acidic in nature (3.2-4.5 pH), hence needs lime application (8-20 kg/m^3).

B. Reed-Sedge Peat

This peat is mainly derived fromremains of grasses, reeds, sedges and other swamp plantsand is reddish brown to almost black in colour. It is moredecomposed than peat moss. Read-Sedge peat has lesswater holding capacity and high in salt content and its pHvaries from 4-7.5.

C. Peat Humus

It is an advanced state of decomposition ofeither reed-sedge peat or hypnum mss. It is dark brown toblack in colour and has less moisture holding capacity. Itcontains 2-3.5 per cent nitrogen and pH varies from 5-7.5.

D. Sphagnum Moss

Sphagnum moss is dehydrated remainsof acid-bog plants from the genus Sphagnum and iscommercially used horticulture peat. It is light in weightand has the ability to absorb 10 to 20 times its weight inwater and pH ranges from 3.5-4.0. Sphagnum moss hashigh water holding capacity and is most desirable form oforganic matter for the preparation of growing media. It possess fungi static substances which accounts for itsability to inhibit damping-off of seedlings.

2. Coco Peat

Cocopeat is also known as Coir pith/Coco peat/Coir dust.Coir is a versatile natural fiber extracted from mesocarp tissue, or husk of the coconut fruit. The husk contains 20 per cent to0 per cent fiber of varying length. After grinding the husk, the fibers are removed and used for preparation of coco-pith which is commonly used as medium. Cocopeat is the protective fibrous coating of the seed of the coconut palm. The coir fibre used in horticulture is the residue remainingafter washing the long coir fibres which are used in rope andmat making. Coir fibre has very good aeration capacity and is used in composts as a bulking agent to increase moisture retention and porosity. Coir fibre is imported from tropicalcountries where it is considered a waste product. Coco Peat holds 8-9 times its weight in water. It can be reusedor up to 4 years. The properties of Coco Peat make it resistant to bacterial and fungal growth. It also extends the life of compost. Cocopeat provides the plants easy porosity for growing roots and good aeration for the healthy plant growth.

3. Compost

Compost is the well decomposed organic matter obtained byaerobic/anaerobic decomposition. Composting is a process of biodegradable organic wastes to stable humus by indigenous microflora including bacteria, fungi and actinomycetes. Compost provides a rich growing medium, or a porous, absorbent material that holds moisture and soluble minerals. It also provides the support and nutrients to the plants.

4. Vermicompost

Vermicompost is also known as worm castings, worm humusor worm manure. Vermicompost is defined as the end-product Compost provides a rich growingmedium, or a porous, absorbent material that holds moisture and soluble minerals. Italso provides the support and nutrients to the plantsof the breakdown of organic matter by the earthworm. Theprocess of producing vermicompost by earthworms is calledvermicomposting. This compost is an odorless, clean, organicmaterial containing adequate quantities of N, P, K and severalmicronutrients essential for plant growth. (Sapna and Poonam, 2020)

Melero *et al.* (2011) used poultry manure plus soil solarization as alternatives to methyl bromide for controlling carnation vascular wilt caused by *Fusarium oxysporum* f. sp. *dianthi* (Fod) under greenhouse conditions. Carnations planted in plots treated either with poultry manure, methyl bromide or soil solarization had lower final disease incidences, smaller areas under their disease progress curves and higher yields in comparison with untreated plots. Soil solarization was performed under suboptimal conditions, and it provided disease levels and yields similar to those of the untreated control plots. Nevertheless, under the same conditions, previous amendment of Fod-infested soil with poultry manure increased disease control over soil solarization alone, improved carnation yield and quality and also increased plant vigor, thus providing a satisfactory alternative to methyl bromide. The application of organic amendment to the same plot before every crop cycle is recommended to ensure continuous disease control, but the rates of application could be reduced to half for the third and fourth crop cycles, thereby reducing undesirable environmental effects.

Chandel (2011) found that soil amended with pine needles recorded maximum reduction in disease (82 per cent), which was followed by neem cake (71 per cent) and mustard cake (59 per cent). The least disease reduction was registered with poultry manure. The results of the four plant spacing treatments (15 x15, 20 x 20, 25 x 25 and 30 x 30 cm) indicated that the wider spacing had lowest incidence (22 per cent at 30 x 30 cm). Only three cultivars, White Candy, Regina and Dona Precas showed resistant reaction. Three others namely, Fantasia, Bologna and Macarena were moderately resistance against *F. oxysporum* f. sp. *dianthi* whereas the rest were susceptible. Disease control was high with *Trichoderma harzianum* (68 per cent) followed by *T. viride* (65 per cent), *T. hamatum* (53 per cent) while the least control was with *Bacillus subtilis*. There were increases in the levels of total sugars, total phenols and ortho-dihydric phenols, which were directly proportional to the disease control.

8.1.2 Media Sterilization

Before starting cultivation, disinfection of the media is absolutely necessary to mini-mize the infestation of soil-borne pathogens like *Phytophthora, Fusarium* and *Pythium* which could otherwise destroy the crop completely. The beds should be drenched/fumigated with 2 per cent formaldehyde (100 ml formalin in 5 litres water/m^2) or methyl bromide (30 g/m^2) solution and then covered with a plastic sheet for a minimum period of 2 to 3 days. Then beds should be watered thoroughly to drain the chemicals before planting. Basile and Lamberti (1977) fumigated loamy soil with 20, 40, 60 or 80 g/m^2 methyl bromide; plots were then aerated by cultivation to 20 cm or were irrigated with 1100 or 2200 m^3 water/ha. Bromine residues were found in the unirrigated plots only, and where they exceeded 2.5µmg/g dry soil many carnation transplants died. Bromine accumulation in carnation 60 days after transplanting was directly correlated with the bromine content of the soil at transplanting. It is important to note that carnations are highly susceptible to bromide toxicity. So methyl bromide should be used only on sandy or very light soil.

8.2.0 Growing Environment

Carnations require sufficient amount of light and proper ventilation and therefore, the design and orientation of the greenhouses are of paramount importance. Ventilation plays an important part at all times of the year and especially during warm summer weather in order to reduce temperature; it is necessary to fit the greenhouses with adequate number of ventilators. It should be equipped with heating and cooling systems that will maintain the required interior temperature.

Pergola *et al.* (1983) used hot air to maintain the temperature at about 2°C in a single span greenhouse with polythene cover. This reduces cost of cultivation, and increasing the flowering. Evaporative cooling is a standard method of lowering greenhouse temperature. Exhaust fans at the ends or sides of the greenhouse pull air into greenhouse through moist excelsior pads on the opposite sides or ends. Evaporative moisture reduces the temperature of the air. The system is very effective, when the outdoor humidity is low. Vidalie (1982) grew two carnation cvs. Le Reve and Scania, in a glasshouse and in a polyethylene tunnel with minimum winter temperatures of 13°and 4-5°C, respectively. The two cvs. yielded about 12 flowers/ plant in the glasshouse and 17 in the tunnel (18 months of production). Percentages of flowers with a split calyx were 14.60 (glasshouse) and 39.25 in tunnel for Le Reve and 26.65 and 47.45 for Scania. Corresponding figures for flower stem length were 72.36 and 68.66 cm for Le Reve and 76.55 and 70.16 cm for Scania. Flowers in the glasshouse were less susceptible to *Botrytis cinerea.* Naveen Kumar *et al.* (1999a) planted rooted cuttings in greenhouse (fitted with evaporative cooling system) or in the open field. Flowering was advanced by 10-15 days in greenhouse. Flower stems were longer and number of flower per plants was higher (Table 4). Photosynthetic activity at both vegetative and flower stage was lower in the greenhouse probably as a result of lower light intensity.

Table 4: Effect of Growing Environment on Flowering of Carnation**

Character	*Red Corso*			*Cabaret*		
	Polyhouse	*Open*	*t. Cal.*	*Polyhouse*	*Open*	*t. Cal.*
Days to flowering	138.00	151.50	2.98*	162.00	177.00	3.41*
Days to 50 per cent flowering	161.00	178.75	4.41*	181.00	195.50	4.01*
Flower stem length (cm)	54.00	42.00	2.68*	38.00	28.00	3.78*
Leaf pairs at flowering	27.00	31.00	1.41	23.00	27.00	1.25
Lateral shoots/plant	8.00	8.50	0.69	7.50	8.00	0.66
Flower diameter (cm)	6.50	4.00	3.06*	7.20	4.80	6.74*
Flowers/plant	7.75	4.50	3.28*	6.25	3.00	3.81*
Petals/flower	64.50	45.50	3.99*	85.50	49.00	7.36*

Character	Red Corso			Cabaret		
	Polyhouse	Open	t. Cal.	Polyhouse	Open	t. Cal.
Photosynthetic activity (mole m^{-2} s^{-2})						
a. Vegetative stage	17.95	22.28	1.53	16.69	23.50	2.06
b. Flowering stage	5.26	18.80	2.37*	6.27	10.68	1.54
Nitrate reductase activity (NRA) at flowering (in moles nitrite formed g^{-1} fr.st.b^{-1})	11.25	29.25	4.03*	1507	3150	5.97*

* Significant at 5 per cent level.

** Naveen Kumar *et al.* (1999a)

8.3.0 Planting

The main objective of a grower is to produce the maximum number of excellent quality flowers at the time when the prices are higher. This requires proper decision regarding planting time, plant density, *etc.*, which is very important factors influencing time of flowering and yield. Deep planting of cuttings should be avoided. After making hole in the bed, rooted cuttings are planted as shallow as possible. The soil temperature should be in the range of 15°C for the first few weeks.

Masaru *et al.* (2020) developed an efficient night cooling technology to produce high-quality carnation cut flowers. The effects of temperature and timing of night cooling treatments (applied using a heat pump) on the flowering time and quality of cut flowers were investigated for standard type carnation flowers grown in a greenhouse in hot conditions. In the first experiment, rooted cuttings of the carnation 'Exceria' were plantedin a greenhouse on July 6, 2012. Night cooling treatments at 18°C, 21°C, and 24°C were carried out, and the harvested cut flowers were compared with those of a non-cooled control group. For flowers harvested in November, those from plants subjected to the night-cooling treatments had harder stems compared with those of the non-cooled control. In the next experiment, to determine the best timing and duration of the nightcooling treatment, we applied three treatments to rooted cuttings planted on June 18, 2013; overnight coolingfrom sunset to sunrise, cooling for 4 h from sunset (End of Day cooling: EOD-cooling), and cooling for 4 h before sunrise (End of Night cooling: EON-cooling), and compared them with a non-cooled control. For flowers harvested in December, the node order of flowering was the same in the EOD-cooling treatment and the overnight cooling treatment, and was lower than those in the EON-cooling treatment and the non-cooled control. The number of days to flowering was shorter in the EOD-cooling treatment than in the non-cooled control. In October and November in the harvest season when the high temperature influenced the earlygrowth stage, the stem quality of cut flowers was improved by night-cooling treatments, as indicated by a significantly smaller stem weeping angle than that in the non-cooled control group. These results suggested that an EOD-cooling treatment at 21°C for 4 h from sunset was as effective as an overnight cooling treatment to improve the quality of cut flowers by promoting stem hardness and flowering. This method is a cost-effective and efficient thermal management strategy to produce high-quality carnation cut flowers in hot.

8.3.1 Support Nets

Before planting, 4-5 layers of nets are laid out. The carnation plants are then planted within the netting with appropriate spacing between them. The nets are then gradually raised with plant growth. Good support material is metal wire with a mesh of 7.5 x 7.5 cm. The cheaper support material is net made of nylon. For every 2.5 to 3.0 m, the wires should be supported with poles. The first net should be fixed at 12.0 cm above the soil. The remaining nets, whose squares should be 12.5 x12.5 cm or 15.0 x 15.0 cm are placed over the first net. These nets are placed 15 cm apart.

8.3.2 Planting Time

Planting schedules are very important to regulate the production and meet market demands without compromising with the profit margin. With the help of staggered planting it is possible to regulate the flowering in this crop. Planting time varies according to light intensity, photoperiod and temperature. With the use ofsupplemental light and CO_2 in heated greenhouses, late planting is possible in August or September, (Besemer, 1980).

Manios and Tsikalas (1982) found that planting dates had a significant effect on the production and its distribution throughout the cropping period. The production of the plots planted on 20th of April was better fitted to the demand followed by those planted on 10th of May. In India, field trials carried out at Ludhiana have established that good qualituy blooms of both standard and spray carnartion can be produced in the open fields (Gill, 1984). Based on experimental result using marguerite cv. Crimson it was found possible to raise carnations at Bangalore from seeds during June - January with varying success (Mukhopadhyay, 1984). Hong *et al.* (1986) planted rooted cuttings of cultivar Scania in polyhouse every month for one year. Planting in April or May 20 produced more flowers than the others, and produced top quality flowers. Yields of plants transplanted from April to October were decreased as the pinching time increased, *i.e.*, no, single, and double pinch, while those from November to February were increased as the pinching time increased. Gursan (1999) obtained maximum total yield for the standard variety from January-September planting while May planting for spray variety.

8.3.3 Spacing

Sousa *et al.* (1981) observed maximum yield in a density of 233333 plants/ha. In a trial conducted by Altanand Altan (1982) lasting from 9 December to 15 July with carnation grown at 20 x 20, 15 x 15 and 15 x 20 cm spacing the high planting density produced the maximum number of flowers (75.5/m^2) but quality was better at low density. Mynett (1982) in his experiment with standard carnation Scania 3C, reported that planting of 100 plants/m^2 gave the highest yield (197 flower stems/m^2 and planting of 50 plants/m^2 showed the lowest yield of 131 flower stem/m^2 (Table 5).

Table 5: Number of Flowers at different Planting Density and Dates of Planting in cv. Scania 3C*

Planting density/m²	*Date of Planting*		
	May 10	*May 20*	*May 30*
50	131	119	103
75	176	152	128
100	197	168	137

* Mynett (1982).

Bunt and Powell (1982) planted carnations, cultivar White Sim, in different periods and different densities of 12.9, 25.8, 51.7 and 103.3 plants/m^2. At high density, a major part of the year's yield came from the first flush, accounting for 48 per cent of the total yield. After the first flush, crop yields were suppressed more at higher densities. At the low density, the first flush was only 26-31 per cent of the year's yield. Densities of 25.8 and 51.7 plants/m^2 gave the best continuity of production per unit area while the lowest density gave the best continuity of production per plant. Sakai and Asano (1990) conducted a trial with different planting densities of 49, 65 or 75 plants/m^2 with different number of shoots (4, 5 and 6/plant), flower quality decreased with increasing density and increasing number of shoots/plant. Optimum shoot number and plant density were 4 and 49 for March-April flowering, 4 and 65 for June-July flowering and 3 and 49 for October-November flowering, respectively. High plant density gave weak stems in cv. Lilac but decreased flower weight in Barbara and Ronja. Sawwan (1998) recorded variation in 3 Sim cultivars with respect to planting density. While flower diameter, stem length, node number, internode length and grade were not significantly affected by plant density. The highest flower production (124.26 flowers/m^2) was obtained from William White at 32 plants/m^2, followed by Manon at 32 plants/m^2 (122.76 flowers/m^2). In a greenhouse trial, cuttings of carnations cv. Dallas, Lereve, Scania, Triunph and White Sim were planted at densities of 35.70 or 100 plants/m^2 (Klasman *et al.*, 1999). After 6 months there were no significant differences in stem length among densities, but the response of dry weight (usually associated with vase life) differed among cultivars.

8.4.0 Irrigation

Rooted cuttings should be watered immediately after planting. Spraying of water through misting has been found to be more effective than the surface watering for ready establishment of the cuttings. After 3 weeks of planting, drip irrigation has to be adopted. The crop requires 4-5 l of water/m^2/day. It is advisable to check the irrigation water quality in terms of EC, pH, chloride and pathogens. Stagnation of water should be avoided to minimise the incidence of diseases. Frequency of watering depends on the weather conditions and the soil type.

Farina *et al.* (1994) studied the water requirement of carnations grown on raised benches using water-balance lysimeters. As the growing cycle progressed, there was a shift in water loss from evaporation to transpiration. The rate of transpiration was affected by DM production and plant developmental stage, being particularly high during bud differentiation and flower growth. The use of a crop co-efficient was not practical due to its variability during the growing cycle. Hence the management of irrigation will have to depend on data for plant and/or substrate water status (Table 6).

Table 6: Cumulate Water Consumption for Transpiration and Retention in Carnation Tissue**

Date	*Pinching on 23 July*			*Pinching on 23 July and on the Period 15-30 August*		
	Trans-piration (l/m²)*	*Plant Tissue Water (l/m²)*	*Water (Per cent)*	*Trans-piration* (l/m²)*	*Plant Tissue water (l/m²)*	*Water (Per cent)*
31 Oct. 1991	235.6	6.07	2.58	110.2	4.70	4.27
31 Jan. 1992	498.5	13.68	2.74	416.9	15.39	3.69
19 May, 1992	1048.8	37.03	3.53	707.2	25.18	3.56

*Calculated by difference between the water losses in the cultivated boxes and water losses in the boxes without plants.

** Farina *et al.* (1994).

Byrne (1974) planted rooted cuttings in large containers filled with 15 cm of relatively porous medium. Half were irrigated daily and half only when needed. Those with daily watering produced far more second-crop flowers than those irrigated less frequently. There was no significant variation in vase life between flowers from daily-irrigated plants and normally-irrigated plants. Shirasaki (1990) reviewed the effects of soil factors on cut flowers and found that flower quality is decreased if soil moisture is low and is also dependent on soil N and electrical conductivity.

Baas *et al.* (1990) studied the consequences of using saline irrigation water in closed nutrient systems in carnation. They observed decrease in peduncle length at 23 mM NaCl. Postharvest vase-life was not affected by salinity. As a similar yield decrease occurred in a treatment in which salinity was increased using increased concentra-tions of major elements, it is concluded that the yield decrease was not related to specific ion-toxic effects of Na^+ and/or Cl^-. Chemical analysis of peduncles of carnation flowers showed that NaCl concentration had no consistent effect on peduncle concentrations of major elements. Influx concentrations of Na^+ were linearly related to the concentration in the nutrient solution. On an average, influx was 2.5 per cent of the Na^+ concentration in the nutrient solution for carnation. Influx concentrations of Cl^- were higher, and not related linearly to the Cl^- concentration in the nutrient solution.

Heins (1973) planted carnation in gravel, a 1:1:1 mixture of soil, sand and peat or a 1:1 mixture of clay loam soil and wood chips. The hydrophilic polymer hydrogel was added at 7.5 lb/yd³ to some pots in each treatment. In addition 3 levels of automatic trickle irrigation were compared, *viz.*, frequent, intermediate and infrequent. The exact frequency of each level of watering depended upon the season, being greater for the second trial than for the first. The best growth in both trials was produced with the most frequent irrigation (daily for the first trial and 3 times daily for the second). The addition of Hydrogel increased fresh and dry weight of plant in the first trial, but had little effect in the second. It was most beneficial when combined with frequent irrigation and gravel as the growing medium. Kong Deheng and Li Wen (1999) found that pH of the growth medium decreases during growth for all irrigation rates. The decrease was maximum with the highest irrigation rate, minimum with the lowest rate.

8.5.0 Fertilization

Carnation plants need regular supply of nutrients throughout their life cycle. Thus, the application of nutrients in small doses but more frequently favours better growth and flower production. Low level of nitrogen and boron may cause calyx splitting. In the first three weeks after planting, no chemical fertilizers are applied. A total nutrition of 250 g N, 80 g P_2O_5, 200 g K_2O, 125 g Ca and 400 Mg per m^2 per annum in 20-24 splits (at fortnightly intervals) will give good plant growth and production of quality flowers.

In carnation, the tissue analysis reflects the uptake and distribution of minerals by plants and a quantitative relationship between absorbed nutrients and the growth. Fifth pair of leaves from terminal end of primary shoots were considered best for sampling when flower bud becomes visible on these shoots. The N:P:K ratio throughout the cycle is approximated to be 1:0.5:1.5.

Huett (1994) studied the effects of 3 nutrient solution formulations on the production and quality of the Sim carnation cultivars grown in rockwool substrate. The solutions were: a standard (Std) formulation with a K:N ratio of 1.29:1.00 (approximately that of carnation leaf standards), with N supplied as NO_3^-; the standard solution with 10 per cent of the N as NH_4 (Std + NH4); and a formulation with increased Ca and reduced K (high Ca). Total flower production from the Std + NH_4 (9.9/plant) and high Ca (9.7/plant) treatments exceeded that of Std solution (8.4/plant). Stem strength, stem diameter and vase life were unaffected by treatments.

Gopinath and Chanshekar (2009) found an application of 80 per cent of the recommenced dose of fertilizers *i.e.* 200:48:160g N, P_2O_5 and K_2O/m^2/annum in water soluble form not only resulted in producing higher cumulative yield of first grade (265.36/m^2) and a lowest cumulative yield of second grade (3.02/m^2) cut flowers coupled with lowest per cent yield of first and second grade cut flowers during winter and summer and highest cumulative yield of first grade cut flowers (11.06/plant/annum and 265.36 m^2). Eighty per cent recommended dose of fertilizers applied in water soluble form produced higher number of petals (67.25/flower head) and increased the diameter of cut flowers (5.97 cm) during winter season. Madhuri and Barad (2018) found better vegetative growth parameters with nutrient level 6000 N: 4000 P_2O_5: 2000 K_2O and recorded highest plant height, fresh and dry weight of plant, internodes per stem andinternodal length under polyhouse.

8.5.1 Role of Nutrients

8.5.1.1 Nitrogen

Nitrogen is probably the greatest limiting factor in the nutrition of carnation and it is needed for vegetative as well as reproductive stages of growth. The deficiency of this nutrient is very commonly observed in this crop. N deficiency symptoms consisted of generalized chlorosis with poor growth and short internodes, limited growth of erect plants, thin stems and stiff leaves which curve at the tip particularly during winter and general yellowing of mature or lower leaves. Blake and Harris (1960) reported that at low level of N, successive leaf pairs developed more slowly and more leaves were formed below the flower so that flower initiation was delayed. Bik (1962) recorded that the leaf N content was markedly affected by N fertilization and there was a close relationship between the leaf N content and the condition of the crop. Research evidences suggested a direct correlation between calyx splitting in flowers and sub optimal levels of nitrogen in plant tissues. Excess nitrogen, on the other hand, delays flowering and drastically reduces the postharvest quality of flowers (Winsor, 1968).

Green and Holley (1974) reported that productivity of carnation can be increased to levels attained by C_4 plants, such as corn, by adjusting the ratio of ammonium to nitrate N supplied according to seasonal solar radiation levels. During periods of low solar radiation, the optimum ratio was one third ammonium and two thirds nitrate N. During periods of highest solar radiation, 100 per cent nitrate was best. At high radiant intensity, reduction of nitrate N in carnation, a C_3 plant, apparently increased net CO_2 assimilation (growth) by decreasing photorespiration. Nitrate reduction served as an alternate sink for excess photosynthetic energy. Pergola *et al.* (1979) found that NH_4–N reduced the severity of wilt compared with NH_3–N. Combination of high pH and the use of NH_4–N drastically reduced the wilt. The different recommendations included 131 ppm (Holley and Woodwards, 1958), 163 ppm (Holley and Baker, 1963) and 113-150 ppm (Winsor, 1968) nitrogen in solution which resulted in high quality carnations.

Nitrogen (N) is said to be an important limiting factor in carnation. A low level of N causes weak stems and stunted plants whereas excessN causes delayed flowering and reduces postharvest quality of flowers. Thakulla *et al.* (2018) evaluated the effects of different doses of N on growth and flowering of exotic variety of carnationvar. chabaud under pot culture with five treatments (0 kg, 100 kg, 200 kg, 300 kg and 400 kg N/ha). Results revealed that maximum flower number (9.83) was recorded at 400 kg N/ha. Early flowering was recorded with 300 kg N/ha and 0kg, 100kg and 200kg. Maximum plant height, length of flower stalk, number of branches andplant width were recorded at the rate of 400 kg N/ha except flower width which was maximum at the rate of 300 kg N/ha and the minimum was recorded in controlled treatment for all parameters. These results suggested that increasing the doses of N has positive effect on most of the growth and flowering attributing parameters while excess N causesdelay in flowering (Thakulla *et al.*, 2018).

8.5.1.2 Phosphorous

Phosphorous is another macronutrient which is required for normal growth of carnation plants. Winsor (1962) recorded that phosphate fertilization had the most marked effect on flowering and produced an average increase of 20 per cent in flower number. Its deficiency also restricts the growth of plants and the leaves become narrower as compared to those from healthy ones. The tips of lower leaves that have suboptimal levels of phosphorous die off and turn yellow particularly during the second year in the field. In severe cases, the whole leaf turns yellow or pale straw colour and die off prematurely. Roelants (1973a) reported that higher level of P nutrition increased the number, diameter and weight of the flower and resulted in more number of petals per flower. In spectophotometric and chromatographic studies, Roelants (1973b) noted that P nutrition increased the quantity of pigmentation and improved colour in carnation flower. Excessive amount is rarely harmful since element does not stimulate any one part unduly. Chan *et al.* (1958) however, did not find any beneficial effect of P on flower production.

8.5.1.3 Potassium

Potassium improves the general vigour of the plant and is essential for the nitrate changes which occur in the plant. It acts as a catalyst in the manufacture and translocation of carbohydrates and also increases resistance to certain diseases. The deficiency is characterized by scorching of leaf blade. In severe cases white spots on foliage appear below the flower, disfigured or malformed flowers, loss of petalcolor and yellowing of the calyx are observed. Chan *et al.* (1958) found that application of potassium markedly increased the yield of flowers with optimum level beyond 90 ppm. Mayak *et al.* (1978) studied the effect of KNO_3. There was an osmotic adjustment in response to KNO_3 treatment, which occurred in the external as well as in the internal compartments. The change in osmotic concentration in the external compartment was well correlated with extension of longevity.

8.5.1.4 Calcium

Calcium deficiency produced thin and weak stems and small flowers (Mantrova, 1976). Production of many sleepy flowers (not opening properly) and consistent foliage scorching were also recorded in plants grown in medium deficiency of this nutrient (Peterson, 1960). Application of lime stone at the rate of 2-4 oz per 100 sq.ft per week starting from flowering time was recommended by Strydom (1962). Incarnation Ca nutrition reduces the incidence of *F. oxysporum* f.sp. *dianthi* (Blanc and Tramier, 1980).

8.5.1.5 Magnesium

Magnesium deficiency produces weak stem and small flowers. Application of ferti-lizers containing Mg removed the depressing effect of K on Mg and improved magnesium uptake by plants (Mantrova, 1976). Lyakh (1986) reported that < 80 mg/l Mg fertilization markedly increased flower yield and also increased the number of rooted cuttings, especially in the presence of ammonium or amide fertilizers. A basal dressing with Mg at 50-150 mg/l increased resistance to *F. oxysporum*. It is suggested that when growing carnations on calcareous soils, with the addition of peat and sand, the Mg content of the substrate should be raised to 150-180 mg/l by applying magnesium sulphate.

8.5.1.6 Boron

Deficiency of B causes pale yellow zone along the mid-rib which later turns purple, young emerging leaves curved, twisted, the internode of the upper part of the shoot is curved (Overtli, 1960). The accumulation of 3000-5000 ppm B was toxic to the plants accompanied by leaf-tip burn symptoms. According to Mastalerz (1978) the deficiency of B caused reduction in internodal length of flower stem, failure of flower bud to reach anthesis stage and development of purple colour in leaves. Adams *et al.* (1980) reported that deficiency of B reduced flower production and average bloom diameter and increased the incidence of calyx splitting. Dabeda *et al.* (1975) recorded that optimum level of B was about 50 ppm, whereas Puustjarvi (1981) recommended that 0.04 mg/l B should be applied continuously through irrigation systems if the pH approaches 6.5.

Adams *et al.* (1979) observed the effects of supplying B at 0 and 0.5 ppm and N at 100-260 ppm in liquid feeds. Total bloom production increased from 966 to 1002/m^2 when B was supplied and decreased by 3 per cent at pH 6.3. The greatest increase (11.4 per cent) resulted from the addition of B at pH >6.8. Calyx splitting decreased markedly when B was supplied to plants receiving 100 ppm N, but only slightly at 250 ppm N. Increasing N levels depressed the soil pH and increased the availability of B, thus reducing calyx splitting. Splitting was related independently to both N and B status of the plants. The incidence of B deficiency symptoms was related to the B content of the leaves.

8.5.1.7 Chlorine

Ishida *et al.* (1979) found that salt injury did not occur with solutions containing 0 or 100 ppm Cl. Both height and top fresh weight of the flowering plants grown in both sand and soil decreased above 250 ppm Cl, and flowering was delayed above 500 ppm Cl. The degree of injury increased as the Cl concentration was raised from 250 to 3000 ppm. Leaf dieback occurred at above 500 ppm Cl in sand culture and at above 1000 ppm in soil culture about 45 days after the start of treatment. Injury was generally more severe in sand than in soil culture, and was highly correlated with Mg and Na levels in the leaves and with Cl level in the leaves, stems and petals. With increasing sea water concentrations, the total N, P and K in the leaves decreased and the Cl, exchangeable Mg and Na and the EC values in the sand and soil increased.

Other nutrients like Mn, Fe, Zn, Mo and Cu, *etc.*, are also needed in small amount as they are catalysts or activators for many plant processes. Mostafa (1996) studied the effects of Mn (0, 0.4 or 1.12 ppm) on the growth of carnations cv. Yellow Dusty Sim, planted in sandy soil. Mn increased length, diameter and DW of stem and also the number of leaves and leaf DW and promoted earlier flowering. The highest content of Mn in leaves was observed at the highest rate.

8.5.1.8 Nitrogen, Phosphorus and Potassium

The balance of different nutrients is very essential for the plants to produce high biomass and better yield. Lindemann and Schwenke (1972) observed the best yields of top grade flowers with 0.2 per cent of 14:10:14 NPK, Where as in 0.1 per cent, 0.2 per cent or 0.3 per cent of 20:5:10 NPK most soil N values were optimal, and the P_2O_5 values were lower than the earlier one. In a trial with cv. Marguerite Scarlet with 25 plants/m^2, Arora and Saini (1976) reported that the maximum number of flowers was obtained with the application of N and P at 40 and 10 g/m^2, respectively. The largest flowers were obtained with N and K at 20 and 60 g, respectively. The incidence of calyx splitting was lowest with N and P at 40 and 10 g, respectively. El Shafie (1977) found that plants which received moderate levels of fertilizers like 56 g ammonium nitrate + 48 g superphosphate + 42 g potassium chloride per 150 x 150 cm bed compared with levels twice as high, and were left with 4 main branches, rather than 2 or 3, were the heaviest plants and produced the largest number of top quality flowers. In an investigation, Magnifico *et al.* (1985) reported highest fresh weight and dry weight of whole plants, flowers and other plant components where fertilizers were applied every 2 weeks to give a total of 160, 40 and 80 kg/1000 m^2 of N, P_2O_5 and K_2O, respectively. Lower rates of N were significantly less effective. Kaptan (1985) used 3:1:1:1 ratio of alluvial soil, peat, manure and sand as growing medium for greenhouse cut flower production. The application of 30 g N, 20 g P, and 20 g K/m^2 at two-month intervals in winter and at 5-week intervals in summer was suggested for this growing medium.

In a greenhouse trial with miniature carnation Red Baron, Burbano *et al.* (1990) used ammonium phosphate, potassium nitrate and ammonium salt (Nitron 26) as nitrogen sources and potassium sulphate as a control. They found that potassium nitrate and potassium sulphate increased soil pH while Nitron 26 and ammonium sulphate decreased soil pH values. The number of *Fusarium oxysporum* f. sp. *dianthi* colonies in the soil decreased following treatments with potassium nitrate and potassium sulphate. The number of nematode cysts and viability of *Heterodera trifolii* were not effected by the treatments. The greatest number of floral stems was obtained with potassium sulphate and the lowest number with ammonium sulphate. Hwang and Yoon (1994) in an experiment in pot containing 4 types of soil which varied in electrical conductivity (0.9-6.3 dS/m), and in the contents of nitrate-N (42-69 ppm), available P (354-644 ppm) and exchangeable K (0.41-1.19 meq/100g). The critical electrical conductivity values were 3.9 dS/m. The optimum concentrations of N, P_2O_5, K_2O, CaO, MgO and Na_2O in shoots for maximum biomass yields were 3.10, 0.50, 4.60, 2.20, 0.60 and 0.40 per cent respectively, at 60 days after treatement.

Hosni and Shoura (1996) studied the effects of N (0.3 g/pot) + P (0.3 g P_2O_5/pot) + K (0.3, 0.6 or 0.9 g K_2O/pot) on carnation. Flower number, stem diameter and fresh weight were significantly increased with the highest rate of K. Flower stem length decreased with increasing K. N and P contents of aerial parts were not affected by different rates of K. However, K contents of aerial parts increased with increasing rates of K.

In an investigation on the role of different rates of N and K, Starck *et al.* (1991) reported that the higher N rate produced bigger flowers in plants grown in sawdust or 25 per cent peat + 75 per cent sawdust. Increase in K rates promoted stem elongation in plants grown in 50 per cent peat + 50 per cent sawdust. In sawdust alone the plants had the lowest total and soluble N contents in leaves. Leaf P contents were broadly similar throughout. The higher K rate increased leaf K content slightly. Leaf Ca content increased as plants aged and was greater at the higher K rate at all sampling dates. Boztok *et al.* (1996) investigated the effect of N:K ratios of 1:1, 1:2 and 2:1 in cv. Scorpio grown in a 10:1 ratio of pumice:peat. The maximum yields and quality (11 stems/plant, 5 flowers/plant with flower diameters of 3.6 cm) were obtained using an N:K ratio of 1:1.

8.5.1.9 Nutrient Film Technique

Bowe and Reinelt (1991) compared nutrient film technique (NFT) procedures and raised bed culture for growing carnations. NFT resulted in earlier flowering and higher flower yields/plant without loss of quality (80 per cent of the harvested flowers were of A1 quality) and with a considerable reduction in water used during production. Some increase in stem brittleness was also noted. Guba *et al.* (1993) adjusted on rockwool slabs, the pH of the circulating nutrient solution reported that on to 5.0, 6.0 or 7.0 had no effect on yield of flowers/m^2 or average length of flower shoots or on cut flower shoot fresh weight.

8.5.1.10 Soluble Fertilizers

Indiscriminate supply of soluble fertilizers not only increases the yield but also increases the disease incidence. Stack *et al.* (1977) planted rooted carnation cuttings on a greenhouse bench in soil which had received one of 5 different fertilizer solutions weekly, comprising N and K at low, intermediate or high levels (each at 40, 200 and 400 mg/l, respectively) in various combinations. When flowering stems appeared all plants were cut back and the stubs inoculated with *Gibberellazeae* (the causal agent of stem rot and stub dieback) were recorded. The incidence of infection rose from 25 per cent with low N and K to 48 per cent with high N and/or K. There was a severe infection (19 per cent) in the high-N + low-K treatment than in any other treatment (2.3-3.5 per cent).

Uri *et al.* (1990) compared three levels of K (93, 252 and 378 g/m^3) as K_2SO_4 and two levels of NO_3-N:NH_4-N ratios (1:0 and 4.3:3) in combinations, via trickle irrigation, supplying a constant total N level of 126 g/m^3. Conventional cultivation was used as the control. Throughout the season, all treatments containing NH_4^+-N lowered the soil pH and it fell from 6.9 at the beginning of the season to 5.8 after 40 weeks. Treatment with KNO_3 treatment kept the pH constant at 6.6-6.9, gave the lowest EC value in the soil 280 days after planting. The highest flower yield was obtained with the KNO_3 treatment (105 flowers/m^2, 17 per cent more than the other treatments). The yield was thus influenced by the form of N supplied. The flower quality, as measured by the incidence of calyx splitting and brittle stem, was improved by supplying the highest K level. A total plant content of > 4 per cent K was observed after 162 days of planting, it was asso-ciated with a low incidence of brittle stem (18 per cent). A content of < 4 per cent K in the 5th leaf from the apex was associated with 31 per cent brittle stem.

Strojny *et al.* (1992) tried different liquid fertilizers on carnation cvs. Tanga and Pallas grown in a mixture of 40 per cent mineral soil + 20 per cent composted pine bark + 20 per cent sphagnum peat + 20 per cent sand. One month after planting the plants were fed with a 0.3 per cent Azofoska solution at every second watering, with Brun's nutrient solution (200-25-205 ppm NPK in summer and 100-20-250 ppm NPK in winter), or with solutions containing 100-25-150 mg NPK or 200-25-400 mg NPK or 300-25-650 mg NPK/l at every watering. First year yields were highest with 100-25-150 mg NPK/l for both cultivars. This treatment also gave the highest 2-year yield of Tanga but treatment with 200-25-400 mg NPK/l gave the highest 2-year yield of Pallas Orange.

Kaplan (1993) studied the effects of ammonium N rate (6, 15 and 24 per cent in the nutrient solution) and 15 per cent ammonium N + a nitrification inhibitor (N-Serve) on the yield and quality of spray carnation cv. Elsy and standard carnation cv. Pallas grown on rockwool in a greenhouse. Increased ammonium-N rates did not significantly affect flower yield of Elsy but increased individual flower FW, stem length and total leaf N. Increasing ammonium-N had no effect on flower yield or other characteristics of Pallas carnation. The pH of nutrient solutions in the root zone decreased to <5.0 during some stages of growth and it is recommended that the pH of the nutrient solution should increase during these stages.

8.5.1.11 Micronutrients

Lakshmi *et al.* (2016) found that fertigation of 0.4 per cent MN-EDTA+ foliar application of 0.2 per cent MN-EDTA at fortnightly intervals, beginning from one month after planting up to harvesting stage significantly improved plant height, number of leaves, number of laterals, earliness in flower bud appearance, stalk length, yield and vaselifeduring the two consecutive flushes, respectively. Nirala *et al.* (2018) found that the fertigation treatment (120 per cent fertigation of RDF of NPK) was found to be very effective, might be due to application of higher doses, resulted maximum plant heights, number of branches early bud formation and flowering, maximum stalk length and girthas well as maximum number of flower per m^2 area. The b/c ratio was also found highest.

8.5.1.12 Hydroponics

Hydroponics or soil-less culture is a technology forgrowing plants in nutrient solutions that supply allnutrient elements needed for optimum plant growth withor without the use of an inert medium such as gravel,vermiculite, rockwool, peat moss, sawdust, coir dust,coconutfibre, *etc.* (Hasan *et al.*, 2018).

Modern hydroponic systems rely on irrigation of plants with balanced nutrient solutions throughout the life of the crop. Huett (1994) recorded the effects of 3 nutrient solution formulations. The solutions were a standard (Std) formulation with a K:N ratio of 1.29:1.00, with N supplied as NO_3^-; the standard solution with 10 per cent of the N as NH_4^+ (Std + NH_4^+); and a formulation with increased Ca and reduced K (high Ca). Production from the Std + NH_4^+ treatment (20.8 flowers/plant) exceeded that for the other formulations and more than 75 per cent of these flowers were first grade. The pH was lower in the Std + NH_4^+ (6.13) and high Ca (6.76) treatments than in the Std treatment (7.17). The high Ca formulation increased Ca and reduced K concentrations in leaves, stems and flowers. Nutrient concentrations in youngest fully expanded leaves for all treatments were within the adequate range. The mean K:N ratio in mature carnations (leaf + stem + flower) was 1.39:1.00 and this ratio with 10 per cent of the N as NH4 was recommended for commercial hydroponic carnation production.

8.5.1.13 Tissue Analysis

Blanc (1962) obtained maximum uniformity in nutrient content in samples from axillary shoots at 5th and 6th nodes from the apex of the floral stem. He also reported that total N of the axillary shoot is directly related to flower production but there was no such relation with P and K. Hasegawa (1992) described a technique for nutritional diagnosis of carnation using small amounts of exudate from the leaf base. In 2 culti-vars, the levels of N in the exudate showed a relationship with the soil N level and decreased as the plant increased in weight. The level of K in the exudate tended to increase as plant weight increased. In cut flowers, on various dates, the variation in the level of K was greater than that in the level of N.

8.6.0. Mulching

Huh and Kim (1995) studied the effect of reflective film mulching (Neopolyshine) on the growth, productivity and quality of carnation flowers cv. Nora under protected cultivation. Hundred per cent flower bud development was observed 16 weeks after planting, compared with 50 per cent bud development in controls. Mulched plants also produced more cut flowers of better quality (lower percentage of calyx splitting), than unmulched plants. More flowers with long stems (46-65 cm) were produced by mulched plants than unmulched plants. Mulching with reflective film (RFM) improved the light environment amongst the plants (73 per cent of incident radiation was reflected back); over 85 per cent of the reflected light was in the range of photosynthetically active radiation.

Ito *et al.* (1996a) also investigated the influence of reflective film mulching (RFM) on the photosynthetic activity of carnations.The amount of downward light reflection was 32 per cent in plots mulched with Polyshine cloth, compared with 3 per cent in plots mulched with black film. The photosynthetic rate on a plant basis increased by 32 per cent in the Polyshine mulched plots compared with the black-mulch plots. They also observed photosynthetic rate on a single leaf basis was increased by only 16 per cent by supplemental illumination of the abaxial leaf surface and larger amount of CO_2 assimilation products was translocated to roots and folded leaves in RFM plots than in black-mulch plots. It was concluded that the higher photosynthetic activity in RFM plots was caused by the activation of CO_2 fixation in lower leaves rather than by the increase of CO_2 fixation in all leaves. In another trial, Ito *et al.* (1996b) investigated the interaction between RFM and shading treatments on carnation growth. The combination of RFM and shading treatments reduced leaf temperature by 5.3°C compared with untreated controls. It was concluded that the shading treatments reduced air and soil temperatures, alleviating high temperature stress and activating nutrient uptake, resulting in increased plant size.

Yamaguchi *et al.* (1996a) examined the light reflectance, transmission and absorbance properties of several shading and mulching materials. They concluded that the most important characteristics of the shading materials were transmittance and light diffusivity (as a percentage of total transmitted light). The optimum transmission percentage during summer ranged from 40 to 70 per cent. Transmission and light diffusivity of shading materials exhibited an interactive effect on carnation growth. For instance, shading materials with a low transmittance of around 40 per cent required a high light diffusivity of >70 per cent to be effective. On the other hand, with moderate shading of about 70 per cent transmittance, light diffusivity did not appreciably affect the acceleration of early growth and flowering. Mulching materials were consistently effective when they reflected >80 per cent of radiation. Suitable combinations of mulching and shading materials increased cut flower yields by up to 50 per cent compared with non-shaded and non-mulched plots. In another study, Yamaguchi *et al.* (1996b) used the combination of RFM and summer or all-season shading (using cheesecloth with 84 per cent permeability to PAR) in a greenhouse trial. They found increased plant height and length of primary and secondary branches in cv. Tanga on the RFM treatments compared with controls. The number of cut flowers harvested in the first 3 months of the harvest period was maximum in the RFM treatment (82.5 flower/m^2, compared with 32.5/m^2 in controls) and total yield was 15 per cent more in the RFM treatment than in controls. They also noticed soil temperature 0.5 and 5 cm under the RFM was lower than that of bare soil during the day, but higher at night, the effect being more pronounced near the soil surface.

8.7.0 Pinching

For successful production of top quality carnations, pinching is a must. During pinching, the tip of the stem is removed to encourage the growth of the lateral shoots. Pinching is very much essential for the standard or Sim carnations. Only 4-6 well grown lateral shoots will be allowed to grow. Pinching is done leaving 6 pairs of basal leaves which are inductive. In single pinch, apical portion measuring 5-7 cm is pinched off. In single and half pinch, after flowering single pinch of main shoot, half of the side branches are pinched off. The plants are usually pinched for the first time about 3-4 weeks after planting. With one and a half pinch, 2-3 shoots are pinched again at 4 well developed pairs of leaves about 3-4 weeks after the first pinch. The procedure depends largely upon the time of flower requirement and cultivars grown. This operation is generally done in the morning, when the plants are fully turgid.

In India, uniform pinching of all shoots is done twice, 40 or 60 days after planting. In contrast, in Western countries 4 systems of pinching are followed. Manios and Tsikalas (1982) stated that pinching had no significant effect on the flower production or its distribution. Ubikata (1999) pinched out flower buds to replace primary with secondary buds thereby increasing the number of flowers in spray. Pinching increased the duration of flower production but decreased the flower at the peak flower production stage. Pinching treatment increased the flower weight but decreased flower length.

8.7.1 Single Pinch

In this method, the top of the main or leader shoot is removed to give about 4-5 lateral shoots which will produce flowers at the same time. This pinched plant produces 2 crops in 38-47 weeks depending on the time of the year. Cardus and Lasala (1962) reported that the highest yield of flowers without reduction of quality was obtained by pinching at 5th node. Volpi and Accati (1974) found in their glasshouse experiments on cv. Alice the highest yield of winter blooms from the May planting with 2 pinchings and close spacing; from the July planting a single pinching and medium spacing was more productive, and from the August planting no pinching and medium spacing. Namikawa (1980) pinched young plants of carnation once or twice after transplanting. The number of flowers was greater for plants pinched only once, but there was no significant difference in the flower quality of plants pinched once or twice. Groshkov and Angelov (1981) recorded that pinching once at the 4th node gave higher yields and returns for both Scania and Arthur Sim cultivars.

8.7.2 Pinch and-a-Half

This method involves removing the tip of the main stem to give 4-5 side shoots. Only half the total number of lateral shoots is pinched when they are about 5 cm long or have 5-6 pairs of leaves. In this method the plants receive 2 or 3 half pinching. This method reduces the size of the first crop but provides a steady supply of flowers without any peak time. This method is followed to regulate the supply of flowers throughout the year. Pathania *et al.* (2000) found the best treatment appeared to be pinch and a half method (with 6 laterals) which staggered the flower production in two flushes (3rd week of May to July end) and yielded 66.66 flowers/m^2 of good quality in a cropping season (Table 7).

8.7.3 Double Pinch

In this method, the main shoot is pinched once followed by pinching of all the lateral shoots arising from the first pinch when they are of desired length. Besemer (1980) reported that this method resulted in a large number of flower bearing shoots and it is seldom practiced in high density planting. The quality of the flowering shoots, particularly after the first flower is poor and hence not preferred in the markets. Farina *et al.* (1985) observed plants of particular ecotype which were small with small flowers and they responded well to continuous pinching (rather than the usual one or two pinchings). Early planting (30th June) and a high planting density gave high yields in Eolo, Gloria, Mei Ling and Mei Wong, and in these cultivars flower production per m^2 was much higher than for Mediterranean carnations.

8.7.4 Single Pinch Plus Pull-Pinch

It is also not a very common method of pinching carnation. This method involves a single pinch of main shoot followed by pulling of growing tips when the lateral shoots are long enough (longer than for normal pinching). Pull-pinching of shoots is followed for about 2 months to provide constant flower production over longer period of time. It has been found that this pinch should be practiced only in high density planting. Hilliard and Hanan (1976) observed the highest shoot break

Table 7: Effect of Various Pinching Methods on Flower Regulation in Sim Carnation cv. Lena*

Treatments	*Plant Height (cm)*	*Days to First Flowering*	*Flower Duration (Days)*	*Stem Length (cm)*	*Flower Size (cm)*	*Calyx Splitting (Per cent)*	*Vase Life (Days)*	*Flower Yield/ m^2*	*Flower Quality***
No pinching (unstopped)	62.10	100.53	34.33	48.77	6.98	38.07	13.13	11.11	+ + + +
Single pinching	48.00	147.20	31.03	37.10	6.84	30.53	9.97	44.44	+ + +
PINCH and 2+4									
HALF									
Ist flush	51.30	137.80	35.83	38.17	6.83	36.63	12.13	22.22	+ + + +
2nd flush	42.67	168.57	29.47	34.50	6.76	34.63	10.10	44.44	+ + +
PINCH and 2+6									
HALF									
Ist flush	53.13	140.00	35.50	38.90	6.83	37.33	12.20	22.22	+ + + +
2nd flush	42.23	172.17	27.83	32.00	6.64	32.06	9.47	66.00	+ + +
Double pinching (16 laterals)	44.50	225.03	23.17	21.10	5.52	19.00	6.10	177.77	+
Double pinching (18 laterals)	39.27	184.43	18.13	25.97	5.60	24.00	5.10	88.88	+ + +
C.D. at 5 per cent	2.33	6.85	2.07	2.47	0.06	2.73	0.45	9.25	

** + + + + = First grade, + + + = Second grade, + + = Third grade, + = Fourth grade.

* Pathania *et al.* (2000).

(4.7) for pinching to leave 10-leaf pairs and the lowest (4.0) for pinching to leave 4-leaf pairs. Pull-pinching 10 days after planting resulted in 4.6 shoot breaks but it is time consuming.

Hong *et al.* (1986) observed yields of plants transplanted from April to October were decreased as the pinching time increased, *i.e.*, no, single, and double pinch, while those from November to February were increased as the pinching time increased. Carnation plants with 3 main shoots on which all the secondary shoots were retained had heavier fresh and dry weights and more flowers than plants which were pruned to leave only 2 secondary shoots on each of the 3 main shoots; however, the plants with 2 secondary shoots produced larger flowers with thicker and stronger stems (El Shafie, 1977). Five greenhouse grown cultivars were pruned above or below the 6th wire 69 and 91 weeks after planting. Total flower production was greater in control plants than in pruned plants. High pruning resulted in greater flower production than low pruning (Martinez and Gonzalez, 1983).

8.8.0 Disbudding

To obtain good quality flowers, in case of standard carnation, the lateral flower buds are to be removed at the appropriate time leaving only the terminal or main flower bud. On the contrary, in spray carnations, the terminal or main flower buds are removed to encourage the lateral flower buds to develop. The best time for disbudding is when the apical flower bud is about 15 mm in diameter. Unlike standard carnation, the spray or miniature types need different kinds of treatment. In these types the main flower bud (terminal) is removed to encourage lateral flower buds to develop (Besemer, 1980). In marguerite carnation all the axillary or secondary flower buds borne on the main stem or terminal shoots are disbudded as soon as they are large enough to handle. Results of the trial indicate that for Bangalore conditions in India, the disbudding operation or removal of axillary flower buds should be done within 7 days from the appearance of these buds (Mukhopadhyay, 1984).

8.9.0 Calyx Banding

Calyx splitting is a serious malady affecting the quality of flowers in carnation. The problem of calyx splitting can be reduced by placing a band around the calyx of the flower bud when they have just started opening. In recent years, a 6 mm wide clear plastic tape is used instead of rubber bands or wire paper twistems. Unlike bands, these plastic tapes can be left in place even after harvest. Care should be taken not to put bands when the buds are too small otherwise it results in misshaped flower buds. These bands should be placed at a point about half the way on the calyx or where the diameter of the bud is the largest (Besemer, 1980).

8.10.0 Economic Feasibility of Carnation Cultivation under Polyhouses

Arshdeep *et al.* (2020) found that average cost of production of carnation per cut stem varied between Rs. 0.53 to Rs. 0.55. The net returns from the production of carnation cut stems were Rs. 3,11,348.30 under 500 m^2 polyhouse, whereas, under 2000 m^2 it was Rs. 12,79,504.72. The net present value of carnation was found to vary between Rs. 9,21,328 to Rs. 42,91,301 under different sizes of non-subsidized polyhouses. Whereas, under subsidized polyhouses, it ranged from Rs. 10,31,093 to Rs.46,80,471 under 20 years life of polyhouse. Benefit cost ratio under subsidized polyhouses of 500 m^2 to 2000 m^2 varied from 1.54 to 1.66 with an internal rate of return 31 to 37 per cent. Similarly BCR varied from 1.45 to 1.57 under different sizes of non-subsidized polyhouses with an internal rate of 27 to 33 per cent.

9.0 Diseases, Pests and Disorders

9.1.0 Diseases

Fungal Diseases

Diseases caused by fungi, bacteria and virus result in considerable loss of flowers. Though carnations are attacked by number of diseases, fusarium wilt which usually invade the plants from contaminated soil is the most serious disease. Now-a-days integrated approach by altering cultural practices along with use of biological control and safe fungicides are the possible methods for the control of most of the diseases.

9.1.1 Wilt

Wilt is caused by *Fusarium oxysporum* f.sp. *dianthi*, also known as vascular wilt. This fusarium wilt is widely prevalent in carnations all over the world. The initial symptoms of the disease are yellowing of leaves and crook neck shoots. The leaves and shoots wither and become brownish. Stems when cut open show brown discolouration at the vascular region. Sometimes the pith and cortex also get discoloured. Finally the stems show shredding of the internal tissue. First the symptoms occur on one side of the plant but ultimately the entire plant shows the symptoms. The chief diagnostic symptoms are abnormal growth and stunting of young shoots. The entire plant wilts and collapses in a very short time.

N fertilizers increased the disease incidence, however it was found that N deficiency did not provide a lasting protection against *F. oxysporum* f. sp. *dianthi* but delayed disease development. Application of potassium nitrate and potassium sulphate in the soil reduced the disease incidence (Burbano *et al.*, 1990). Sezgin and Onan (1987) found virulence of fusarium wilt was increased by urea. Blanc and Tramier (1980) found that the disease injury was related to the initial Ca content of the cutting; higher Ca content at initial stage reduced the injury. Blanc *et al.* (1983) found that Sacha had a higher Ca content than Scania and that may be the probable reason for the resistance.

Duskova *et al.* (1989) reported that the disease spread most rapidly in top soil and peat (pH 5.0) and most slowly in a mix of topsoil and manure. The effect of higher pH resulted in a reduced number of diseased plants.

Kishore and Srikant (2008) found that under *in vivo*, wilt disease was completely reduced in carbendazim alone and carbendazim + mancozeb next best treatments were mancozeb (16.67 per cent), carbendazim + iprodione (16.67 per cent) and garlic extract (16.67 per cent) followed by thiophanate methyl (33.33 per cent).

9.1.1.1 *In vitro* Selection

The *in vitro* selection using culture filtrates may provide information on methods of selection of new wilt resistant carnation varieties. Arai and Takeuchi (1993) observed the influence of culture filtrates (CF) of *F. oxysporum* f. sp. *dianthi* on growth and viability of *in vitro* leaf segments. However, CF of virulent race 2 decreased viability and suppressed growth of callus of the susceptible cultivar. Pollen assay is one of the feasible methods for screening resistant cultivars. Mercuri *et al.* (1992) cultured pollen grains of 8 commercial carnation varieties *in vitro* on germination medium containing increasing concentration (0, 7.5, 15 and 30 per cent) of axenic culture filtrate of *F. oxysporum* f. sp. *dianthi*. Germination and tube growth of pollen was monitored in response to the filtrate. They found that pollen assay is a feasible method for screening resistant cultivars.

9.1.1.2 Chemical Control

Nash and Baker (1972) found that thiabendazole (TBZ) and benomyl gave 98 per cent and 95 per cent controlled stem rot respectively, whereas untreated plants showed 59 per cent infection. No phytotoxicity was observed with either fungicide and rooting was stimulated. Costache *et al.* (1979) reported carnation wilt in greenhouses, caused by *F. oxysporum* f.sp. *dianthi* was successfully controlled with a sequence of treatments. Cuttings (cv. Florance) were dipped for 10 h in a 0.2-0.3 per cent benlate solution, laterbenlate or topsin-M [thiophanate-methyl] each at 0.05 per cent or mycodifol [captafol + folpet] at 0.2 per cent were applied to the rooting medium and the soil. Dalla Guda *et al.* (1982) found that application of vapam (Metam sodium 32.7 per cent) twice in 10 days at 500 cc/m^2 was most effective. The treatments with formalin (30 per cent) were not effective. Garibaldi *et al.* (1988) found that prochloraz was most effective (74 per cent reduction of disease severity) when used at 2 g/m^2 during transplanting of rooted cuttings and then at 10-d intervals. Prochloraz wettable powder at high dosages (10-20 g/m^2) was phytotoxic but granular formulation did not show phytotoxic effect when applied once, at transplanting, at 10 g/m^2. The effectiveness of benomyl at 10 g/m^2 (1-2 applications) was reduced after the 1st year. Orozco de Amezquits *et al.* (1993) found that Vapour + dazomet and lower K doses effectively controlled the diseases. Acid conditions favoured the pathogen development in soil. Garcia *et al.* (1994) compared the effect of fumigant 1,3-D chloropicrin with soil steaming. Both treatments reduced disease incidence and fungal populations in soil. Yephet *et al.* (1994) reported that Methyl bromide reduced the *F. oxysporum* f. sp. *dianthi* propagule count at all soil depths between 0 and 60 cm, but was most effective in the upper 30 cm. In case of monocultures of susceptible carnation cultivars, methyl bromide did not provide total protection, even for one growth season. Elena and Tjamos (1997b) found that soil disinfestation with 60 g/m^2 dazomet was equally effective to solarization. Penaranda and Bolton (1984) observed that drenching the growing medium with ethazol + benomyl gave good control of pathogens.

Manasa *et al.* (2017) found that six fungicides which were applied to the soil by drenching method propiconazole @0.1 per cent (1.66 x10^3 cfu per gram of soil and 9.33 per cent wilt incidence) followed by carbendazim @ 0.2 per cent (2.66 x10^3

cfu per gram of soil and 11.19 per cent wilt incidence) were significantly found effective in reduction of soil population of *Fusarium* and wilt incidence in polyhouse condition. Savita and Harender (2019a) found, mong non-systemic fungicides, Saaf (carbendazim 12 per cent + mancozeb 63 per cent) was most effective with average inhibition of 60.19 per cent in mycelial growth of the wilt pathogen. Among systemic fungicides, Bavistin was found most effective with 92.04 per cent average inhibition in mycelial growth while Systhane was found least effective treatment with 25.68 per cent average inhibition in mycelial growth of the pathogen.

9.1.1.3 Biological Control

Garibaldi *et al.* (1990) found that combination treatments of benomyl with antagonists fusarium isolates from the rhizosphere of carnations controlled fusarium wilt of carnation. In a greenhouse experiment, Weber *et al.* (1998) tested the efficacy of 10 isolates of soil fungi (*Actinomucor elegans, Absidia californica, Penicillium funiculosum, Trichoderma hamatum, T. harzianum, T. longibrachiatum, T. viride* and *Zygorhynchus moelleri*) as biological control agents for *F. oxysporum* f.sp. *dianthi*. The fungal isolates gave greatest protection. Rattink (1992) found that antagonists *F. oxysporum* 618-12 reduced disease incidence on the susceptible cultivar often by >70 per cent. On moderately resistant cultivar, *Trichoderma harzianum, Streptomyces griseoviridis* were ineffective (Table 8).

Table 8: Percentage Disease Reduction in Susceptible (S) and Moderately Resistant (MR) Carnations in Natural Substrate after 26 Weeks after Pre-inoculation with Antagonists*

Experimental Cultivar	*1*		*2*		*3*	
	S	*MR*	*S*	*MR*	*S*	*MR*
Trichoderma	–5	42	20	7	65	97
Streptomyces	5	59	4	–128	0	–80
npF (618-12)	84	100	66	50	12	57

* Rattink (1992).

Duijff *et al.* (1993) reported that treatment of carnation roots with *Pseudomonas* sp. str. WCS417r significantly reduced wilt and consistent suppression of fusarium wilt of carnation by WCS417r involved multiple mechanisms (induced resistance, siderophore-mediated competition for iron and possibly antibiosis).

Dolors *et al.* (2010) found that plant growth medium based on grape marc compost (compost:peat 1:1, v/v) amended with the biological control agent *Trichoderma asperellum* strain T34, restores composts' suppressive capacity against *Fusarium* wilt of carnation compared with the non-amended medium under greenhouse conditions. Standard chemicals (tocloflos-methyl 50 per cent and captan 85 per cent) used in the cultivation of carnation crops did not improve the suppressive capacity of the growth media or the T34 populations. Plants in T34-amended growth media had higher water demand, regardless of the treatment, which implied their growth could be enhanced by increasing fertirrigation, compared to plants non-amended with T34. In conclusion, the amendment of formulated grape marc compost with *T. asperellum* T34 improves the suppressive capacity of this growth medium against Fusarium wilt in carnation.

9.1.1.4 Solar Radiation Intensity

Ben Yephet and Shtienberg (1994) reported that the most severe epidemics developed at low radiation intensities (200-300 mE/m²/s) and at temperature close to 25-26°C. Plants remained symptomless at solar radiation intensities of >1000 mE/m²/s and temperature <18°C. Elena and Tjamos *et al.* (1997a) found soil solarization for nearly 7 weeks significantly reduced or eliminated inoculums. Baayen and Schoffelmeer (1997) found that the effect of air humidity was negligible, disease indices were slightly lower at 45 per cent compared with 80 per cent humidity. High light intensities led to much lower percentages of diseased plants than low light intensities.

Negi and Harender Raj (2013) observed that soil solarization with transparent polyethylene sheet and biodegradable plastic sheet (both of 25µm thickness) for 40 days resulted in an increase of 5.6 and 3.0°C, respectively in average maximum soil temperature inside polyhouse. Soil solarization of the culture of wilt pathogen with same sheets at 5 cm soil depth for 40 days increased the incubation period of the disease by 47.10 and 33.86 per cent and reduced the disease incidence by 81.82 and 63.63 per cent, respectively in comparison to pathogen culture retrieved from unsolarized plots inside polyhouse at same depth after same duration. Soil solarization of the culture of wilt pathogen with transparent polyethylene sheet and biodegradable plastic sheet at 5 cm soil depth for 40 days also resulted in 84.43 and 68.84 per cent reduction, respectively.

9.1.1.5 Integrated Control

Jacob *et al.* (1988) tested bacterial antagonists *Streptomyces* sp. str. G 43 and *Bacillus subtilis* str. T 99 to control *F. oxysporum* f. sp. *dianthi* in carnation. At low infection pressure, post-inoculation treatment with *B. subtilis* was 50 per cent more effective than with streptomyces. Similarly, pre- and post-infection treatment with *B. subtilis* alone or in combination with zineb gave best disease control; fungicide treatment alone was almost completely ineffective. Further tests demonstrated a synergistic interaction in disease control between *B. subtilis* and suboptimal doses of either zineb or benomyl. Elena and Tjamos (1997a) found that integrated control was highly effective by using soil solarization for 4 weeks alone or in combination with biological factors, such as the antagonistic fungus *Trichoderma harzianum* and dazomet at a rate of 60 g/m² alone. The best way to control wilt of carnation is an integrated approach consisting of clean stock, soil treatment and phytosanitation. Chemical treatment of soil with benomyl @ 1.15 kg a.i/93.03 m² or with thiophanate methyl @ 1.6 kg a.i/93.03 m² before planting is reported to give good control of the disease. Biological agents like *Bacillus subtilis* and *Streptomyces* are also reported to have given good control of fusarium wilt. In the investigation on integrated disease management, Martinez and Pinzon (1999) suggested that to succeed in the biological control programme of *F.oxysporum* f.sp. *dianthi*, it was necessary to implement a programme for the control of the nematode. When the nematode was controlled along with the use of the most promising isolates of *Trichoderma* spp., it was possible

Diseases and Disorder of Carnation

Septoria leaf spots

Flower blight

Fairy ring spots

Bacterial wilt

Alternaria leaf spots

Fusarium wilt

Calyx splitting in carnation

to grow susceptible varieties of spray carnation in areas heavily infested with the *F. oxysporum* f.sp. *dianthi* without the need of chemical disinfestation of the soil.

Basallote *et al.* (2016)found that Soil treatments combining organic amendments and soil solarization significantly reduced disease incidence (86–99 per cent) and increased the number of commercial carnation stems by 5–9 times compared to non-treated plots. *Fusarium* species associated with carnation cultivation were identified as *Fusarium oxysporum* (43 per cent), *Fusarium proliferatum* (24 per cent), and *Fusarium solani* (33 per cent). The phylogenetic analysis of the translation elongation factor 1 alpha (EF-1a) region distinguished highly aggressive isolates of *F. oxysporum* f.sp. *dianthi*, from low aggressive isolates. The pathogenicity tests showed that FWC has a complex etiology, with several *Fusarium* spp. identified as causal agents. *F. proliferatum* and *F. solani* are associated with carnation wiltfor the first time in Spain.

Lingan *et al.* (2014) found the *Pseudomonas fluorescens* and *Trichoderma viride* enriched Bioconsortium Pf1 + Tv (as seedling dip + soil application enriched with FYM + foliar spray @ monthly interval) recorded the disease incidence of 7.8 per cent in in the first year and 9.5 per cent in the second year which was at par with carbendazim treatment. Simultaneously, when compared to chemical treatment, the stem length and number of flowers of carnation (86.0 cm and 47.5 per m^2) were maximum in beds treated with Pf1 + Tv (seedling dip + soil application enriched with FYM + foliar spray @ monthly interval) followed by individual bioagent application.

Harinder *et al.* (2016) found combination of treatment of soil solarisation with transparent polyethylene mulch for 40 days with soil amendment of neem cake followed by cabbage leaves was found at par with combination of soil solarization followed by three drenching with combi-fungicide comprising of carbendazim 25 per cent + iprodione 25 per cent which resulted in reducing the wilt incidence by 89.0 to 100 per cent in comparison to unsolarized and unamended control. Combination of soil solarization with soil amendment with neem oil cakes/granules or *T. viride* formulation was more effective than soil drenching with chemical fungicides in improving the plant growth and quality of the flowers. Integration of soil solarization with soil application of *Trichoderma viride* followed by amendment of neem cake was found most effective in improving the shoot length by 79.8 to 84.6 per cent with 48 to 55 times increase in number of A grade flowers in comparison to unsolarized and unamended control. Savita *et al.* (2019b) found among amendments, neem cake was found most effective with 75.0 per cent reduction in the wilt incidence. Bavistin among fungicides, Neemazal among botanicals and *Trichoderma viride* among bio-control agents were found effective with 100.0, 71.1 and 93.6 per cent reduction in the wilt incidence. Among different treatments, root dip of cuttings in Bavistin (0.1 per cent), soil amendment with Neemcake (1kg/m^2), root inoculation with culture of AM fungi and *A. chroococcum* (5g culture/plant) and soil application of *T. viride* formulation (10g/m^2) in solarized soil was found most effective with 97.1 per cent reduction in the wilt incidence. This treatment combination also resulted in maximum increase of 50.97, 100.4, 39.2, 57.3, per cent in plant height, number of flowers per plant, flower size and length of flowering stem, respectively in comparison to control.

Arici *et al.* (2019) evaluated the effect of somebiological preparations [*Trichoderma harzianum, Pseudomonas fluorescens; Bacillus subtilis* QST 713(SERENADE), *Mycorrhiza* spp., *Trichoderma* spp., *Bacillus* spp. (PANORAMIX), *Lactobacillus acidophilus* + *Lactobacillus paracasei* (VITANAL), tea tree oil extract (TIMOREX GOLD), orange oilextract (PREV-AM), plant extracts of Reynoutria spp. (REGALIA) against Fod were investigated.After 30 days, results were obtained with *T. harzianum* (96 per cent), SERENADE (96 per cent) and PANORAMIX (88 per cent) against *Fusarium oxysporum* f. sp.*dianthi*. All of the control plants (+) were completely dead due to the disease.

9.1.2 *Phialophora cinerescens*

In case of *Phialophora cinerescens* infections the plants show gradual wilting. Leaves of the infected plants fade and turn into straw colour. The vascular system gets discoloured into chocolate brown and the colour is much darker compared to fusarium wilt. Rumine and Parrini (1982) used benzimidazole fungicides like Benomyl, carbendazim, Thiophanate-methyl, Thiabendazole to control carnation wilt caused by *P. cinerescens*, but they were not effective in preventing or curing the infection. In carnation, few varieties are resistant to both *Fusarium* and *Phialophora*. Among different colours, pink flower producing cultivars are susceptible to wilt whereas red flower cultivars are resistant. The cultivars with white flowers are known to be intermediate or variable in their reaction.

9.1.3 Foot Rot

Several soil borne fungi *Phytophthora* sp., *Phythium* sp., *Rhizoctonia solani*, *Sclerotinia sclerotiorum* attack the roots and collar portion of carnation especially under high moisture conditions. The infection results in rotting of the stem at soil level resulting in wilt. In case of root infections, the leaves gradually get discoloured and start drying up from bottom upwards. Reduction in soil moisture is an important step in the control strategy.

Vinod *et al.* (2015) reported the occurrence of stem rot incited by *Sclerotinia sclerotiorum*, for the first time in India. The pathogen was identified as *Sclerotinia sclerotiorum* on the basis of phenotypic and genotypic characteristics. Carpogenic germination was induced artificially, and life cycle of the fungi was studied. Microscopic studies of the apothecium revealed the presence of spermatia, croziers, paraphyses as well as monomorphic ascospores specific to *S. sclerotiorum*.

9.1.4 Phytophthora

The *Phytophthora* sp. associated with this disease is *P. nicotianae* var. *parasitica*. Migheli *et al.* (1993) isolated ten isolates of antagonistic fungi, from soils and substrates suppressive against *Phytophthora cryptogea* and *Pythium ultimum*. The fungal species involved were *Trichoderma harzianum, Fusarium oxysporum, Penicillium viridicatum, Aspergillus fumigatus, Mycelia sterilia, Penicillium canescens* and *Penicillium puberulum*. In 3 trials, carnation cuttings were transplanted successively in an artificially infested potting mix in order to evaluate the effect of the antagonists on disease severity over a period of 4-10 months. Some of the isolates *e.g.*, F2 (*T. harzianum*), F34 (*P. canescens*), C16 (*A. fumigatus*), F60 (*P. puberulum*) and F69 (*M.*

sterilia) effectively reduced the incidence of *P. Cryptogea var. parasitica* on carnation, most of the tested strains progressively lost their antagonistic ability after the first transplant. High variability observed in the biological control activity of the tested isolates indicates the necessity to reduce the influence of different environmen-tal and cultural conditions on the antagonists by developing alternative methods for strain improvement, formulation and delivery systems of antagonistic microorganisms. The common fungicide used against this disease is Fosetyl-Al and Metalaxyl.

Sunita and Nandita (2004) tried six different species of *Trichoderma viz., T. harzianum, T. viride, T. hamatum, T. virens, T. koningii* and *T. iongibrachiaturn* were assessed both in vitro and in vivo for their antagonistic activity against *Phytophthora. Trichoderma virens* suppressed maximum growth of *Fusarium oxysporum f.sp. dianthi* in dual culture technique while *T. harzianum and T. viride* were found most effective in inhibiting the mycelial growth of Phytophthora nicotianae var. parasitica and Sclerotium rolfsii respectively.

9.1.5 Rhizoctonia

Raabe *et al.* (1977) incorporated benomyl, thiabendazole or thiophanate-methyl into the soil one day before planting. All fungicides, particularly benomyl, gave good preventive control. Cultivars Elegance and Caribe were most susceptible to infection and Atlantis least susceptible. Hadar *et al.* (1982) kept the soil wet for 45 days, resulted in a sharp decline in inoculum potential. The use of either methyl bromide or vapam resulted in complete control of the disease and reduced inoculum potential. The fungi-cides useful are benomyl or ipridione against *Rhizoctonia solani.*

Gupta *et al.* (1995) evaluated 12 cultivars of carnation for resistance against *R. solani* stem rot under natural epiphytotic conditions, none was found to be resistant. Disease incidence was greatest on cultivars Scania and Sam's Pride and least on William Sim and Vermilion Protruding.

Sunita and Nandita (2004) tried six different species of *Trichoderma viz. T. harzianum, T. viride, T. hamatum, T. virens, T. koningii* and *T. iongibrachiaturn* were assessed both *in vitro* and *in vivo* for their antagonistic activity against *Rhizoctonia*. Minimum per cent mycelial reduction in *Rhizoctonia solani* was observed highest with *T. harzianum* followed by *T. viride*. In pot culture experiment, *T. harzianurn* gave excellent control of wilt, root and basal rot diseases by giving maximum disease reduction when applied twice as soil drench whereas stem rot incited by *Rhizoctonia solani* was minimized to great extent with addition of *T. hamatum* in soil a week ahead of transplantation of cuttings and 45 days after establishment of cuttings.

9.1.6 Pythium

In greenhouse experiment, Baker and Harman (1981) found that soil steaming + ethazol [etridiazole] application in soil at 18 oz/1000 ft^2 successfully controlled pythium rot and increased flower production in several carnation cultivars. Hanan and Baker (1983) planted the cv. White No.1 in steamed, raised soil beds in a greenhouse with different treatments. Etridiazole (which is specific to pythium-like moulds and also a nitrification inhibitor) raked into the top soil layer at 15 ml/plot (27.3 ft^2); metalaxyl at 10.5 ml/plot; or N-Serve (a nitrification inhibitor possibly fungicidal to phytophthora) at 1.22 g/plot control. Plants in the etridiazole-treated plots produced more number of flowers and also peaked 3 weeks ahead of the other treatments including control.

9.1.7 Sclerotium Rot (*Sclerotium rolfsii*)

Madhu Meeta *et al.* (1995) found Folicur (tebuconazole) was the most effective in preventing germination of soaked sclerotia of *Sclerotia rolfsii*. None of the sclerotia germinated following immersion in 0.01 per cent tebuconazole for 2 h. In pot tests, inoculated plants wilted within 7 d but 100 per cent survival was obtained when plants were drenched with 0.01 per cent tebuconazole after inoculation.

Sunita and Nandita (2004) tried six different species of *Trichoderma viz. T. harzianum, T. viride, T. hamatum, T. virens, T. koningii* and *T. iongibrachiaturn* were assessed both *in vitro* and *in vivo* for their antagonistic activity against *Rhizoctonia*. Minimum per cent mycelial reduction in *Rhizoctonia solani* was observed highest with *T. harzianum* followed by *T. viride*. In pot culture experiment, *T. harzianurn* gave excellent control of wilt, root and basal rot diseases by giving maximum disease reduction when applied twice as soil drench whereas stem rot incited by *Rhizoctonia solani* was minimized to great extent with addition of *T. hamatum* in soil a week ahead of transplantation of cuttings and 45 days after establishment of cuttings.

9.1.8 Stem Rot (*Fusarium roseum*)

Sometime, due to the injuries caused by continuous harvesting in the mother blocks, stem rot of carnation occurs. Rotting at the base of the plants is seen quite often.

In the main field, a dry rot at the base makes the plant wilt. The disease is very common in neglected or badly managed gardens, especially after a very high flower yields. Application of benomyl, captan and chlorothalonil to foliage and stem controls the disease.

Sharma and Chandel (2016) found Integration of *Melia azedarach* seed (powder) along with *Trichoderma viride* as well as Neem cake with *T. viride* and *T. harzianum* gave maximum reduction (75.67 per cent) in disease incidence of stem rot with minimum disease incidence (6 per cent), in comparison to 24.67 per cent in control followed by Bacillus subtilis + *Melia azedarach* (8 per cent). These treatment also improved the plant growth and flower quality parameters by giving the average plant height to 82.32 cm, 81.05 cm, stem length (76.64 cm, 75.91 cm), number of flowers per plant (3.79, 3.68), flower size (7.20 cm, 7.17 cm) and took less number of days to first flowering. The integration of effective management practices such as plant oils, botanicals and antagonists would be important for the successful management of plant diseases, thus showing its practical applicability in field in mitigating the losses caused by disease in carnation.

9.1.9 Alternaria

In glasshouse tests, Strider (1978) found a direct relationship between symptom expression and length of time foliage was kept moist after inoculation. Some infection occurred, when inoculated plants were placed immediately on the glasshouse bench and not in a moist chamber. All cultivars examined were susceptible except Dusty, New Pink Sim, Light Pink Barbi, Maj Britt, Pink Ice and Red Gayety were significantly less susceptible than most. Hilal and Kamel (1990) found that the cultivars Leza, Amber Rose, Lena and Yellow were resistant to *Alternaria dianthi* while New Arthur was very susceptible. In field experiments on 4 cultivars (Arthur Sim, Exquisite, Sam's Pride and White Sim), Dhiman *et al.* (1991) reported that the level of disease decreased gradually with subsequent delayed planting. Exquisite was the most resistant cultivar. Madhu Meeta *et al.* (1996) found that cultivars Yellow Dusty, Harvest Moon, Leena and Shocking Pink were highly susceptible and cultivar Scania was the least susceptible. Wang Kuei Mei *et al.* (1994) reported that leaf spot caused by *Alternaria dianthi* is the most serious air-borne disease of carnation. Infection was much more severe on field plants than in those grown under protection from rain. Cultivars with red flowers were the most resistant.

Different workers used different fungicides for the control of this disease. Strider (1978) found that anilizine afforded complete and excellent protection, they further observed that captafol, chlorothalonil, captan, zinc ion-maneb complex and zineb, all at one-third the recommended doses gave good protection. During the storage, temperature fluctuation should be avoided. Krzyzanowska (1982) found that spraying of mother plants with rowral at 0.1 per cent prior to cutting is the best control of alternaria blight. Dusting mother stock with captan 5 per cent or Dithane 5 per cent was also effective. Chlorothalonil, mancozeb and cuprosan 311 SD [copper oxychloride + maneb + zineb], applied 5 times at 10-days interval, gave good control of *A. dianthi* (Hilal and Kamel, 1990). Wang KueiMei *et al.* (1994) recorded good control by spraying 75 per cent chlorothalonil or 24.9 per cent difenoconazole before or immediately after infection in the field and improved carnation quality. Tripathi and Lal (1992) tested the efficacy of 6 fungicides and reported that difolatan, dithane, 0.2 per cent defolatan and 0.2 per cent dithaneM-45 gave the best control of leaf blight of carnations caused by *Alternaria dianthi* in Uttar Pradesh, India.

9.1.10 Leaf Spot (*Heterosporium echinulatum*)

Penaranda *et al.* (1992) found that the protectant (dichlofluanid and propineb) fungicides gave better control of the disease than the systemic fungicides (penconazole and triforine). Propineb was the most effective in miniature carnation variety Sam's Pride. The systemic fungicides were, however, more effective in inhibiting spore germination in *Heterosporium echinulatum*, particularly triforine. It was suggested that a combination or a simultaneous application of both protective and systemic fungicides could be the best strategy for the control of the disease. Ortiz Borrero and Arbelaez Torrez (1992) found, out of 19 fungicides tested *in vitro*, the mixture of triforine + propineb was the most effective against *H. echinulatum* on carnations. A combination of leaf pruning with sprays of triforine + propineb was recommended for disease control. Abe *et al.* (1997) found that fungicides (tetraconazole and fluazinam) were most effective in controlling leaf spot disease caused by *H. echinulatum* in carnation variety Light Pink Barbara.

Misawa *et al.* (2018) reported leaf spot symptoms which appeared as brown or purple spots with concentric rings c. 5-10 mm in diameter on the upper leaves.

Disease incidence reached approximately 20 per cent in the greenhouse (230 m^2) in early May. Three fungal isolates (CAR4, CAR6, and CAR8) were obtained from the leaf lesions. Conidiophores of the isolates grown on potato carrot agar medium were erect, pale to mid brown, with swellings at the distal ends, bearing a single conidium at the apex.

9.1.11 Cladosporium

Cladosporium echinulatum infection on leaves begins as purple specks which later enlarge and show a tan to grey centre enclosed by a purple border. On the petals the pathogen produced light-brown lesions (Cedeno and Carrero, 1997). Lo Chaur Tsuen *et al.* (1997) reported that disease severity was enhanced under conditions of high moisture and could be controlled by reducing the relative humidity. Fungicides chlorothalonil and difenoconazole also reduced the severity of the disease but difenoconazole was most effective.

9.1.12 Rust

Rust is caused by *Uromyces dianthi*. Semina and Shestachenko (1981) reported that in Sim group, cvs. Rosa and Arthur Sim were highly resistant to rust, but many hybrids were susceptible. Aloj *et al.* (1990) reported that white rust, caused by *Albugo* sp. (cultivars Fede and Argallo) affected only floral calyces.

Hill (1974) reported that weekly applications of zineb (0.14 per cent) or maneb + nickel sulphate (0.2 per cent) gave good control of the disease but did not eradicate it. Fortnightly applications of oxycarboxin were not so effective but gave some control against carnation rust. Tatarynowicz (1981) reported that plantvax [oxycarboxin] at 0.1 per cent + dithane M-45 [mancozeb] at 0.2 per cent gave 100 per cent control of *U. dianthi* when applied every 7 or 14 days and 99.5 per cent control when applied every 21 days. Mitrofanob *et al.* (1982) found that fundazol [benomyl] at 0.2 per cent, cuprosan at 0.5 per cent or zineb at 0.5 per cent were less effective to control rust on rooted perpetual carnation cuttings than Bordeaux mixture at 1 per cent. However, when Fundazol plus cuprosan or zineb was applied the effectiveness improved markedly. Dalla Guda and D'Aquila (1982) reported that weekly application of biloxazol at 0.28 ml/l gave very good results; treatment with propinconazol was the most effective but stunted the plants as treatment with triadimefon. Triadimefon did not give good control. Captafol + folpet is not to be used on flowers in anthesis (Rumine and Bonifacio, 1983). In absence of chemical treatment, heavy rust infections appeared only on wet leaves whereas the dry ones had a level of infection as low as that of the treated plants. Therefore, protected cultivation, prevention of moisture on the plants and a ground-localised irrigation system are safer and less expensive alternatives to the chemical treatments. In a trial with triforine, triadimefon and bitertanol, applied at various concentrations, either by watering on or by spraying, as a single application, just before or just after inoculation. Gjaerum and Langnes (1984) suggested that when watered on, both triadimefon and bitertanol were effective but the latter caused leaf damage. When sprayed on, only bitertanol at the higher concentrations was effective, although not for an already visible attack, and it was less phytotoxic when applied in this way. Cifuentes and Arbelaez (1984) reported that soil application of oxycarboxin at

375-7300 ppm gave the best control (>99 per cent). Triforine also gave good control but was phytotoxic at soil application rates at and above 237.5 ppm for control of *U. caryophyllinus* [*dianthi*].

9.1.13 *Peronospora dianthicola*

Quintero (1992) reported that the most susceptible carnation varieties were Alice and Redalice on which fungal spots of *Peronospora dianthicola* were observed on the leaf and stem. Several varieties (Rodeo, Hellas, Tango bambi, Peterson, Pallas, Yellow Dusty, Improved White and Rayo de sole) were only affected on leaves. A reduction in temperature and RH inside greenhouses, implementation of a good ventilation system, improved soil drainage and periodical spraying with mancozeb and maneb are recommended for the control of the pathogen.

9.1.14 Stub Dieback (*Gibberella zeae*)

Stack *et al.* (1978a) examined several cultural practices used in glasshouse production of carnations for their effect on the incidence of infection by *G. zeae*. Cold storage of rooted cuttings prior to planting increased the incidence of infection when the cuttings were inoculated after planting. Vegetative stems were less susceptible to infection than reproductive ones. Shorter stubs were more susceptible to infection and infections in shorter stubs were more likely to expand into severe lesions. Fresh stubs were infected at higher rates than older stubs. Manipulation of these practices may reduce the incidence of stub dieback. Stack *et al.* (1978b) found that combination of high temperature (24°C) and intermittent high RH resulted in high incidence and severity of infection. Intermittent wetting or presence of a saturated atmosphere for a short period following inoculation increased infection frequency. Growing plants at a high light intensity prior to inoculation resulted in more diseases than in plants grown at a lower intensity.

9.1.15 Powdery Mildew (*Oidium dianthi*)

Saenz *et al.* (1995) found that Cerise, Vanessa, President and Katia were susceptible to powdery mildew (*Oidium dianthi*).

Bacterial Diseases

9.1.16 Bacterial Wilt

Zoina and Aloj (1979) found that all the carnation cultivars belonging to the Sim group and Mediterranean group were susceptible to *Pseudomonas woodsii*, although at different levels. *Pseudomonas caryophylli* causing wilt is prevalent at high temperatures. The symptoms are sudden wilting and production of greyish green foliage. Vascular discolouration and rotting of roots are the other symptoms. Deep and longitudinal cracks are seen on the lower most internode of the plant. This symptom is commonly produced during winter when the soil temperature is low. Soil sterilization and using disease free plants and cuttings reduce the disease risk.

9.1.17 Bacterial Spot

Diatloff and Rochecouste (1991) mapped the spread of bacterial spot in 2 commercial outdoor plantings of standard Sim carnations and multiflowered Spray carnations and related to temperature and rainfall. Disease in Spray carnations began as marginal 'hot spots' traced to be infected with *Pseudomonas andropogonis*. In Sim carnations, the initial infections occurred later and were randomly distributed. Mechanical transfer of inoculum by workers servicing both crops and wind-blown rain were most probably responsible. The subsequent disease spread was similar in both types of carnation and had 3 distinct phases slow peripheral spread, rapid extensive spread and declining vertical spread which was related to rainfall and temperature.

Rodoni *et al.* (1994) reported that among viruses *viz.*, carnation mottle carmovirus (CaMV), carnation *etc.*hed ring caulimovirus (CERV), carnation vein mottle potyvirus (CarVMV), carnation latent carlavirus (CarLV) and carnation ringspot dianthovirus (CarRSV) CarMV was the most prevalent virus (9.9 per cent), and lower levels of CERV (0.2 per cent), CarVMV (0.1 per cent) and CarLV (0.1 per cent) were detected. Singh and Singh (1989) reported that aphid (*Aphis gossypii*) as the vector of virus 10 per cent transmission of the virus was recorded after a 1-h test feed and 100 per cent transmission was recorded after a 12-h test feed. The results of test feeds for all other insect species were negative. Further they found that *Aphis gossypii* can acquire sufficient virus in a 60-min acquisition feed and a 6-h test feed to give 100 per cent transmission of the virus.

De La *et al.* (2016) observed several viral symptoms, such as leaf mottle and severe plant deformation, in many young carnation plants. Carnation mottle virus (CarMV) and Carnation *etc.*hed ring virus (CERV) were detected in single and mixed infections by DAS-ELISA, electrophoretic dsRNA assay, and dot-blot analysis using digoxigenin labeled riboprobes. The identity of both viruses was confirmed by sequencing the coat protein gene (CP) amplicons obtained by RT-PCR.

9.1.18 Carnation Mottle Virus and Carnation Ringspot Dianthovirus

Kowalska (1972) isolated both carnation mottle virus and ringspot dianthovirus from glasshouse carnations. The mottle virus was symptomless, whereas ring-spot dianthovirus induced chlorotic and necrotic spots and leaf distortion. Goethals and Hoof (1971) found that meristem culture of carnations infected with carnation mottle virus resulted in 28 per cent of the plants free from the virus. Two months' heat treatment of the infected mother plants at around 38°C before taking the meristems raised the level of healthy plants to 100 per cent. This combined technique was also suitable for eliminating *etc.*hed ring and streak viruses. Kowalska (1974) observed the largest number of carnation mottle virus free plants from tips 0.4-0.7 mm long. Short (4 weeks, 37°C) thermotherapy before excision of the meristem tips increased the number of virus-free plants. Goethals *et al.* (1973) outlined the elementary precautions to limit the spread of carnation mottle virus. They found that virus was transmitted 2 per cent by contact between infected and healthy leaves, 100 per cent by applying extracts of infected plants to wounds, and 33 per cent by rubbing the leaves with an abrasive. Cuttings taken by 4 different methods, *viz.*, nipping off with the fingernails, using an unsterilized knife, using a sterile knife, and breaking stems at an internode resulted in infection rates of 61, 75, 8 and 1 per cent, respectively. Toth *et al.* (1984) reported that carnation mottle virus is the most widespread carnation virus, causing up to 20 per cent yield losses. The most effective procedures suggested to minimise the infection were (i) heat treatment (for cultivars difficult to immunize); (ii) meristem tissue culture (tips cut < 0.2 mm in length); (iii) testing of *in vitro* plants by ELISA; (iv) planting out into isolation greenhouses, where each plant is individually recorded and treated; and (v) 3 tests for virus infection, after 1, 4 and 7 months. Ko Jae Young *et al.* (1997) found that all meristem-cultured and seed-propagated plants were infected with carnation ring spot virus. Seed-propagated plants were free from carnation mottle virus.

9.1.19 Carnation Mottle Carmovirus (CaMV)

Hollings *et al.* (1977) observed that Sim carnations were free from other viruses, CarVMV caused slight diffuse chlorotic mottling in the younger leaves and decreased flower yield by about 22 per cent, (due to flower breaking). In the non-Sim cultivars leaf symptoms and flower breaking were more pronounced. In mixed infections with carnation mottle virus, symptoms were much more severe. CarVMV was not eliminated from carnation or *D. barbatus* plants grown for 4 weeks at 37°C, and only rarely from cuttings taken from them, but it was readily eliminated by meristem-tip culture. Mokra *et al.* (1988) found the symptoms to be yellow or purple and partially necrotic streaks or flecks in older leaves which dried later. Carnation mottle carmavirus (CarMV) affected plants were mottled, yellow and dwarfed. The flowers were small and discoloured. Wang *et al.* (1990) found that the shoots from callus cultures of petals yielded 78.26 per cent plants free from carnation mottle carmovirus while infection was eliminated from shoots obtained by ovule culture. When meristem-tip culture was combined with use of an antiviral agent the percentage of virus-free shoots increased according to the time of treatment with ribavirin (5 mg/litre). Using this virus-free shoots were produced effectively from meristem tips up to 3 mm long. The virus was completely eliminated from the meristem tip after 5 weeks treatment. Fawzy *et al.* (1991) reported that nematode was not involved in virus transmission.

9.1.20 Carnation Yellow Fleck Closterovirus

Carnation yellow fleck closterovirus (CYFV) symptoms (streak) are more severe at the flowering stage. Plants infected by CYFV alone often remain symptomless. CYFV is not transmitted mechanically but only by grafting or by aphids (Poncet and Devergne, 1988). Batlle and Lavina (1994) identified carnation necrotic fleck closterovirus (CNFV) in a commercial carnation cultivation in Spain, its presence was confirmed by mechanical inoculation of *Saponaria vaccaria*.

9.1.21 Carnation Vein Mottle

Oertel (1977) reported that 70-90 per cent infection of planting material infected by ringspot, mottle, vein mottle, *etc.*hed ring and latent viruses and Arabis mosaic virus was reduced by preselection followed by extended heat treatment and meristem culture to only 0.7 per cent. Heat treatment also increased meristem rooting by 22.6 per cent and the rate of successful transplants by 25 per cent.

9.2.0 Pests

9.2.1 Thrips

Thrips (*Thrips tabaci*) suck the sap from the leaves, causing them to turn yellow and patchyoften with black specks and slight crinkling. A severe attack adversely affects the growth, they also cause streaks in the flower making them unmarketable. The dark and pink cultivars normally suffer most. Tkachuk (1978) found that rogor [dimethoate], metathion [fenitrothion], fosphothion [malathion], cidial [phenthoate] or DDVF [dichlorvos], all at 0.1-0.2 per cent gave the best control of carnation thrips (*Taeniothrips dianthi* and *Thrips tabaci*). Tjosvold and Ali (1989) reported that wrapping cut carnation flowers in polyethylene bags containing 1 per cent chlorpyrifos as a fumigant for a period of 3 days during refrigeration at about 10°C reduced infestation by about 50 per cent, but complete control was not achieved. Riudavets *et al.* (1992) sampled carnation crops for predators of western flower thrips (*Frankliniella occidentalis*). The most common predators were *Orius majusculus, O. laevigatus, Dicyphus tamaninii and Macrolophus caliginosus* and mean development time was 25°C, 70 per cent RH and LD 16:8 ranged from 12.9 days for *O. majusculus* to 16 days for *M. caliginosus*. The mean number of *F. occidentalis* consumedranged from 4.9/days for *O. majusculus* to 3.1/days for *M. caliginosus*. In another trial Riudavets *et al.* (1995) found that *Orius laevigatus* and *O. majusculus* were scarce on carnation. *Dicyphus tamaninii* and *Macrolophus caliginosus* were absent on carnation. All species completed nymphal development, laid eggs and survived as adults for more than 20 days when fed larvae of *F. occidentalis* at 25°C. Tjosvold *et al.* (1995) harvested carnations infested with *F. occidentalis* from a commercial greenhouse and prepared for shipment. The cut flowers were fumigated at 23°C with Naled Dibrom 8 EC at 4.49, 8.98 and 17.96 g a.i./100 mfor 1 h and with 35.92 and 89.80 g a.i./100 m^3 for 2 h. Naled Dibrom reduced the incidence of thrips on flowers by 78.4-89.2 per cent. Sulfotep (plantfume 103) was applied at 13.77 g a.i./100 m^3 for 2 h and 15 h, and reduced the population of thrips in flowers by 52.4 and 64.35 per cent, respectively. Phytotoxic effect was noted with NaledDirbom at 8.98 g a.i./100 m^3 and sulfotep on cv. Chianti. Application of NaledDibrom at 4.49 g a.i./100 m^3 for 1 h resulted in a significant reduction of thrips population without phytotoxic effects. Increased fumigant rates and fumigation time did not eliminate *F. occidentalis* completely.

9.2.2 Bud Borer

Carnation is one of the most preferred hosts for *Heliothis armigera*. The pest causes 50 per cent damage to flower buds. Bud borer bore into buds and feed on developing petals. Affected buds bear holes and excreta. Sood and Kakar (1990) studied the control of bud borer on 5 cultivars of carnations in the laboratory at 22°C and in the field. Infestations of 40.9-91.0 per cent were observed, with damage being greatest on White Sim (81.3 per cent) and lowest on Yellow Dusty (63.8 per cent).

Pal *et al.* (2014) found that bud borer infestation varied significantly amongst different carnation varieties during peak incidence period of the pest for both the seasons. The mean bud borer infestation ranged from 8.18 per cent in the variety Dover to 17.61 per cent in the variety Charment. The varieties *viz.*, Dover, Lavender

Pests of Carnation

Aphid infestation

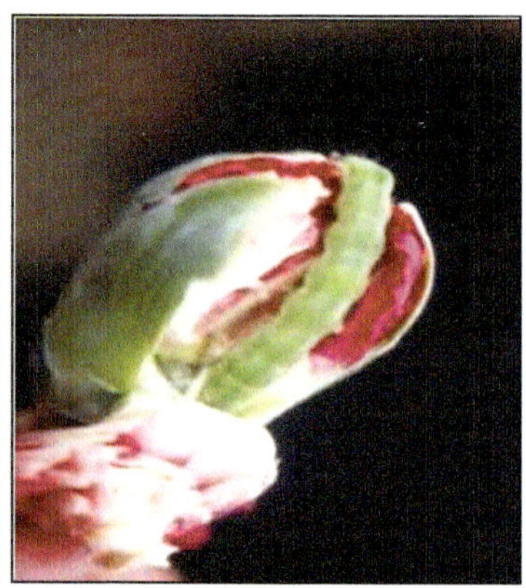

Ball worm infestation

Red spider mite attack

Thrips infestation

Lace, Tikar and Madras were found to be the least susceptible varieties showing less than 10 per cent bud infestation. There was significant difference in various vegetative and floral parameters among different varieties of carnation. The bud borer infestation was found to be significantly and positively correlated with the plant height, number of branches per plant and number of flowers per plant. A negative correlation was found between bud borer infestation and thickness of calyx.

9.2.3 Red Spider Mite (*Tetranychus urticae*)

Red spider mite is most serious pest of carnation. They are minute red insects, feed on the under sides of leaf, suck the sap and eventually leaves turn pale and have a dusty coating and fine webs. In severe attacks plants become stunted, wilt and dry, and the flowers were also invaded. The varieties with curl leaves were more affected than flat leaves. Poe and Wilfret (1973) studied the relative susceptibility of 18 different cultivars and reported that cultivars with low borne leaves, leaves that are curled, bent or reflexed, had larger mite population than those with straight flat leaves. Price (1983) observed that dipping of harvested carnations in dicofol, an emulsifiable concentrate of oxydemeton-methyl at 1.88 ml/l did not eliminate the infestation of *Tetranychus urticae*. He also tried with the release of 16 females of *P. macropilis* per vase which reduced spider mite populations after 6 days, population number of spider mites was found after 13 days. However, it did not eliminate infestation by *T. urticae* completely. Pandolfo (1984) found that the most effective compound against *T. urticae* was propargite, which was effective against adults, eggs and larvae; followed by danitolfiori (fenpropathrin) alone at 150 g/l or combined at 100 g/l with torque [fenbutatin oxide] at 60 g, and also the dicofol-tetradifon mixture. Shelton *et al.* (1997) reported that mortality of diapausing spider miteswas low, in sequential low O_2 (0.1 per cent), high CO_2 (60 per cent) controlled atmosphere (CA), than air. Walung *et al.* (1999) compared the efficacy of 25-100 g/ha chlorofenapyrote, 50 g fluvalinate, 231 g dicofol and 50 g of flufenoxurn to control *T. cinnabarineeus*.Chlorofenapyrote at 50-100 g/ha significantly reduced mite population for 21 days. Valunj *et al.* (1999) also observed that chlorfenapyrote at 500 ml/ha was most effective treatment to suppress the pest for three weeks with 90-95 per cent efficacy. Fluvalinate at 500 ml/ha effectively suppressed the population for 1 week (Table 9).

9.2.4 Aphids (*Myzus persicae*)

Aphids suck the sap from the leaves and disfigure the young growth. In severe attacks, they leave sticky deposits on the leaves and flower buds. On young plants, a single drench with 0.35 per cent rogor or 0.25 per cent demeton-methyl gave good control over 4-6 weeks whereas 2 or 3 sprays of these insecticides were necessary. On older plants rogor and demeton-methyl drenches were not more effective than sprays and had a residual action only for 1 or 2 weeks (Gould, 1960). Bivins (1976) did not find phytotoxicity symptoms on 6 carnation cultivars sprayed with primor (primicarb) 50 w.p. at 2 or 4 oz a.i./100 gal, 3 times at fortnightly intervals. No aphid was found in any plot other than the control plots of the cultivar Shibuya Light and Pink Sim, which was infested with aphid. Shelton *et al.* (1997) investigated the efficacy of sequential low O_2, high CO_2 controlled atmosphere (CA) treatments to

Table 9: Bioefficacy of AC-303-630 (pyrole 10 EC) against Carnation Mite*

Treatment	*Dose (ml/ha)*	*Precount 3 Leaf/ Plant*	*Average Number of Two Sprays*								
			Average/Mean Mite Population (3 leaves/plant)								
			3 day	*Per cent eff*	*7 day*	*Per cent eff*	*14 day*	*Per cent eff*	*21 day*	*Per cent eff*	*28 day*
AC-303-630 (Pyrole 10 EC)	250	11.45	0.58	95.13	0.13	98.91	1.98	84.52	2.85	77.02	19.08
AC-3-3-630 (Pyrole 10 EC)	500	11.80	0.28	97.09	0.05	99.00	1.45	90.05	1.48	89.50	11.26
AC-3-3-630 (Pyrole 10 EC)	1000	13.00	0.25	98.20	0.00	100.00	0.69	95.00	0.76	94.77	10.15
Fluvalinate (Mavrik 25 EC)	500	14.21	2.09	86.17	2.03	84.83	6.90	57.53	2.18	86.15	10.78
Kelthane 18.5 EC	150	11.85	4.03	68.85	4.03	68.85	7.95	46.44	3.55	75.32	14.83
Flufenxuron (Cascade 10 DC)	500	12.15	5.91	64.44	5.13	64.44	9.35	47.65	3.78	78.16	15.95
Untreated control		11.68	15.05		15.10		16.18		15.68		15.80
S.Em		0.63	0.319		0.34		0.33		0.226		0.976
C.D. at P=0.05		N.S.	0.948		1.016		0.97		0.672		N.S.

* Valunj *et al.* (1994).

kill green peach aphids (*Myzus persicae*) and melon aphids (*Aphis gossypii*). Aphid mortality was high under a treatment of 0.1 per cent O_2 followed by 60 per cent CO_2 than air. Oettinget *al.* (1978) found that pirimor [pirimicarb] at 4 oz a.i./100 gal and orthene [acephate] at 8 or 12 oz a.i./100 gal were most effective to control aphid on carnation grown in polyethylene-clad greenhouses for up to 5 weeks. Metasystox-R [oxydemeton-methyl] at 8 or 12 oz a.i./100 was less effective than pirimor or orthene. Furadan [carbofuran] at 8 oz a.i./100 gal also gave good initial control but only remained active for 2 weeks.

9.2.5 Carnation Tortrix Moth [*Cacoecimorpha pronubana* (*Tortrix pronubana*) *Epichoristodes acerbella*]

The pest is found on carnation in the summer and early autumn and it is rather difficult to control once it is sheltered in the united leaves or in the flower buds, as spray cannot reach it. The most effective chemical treatment was found to be spraying at 10 day intervals from July until the final harvesting with a 0.5 per cent emulsion of 10 per cent parathion (Cuscianna, 1963). A mixture of sevin and kelthane was also recommended for control of tortrix moth in nurseries and beds (Arroyo,1963). Akbulut and Zumreoglu (1992) in surveys on carnations in Turkey identified *Liriomyza trifolii* as a pest. The infestation rate in greenhouses was 37.14-58.33 per cent.

9.2.6 Nematode

Meloidogyne arenaria, M. hapla, M. javanica and *M. incognita* are the limiting factors in the cultivation of carnation causing severe damage. Lamberti *et al.* (1987) also reported *Ditylenchus dipsaci, Heterodera trifolii, Paratylenchus Dianthus* and *Criconemoides curvatum* cause damage to carnation.

According to Streu *et al.* (1961) carnation is the host of 21 species of nematodes. *Criconemoides curvatum* feed on the roots as endoparasites as well as ectoparasites, and the nematodes were observed embedded in the cortical tissue. Soil steaming methods in current use generally caused little reduction in nematode population. Fawzy *et al.* (1991) found that *M. incognita* infected carnation plants were mottled, yellow and dwarfed. The flowers were small and discoloured. Cho *et al.* (1996) found variation in resistance against *M. incognita* between cultivars.

Soil fumigants like 1,3 dichloropropene alone or in mixture with methyl isothiocyanate or non-volatile systemic nematicides such as aldicarb, carbofuran, fenamiphos and thionazin are usually very effective. The application of methyl bromide is limited due to toxic effect of the bromine residues accumulated in the soil after treatment. These residues can be however, removed by abundant irrigations. Cuany *et al.* (1973) suggested that application of aldicarb 4 kg/ha at planting and 8 kg a.i./ha in autumn controlled *M. incognita*. Cuany and Ritter (1977) recommended the use of nematicide like carbamates to control *Meloidogyne* spp., *Heterodera trifolii, Paratylenchus Dianthus* and *Macroposthonia curvata* on carnation. High levels of vapam, di-trapex, di-trapex-CP and normal levels of nemacur were also very effective, with low levels of nemagon and DD being less effective.

Pal and Sarkar (2010) tested germplasms and found significant variation in their reaction to different pests of carnation. The germplasms namely Acc. No. 19, Danubio, Atletic Schubert and Lolita were found least susceptible to green peach aphid, *Myzus Persicae* Sulz. (<3 aphids/apical twig). More number of mites was observed on buds/flowers as compared to the leaves. The germplasms Acc. No. 19 and Atletic Schubert were found as least susceptible to red spider mite infestation. The germplasms *viz.* Acc. No. 19, Acc. No. 15 and Atletic Schubert were found to be least susceptible to bud borer infestation. The germplasms Acc. No. 19 and Atletic Schubert can be commercially cultivated in this region due to multiple tolerance to all the pests.

9.3.0 Disorders

9.3.1 Calyx Splitting

The splitting of calyces is a function of quality in carnation. The sepals beneath the flower are unlike to form a cylindrical calyx tube which supports the base of the petals. As the bud opens and the petals approach their full size the calyx may split down either half or completely. The primary cause of the splitting of calyx is unknown but may be due to factors like genetical, environmental and nutritional (Arora and John, 1976). Calyx splitting affected from 3.64 per cent (charmeur) to 36.1 per cent (mataro), out of 25 cultivars tested. Kim *et al.* (1992) observed that Falip, Rimon and Beton had longer flower stems and less calyx splitting compared to other cultivars. Baily (1963) reported that calyx splitting is mainly due to fluctuation of temperature at the time of flowering. Huy and Kim (1995) found that the use of reflective film mulch lowered the percentage of calyx splitting. Saini and Arora (1975) stated that increasing level of nitrogen reduced the percentage of calyx splitting, while increased potassium rate enhanced it. In contrast, Uri *et al.* (1990) and Medina (1992) reported that increased potassium content in tissues reduced calyx splitting. Boron availability to plants also lowered the percentage of calyx splitting (Adams *et al.*, 1979).

9.3.2 Curly Tip

This disorder affects the growing tips which curl and become distorted and is usually interpreted as some form of disease. The tips of the young shoots fail to separate and continuation of growth results in a characteristic curvature. The symptoms usually present themselves during low light intensity and low temperature. Thus poor light and other adverse conditions are thought to be the cause of the disorder.

10.0 Harvesting and Yield

10.1.0 Harvesting

The term maturity refers to that point of time when a flower can be harvested irres-pective of their physiological maturity and will continue to grow and develop the maximum quality. The standard flowers should be harvested at a large bud (tight bud) or at cross bud (petals visible) stage that is when the outer petals are unfolded nearly perpendicular to the stem. In case of spray carnations, harvesting should be done when at least two flowers have opened and the other buds are showing colour. Stems are usually cut with a sharp knife. It takes 15 to 20 weeks for the single pinched plant to yield cut blooms. The cut flowers should be immediately placed in clear water. Nowak *et al.* (1983) observed that the quality of flowers cut at the stages of large tight buds or cross petals is higher than the quality of flowers cut as in commercial practice paintbrush stage (petals enlarged, 20-25 mm).

In recent years, the trend is to harvest flowers in a tighter bud stage because it can be kept at low temperature for even 20 weeks, and opened according to the demands of the market. The success of the bud harvest greatly depends on the ability of the cut bud flowers open to their normal size. Number of commercial chemical bud opening solutions like 'Everbloom', 'Cornell' and 'Kagawa solution' are available in the market. The length of the stem cut with the flower depends on the time of the year. During fall, the cut should be made enough to encourage more lateral shoots while the flowers can be cut with longer stems when the plants are to be discarded. The best position of cutting stem is the area where leaves are well-spaced and where atleast 2 axillary shoots appear.

10.2.0 Yield

Flower yield depends on type and cultivar, growing region and environment, planting time, planting density, pinching and other management practices. In general, spray carnation produces more number of flowers than standard type. Sheared and Bunt (1963) obtained 15-24 blooms per square foot in the year cropping with a spacing of 15 x 15 cm. Altan and Altan(1982) recorded 75.5 flower/m^2. In an investigation, Mynett (1982) planted 100 plants/m^2 of cv. Scania 3C and recorded 197 flower stems. Swan (1998) reported the highest flower production (124.26 flowers/m^2) from William White at 32 plants/m^2 followed by Manon (122.76 flowers/m^2) at the same spains.

Kumar *et al.* (2019) found cocopeat + river sand (1:1) and 200 ppm IBA was found the best for rooting of cuttings. Within the interaction treatments, the best combination was found with the cocopeat + river sand (1:1) rooting media with 200 ppm IBA concentration for different flowering parameters like days to initiation of first flower bud (20.47 days), days to 50 percent flowering, number of floral buds per plant (6.20), number of flowers per plant (5.80), diameter of flower (5.61 cm) and stalk length of flower (58.12 cm).

Harshavardhan *et al.* (2016) found that 75 per cent recommended dose of nitrogen and phosphorus and 100 per cent potassium + *Azospirillum brasilense* + *Bacillus megaterium* + *Glomus fasciculatum* (VAM Fungi) + *Trichoderma harzianum* + vermicompost + panchagavya + jeevamrutha was best in improving flower quality and yieldwith flower bud diameter (1.67 cm), stalk length (75.45 cm), yield per plant (22.19) and yield per m^2 (634.27). Over all net returns (Rs 13,98,502 per 1000 m^2) was found to be maximum in this treatment with a B:C ratio of 1:1.70.

11.0 Postharvest Management

11.1.0 Grading

Grading is essential for selection of flowers according to the quality parameters which brings better price than flowers of equal quality but of poor grade. Various standardized grades based on stem length, flower diameter and physical conditions of the flower have been proposed over the years. Free from diseases and insects, other blemishes on flower and foliage, sleepiness of flowers, stem crooks, slabslides and bullheads, split calyx, faded colours are other factors to be considered while grading flowers. The Society of American Florists (SAF) suggested the following inspection standards for cut-carnations (Staby *et al.*, 1978):

1. Bright, clean, firm flowers and leaves
2. Fairly tight petals near the centre of the unopened flowers
3. Symmetrical flowershape and size characteristics of the cultivar
4. No split or mended calyx, flowers with split and mended calyx must not carry SAF emblem
5. No lateral buds or suckers
6. No decay or damage
7. Straight stem and normal growth.

The grades and standards developed by the European Economic Community (EEC) as cited by Staby *et al.* (1978) are given below:

Code	*Stem length (cm)*
0	5 or less
5	5-10
10	10-15
15	15-20
20	20-30
30	30-40
40	40-50
50	50-60
60	60-80
80	80-100
100	100-120
120	120 or above

The present standard carnation grades developed by the SAF are presented below:

Factors	Blue (fancy)	Red (standard)	Green (short)
	1	2	3
Minimum flower	Tight–50	Tight–44	Tight–none
diameter(mm)	Fairly tight–62	Fairly tight–56	Fairly tight–none
Open–75	Open–69	Open–none	
Minimum length (cm)	55	43	30

Overall.

Stem strength: Shall be determined by holding the stem horizontally at a point 25 mm above the minimum length for the grade, and the devia- tion of the flower head shall not be more than 30° below the horizontal place with the natural curvature of the stem down.

Defects: Blue grade shall consist of full, symmetrical flowers reasonably free from slabsides, bullheads, blown heads, singles, sleepy appearance, splits, discolouration, insects, diseases, and other damages. Blue grade shall have an essentially straight stem free from disbuds damage or discolouration by insects, disease and other means. Red grade, other than diameter of flower and length of stem, shall have the same requirements as the blue grade with moderatevariations.

All these methods do not take into consideration the potential lasting quality of cut-flowers. Several workers have employed different qualitative (flower and bud development, retention of flower colour) and quantitative (vase-life, storage life, water uptake, *etc.*) parameters for the evaluation of postharvest qualities of cut-carnation (Salingar, 1975; Sytsema, 1975; Halevy and Kofranek, 1977). In all these methods the vase-life and ornamental (aesthetic) value of flowers are determined based on subjective evaluation.

Cut flowers of the spray type carnation Light Pink Barbara graded as: high (H), middle (M) and the low (L) in a flower market were investigated with respect to their vase life, fresh weight and water and sugar contents in the floral organs (Miura *et al.*, 2000). Total fresh weight, per cent water content and water absorption were highest in H and lowest in L throughout the experiment. Buds of H and M opened normally within 10 days, whereas those of L blasted; vase life in H and L were 15 and 10 days, respectively.

Cut flowers of the spray type carnation Light Pink Barbara graded as: high (H), middle (M) and the low (L) in a flower market were investigated with respect to their vase life, fresh weight and water and sugar contents in the floral organs (Miura *et al.*, 2000). Total fresh weight, per cent water content and water absorption were highest in H and lowest in L throughout the experiment. Buds of H and M opened normally within 10 days, whereas those of L blasted; vase life in H and L were 15 and 10 days, respectively.

11.2.0 Storage

With appropriate chemical pretreatments, carnations can withstand prolonged storage at 0-1°C in bud form. The buds can be opened with sucrose solutions after storage, and acceptable vase-life periods achieved. However, in Australia, growers have reported erratic results related to condition of the stock plants, developmental stage of buds stored, and conditions applied during the opening period. Number of trials was established on the storage of carnations to ease depressed markets because of over-production.

11.2.1 Low Temperature Storage

Barth and Frensham (1989) harvested buds in 3 stages of development and stored for 2- and 4-month periods. After storage, the buds were opened at 2 shed temperatures in 2 bud-opening solutions. Opening at 25°C using a 10 per cent sucrose solution was more successful than opening at 18°C using a Chrysal solution. Buds stored for 4 months opened more rapidly than those stored for 2 months, although total vaselife was less. Tighter bud stages are recommended for longer storage periods as some maturation occurs during storage. Growers can safely use 2- or 4-month storage periods to target specific markets, if shed conditions are optimal during bud opening. Menguc *et al.* (1994) also investigated the maturity stages *viz.* (i) tight bud, (ii) cross, (iii) paint brush, and (iv) fully open on storage life. Half of the samples were held in 550 mg silver thiosulfate (STS)/l + 100 g sucrose for 20 h. Treated and non-treated flowers were then stored in plastic bags at 0-0.5°C and vase life was subsequently evaluated at 17-22°C. STS treatment had no effect on storage life which averaged 6.0 days for treated and untreated flowers. However, storage life was significantly affected by harvest maturity and it declined as harvest maturity advanced. The flowers at harvest stage 4 showed the longest vase life followed by those from stage 3 and stage 2 (Table 10). The number of days to bud opening decreased as harvest maturity advanced and was reduced by STS treatment. STS treatment increased flower diameter.

Table 10: Effect of different Harvesting Stages on the Vase Life of Flowers after Storage*

Harvesting Stages	*Vase Life (Day)*
Stage 1	7.4 a
Stage 2	6.8 b
Stage 3	4.6 c
Stage 4	3.8 d

P = 0.01. *Menguc and Usta (1994).

11.2.1.1 Pre-treatment

Pre-treatment with growth regulators, STS and sucrose showed good effect on quality and vase life. Goszczynska and Nowak (1979) observed good quality and more vase-life of tight carnation buds pretreated before cold storage with growth regulators. Dry storage of tight carnation buds for 5 weeks resulted in flowers of good quality and longevity. Gibberellic acid treatment accelerated stored buds' opening and increased the diameters of fully open flowers. A mixture of 5 mg/l BA with 20 mg/l of NAA accelerated bud opening after storage, increased the diameters

Harvesting of Carnation

Paint brush stage

Full bloom stage

Freshly harvested carnation

Carnation harvested in open

Harvested flowers ready for marketing

Packaging of Carnation

Packing in polythene

Packing in bags

Individual flower in paper boxes

Packaging in card board boxes

of fully open flowers and their longevity. Treatment with low concentrations of ethephon hastened bud-opening after storage. Heins*et al.* (1981) observed tight bud and open flowers survived –4°C for 5 days without injury if flowers were pulsed with 20 per cent sucrose at 25°C for 24 hours prior to storage. Between the 8th and 10th day of storage, both stem and floral tissue froze. Goszczynska and Rudnicki (1982) stored carnation successfully up to 16 weeks. They treated cut flowers of the carnation cv. Scania either as tight buds or when colouredpetals began to be visible (as a red cross) with various chemical solutions prior to storage for 14-24 weeks at 0 to 1°C. Treatment in a solution containing silver thiosulphate (STS) at 550 mg/l and sucrose (S) at 100 mg/l improved flower quality assessed after 14 weeks of storage. The use of this preser-vative combined with dipping in a 0.1 per cent solution of the fungicides Rovral [iprodione] or Sumilex [procymidone] before storage, improved flower quality after 16-24 weeks and gave a vase life of 7-8 days.

Lee (1988) found that flowers treated with STS alone at 4 mM or STS at 4 mM + 3 per cent sucrose and wrapped in film could be stored for 5 weeks with no significant reduction in vase life, compared with freshly-harvested non-stored flowers. Vase life of buds, after cold storage, was longer than that of open flowers, but the buds often failed to open properly; STS reduced FW loss and flower decay during storage and also markedly reduced ethylene production after storage. Koyama *et al.* (1995) pre-conditioned carnation flowers with STS and/or sucrose and stored at 1°C for 12 weeks. The vase life of flowers was extended after 12 weeks of cold storage by STS treatment. After cold storage, petals of shoots pre-conditioned with 0.3 mM STS + 10 per cent sucrose had the highest sugar concentrations but no promotive effect on vase life was observed. They concluded that high-grade cut carnation flowers can be obtained in mid-summer by pre-treating budding shoots with STS and cold-storing them at 1°C for 12 weeks.

11.2.1.2 Post-storage Treatment

Goszczynska and Rudnicki (1983) observed that bud fresh weight fell by 8 per cent when sealed in polyethylene bags. After storage the buds were opened in a solution containing 200 mg/l 8-hydroxyquinoline citrate, 25 mg/l $AgNO_3$ and 70 g/l sucrose. After 20 and 24 weeks' storage the flowers had a vase life of about 12 and 6 days, respectively. Koyama and Uda (1994) found storage period longer than 16 weeks delayed the onset of anthesis and is therefore not recommended. The longest practical storage period was 12 weeks. Shoots stored for 4 weeks at 1°C were successfully forced in a greenhouse in early May and petal colour was nearly normal. Dipping the cut ends of the shoots in 1 mM STS for 2 h after storage for 4 weeks at 1°C extended vase life from 8.6 days in controls to 26.8 days. High-grade cut flowers were obtained by using a greenhouse as a forcing room together with the application of STS dip after storage and treatment with a commercial forcing solution.

11.2.1.3 Dry Storage

Maxie *et al.* (1973) stored cut flowers of carnation cv. Scania for 0, 7, 14, 21 or 28 days at 32o or 41°F, dry or in tap or distilled water. After storage the vase life at room temperature was then determined. Flowers stored at 32°F for 2-3 weeks showed little loss in quality, but storage at 41°F for any period of time adversely affected vase life. Storing in water showed no obvious advantage over dry storage.

11.2.1.4 Control Atmospheric (CA) Storage

Controlled atmospheric (CA) storage of carnation involves (i) increasing the levels of carbon dioxide, (ii) decreasing level of oxygen, (iii) low storage temperature, and (iv) prevention of build up of endogenous ethylene levels to threshold that could trigger senescence reaction. Tinga (1956) indicated harmful effects of storing cut carnation under high CO_2 conditions. In toxic levels colour change, petaland pedicel necrosis, and desiccation of flowers occur during the storage. Red cultivars seem to be more susceptible for this storage. Smith and Parker (1966) found that carbon dioxide levels of 0.35 per cent or higher were enough to offset the effects of ethylene at or highly above threshold levels of injury on carnation (0.05 ppm for 3-4 days), while Uota and Garazi (1967) found 7-20 per cent of CO_2 to be effective in suppressing sleepiness in carnation caused by ethylene for 24 hours at 20°C. Johnston *et al.* (1997) investigated the effects of CA storage in (1) air, (2) 0.1 per cent O_2 (48 h), 60 per cent CO_2 (48 h), air, (3) 60 per cent CO_2 (48 h), 0.1 per cent O_2 (48 h), air, or (4) 60 per cent CO_2 + 0.1 per cent O_2 (48 h), air. Hypercarbic and anoxic controlled atmosphere (CA) storage accelerated pigment loss and p*et al*wilting, and reduced flower longevity. Shelton *et al.* (1997) investigated the efficacy of sequential low O_2, high CO_2 controlled atmosphere (CA) treatments to kill aphids. This experiment tested the feasibility of a series of 6-day sequential CA treatments to disinfest shipped cut flowers. Aphid mortality was high though not significantly different across treatments, while spider mite mortality was low, reaching a maximum of only 25 per cent under a treatment of 0.1 per cent O_2 followed by 60 per cent CO_2, then air. Vase life was generally reduced by CA treatments. The sequence of O_2 and CO_2 exposure significantly affected floral tolerance of several CA treatments.

11.2.1.5 Modified Atmospheric Storage

Vase life of cut carnation at the room temperature was extended by storing the flowers under modified atmospheric condition created by sealing the flowers in plastic bags. The success of MA storage depend largely upon the the permeability of films to CO_2 and O_2, temperature and length of storage and cultivar (Hardenburg *et al.*, 1969).

11.2.1.6 Low Pressure Storage

The advantage of hypobaric storage of carnation is well established (Staby *et al.*, 1978). Goszczynska and Rudnicki (1982) observed that storage of buds for 20 weeks at low pressure (LP) considerably improved flower quality compared with 20 weeks in standard air-storage. Buds stored under LP developed into fully opened flowers, irrespective of pre-conditioning, although acceptable quality and vase-life were only achieved by those which were pre-conditioned with STS + sucrose prior to storage.

11.3.0 Packaging

Carnations are packed in corrugated cardboard containers. About 800 carnations are packed in a standard size carton 30 cm high, 50 cm wide and 122 cm long. The boxes or containers should be well insulated and should have a lining of vapour barrier like polythene lining to help in maintaining high relative humidity inside the packages. Bunches of 25 flowers are then placed in these boxes with half of the total number of bunches oriented on each end of the container. Close packing of flower helps to minimize the distance for heat exchange between flower layers. In order to maintain high humidity insulating paper is put across the box to cover the flowers completely. The lids are fixed to the box by taping or twing it. Sometimes the top lid is made partially or completely transparent for viewing the flowers without disturbing them. Nichols (1979) presented results to show that the survival of carnation cut flowers can be affected by enclosing them with pollinated flowers. Unpollinated flowers were enclosed in imperforated but unsealed cardboard boxes (38 x 24 x 8 cm) with one or two pollinated flowers for one or two days. The survival of the unpollinated flowers after transfer to water was often, but not always, shortened and the variability of the results could be explained by the amounts of ethylene sometimes accumulated inside the boxes.

11.4.0 Vase Life

The physiological aspects have been separately discussed under physiology. The senescence of flower petals is a highly regulated developmental process which requires active gene expression and protein synthesis. The biochemical changes associated with petal senescence in carnation flowers include an increase in hydrolytic enzymes, degradation of macromolecules, increased respiratory activity and a climacteric-like increase in ethylene production. Ethylene plays a critical role in the regulation and coordination of senescence processes.

11.4.1 Pre-harvest

Accati and Jona (1982) found that vase life is longer in the cultivar whose initial stylar development is slower. Once the style has reached maturity, thus protruding stigmatic papillae, the flower wilts within 3.5-4 days. Celikel *et al.* (1995) reported that as the plants aged, flower head and stem diameter decreased, but vase life slightly increased. Flowers grown under glass had a longer vase life but a lower FW than those grown in plastic-clad greenhouses. Vase life was longest in autumn. There were considerable differences in longevity and flower quality among the cultivars. In addition, season, method of growing and cultivar markedly influenced the response of flowers to STS pulsing and storage at 1°C for 1 week. They concluded that preharvest conditions have a considerable effect not only on quality and longevity of cut carna-tions at harvest, but also on their response to postharvest treatments. Lukaszewska (1995) also found vase life was longer for flowers raised in greenhouses. Removing the main bud from the inflorescence extended the longevity of the remaining florets.

11.4.2 Postharvest Treatment

Reid *et al.* (1980) reported that carnations pretreated with 4 mM STS for 20 minutes and subsequently held in commercial preservative solutions lasted up to 4 times longer than untreated carnations (11.9 vs. 2.9 days). Lee *et al.* (1980) found that pretreatment with $AgNO_3$ followed by holding in preservative solution extended vase life 3 times longer than control. Le Masson and Nowak (1981) found twenty-four hours of pretreatment with $AgNO_3$ increased the vase life of flowers from 3 to 4 days for carnations kept in water or in preservative solutions after simulated transport conditions. A mixture of $AgNO_3$ with sodium thiosulphate or with EDTA was considerably better than $AgNO_3$ alone. Mor *et al.* (1981) recorded that flower buds pretreated with STS (containing 4 mM STS) plus the biocide physan (a quaternary ammonium compound) at 200 ppm markedly showed more vase life in deionized water by 11.5 and 5.2 days, respectively, over the controls.

Singh and Singh (1994) observed pretreatment with 100 µM spermidine for 24 h slightly improved longevity of ACC and ethylene treated petals, but completely overcame the effect of IAA, increasing the longevity of petals from 4.2 days (IAA only) to 9.5 days (spermidine followed by IAA). Yamamoto *et al.* (1995) studied the effects of cis-propenylphosphonic acid (PPOH) and alpha-aminoisobutyric acid (AIB) as a pulse treatment, the 20 mM equimolar combination was comparable with 0.2 mM silver thiosulfate, which is used commercially to prolong vase life. These results show that the combination of PPOH and AIB may be useful. Onozaki *et al.* (1998) in their pretreatment trials found that exposure to 60 mM AIB or 60 mM AIB + 10 mM calcium nitrate for 21 or 24 h significantly extended vase life by 20 and 14 per cent, respectively, over the AIB treatment.

11.4.2.1 Pulsing

Nityasiri Suisuwan *et al.* (1985) showed that pulsing the blossoms with $AgNO_3$ at 1000 ppm for 15 minutes and repulsing with 10 per cent sucrose + 150 ppm citric acid for one hour would give the best results compared to blossoms that were untreated or pulsed with other solutions. Ketsa *et al.* (1987) found flowers pulsed in the solution containing 68 mg/l $AgNO_3$ + 794 mg/l $Na_2S_2O_3.5H_2O$ +10 per cent sucrose for the 18 hours then packed dry and shipped by non-refrigerated truck had the longest vase life (4.03 days) while the control vase life was 2.05 days only. Improvement of vase life was also obtained with 50 mg/l BA included in the pulsing solution plus packing wet with stem bases wrapped in cotton during transit. Bichara *et al.* (1993) investigated the longevity of flowers using a 50 per cent senescence value (S50 at which 50 per cent of flowers senesced). A 22-h pulsed treatment using a combination of 4.6 mM AOA, 4.2 x 10^2 mM kinetin and Triton X-100 at 500 mg/l enhanced flower vase life comparative to a water control. However, a single AOA pulse treatment was as effective as the mixture is increasing longevity. With the exception of dihydrozeatin, which greatly enhanced longevity (S50 approx. 12 days). S50 values were more useful for expressing longevity than S100 values.

Staden *et al.* (1994) pulsed buds with silver thiosulphate (0.2 mM $AgNO_3$ and 2.5 mM $Na_2S_2O_3.5H_2O$) for 2 hours, amino oxyacetic acid (4.6 mM) for 22-h and 5 per cent ethanol. The treatments did not prevent stem bending but extended the longarly in silver thiosulphate treated flowers. Onozaki *et al.* (1998) found that cut flowers pretreated with 60 mM AIB + 10 mM calcium nitrate for 20 or 24 hours significantly extended vase life by 20 or 14 per cent, respectively over AIB treatment alone.

The longevity of cut carnation flowers can be improved by pulse treatment with inhibitors of ethylene biosynthesis or action such as silver thiosulfate (STS), aminooxyacetic acid (AOA), ∞-aminoisobutyric acid (AIB), 3-amino-1,2,4-triazole (ATA), 2,4-dinitrophenol (DNP) and sodium benzoate (SB). Among all the inhibitors tested, STS and ATA were most effective. The longevity of flowers of cv. Charlotte pulse-treated with 0.2 mM silver thiosulfate (STS) or 100 mM 3-amino-12,4-triazole (ATA) was extended to 20 days as compared to 11 days in the water control. 2,4-dinitrophenol (DNP) (0.8 mM), sodium benzoate (SB) (2 mM), ∞-aminoisobutyric acid (AIB) (4 mM), aminooxyacetic acid (AOA) (2 mM) increased flower vase life by 53, 43, 31 and 23 per cent, respectively, over the water control. Solutions containing STS, AOA or SB also maintained the flower diameter at higher levels than distilled water (Wawrzynczak and Goszczynska, 1999).

Nahid *et al.* (2020) investigated the effects of silver nanoparticles (AgNPs) on ethylene-sensitive (carnation) and ethylene-insensitive (chrysanthemum) cut flowers. AgNPs were synthesized by chemical methods and then applied on both flowers by a pulsing method. The treatments involved two concentrations of AgNPs (0.04 and 0.08g/L) along with the control (deionized water), and the duration of exposure lasted for 24 h. There were increases in hydrogen peroxide content (H_2O_2) and peroxidase (POD) activity. The bacterial population of the stem end and total soluble protein (TSP) increased in carnation petals, but decreased in chrysanthemum petals. The activity of superoxide dismutase (SOD) dropped in carnation petals, whereas it rose in chrysanthemum petals. Using AgNPs at concentrations of 0.08 and 0.04g/Lcan optimally extend the vase life of carnation and chrysanthemum, respectively.

11.4.2.2 Common Preservatives

Sucrose

Water content is generally higher in sucrose treated leaves. The rate of chlorophyll degradation with time was significantly lower in sucrose treated leaves than in water treated leaves. Starch and total sugar contents decreased steadily with incubation time, both in leaves and petals; the decline was more rapid in water-treated leaves. Reducing sugars initially increased, then decreased sharply (Behera, 1993). In the past some work was carried out using only sucrose. Mayak and Dilley (1976) reported that flowers developing during periods of low light intensity senesced rapidly and responded to kinetin only when supplemented with sucrose. Flowers supplemented with sucrose senesced less rapidly when treated with ethylene or ABA than those not treated with sucrose.

Lukaszewska (1995) placed cut flowers of 9 carnation cultivars in solutions containing 5 per cent sucrose and 30 ppm $AgNO_3$, 200 ppm 8HQC or 50 ppm $NaNO_3$. All holding solutions increased the number of open florets and the number of days until the last floret wilted. Naveen Kumar *et al.* (1999b) found carnation grown under polyhouse showed better vase life than open field, further most of the carnation preservatives were on par with distilled water.

Silver Thiosulphate

It is a very common chemical used by the most of the workers. Mayak *et al.* (1977) found pretreated stems with 1000 ppm $AgNO_3$ for 30 min greatly reduced bacterial counts in the holding solution and extended flower. Goszczynska and Rudnicki (1978) found the effect produced by the 5 and 10 ppm of ethephon (delay in opening of tight buds) was counteracted by a simultaneous application of $AgNO_3$ at 50 ppm. Veen (1979) found vase life was nearly doubled by pretreatment with silver thiosulphate of 0.2 mM. Goszczynska and Nowak (1980) obtained best results with regard to days to open (2-3), flower diameter (73.2 mm), and vase life (11.4 days) when cut flowers, at a stage with petals emerged about 0.5 cm above the calyx, were treated with 1000 mg/l $AgNO_3$ + 10 per cent sucrose and stored for 8 weeks. Reid *et al.* (1980) observed that combining silver nitrate with sodium thiosulphate (molar ratio 1:4) doubled the vase life (from 5 to more than 10 days). The effect could be achieved by treating stems with solutions containing as little as 1.0 mM Ag with a pulse as short as 10 min. Silver uptake estimations indicated that a minimum of 0.5 m Mol Ag was required per stem for maximum vase life and that more than 5 m Mol Ag/stem was toxic. Choi and Roh (1980) found that the longevity of carnation extended when the stem base of the flower was treated with 1,000 ppm $AgNO_3$ solution for 5 minutes and then preserved in the solution containing 2 per cent sucrose. Dimalla and Staden (1980) reported an increase in postharvest life when the cut stems were immersed in silver thiosulphate for 10 minutes. The preservative prevented gynaecium enlargement and apparently ensured a continuous transport towards and utilization of nutrients in the petals. Aldrufeu *et al.* (1981) showed that the distilled or tap-water solutions containing 68 mg $AgNO_3$ + 794 mg $Na_2S_2O_3$/l increased the longevity of cut bud carnations, by 8 or 9 days when they were treated just after harvest. Han and Lee (1992) found that aminooxyacetic acid (AOA) at 8 mM and STS at 1 mM extended vase life by 7-8 days, compared to distilled water. Further they observed AOA was superior to STS for extending vase life when the stems were harvested at stage II (petals 2.22-3.18 cm long, showing colour and opened 40-50 degree from the vertical), which was considered as the optimum stage of maturity for longest vase life and flower quality. Uda *et al.* (1995) found the ratios of Ag:thiosulfate ratios 1:4-1:32 doubled vase life compared with flowers treated with deionized water, while ratios of 1:0-1:2 extended vase life only slightly. Injury to upper leaves increased proportionally with the increase in Ag accumulation and with the increasing ratios of thiosulfate to Ag. They recommended 1:8 molar ratio of $AgNO_3$: $Na_2S_2O_3.5H_2O$ in the STS solution to extend the vase life of carnations. In another trial, Uda *et al.* (1996) suggested to measure the amount of Ag in STS solution absorbed by a cut flower. Absorption of 0.5-1.0 mM Ag per cut carnation stem optimized vase life without causing toxicity.

Novel Preservatives

Altman and Solomos (1993) found continuous postharvest treatment of flowers with 50 or 100 mM aminotriazole [amitrole] significantly extended vase life to 18 days compared to water. Flowers treated with 50 mM aminotriazole for 2 days senesced in response to application of 10 µl exogenous ethylene/l, whereas flowers treated for 4 days exhibited no morphological response to ethylene treatment. Singh and Singh (1994) treated the basal 1 cm of petals with spermidine (50 or 100 µM), IAA or ACC (both at 50 µM). Spermidine application caused the greatest increase in petal FW while, ACC treatment resulted in the lowest FW increase. Spermidine increased longevity to 11 days, compared to 8 days in control; IAA and ACC reduced longevity compared to control. Drory *et al.* (1995) found that 24h treatment of non-stressed cut flowers and isolated petals with 60 μM cycloheximide (CHI), resulted in a delay in wilting, reduction of membrane permeability, inhibition of protein synthesis and ethylene production and an increase in electrolyte content (Table 11). The results indicate that the response of cut flowers to transient water stress may be dependent on protein synthesis.

Table 11: Effects of Water Stress and CHI Pretreatments on Longevity of Cut Carnation Flowers*

Pretreatment	*Water Stress*	*Longevity[a] (days)*
Water	–	5.2 b
Water	+	4.0 a
CHI	–	>9.0 c
CHI	+	>9.0 c

Flowers were held for 24 h in water or 60 M CHI, then exposed to water stress for 24 h, followed by recovery in water and observation for wilting at 20°C, RH 70-80 per cent. Longevity was measured from the end of the water stress. Results are means of five replicates.

a Values followed by the same letter are not significantly different at *P=0.05*.

* Drory *et al.* (1995).

Yamamoto *et al.* (1995) studied the effects of cis-propenylphosphonic acid (PPOH) and ∞-aminoisobutyric acid (AIB) on the vase life. It increased markedly with PPOH + AIB (up to 12.2 days with continuous treatment) compared with treatment with each compound alone (7.6 and 5.0 days, respectively). Midoh *et al.* (1996) investigated the extension of vase life by a novel preservative like 1,1-dimethyl-4-(phenyl-sulfonyl) semicarbazide (DPSS). It appeared DPSS provided no protection from the effects of ethylene treatment (10 μl/l for 20 h) nor did it inhibit ACC synthase, however, it inhibited ACC-dependent ethylene production in petal discs, suggesting possible potential for inhibiting ACC oxidase. Onozaki *et al.* (1998) found that the addition of 2.5 mM calcium nitrate + 10 mM AIB significantly prolonged vase life compared to control and the single application of AIB or calcium nitrate. The combined treatment was as effective as 20 mM AIB solution in extending vase life.

Cut carnations, cultivars Buffalo and Figaro were trimmed to 30 cm and their bases immersed in various test solutions (Petridon *et al.*, 1999). Ethanol, industrial alcohol, methanol, acetylsalicylic acid (aspirin), salicylic acid, and hydroxyquinoline sulfate (HQS), inhibited climacteric ethylene synthesis, and significantly delayed senes-cence, except for methanol which was effective only in Figaro. Vase life as extended from 7.0 days (in distilled water) to >16 days (2-5 per cent ethanol) in Buffalo, and from 7.5 days to >16 days (2-3 per cent methanol, 3 per cent ethanol or 120 ppm acetylsalicylic acid) in Figaro.

Other Chemicals

Camprubi and Aquila (1973) observed cupferron and mandelic acid, which act as metal-complexing agents, had the same beneficial effects on the fresh weight, diameter, petalgrowth and vase life of flowers as did 8-HQS, also a complexing agent. The results supported the hypothesis that blockage of the vessels may be an enzymatic process that could be inhibited by removing the microelements of the coenzyme.

Temperature/Light Forcing

Systema (1971) stated that a temperature above 20°C and 70 per cent RH is necessary to force good quality flowers within 1 week. Some light is needed for good colourdeve-lopment in red cultivars. The vase life of flowers cut in bud is the same as that of flowers cut at the usual stage; the incidence of calyx splitting does not appear to be affected by cutting in bud. Cut buds could be stored in plastic wrapping at 0-2°C for 14 days with no ill effects before forcing them into flower. Neumannova and Paulin (1974) found nutrient solution used at 27°C increased the rate of bud opening. It also extended the vase life of both cold-stored and freshly cut blooms. Varying RH at 27°C had no major effect on the rate or quality of bud opening or vase life of either cold-stored or freshly cut carnations.

Koyama and Uda (1994) observed as the sucrose concentration of the forcing solution increased, flower diameter increased and vase life was extended, but at 7 per cent sucrose, dehydration symptoms were observed at the tips of the leaves and calyxes. The best forcing conditions for obtaining optimum-quality flowers with normal petalcolour were 20-25°C, 1-2 klx for Nora (pale pink) and 2-3 klx for Coral (red) and 3 per cent sucrose in the forcing solution. Irving and Honnor (1994) applied elevated CO_2 atmosphere treatments (60 per cent CO_2 in air) for 1 or 30 days at 0 or 20°C to cut flowers as a potential disinfestation treatment. CO_2 did not affect longevity at 0 or 20°C. The higher temperature was more detrimental to the flowers than 60 per cent CO_2. There was a transient yellowing of the petals upon removal from the 60 per cent CO_2 treatments at 20°C. They concluded that carnation flowers can withstand 60 per cent CO_2 at 0°C without detriment to vase life conditions that could constitute a disinfestation treatment against some insects.

Effect of Microbes

Doorn *et al.* (1991) observed that the bacterial population developing in the stems of cut carnation flowers during vase life leads to vascular occlusion but this apparently has little effect on flower longevity. Jones and Hill (1993) analysedthe microbial concentrations in vase water and found that proliferation was effectively controlled by DICA and BCDMH, but not by HQC. Populations of up to 106 cfu/ml were detected in vase water, indicating that carnation can tolerate these levels. Longevity was significantly reduced when flowers placed in solutions containing high micro-organism counts (>10^8 cfu/ml) isolated from carnations. Doorn *et al.* (1995) evaluated the vase life after keeping the cut flower in high bacterial concentration. The vase life of flowers was reduced in a suspension of bacteria with an initial exogenous concentration of 10^8 colony-forming units (cfu) per ml. The composition of the bacterial flora had no effect on longevity.

11.5.0 Physiology of Senescence

Westhuizen *et al.* (1978) observed that the concentrations of glutamine, glycine, histidine, methionine, phenylalanine, proline, serine, threonine and valine increased throughout the vase life of the flowers. The concentrations of arginine, aspartic acid, isoleucine, leucine, lysine and tyrosine increased initially but decreased thereafter. There was a noticeable decline in the concentration of ∞-alanine, ∞-aminobutyric acid and asparagine immediately before the climacteric maximum of respiration, whereas at other times an increase was evident. The concentration of all other free amino acids fluctuated. Ferraris *et al.* (1982) observed that flowers of long-lived cultivars Alice, Biancaneve, Giallo 9942 and Champagne 9538 accumulate large amounts of polymethylgalacturonase (PMG), those of the short-lived cultivars Astor, Manon and Maya, on the contrary, contain polygalacturonase (PG) and polygalacturonate transe-liminase (PGTE). These enzymes were partially purified. Paulin *et al.* (1982) placed cut flowers of carnation in a preservative solution (containing an antiseptic, an acidifying agent and an agent for precipitating calcareous compounds) with or without the addition of glucose at 30 g/l. When glucose was present, the activities of sucrose-synthetase, sucrose-P-synthase and glucose-6P-isomerase in the flowers remained high from bud opening to wilting. In the absence of glucose, enzyme activity decreased as the flower aged. The decrease in activity appeared to be related to very low protein synthesis. Eisenberg and Staby (1985) monitored senescence of carnation cv. Improved White Sim flowers after placing them in either a preservative solution or deionized water. Flower fresh weight, ethylene, and CO_2 levels increased as the flowers aged, but fresh weight and CO_2 levels fell once flowers began to senesce regardless of holding solution. Preservative-treated flowers senesced at a slower rate than deionized water-treated flowers. The amount of ADP phosphorylated to ATP per oxygen atom consumed, using mitochondria isolated from petal tissue provided with either succinate or malate as substrates, increased as flowers senesced and then decreased in the later stages of senescence. Respiratory control ratios with succinate as the substrate did not change appreciably until the final stages of senescence while respiratory control values using malate showed greater variation but no consistent pattern relative to the progress of senescence. Cyanide-resistant respiration was noted with isolated mitochondria oxidizing either substrate, but no correlation between cyanide-resistant respiration and senescence was found.

Paulin *et al.* (1985) placed flowers at usual marketable stage (when the outer petals were at right angles to the stem) in a solution of the basic vase-life preservative SB alone or supplemented with 10^{-3}M oxyaminoacetic acid. They were then held at 23°C, 40 per cent RH and 12 h light/day at 12 W/m^2 for 18 days. Ethylene emission was almost completely inhibited by oxyaminoacetic acid, which also reduced the breakdown of phospholipids and fatty acids (particularly linoleic, linolenic and oleic acids). Vase life was thus extended, complete senescence being observed on day 18 compared with day 8 for the controls. Kapchina and Yakimova (1989) treated buds of spray carnation with different solutions like Ronilan, silver nitrate and sucrose. All the preservatives extended the phase of increasing invertase activity in developing buds but there was no correlation between the level of invertase activity and the effectiveness of any particular preservative treatment. Changes in the ratio of petal sucrose: reducing sugar concentrations reflected the changes in invertase activity. Petal protein content increased during withering in all treatments.

11.5.1 Role of Ethylene

The senescence of flower petals is a highly regulated developmental process which requires active gene expression and protein synthesis. The biochemical changes associated with petal senescence in carnation flowers include an increase in hydrolytic enzymes, degradation of macro-molecules, increased respiratory activity and a climacteric-like increase in ethylene production. It is clear that the gaseous phytohormone ethylene plays a critical role in the regulation and coordination of senescence processes. Many reviews on physiology and mode of action of ethylene are available. Molecular cloning led to the isolation of genes involved in ethylene biosynthesis and action (Altvorst and Bovy, 1995). Larsen *et al.* (1995) investigated the pollen-pistil interactions that result in the stimulation of ethylene production and peal senescence in carnation flowers. The results suggested that physical growth of pollen tubes is insufficient to elicit a sustained increase in ethylene production or to lead to the production of signals necessary for elicitation of petal ethylene production and senescence. Rather, the cell-cell recognition reactions leading to sexual compatibility in *Dianthus* appear to play a role in this inter-organ signalling after pollination.

11.5.1.1 Ehylene vs Temperature

Barden (1972) exposed open flowers or buds of the cv. White Sim to four ethylene levels. Flower life decreased with rising ethylene concentration. Temperature had a greater effect than ethylene concentration on vase life. Similar observation was made by Maxie *et al.* (1973) and they found that in ethylene-free air the flowers tolerated high temperatures for a short period (*e.g.* 40°C 4 h), but their vase life was shortened as the ethylene concentration increased. Woltering and Harkema (1987) determined the response of cv. Scania flowers to various concentrations of exogenous ethylene. The ethylene threshold concentration depended on exposure time and the threshold time depended on temperature.

Buds and open flowers behave differently for different ethylene treatments. It is gene-rally felt that buds withstand higher ethylene concentrations, longer exposures and higher temperatures better than open flowers. Camprubi and Nichols (1978) observed that flowers cut at various stages of development became increasingly sensitive to ethylene as they aged. Ethylene (*e.g.*, 1 v.p.m. for 24 h) caused irreversible wilting of open flowers but had little or no effect on the youngest flowers (buds), provided they were opened in a sucrose-germicide solution.

11.5.1.2 Ethylene Inhibitors

Veen and Geijn (1978) reported that when a solution of silver nitrate was sprayed over cut carnation flowers, the keeping quality was improved by suppressing ethylene production, but it caused black spots on the petals. When added to the vase water, it was not absorbed. When used as a silver-thiosulphate complex it was rapidly absorbed and suppressed ethylene production. Heins (1980) found that ethanol in the holding solution inhibited climacteric ethylene synthesis and delayed senescence of carnation flowers up to 230 per cent over distilled water controls. Once climacteric ethylene synthesis had started ethanol was not effective in inhibiting ethylene synthesis or delaying senescence. Further, Heins and Blakely (1980) found that conversion of ACC to climacteric ethylene was inhibited by adding ethanol to the holding-solution. Ethanol was the most effective alcohol in delaying carnation flower senescence of the tested series methanol, ethanol, propanol, tert-butanol and n-butanol. Ovary development was also inhibited in carnation flowers by ethanol. Wu *et al.* (1992) also found that continuous treatment with 8 per cent ethanol doubled the vase life. When carnation flowers were cut in bud and held in a solution of 500 ppm sodium borohydride ($NaBH_4$), ethy-lene production was inhibited and senescence delayed by 3-5 days compared to control in distilled water (Camprubi *et al.*, 1981). Ronen and Mayak (1981) observed that the rate of ethylene production rose after 2 and 4 days in cut carnations, in a 0.05 mM ABA solution and in water (control), respectively. Continuous uptake of 0.17 mM AVG (aminoethoxyvinylglycine) delayed ethylene production in ABA-treated anduntreated flowers. Broun and Mayak (1981) observed that ∞-aminooxyacetic acid delayed ethylene production in cut carnation flowers at a concentration of 3×10^{-4} M. Wu *et al.* (1989) reported that the physan-20 repressed ethylene biosynthesis. The EFE [ethylene forming enzyme] activity and ACC content of the flowers remained unchanged for 6 days in the vase. Exogenous application of ethylene did, however, induce 'sleepiness'. Michalczuk *et al.* (1989) founddiazocyclopentadiene (DACP) has been shown tobe an effective ethylene action inhibitor after illumination with fluorescent light and the treatment delayed and reduced endogenous ethylene production.

Saks *et al.* (1992) observed GA_3 at 10^{-4} M delayed the onset and reduced the climacteric peak of ethylene. Pretreatment with GA_3 at 10^{-4} M for 48 h reduced the sensitivity of flowers to both ethrel or 2,4-D. It also affected the respiratory climacteric. Application of 0.1 mM GA_3 for 48 h to preclimacteric flowers suppressed style development and inhibited ethylene production (Saks *et al.*, 1993). Son and Chae (1993) recorded that ethylene production reached peak after 3 days in control, but treatment with 10 mM putrescine, spermidine or spermine delayed the peak until day 13. These treatments also delayed petalinrolling. Lee and Kim (1993) found both AOA and STS doubled the vase life as a result of inhibition of ethylene production. STS inhibited ethylene production by completely inhibiting ACC accumulation and stopping the activity of ACC synthase and ethylene forming enzyme (EFE) in the petals.

Sisler *et al.* (1993) found that DACP inhibited ethylene production after illumination with fluorescent light. When flowers were preincubated with DACP in the dark, then exposed to fluorescent light in the air, the lowest concentration of DACP required for protection against 10 mµl/l of ethylene was >700 µM/l. Son *et al.* (1994) also observed that addition of AOA or AVG (ACC synthase inhibitors) greatly reduced ethylene production. Further they reported that solutions of aspirin, DNP and a polar fraction of *Allium sativum* were all effective inhibitors of ethylene forming enzyme (EFE) activity. A number of gaseous compounds that appear to block the ethylene receptor have been discovered recently. They inhibit a ripening of fruits, and senescence or abscission of flowers, buds or leaves. The compounds block the receptor, preventing the physiological action of ethylene for up to 12 days at 25°C when provided in a single exposure. Some of the inhibitors are active in very low concentrations. For example as low a concentration as 0.5 ml/l of 1-methylcyclopropene (1-MCP) is sufficient to protect carnation flowers for several days against ethylene (Sisler and Serek, 2001).

Son and Byoun (1995) observed that although methionine inhibited ethylene production but it did not extend the vase life of carnations. Phytotoxicity was exhibited at the highest doses of methionine and ethionine. Altman and Solomos (1994) found that although treatment with 50 mM aminotriazole for 4 days did not completely inhibit components of the ethylene biosynthetic pathway, no morphological or respiratory responses to the application of exogenous ethylene at 10 ml/l were observed, indica-ting that prolonged aminotriazole treatment inhibited ethylene action. Altman and Solomos (1995) suggested that the interaction between silver ions and ethylene is competitive. The differential distribution of silver ions between basal and apical regions of pulsed or continuously treated petals showed that the petal base may play an important role in regulating carnation petal senescence. In another experiment, Lee and Yourga (1999) found that treatment with 6 per cent ethanol induced extended vase life by 2-4 times compared to control. High concentration of ethanol caused bent stems, 4-6 per cent ethanol treatment was most effective that improves the quality and extended vase life. They suggested that ethanol could delay the senescence of cut carnation by inhibiting ethylene synthesis.

Satoh *et al.* (1999) also reported that 1,1-dimethyl-4-(phenylsulfonyl) semicarbazide (0.08 mM) prevented increases in ACC synthase activity, the content of ACC and ethylene production in cut flowers of carnation cv. Nora, and increased the vase life of fully open flowers from 5 days to > 2 weeks.

11.5.1.3 Ethylene Concentration

Staby *et al.* (1993) reported that treatment with STS resulted in the longest vase life at lower ethylene concentrations (0.01-0.35 ppm), however, at higher ethylene concentrations, STS had little or no effect. Neither EVB (which contain AOA) nor florish (contain AVG) improved vase life, even at low ethylene concentrations. Acetaldehyde is the causal agent of ethanol and it increased the vase life by 50 per cent. Ethylene formation was reduced or inhibited as a result of acetaldehyde application but no prevention of ethylene concentration (Podd and Staden, 1999).

11.5.1.4 Genotype difference on Ethylene

Vase life, ethylene production and ethylene forming enzyme (EFE) activity, ion leakage, ethylene sensitivity and ACC synthase activity differ from variety to variety. Treatment of White Sim flowers at anthesis with 1.0 ml ethylene/l resulted in increased ethylene biosynthesis and premature petal senescence (Snyder and Rae, 1973).

Wu *et al.* (1991) found that cut flowers of cultivars Sandra and Chinera lasted about twice as long (approximately 14 days) as those of White Sim. The vase life of

White Sim was unaffected by the age of the flower at harvest, but the life of Sandra and Chinera declined as flower age at harvest increased. The senescence of White Sim and Chinera indicated by in-rolling of petals and wilting, was accompanied by a marked increase in respiration, ACC content, ethylene forming enzyme (EFE) activity, and ethylene production. In contrast, Sandra flowers showed neither in-rolling of petals nor climacteric respiration and ethylene production during their eventual senescence. Similar observation was recorded by Brandt and Woodson (1992) where in all varieties; the extended vase-life genotypes did not exhibit large increase in ethylene production that typically accompanies petal senescence in carnation. It is concluded that the low level of ethylene produced by these flowers during postharvest aging is the result of low activities of both ACC synthase and the ethylene forming enzyme. Woltering *et al.* (1993) observed that the longevity was markedly greater in cv. Chinera than in White Sim. In Chinera, the increase in ACC synthase and EFE activity, ethylene production and membrane permeability (which are associated with the start of senescence) occurred later and the maximum levels during senescence were much lower than in White Sim. Doorn *et al.* (1993) found that in Epomeo the maximum ethylene production during natural senescence was about one third of that in White Sim, and in Chinera about one third of that in Epomeo. When treated with a high dose of ethylene for a short period, Chinera and Epomeo responded by producing high amounts of ethylene, indicating that all necessary steps for autocatalytic ethylene production are present or can be rapidly induced.

In an investigation on expression of genes responsible for ethylene production and wilting in carnation, Kosugi *et al.* (2000) analyzed the expression of these genes in petals of a transgenic carnation harbouring a sense ACC oxidase transgene and in petals of carnation flowers treated with 1,1-dimethyl-4-(phenylsulfonyl) semicarbazide (DPSS). In petals of the transgenic carnation flowers, treatment with exogenous ethylene caused accumulation of the transcript for cysteine proteinase (CPase) and in-rolling (wilting), whereas it caused no or little accumulation of the transcripts for ACC oxidase and ACC synthase and negligible ethylene production. In petals of the flowers treated with DPSS, the transcripts for ACC synthase and ACC oxidase were accumulated but no significant change in the level of the transcript for CPase was observed. These results suggest that the expression of ACC synthase and ACC oxidase genes, which leads to ethylene production, is differentially regulated from the expression of CPase, which leads to wilting, in carnation petals.

Takashi (2008) developed two cultivars, 'Miracle Rouge' and 'Miracle Symphony', with genetically determined long vase lives of 17.7 to 20.7 days (3.2 to 3.6 times that of 'White Sim') under standard conditions (23°C, 70 per cent RH, 12-h photoperiod). Line 532-6 showed an ultra-long vase life averaging 27.8 to 32.7 days (4.6 to 5.4 times that of 'White Sim'). Another line 806-46b with both ultra-long vase life (27.1 days, 4.4 times that of 'White Sim') and ethylene resistance. Analyses using six cultivars and 123 selected lines from the 1st to the 7th generations revealed that the long vase life was strongly associated with a decrease in ethylene production.

12.0 Biotechnology

During the last two decades, significant progress has been made for *in vitro* multi-plication and transformation of carnation. Protocols have been standardized for regeneration and transformation of colour and extended vase-life. However, further emphasis has to be laid to develop varieties/hybrids with flower pigment gene, long vase-life by delayed senescence and resistance to biotic and abiotic stress besides development of markers and protocol for cryopreservation (Nagaraju and Palai, 2000).

12.1.0 Tissue Culture

Micropropagation of carnation has been summarized in propagation.

12.1.1 Anther Culture

Mosquera *et al.* (1999a) reported maximum callus induction in cv. Tanga when the anthers were incubated in total dark conditions for 5 weeks on MS basal medium supplemented with 1 mg/l each of 2,4-D and BA. Thirty proembryos were obtained but only one plantlet regenerated.

12.1.2 Protoplast Culture

Very little work has been done on protoplast isolation and culture of carnation. Miiand Cheng (1982) isolated mesophyll protoplast. They found that only 5 to 10 per cent of the protoplasts divided to form cell colonies in MS liquid medium containing 0.7 µM NAA and 2.2 µM BA, $5x10^4$ cells/ml were optimum plating density. Protoplast division was induced when isolated from young leaves, while from manure leaves, they failed to induce division under similar culture conditions. Kim and Lee (1996) also isolated leaf mesophyll protoplasts of *Dianthussuperbus* cultured at a density of $5x10^4$ protoplasts/ml and divided at about 18 per cent planting efficiency in MS liquid medium supplemented with 0.5 mg/l BAP, 2.0 mg/l NAA and 9 per cent mannitol after 2 weeks. Subsequently, green microcalli were transferred to MS solidified medium with 2.0 mg/l 2,4-D that induced shoot-forming calli after 4 weeks. These calli were transferred onto N6-2 medium containing 0.1 mg/l 2,4-D, 0.1 mg/l NAA, 2.0 mg/l Kin and 2.0 mg/l casein hydrolysate and were cultured under light. After 5 weeks the calli gave rise to multiple shoots. Upon transfer to MS medium containing 2.0 mg/l NAA, individual shoots were rooted in 4 weeks. The regenerants were successfully transferred to soil.

12.1.3 Somatic Embryogenesis

Callus has been induced from flower buds, leaf, hypocotyls, internodal sections, shoot-tips without apex and meristems of various carnation cultivars. MS medium supplemented with 2,4-D was found to be the most suitable medium for callus induction. The balance between auxin and cytokinin is the essential factor for callus induction and regeneration (Murashige, 1974).

Lee and Ahn (1997) reported that embryogenic callus induced from flower buds of carnation cv. Desio proliferated over 120 weeks without loss of embryogenic compe-tence through periodic subculture in MS medium with 2,4-D (0.5 mg/l).

When liquid agitation-cultured embryogenic callus was transferred to a regeneration medium (MS with 3 per cent sucrose and 0.5 per cent phytagel), numerous immature embryos were formed. These embryos matured with high efficiency on the same medium containing 7.5 per cent polyethylene glycol (PEG) and 0.03 mg/l ABA. Yantcheva *et al.* (1998) found direct somatic embryogenesis and plant regeneration from leaf explants of carnation cv. Lena (Sim group) and Bulgarian spray cultivars, Yanita, Regina and Line 84, MS liquid medium supplemented with 1 mg, 2,4-D and 0.2 mg BAP/l was used for direct induction of embryoids without an additional callus phase. Somatic embryo maturation was promoted by casein hydrolysate (1000 mg/l) in MS liquid media. The percentage conversion of embryos and polyembryos to whole plants varied between 10 and 75 per cent among the cultivars.

Shankhla *et al.* (1995) studied regeneration potential of petals, sepals, leaves and nodes of carnation cv. German Red using TDZ or BA and found that TDZ was more effective in inducing shoot formation. Prolonged growth in TDZ resulted in hyperhydricity. Petaland sepal explants exhibited good regeneration potential, but the shoots produced from these explants invariably developed flower buds *in vitro*. Explants from the second or third pair of leaves down from the apex exhibited better regeneration potential.

Carnation leaf explants derived from micropropagated plantlets of cv. Nora Barlo were used by Mosquera *et al.* (1999b) to select plantlets resistant to *Fusarium oxysporum* f.sp. *dianthi*. Callus induction occurred on medium containing up to 10 per cent fungus homogenate and adventitious regeneration of plantlets occurred on 5 per cent fungus medium. The plantlets showed good morphological and physiological characteristics such as vegetative vigour and fast regeneration.

Omid *et al.* (2008) described the culture conditions necessary to induce secondary embryogenesis in two carnation cultivars (Nelson and Spirit). In the first step, embryogenic calli were induced on petal explants followed by development of primary somatic embryos from the calli. In the second stage, secondary somatic embryos were obtained when precotyledonary and cotyledonary primary embryos were isolated and transferred onto a series of culture media all containing MS basal salt mixture, and supplemented with different concentrations of 2,4-D,BA, sucrose and mannitol. The highest rate of secondary embryogenesis occurred on mannitol containing media. Secondary somatic embryos were converted into plantlets when they were transferred onto growth regulator-free half-strength MS medium and successfully acclimated in the greenhouse.

Mostafa *et al.* (2019) achieved best results with 2,4-dichlorophenoxyacetic acid (2,4-D) + N-(2-chloro-4-pyridyl)-N2-phenylurea (4-CPPU) for ISE and picloram + 4-CPPU or naphthoxyacetic acid (NOA) + 6-benzylaminopurine (BAP) for DSE. Highest DSE, formation of embryogenic callus, embryo maturation (generation of globular, heart and torpedo shapes) and ISE rate was observed in carnation explants exposed to blue light (450–495 nm). In contrast, green (495–570 nm), red (610–700) and far red (710–730 nm) lights caused negative effects on embryogenesis compared to white light controls (380–750 nm). For the first time, genetic stability of regenerated carnation plants was estimated using inter-simple sequence repeat (ISSR) markers. The amplified products showed 75 distinct and scorable bands, and regenerants

Somatic Embryogenesis in Carnation

(a) Embryogenic callus originated from petal explants on MS medium containing 2.0 mg/l 2,4-D and 0.2 mg/l BA. (b) Somatic embryogenesis at different developmental stages on medium containing 3 per cent of sucrose and 6 per cent of mannitol, after 4 weeks of culture. (c) Direct secondary embryogenesis from cotyledons on medium containing 0.2 mg/l 2,4-D after 3 weeks of culture. (d) Indirect secondary embryogenesis from hypocotyls on medium containing 0.2 mg/l 2,4-D after 2 weeks of culture. (e) Plantlet regenerated from a somatic embryo cultured on half-strength MS medium after 2 weeks. (f) A potted plant ingreenhouse (Omid *et al.,* 2008).

[plants obtained by primary (PSE) and secondary SE (SSE)] were completely identical to the mother plant. Similarly, flow cytometric analysis confirmed that somatic embryo-derived plants had on average 1.53 pg nuclear DNA (2C), and all plants maintained their ploidy. In conclusion, obtained embryos under blue light were big in size and torpedo-shaped and their germination was highest compared to other light spectra. Moreover, blue light was effective for direct and indirect somatic embryogenesis in carnation without induction of somaclonal variation.

12.2.0 Biochemical and Molecular Markers

12.2.1 Isozymes

Though carnation is an important cut flower, little work has been done on biochemical and molecular markers. Messeguer and Arus (1996) studied ten enzyme systems for polymorphism in the parents and the progeny of crosses between 12 carnation cultivars. The inheritance of nine isozyme loci was determined — 3 code for alanyl aminopeptidase (Aap-1, Aap-2 and Aap-3), 2 of glutamate-oxalacetate transaminase [aspartate amino-transferase] (Got-1 and Got-3) and 1 for each of esterase (Est-2), phosphoglucoisome-rase [glucose-6-phosphate isomerase] (Pgi-2), phosphoglucomutase (Pgm-2) and shikimate dehydrogenase (Sdh-1).

Droillard *et al.* (1989) found that the total superoxide dismutase activity decreased between the end of flowering and the withering of petals of cut carnation. The three forms of superoxide dismutase (Cu/Zn-SOD, Fe-SOD and Mn-SOD) identified on polyacrylamide gel exhibited several isozymes, the activities of which all decreased some time during aging.

Droillard and Paulin (1990) reported the balance between reactions involving free radicals and processes which ameliorate their effect and plays an important role in the regulation of plant senescence. They developed a method to isolate peroxisomes and mitochondria from carnation petals. Based on electron microscopy and marker enzyme levels, the proportion of peroxisomes to mitochondria increased during senescence.

Hitendra *et al.* (2007) conducted isozyme analysis for nine enzyme systems and seven of them showed significant activities providing typical banding patterns. Four enzyme systems, NADH dehydrogenase, peroxidase, a-esterase and aspartate amino transferase showed considerable and veryuseful variation among 13 carnation (*Dianthus caryophyllus* Linn.) cultivars. Allele frequency in the present set of cultivars ranged from 0.5 to 1.0 and at polymorphic loci only two alleles could be identified. Polymorphism in the selected 13 cultivars was found to be 13.8 per cent. A dendrogram showed grouping of cultivarsin two clusters on the basis of genetic distance values and there were three sub-clusters.

The findings of Shigeru *et al.* (2020) suggested that the 'Mini-tiara' carnation has two genes from a virtual group of at least five ACS1 gene homologs of carnation and *D./knappii*. The discovery of a DcACS1b homozygotic carnation and inheritance of ACS1 homologous genes in the 'Mini-tiara' carnation from carnation and *D./knappii* suggests the allelic nature of carnation DcACS1a and DcACS1b genes.

12.2.2 Molecular Markers

DNA fingerprinting (DFP) using mini- and micro-satellite sequences to the identification of genotypes and to the establishment of genetic distances using progeny from the selfing of breeding loine 2217 was reported by Vainstein *et al.* (1995) and Ben *et al.* (1997). The probability of 2 offsprings from the crossing of similar genotypes having identical DNA fingerprints was $1.8x10^{-6}$. A comparison of genetic relationships within and between phenotypic categories, based on known genetic history, with genetic relationships derived from DFPs revealed a perfect match for flower types. Correlation analysis of the results yielded by each of the 3 probes (2 mini- and micro-satellite) revealed a significant correlation. A marker linked to bullheaded flowers in carnation was identified using RAPD analysis.

The RAPD technique was used to identify genetic markers useful in the development of a diagnostic method for *F. oxysporum* f. sp. *dianthi*, the causal agent of the carnation wilt disease. A total of 18 different strains isolated from different locations around the world, as well as 17 *F. oxysporum* isolates corresponding to f. sp. other than *dianthi* (non-*dianthi*) were amplified using 15 primers. No direct correlation was observed between the RAPD pattern and the race of an isolate (Hernandez *et al.*, 1999).

Scovel *et al.* (1998) investigated on flower doubleness as a breeding characteristic in carnation. Flowers of standard and spray varieties, which constitute the largest market share, are usually of the double and semi-double type respectively. These flower types are not easily distinguishable due to phenotypic overlaps caused by environmental conditions. Using random decamer primers, RAPD marker which is tightly linked to this recessive allele was identified. The RAPD marker was cloned and used to generate a RFLP marker. This RFLP marker discriminated with 100 per cent accuracy between the semi-double and double-flower phenotypes in carnations of both Mediterranean and American groups.

Masafumi *et al.* (2017) detected the STS-WG44 marker tightly linked to the major resistant gene for bacterial wiltin all backcross lines with resistance selected by the root-soaking method. Quantities of markers OQ12 and STS-WB66 close to QTLs with slight effects in resistant lines were lower in succeeding generations of backcrossing. These findings suggested that STS-WG44 is available for selecting resistant lines. In practical breeding populations, the difference of mean disease incidence between two groups categorized as having or lacking STS-WG44 was 62.6 per cent and STS-WG44 was possessed in most lines showing resistance less than 20 per cent in disease incidence. These findings suggested that STS-WG44 as a selective marker facilitated narrowing of populations to those that are highly resistant for practical breeding. A high-density genetic map, an essential tool for comparative genomic studies and quantitative trait locus fine mapping, can also facilitate genome sequence assembly. The sequence-based marker technology known as restriction site associated DNA (RAD) enables synchronous, single nucleotide polymorphism marker discovery, and genotyping using massively parallel sequencing. Masafumi *et al.* (2017) constructed high-density linkage map for carnation (*Dianthus caryophyllus* L.) based on simple sequence repeat (SSR) markers in combination with RAD markers developed by double-digest RAD sequencing (ddRAD-seq). A total of 2404

(285 SSR and 2119 RAD) markers could be assigned to 15 linkage groups spanning 971.5 cM, with an average marker interval of 0.4 cM. The total length of scaffolds with identified map positions was 95.6 Mb, which is equivalent to 15.4 per cent of the estimated genome size. The generated map is the first SSR and RAD marker-based high-density linkage map reported for carnation.

Masafumi (2014) conducted breeding using CBW resistance derived from *Dianthus capitatus* ssp. and *rzejowskianus*. To map the genetic loci involved in resistance to CBW and develop the linked markers, the first molecular linkage map for carnation was constructed by using 134 progeny derived from a cross between 'Carnation Nou No. 1' (an interspecific hybrid of carnation and *D. capitatus*) and 'Pretty Favvare', a susceptible cultivar. The map consisted of 146 DNA markers and covered 16 linkage groups. QTL analysis identified a QTL with a significant effect and two QTLs with small effects. Evaluation of disease incidence in relation to the presence of the STS-WG44 marker, which is linked to a QTL with a large effect, revealed that marker-assisted selection (MAS) using STS-WG44 enables the tested population to be narrowed down by half. Repeated crossing and selection viaboth conventional disease screening and MAS led to successful development of the first CBW-resistant carnation cultivar, 'Rouge'.

Masafumi *et al.* (2020) reported that flowering time showed large phenotypic segregation in two F_2 populations and in different years. Despite the different populations and different years, one major common QTL for flowering time was detected in linkage group 10 with the effect explaining from 18.2 per cent –22.5 per cent of the overall phenotypic variance. They developed a DNA marker, qD1Flw1-sc43-4, that was located close to the detected QTL for flowering time. They could distinguish the flowering time and categorize the genotypes of an F_1 population derived from a cross between late flowering 'Light Pink Barbara' and early flowering 'Kaneainou 1 go' using the qD1Flw1-sc43-4 marker.The results suggested that flowering time in carnation involves several genetic factors.

The whole-genome sequence of carnation (*Dianthus caryophyllus* L.) cv. 'Francesco' was determined using a combination of different new-generation multiplex sequencing platforms by Masafumi *et al.* (2014). The total length of the nonredundant sequences was 568 887 315 bp, consisting of 45 088 scaffolds, which covered 91 per cent of the 622 Mb carnation genome estimated by k-mer analysis. The N50 values of contigs and scaffolds were 16644 bp and 60 737 bp, respectively, and the longest scaffold was 1 287 144 bp. The average GC content of the contig sequences was 36 per cent. A total of 1050, 13, 92 and 143 genes for tRNAs, rRNAs, snoRNA and miRNA, respectively, were identified in the assembled genomic sequences. For protein-encoding genes, 43266 complete and partial gene structures excluding those in transposable elements were deduced. Gene coverage was <"98 per cent, as deduced from the coverage of the core eukaryotic genes. Intensive characterization of the assigned carnation genes and comparison with those of other plant species revealed characteristic features of the carnation genome. The results of this study will serve as a valuable resource for fundamental and applied research of carnation, especially for breeding new carnation varieties.

12.3.0 Genetic Transformation

Firoozabady *et al.* (1991) reported that *A. tumefaciens* strain A 281 was the most virulent in *in vitro* stems of carnation and regenerated transformed plants from leaf and petal bases. However, transformation was most successful using *in vitro* leaf bases cocultivated with *A. tumefaciens* EHA101 containing the binary vectors pSLJ191 (p35S NPTII/p35SGUS) and pWTT2084 (p35SALS/p35SGUS). The frequency of GUS-positive explants is 40-60 per cent, however, only a low percentage of these produced fully transformed shoot clusters (3-10 per cent). The regeneration was localized to only the basal portion of the leaf, only 5-50 per cent of the explants regenerated shoots depending on the selection (genticin or chlorsulfuron), some transformed sectors regenerated chimeric shoots suggesting that the origin of the regenerated plants is multicellular. Transformed plants after hardening when transferred to the greenhouse were very vigorous and flowered normally with no evidence of chimeras. This technique for production of transgenic plants was successful for Improved White Sim and two other commercial cultivars Manon and Nathalie. Five cultivars of carnation were transformed by cocultivation of leaves, petals or stem sections with *Agrobacterium* followed by direct or indirect organogenesis (Robinson and Firoozabady, 1993). Firoozabady *et al.* (1995) further improved transformation and regeneration system for carnation.

Petal explants prepared from two cultivars of carnation and one interspecific hybrid cultivar (*D. chinensis* x *D. caryophyllus*) were cocultivated with either disarmed *Agrobacterium tumefaciens* strain EHA 101 or virulent *A. rhizogenes* strain A 13, both of which harboured the binary vector pIG121Hm carrying the *nptII*, hygromycin phosphotransferase (*hpt*) and intron containing *gus* genes. Transient expression assay of GUS immediately after cocultivation indicated a large difference in the transformation frequency among different *Dianthus-Agrobacterium* combinations. Cocultivated explants were cultured on a shoot induction medium containing 250 mg kanamycin per litre, which severely inhibited the regeneration of green shoots from control explants. Two months after cocultivation, several green shoots were regenerated. However, only a few transgenic plantlets proved to be stably transformed by Southern blot analysis. In addition, some of the transgenic plantlets exhibited an abnormal phenotype including dwarfness and precocious flowering (Nakano *et al.*, 1999).

Van Altvorst *et al.* (1996) obtained transgenic carnation plants after infection of petalexplants with the supervirulent *Agrobacterium tumefaciens* strain AGLO. The transgenic nature of the plants were confirmed using Southern blot techniques and the expression of *gus* gene was verified by histochemical assays on selected shoots. But they faced difficulty in transferring the transgenic plants to the greenhousedueto vitrification and premature flowering. Zuker *et al.* (1995) developed transgenic carnation plants by microprojectile bombardment of highly regenerative stem segments. Two-step regeneration procedure based on the use of two different cytokinins (BAP and TDZ) was employed for the production of adventitious shoots from stem segments. The size of original stem was found to affect the regeneration efficiency of stem segments, the highest efficiency of adventitious shoot regeneration was obtained with segments originating from stems with two mature leaves, as compared to those with four, six or eight mature leaves. Bombardment of cultivar White Sim stem segments was performed with a plasmid containing *uidA* and *bar* genes encoding ∞-glucuronidase and phosphinothricin-acetyltransferase respectively. Presence of the *bar* gene in plants stably expressing *uidA*was confirmed by Southern blot analysis.

Cornish *et al.* (1991) obtained transgenic carnation plants with some reduced senes-cence response by introduction of *carn 363,* a carnation gene, this homologous to the tomato ethylene forming enzyme (pTOM13). Bovy *et al.* (1995) studied the role of ethylene in the senescence process of carnation. The senescence and ripening process is characterized by a climacteric increase in the production of ethylene, leading to the induction of genes involved in programmed cell death and ripening.

Savin *et al.* (1995) studied ethylene biosynthesis and reported that ACC synthase (ACS) catalyzes the conversion of S-adenosylmethionine to 1-aminocyclopropane-1-carboxylic acid (ACC), while ACC oxidase (ACO) catalyzes the subsequent step, the conversion of ACC to ethylene. An ACO cDNA isolated from carnation cv. Scania was used to produce transgenic carnation plants containing an antisense *ACO* gene. Flower of one transformant from each of 2 cultivars (Scania and White Sim) containing this antisense gene exhibited low climacteric ethylene production and a markedly delayed petalsenescence.

The genetically modified flowers of carnation (*Dianthus caryophyllus*) have been sold to the floristry industry. The transgenic carnation carries a herbicide tolerance gene (a mutant gene encoding acetolactate synthase (ALS) and has been modified to produce delphinidin-based anthocyanins in flowers, which conventionally bred carnation cannot produce. The modified flower color has been achieved by introduction of a gene encoding flavonoid 32,52-hydroxylase (F3252H). Transgenic carnation flowers are produced in South America and are primarily distributed to North America, Europe, and Japan. Although a nonfood crop, the release of the genetically modified carnation varieties required an environmental risk impact assessment and an assessment of the potential for any increased risk of harm to human or animal health compared to conventionally bred carnation. The results of the health safety assessment and the experimental studies that accompanied them are described by Stephan *et al.* (2013). The conclusion from the assessments has been that the release of genetically modified carnation varieties which express *F3252H* and *ALS* genes and which accumulate delphinidin-based anthocyanins do not pose an increased risk of harm to human or animal health.

Potted carnation (*Dianthus caryophyllus* L. cv. Lillipot) plants were transformed by Tomoyuki *et al.* (2006) with cDNAsfor carnation 1-aminocyclopropane-1-carboxylate(ACC) synthase (DC-ACS1, s/aACS transgenes) orACC oxidase (DC-ACO1, s/aACO transgenes) insense or antisense orientation or mutated carnationethylene receptor cDNA (DC-ERS2¢) byAgrobacterium-mediatedgene transfer. The presence of acetosyringoneat 100 lM in media for shoot culture priorto leaf explant preparation and preculture of Agrobacteriumin addition to the conventional method ofaddition to media for infection and coculture, and theuse of water instead of nutrient media for infectionand coculture increased the transformation efficiencyto 4.0 per cent compared to the 0.1 per cent obtained by the conventionalmethod. PCR analysis as well as Southernblot analysis confirmed the integration of the ethylene-relatedtransgenes. Leact segments of culturedshoots of some lines transformed with s/aACOtransgenes had less activity to convert ACC to ethylenethan that of the non-transformed control plant,indicating that the integrated s/aACO transgenes.

Shigeru (2011) reported that senescence of carnation flowers is characterized by autocatalytic ethylene production from petals and subsequent wilting of the petals. Recent studies on the regulation of ethylene production and wilting in senescing carnation petals revealed that (1) petal senescence is triggered by ethylene evolved from the gynoecium during natural senescence, (2) ethylene production in the gynoecium is induced by a factor(s) other than pollination signals in carnation flowers lacking anthers, (3) there are two subsets of ethylene responses in the petals, one responsible for autocatalytic ethylene production and the other for wilting, (4) expression of genes involved in the execution of petalwithering is differently regulated between ethylene-dependent or -independent senescence. Furthermore, it was revealed that the generation of transgenic carnation without detectable ethylene production is useful to prolong the vase life of cut carnation flowers.Roses, carnations and chrysanthemums lack a blue/violet flower colour because of their inability to synthesize the delphinidin-based anthocyanins, which most plants with blue/violet flowers produce. Flavonoid 3′,5′-hyroxylase is the key enzyme on the pathway to the synthesis of delphinidin-based anthocyanins. After countless trial-and-error experiments, introduction and expression of the flavonoid 3′,5′-hyroxylase gene in these important flower species has been achieved, resulting in significant colour change toward blue in flowers of transgenic plants. After clearing the complicated regulatory procedures for GM plants, the colour-modified carnations (Florigene 'Moon' series™) have been sold in North America, Japan, Europe and other countries since 1996.Quantitative real-time RT-PCR (qRT-PCR) analysis showed that the expressions of DcaAG genes in the flower bud were significantly different from those in the root, stem, and leaf between the single and double flower phenotype carnations, and that DcaAG genes were specifically expressed in the stamen and carpel of carnation. Moreover, the expression of other floral organ identity genes (AP1 and AP2, PI and AP3, SEP1 and SEP3 corresponding to the A-, B-, and E-class of genes, respectively) showed no significant difference in all floral organs between the single and double flower phenotype carnations, suggesting that C-class (DcaAG) genes might play an important role in the double flower phenotype in carnation.

12.4.0 *In vitro* Conservation

Fukai (1992) evaluated a number of protocols with a view to standardize *in vitro* storage of germplasm of carnation. Shoot-tips from 29 *Dianthus* species and hybrids were cooled at a rate of 0.5°C/min down to –4°C in the presence of 10 per cent dimethyl sulphoxide (DMSO) and 3 per cent glucose prior to immersion in liquid nitrogen and cryopreserved (Fukai *et al.*, 1991). The shoot-tips of all species and cultivars were viable after thawing. Fukai *et al.* (1994) also reported that nodal segments of *in vitro* grown *D. hybridus* cv. Sakuranadesiko were cryopreserved by the 2-step freezing and encapsulation-dehydration methods. Drought tolerance was

Marketing of Carnation

Loose flowers in local market

Cut flowers in world market

induced in alginate encapsulated explants by preculturing on a medium enriched with 0.2-1 M sucrose. The encapsulated explants, precultured and dried to 20 per cent water by weight, showed 100 per cent survival and shoot regeneration following immersion in liquid nitrogen. Explants cryopreserved by conventional 2-step freezing had lower shoot regeneration frequen-cies (49 per cent).

Tannoury and Vintejoux (1997) also encapsulated terminal or axillary buds in alginate beads, subjected to dehydration or vitrification, then immersed in liquid nitrogen. The shoot-tip cells remained viable after freezing in liquid nitrogen. Cytological studies indicated that alterations occurred mostly in basal tissues and leaf primordia. Some ultrastructural changes were detected in cell walls, the nucleus, the plastide and the vacuoles.

12.4.1 Gene Anologues

Flowers with more petals are of more ornamental value. It is well known that AGAMOUS (AG) is the core member of the C-class gene which plays an essential role in double flower formationand identification of stamens and carpels in Arabidopsis thaliana. Qijian *et al.* (2020) searched C-class genes in thegenome of the carnation, and found two AGorthologs (*DcaAGa, DcaAGb*). Phylogenetic analysis showed that the two genes were closely related to the euAG subclade. Then they searched the genomesof other Caryophyllales plants (*Beta vulgaris, Spinacia oleracea, Chenopodium quinoa*) for C-class genes,and found that their C-class genes all belonged to the euAG subclade. Semi-quantitative PCR (sq-PCR) analysis indicated that the expression of *DcaAG* genes in the single flower phenotype was higher than that in the double flower phenotype. Quantitative real-time RT-PCR (qRT-PCR) analysis showed that the expressions of *DcaAG* genes in the flower bud were significantly different from those in the root, stem, and leaf between the single and double flower phenotype carnations, and that DcaAG genes were specifically expressed in the stamen and carpel of carnation. Moreover, the expression ofother floral organ identity genes (*AP1* and *AP2, API* and *AP3, SEP1* and *SEP3*) corresponding to the A-,B-, and E-class of genes, respectively) showed no significant difference in all floral organs between the single and double flower phenotype carnations, suggesting that C-class (*DcaAG*) genes might play an important role in the double flower phenotype in carnation. Petalloss or decrease, precocious flowering, silique shortening, and seed sterility were observed in 35S::*DcaAGa* and 35S::*DcaAGb* transgenicArabidopsis plants. All these results show that *DcaAG* genes might affect the petal number negatively and have a specific function in stamen and carpel development in carnation.

Endogenous signals in response to exogenous factors determine the senescence offlowers. Interactions amongphytohormones especially abscisic acid (ABA) and ethylene are the major determinant of the senescence. Mostafa *et al.* (2020) found complex expression patterns of the genes related to ABA and ethylene as endogenous signals wereinvestigated on cut carnations (*Dianthus caryophyllus* L.) that were exposed to different light spectra. Expression of ethylene biosynthetic (*DcACS* and *DcACO*), and signaling (*DcETR* and *DcEin2*) genes and also genes involvedin ABA biosynthesis (*DcZEP1* and *DcNCED1*), transport (*DcABCG25* and *DcABCG40*) and catabolism (*DcCYP707A1*) were evaluated in petals of carnations exposed to three light spectra [white, blue and red]. Lowest relative membrane permeability (RMP) was detected inflowers that exposed to Blue light (BLFs), as aconsequence, the longest vase life was found in BLFs. The Red and White lights markedly accelerated flower senescence and increased expression of *DcACS* and *DcACO* on day 6 and 10 of vase life assessment respectively;while Blue light inhibited the expression of ethylene biosynthetic genes. Expression of the genes involved in theproduction and transport of ABA and in signal transduction of ethylene was elevated during vase life offlowersirrespective of exposure to different light spectra. In conclusion, Blue light can be an effective environmentalfactor to extend the vase life of carnationflowers by delaying the petalsenescence through down-regulation ofethylene biosynthetic genes and up-regulation of ABA biosynthetic genes.

13.0 References

Abe, H., Shinmura, A. and Shiba, H. (1997) *Ann. Rep. Soc. Plant Protection of North Japan.*, **48:** 122-125.

Accati, E.G., Basoccu, L., Sulis, S. and Garibaldi Accati, E. (1979) *Acta Hortic.*, **91:** 115-119.

Adams, P., Graves, C.J. and Winsor, G.W. (1980) *J. Hort. Sci.*, **55**: 89-96.

Adams, P., Hart, B.M.A. and Winsor, G.W. (1979) *J. Hort. Sci.*, **54**: 149-154.

Akbulut, N.and Zumreoglu, S. (1992) *Proc.II Turkish Natl. Cong. Entomol.*, pp. 549-557.

Aldrufeu, A., Folch, I. and Camprubi, P. (1981) *HortScience*, **16:** 224-225.

Aloj, B., Nanni, B. and Marziano, F. (1990) *Informatore Fitopatologico.*, **40:** 10, 55-56.

Alpi, A. and Buiatti, M. (1972) *Rivistadella Ortoflorofrutticoltura Italiana.* **56:** 119-130.

Altan, T. and Altan, S. (1982) *Cukurova Univereseti Ziraat Fakulltesi Yilligi*, **13**: 47-57.

Altman, S.A. and Solomos, T. (1994) *J. Amer. Soc. Hort. Sci.*, **119:** 282-287.

Altman, S.A. and Solomos, T. (1995) *Postharvest Bio.Technol.*, **5:** 331-343.

Altman, S.A. and Solomos, T. (1993) *HortScience*, **28:** 201-203.

Altvorst, van. A.C. and Bovy, A.G. (1995) *Plant Growth Regul.*, **15:** 43-53.

Amar, B.P.M. (2019) *Asian Pl. Res. J.*, **2**(1): 1-6.

Andersen, H. and Aabrandt, Z. (1989) *Gartner-Tidende.* **105:** 38, 941-943.

Andreucci, E. (1964) *Inf. Fitopat.*, **14**: 410-412.

Annonymous (2000) *Pl. Var. J.*, **13**: 27.

Arai, M. and Takeuchi, M. (1993) *Plant Cell, Tissue and Organ Cult.*, **34:** 287-293.

Arici, S.E., Erdogan, O. and Tuncel, Z.N. (2019) *Appl. Ecol. Environ. Res.*, **17**(6): 15255-15265.

Arora, J.S. and Saini, S.S. (1976) *J. Res. (PAU)* Ludhiana, India. **13:** 4, 362-366.

Arora, J.S. and John, A.G. (1976) *Indian J. Orn. Hort.*, **7:** 21-25.

Arroyo, M. (1963) *Agriculture (Madrid)*, **32**: 283-285.

Arshdeep, S.C., Ravinder, S. and Dev, D. (2020) *Indian J. Ecol.*, **47**(2): 369-377.

Arus, P., Llaurado, M. and Pera, J. (1992) *Acta Hortic.*, **307**: 57-64.

Ashok, A.D., Arun, D.S. and Rengasamy, P. (1999) *South Indian Hort.*, **47**: 115-118.

Atanasova, B. (1998) *Rasteniev"dniNauki.*, **35:** 200-205.

Baas, R., Nijssen, H.M.C., Berg, T.J.M van den., Warmenhoven, M.G. and Weel, P. van. (1990) *Vakbladvoor de Bloemisterij.*, **45:** 40, 56-57.

Baayen, R.P., Sparnaaij, L.D., Jansen, J. and Niemann, G.J. (1991) *Netherlands J. Plant Pathol.*, **97:** 2, 73-86.

Baayen, R.P. and Schoffelmeer, E.A.M. (1997) *Proc. 49th Internatl. Symp. Crop Prot., Gent, Belgium*, 6 May, Part IV.

Bagaturiya, B.Y. and Tavberidze, D.M. (1976) *Vopr. Intro. Rast. Zelen. Str-va.Tbilisi*, Georgian, SSR, Mecniereba, **10**: 52-56.

Bailey, J.H. (1963) *The Standard Cyclopaedia of Horticulture*, Macmillan and Co., New York, Vol. 1, p. 167.

Baker, R. and Harman, G. (1981) *Res.Bull., Colo. Greenhouse Grrs' Ass.*, **376:** 4.

Barden, L.E. (1972) *Bull. Colo. Flower Grs' Asso.*, **263:** 1-3.

Barth, G.E. and Frensham, A. (1989) *Australian J. Expt. Agric.*, **29:** 735-739.

Basallote, M.J., Delgado, M.D.V., Capote, N., Vara, J.M.M., Herrera, C.J.L., Ligero, A.M.P. and Rubia, M.F.T. (2016) *Crop Prot.*, **89**: 184-192.

Basile, M. and Lamberti, F. (1977) *Rivistadella Ortoflorofrutticoltura Italiana.*, **61:** 149-156.

Batlle, A.andLavina, A. (1994) *Plant Dis.*, **78:** 11, 1122.

Behera, P.K. (1993) *Plant Physiol. Biochem.*, **20**: 59-62.

Ben Yephet, Y. and Shtienberg, D. (1994) *Phytopathol.*, **84:** 1416-1421.

Ben Yephet, Y., Shtienberg, D., Reuven, M. and Mor, Y. (1993) *Netherlands J. Plant Pathol.*, **99:** 3-12.

Ben, M.H., Scovel, G., Ovadis, M., Vainstein, A. and Altman, A. (1997) *Acta Hortic.*, **447**: 599-601.

Bichara, A.E. and Van Staden, J. (1993) *Plant Growth Reg.*, **13:** 161-167.

Bierman, E.P.B., Brouwer, D.H. and Van Hemmen, J.J. (1997) *Bull. Env.l Contamination and Toxicology*, **58:** 1006-1013.

Bik, R.A. (1962) *Stikstof.*, **35**: 502-520.

Bivins, J.L. (1976) *Flower and Nursery Rept., Calif. Univ.* Jan.-Feb., pp. 9-10.

Blake, J. and Harris, G.P. (1960) *Ann. Bot.*, **24**: 247-256.

Blanc, D. (1962) *C.R. Acad. Agril. France*, **48:** 791-797.

Blanc, D. and Tramier, R. (1980) *C.R. Acad. Agric., France*, **6**: 532-538.

Blanc, D., Tramier, R. and Pallot, C. (1983) *Acta Hortic.*, **141:** 115-123.

Bovy, A. G., Altvorst, A. C. van., Angenent, G. C., Dons, J. J. M. and Fjeld, T. (1995) *Acta Hortic.*, **405**: 179-189.

Bowe, R. and Reinelt, J. (1991) *Gartenbau Magazin.*, **38:** 4, 42-44.

Boztok, S., Cokuysal, B. and Yildirim, T.B. (1995) *Ege Universitesi Ziraat Fakultesi Dergisi.*, **32:** 111-118.

Boztok, S., Cokuysal, B., Hatipoglu, A., Colakoglu, H. and Gulgun, B. (1996) *Bahce*, **25:** 95-102.

Brandt, A.S. and Woodson, W.R. (1992) *HortScience*, **27:** 1100-1102.

Broun, R. and Mayak, S. (1981) *Sci. Hort.*, **15:** 277-282.

Buiatti, M., Gimelli, F., Venturo, R., Bogani, P. and Picconi, T. (1986) *Proc. Sem. Somaclonal variations and crop improvement*, Gembloux, Belgium, 3-5 September, 1985, pp. 251-256.

Bunt, A.C. (1974) *Commercial Grower*, **4081**: 459-464.

Bunt, A.C. and Powell, M.C. (1982) *Sci. Hort.*, **17**: 177-186.

Burbano, L.E., Erazo, A., Orozco de Amezquita, M., Garces de Granada, E. (1990) *Agronomia Colombiana*, **7:** 61-69.

Burchi, G., Bianchini, C., Mercuri, A., Foglia, G., Rosellini, D. and Schiva, T. (1999) *J. Genet. Breed.*, **53**: 301-306.

Bylov, V.N. and Smirnova, A.J. (1984) *Byulleten'GlavnogoBotanicheskogoSada*, **130**: 55-58.

Byrne, T.G. (1974) *Flower and Nursery Rep.* (July), Calif. Univ., p. 7-8.

Byrne, T.G. (1982) *Flower and Nursery Rep.*, Calif. Univ. (Summer), p. 3-4.

Camprubi, P. and Aquila, J.F. (1973) *Acta Horticulturae*, **43**: 307-318.

Camprubi, P. and Nichols, R. (1978) *J.Hort. Sci.*, **53:** 17-22.

Camprubi, P., Aldrufeu, A., Pages, M., Bargallo, R. and Lopez, J. (1981) *Acta Hortic*, **113**: 51-57.

Cardus, J. and Lasala, M. (1962) *Proc. 16th Int. Hort. Cong.*, Brussels, p. 314.

Cassells, A.C., Walsh, C. and Periappuram, C. (1993) *Euphytica*, **70:** 167-174.

Cecilia, P., Hoesel, S. and Crutcher, F. (2020) *J. Microbiol. Methods*, **170.**, doi10.1016/j.mimet.2020.105841.

Cedeno, L. and Carrero, C. (1997) *RevistaForestalVenezolana*. **31:** 41-1, 91-92.

Celikel, F.G., Karaaly, Y. and Stromme, E. (1995) *Acta Hortic.*, **405:** 156-163.

Ceron Madrigal, J.J., Hernandez Ruiperez, F. and Bernal Gambin, L.J. (1995) *AvancesenAlimentacion-y-Mejora-Animal.*, **35:** 1, 11-14.

Champeroux, A. (1983) *Agronomie.*, **3:** 707-715.

Chan, A.P., Heeney, H.B., Maginess, E.A. and Cannon, H.B. (1958) *Proc. Amer. Soc. Hort.Sci.*, **72**: 473-480.

Chandel, S. (2011) ***J. Mycol. Plant Pathol.*, 41**(1): 111-116.

Cheever, D.W. (1967b) *Bull. Colo. Flower Grs.' Asso.*, No. 208. pp. 1-3.

Cho, M.R., Kim, J.Y., Song, C., Ko, J.Y., Na, S.Y. and Yiem, M.S. (1996) *J.Nematol.*, **28:** 639-642.

Choi, H.R. and Roh, S.M. (1980) *J.Korean Soc. Hort. Sci.*, **21:** 176-184.

Cifuentes, Y. and Arbelaez, G. (1984) *Agronomia Colombiana*, **2**: 101-108.

Coats, A.M. (1956) *Flowers and their Histories*, Pitman Publishing Corporation, New York.

Cornish, E.C., Baudinette, S.C., Graham, M. H., Stevenson, K. R., White, E. L. J., Michael, M. Z., Lu, C., Chandler, S. F. and Savin, K. W. (1991) 31st*Annual Meeting of the Australian Soc. plant Physiol., Australian National University*, pp. 2-4.

Cox, R.J. (1967) *Sparkes Tech. Rept.*, A. G.Sparks Ltd., Sussex, UK.

Cuany, A. and Ritter, M. (1977) *Acta Hortic.*, **71:** 201-207.

Cuany, A., Berge, J.B., Scotto La Massese, C. (1973) *Bull. OEPP*, **3**: 7 5-87.

Cuscianna, N. (1963) *Progr. Agric. Bologna*, **9**: 662-675.

Dabeda, R., Muller, T.C. and Silva, V.R. (1975) *Annuals*, **26:** 327-336.

Dalbir Singh, Gill, A.P.S., Kumar, R., and Singh, D. (1994) *J.Ornamental Hort.*, **2:** 51-53.

Dalla Guda, C. and D'Aquila, F. (1982) *Annalidell'IstitutoSperimentale per la Floricoltura, Sanremo*(Italy), **13**: 13-22.

Debergh, P. (1972) *Meded. Fac. Landb. Gent.*, **37:** 41-46.

Debergh, P. (1973) *Meded. Fac. Landb. Gent.*, **38**: 402-405.

Demmink, J.F., Custers, J.B.M. and Bergervoet, J.H.W. (1987a) *Acta Horticulturae*, **216**: 343-344.

Demmink, J.F., Karsten, M., Giessen, A.C. and Vander, R. (1982) *Vakbladvoor de Bloemisterij.*, **37:** 24-27, 29.

Dereuddre, J., Fabre, J. and Basaglia, C. (1988) *Plant Cell Rep.*, **7**: 170-173.

De, La, T.A.R., Pallás, V. and Navarro, J.A.S. (2016) *Plant Dis.*, **100**(7): 1509.

Dhiman, J.S., Arora, J.S. and Gill, P.S. (1991) *Indian Phytopathol.*, **44:** 356-359.

Diatloff, A. and Rochecouste, J. (1991) *Australian Plant Pathol.*, **20:** 27-30.

Dimalla, G.G. and Staden, J.Van (1980) *Zeitschrift fur Pflanzenphysiol.*, **99:** 9-17.

Dolors, S., Casanova, E., Segarra, G., Aviles, M., Reis, M. and Trillas, M.I. (2010) *Biol. Control.*, **53**(3): 291-296.

Doorn, W.G. van., Woltering, E.J., Reid, M.S., Wu, M.J., Van Doorn, W.G., Pech, J.C., Latche, A. and Balague, C. (1993) *Proc.Internatl. Symp. Cell Mol. Aspects of biosynthesis and action of plan hormone ethylene.* Kluwer Academic Publishers Dordrecht Netherland, pp. 188-194.

Doorn, W.G.van., Witte, Y.de., and Harkema, H. (1995) *Postharvest Biol.Technol.*, **6:** 111-119.

Doorn, W.G.van., Zagory, D., Witte, Y.-de., and Harkema, H. (1991) *Postharvest Biol.Technol.*, **1:** 161-168.

Droillard, M. J. and Paulin, A. (1990) *Plant Physiol.*, **94**: 1187-1192.

Droillard, M.J., Bureau, D. and Paulin, A. (1989) *Physiol. Plant.*, **76**: 149-154.

Drory, A., Beja Tal, S., Borochov, A., Gindin, E. and Mayak, S. (1995) *Sci. Hort.*, **64**: 167-175.

Dubey, K.C. (1972) *Plant Sci.*, **4**: 121-123.

Duijff, B.J., Meijer, J.W., Bakker, P.A.H.M. and Schippers, B. (1993) *Netherlands J. Plant Pathol.*, **99:** 277-289.

Dumlao, R.C. (1969) *Diss. Abstr. B.*, **29**: 2771.

Duskova, E., Prokinova, E. and Vancura, V. (1989) *Proc. Internatl. Symp. Interrelationships between microorganisms and plants in soil.* Elsevier Science Publishing Co. Amsterdam, Netherland, pp. 403-410.

Eisenberg, B.A. and Staby, G.L. (1985) *Plant and Cell Physiol.*, **26:** 829-837.

El Shafie, S.A. (1977) *Archiv fur Gartenbau.*, **25:** 337-345.

Elena, K. and Tjamos, E.C. (1997a) *Annales-de-l'Institut-Phytopathologique-Benaki.*, **18:** 13-24.

Elena, K.andTjamos, E.C. (1997b) *Phytopathologia Mediter.*, **36:** 2, 87-93.

El-Fouly, M.M., Mohamed, B.R. and Fawzi, A.F.A. (1977) *Sci. Hort.*, **6:** 241-249.

Engvild, K.C. (1972) *Physiol. Plant.*, **26:** 62-66.

Farina, E. (1984) *Rivistadella Ortoflorofrutticoltura Italiana*, **68:** 307-315.

Farina, E., Cervelli, C. and Leoni, S. (1994) *Acta Hortic.*, **361:** 478-485.

Farina, E., Garibaldi, E.A. and Accati Garibaldi, E. (1985) *Annalidell'IstitutoSperimentale per la Floricoltura,* **14:** 1, 19-36.

Farnham, D.S., Ueda, T., Kofranek, A.M., Halevy, A.H., McCain, A.H. and Kubota, J. (1978) *Flower and Nursery Report.* Autumn, 4-7.

Fawzi, A.F.A. and El Fouly, M.M. (1979) *Physiol. Plant.*, **47:** 245-249.

Fawzy, R.N., Riad, F.W. and Rizhalih, L.R. (1991) *Ann.Agri.Sci., Moshtohor,* **29:** 797-810.

Ferraris, L., Abbattista Gentile, T. and Accati, E. (1982) *Rivista di Ortoflorofrutticoltura,* **66:** 135-145.

Ferrero, F., Silvy, A., Jay, M. and Ledeme, P. (1987) *Acta Hortic.*, **216:** 205-214.

Firoozabady, E., Lemieux, C. S., Moy, Y.S., Mall, B., Nioholas, J. A. and Robinson, K. E. P. (1991) *In vitro*, **27**: 96A.

Firoozabady, E., Moy, Y., Tucker, W., Robinson, K. and Gutterson, N. (1995) *Mol. Breed.*, **1:** 283-293.

Forkmann, G., Dedio, J., Henkel, J., Min, B.W., Wassenegger, M., Vainstein and Weiss, D. (1995)*Acta Hortic.*, **420:** 29-31.

Forsberg, J.L. (1963) *Diseases of Ornamental Plants. Special Publ. No.3*, Univ. of Illionis, Coll. of Agric., Urbana III.

Fukai, S. (1992) *Mem. Fac. Agric. Kagawa Univ.*, **56**: 79.

Fukai, S., Goi, M. and Tanaka, M. (1991) *Euphytica*, **56**: 149-153.

Fukai, S., Togashi, M. and Goi, M. (1994) *Tech. Bull. Fac. Agric. Kagawa Univ.*, **46**: 101-107.

Garcia, D.L., Arbelaez, M.G. and Arbelaez Torres, G. (1994) *Agronomia-Colombiana.* **11:** 131-146.

Garibaldi, A. (1983) *RivistadellaOrtoflorofrutticolturaItaliana*, **67:** 261-270.

Garibaldi, A. and Garibaldi, E.A. (1982) *ColtureProtette.* **11:** 19-22.

Garibaldi, A., Aloi, C. and Gullino, M.L. (1988) *Atti-Giornate-Fitopatologiche.*, **1:** 469-480;

Garibaldi, A., Gullino, M.L. and Aloi, C. (1990) *Brighton Crop Protection Conf. on Pests and Diseases* - 1990. 1. 89-95.

Garrido, G., Cano, E.A., Acosta, M. and Sanchez Bravo, J. (1998) *Sci. Hort.*, **74:** 219-231.

Gatt, M.K., Hammett, K.R.W., Markham, K.P. and Murray, B.G. (1998) *Sci. Hort.*, **77:** 207-218.

Gill, A.P.S. (1984) *Indian Hort.*, **29**: 37-42.

Gill, A.S. and Grewal, N.S. (1987) *Punjab Hort. J.*, **27:** 1-2,

Gimelli, F., Ginatta, G., Venturo, R., Panizzi, A. and Buiatti, M. (1982) *Rivista di Ortoflorofrutticoltura*, **66:** 97-102.

Gimelli, F., Ginatta, G., Venturo, R., Positano, S. and Buiatti, M. (1984) *Rivista dell ortoflorofrutticolturaItaliana*, **68:** 107-121.

Gjaerum, H.B. and Langnes, R. (1984) *Gartneryrket.*, **74:** 122-123.

Goethals, M. and Hoof P. Van (1971) *Parasitica*, **27:** 36-41.

Goethals, M., Weerts, J. and Verhoyen, M. (1973) *Parasitica.*, **29:** 29-34

Gopinath, G. and Chanshekar, S.Y. (2009) *J. Ornam.Hortic.*, **12**: 56-58.

Goszczynska, D. and Nowak, J. (1980) *PraceInstytutuSadownictwa I Kwiaciarstwa w Skierniewicach.* **5:** 73-81.

Goszczynska, D. and Nowak, J. (1979) *Acta Hortic.*, **91:** 143-146.

Goszczynska, D. and Rudnicki, R.M. (1982) *Scientia Hort.*, **17**: 289-297.

Goszczynska, D. and Rudnicki, R.M. (1983) *Acta Hortic.*, **141:** 203-212.

Goszczynska, D.andRudnicki, R.M. (1978) *Bulletin de l'Academie Polonaise des Sciences.* **26:** 2, 137-140.

Goto, T., Masaoka, K., Kageyama, Y. and Konishi, K. (1998) *Scientific Rep.Fac.Agrl., Okayama University*, No. 87, 141-148.

Gould, H.J. (1960) *Plant Pathol.*, **9:** 106-109.

Green, J.L. and Holley, W.D. (1974) *J. Amer. Soc. Hort. Sci.*, **99:** 5, 420-424.

Guba, W., Starck, J.R.andStrojny, Z. (1993) *J. Fruit and Ornamental Plant Res.*, **1:** 1, 27-33.

Guda, C.D. and D'Aquila, F. (1980) *Annali dell Istitutosperimentale per la Floricoltura*, **11:** 13-18.

Gupta, D., Madhu Meeta and Pathania, N.S. (1995) *J.Orn. Hort.*, **3:** 1-2, 1-3.

Gursan, K. (1999) *Acta Hortic.*, **491:** 415-420.

Hackett, W.P. and Anderson, M.J. (1967) *Proc. Amer. Soc. Hort. Sci.*, **90:** 365-369.

Hadar, E., Elad, Y., Hadar, Y. and Chet, I. (1982) *Plant and Soil*, **65:** 303-307.

Han, bK. and Lee, J.S. (1992) *J. Korean Soc. Hort. Sci.*, **33b:** 343-350.

Hanan, J.J. (1959) *Proc. Amer. Soc. Hort. Sci.*, **74:** 692-703.

Hanan, J.J. and Baker, R.R. (1983) *Research Bulletin, Colorado Greenhouse Growers' Assn.*, **398:** 1-2.

Hanic, E. (1990) *Proc.Yugoslavsymp.Intensive veg. breeding and greenhouse production*, Yuygoslavia, pp. 229-236.

Hardenburg, R.E., Uota, M. and Parson, S.S. (1969) *Proc. 12th Int. Cong. Refrig.* (Madrid), **3:** 339-347.

Harinder, R., Praneet, C., and Shalini, C. (2016) *Indian Phytopathol.*, **69**(4): 396-399.

Harshavardhan, M., Kumar, D.P., Rajesh, A.M., Yathindra, H.A. and Shivanand, H. (2016) *Ecol. Environ. Conserv.*, **22:** 29-33.

Hasan, M., Sabir, N., Singh, A.K., Singh, M.C., Patel, N., Khanna, M., Rai, T., and Pragnya, P. (2018) *Tech. Bull.*, TB-ICN 188/2018.Publ. by I.A.R.I., New Delhi-110012 India p. 30.

Hasegawa, K. (1992) *Agriculture and Horticulture*, **67:** 301-305.

Hauzinska, E. (1975) *I Hod. Rosl. Aki. Easi.*, **19:** 363-376.

Heins, R.D. (1973) *Bull. Colo. Flower Grs.' Ass.*, No. 281, p. 4.

Heins, R.D. and Blakely, N. (1980) *Sci. Hort.*, **13:** 361-369.

Heins, R.D., Howell, G.S. and Wilkins, H.F. (1981) *Sci. Hort.*, **14:** 269-275.

Hempel, M. (1977) *Ph.D. Thesis*, Academy of Agriculture Lublin, p. 81.

Hentrich, W. and Glawe, M. (1982) *Archiv fur Zuchtungsforschung*, **12:** 197-207.

Hernandez, J.F., Posada, M.A., Portillo, P. del, Arbelaez, G. and Fischer, G. (1999) *Acta Horticulturae*, **482:** 123-131.

Hilal, A.A. and Kamel, B.K.M. (1990) *Agricultural Research Review*, **68:** 3, 543-554.

Hill, S.A. (1974) *Plant Pathol.*, **23:** 4, 151-155.

Hilliard, D. and Hanan, J.J. (1976) *Bull. Colo.Flower Grs.' Ass.*, **318:** 1.

Hitendra, P., Poonam, S., Rajesh, B. and Gaurav, Z. (2007) *J. Ornam. Hortic.*, **10**(2): 78-84.

Holley, W.D. (1960) *Bull. Colo.Flower Grs.' Ass.*, **134:** 9-12.

Holley, W.D. and Baker, R. (1963) *Carnation Production*, W.C. Brown Co., Iowa (USA), pp. 1-142.

Holley, W.D. and Meyers, T.A. (1970) *Bull. Colo. Flower Grs.' Ass.*, **241:** 5.

Holley, W.D. and Woodwards, C.W. (1958) *Bull. Colo.Flower Grs.' Ass.*, **104.**

Holley, W.D., Goldsberry, K.L. and Juengling, C. (1964) *Bull. Colo. Flower Grs. Ass.*, No. 174, pp. 1-5.

Hollings, M., Stone, O.M., Atkey, P.T. and Barton, R.J. (1977) *Ann. App. Biol.*, **85:** 59-70.

Hong, Y.P., Kim, J.Y and Kim, E.Y. (1986) *Res. Rep.Rural Development Administration Hort.*, **28**: 108-116.

Hopkins, H.W. and Holley, W.D. (1962) *Bull. Colo.Flower Grs.' Ass.*, No.153, pp. 1-4.

Hosni, A.M. and El Shoura, H.A.S. (1996) *Ann. Agrl. Sci., Cairo*, **41:** 351-365.

Huett, D.O. (1994) *Australian J. Expt. Agril.*, **34:** 691-697.

Huy, K.Y and Kim, K.S. (1995) *J. Korean Soc. Hort. Sci.*, **36:** 265-272.

Hwang, K.S and Yoon, J.H. (1994) *RDA J. Agrl. Sci. Soil and Fertilizer*, **36:** 268-272.

Irving, D.E. and Honnor, L. (1994) *Postharvest Biol.Technol.*, **4:** 281-287.

Ishida, A., Masui, M., Nukaya, A, and Ogura, T. (1979) *J. Japanese Soc. Hort. Sci.*, **48:** 322-326.

Ito, A., Yamaguchi, T. and Koshioka, M. (1996a).*JapaneseAgrl. Res. Quart.*, **30**: 255-261.

Ito, A., Yamaguchi, T. and Koshioka, M. (1996b) *Japanese Agrl. Res. Quart.*, **30:** 263-267.

Jacob, M. and Hamdan, I. (1992) *GartenbauMagazin*, **1:** 62-63.

Jacob, M., Junge, H., Werner, E. and Obieglo, U. (1988) *Nachrichtenblatt fur den Pflanzenschutz in der DDR.*, **42:** 246-250.

Jayashree, C. H., Anand, N.M. and Hegde, R.V. (2004) *J. Ornam. Hortic.*, **7**: 271-275.

Jensen, H.Ek. (1997) *Acta Horticulturae*, **435:** 77-86.

Johnson, R.T. (1980) *HortScience*, **15**: 605-606.

Johnston, J.W., Carpenter, A., and Lill, R.E. (1997) *VII internatl. Controlled atmos. Res. Conf., Calif. Univ., Calif.* 1997. No.**15:** 114-122.

Jones, R.B. and Hill, M. (1993) *J. Amer. Soc. Hort. Sci.*, **118:** 350-354.

Joung, H.Y., Kim, H.D., Choi, S.R., Kim, T.I., Sim, H.K., Kim, J.Y., Ko, J.Y., Keun, O.K. and Choi, X.H. (1998) *J. Hort. Sci.*, **40**: 120-122.

Juan, M.S.D., Silvia, J.B. and Manuel, J. (2017) *Hort. Sci.*, **44**(1): 14–20.

Juga, M. (1963) *Grad. Via. Liv.*, **21**: 53-54.

KadmanZahavi, A. and Ephrat, E. (1982) *Scientia Hort.*, **18:** 159-167.

Kageyama, Y., Okamoto, N. and Konishi, K. (1985) *Scientific Rep. Fac. Agr.*, Okanyama Univ., **65:** 15-21.

Kakehi, M. (1970) *Bull. Hiroshima Agric. College*, **4:** 40-49.

Kallak, H., Hilpus, I. and Virumae, K. (1996) *Scientia Hort*, **65:** 181-189.

KanyaratChaisut (1995) *GradenarskaiLozarskaNauka*, **37**: 94-97.

Kapchina, V.M. and Yakimova, E.T. (1989) *Acta Horticulturae*, **251:** 53-57.

Kaviarasu, A. and Selvaraj, N. (2015) *Society for Advancement of Science and Rural Development*, **8**(20): 5541-5548.

Kazaz, S., Yilmaz, S. and Askin, M.A. (2010) *Acta Hortic.*, **883**: 78-79. doi: 10.17660/ActaHortic.2010.883.51.

Ketsa, S., Tongumpai, P. and Siripanich, T. (1987) *Kasetsart J. Natural Sci.*, **21:** 207-212.

Keun, K., Manjulatha, M. and Kim, A.K. (2019) *Hortic.Environ. Biotech.*, **60**(6): 123-27.

Khalighy, A. (1976) *Iranian J. Agrl.Res.*, **4:** 35-39.

Kho, Y.O. and Baer, J. (1973) *Euphytica*, **22**: 467-470.

Kim, J.C. and Lee, E.A. (1996) *Plant Cell Rep.*, **16**: 18-21.

Kim, J.Y., Kim, K.H., Ko, J.Y. and Choi, J.K. (1992) *Res. Rep. Rural Development AdminisHort*, **34**: 27-35.

Kim, K.W. and Byun, M.S. (1985) *J. Korean Soc. Hort. Sci.*, **26:** 76-82.

Kim, K.W., Byun, M.S. and Kang, M.S. (1988) *J. Korean Soc. Hort. Sci.*, **29**: 208-215.

Kishore, C. and Srikant, K. (2008) *J. Plant Dis. Sci.*, **3**(1): 43-47.

Klasman, R., Laino, M. and Benedetto, A. (1999) *Horticultura Argentina*, **18**: 66-69.

Ko, J. Y., Kim Jae Young and Ahu Byung Joon (1997) *J. Korean Soc. Hort. Sci.* **38:** 783-788.

Koganezawa, H., Sasaya, T., Sato, T., Takezaki, A., Tomioka, K. and Morin, M. (1998) *Bull. Shikiky Natl. Agrl. Expt. Stn.***62:** 9-15.

Kohl, H.C. and Rundle, D.L. (1974) *Flower and Nursery Report*, (1974), pp. 9-11.

Kong Dezheng and Li Wen (1999) *J. Henan Ag. Sci.*, **8:** 29.

Kosugi, Y., Shibuya, K., Tsuruno, N., Iwazaki, Y., Mochizuki, A., Yoshioka, T., Hashiba, T. and Satoh, S. (2000) *Plant Sci.*, **158**: 139-145.

Kowalska, A. (1972) *PhytopathologischeZeits.*, **74:** 329-341.

Kowalska, A. (1974) *PhytopathologischeZeits.*, **79:** 301-309.

Koyama, Y. and Uda, A. (1994) *J. Japanese Soc. Hort. Sci.*, **63:** 203-209.

Koyama, Y., Uda, A., Wada, O. and Fujino, M. (1995) *J. Japanese Soc.Hort. Sci.*, **63:** 835-842.

Krzyzanowska, E. (1982) *Ogrodnictwo* (Poland), **19:** 43-44.

Kumar, P.N., Singh, B., Voleti, S.R. and Sindhu, S.S. (1999) *Indian J. Pl. Physiol.*, **4**: 185-188.

Kumar, S., Kumari, R., Baheti, T., Thakur, M. and Ghani, M. (2016) *Indian J. Agric. Sci.*, **86**(11): 74-77.

Kumar, S., Malik, A. and Yadav, R. (2019) *Int. J. Curr. Microbiol. Appl. Sci.*, **8**(10): 78-83.

Lakshmi, D.M., Jawaharlal, M. and Kannan, M. (2016) *Indian J. Agric. Sci.*, **86**(10): 44-49.

Lal, S.D. and Danu, N.S. (1985) *Progressive Hort.*, **17:** 145-147.

Lamberti, F., Tacconi, R., Marinari, A., D' Errico, F.P., Basile, M. and Errico, F.P.d' (1987) *DifesadellePiante.*, **10:** 77-84.

Larsen, P.B., Ashworth, E.N., Jones, M.L. and Woodson, W.R. (1995) *Plant Physiol.*, **108:** 1405-1412.

Laurie, A., Kiplinger, D.C. and Nelson, K.S. (1968) *Commercial Flower Forcing*, McGraw Hill BookCompany.

Le Masson, B. and Nowak, J. (1981) *Scientia Hort.*, **15:** 383-390.

Lee, H.S. and Ahn, B.J. (1997) *J. Korean Soc. Hort. Sci.*, **38:** 771-775.

Lee, J. S. and Kim, Y. (1999) *J. Korean Soc. Hort. Sci.*, **40:** 245-247.

Lee, J.S. (1988) *J. Korean Soc. Hort. Sci.*, **29:** 145-154.

Lee, J.S. and Kim, Y.A. (1993) *J. Korean Soc. Hort. Sci.*, **34:** 463-473.

Lindemann, A. and Schwenker, K.P. (1972) *Deutsche Gartnerborse*, **72:** 504-506.

Lingan, R., Raja, P., Jegadeeswari, V., Santhi, V.P. and Selvaraj, N. (2014) *Indian Phytopathol.*, **67**(1): 77-81.

Lo ChaurTsuen., Huang ShiouHwa., Lo, C.T. and Huang, S.H. (1997) *Plant Pathol. Bull.*, **6:** 83-87.

Luc, T.T., Park, Y.G. and Jeong, B.R. (2019) *In Vitro Cell. Dev. Biol.*, **55**(5): 1475-2689.

Lukaszewska, A.J. (1995) *Ann. Warsaw Agril. Univ. SGGW, Horticulture*, **17:** 25-31.

Lundqvist, A. (1995) *HereditasLandskrona.*, **122**: 85-89.

Lyakh, V.M. (1986) *Nauchnye Trudy Nauchno Issledovatel'skogo Instituta Gornogo Sadovodstva I Tsvetovodstva*, **33:** 24-29.

Madhuri, G and Barad, A.V. (2018) J. *Pharma Innovation*, 7(5): 559-561

Madhu Meeta, Gupta, V.K. and Meeta, M. (1995) *Plant Disease Res.*, **10:** 170-173.

Madhu Meeta, Shyam, K.R., Pathania, N.S. and Meeta, M. (1996) *Plant Disease Res.*, **11:** 75-78.

Magnifico, V., Talia, M.A.C., Mininni, M. and Cordella, S. (1985) *Colture-Protette*, **14:** 47-54.

Mahmoud, A. (2011) *J. Hortic. Sci. Biotech.*, **86**(2): 135-140.

Maike, B., Jens, M. and Hans, P. (2018) *Postharvest Biol.*, **143:** 129-136.

Malczewska, E., Molas, R., Skrzyczack, C.Z. and Hempel, M. (1979) *Acta Horticulturae*, **91:** 345-351.

Manasa, B.G., Somashekara, Y.M., Shankara, K and Chikkanna, S. (2017) *Int. J. Curr. Microbiol. Appl. Sci.*, **6**(10): 2559-2565.

Manios, V. and Tsikalas, P. (1982) *Georgiki Erevna*, **6:** 93-102.

Manoj, K.Y., Rajesh, K.S., Dinesh, Y., Tripathi, M.K., Bhakti, R. Anil, Si., Rajendra, K. and Garg, G.K. (2012) *Prog. Agric.*, **12**(1): 11-13.

Mantrova, E.Z. (1976) *Agrokhimiya*, **4**: 99-103.

Maria, A.P.L., Jose, M.B.U., Jose, C.L.H. and Jose, M.M. (2007) *HortScience*, **42**(3): 596-599.

Martinez, B.R. and Gonzalez, M.E. (1983) *Agronomia Colombia*, **1**: 21-34.

Martinez, F.X., Fabregas, X., Tapia, L., Caballero, A. and Tattini, M. (1993) *Acta Horticulturae*, **342:** 77-84.

Martinez, G. and Pinzon, L. (1999) *Acta Hortic.*, **482**: 139-144.

Marzieh, A., Alireza, B., Saber, S., Shahriar, H. (2017)*J. Genet. Eng. Biotechnol.*, **15**(2): 309-315.

Masafumi, Y., Shirasawa, K., Hirakawa, H., Isobe, S., Matsuno, J., Uno, Y., Tanase, K., Onozaki, T., Yamaguchi, H. (2020) *Sci. Hortic.*, **262:** 333-337. doi 10.1016/j.scienta.2019.109053.

Masafumi, Y., Kyutaro, K., Shunichi, K., Hideki, H., Akemi, O., Koji, T., Taro, H., Nobuhiro, S., Masayoshi, N., Kazuo, I., Takashi, O., Hiroyasu, Y., Takamasa, S., Taira, M., Yuzo, N., Yoshihiro, O., Noriko, N., Akiko, W., Yoshikazu, T., Shusei, S., Kenta, S., Sachiko, I., Yoshinori, M., Shinobu, N., Yoshie, K., Mitsuyo, K. and Satoshi, T. (2014) *DNARes.*, **21:** 231–241.

Masafumi, Y., Kenta, S., Takamitsu, W., Takashi, K., Sachiko, I., Koji, T. and Hiroyasu, Y. (2017) *Plant Mol. Biol. Rep.*, **35**: 110–117. doi10.1007/s11105-016-1010-2.

Masaru, H., Shinji, K., Yuichi, U. and Masahito, Y. (2020)*Hortic. J.*, **89**(1): 61–68. doi: 10.2503/hortj.UTD-101.

Mastalerz, J.W. (1978) *Greenhouse Environment*, John Willey and Sons, New York. pp. 629.

Matzke, Q. (1965) *Bodenkultur*, **16**: 392-396.

Maxie, E.C., Farnham, D.S., Mitchell, F.G.and Sommer, N.F. (1973) *Flower and Nursery Rep.* (Nov.4).

Mayak, S. and Dilley, D.R. (1976) *J. Amer. Soc. Hort. Sci.*, **101**: 583-585.

Mayak, S., Garibaldi, E.A. and Kofranek, A.M. (1977) *J. Amer. Soc. Hort. Sci.*, **102:** 637-639.

Medina, T LA. (1992) *Acta Horticulturae*, **307***:* 203-212.

Mehlquist, G.A.L. (1940) *Proc. Amer. Soc. Hort. Sci.*, **37**: 1019-1021.

Mehlquist, G.A.L. (1945) *Proc. Amer. Soc. Hort. Sci.*, **46**: 397-406.

Mehlquist, G.A.L. (1957) *Flor.Exch.*, **129:** 15, 50-51.

Mehlquist, G.A.L. (1941) *Proc Amer. Soc. Hort. Sci.*, **38**: 699-704.

Melero, J.M., Herrera, C.J.L., -Ligero, A.M.P., M D Vela-Delgado, M.D., Navas-Becerra, J.A. and Basallote-Ureba, M.J. (2011) *Crop Prot.*, **30**(8): 970-976.

Menemen, Y. and Hamzaoglu, E. (2000) *Ann. Bot. Fenn.*, **37**: 285-287.

Menguc, A., Usta, E. and Sass, P. (1994) *Acta Horticulturae.*, **368:** 802-807.

Mercuri, A., Schiva, T., Baratta, G., Fenoglio, G. and Burchi, G. (1992) *Acta Horticulturae*, **307:** 225-232.

Messenguer, R. and Arus, P. (1996) *J. Heredity*, **87**: 112-118.

Midoh, N., Saijou, Y., Matsumoto, K. and Iwata, M. (1996) *Plant Growth Reg.*, **20:** 195-199.

Migheli, Q., Mocioni, M., Varese, C. and Garibaldi, A. (1993) *DifesadellePiante.*, **16:** 3-10.

Mii, M. and Cheng, S.M. (1982) In: *Plant Tissue Culture* (Ed. A. Fujiwara) (Maruzen: Tokyo),

Miroshnichenko, D.N. and Folgov, S.V. (1999) *Genetika (Moskva)*, **35**: 1208-1213.

Miroshnichenko, D.N. and Folgov, S.V. (2000) *Acta Hortic.*, **508**: 255-258.

Mirzaev, M.M. (1988) *Biotekhnologiya v Sadovodstve I Vinogradarstve.*, pp. 31-34.

Misawa, T., Kubo, C., Kurose, D. and Uematsu, S. (2018) *New Dis. Rep.*, **38**. doi: 10.5197/j.2044-0588.2018.038.012.

Mitrofanob, V.I., Evmenenko, A.F. and Danilenko, V.F. (1982) *Byul. GosudarstvennogoKiktskogo Botanicheskogo*, Sada, **47**: 73-76.

Miura, Y., Sekuima, K., Hikai, T., Igarashi, D., Inoue, T., Uematsu, H., Kuboi, S., Matuyama, A., Yoshida, M. and Suzuki, A. (2000) *J. Japanese Soc. Hort. Sci.*, **69:** 497-504.

Mokra., Mertelik, J and Gotzova, B. (1988) *Acta Horticulturae.*, **234**: 227-233.

Montero, F.P., Aguera, E. and Jimenez, F. (1985) *Agrochimica*, **29:** 356-363.

Montes, S., Ramivez, L., Hernandez, M.M., Santana, N., Martinez, M. and Lara, D. (1997) *CultivosTropicales*, **18:** 75-81.

Mor, Y., Hardenburg, R.E., Kofranek, A.M. and Reid, M.S. (1981) *Hort Science*, **16:** 766-768.

Mor, Y., Reid, M.S and Kofranek, A.M. (1980) *Scientia Hort.*, **13:** 377-383.

Morton, R.J. (1953) *Perpetual Flowering Carnations*, W.H. and L. Collinridge Ltd., London.

Moshari, A. and Danesh, D. (1982) *Iranian J. Plant Pathol.*, **18:** 43-51.

Mosquera, T., Rodriguez, L.E., Bernal, C. and Lopez, F. (1999b) *Acta Horticulturae*, **482**: 309-311.

Mosquera, T., Rodriguez, L.E., Parra, A. and Rodriguez, M. (1999a) *Acta Horticulturae*, **482**: 305-308.

Mostafa, A., Sasan, A., Shirin, D., Mahboobeh, Z.M., Erik, L., and Margrethe, M. (2019) *Plant Cell, Tissue and Organ Cult.*, **139**(3): 1573-5044.

Mostafa, M.M. (1996) *Alexandria J. Agril. Res.*, **41:** 109-122.

Muench, J. and Fritzxche, G. (1974) *Taspo,* **108:** 4-5.

Mukhopadhyay, A. (1984) *Indian Hort.*, **29**: 43-46.

Murashige, T. (1974) *Annual Review of Plant Physiology*, **25:** 135-166.

Mynett, K. (1979) *Symp. Growth regulators in floriculture*, pp. 461-465.

Mynett, K.J. (1982) *XXI Internatl. Hort. Cong.*, 2: Abstract No.1660.

Naglaa, M.E., Abdullah, A. and Mohamed, N.B. (2013) *J. Food* ***Agric.*** *Environ.*, **11**(1): 1113-1117.

Nahid, R., Nazari, F., Javadi, T and Samadi, S. (2020) Advances in Horticultural Science, 34(2): 45-49

Nakano, M., Koike, Y., Watanabe, Y. and Suzuki, S. (1999) *Bull. Fac. Agric., Niigata University*, **51**: 105-114.

Nakasone, H.Y. and Kamemoto, H. (1957) *Bull. Hawaii Agric. Expt. Stat.*, No. 112, p. 15.

Namikawa, O. (1980) *Bull.Kanagawa.Hort.Expt.Stn.*, **27:** 45-51.

Narayan, R., Ahmed, N., Kumar, A. and Mer, M.S. (2015) *Int. J. Appl. Agric. Sci.*, **13**: 23-27.

Nash, C.H.and Baker, R. (1972) *Bull., Colorado Flower Growers ' Assn.***271:** 1-3.

Naveen Kumar, P., Singh, B. and Voleti, S.R. (1999b) *J. Orn. Hort.*, **2**: 139.

Naveen Kumar, P., Singh, B., Sindhu, S.S. and Voleti, S.R. (1991) *J. Orn. Hort.*, **2**: 137.

Negi, H.S., and Raj, H. (2013) *Indian Phytopathol.*, **66**(3): 224-227.

Neumannova, B. and Paulin, A. (1974) *Annales de Technologie Agricole.* **23:** 467-480.

Nichols, R. (1979) *Acta Horticulturae,* **91:** 93-98.

Nimura, M., Yamaguchi, N. and Itoh, K. (1996) *Res.Bul. Aichi ken Agril. Res.Center,* **28:** 227-232.

Nirala, S.K., Suresh, R. and Kumar, S. (2018) *Int. J. Curr. Microbiol. Appl. Sci.*, **7**(8): 3746-3755.

Nityasiri, S., Ch., Phanchana. M. and Manot, R. (1985) *Proc.23rd Natl.Conf.*, pp. 100-106.

Nowak, J. and Rudnicki, R.M. (1975) *PraceInstyt.Sadownic.IKwiaciarstwa w Skiernie.*, *B***1:** 173-179.

Nowak, J., Goszczynska, D. and Rudnicki, R. (1983) *Ogrodnictwo.v.* 20(8) pp. 18-19.

Obieglo, U. (1992) *Bul. OILB SROP.* **15:** 1, 127-129.

Odom, R.E. (1953) *Bull. Colo.Flower Grs.' Assn.*, **46:** 1-3.

Oertel, C. (1977) *Archiv fur Gartenbau*. **25:** 11-25.

Oetting, R.D., Morishita, F.S., Jefferson, R.N., Humphrey, W.A.andBesemer, S.T. (1978) *Flower and Nursery Rep.* Spring, 4-5.

Omid, K., Ali., D. Gona., K. and Kordestani (2008) *Plant Cell Tissue Organ Cult.*, **92:** 273–280.

Onozaki, T., Ikeda, H. and Yamaguchi, T. (1998) *J. Japanese Soc. Hort. Sci.*, **67**: 198-203.

Onozaki, T., Ikeda, H. and Yamaguchi, T. (2001) *Sci. Hortic.*, **87**: 107-120.

Onozaki, T., Ikeda, H., Yamaguchi, T. and Himeno, M. (1998) *Acta Horticulture*, **454:** 127-132.

Oradee, S. and Chaichoompon, S. (1985) *Proc. 23Natl.Conf.: Poster session*, volume 1. Bangkok (Thailand), pp. 205-214.

Orozco de Amezquita, M., Garces de Granada, E. and Arbelaez, G. (1993) *Agronomia Colombiana*. **10:** 90-102.

Ortiz Borrero, M. and Arbelaez Torrez, G. (1992) *ASCOLFI Informa*. **18:** 5-6.

Overtli, J.J. (1960) *Gartenbauwiss.*, **25**: 287-292.

Ozawa, H. and Kikugawa, I. (1953) *Bull. Kagawa Hort. Expt. Stat*, **1:** 35-39.

Pal, S. and Sarkar, I. (2010) *J. Ornam. Hortic.*, **13**(2): 16-20.

Pal, S., Sarkar, I., Chatterjee, H. and Senapati, S.K. (2014) *J.Ornam.Hortic.*, **17**(3-4): 34-37.

Pandolfo, F.M. (1984) *InformatoreFitopatologico* (Italy), **34**: 45-46.

Paolo, C., Marcello, D., Paola, D., Virginia, L., Cooman, L. D. (2003)*Phytochem. Anal.*, 14: 8–12.

Pathania, N.S., Sehgal, O.P. and Gupta, Y.C. (2000) *J. Ornam. Hort.*, **3**: 114-117.

Paulin, A., Bureau, J.M. and Droillard, M.J. (1985) *C.R des Seances del'Academie des Sciences, III Sciences de la Vie*. **300**: 7, 301-304.

Paulin, A., Pages, A. and Elavummoottil, O. (1982) *C. R. de Seances de l'Academie des Sciences, III Sciences de la Vie*. **295**: 13, 803-806.

Pawliczuk, Z. and Orlikowski, L.B. (1994) *ZeszytyNaukowePraceInstyt.Sadownic. IKwiaciarstwa w Skiernie.*, B **1**: 105-109.

Penaranda, Bolton, A.T. (1984) *Canadian J., Plant Pathol*, **6:** 321-324.

Penaranda, L.A., Torres, H.J. and Arbelaez, G. (1992) *Acta Horticulturae*, **307**: 247-250.

Pennazio, S. (1975) *J. Hort. Sci.*, **50:** 161-164.

Pergola, G., Ranieri, M.De and Grassotti, A. (1983) *ColtureProlette*, **12**: 37-42.

Pergola. G., Guda. C. dalla. and Garibaldi, cA. (1979)*Jornadastecnicas*, pp. 180-183.

Peterson, R. (1960) *Bull. Colo.Flower Grs.' Ass.*, **134**: 58.

Petridou, M., Voyiatzi, C. and Voyiatzis, D. (1999) *Adv. Hort. Sci.*, **13**: 161-164.

Phillips, D.J. and Matthews, G.J. (1964) *Bot. Gaz.*, **125:** 7-12.

Pirone, P.P. (1970) *Diseases and Pests of Ornamental Plants*, 4th ed. Ronald Press, New Yord.

Pobudkiewicz, A.K. and Goldsberry, K.L. (1989a) *Res. Bull. Colo. Greenhouse Grs' Ass.*, **468**.

Pobudkiewicz, A.K.and Goldsberry, K.L. (1989b) *Res. Bull. Colorado Greenhouse Grs' Ass.* No. 469, pp. 1-2.

Podd, L.A. and Staden, J. Van (1999) *Plant Growth Reg.*, **28:** 175-178.

Poe, S.L. and Wilfret, G.J. (1973) *Proc. 85th Ann. Meet. Fla. St. Hort. Soc.*, pp. 384-387.

Polyanista, G.I. (1971) *Bull. Gosuda. Nikits. Botanich. Sada*, **1:** 37-40.

Polyanista, G.I. (1972) *Trudy Gosuda. Nikits. Botanich. Sada*, **59**: 91-96.

Poncet, C and Devergne, J.C. (1988) *Acta Horticulturae*, **234:** 209-220.

Ponzi, R. (1982) *AnnalidellaFacolta di ScienzeAgrariedell'UniversitadegliStudi di Napoli Portici*, **16**: 59-64

Pop, I.V., Sandu, C. and Constantinovici, D. (1994) *Acta Horticulturae*, **377:** 335-340.

Poucher, W.A. (1950) *Perfumes, Cosmetics and Soaps*, Chapman and Hall Ltd., London.

Powell, M.P. and Bunt, A.C. (1983) *Scientia Hort.*, **20**: 193-202.

Pratibha, C., Dhiman, S.R., Bharti, K., Gupta, Y.C., Gupta, R.K. and Dogra, R.K. (2017) *J.Ornam. Hortic.*, **20**(3-4): 120-125.

Pratibha. C., Dhiman, S.R., Gupta, Y.C., Bharti, K., Gupta, R K. and Dogr, R.K (2017) *Journal of Ornamental Horticulture*, **20**(1-2): 25-27.

Puustjarvi, V. (1981) *Peat and Plant Yearbook*, Helsinki, Finland, pp. 3-12.

Puustjarvi, V. (1983) *Peat and Plant Yearbook*, Helsinki, Finland, pp. 38-43.

Qijian, W., Naizhen, D., Xiaoni, Z., Shengnan, L., Manzhu, B. and Xiaopeng, F. (2020) *Plants.*, **9**(1): 75-80.

Quak, F. (1977) In: *Plant Cell Tissue and Organ Culture* (Eds. J. Reinhart and Y.P.S. Bajaj), Springer-verlag, Berlin, pp. 598-615.

Quintero, G.H. (1992) *Acta Horticulturae*, **307:** 241-245.

Raabe, R.D. and Hurlimann, J.H. (1972) *California Agric.*, **26:** 4 -5.

Raabe, R.D., Hurlimann, J.H. and Byrne, T.G. (1977) *Flower and Nursery Report*. March-April, 3.

Rajib and Jagatpati (2011) *Res. Plant Biol.*, **1**(4): 23-32.

Rattink, H. (1983) *Mededelingen van de FaculteitLandbouwwetenschappen, Rijksuniversiteit Gent.*, **48**: 3, 699-703.

Rattink, H. (1992) *Acta Horticulture*, **307:** 37-42.

Read, P.E. and Hoysler, V. (1971) *HortScience*, **6:** 350-351.

Reid, M.S., Farnham, D.S. and McEnroe, E.P. (1980) *HortScience*, **1**: 807-808.

Reid, M.S., Paul, J.L., Farhoomand, M.B., Kofranek, A.M. and Staby, G.L. (1980) *J. Amer. Soc. Hort. Sci.*, **105**: 25-27.

Riudavets, J., Castane, C., Gabarra, R., Parker, B., Skinner, M. and Lewis, T. (1995) *Thrips Biology and management: Proc. Internatl.Conf.* Thysanoptera., 1993, pp. 255-258.

Riudavets, J., Gabarra, R. and Castane, C. (1992) *Bul.OILB SROP*. **16:** 137-140.

Robinson, K. E. P. and Firoozabady, E. (1993) *Scientia Hort.*, **55**: 83-99

Rodoni, B.C., Hepworth, G., Richardson, C.and Moran, J.R. (1994) *Australian J. Agric. Res.*, **45**: 223-230.

Roelants, A. (1973a) *Revue de L' Agriculture*, **26:** 1329-1339.

Roelants, A. (1973b) *Revue de L' Agriculture*, **26**: 1011-1026.

Ronen, M. and Mayak, S. (1981) *J. Expt. Bot.*, **32**: 759-765.

Rumine, P. and Bonifacio, A. (1983) *InformatoreFitopatologico* (Italy), **33:** 39-44.

Rumine, P. and Parrini, C. (1982) *Annalidell'IstitutoSperimentaleper la Floricoltura, Sanremo* (Italy)., **13**: . 79-91.

Rupali, S. and Ranjan, S. (2008) *J.Ornam.Hortic.*, **11**(4): 251-256.

Rymon, D., Nakar, R., Wegner, J. and Umiel, N. (1987) *Acta Horticulture*, **203:** 221-225.

Saenz, G.S., Koike, S.T., Tjosvold, S.A. and Greene, I.D. (1995) *Plant Disease.*, **79:** 320.

Sagawa, Y. and Mehlquist, G.A.L. (1957) *Amer. J. Bot.*, **44:** 397-403.

Saini, S.S. and Arora, J.S. (1975) *J. Res. PAU*, **12**: 78-81.

Sakai, K and Asano, M. (1990) *Res. Bull. AichikenAgrl. Res.Center*, **22**: 191-198.

Sakai, K., Takamatsu, M. and Yamaguchi, N. (1993) *Res.Bul.Aichi ken Agrl.Res. Center*, **25:** 243-249.

Saks, Y. and Staden, J. van (1992) *J. Plant Physiol.*, **139:** 484 - 488.

Saks, Y., and Staden J. van (1993) *Plant Growth Reg.*, **12:** 99-104.

Samata, Y., Tsuyuki, Y., Inazu, K. and Iizuka, K. (1979) *Bul.Fac.Agric., Tamagawa Univ.*, **19:** 29-41.

Sankhla, D., Davis, T.D., Sankhla, N. and Upadhyay, A. (1995) *Acta Horticulturae*, **280**: 173-176.

Sankhla, D., Davis, T.D., sankhla, N. and Upadhyaya, A. (1994) *Plant Cell Rep.*, **13**: 514-518.

Sapna, K. and Poonam Kumari(2020)*J. Pharmacogn. Phytochem.*, **9**(2): 1056-1061.

Sarafi, A. and Mahdavian. M. (1978) *VII Internatl. Conf. Controlled Atmosphere Res. on Postharvest Hort.*, Series No. 15, 114-122.

Sato, S., Hagimori, M. and Iwai, S. (1993) *Plant Cell Rep.*, **12:** 370-374.

Satoh, S., Mikami, M., Kiryu, S., Yoshioka, T. and Midoh, S. (1999) In: *Biology and biotechnology of the plant hormone ethylene II, Proceedings of the EU-TMR-Euroconferene Symposium, Greece, 5-8 September, 1998*, pp. 441-442.

Savin, K. W., Baudinette, S. C., Graham, M. W., Michael, M. Z., Nugent, G.D., Chandler, S. F., Cornish, E. C. and Lu, C.Y. (1995) *HortScience*, **30**: 970-972.

Savita, S, Harender, R. and Neeraj, S. (2019a) *Int J Curr Microbiol Appl Sci.*, **8**(1): 2484-2499.

Savita, S. and Harender, R. (2019b) *Int. J. Chem. Stud.*, **7**(1): 1416-1418.

Sawwan, J. (1998) *Dirasat Agri. Sciences*, **25:** 375-378.

Scovel, G., Altshuler, T., Liu, Z. and Vainstein, A. (2000) *J. Hered.*, **91**: 487-491.

Scovel, G., Ben-Meir, H., Ovadis, M., Itzhaki, H. and Vainstein, A. (1998) *Theor. Appl. Genet.*, **96**: 117-122.

Semina, S.N and Shestachenko, G.N. (1981) *Mikologiya I Fitopatologiya*. **15:** 238-240.

Sezgin, E. and Onan, E. (1987) *Doga, Turk TarimveOrmancilikDergisi.*, **11:** 435-447.

Shafi Bhat, M., Jhon, A.Q. and Abdullah, M. (1991) *Progressive Hort.*, **23**: 60 - 67.

Sharma, B.P., Gautam, A., Gupta, Y.C., Dhiman, S.R. and Bhalla, R. (2010) *J. Ornam. Hortic.*, **13**(2): 23-26.

Sharma, S. and Chandel, S. (2016) *J. Appl. Nat. Sci.*, **8**(3): 1680-1684.

Sheard, G.E. and Bunt, A.C. (1963) *Ann. Rep. Glasshour Crops Res. Institute*, pp. 133-171.

Shelton, M., Carpenter, A., Van Epenhuijsen, C.W., Thompson, J.F. and Mitcham, E.J. (1997) *Postharvest Horticulture Series*, Department of Pomology, University of California, No. 15, 127-131.

Shigeru, S. (2011) *J. Jpn. Soc. Hortic. Sci.*, **80**(2): 127-135.

Shigeru, S., Hideki, N., Takayuki, I. (2020) *J. Appl. Hortic.*, **7**(1): 8-10.

Shirasaki, T. (1990) *Agriculture and Horticulture*, **65:** 626-630.

Simard, M.H., Michaux-Ferrier, N. and Silvy, A. (1992) *Plant Cell, Tissue and Organ Cult.*, **29:** 37-42.

Singh, K. and Singh, K. (1994) *Phyton Horn.*, **34:** 2, 309-313.

Singh, R. and Singh, A. (1989) *Current Sci.*, **.58:** 931.

Sisler, E.C. and Serek, M. (2001) *Acta Horticulturae*, 543: 33-37.

Sisler, E.C., Blankenship, S.M., Fearn, J.C. Haynes, R. (1993) In: *Current Plant Science and Biotechnology in Agriculture*, (Eds. Pech, J.C., Latche, A. and Balague, C.), **16**: 182-187.

Skalska, E. (1977) *Sbornik UVTIZ, Zahradnictvi.*, **4**: 259-267.

Skipina, K.P. (1983) *Nauchnye Trudy NII GornogoSadovodstva I Tsvetovodstva*, **30:** 127-133.

Smith, W.H. and Parkar, J.C. (1966) *Nature*, **211:** 100-101.

Snyder, R.G. and Rae, H.L. (1973) *J. Amer. Soc. Hort. Sci.*, **98:** 568-572.

Son, K.C. and Byoun, H.J. (1995) *J. Korean Soc. Hort. Sci.*, **36:** 420-425.

Son, K.C. and Chae, Y. (1993) *J. Korean Soc. Hort. Sci.*, **34:** 75-80.

Son, K.C., Lee, I.K. and KO, J.Y. (1994) *J. Korean Soc. Hort. Sci.*, **35:** 57-65.

Sood, A.K.andKakar, K.L. (1990) *Indian J.Agric. Sci.*, **60:** 367-368

Soto, S.J.C., Clavijo, O.M.J., and Filgueira, D.J.J. (2012) *Agronomia Colombiana.*, **30**(2): 172-178.

Sousa, M.A., De, L.B. and Simao, S. (1981) *Anais da Escola superior de agricultura Luiz de Qweiroz*, **38**: 623-641.

Sparnaaij, L.D. (1979) *HortScience*, **14**: 496-499.

Sparnaaij, L.D. and Beeger, G.W. (1973) *Euphytica*, **22:** 274-278.

Sparnaaij, L.D. and Demmink, J.F. (1970) *Euphytica*, **19**: 310-316.

Sparnaaij, L.D. and Demmink, J.F. (1985) *Vakbladvoor de Bloemisterij*, 40: 112-113.

Sparnaaij, L.D., Demmink, J.F. and Koehorst van Putten, H.J.J. (1990a) *Euphytica*, **50:** 35-42.

Sparnaaij, L.D., Demmink, J.F. and Koehorst van Putten, H.J.J. (1987) *Acta Horticulturae*, **216:** 303-311.

Sparnaaij, L.D., Koehorst van Putten, H.J.J. and Bos, I. (1996) *Euphytica*, **90:** 183-194.

Sparnaaij, L.D., Koehorst, H.J.J., Segers, T.A. and Jong, J.de (1990c) *Proc. EUCARPIASymp. Integration of in vitro techniques in ornamental plant breeding*, 10-14 November 1990, pp. 68-73.

Sparnaaij, S.D. and Demmink, J.F. (1976) *Euphytica*, **25:** 329-338.

Spinsky, P., Beck, G.E. and Mc.Cown, B.H. (1974) *HortScience*, **9:** 270.

Staby, G.L., Basel, R.M., Reid, M.S. and Dodge, L.L. (1993) *HortTechnol.*, **3:** 199-202.

Staby, G.L., Robertson, J.L., Kipliger, D.C. and Conover, C.A. (1978) *Chain of Life*, Ohio Florists' Assoc., Ohio State Univ., Columbus.

Stack, R.W., Horst, R.K and Kawamoto, S.O. (1977) *Proc. Amer. Phytopathological Soc.*, **4:** 125.

Stack, R.W., Horst, R.K., Nelson, P.E. and Langhans, R.W. (1978a) *Phytopathol.*, **68:** 423-428.

Stack, R.W., Horst, R.K., Nelson, P.E. and Langhans, R.W. (1978b) *Phytopathol.*, **68:** 429-433.

Staden, J.van., Armitage, A.E. and Van Staden, J. (1994) *J. Southern African Soc. Hort. Sci.*, **4**: 19-20.

Stoe, O.M. (1963) *Ann. Appl. Biol.*, **52:** 199-209.

Streu, H.T., Jenkins, W.R. and Hutchinson, M.T. (1961) *N. J. Agric. Expt. Sta.*, No. 800, pp. 32.

Strider, D.L. (1978) *Plant Disease Rep.*, **62:** 24-28.

Strojny, Z., Kowalczyk, W. and Marynowska, M. (1992) *PraceInstytutuSadownictwa I Kwiac. w Skiernie.*, **17:** 15-21.

Suh JeungKeun., Lee SunHwa., Suh, J.K. and Lee, S.H. (1997)*J. Korean Soc. Hort. Sci.*, **38:** 292-296.

Suman, K., Sumita, K. and Shanker, L.K. (2013) *Propag. Ornam. Plants.*, **13**(1): 3-11.

Sunita, C. (2015) *Int. J. Plant Prot.*, **8**(1): 130-133.

Sunita S. Chandel and Nandita Pathanie (2004) *Ecol. Environ. Conserv.*, **10**(4): 475-478.

Sutter, E. (1988) *J. Amer. Soc. Hort. Sci.*, **113:** 234-238.

Sweet, D.V. (1952) *Quart. Bull. Mich. Agric. Expt. Stat.*, No. 342: p.24.

Sytsema, W. (1971) *Vakbladvoor de Bloemisterij*. **26:** 9.

Takanori H., Kasumi, H., Mio, K., Yuuki, K., Yuuka. and Maho. (2020) *Environ. Control Biol.*, **58**(1): 15-20. doi: 10.2525/ecb.58.15.

Takashi, O. (2008) *Jpn.Agric. Res. Q.*, **42**(3): 137-144.

Takashi, O., Natsu, T., Masafumi, Y., Hiroshi, I., Katsuhiko, S., and Michio, S. (2006) *J. Jpn. Soc. Hortic. Sci.*, **75**(3): 256-263.

Takeda, Y. (1974) *Special Rept. Shiga Pref. Agric. Exp. Sta.*, **11:** 1-124.

Tannoury, M., Vintejoux, C. (1997) *Acta Botanica Gallica*, **144**: 107-118.

Tannoury, M., Vintejoux, C. and Dereuddre, J. (1995) *Acta Botanica Gallica*, **142**: 415-424.

Tatarynowicz, B. (1981) *RocznikiAkademiiRolniczej w PoznaniuOgrodnictwo* (Poland), **129:** 261-274.

Thakulla, D., Khanal, A. and Bhatta, L.R. (2018)www.symbiosisonlinepublishing.com.

Thu, H.T.M., Naing, A.H., Jeong, H.Y. and Kim, C.K. (2020)*Plants*, **9**: 950., doi: 10.3390/plants9080950.

Tinga, J.H. (1956) *Ph.D. thesis*, Cornell Univ., pp. 189.

Tjosvold, S.A. and Ali, A.D. (1989) *Flower and Nursery Rep.*, summer, 3-4.

Tjosvold, S.A., Ali, A.D. Parker, B.L., Skinner, M. and Lewis, T. (1995) *Thrips Biology and management: proc. Internatl.Conf.* Thysanoptera., 1993, pp. 351-353.

Tkachuk, V.K. (1978) *Khimiya v Sel'skomKhozyaistve*, **16:** 1, 43-45.

Tomoyuki, K., Reiko, E.A., Yamashita, and Shigeru. (2006) *Plant Cell Tissue Organ Cult.*, **86**: 27–35, doi10.1007/s11240-006-9093-3.

Toth, E., Gyertyanos, K. and Gyorgy, K. (1984) *Kertgazdasag*, **16:** 81-84.

Tripathi, K.C. and Lal, S.D. (1992) *Progressive Hort.*, **24:** 115-117.

Ubikata, M. (1999) *Bulletin of HokkardoPrefec. Agri. Expt. Stat.*, No.77, pp. 39-43.

Uda, A., Koyama, Y. and Fukushima, K. (1995)*J. Japanese Soc. Hort. Sci.*, **64:** 185-191.

Uda, A., Yamanaka, M., Fukushima, K. and Koyama, Y. (1996) *J. Japanese Soc. Hort. Sci.*, **64:** 927-933.

Uitermark, C.H.T. (1984) *Vakbladvoor de Bloemisterij*, **39**: 51.

Umiel, N., Dehan, K. and Kagan, S. (1987) *Acta Horticulture*, **216:** 355-358.

Uota, M. and Garazsi, (1967) *USDA Marketing Res. Rep.*, **796**: 9.

Uri, Y., Kafkafi, U. and Kalo, H. (1990) *Hassadeh.*, **70**: 5, 742-746.

Vainstein, A., Ben, M.H., Zuker, A., Watad, A.A., scovel, G., Ahroni, A., Ovadis, M. and Vainstein, A. (1995) *Acta Horticulturae*, **420**: 65-67.

Valunj, A.R., Pawar, S.A., Kadam, U.K. and Khoire, V.M. (1999) *Pest Management in Hort Eco System.*, **5**: 21-23.

Van Altvorst, A.C., Koehorst, H.J.J., Dejong, J. and Dons, J.J.M. (1996) *Plant Cell, Tissue and Organ Cult.*, **45:** 169-173.

Van Os, H. (1961) *Tijdschr. Plziekt.*, **67:** 59.

Veen, H. (1979) *Acta Horticulture*, **91:** 99-103.

Veen, H. and Geijn, S.C. van de (1978) *Planta*, **140:** 93-96.

Vidalie, H. (1982) *Revue Horticole*, **228:** 33-38.

Vigodsky-Hass, H., Hada, E. Sharoni, M., Kirschner, B., and Sagi, Y. (1982) *Hassadeh*, **62**: 1442-1445.

Vinod, K.S., Rajeshkumar, P., Nakkeeran, S. and Eraivan, A. K. (2015) *Afr. J. Microbiol. Res.*, **9**(49): 2328-2336.

Voogt, W., Adams, P., Hidding, A.P., Kpp, J.A., Sonneveld, C. and Kreij, C.de (1995) *Acta Horticulture*, **401:** 327-336.

Vosamae, H. and Kask, K. (1972) *Biologia*, **21**: 178-180.

Wahome, P.K., Jesch, H.H. and Grittner, I. (2001) *Scientia Hort.*, **87**: 207-216.

Walunj, A.R., Pawar, S.A. and Kadam, U.K. (1999) *Tests of Agrochemicals and Cultivars*, **20:** 10-11.

Wang Kuei Mei., Lo ChaurTseun., Tu ChisnChyn., Wang.K.M., Lo.C.T.andTu.C.C. (1994) *J. Agric. Res. China.*, **43:** 195-207.

Wang, B., Lu, M.K., Yu, S.Q. and Zhang, R.P. (1990) *Chinese J. Virol.*, **6**: 341-346.

Wang, M., Jie, Xiao., Wei, H. and Jeong, B.R. (2020) *Agron.*, **10:** 1217. doi: 10.3390/agronomy10081217.

Wawrzynczak, A. and Goszczynska, D.M. (1999) *J. Fruit and Ornamental Plant Res.*, **7**: 105-117.

Weber, Z., Werner, M. and FruzynskaJozwiak, D. (1998) *Phytopathologia Polonica.*, **16:** 37-43;

Wen, G.B., Yu, J.P., Zhao, X.M., Fan, G.C., Lin, D.W., Yang, Z. and Zhu DeWei. (1995) *Acta Horticulture*, **404:** 82-90.

Westhuizen, A.J.van der and Swardt, G.H.de (1978) *Zeitschrift fur Pflanzenphysiol.*, **86:** 125-134.

Wickel, H. and Borner, R. (1979) *Acta Horticulturae*, **91:** 111-114.

Winsor, G.W. (1962) *Grower*, **58**: 528-529.

Winsor, G.W. (1968) *Scientia Hort.*, **20**: 26-40.

Woltering, E.J. and Harkema, H. (1987) *Acta Horticulture*, **216:** 255-262.

Woltering, E.J., Somhorst, D., Beer, C.A.de and De Beer, C.A. (1993) *J. Plant Physiol.*, **141:** 329-335.

Wu, M.J., Doorn, W.van., Mayak, S., Reid, M.S. and Van Doorn, W. (1989) *Acta Horticulture*, **261:** 221-225.

Wu, M.J., Zacarias, L., Saltveit, M.E. and Reid, M.S. (1992) *HortScience*, **27:** 136-138.

Wu, M.J., Zacarias, L. and Reid, M.S. (1991) *Scientia Hort.*, **48:** 109-116.

Wurr, D.C.E., Fellows, J.R. and Andrews, L. (2000) *Sci. Hortic.*, **86**: 57-70.

Yagi, M., Tanikawa, N. Shibata, M. and Onozaki, T. (2012) *Acta Hortic.*, **937**: 7 doi: 10.17660/ActaHortic.2012.937.52.

Yamaguchi, T., Ito, A. and Koshioka, M. (1996a) *Japan Agrl. Res. Quarterly*, **30**: 189-196.

Yamaguchi, T., Ito, A. and Koshioka, M. (1996b) *Japan Agrl. Res. Quarterly*, **30:** 181-188.

Yamamoto, K., Yokoo, Y., Makimoto, Y., Arima, Y., Oshima, K. and Halevy, A.H. (1995) *Acta Horticulturae*, **394**: 289-295.

Yantcheva, A., Vlahova, M. and Antanassov, A. (1998) *Plant Cell Rep.*, **18**: 148-153.

Yephet, Y.B., Reuven, M., Szmulewich, Y. and Mor, Y. (1994) *Crop Protection*, **13:** 357-361.

Zhola, I. and Ushkaura, A. (1983) *Trudy LatviiskoiSelskokhozyaistvennoiAcademii*, **207:** 36-38.

Zia, M., Yaqoob, K., Mannan, A., Nisa, S., Raza, G., and Rehman, R. (2020) *Vegetos: Intl. J. Plant Res. Biotechnol.*, **33**(1): 222-229.

Zoina, A. and Aloj, B. (1979) *AnnalidellaFacolta di ScienzeAgrarie, dell'UniversitadegliStudi di Napoli Portici*, **13:** 101-108.

Zuker, A., Chang, P.F.L., Ahroni, A., Cheah, K. and Woodson, W.R. (1995) *Scientia Hort.*, **64:** 177-185.

Zuker, A., Tzfira, T., Scovel, G., Ovadis, M., Shklarman, E., Itzhaki, H. and Vainstein, A. (2001) *J. Amer. Soc. Hort. Sci.*, **126**: 13-18.

CRYSANTHEMUM

Rajiv Kumar, M.L Chaithra, M.A. Kher,
A. Mahapatra, S.K. Palai and P. Das

1.0 Introduction

Chrysanthemum (*chrysos*, golden; *anthos*, flower) is a popular flower crop of commercial importance. It stands third in the world cut flower trade and first in Japan and China, 3rd position in Germany, 2nd position in United Kingdom (Datta and Janakiram, 2015). In India, chrysanthemum is being grown in 20.55 thousand heactare with 188.81 thousand metric ton of loose flower and 15.38 lakh cut flowers (Anon., 2016-17). In the Netherlands, during 2019, chrysanthemum was imported with a value of 2754 thousand euros, while, export value of 365972 thousand euros (Anon., 2020). As a result of control over the growing conditions in the polyhouse environment, fertilization, use of growth regulating chemicals, chrysanthemum can now be produced any time of the year.

2.0 Importance and Uses

It is known as the 'Queen of the East', chrysanthemum is admired by enthusiasts all over the world. In several countries the membership of the National Chrysanthemum Society runs into thousands of active members. This society classifies cultivars, conducts shows and publishes journals regularly. The fact that chrysanthemum shows draw biggest crowds everywhere is an ample proof of the popularity of this flower. In India too, chrysanthemum occupies a place of pride both as a commercial crop and as a popular exhibition flower. The wide variations exhibited by the large number of cultivars in respect of growth habit, size, colour and shape of blooms make the chrysanthemum suitable for every purpose conceivable of a flower. The erect and tall growing cultivars are suitable for background planting in borders or for use as cut flowers. The cultivars with the dwarf and compact growing habit, on the other hand, are suitable for front row plantation or pot culture. The decorative and fluffy bloomed small-flowered cultivars are ideal for garland making and hair decoration. The extra large-bloomed cultivars are prized for their exhibition value. The simultaneous blooming habit and the ease with which the plants can be trained into any desirable shape offer a great scope for a skilled gardener to display his or her skill. The utility and popularity of chrysanthemum have increased immensely with the introduction of the techniques for year-round flowering based on scientific research in the field of photoperiodism and genetics.

Chrysanthemum morifolium is also an important source of essential oil and sesquiterpenoid alcohol (Hu and Chen, 1997 and Yang *et al.*, 1997). Certain species like *Chrysanthemum cinerariifolium* and *C. coccineum* are also cultivated as sources of pyrethrum, an important insecticide (Chittenden, 1956; Carter, 1980). Apart fromthis, seeds of *Dendranthema tanacetum* (*Chrysanthemum corymbosum*) is an important source of acetylenic acid (Tsevegsuren *et al.*, 1998). Ryori Giku is a yellow flowering culinary type which is eaten as delicacy in Japan after frying. Chrysanthemum has also figured prominently in philately in China and Japan.

Flower extracts of *C. fuscatum* and *C. grandiflorum* has insecticidal activity on confused flour beetle (*Tribolium confusum*) as an alternative pest control method for minimizing the noxious effect of pesticides on the environment (Haouas *et al.*,

2008). Essential oil from *C. indicum* is an effective inhibitor of oral bacteria *e.g.*, a component of toothpaste and/or gargling solution (Jung, 2009), and its flowers are used in traditional Chinese medicine with strong aroma. It flowers have abundant flavonoids such as Quercitrin, myricetin and luteolin7-glucoside, which is significant for the development of potential pharmaceuticals (Liang *et al.*, 2010). *Chrysanthemum morifolium*is used in a tea infusion and as a popular beverage due to its health benefits, such as antioxidant effect (Zheng *et al.*, 2015; Han *et al.*, 2019).

Chrysanthemum is a medicinal and edible plant. The ancient book "Shen Nong's Herbal Classic" described chrysanthemum as body refreshing and life prolonging. Chinese Pharmacopoeia (2015) describes its medicinal properties as heat-clearing and detoxifying. As an herbal medicine, it has antibacterial, anti-inflammatory, anti-HIV and antitumor properties. The functional efficacies attribute to the constituents, caffeoylquinic acids, flavonoids and triterpenes (Akihisa *et al.*, 1996; Ukiya *et al.*, 2002; Lee *et al.*, 2003; Chen *et al.*, 2012).

The flowering head of *C. morifolium* has been documented for "scattering cold", "cleaning heat and toxin" and "brightening eyes" in the traditional Chinese medicine book of Compendium of Materia Medica (Zhang, 2019). More recently, it has been found to exhibit various potential health benefits including antioxidant activities (Wang *et al.*, 2017), cardiovascular protective effects (Jiang *et al.*, 2004), anticarcinogenesis activities (Miyazawa and Hisama, 2003), anti-inflammatory effects (Su *et al.*, 2015), and anti-HBV activities (Gu *et al.*, 2013), due to significant amount of bioactive compounds including flavonoids (Su *et al.*, 2015), phenolic acids (Kim and Lee, 2005), triterpenes, and polysaccharides (Zheng *et al.*, 2006).

3.0 Origin and History

Chrysanthemum, believed to be the native to the northern hemisphere chiefly Europe and Asia, was claimed to have been originated in China (Carter, 1980). Wang *et al.* (1999) opined that *Dendranthema vestitum* was one of the ancestors of *D* x *grandiflora* (*D. grandiflorum* = *D. morifolium*). The numerous species involved in the development were mentioned by various workers as *C. sinense*, *C. indicum*, *C. japonicum*, *C. ornatum*, *C. satsumense* and *C. boreale* (Ramatuelle, 1792; Sabine, 1822; Hemsly,1889; Makino, 1897; Henry, 1902; Nakai, 1928; Niwa, 1936). Nevertheless, it is now generally accepted that modern, large, double and exquisitely flowered cultivars owe their origin to relatively small, single and not so attractive types. This great transformation is the result of natural cross pollination and selection by keen growers and also hybridization and natural and induced mutations. The important mechanism in the cytogenetical evolution of garden chrysanthemum has been enumerated by Nazeer and Khoshoo (1982) as outbreeding, spontaneous and intentional hybridization coupled with mutation, chromosomal differentiation and repatterning and polyploidy.

The history of origin, distribution and evolution from wild daisy-like flowers to present day magnificent cultivars is fascinating.

3.1.0 In China

The earliest mention of chrysanthemum in recorded history is found in the writings of Confucius, the great Chinese philosopher, who described its yellow glory as early as in 550 BC. The early remoulding of the wild ancestors into much superior diverse forms is also credited to the Chinese. The first white blooming chrysanthemum was raised by Tao Hungching during 452-532 AD (Cumming, 1964).

3.2.0 In Japan

Cumming (1964) reported that seeds of chrysanthemum reached Japan from China via Korea as early as 386 AD. Niwa (1936), however, believed that the cultivated chrysanthemum was brought from China to Japan after some development in the era of Tempyo of the Nara period (724-749 AD).

Chrysanthemum was proclaimed as the National Flower by emperor Uda in the year 910 AD and a sixteen petalled single large chrysanthemum called Ichimonji was adopted as the imperial crest during 12th century. However, at present cherry blossom is the National Flower and chrysanthemum the symbol of Royalty in Japan. After reaching Japan, chrysanthemum underwent tremendous transformation and Japan had the largest number of cultivars, representing widest range of variability (Nakajima and Young, 1965). Chrysanthemum reached its most refined stage of cultivation, a heritage of which any country would be justifiably proud of. In China, the emphasis being on regularity, the course of development in Japan was much wider and types varying from perfectly regular to most irregular were evolved and appreciated.

3.3.0 In Europe

The first attempt to grow chrysanthemum in Europe was made in Holland in 1688 according to Niwa (1936), who stated that a Japanese cultivar was introduced there and perished. In 1689, Jacob Breynius, a Dutch botanist described 6 Chinese cultivars grown in Holland (Cumming, 1964). In 1690, H. van Rheede described a Holland-grown type from India called Gool-doodi. The first chrysanthemum received in England died, and both England and France imported Old Purple cultivar and cultivated at Kew Gardens and Marseilles. The fate of chrysanthemum in these three European countries was strikingly different. In Holland, it never became popular. In France, though its value was appreciated, it primarily remained as cemetery flower with a preference for white and purple. In England, the emphasis was on exhibitions, the first being held in 1843 at Norwich.

Its popularity is steadily increasing and the National Register of Chrysanthemum issued by the National Chrysanthemum Society of England lists more than 5000 cultivars. Skilled growers present exquisite forms grown with utmost care to horticultural perfection in the shows held in England. The cut-flower as well as pot-mum trade has been organized on scientific lines.

3.4.0 In USA

The first chrysanthemum known to the USA was 'Dark Purple' imported by a scientist John Stevens in 1798 and it was followed by many more cultivars from China. Unlike any other country, the chrysanthemum growing became commercialized in America in the shortest period. Although shows were held at many places, the main emphasis was on commerce and spray types representing different forms, such as pompon, decorative, anemone and spider received greater attention here. Instead of big size and novelty in colour the main qualities sought after were sturdy growth, long vase-life and higher yield. The National Chrysanthemum Society of America was founded in 1890, consisting exclusively of commercial growers, at New York. The society organized the first exhibition in 1902 at Chicago.

3.5.0 In India

In the absence of any chronological evidence, the history of first cultivation of chrysanthemum in India was in Marathi literature, however, the first written mention was in Jnyaneshwari (1290 AD) (Chapter 15, *Shloka* 20). As mentioned earlier, a cultivar from India was reported to be grown in Holland in 1690 under the name *gool-doodi*. It was interesting to note that chrysanthemum was still called *gul-daudi* in Hindi.

In North India, chrysanthemum is primarily grown for landscape gardening either in the ground or in pots. On the contrary, in southern part of the country, it is mostly grown in farmers' fields for supply to the market as loose flowers for garland, hair decoration by women and offerings to God. While yellow coloured flowers are preferred in the South, in the North various hues of red, purple, yellow and white has the common appeal. The use of chrysanthemum as cut flower and pot plant is increasing steadily and it is being extensively grown by commercial growers in almost all parts of the country, particularly near the marketing centres in cities and towns.

4.0 Classification

The species of the genus *Chrysanthemum* are annual and perennial herbs, sometimes partly woody. Leaves alternate, from nearly entire to much dissected. The growth pattern of chrysanthemum was divided in different stages. The nine developmental stages identified included a vegetative stage, a domming stage, early and late stages of involucre and floret formation, and early, mid and late stages of corolla formation. Inflorescence-heads many flowered; disc florets perfect and mostly fertile; ray florets pistillate, mostly fertile; receptacle naked, flat or convex; involucre scales imbricated and angled or terete, or more or less ribbed; pappus absent or a scale-like cup or raised border.The time of capitulum initiation, its development rate and the time of anthesis differed depending on species. Two types of disc floret development *i.e.*, petal closed before or after stamen primordial initiation, were observed (Zhang *et al.*, 1998).

The name for garden chrysanthemum, *Dendranthema* was given in 1961 and was brought under separate genus from the corn marigold (*Chrysanthemum coronarium*). However, chrysanthemum is widely being used as a common name; contrary to the term *Dendranthema*. Hence, a proposal was submitted in 1995 to change back to the original generic name and the proposal was ultimately accepted to overcome such problems (Brummitt, 1997). Song *et al.* (2018) classified the ray floret into nine types, including flat-straight, flat-curve, flat-atypical, spoon-straight, spoon-curve, spoon-atypical, tubular-straight, tubular-curve, and tubular-atypical, which can be

Uses of Chrysanthemum

In pot

In border of a home garden

In flower arrangement

Oil from chrysanthemum

In a garden

Chrysanthemum bonsai

used in the classification of capitulum shape and for the identification, breeding, and international standardization of chrysanthemum cultivars.

The genus *Chrysanthemum* belongs to the family Asteraceae. Many other commonly grown flowers like marigold, dahlia, aster, cornflower, zinnia and sunflower also belong to this family. The Chrysanthemum Societies in various countries have classified the chrysanthemums grown in their respective countries on the basis of number, size, form and arrangement of disc and ray florets. The following is the classification adopted by the National Chrysanthemum Society (NCS) of America.

4.1.0 Classification of the NCS of America

4.1.1 Division A (Classes 1-8)

Ray florets flattened to concave or convex, visible portions never tubular.

4.1.1.1 Section 1 (Classes 1-3)

Disc prominent, circular in outline, composed of many disc florets.

Class 1. Single

Ray florets in a single row at right angles to the stem. Disc flat to slightly rounded and may be of contrasting colours, *e.g.*, Potomac.

Class 2. Semi-double

Ray florets in more than one row at right angles to the stem but may curve downward at the tips. Disc as in class 1.

Class 3. Anemone

Ray florets variable, from flattened, broad and equal in length to reflexing, pointed at tip and unequal in length. Disc florets are numerous tube-like and elongated so as to form a prominent disc which may range from flat to hemispherical in form.

4.1.1.2 Section 2 (Classes 4-8)

Disc must not be apparent. Disc florets, if present in the centre or scattered over the receptacle, must be concealed. Disc florets may be entirely absent.

Class 4. Pompon

Bloom globular, somewhat flat in young stage or small button type. Ray florets broad, incurved, smooth and firm with good substance.

Class 5. Incurve

Breadth and depth should be equal to produce a globular bloom. Ray florets narrow to broad, smooth and incurve in a regular to an irregular manner without producing an open centre, *e.g.*, Snow Ball, Mountaineer, Nob Hill.

Different Classes of Chrysanthemum

Single and semi double

Intermediate incurve

Anemone

Spider

Spoon

Reflex

Classes of Chrysanthemum

Regular incurve — Pompon

Decorative — Brush

Quill — Unclassified

Class 6. Reflexing Incurve

Ray florets usually broad and smooth. Breadth and depth nearly equal to form a globular bloom, sometimes flattened, may be less compact than incurve. All mature florets not completely incurving and not all completely reflexed. The lower florets sometimes reflexing to give a skirted effect, *e.g.*, Dream Castle, Indianapolis.

Class 7. Decorative

Ray florets from short and broad to narrow, long and pointed; they generally reflex, although upper florets may tend to incurve. Blooms more flattened than globular, *e.g.*, Otome Pink, Princess Anne.

Class 8. Reflex

Bloom globular with equal depth and breadth and a full centre, or somewhat flattened. Ray florets narrow to broad, gracefully overlapping in either a regular or in an irregular manner and reflexed, *e.g.*, Coronation Pink.

4.1.2 Division B (Classes 9-11)

Ray florets tubular, coiled or straight, thread-like to coarse, closed, spatulate, flattened, or hooked at the distal end.

Class 9. Spoon

Ray florets regular and tubular, usually straight, distal portion open, flattened and spoon-like. Disc must be apparent with form as in class 1 or 3.

Class 10. Quill

Ray florets tubular, straight and not coiled; may be either closed to tip and pointed or open and spatulate. The bloom is fully double; disc must not be visible.

Class 11. Spider

Ray florets long and tubular, very fine to coarse and may assume a wider variety of direction. Distal portion of the florets may be closed or open, and spatulate tips show definite coils or hooks. Disc must not be obvious. Four sub-classes based on fineness of tube have been suggested as thread, fine, medium and coarse-tubed.

4.1.3 Division C (Classes 12-13)

Ray florets flattened or tubular. Disc may or may not be obvious.

Class 12. Lacinated

Ray florets may be lacinated or feathered at tips. Bloom form may be any of the classes 1 to 11, *e.g.*, Jack Straw.

Class 13. Brush or thistle-like

Ray florets are fine tubes grown almost parallel to the stem in a brush or thistle like manner. Disc nearly concealed, *e.g.*, Saga, White Ise.

In India, the chrysanthemums have been classified into two broad groups, namely large-flowered (which are usually grown as standards with 1-3 stems, bearing a single flower each) and small-flowered (which are grown as bushes with multiple branches, bearing a very large number of blooms per plant). The sub-classes of large flowered are almost the same as described from classes 1 to 12 of the NCS of America. In addition, there is a class called ball or rayonante in which the ray florets are broad and radiate in all directions from centre like sun rays; cultivar Pride of Madford is a typical example. The small-flowered types are much more thoroughly classified as given below:

Anemone: Same as class 3 above.

Button: Blooms 2-3 cm in diameter with florets either like ball or pompon described above.

Charm or cineraria: Like single or semi-double but size is not more than 3 cm across.

Decorative: Same as class 7 of the NCS.

Korean: Same as classes 1 and 2 of the NCS.

Pompon: Same as class 4 of the NCS.

Spoon: Same as class 9 of the NCS.

Quill: Same as class 10 of the NCS.

Stellate: Like Korean but the ray florets are twisted to give the bloom a star-like appearance.

5.0 Species and Cultivars

5.1.0 Species

The number of species under the genus *Chrysanthemum* varied from 100 to 200 (Carter, 1980). Description of some important species of chrysanthemum are:

Chrysanthemum boreale **(Abura-Giku)**

Believed to have also been involved in the evolution of florists' chrysanthemum by Niwa (1936) due to the morphological resemblance. This species was mentioned in the Index KewensisCz.

C. carinatum **(Tricolour chrysanthemum)**

It is a winter season annual bearing tricoloured blooms, 5 cm across.

C. coronarium **(Garland chrysanthemum)**

It is another winter season annual producing yellow and white blooms, often used for garland making.

C. cinerariifolium **(Dalmatian pyrethrum)**

Grown in temperate regions for making an insecticide called pyrethrum.

C. coccineum **(Painted daisy)**

Grown in temperate countries from seeds. Plants are 30-90 cm tall and bear

Species of Chrysanthemum

Chrysanthemum coccineus

Chrysanthemum cinerariaefolium

Chrysanthemum coronarium

Chrysanthemum japonicum

Chrysanthemum indicum

Chrysanthemum frutescens

Species of Chrysanthemum

Chrysanthemum carinatum

Chrysanthemum maximum

Chrysanthemum morifolium

Chrysanthemum ornatum

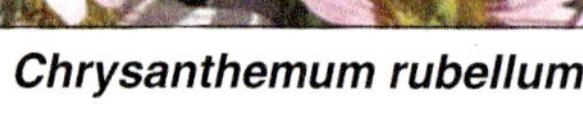

Chrysanthemum rubellum

Chrysanthemum sibiricum

blooms about 5 cm across with white, pink and red colours and look like anemone chrysanthemum. It is a perennial.

C. *frutescens* (Marguerite or Paris daisy)

It is a shrub growing 60-90 cm tall, bearing white and soft yellow flowers and is quite popular as pot plant also.

C. *indicum*

Although native of China and Japan, it was inappropriately named as indicum. It is supposed to be one of the ancestral species involved in the evolution of modern florists' chrysanthemum. It bears yellow flowers.

C. *japonicum* (syn. C. *nipponicum*)

It is found growing in the Pacific coastal region of Japan and widely used as an ornamental plant.

C. *maximum* (Shasta daisy)

It is grown as a popular cut flower bearing white and yellowish blooms. It is perennial.

C. *morifolium* (Florists' chrysanthemum)

This is a hybrid species which is the result of repeated cycles of inter-specific crossing among elemental species extending over a period of more than 2500 years. Its plants are perennial, bear blooms which are actually composite inflorescence, with two types of florets arranged on a flattened axis called capitulum. The outer florets are called ray florets and the inner ones with different morphology are called disc florets. The ray florets have only female parts while the disc floret may have both male and female parts, and a tubular corolla made up of five united petals.

C. *ornatum* (syn. C. *marginatum*)

Allied to *C. indicum* and *C. sinense,* it is also supposed to have involved in the evolution of florists' chrysanthemum.

C. *rubellum*

It is exceptionally sturdy and used by breeders for its hardiness.

C. *satsumense* (Satsuma-nogiku)

Yet another possible ancestor of the present-day cultivars mentioned by Japanese workers. The species is not mentioned in the Index Kewensis.

C. *sibiricum*

This is said to be one of the parents of Korean hybrids evolved in early 1930.

C. *sinense*

This is also supposed to be one of the sources (with *indicum*) of today's florists' chrysanthemum. Native of China and bears blooms with white ray florets.

5.2.0 Cultivars

The number of chrysanthemum cultivars is incredibly large with more than 15000 listed in Japan only. The National Chrysanthemum Society of Britain lists over 6000 cultivars (Machin and Scopes, 1978). In India, too, the number would easily cross 500 mark. Given below is a list of some selected cultivars grown widely throughout the world.

5.2.1 Spray Cultivars

Varieties: Bronze Dee Gem, Chatsworth, Dee Gem, Heide, Karen Taylor, Linda, Myss Debbie, Myss Dorothy, Myss Rihanna, Myss Saffron, Pat Bahn, Pennine Bullion, Pennine Point, Pennine Swan, Purple Dee Gem, Roen Sarah, Rosie Lyttle, Yellow Heide, Alyece Shaw, Bronze Enbee Wedding, Cerisa, Edina, Enbee Wedding, Finn Lyttle, Frizbee, Gold Hoagy, Hoagy, Honey Enbee Wedding, Natalie Sarah, Orange Enbee Wedding, Pacific Lady, Paloma Redeye, Paloma Regent, Peach Enbee Wedding, Peach Southway Sheeba, Pennine Polo, Primrose Nbee Wedding, Rita Mcmahon, Rose Enbee Wedding, Salmon Enbee Wedding, Southway Sheeba, Southway Shimmer, Southway Sloe, Southway Strontium, Tonto, White Enbee Wedding, Yellow Enbee Wedding, Barbara, Bravo, Cassablanca, Gladys, Golden Chalice, Holly, Jenny Wren, Lighthouse, Lynn, Morning Star, Nicole, Redbreast, Ringdove, Robin, Triumph, Aunt Millicent, Daphne Davies, Emperor Of China, Grandchild, Mary, Mary Stoker, Mauve Gem, Nell Gwynn, Percy Salter, Perry's Peach, Ruby Mound, Ruby Raynor, Sea Urchin, Starlet, Venus 1, White Gem.

5.2.2 Standard Cultivars

Varieties: Bronze Max Riley, Egret, Max Riley, Music, Peter Rowe, Anne Jones, Apricot Courtier, Bronze Matlock, Christopher Lawson, Courtier, Dorridge Crystal, Gambit, George Griffiths, Golden Courtier, Matlock, Maxine Charlton, Membury, Migdale, Peach Courtier, Pearl Dorridge Crystal, Thoroughbred, Venice, Yellow Courtier, Clapham Delight Cornetto Dana Doris Ozall Dorothy Stone, Etta Dakin, Hanenburg, John Harrison, Judy Dakin, Malcolm Perkins, Marion Couchman, Millenium, Lilly Emily, Orange Allouise, Primrose Allouise, Promise, Red Regal Mist, Regal Mist, Salmon Allouise, Savanah Charlton, White Allouise, Yellow Clapham Delight, Alec Bedser, Allouise, Astro, Barbara Dakin, Bill Wade.

5.2.3 Pot cultivars

Sl. No.	*Colour*	*Cultivars*
1	Yellow gold	Bethany™ Yellow, Beverly™ Gold, Brittany™ Yellow, Chelsey™ Yellow, Cheryl™ Golden, Cheryl™ Sparkling Yellow, Chloe™ Yellow, Dawn™ Yellow, Elena™ Gold, Gigi™ Gold, Gigi™ Yellow, Hailey™ Gold, Hankie™ Yellow, Jacqueline™ Yellow, Makayla™ Yellow, Mary™ Yellow, Michelle™ Gold, Nikki™ Yellow, Okra™ Yellow, Rhonda™ Yellow, Stephany™ Yellow, Tanya™ Yellow, Ursula™ Sunny Yellow, Wilma™ Yellow, Yolanda™ Yellow, Yvette™ Yellow

Cultivars of Chrysanthemum

Annie Girl

Autumn Sunset

Billy Bell

Bola-de-Oro

Crimson Tide

Descanso

Fortsmith

Gambit

Cultivars of Chrysanthemum

Hagoromo

Honey Glow

Ja Dank

Lavender Pixie

Lily Galon

Paint Box

Moira

Red Win

Spray Type Cultivars of Chrysanthemum

White Marble — Elegance — Anastasia pink

Feeling Green — White Spider — Polaris

Cerisa — Chatsworth — Karen Taylor

Linda — Myss Saffron — Ruby Mound

Source: W.N.Hall and Sons Ltd (T/A Halls of Heddon, Registered Office: West Heddon Nursery Centre, Heddon on the Wall, Northumberland, NE15 0JS).

Sl. No.	*Colour*	*Cultivars*
2	Orange-Bronze	Aubrey™ Orange, Beverly™ Bronze, Beverly™ Dark Bronze, Beverly™ Orange, Cheryl™ Spicy Orange, Electra™ Amber, Gigi™ Orange, Hailey™ Orange, Hannah™ Orange, Jacqueline™ Orange Fusion, Keeley™ Orange, Makenna™ Orange, Miranda™ Orange, Nikki™ Bronze, Nikki™ Orange, Rhonda™ Bronze, Stacy™ Dazzling Orange, Stephany™ Bronze, Ursula™ Fancy Orange, Ursula™ Jazzy Coral, Yvette™ Orange Bicolor
3	Red	Aideen™ Red Fire, Ashley™ Red, Autumn Sunset™, Bonnie™ Red, Brandi™ Burgundy, Cheryl™ Jolly Red, Christina™ Red, Cynthia™ Scarlet, Danielle™ Red, Edana™ Red, Hailey™ Red Bronze, Hestia™ Hot Red, Kathleen™, Dark Red, Mila™ Red, Patty™ Pomegranate, Rhonda™ Red, Samantha™ Red, Tabitha™ Scarlet, Wanda™ Red
4	Purple-Dark Pink	Arlette™ Purple, Brooke™ Dark Pink, Cheryl™ Regal Purple, Danielle™ Purple, Emelda™ Purple, Gigi™ Dark Pink, Jacqueline™ Pink, Nikki™ Dark Pink, Patty™ Purple, Rhonda™ Purple, Veronica™ Dark Pink, Wanda™ Purple.
5	Pink-Lavender	Chelsey™ Coral, Chelsey™ Pink, Cheryl™ Pink, Debbie™ Hot Pink, Gigi™ Coral, Jacqueline™ Peach Fusion, Jacqueline™ Rose, Marsha™ Pink, Nikki™ Pink Bicolor, Rhonda™ Pink, Stacy™ Pink, Ursula™ Lavender, Wanda™, Lavender.
6	White	Addison™ White, Bertha™ White, Chelsey™ White, Cheryl™ Frosty White, Edith™ White, Gigi™ Snow, Jacqueline™ Pearl, Katie™ White, Makenzie™ White, Mildred™ White, Rhonda™ White, Shannon™ White, Vanna™ Snow, Wilma™ White

Source: Syngenta flowers, 2280 Hecker Highway, Gilroy, California 95020 USA, Syngentaflowers-us.com.

Cut Flower Cultivars

Arctic Queen, Pina Colada, Champagne Golden, Champagne Yellow, Feeling Green Park, Champagne Orange, Lexy Red, Candor Pink

(Source: https://www.risenshine.in)

5.2.4 Cultivars in India

Given below is a list of some attractive and popular cultivars grown in India.

Cultivars	*Year of Release*	*Breeding Method*	*Pedigree/ Parentage*	*Important Traits*	*Institutes/ Universities*
Arka Chandrakant	1990	Hybridi-zation	Angela x G.P.67	White, garland, floral decoration, pot plant, bedding plant	IIHR Bangalore
Arka Chandrika	1994	Hybridi-zation	Angela x G.P.1	White, white, garland, floral decoration, making bouquet	IIHR Bangalore
Arka Indira	1980	Hybridi-zation	Lord Doonex -14x Flirt x Valentine-1	Golden yellow, cut flower, pot plant	IIHR Bangalore
Arka Kirti	1994	Hybridi-zation	Angela x G.P.1	White, floral decoration, pot plant, bedding plant	IIHR Bangalore
Arka Nilima	1992	Hybridi-zation	Flirt x Valentine	Red-purple (70.8) fades to red-purple(62.C), cut flower, floral decoration, making bouquet	IIHR Bangalore
Arka Pankaj	1990	Hybridi-zation	Lord Doonex -14x Flirt x valentine-1	Light pink fades to pale pink, cut flower, floral decoration, making bouquet	IIHR Bangalore
Arka Rakhee	1980	Hybridi-zation	Open pollinated seedling of Lord Doonex	Primrose yellow with red stripes on ray florets, pot plant	IIHR Bangalore
Arka Ravikiran	1992	Hybridi-zation	Flirt x Valentine	Greyed-red (178.B) fades to greyed-orange (174.C), cut flower, pot plant	IIHR Bangalore
Arka Red Gold	1980	Hybridi-zation	Flirt x Valentine	Greyish orange fades to golden yellow, Cut flower, making bouquet	IIHR Bangalore
Arka Yellow Gold	1992	Hybridi-zation	Induced mutant of 'Flirt'	Yellow (33.A) having greyed-orange (173.B) stripes fades to yellow (2.B), Garland, cut flower, making bouquet	IIHR Bangalore
Arka Yellow star	1990	Hybridi-zation	Flirt xValentine-22	Yellow, garland, floral decoration, making bouquet	IIHR Bangalore
Arka Usha Kiran	2001	Induced mutation	Kirti	Yellow, garland, floral decoration	IIHR Bangalore
Pusa Anmol	2005	Mutation	An induced radio mutant of hybrid variety Ajay (evolved at NBRI, Lucknow).	Highly floriferous, bushy variety, spray group, relatively thermo and photo insensitive	IARI, New Delhi
Pusa Centenary	2005	Mutation	An induced radio mutant of a very popular commercial exotic cut flower variety Thai Chen Queen that produces light orange flowers, which are suitable for cut flower purpose.	Vigorous growing, flowers possess brilliant yellow incurving ray and disc florets which are fleshy and delicate, suitable for cut flower and exhibition purpose	IARI, New Delhi

Cultivars	Year of Release	Breeding Method	Pedigree/ Parentage	Important Traits	Institutes/ Universities
Shanti	2000	Selection	Open-pollinated seedling selection	Small flowered, white decorative type, cut flower and garland variety	NBRI, Lucknow
Y2K	2000	Selection	Open-pollinated seedling selection	Samll flowered, white Anemone type mini chrysanthemum, plants dwarf, bushy, profuse blooming	NBRI, Lucknow
Sadbhavana	2000	Selection	Open-pollinated seedling selection	Samll flowered, red yellow, open disc double Korean type mini-chrysanthemum	NBRI, Lucknow
Kargil 99	2000	Selection	Open-pollinated seedling selection	Samll flowered, spoon-shaped, purplish mauve florets, leaves variegated	NBRI, Lucknow

6.0 Genetics and Breeding

6.1.0 Genetics

Nazeer and Khoshoo (1982) revealed that while the basic number for the genus is 9, ploidy level varies widely in different cultivars of the species with 2n=36, 45, 47, 51-75. Li Maoxue *et al.* (1983) found most Chinese cultivars hexaploids and aneuploids, showing three groups - 541 chromosomes in early and some large-flowered cultivars; 52-57 chromosomes and 58-71 chromosomes in several large-flowered cultivars. Endo (1990) reported the occurrence of B chromosome in *C. morifolium*. This chromosome was found to occur more in aneuploid than euploid cultivars. These chromosomes were seldom found in hybrids from crosses or self pollinations involving a parent containing B chromosome.

Genetic studies on self incompatibility by Fryxwell (1957) and Drewlow *et al.* (1973) showed that the self incompatibility reaction was sporophytic and involved more than one locus. They also found that nine sibling clones selected from a highly compatible cross were all self incompatible. Those nine sibs could be separated into 9 distinct incompatibility patterns from the pollinations made in that study. Zagarski *et al.* (1983) reported that 11 garden chrysanthemum siblings obtained through 2 generations of single plant self pollination, were selfed and crossed in a complete diallel outcrosses to unrelated individuals and were tested for the performance of male and female fertility. Since 11 crossing patterns were found in the diallel, at least 3 genes were assumed to govern self incompatibility. The data suggested gene independent action with intra-genic dominance. One inbred sibling 77-AM3-3 expressed high level of pseudo-self incompatibility (PSC) by giving some very low cross seed sets together with high set when selfed. Seasonal variation in PSC was also observed.

The variation in ovule number in the disc floret was found to be responsible for self incompatibility among cultivars. Ovule number was significant between and within genotypes for self (disc and ray florets) but not cross (ray floret only) pollinations. The disc floret in self pollinations was found to be responsible for increasing the variability of ovule number, which in turn resulted in incompatibility (Anderson *et al.*, 1988).

Inheritance of different characters of chrysanthemum was studied by many workers. The somatic-genetic analysis of the apical layers of 16 chimeral sports of the Indian-apolis cultivar led to the understanding of complex nature of colour inheritance (Stewart and Derman, 1970).

It was shown that a cultivar could be genetically of one colour in L1 and of another colour in L2. As the sex cells arise from L2, a cultivar which is white in L1 and pink in L2 would have white colour but breed as pink and *vice versa.* Jordan *et al.* (1983) studied the inheritance of characters, cell sap anthocyanin and chromoplast carotene in petal cells in a wide range of progenies from crosses. The presence of one gene *A* ensured anthocyanin production while gene *I* inhibited carotene production. Yellow flowers were formed in the absence of both *A* and *I*, while white flowers in the absence of *A* alone. The combination of *A* and *I* resulted in pink, carmine, bluish red flowers, whereas *A* without *I* resulted in bronze and brownish red flowers. Segregation of flower colour was suggested by them to help in the identification of parental genotype.

The flower colour is determined by the presence or absence of carotenoids, anthocyanins, flavonols or other glucosides in the plant. Anthocyanin, cyanidin 3-O-(6-O- malonyl-β-D-glucopyranoside) was found to be responsible for the red colour in chrysanthemum cultivars (Saito *et al.*, 1988). Teynor *et al.* (1989a) reported that disomic inheritance of a dominant gene inhibited plastid pigmentation. Under greenhouse conditions, plants contained more colourless plastid pigments, while under open conditions, yellow plastids were more in number indicating the effect of environment on the expression of the pigments. Studies on the intensity of the vacoular pigmentation in crosses between 2 chrysanthemum cultivars with 16 inbred clones revealed that the progenies exhibited red and orange colour when grown in green-houses. The colour variation between red and orange and also yellow was noticed under field conditions. The red colour expression might be due to heterozygous genotype and the orange colour due to a single dominant gene causing mosaic like distribution of red vacoular pigmentation over yellow plastid pigmentation (Teynor *et al.*, 1989b). In some varieties a heterozygous dominant inhibitor gene controlled the carotenoid pigment while in other varieties, a heterozygous dominant inhibitor gene as well as a heterozygous dominant gene were responsible for this (Hattori, 1991). The crossing of *C. morifolium* and *C. japonense* did not produce any yellow colour flowers suggesting that the wild species contained a homogygous dominant inhibitor gene. The yellow colour of *C. segetum* petals was due to the presence of the 7-O-glucosides of quertin and gossypetin (Stich *et al.*, 1997). Cyanidin 3-dimatonyl glucoside is responsible for the purplish red flower colour (Nakayama *et al.*, 1997). Besides these, other compounds like flavonols have been found to be responsible for different colours in chrysanthemum (Stich *et al.*, 1994; Schwinn *et al.*, 1994).

Sun *et al.* (2019) reported that genetic basis of branching was revealed through a quantitative trait locus (QTL) analysis of an F_1 population generated by crossing the varieties 'Nannong Xuefeng' and 'Monalisa' grown under contrasting planting densities. Under the low planting density regime (E1), 12 additive QTLs involving seven branching-associated traits were detected, while under the high planting density regime (E2), the number of QTL detected was only eight. A joint QTL analysis across the two environments identified two QTLs which were separately detected in E1. A set of four QTLs exhibiting additive x additive epistasis was identified, few of which had interaction with environments.

Jong (1984) found 70 per cent heritability of characters, like number of days to flower and number of flowers per plant, in 79 F1 populations from 15 parents. A significant general combining ability for days to flowering was noted. The progenies of parents that flowered early at one temperature were generally early at all temperatures at which plants were studied (12°, 13°, 15° and 17°C at night). The inheritance of flower doubleness and floret corolla shape in 70 F1 populations from 16 parents were studied. Singleness was partially dominant over doubleness of type. The percentage varied with parents.

The anemone type was inherited as if it was single. Crosses between ligulate the tubular corollas produced ligulate, intermediate and tubular types in various ratios depending on parents used. Flower doubleness and corolla shape were not linked. Hemlata *et al.* (1992) indicated the existence of high heritability for several characters. High heritability was accompanied by high genetic advance for number of branches, leaves, leaf area and number of flowers/plant, showing an additive gene effect.

High heritability coupled with high genetic advance was observed for number of suckers/plant, disc diameter and number of flowers/plant (Baskaran *et al.*, 2009), number of leaves/plant, number of ray florets/flower, branches/plant, flowers/plant, flower weight, floret length, flower diameter, plant height, plant spread and days to 1st bud appearance (Sahu and Sharma, 2014), number of flowers per plant, flower size and plant height (Kumar *et al.*, 2015a), carotenoid andanthocyanin content, weight and number of ray floret/flower head, ray floret width (Verma and Sahukla, 2018), and most of the vegetative and flowering traits (Negi *et al.*, 2020), indicating that these traits are governed by additive genes which could be effectively improved through selection. Imai *et al.* (1995) observed that cebrosides containing unsaturated hydroxy fatty acids were responsible for cold resistance in chrysanthemum.

6.2.0 Breeding

All the present day colourful chrysanthemum varieties have been developed through complex interspecific crosses among elemental species, open pollination, indiscriminate intervarietal hybridization, spontaneous and induced mutation, selection and management of chimera. The main agencies responsible for varietal improvement are individual enthusiasts, nurserymen and breeders working in research institutes and agricultural universities. Being a cross pollinated crop, new varieties arise mainly as seedlings obtained either by natural or conscious cross-pollination (Datta and Janakiram, 2015). New varieties of chrysanthemum are being developed throughout the world mainly through hybridization and induced mutagenesis. New varieties are also developed through sports. International Atomic Energy Agency, Vienna is maintaining a database for new varieties developed through induced mutagenesis of all crops including chrysanthemum.

Extensive research is going on chrysanthemum breeding at CSIR and ICAR institutions in India. National Botanical Research Institute (NBRI), Lucknow, India has done a commendable work on chrysanthemum on both basic and applied aspects. Interdisciplinary research on chrysanthemum is going on under All India Coordinated Research Project on Floriculture at ICAR institutes with State Agricultural Universities.

The chrysanthemum has developed considerable heterozygosity and the variability in habit, height, vigour, period and quality of bloom, colour, size and shape of flowers, and fertility is expressed under cultivation. Crossbreeding has been utilized as one of the main method to increase further genetic variability. Systematic efforts were made at CSIR-NBRI and ICAR to develop high yielding variety, pot culture variety, cut flower variety, garland purpose, exhibition type by selection and incorporating desirable genes which had been missing among cultivars grown in India but found elsewhere in the world through natural crossing or conscious selective crossing.

Ueno *et al.* (2013) introduced termed 'Ion Beam Breeding' as a novel approach which facilitated the mutation-linked induction of genetic changes plants. They used ion beam irradiation and introduced two new cultivars *viz.*, 'Aladdin' and 'Imagine' with reduced axillary buds, and ion beam re-irradiation of 'Aladdin' yielded cultivar 'Aladdin 2' with combined traits of few axillary buds and early flowering at a low temperature.

Hui *et al.* (2014) reported that with the combination of molecular marker technology and traditional morphological approaches, breeding success will be improved. It will clarify the cross-compatibility among chrysanthemum cultivars and expand the range of breeding resources for genetic improvement. The breeding period will be shortened with the directional transfer of favorable genes and the accumulation of target genes. It will provide an accurate reference library for the protection of chrysanthemum germplasm resources and the innovation of taxonomy.

Zhang *et al.* (2018) developed an effective mathematical analysis method to breed new cultivars of early-flowering multiflora chrysanthemum. On the basis of flowering time, 505 multiflora chrysanthemum germplasm divided into five groups using a grading analysis method, including extremely early group (daylength longer than 13.5 hours), early group (daylength 13.5–12.0 hours), medium group (daylength 12.0–11.0 hours), late group (daylength 11.0–10.0 hours), and extremely late group (daylength shorter than 10.0 hours). Thirty seven genotypes and 367 progenies were were evaluated combining AHP and the gray relational analysis (GRA).

Nedoluzhko (2018) carried out intervarietal, interspecific, and closely related crosses, as well as radiation mutagenesis, were used. Interspecific hybridization based on adaptive Manchurian and Korean high-mountain species of the Chrysanthemum genus made it possible to improve the genetic basis of existing varieties and solve the problem of individual and complex economic features. The genotypes that combine the main target features of immunity to white rust, winter hardiness, and optimal flowering period are created for the first time in the native breeding of the garden chrysanthemum. Hybrids and varieties of a new generation are obtained, which make it possible to expand the domestic assortment.

Ploeg *et al.* (2009) suggested that energy efficiency of greenhouse cut chrysanthemum may be increased by breeding. In addition to breeding for cultivars with a shorter reaction time at suboptimal temperatures, an alternative approach would be to develop cultivars that are heavier at suboptimal temperatures so that they could be grown at a higher plant density, enhancing the production per unit area. At suboptimal temperatures, growth rate early in the cultivation period decreased as a consequence of a lower light interception resulting from a lower specific leaf area. There are possibilities for breeding for suboptimal temperature-tolerant cultivars. However, the substantial increase in the demand of chrysanthemum both as pot plant and cut flower crop is largely due to the basic role played by breeders in evolving cultivars more suited to newer techniques and trends. About one-third of commercial cultivars of chrysanthemum have arisen as sports or somatic mutants as a result of loss or gain of chromosomal material during mitosis (Dowrick and El-Bayoumi, 1966). It was suggested that most chrysanthemum cultivars were periclinal chimeras and that sorting out of different layers might have resulted in formation of new colour forms. Hybridizing and induction of mutations were responsible for widening the genetic gap and consequently varied plant forms and new colour forms in chrysanthemum.

Protection of Plant Vareiety and Farmers Rights Authority, New Delhi identificed two centres *viz.*, IARI, New Delhi and ICAR-IIHR, Bangalore for developing DUS (distinctiveness, uniformity and stability) test guidelines for protection of Chrysanthemum genotypes. The DUS test guidelines have been developed for 78 vegetative and floral characters for DUS characterization (Anon., 2020b).

6.2.1 Objectives

The aims of breeding change with time and place. For early breeders the main aim was development of novelty types for the shows. With emphasis on shifting to commercial production, other qualities, *e.g.*, winter hardiness, earliness, floriferousness and resistance to diseases became more important (Sahlstrom, 1966). Cultivars with low temperature requirement, having rapid growth following a short period of long days, small to medium leaves to enable close planting and with stem length of at least 80 cm are preferable. Pollen-free cultivars have been reported to be desirable by Broertjes and Jong (1983) as pollens spoil the appearance and also induce allergic reactions during handling.

Promising varieties comprising novel commercial characters like attractive flower colour and shape, no pinch no stake dwarfness, out- of- season blooming, cut flowers (attractive colour, long erect stem, uniform bloom opening, tough florets, long vase life and healthy leaves), pot culture (dwarf and compactness, profuse branching, uniform spreading of branches, simultaneous blooming habit, attractive colour and good colour retention quality and healthy leaves), high yielding, garland purpose, exhibition type, chlorophyll variegation in leaves, showy decorative leaves, *etc.* have come out from systematic efforts of all above mentioned methods (Datta and Janakiram, 2015).

6.2.2 Selection

Selection from seedlings raised from seeds obtained from open pollinated flowers is the oldest method for developing new cultivars with improved characters. The transformation of this once wild single daisy-like flower species into today's large, gorgeous double blooming type was achieved through the method of selection by enthusiastic growers and breeders during last several centuries. The drawback of this method is that the parentage of the selected one is not known, and a very large number of seedlings need to be raised for selection. Moreover, many of the improved types with double, long and tubular ray florets do not set seed due to mechanical barriers and hence, cannot be used as parents in the conventional breeding method.

Unique Chrysanthemum Germplasm Registered with ICAR-NBPGR, New Delhi

Sl. No.	*IC Number*	*INGR Number*	*Year*	*Pedigree*	*Unique Traits*	*Source*
1	IC0568708	10114	2010	Selection from Punjab Gold	Early and off-season flowering, suitable for pot and garden use	Anon., 2020a
2	IC0568709	10115	2010	Selection from Punjab Gold	Early and off-season flowering, suitable for pot and garden use	Anon., 2020a
3	IC0623437	17059	2017	Half-sib progency from cv. Pink cloud	Flower colour: 77.B, Purple group, Fan 2. Stellate ray florets (Cylindrical shape)	Kumar *et al.* (2019)
4	IC0623438	17060	2017	Half-sib progency from cv. Red stone	Flower colour: 162.D, Gray yellow group, Fan 4. Stellate ray florets (Cylindrical shape)	Kumar *et al.* (2019)

Nevertheless, this method is being practised even today and most of the outstanding spray and loose flower-type cultivars evolved in India, *viz.*, Birbal Sahni, Apsara, Kundan, Jaya, Jayanti, Sharad Singar, *etc.*, and those out of season blooming-types evolved at the National Botanical Research Institute, Lucknow, which have enabled year round blooming without environment control, were evolved by the selection method only (Kher, 1986). Ohishi (2000) studied various characteristics of fifty-two clones of chrysanthemum cv. Shuho-no-chikara collected in Aichi, Japan, to determine the superior clone. There were significant differences in the earliness in flowering, degree of appearance of red-purple stripes on white petals, frequency of change from incurving floret to tubular floret, stem length and the diameter of the head among the collected clones grown in an open field, a plastic house without heating and a glasshouse with heating. The mutation of flower colour and the appearance of thin red-purple stripes on white petals were observed in 38 clones grown in the field. This mutation decreased in those clones grown under structures.

Zalewska *et al.* (2007) reported that altered inflorescence colour and altered number of ligulate florets when propagated from adventitious buds technique. The changes might be an effect of either chimeral structure or somaclonal variation. The adventitious buds technique might be useful in chrysanthemum breeding as a source of a new variability.

6.2.3 Hybridization

Hybridization method allows choosing of male and female parents to cross pollinate them and select desirable types out of the resulting seedling population. This method enables breeders to obtain seeds even from those cultivars which naturally do not set seed due to either enormous length of ray florets which act as barriers for pollination or due to condition of protandry and sporophytic self incompatibility (Fryxwell, 1957; Drewlow *et al.*, 1973).

Following new varieties of chrysanthemum have been developed through conventional breeding.

Bala (2019) reported two releases of two Chrysanthemum varieties, *viz.*, Punjab Shingar and Punjab Mohini from PAU, Ludhiana in 2018. 'Punjab Shingar' is floriferous, compact and upright growth, flowers are white, decorative type, suitable for loose flower. 'Punjab Mohini' is dwarf, compact growth, no pinch no stake, flowers are white with yellow center, single Korean type and suitable for pot culture.

A new vareity Punjab Gold found suitable for pot culture (Arora *et al.*, 2002). New varieties 'Yellow Delight' for cut flower production (Kumar and Poonam (2008a), Anmol and Royal Purple for for pot culture (Kumar and Poonam, 2008b) have been released from PAU, Ludhiana. 'Anmol' has bright yellow anemone type of flowers; 'Royal Purple' has purplish pink anemone type of flowers. Both varieties are suitable for pot culture. Another variety Yellow Delight is an early flowering with bright yellow pompon type of flowers and is suitable as cut flower.

Kumar and Poonam (2008c) reported release of three new varieties from PAU Ludhiana such asGarden Beauty, Winter Queen and Autumn Joy suitable for garden purpose. The variety 'Garden Beauty' has maroon spoon type of flowers; 'Winter Queen' has pink spoon type of flowers; 'Autumn Joy' has pink decorative flowers, early flowering, highly floriferous. Han *et al.* (2019) reported the purple chrysanthemum cultivar, ARTI-Dark Chocolate, is worthy of breeding as a new tea cultivar.

6.2.3.1 Hybridization Technique

1. The anthers of disc florets are clipped before dehiscence in seed parent and the heads are covered with the wax paper bags to prevent natural pollination.

2. The long ray florets in seed parent are trimmed to expose the stigmas. This isdone in stages as the florets open centripetally starting from the outermost whorl, to expose only mature stigmas (as indicated by sticky substance on lobes).
3. The pollens from dehiscing anthers of male parent are collected by tapping the bloom on a petridish or with the help of a soft brush directly.
4. The emasculated flowers are cross pollinated by dusting the pollen on the receptive stigma of seed parent after opening the wax paper bag temporarily and then placing back the bag after pollination.
5. A tag is tied giving details of female and male parents.
6. The paper bag is removed after all the stigmas wilt.
7. The seeds are collected after the blooms completely dry up (it takes 1 to 2 months) and stored in dry place till sowing time.

It requires at least 5 years releasing a new cultivar. Only commercial nurseries and institutes with adequate resources and expertise should take up planned breeding programme. Langton and Cockshull (1976) adovacted the concept of deotype breeding. An ideotype is a model of a plant expected to react in a predictable manner in a defined environment (Donald, 1968). Thus a summer blooming ideotype was different from that for a winter blooming one.

Shibata *et al.* (1986) made crosses between 'Pink Pearl' and 'Gem' and obtained 'Spring Song', that required relatively low temperature to grow. They (1988) produced summer blooming cultivar 'Summer Queen' by crossing Japanese summer-flowering cultivars for introduction of the year round types. The interspecific hybridization between *C. morifolium* and *C. pacificum* was done by Shibata *et al.* (1988) which finally led to the evolution of 'Moon light'. The cultivar was found to be heptaploid with chromosome number 2n=64.

Xu *et al.* (2000) studied some characters of flowers, stems and leaves in F_1 generation by using multiple combinations of fourteen varieties of *D. morifolium* and the selfed combination of two varieties. The results showed that not all of the colours of flowers resulted from matroclinal inheritance. Inflorescences showed paternal and matroclinal inheritance, and flower type, and plant height showed matroclinal inheritance to a certain degree. They also suggested that because of the wide variation of characters of the F_1 generation and the existence of special parental variants, it should be possible to select new varieties.

Reciprocal crosses between Chrysanthemum (*D. grandiflorum*) and ten other *Dendranthema* species (*D. arcticum* subsp. *maekawanum, D. boreale, D. crassum, D. indicum, D. japonicum, D. ornatum, D. occidentalijaponenmse* var. *ashizuriense, D. pacificum, D. yoshinaganthum* and *D. zawadskii* var. *latilobum*) native to Japan were performed (Fukai *et al.*, 2000). The results showed that the reciprocal crosses between chrysanthemum and other species are possible when the appropriate genotype is carefully chosen. Hybrid progenies showed a wide range of variations in flowers. The hybrids required shorter days to flower than native species.

6.2.4 Mutation Breeding

Chrysanthemums sometimes mutate to produce new cultivars called sports. The frequency of such mutations in nature is low. The rate of mutations has been found to increase manifold by irradiating the planting material with X-rays or gamma rays (Broertjes, 1966; Dowrick and El-Boyoumi; 1966, Cowse, 1966; Chan, 1966). Broertjes (1966) compared various radiations to find out the best mutation frequency and/or mutation spectrum. X-rays showed a good dosage relationship (Table1).

The mutation rate increased almost linearly with increasing dosage. In spite of lower mutation rate the optimal dosage proved to be 1500 R. Electrons were found to be fairly ineffective. With fast neutrons there was a marked increase in mutation frequency with increasing dosage (Table 2); the survival percentage for the dosages applied being 100. With thermal neutrons the dosage response curve proved to be irregular.

Gamma irradiation in cv. Lalima at 25 Gy found mutation in flower head shape in M1 generation and established in pure form. Significant reduction in plant height, flower head diameter and guard cell width were recorded, however, no change in chromosome number was recorded in the mutant. Gamma ray-induced flower head shape mutant of 'Lalima' was found very attractive and ideal for pot culture (Banerji and Datta, 2002).

The *in vitro* gamma ray mutagenesis in cv. 'Delistar White' in ray florets at 0.5 Gy dose found most effective in inducing mutations in flower shape and number of florets per flower head, although no change in flower colour was observed, but conversion from tubular florets to spoon-shaped florets indicating no chimera formation, which concludes that flower colour/shape mutations can be developed through direct *in vitro* mutagenesis by avoiding chimeric phase (Barakat *et al.*, 2010).

Irradiated rooted cuttings of cv. 'Maghi' reseulted into increase in morphological abnormalities in leaves and plants. Leaf abnormality includes asymmetrical development of leaf lamina, reduction in leaf size, notching of different grades, leathery texture, fission and fusion of leaf lamina, light and dark green leaves, sickle shaped leaf, narrow and linear leaf and obtuse apex. Fasciations in stem were also recorded after irradiation. Chimera in flower head colour was observed. Frequency of chimera varies from a narrow yellow stripe on ray floret, half of the ray floret, full ray floret, bunch of ray floret on a flower head to entire flower head. Three types of chimera in flower head colour *viz.*, sectorial, mericlinal and periclinal were observed in treated population (Batra and Dwivedi, 2015).

Reduction in various vegetative and floral traits alongwith increase in morphological abnormalities in flowers and leaves was found after exposure of cuttings to gamma irradiation. The floral abnormalities like fasciation of flower head, asymmetric development of ray florets were increased at 10 Gy dose. The chimera in flower colour was observed in treated population at 10 Gydose of gamma rays. In cultivar 'Purnima ', total three somatic mutants were detected *i.e.* orange (Orange White Group 159 D/4) and white tubular (White Group 155 B/4) at 10 Gy and white with yellow (Yellow Group 2B/1) at 15 Gy dose. One mutant was detected in cv. Otomazakura (Purple Group 75 A/1) at 10 Gy dose of gamma rays (Tables 1 and 2) (Verma *et al.* (2016).

Table 1: Effect of Gamma Irradiation on Floral Traits on Chrysanthemum Cultivar

Dose (Gy)	*Floral Traits*									
	Days to Bud Initiation	*Days to First Flower Colour Shown*	*Days to Full Bloom*	*No. of Flower Heads/ Plant*	*Flower Head Size (cm)*	*Ray Floret Length (cm)*	*Ray Floret Width (cm)*	*No. of Ray Florets/ Flower Head*	*Fresh Weight of Flowers/ Head (g)*	*Flower Abonor-malities*
Control	48.63 ± 1.95	93.00 ± 2.17	11.70 ± 4.65	32.0 ± 2.44	66.59 ± 0.09	3.16 ± 0.02	1.12 ± 0.03	190.00 ± 1.55	6.00 ± 0.05	0.00
10	49.00 ± 2.54	74.00 ± 3.25	98.6 ± 1.12	30.00 ± 1.30	6.20 ± 0.05	3.1 ± 0.08	1.20 ± 0.07	185.00 ± 8.65	6.05 ± 0.01	4.00
15	51.12 ± 1.23	72.70 ± 1.74	97.0 ± 1.74	25.42 ± 3.21	6.34 ± 0.03	3.06 ± 0.05	1.32 ± 0.09	182.00 ± 6.12	5.85 ± 1.00	8.00
20	55.63 ± 2.51	73.50 ± 1.31	101.2 ± 0.94	21.41 ± 1.19	6.00 ± 1.69	2.98 ± 1.01	1.13 ± 1.00	179.00 ± 5.36	5.37 ± 0.97	15.00
Control	44.12 ± 1.00	67.00 ± 2.06	123.00 ± 3.82	20.74 ± 1.21	11.34 ± 1.35	4.22 ± 0.15	1.21 ± 0.67	230.90 ± 8.04	15.43 ± 0.32	0.00
10	42.57 ± 1.32	69.00 ± 1.48	128.00 ± 2.00	18.34 ± 2.64	11.08 ± 0.43	4.21 ± 0.06	1.20 ± 0.18	226.04 ± 3.96	14.00 ± 0.48	8.00
15	45.61 ± 1.00	73.7 ± 1.00	132.16 ± 1.55	14.10 ± 2.17	10.21 ± 0.89	3.19 ± 0.11	1.18 ± 0.17	221.65 ± 2.45	13.63 ± 1.21	5.00
20	47.22 ± 1.30	47.3 ± 2.23	135.11 ± 3.13	10.87 ± 1.14	8.76 ± 1.05	0.18 ± 1.03	1.15 ± 0.43	118.00 ± 2.17	11.74 ± 1.04	6.00

Table 2: Spectrum of Mutations in Flower Colour and Shape Induced after Gamma Irradiation in Chrysanthemum*

Cultivar	*Original Colour*	*Mutant Colour (Floral/foliage)*	*Mutant Type (Flower shape)*	*Doses Used (Gy)*
Purnima	White group 155 B, Fan-4	Orange White group 159 D, Fan-4	Flat	10
	White group 155 B, Fan-4	White group 155 B, Fan-4	Tubular	10
	White group 155 B, Fan-4	Yellow group 2 B, Fan-1	Chimera	15
Otomezakura	Purple group 76 D, Fan-1	Purple group 75 A, Fan-1	Flat	10

* Verma *et al.*, 2016).

Bowen *et al.* (1962) reported that mutations caused by radiation resulted in sectoral or periclinal chimeras.

The colour mutations were found to occur in definite direction, *e.g.*, pink cultivars were made to produce bronze, yellow or white mutants, but reverse was not true. Bronze would generally produce red and yellow sports and white would yield yellow. This directional pattern has been diagrammatically shown in figure below.

	Carotenoid	
	Without	With
With	Pink	→ Bronze
Anthocyanin		↓
Without	White	→ Yellow

The above pattern was based on either the loss of a whole chromosome or a single gene (Dowrick and El-Bayoumi, 1966).

Stewart and Derman (1970) discussed the chimeral nature of mutants of Indianapolis cultivar in context of Ll and L2 and L3 layers. Petals contain only Ll and L2 layers and a cultivar can be of one colour in Ll and of another colour in L2. Thus a cultivar white in Ll and pink in L2 will have white flower but breed as pink since sex cells arise in L2. Deeper yellow cultivars are white in Ll and yellow in L2, while golden yellow ones would have yellow chromoplasts both in Ll and L2. Langton (1976) discussed the aspect of chimeral instability in mutants in great detail.

Mitiouchkina and Dolgov (2000) used the *rol C* gene from *A. rhizogene* to modify plant architecture of chrysanthemum cv. White Snowdon. One of transformed lines demonstrates dramatic changes in plant size, branching, flowers and petal form. These plants have dwarf bush habit with smaller flowers in more compact flower stems. The petals were wider and their form was changed.

Gupta (1971) classified cultivars of chrysanthemums into three groups on the basis of their sensitivity to gamma rays. Cultivars with LD_{50} lower than 1.5 kR were termed highly sensitive and those with more than 2.5 kR as mildly sensitive, while the rest lying between 1.5 and 2.5 kR as sensitive.

Rooted plants, originated from 6 axillary buds from the M1V1 branches of the chrysanthemum cultivar Ingrid (dark pink) were irradiated with 20 Gy of γ-rays and pruned 40 days after planting (Adames *et al.*, 1999). Frequency and spectrum of flower colour mutants were evaluated. No mutants were observed in the control population. In the M1V2 population, 22.1 per cent of the total plants were mutants (white, dark bronze, pale pink, yellow, wine, variegated and cream). Among them, 1.8 per cent were periclinal chimaeras (with only one different colour from the original) and the others showed mutated sectors.

Gamma-ray sensitivity of chrysanthemum Bongwhang shoots grown *in vitro* was investigated by Goo *et al.* (1999). Lethality of the buds increased according to the increasing γ -ray dose. Higher γ -ray dosage resulted in lower shoot proliferation ratios. In comparison with the control, mean proliferation ratios of γ -irradiated shoots was 2.7-fold in the M1V3 generation and 6.4-fold in the M1V3 generation. Nonbranching and flower shape mutants appeared following irradiation with 17.5 Gy, leaf and stem mutants at 15.0 Gy, and early flowering at 25.0 Gy.

The rate of mutations was found to be the highest in pink cultivars (Jank, 1957). Apart from colour changes, other characters have also been reported to change by

X-ray and gamma rays. Broertjes and Jong (1983, 1984) obtained pollen-free male sterile mutants by treating chrysanthemum cuttings with X-rays. Similarly, Broertjes *et al.* (1983) reported low temperature tolerant mutants of chrysanthemum cultivar Miros.

Change in somatic chromosome number was reported as a result of γ-irradiation (Yamakawa and Ichikawa, 1970; Drewlow and Widmer, 1971). Datta and Gupta (1983) recorded significant increase in ray floret length of a γ-ray induced mutants.

In vitro multiplication techniques opened new vistas in the concept of breeding. Ben Jaacov and Langhans (1972) found that in a batch of plants grown from a single meristem tip in tissue culture, 28 per cent plants had sported from white to yellow. Broertjes *et al.* (1976) showed that solid mutants (non-chimeral mutants) could be produced by growing plantlets *in vitro* from single cells of the epidermis following irradiation. Mutations in Ll layer mainly concerned colour while those in L2 layer involved plant height, vigour and flower size. Mutation breeding under *in vitro* conditions have been employed for breeding cultivars tolerant to low temperature. (Zimmer and Preil, 1987; Benetka and Kodytek, 1988; Berninger and Philouze, 1988). *In vitro* cultured plants when irradiated with gamma rays produced mutants having varied height, leaf, flower shape, petal size and flowering time (Ahloowalia, 1992; Nagatomi *et al.*, 1993 and Furuya, 1998). A number of plants were regenerated from petal, bud and leaf explant excised from chronic, acute and unirradiated plants. It was found that the combined methods of irradiation and culture yielded a mutation rate of 10 times higher than the conventional chronic cutting method and also produced non-chimeric mutants. The six flower colour mutant varieties registered were derived from chronic radiation (Nagatomi *et al.*, 2000).

Dwivedi *et al.* (2000) reported morphological abnormalities in leaf and flower in cv. Lilith irradiated with gamma rays. Chlorophyll variegation in leaves and sectorial flower colour mutations were the most prominent. Chimeric yellow ray florets were cultured in MS medium supplemented with NAA (0.2 mg/L), BAP (1.0 mg/L), sucrose (3 per cent) and bactoagar (8 per cent) at pH 5.6. Regenerated plants successfully transplanted in field condition flowered true to the explant flower colour.

Attempt has been made to regenerate chrysanthemum plants from a single cell, *i.e.* through somatic embryogenesis for management of single cell mutation event. An efficient somatic embryogenesis protocol has been standardized in chrysanthemum. This technique will open up a new way for isolating new flower color/shape ornamental cultivars through retrieval of single mutated cell (Mandal and Datta, 2005).

An efficient regeneration system from ray florets of a chrysanthemum mutant was developed to isolate, purify and field establish the novel mutant. The callus induction was maximum on MS medium supplemented with BAP (4.0 mg/l) and NAA (1.0 mg/l). The maximum regeneration of microshoots (95.56 per cent) from the ray floret induced callus was recorded on MS medium fortified with BAP (4.0 mg/l) and NAA (0.1 mg/l). MS medium supplemented with BAP (5.0 mg/l) + NAA (0.1 mg/l) + GA_3 (0.2 mg/l) gave the highest microshoot proliferation (100.0

per cent). *In vitro* rooting ideal (99.0 per cent) on half-strength MS medium fortified with 0.5 mg/l NAA. The isolated mutant produced flowers that were true-to-type to the original red mutant, *i.e.*, solid mutant (Verma *et al.*, 2012).

Kumar *et al.* (2017) developed efficient ray floret regeneration protocols to isolate, purify and establish a novel mutant which spontaneously appeared as chimera in the form of yellow flowers in cv. Lalit (white). MS medium supplemented with BAP (4.0 mg/L) + NAA (0.05 mg/L) + GA_3 (0.1 mg/L) was found to be best for highest micro-shoot proliferation (92.0 per cent). Highest rooting (86.0 per cent) was induced after culturing the micro-shoots individually on half-strength MS medium fortified with 0.5 mg/L NAA and 50 g/L sucrose. The *in vitro* raised plants produced all bright yellow flowers as compared to the original variety Lalit with white flowers.

Yoosumran *et al.* (2018) induced mutation in cv. Vivic by ethyl methasulphonate (EMS) resulted in three different mutations characteristic, first with 0.5 per cent EMS for 60 min resulted in white stacked ray florets with 2 layers, green disk florets; second with 1 per cent EMS for 60 min showed flower turned into yellow petal stacked with 2 layers, green disk florets and also the white and yellow ray florets and third with 0.5 per cent EMS for 20 min showed ray florets slender white stacked with 2 layers and green disk florets (Table 3)

Table 3: The Effect of EMS Concentration and Duration of Soak on Height of Shoot and the Number of Leaves on MS Medium for 8 Weeks*

		Height of Shoot (cm)	*No. of Leaves*
Per cent EMS	0	2.80±0.05b	5.23±0.08b
	0.5	3.92±0.06a	7.03±0.45a
	1.0	3.92±0.05a	7.41±0.19a
F-test		**	**
Duration (min)	60	3.57±0.54a	6.58±1.00a
	120	3.53±0.57a	6.53±1.08a
F-test		ns	ns
Per cent EMS	Duration (min)		
0	60120	2.84±0.01b2.77±0.06b	5.26±0.05b5.20±1.10b
0.5	60120	3.94±0.08a3.91±0.03a	7.13±0.20a6.93±0.66a
1.0	60120	393±0.03a3.91±0.08a	7.36±0.15a7.46±0.25a
F-test		**	**
CV per cent		1.64	4.76

ns: Non significant difference, ** Significant different at P≤0.01; Mean within column followed by the same later are not significant different as determined by Duncan's multiple range test.

* Yoosumran *et al.* (2018).

New cultivars or mutants were produced from cv. Anupam by irradiation of its rooted cuttings with 1.5, 2.0 and 2.5 Krad γ-radiations (Banerji and Datta, 1991). Apart from irradiation, chemical mutagens when added to culture medium also result in mutations producing mutants having varied chromosome number (Endo *et al.*, 1991).

Comparative spectrophotometric and thin layer chromatographic (TLC) studies of original and mutant cultivars by Datta and Gupta (1981) and Datta (1985, 1986) showed both qualitative and quantitative changes in pigments.

The main bottleneck in inducing mutations in vegetatively propagated ornamentals is the development of chimeras. Ephemeral small sector apparent on petals cannot generally be recorded as solid mutant shoots. An efficient technique has been deve-loped for direct shoot regeneration from individual florets of chrysanthemum cv. Colchi Bahar. The regeneration potential varied with the developmental stages of flower heads. The highest percentage of shoot formation and most shoots per responding explants were observed on florets taken from stage II flower heads. The regenerated plants were rooted *in vitro* and successfully transferred to soil where they grew vigorously and flowered true to the floret colour type (Chakrabarty *et al.*, 1999).

Some outstanding chrysanthemum cultivars have originated in India through spontaneous mutations. The most notable varieties are Kasturba Gandhi (white) developed from Mahatma Gandhi (mauve); Sonar Bangla (yellow) from Snow Ball (white); White Cloud (white) form Pink Cloud (pink); Sharda (yellow) from Sharad Shobha (white); Queen of Tamluk (yellow) from Casa Grandi (white); R Venkatraman (yellow) from S S Arnold (white); William Turner (white) from Pink Turner (pink); J S Lloyd (yellow) from William Turner (white); White Ball from Pride of Madford, *etc.*

Anand *et al.* (2020) developed an efficient ray floret regeneration protocol to isolate, purify and establish a novel gamma ray induced mutant of cv. Tata Century (pink). Maximum survival and callus formation in minimum duration were recorded when the ray florets were pre-treated with carbendazim (0.2 per cent) + ridomil (0.2 per cent) + 8-HQC (200 mg/L) for 2 h followed by surface sterilization with $HgCl_2$ (0.1 per cent) for a duration of 4 min and cultured on MS medium supplemented with BAP (4.0 mg/L) and NAA (1.0 mg/L). The maximum regeneration of microshoots from the ray floret induced callus was recorded on MS medium fortified with BAP (4.0 mg/L) + NAA (1 mg/L). MS medium supplemented with BAP (4.0 mg/L) + NAA (0.05 mg/L) + GA_3 (0.1 mg/L) was found to be best for highest micro-shoot proliferation. Highest rooting was induced after inoculating the microshoots individually on half-strength MS medium fortified with 0.5 mg/L NAA and 60 g/L sucrose.

6.2.4.1 Breeding for Abiotic Stress

Geest *et al.* (2012) developed a theoretical approach to assist breeding against adverse leaf deformations that severely reduce the ornamental value of some chrysanthemum genotypes during greenhouse cultivation in winter. Starch accumulation, plant hormones such as auxin and ethylene may play a role in leaf epinasty. Rai *et al.* (2017) grouped varieties into tolerant, moderately tolerant, susceptible and highly susceptible on the basis of modified standard evaluation score (SES) of visual salt injury. The varieties Little Pink, Lalit, Haldighati and Pusa

Aditya grouped under the tolerant category, while, varieties Himanshu and Pusa Sona are grouped under susceptible and highly susceptible category, respectively.

Ullas *et al.* (2018a) screened varieties for total carotenoids, phenolic content and antioxidant activities revealed that the variety 'Jubilee' exhibited high total carotenoids (32.82 mg/100g) and highest antioxidant activities of carotenoids {CUPRAC-Cupric reducing antioxidant capacity (149.44 μmol trolox/g), FRAP-Ferrous reducing power (40.09 per cent), DPPH radical scavenging activity (11.24 per cent)} and highest lutein content (19.90 μg/g), while, cv. Little Orange recorded highest β-carotene content (5.51μg/g). Vanlalruati *et al.* (2019) reported that the genotypes Mother Teresa, Anmol, Basanti and Sweta Singar were found to be tolerant to saline stress due to enhanced photosynthetic pigments, proline concentration and better shoot and root growth with higher salinity. Salt stress had significant effects on plant growth, assimilator pigment content and proline concentration.

6.2.4.2 Evaluation for Flower Quality

Joshi *et al.* (2009) evaluated seven varieties *viz.*, IIHR-6, Shyamal, Mayur, Red Gold, Honey Comb, Panchon and Nilima, revealed that variety Mayur recorded early first bud initiation and initiation of first flower; Nilima recorded maximum size of flower and average weight of flower; Red Gold recorded maximum number of flowers/plant, however, 'Nilima' recorded maximum weight of flowers/plant and flower yield.

Punetha *et al.* (2011) reported that cvs. White Anemone, Shanti and Charming were found to be highly suitable cut and loose flowers under mid hill of Garhwal Himalaya. Vetrivel and Jawaharlal (2014) evaluated twenty varieties indicated the 'Amalfi' recorded highest plant height, flower stalk length, flower bud diameter and number of flower stalks/m^2; 'Calmiro Sunny' recorded early flower bud appearance, more number of flower buds per plant, early harvest and longest vase life. Srilatha *et al.* (2015) evaluated nine varieties and found that 'Yellow Gold' was early first flowering, while, 'Poornima' and 'White Anemone' were late flowering and recorded 253 per cent and 154 per cent higher yields, respectively. 'Yellow Gold' is suitable for catching early market, while, 'Poornima' and 'White Anemone' are suitable for late flowering.

Kumar *et al.* (2015) evaluated varieties for loose flower production, Pompon Rosy Pink recorded earlier appearance of bud, maximum flower weight, 100 flower weight, size of flower; Suneel recorded maximum flower diameter and floret size; Decorative White recorded maximum number of flower per plant and highest yield. Asha *et al.* (2016) characterized 30 genotypes, six characters were monomorphic, five dimorphic and 30 were polymorphic. It is useful for breeders/researchers/farmers to identify distinctiveness in chrysanthemum genotypes and to seek protection under Protection of Plant Varieties and Farmers Rights Act. Kavitha *et al.* (2019) reported that CO1 performed well and recorded highest yield (11.2 t/ha) than the local variety (5.8 t/ha). The major attribute for increase in yield is big flower size (4.0–4.5 cm) in CO 1. It also gives ratoon crop that led to reduction in cost of cultivation.

6.2.4.3 Polyploidy Breeding

Kim *et al.* (2020) successfully demonstrated induction of tetraploid plants by colchicine treatment. A total of 21 tetraploid and 110 chimera plants were identified using the chromosome counting method, and the colchicine 200 ppm (48 h) and 300 ppm (12 and 24 h) were the most effective for inducing tetraploidy. The tetraploid plants exhibited thicker and larger leaves, as well as larger and fewer stomata than the diploid plants.

7.0 Growth and Flowering

The genotype, environmental factors, nutrition, soil moisture and other management practices influence the growth and development in chrysanthemum. Certain operations like pinching, disbudding, dis-shooting, *etc.*, also affect plant growth and flower production. The use of chemicals in the regulation of growth and flowering is also well known for commercial production of chrysanthemum.

7.1.0 Genetic or Inherent Factors

Chrysanthemums today are unparallelled by the degree of variability found in their types and cultivars. There are now tall cultivars growing several metres high for background planting. On the other hand, cushion mums which do not grow taller than 15 cm are ideal for edging of the border.

Size of flowers varies from less than 2 cm, suitable for miniculture, to large exhibition blooms with more than 20 cm diameter. A wide choice of cultivars with long sturdy and erect stems suitable for cut flowers, with compact, profusely branched stems and short internodes for pot culture and fluffy decorative flowers for garland making are available today. Similarly, different cultivars naturally bloom in summer, rainy season, autumn, winter and spring are also available. Breeders have been regularly producing cultivars more suitable for the changing techniques and demands. Genetic method of growth regulation is the most basic and reliable method but requires long term effort.

7.2.0 Environmental Factors

The major factors which constitute the environment of a plant are light, temperature, relative humidity and air composition. The effects of these factors on chrysanthemum plant have been studied in details. The results obtained have been of great use to the commercial growers for improving the quantity and quality of flowers and their year round production.

Chrysanthemum is a short day leading commercial flower. It develops flower buds when days are less than 12 hours long and the blooming period is short under traditional cultural method. For commercial floriculture programmed blooming is necessary for year round availability of flowers. Extensive work has been done to categorize the leading varieties into their response group and their sensitivity to photoperiodism for year round blooming (Datta and Gupta, 2012).

7.2.1 Light

Light affects the growth of chrysanthemum quantitatively by influencing the photosynthetic activity and qualitatively due to the phenomenon of photoperiodism.

The response of chrysanthemum to light at different stages of growth and flowering has been discussed below separately in the context of scientific investigations.

7.2.1.1 *Effect on Vegetative Growth*

Cathey (1974) observed that quality of light influenced internode elongation and light from incandescent lamps of roughly equal red and far-red wave lengths increased internode length.

As the number of leaves and internode length determined the stem length, the influence of quantity and quality of light on stem length was clear. Hicklenton (1984) found that supplementary illumination with high pressure sodium lamps increased the number of flowering stems, mean stem break length and total dry weight. He (1985) also studied the influence of different levels and timing of supplemental irradiation on pot chrysanthemum production. Crop growth efficiency (calculated as the ratio of the observed to potential dry weight gain) was the highest under 77 μmol $m^{-2}s^{-1}$ treatment and decreased markedly at higher radiances.

Schoellhorn *et al.* (1996) observed more number of lateral branches as photosyn-thesis photon flux (PPF) increased. Short-term variation in light quality using high pressure sodium lamp and mercury metal halide lamps with higher net photosynthetic rates at each stage of plant growth was recorded by Walz and Horn (1997).

Table 4: Effect of Supplemental Night Lighting by LED, Compact Florescent 3U (CFL) and Incandescent (IL) Lamps to Plant Height and Stem Diameter in Chrysanthemum*

Week	*Plant Height (cm)*				*Stem Diameter (cm)*			
	LEDs	*CFLs*	*ILs*	*LSD*	*LEDs*	*CFLs*	*ILs*	*LSD*
0	2,02[a]	1,95[a]	1,93[a]	0,195	0,89[a]	0,89[a]	0,88[a]	0,019
1	2,77[a]	2,56[a]	2,93[a]	0,226	0,95[a]	0,95[a]	0,97[a]	0,027
2	3,43[a]	2,93[b]	3,15[c]	0,196	1,36[a]	1,29[a]	1,31[a]	0,081
3	4,16[a]	3,55[b]	3,83[c]	0,145	0,71[a]	1,54[b]	1,60[b]	0,078
4	4,85[a]	4,31[b]	4,59[c]	0,085	2,02[a]	1,87[b]	1,98[a]	0,056
5	5,61[a]	5,04[b]	5,34[c]	0,100	2,43[a]	2,21[b]	2,31[b]	0,151
6	6,87[a]	6,18[b]	6,63[c]	0,133	2,69[a]	2,49[b]	2,55[b]	0,162
7	8,89[a]	7,75[b]	8,09[c]	0,157	2,98[a]	2,63[b]	2,73[b]	0,123
8	11,48[a]	9,88[b]	10,04[b]	0,192	3,21[a]	2,87[b]	2,88[b]	0,221
9	14,32[a]	12,65[b]	12,53[b]	0,276	3,49[a]	2,97[b]	3,03[b]	0,312
10	17,65[a]	15,61[b]	15,60[b]	0,291	3,58[a]	3,09[b]	3,19[b]	0,293
11	20,90[a]	18,70[b]	18,35[c]	0,267	3,74[a]	3,23[b]	3,30[b]	0,341
12	24,50[a]	21,73[b]	21,13[c]	0,440	3,98[a]	3,44[b]	3,57[c]	0,272

Letters a, b, c represents statistically significant differences between variables in different light systems ($P \leq 0.05$).

* Thao *et al.* (2018).

Hisamatsu *et al.* (2008) observed that a brief end-of-day (EOD) far-red (FR) exposure promotes extension growth in plants during the dark period and subsequent light period due to increased responsiveness (sensitivity) to gibberellins. Thao *et al.* (2018) reported the effects of LED lighting source toward plants growing under open-field conditions (Table 4).

7.2.1.2 Effect on Flowering

Post (1942) divided the flower formation in chrysanthemum into three well defined stages (i) flower bud initiation stage up to the formation of first floral primordia, (ii) flower bud development stages lasting till the flower bud is macroscopically visible, and (iii) flower bud elongation stage till the flower was fully open. Microscopical studies by Chan (1950), Andrews and Watson (1952) and Dooreubos and Kofranek (1953) confirmed the existence of these stages.

7.2.1.3 Flower Bud Initiation

Certain amount of growth is necessary before the plants became responsive to photoperiod for the initiation of flower bud and that the cultivars showed variation in the amount or the period of growth needed for the response to the treatments (Stefanis and Langhans, 1983). Machin (1973) suggested that larger plant, measured either by the number of leaves or stem length or total dry weight, responded to short days more rapidly than a small plant. Jong (1989) reported that minimum number of leaves required to flower was higher under low radiation or fluctuations of radiation.

Intensity of light during the day influences the response to short day and higher light intensity during the days induced early flowering under short days (Watson and Andrews, 1953). Poesch (1935) showed that light intensity of 55-110 lux at the tip of the plant was enough to retard flowering, while Post (1953b) found 2 hours of illumi-nance at 86 lux insufficient and exposure for 3 hours completely inhibited bud initiation. Cathey and Borthwick (1964, 1967) showed that fluorescent light of very high intensity given only for one minute every night completely inhibited flowering in chrysanthemum.

The spectral quality of light too has been found to influence the photoperiodic res- ponse in chrysanthemum. Borthwick (1947) found that wave length between 600 and 680 nm (red) was most effective for initiation of flowering in SD plants. Wave length of 400 nm (blue) was less effective. It was found that red light (660 nm) set a particular response in motion reversed by far-red light (730 nm).

The isolation of plant pigment called phytochrome by Butler *et al.* (1959) set way to the mystery of the photoreception responsible for the phenomenon of photoperiodism. Phytochrome in two forms, Pr (P 660) the red-absorbing and Pfr (P 730) the far red-absorbing were recognized. These two forms are reversible. The initiation of flower bud in chrysanthemum was reported by Cathey and Borthwick (1964) to be due to the conversion of phytochrome pigment from the inactive Pr form to active Pfr form.

Borthwick (1947) showed that it was the length of continuous dark period that determined the photoperiodic response. For chrysanthemum, the minimum length of continuous dark period per day, necessary for flower bud initiation, was found to be 9-9½ hours by Post (1947, 1948, 1953 b), depending on the cultivars. Post and Kamemoto (1950) also found that a minimum of 3-5 consecutive short days were necessary for bud initiation. Borthwick and Cathey (1962) showed that radiation from fluorescent or ruby-red incandescent-filament lamps inhibited flowering equally well when used at 5 per cent of the time in cycles of 15 minutes for a period of 4 hours. When cycles were increased to 30 minutes radiation from ruby-red lamps it became ineffective, and at larger cycles that from incandescent-filament ones lost effectiveness to a much greater extent than from fluorescent lamps.

Oda *et al.* (2010) effect of photoperiod on the flowering of small-flowered spray chrysanthemum cultivars with July- and August-flowering. The flowering of 'Hotaru' could be manipulated by lighting without the influence from annual climate changes.

7.2.1.4 Flower Bud Development

Although flower bud initiation may take place even under long days, in most cultivars the development of initiated bud requires a longer and continuous dark period. Post (1947, 1948) found that while day lengths of 14-14½ hours could initiate flower buds, their rapid development occurred only when day lengths were of 13-13½ hours. Furuta (1954) also showed that photoperiod required for bud development was one hour shorter than that required for bud initiation. Crown buds were formed with a favourable day length for bud initiation but not for development. The reproductive morphogenesis has been studied by Chan, 1950; Dooreubos and Kofranek, 1953; Schwabe, 1959; Popham, 1963; Cockshull, 1972. The receptacle was initiated in 4-6 days and leaf production ceased after 7 short days (Cockshull, 1972). The differentiation of bracts from leaf primordia occurred after 12 days (Popham, 1963).

The first floret primordia were found by Chan (1950) after 12 days, by Popham (1963) after 10 short days and by Dooreubos and Kofranek (1953) after 18 days. The variation in the requirement of photoperiod for initiation of florets was mainly due to the cultivars used. Floret primordia covered whole of receptacle after 24-26 days and Popham (1963) found that the diameter of receptacle increased 12 times after 30 short days.

At this stage the receptacle and florets were covered by at least 8 bracts. Schwabe (1959) showed that time required for bract formation under short days was only 9 hours as compared to 2-3 days required for leaf production (plastochron) under long days. Schwabe (1953) also observed that at low light intensity the development of initiated buds was delayed.

Mori *et al.* (2014) reported that the early budding was completely prevented and flowering was adjusted arbitrarily, using interruption treatment (NI) in 'Seikomaki', 'Subaru', 'Haruka', and 'Seichigusa'. NI delayed budding for more than twenty days in 'Seiun', 'Seikomaki,', 'Subaru', 'Haruka', and 'Seichigusa', which are suitable to meet the peak demand using NI.

Mori *et al.* (2016) investigated the effect of interrupted lighting (IL), the intercalation of a period of long days (by night interruption) into the natural day length (NDL) period of growing, in order to regulate spray formations in Summer-to-autumn-flowering small-flowered spray-type (SAFS) chrysanthemum cultivars 'Haruka' and 'Subaru', in which lighting can markedly delay flowering. The effects of the NDL period before IL and those of the IL period were investigated. First, we investigated the effect of 2–12 days of NDL followed by 12 days of IL. Plants subjected to 2–6 days of NDL had longer flower clusters and a greater number of flower buds on the upper lateral flower stems and developed a broom-like spray. Next, we investigated the effect of 4–20 days of IL after 4 days of NDL. Plants subjected to 8–20 days of IL had longer flower clusters, more flower buds on the upper lateral flower stems, and broom-like sprays. These changes were more pronounced when the IL periods were prolonged. Therefore, spray formation in SAFS chrysanthemum cultivars 'Haruka' and 'Subaru' can be regulated by controlling the timing and period of IL (Table 5).

Table 5: Effect of Neutral Day Length (NDL) Period before 12 Days of Interrupted Lighting (IL) on Flowering and Cut Flower Characteristics of Sumer-to-Autumn Flowering Small-flowered Spray Type Chrysanthemums*

Cultivar	*NDL Period (Days)*	*Date of Flowe-ring (m/d)*	*Cut Flower Length (cm)*	*Stem Length (cm)*	*Flower Cluster Length (cm)*	*Neck Length of Terminal Capitulum (cm)*	*Number of Primary Lateral Flower Stems*	*Total Number of Capitula per Cut Flower*	*Flowe Cluster Width (cm)*
Haruka	2	8/28 a[z]	96 a	89 a	7.3 b	2.9 abc	14.7 a	29 ab	8.8 ab
	4	8/25 b	96 a	84 ab	11.5 a	4.4 a	9.5 a	36 a	11.8 a
	6	8/21 c	89 ab	83 ab	6.3 c	4.1 abc	9.4 a	11 b	8.0 b
	8	8/21 c	86 bc	80 ab	5.6 c	3.6 abc	9.8 a	11 b	7.2 b
	12	8/20 c	85 bc	80 b	5.3 c	2.7 bc	14.4 a	18 ab	9.0 ab
	No IL	8/14 d	80 c	76 b	4.6 c	2.4 c	15.5 a	19 ab	8.8 ab
Sabura	2	8/22 b	97 a	93 a	3.6 c	1.5 b	28.1 a	76 ab	10.8 a
	4	8/25 a	94 a	82 b	11.8 ab	2.5 b	15.0 b	85 ab	12.7 a
	6	8/20 bc	93 ab	79 bc	13.3 a	4.3 a	19.2 ab	116 a	15.0 a
	8	8/20 c	90 ab	81 bc	9.7 ab	4.1 a	17.1 ab	66 ab	14.2 a
	12	8/17 d	86 bc	78 bc	8.2 b	4.8 a	16.1 ab	38 b	13.3 a
	No IL	8/9 c	80 c	76 c	4.3 c	2.4 b	21.1 ab	57 ab	12.8 a

z Different letters indicate significant difference at $P<0.05$ by using Tukey-Kramer multiple range test.

* Mori *et al.* (2016).

7.2.1.5 Flower Bud Elongation

Anatomical studies revealed that the floral parts enlarged and started elongation after about 24 short days. However, the position was still reversible and Okada (1951) found that after 40 short days the flower development process reached an irreversible stage. Cockshull and Hughes (1968) reported that after about 42 short days, assimilates from photosynthesis were used up for the development of floral parts and their growth continued irrespective of subsequent day length. Schwabe (1950) suggested that after the stage of ovule formation the floret development could not be suppressed and therefore, continued to develop even under long days.

Investigations by Cockshull and Hughes (1971), Canham (1970) and Machin (1973) showed that daily light levels in which plants were grown in early short days had a bearing on their time of flowering and by providing supplementary light, the response to flowering was hastened by 3-8 days.

7.2.1.6 Effect on Flowering Stem and Flowers

Longer duration of light and low light intensity during early short days gave rise to compound sprays. The pedicel length and floret number were also found to be affected by alternate long and short-day conditions by Post (1947) and Post and Lacey (1951a). Jong (1986) investigated the adaptation of chrysanthemum to low light levels. Fourteen genotypes of *C. morifolium* were induced to flower under a gradient that ran from 3 to 35 W m^{-1} visible light at a day length of 9 hours. The sign of light stress was a reduction in number of flowers, followed by delayed flowering and finally no flowering with more light stress. At lower temperature the minimum level of lights for flowering increased.

The time and duration of supplementary lighting increased the production of flower buds per plant in pot chrysanthemum (Aabrandt, 1987). Horridge and Cockshull (1989) suggested a 4 h night break at higher latitudes where the total night length was greater, provided that no period of interrupted darkness, before or after the night break exceeded 8-9 h in length. In glasshouse, supplementary light under short-day conditions (11 h) influenced the flowering time and the number of flower buds of chrysanthemum cv. Surf. Best performance was recorded at 35 $\mu mol\ m^{-2}s^{-1}$ (Andersson, 1990). Supplementary lighting produced early flowering and increased the number of flowers/plant in pot chrysanthemum (Noble and Ellis, 1990). Interruption of the short-day period by lighting in year round chrysanthemum culture resulted in late flowering, heavier stems, longer internodes and flower stems, larger leaves at the top of the plant, possibly larger flowers and improvement of root function. The longer duration of interruption at early stage of growth caused excessive stem growth and bud desiccation, resulting in uneven trusses of flowers (Roelops, 1990). Warrington and Norton (1991) obtained field phenotypes (leaf area, flowering, *etc.*) under controlled environmental (CE) conditions, providing the field daylength and daily quantum integral conditions were reproduced. Best growth and flowering were recorded at 18 h daylength under highest PPF (highest daily quantum integral). Light which leaked from roofs of greenhouse in which supplementary lighting was used can be reflected by clouds anddiffused over the surroundings. When clouds are low, light levels upto 0.2 m mol $m^{-2}s^{-1}$ may be received by neighbouring greenhouses. This reflected light affects the crops of neighbours. Regulation of flowering in chrysanthemum through supplementary illumination has also been reported by Dutta and Seemanthini (1995). As the daylength increased from 14, 16, to 18 h, appearance of flower buds as well as commencement of flowering were delayed but the flower size, weight, stalk length and yield of flowers were increased. Long-day exposures extended duration of flowering. The longest duration of flowering was registered when plants were exposed to daylengths of 16 h for 15 days to 1 month after planting.

Thao *et al.* (2018) reported that the effects of LED lighting source toward plants growing under open-field conditions. LEDs are an advanced lighting system that emits appropriate light spectrum for the plant (Tables 6, 7 and 8).

Table 6: Effect of Supplemental Lighting by LED, Compact 3U (CFLs) and Incandescent (ILs) Lamps to the Number of Flowering Buds and Number of Qualified Plants for Harvesting*

	LEDs		*CFLs*		*ILs*	
	Plants	*per cent*	*Plants*	*per cent*	*Plants*	*per cent*
Total number of plants	1416	100 per cent	1231	100 per cent	1217	100 per cent
Total nuber of plants having flower(s)	1274	90 per cent	1083	88 per cent	1059	87 per cent
Total number of qualified plants	1062	75 per cent	850	69 per cent	780	64 per cent

* Thao *et al.* (2018).

Table 7: Effect of Supplemental Night Lighting by LEDs, Compact Florescent 3U (CFLs) and Incandescent (ILs) Lamps to the Mean Flower Number per Plant*

Week	*Flower Number/Plants*			*LSD*
	LEDs	*CFLs*	*ILs*	
9	7,05[a]	5,32[c]	6,30[b]	0,637
10	12,72[a]	8,83[b]	8,87[b]	1,355
11	17,88[a]	14,52[b]	14,87[b]	1,211
12	22,22[a]	18,62[b]	18,77[b]	1,163

Letters a, b, c present the level of statistically significant difference between variables in different light systems (P<0.05).

*Thao *et al.* (2018).

Table 8: Effect of Supplemental Lighting by LEDs, Compact 3U (CFLs) and Incandescent (ILs) Lamps to the First Date of Flower Bud, and the Averge Diameter of the Main Flowers*

Flower Related Parameters	*LEDs*	*CFLs*	*ILs*
First emerging date of flower bud	52	59	57
Diameter of the main flowers (cm)	6.63[a]	5.70[b]	5.86[b]

Letters a, b, c represent statistically significant differences (P<0.05).

* Thao *et al.* (2018).

Meyer and Horn (1995) showed that supplementary light from high pressure sodium (HPS) lamps induced better growth and flowering in pot chrysanthemum. Dutta *et al.* (1996) also recorded the highest yield of flowers per plant under exposure to 14h photoperiod for 30 days after 2 months from planting. Hanke (1996) observed that short days decreased the risk of later flowering of chrysanthemum. Plants subjected to short days (19.00-08.00 h dark period) by covering with black Mypex film came to bloom earlier than those grown in natural daylength. Potted plants of chrysanthemum cv. Co 1 when exposed to 18 h artificial light for 30 days after 1 month from planting, resulted in increased yield per plant and weight of cut flowers. The vase life of the flower also increased (Dutta *et al.*, 1998). Increasing daylength (natural daylength 21 h) delayed flowering in different cultivars of chrysanthemum. Flower bud differentiation was inhibited using a night break treatment (Shin *et al.*, 1999a). In order to improve the annual stability and uniformity in the flowering period, 'Kasuga-no-beni', a breeding line, showed less than five days difference in flowering under unheated plastic house and open field conditions (Naka *et al.*, 2018).

7.2.2 Temperature

Effects of temperature on growth and development of chrysanthemum during the different stages were studied by many workers. The results of the investigations are summarized hereunder.

7.2.2.1 Effect on Vegetative Growth

Bonamino and Larson (1980) observed taller plants with greater stem diameter and larger and heavier flowers when grown at 10°C night temperature for 9 to 10 hours and the remaining period at 16°C as compared to those grown at a constant 16°C. Similar observation was made by Parups and Butlar (1982) and 22°C at day and 16°C at night respectively, were found to produce the best overall growth. Tayama and Miller (1963) stated that as the plants aged from 0 to 10 weeks, the optimum temperature for growth dropped from 65°F (18.3°C) to 40°F (4.4°C).

Chrysanthemum cuttings subjected to natural low temperature in an open vinyl house for 30 days from early or mid-November, produced as many plants as were normally obtained after a five-week chilling treatment without chilling injury (Nishio and Fukuda, 1987). In greenhouse chamber, a negative DIF (day/night temperature difference) treatment of – 6°C reduced plant height of chrysanthemum. DIF was found to be more effective than short temperature drop (Cuijpers and Vogelezang, 1992). Average time to flower and plant weight were found to increase by increasing the night temperature but reduced the plant height (Willits and Bailley, 1994). When the difference between positive and negative DIF was as large as –12° as contrast to 4°C, reduction in plant height was observed. A 6 h drop in day temperature (DROP treatment) reduced the shoot length. The effect was more pronounced when DROP treatment was given at the start of the day, rather than the end, and the effectiveness increased with increasing temperature drop up to 8°C (Cockshull *et al.*, 1995). Karlsen and Bertram (1995) observed that both root DIF regime and aerial DIF regime affected the stem elongation. At constant root temperature, the negative aerial DIF reduced stem elongation by 41 per cent. Similarly, when air temperature was constant, root DIF reduced stem elongation by 18 per cent. Combination of water stress (application of 20 rather 40 ml of water/day) and a 6°C negative DIF treatment (*i.e.*, night temperature 6°C lower than day temperature) reduced the height of chrysanthemum cv. Maya (Roeber *et al.*, 1995). Karlsen (1997) also observed elongation of stem due to change in root temperature. Root temperature of 25/15°C or 15/25°C day/night significantly affected stem elongation of plant as compared to plants grown at a constant temperature of 20°C. Langton and Cockshull (1997) found that stem elongation in chrysanthemum was dependent upon the absolute day and night temperatures rather than DIF. Increase in day temperature over the full temperature range of 12-32°C and increase in night temperature up to about 24°C resulted in increased extension growth (Table 9).

Table 9: Length of the Internode (mm) Above the Tagged Leaf after 10 Days of Growth*

Night Temperature (°C)	*Day Temperature (°C)*					
	12	*16*	*20*	*24*	*28*	*32*
12	-	6.2	7.7	-	-	-
16	5.1	7.5	9.4	12.1	13.1	-
20	6.2	8.0	9.5	11.6	11.3	-
24	-	7.9	9.6	11.7	11.7	11.8
28	-	9.5	10.0	10.5	10.7	11.4
32	-	-	-	12.9	13.3	-

* Langton and Cockshull (1997).

Bertram and Karlsen (1994) studied the effect of irradiance and day/night tempera-ture on the stem elongation rate of chrysanthemum plants. The stem elongation rate (SER) was found to be high during the night and low during the day. This was attributed to the fact that high irradiance during the day induced high SER during the following night. Thus, plants grown under a negative day/ night temperature programme (low day/high night temperature) had a lower SER during the night and day than plants grown at constant temperature. Plant height and internode length were reduced by 10 h (artificial light at night) of – 4 and – 8 pulse DIF (difference in day and night temperature). The dry matter content was also found to increase by negative pulse DIF (Jensen, 1994). Under alternating 11 h days and 13 h nights, stem elongation rate (SER) in chrysanthemum was strongly temperature dependent. Total daily elonga-tion was similar at constant 18.3°C and at 11.5°C days and 24°C nights, but was much greater at 25.7°C days and 12°C nights. SER was rhythmic in continuous light with period of slightly less than 24 h. In continuous dark condition, however, SER declined rapidly and the rhythm disappeared within 11 h. Application of low temperature pulses (a rapid decline from 18.3°C to 8.3°C) applied for 2, 4, 6, 8 or 11 h during the day showed declining trend in SER followed by a slow recovery and peak shortly after the end of the pulse (Tutty *et al.*, 1994). Plants at vegetative stage was subjected to a 2 h temperature drop from 18° to 6°C in combination with a short (10 h) or a long (14 h) photoperiod. Average diurnal temperature was 17-18°C. The temperature drop treatment was applied to plants before sunrise (positive DIF), after sunrise (negative DIF), before sunset (negative DIF) or after sunset (positive DIF). Negative DIF after sunrise resulted in shorter internodes in chrysanthemum grown under short-days. Flower, stem elongation in chrysanthemum was not affected by temperature drop under short days, but under long days, stem length increased with a temperature drop after sunset or before sunrise. Khattak and Pearson (1997) observed that plant height was significantly affected by light quality and temperature. Plant height seemed to be regulated by the action of both phytochrome and a blue acting photoreceptor; internodal length decreased as temperature increased.

Effect of temperature on growth and floral development of chrysanthemum was also studied by Whealy *et al.* (1987) who reported that exposure to high temperatures during first 4 weeks of growth under short days increased the number of nodes, leaf area, stem length and dry weight of leaves and stems. From the results of the investigations of Schwabe (1950, 1953, 1959), Vince (1955a) and Okada (1959) it became evident that some cultivars required a certain period (3-4 weeks) of exposure to low temperature (4° to 7°C) for normal stem elongation and inflorescence initiation and thus failed to grow normally and formed rosette in the absence of low temperature. On the other hand, many cultivars did not require low temperature, hence rosettes were not found at any stage.

Table 10: Effect of Temperature and Cultivar on Relative Growth Rate (RGR), Net Assimilation Rate (NAR), Leaf Area Ratio (LAR), Specific Leaf Area (SLA), and Leaf Weight Ratio (LWR) of Chrysanthemum during the Long-day (LD) Period*

	RGR $(g.g^{-1}.d^{-1})$	*NAR* $(g.m^{-2}.d^{-1})$	*LAR* $(cm^2.g^{-1})$	*SLA* $(cm^2.g^{-1})$	*LWR* $(g.g^{-1})$
Temeprature (°C)					
15	0.093 a	3.70 a	223	315	0.712 c
18	0.110 b	4.37 b	224	334	0.675 b
21	0.119 c	4.63 b	231	359	0.646 a
24	0.122	4.62 b	245	373	0.659 ab
F_{pr}^{z} Linear	<0.001	0.006	0.047	0.005	0.004
Quadratic	0.008	0.036	0.267	0.677	0.017
Cultivar					
Annecy	0.110 b	3.91 a	248 c	343 b	0.726 b
Delianne	0.114 b	4.48 b	226 ab	319 a	0.709 b
Reagan	0.114 b	4.93 c	216 a	347 b	0.623 a
Supernova	0.105 a	3.98 a	234 b	371 c	0.633 a
F_{pr}^{z}	0.005	0.001	0.009	0.002	<0.001

[z]F probability; signigicant levels less than 0.05 presented in bold type. Different letters indicate significant differences between treatments based on studen's *t* test (P = 0.05).

* Van der Ploeg *et al.* (2007).

Van der Ploeg *et al.* (2007) observed significant variation in temperature response for reaction time, total dry weight produced, stem length, and flower size and number under greenhouse. Depending on the cultivar, stem length increased with temperature, especially during long day, as a result of both increasing internode number and average internode length. The response of both flower size and number to temperature was also highly cultivar specific (Table 10).

Heinsvig *et al.* (2007) found that plants grown with low night temperature formed fewer leaves and total leaf area decrease and increased leaf dry matter content at the expense of root dry matter content because of starch accumulation as starch synthesised during the photoperiod did not disappear during the dark period. However, low night temperatures had no effect on stem length.

7.2.2.2 Effect on Flowering

Post (1953c) classified chrysanthemum as high temperature plant since flower buds are initiated above a critical temperature. Watson and Andrews (1953) showed that in cultivar Gold Coast, no flower bud initiation occurred at l0°C while 99 per cent initiation was noted at 16°C. Flowering was, however, delayed at high temperature and in cultivar Seagull, temperatures between 24° and 27°C did not delay flowering but at 29°C a delay of 11 days was observed while at 38°C no flowering took place (Post and Kamemoto, 1950). A temperature of 10°C at floret elongation stage did not delay flowering in most cultivars (Vince, 1953, Cathey, 1954). The effect of night temperatures was found to be more important than the day temperatures in regulating the growth and flowering in chrysanthemum (Post and Lacey, 1951b). Cathey (1955a) showed that night temperature was three times more effective than day temperature in influencing the flowering response.

The flower quality is influenced by the prevailing temperature (Post, 1948; Post and Lacey, 1951b; Post, 1953a). Flowers of cultivar Gold Coast, for example, produced incurved florets at 15.6°C but reflexed florets at 10°C. At low temperature Indianapolis Pink developed tubular florets on outer whorl instead of normal flat ones. Flower colour, specially pink and bronze, faded at high temperatures.

Post and Lacey (1951b) suggested that high temperature might increase the number of florets formed on receptacle like long days but the blooms might get distorted. In yellow cultivars, bronzing of petals was recorded due to a drastic fall in temperature (Hendricks, 1984). Cairol (1982) noted the highest percentage of flowering when plants were grown at 20°C throughout and had higher dry matter content compared the ones grown either unheated or heated during LD or SD. In an experiment on day and night temperatures, Parups and Butler (1982) reported that number of blooms and buds in cvs. Accent, Fire Power and Blue Marble increased by split night temperature. In cultivar Divinity, however, the yield decreased significantly. Whealy *et al.* (1987) reported that increasing the exposure time to high temperature (30°C at day/26°C at night) increased the time to flowering.

Nakano *et al.* (2015) investigated heat-sensitive hours for flowering in *C. seticuspe*. Heat treatment at 30°C was applied for 16, 8, or 4 h d^{-1}, at different times of the day, under SD conditions (8 h photoperiod from 08.00 – 16.00h), using 20°C as the optimum temperature. Heating between midnight and dawn (00.00 – 08.00 h) delayed floral initiation for up to 2 d, but delayed flowering for > 25 d, which represented the heat-sensitive hours for capitulum development. Heating during this period suppressed expression of the *FLOWERING LOCUS T-like3* (*CsFTL3*) gene, a promoter of flowering induced in leaves by SDs (Table 11).

Table 11: Effect of Daily Heating Treatment at 30°C on Flowering of *Chrysanthemum seticuspe**

Experiment	*Heating Period**	*No. of Nodes#*	*DVC!*	*DF*
Experiment (16 h heating)	Control (20°C constant)	15.1±0.8 a	19.9±0.9 a	54.9±1.3 a
	ZT 0-16	15.9±1.3 ab	20.7±0.7 a	57.2±1.7 b
	ZT 0-8 and 16-24	16.7±1.0 ab	19.5±0.5 a	>71¶ c
	30°C constant	16.3±0.9 b	20.0±0.9 a	>71 c
Experiment (28 h heating)	Control (20°C constant)	13.4±0.5 a	21.0±1.0 a	54.6±1.8 a
	ZT 0-8	13.6±0.8 a	20.9±0.9 a	58.5±1.6 b
	ZT 8-16	13.4±0.5 a	20.5±0.07 a	60.0±2.4 b
	ZT 16-24	14.4±0.6 b	23.0±1.5 b	>71 c
Experiment (34 h heating)	Control (20°C constant)	14.0±1.2 a	19.5±1.3 a	48.3±1.1 a
	ZT 8-12	14.3±0.9 a	19.3±1.1 a	49.0±1.0 a
	ZT 14-18	14.9±0.6 a	20.5±1.5 ab	61.2±1.8 b
	ZT 20-24	14.7±0.6 a	21.2±1.2 b	65.8±1.6 b

* Photoperiod was 8 h light (ZT0-8) and 16 h dark (ZT8-24).

Number of nodes that developed until floral initiation. The uppermost visible node was marked and set to 0 at the start of each experiment.

! Mean values ± SD (n=18 in experiment 1; n=16 in Experiment 2; n=12 in experiment 3). Mean values followed by different lower cases letters in each column in the same Experiment differed significantly at P≤0.05 by the Steel-Dwass test.

¶ None of the plants flowered within the 70 d experimental period.

! DVC, days to visible capitulum; DF, days to flowering.

* Nakano *et al.* (2015).

A split night temperature regime involving a gradual decrease of temperature reduction from 65 (18.3°C) to 50°F (10°C) for about 4 h and again increased gradually to 65°F (18.3°C), recorded significant increase in plant height with more number of flowers/plant (Shabot and Koths, 1987, 1988). Hendriks (1988) stated that constant temperature (20°C) produced larger plants with more leaves while frequent fluctuations (alternating 24/16°C) in temperature might sometimes accelerate flowering.

Karlsson *et al.* (1989a) noted that the optimum temperature requirement for different developmental phases *viz.*, stage: I - start of short days to visible bud (2 mm diameter terminal bud); II-visible bud to disbud (10 mm diameter terminal bud); III-disbud to colour (flower bud showing colour); IV-colour to flower were 21.3, 20.3, 23.1 and 19.3 °C, respectively. Delaying effect on flowering at low night temperature 5° or 13° could be compensated by applying 27, 21, 17 or 13°C during the 12 h light period at appropriate growth stages (Wilkins *et al.*, 1990). It was found that at a night temperature of 32°C, transfer for 7 nights at the start of short-day treatment (SD) delayed the onset of flower bud initiation, while transfer for longer period, delayed the completion of flower bud formation. Delayed anthesis was also recorded. The maximum delay in anthesis occurred when high night temperature was given at the onset of floret initiation stage, (between 14 and 21 nights of SD). Night temperature was reduced (between 13 and 21 nights of SD) under black-out covers in summer as it affected the flowering (Cockshull and Kofranek, 1993). Flower diameter was found to decrease while the number of blooms per plant increased with the rise in night temperature (Willits and Bailey, 1994). An increase in stem length and earlier flowering was observed. Plants grown at 15°C night temperature and long day condi-tions were found to produce taller plants with early flowering as the temperature increased from 15 to 25°C. In an experiment, Shibata (1997) observed that high temperature during the summer delayed flowering and produced poor quality flower while summer-to-autumn flowering chrysanthemum showed smaller heat delay and were able to flower normally even under high temperature conditions (30/25°C, day/night). Adams *et al.* (1998) observed that from pinching, flowering occurred most rapidly in plants grown at a mean temperature of 20.4°C, while those grown at 10.9°C had the lowest leaf number at flowering. The final leaf number from the pinch to the inflorescence increased with increasing temperature above 10.9°C. This could be attributed to a marked sigmoidal increase in the rate of leaf initiation with temperature, not delayed initiation. Temperature below 20°C led to a considerable delay in floral initiation and once initiated, the inflorescence developed most rapidly at 20.2°C (Table 12). The number of axillary bud set increased significantly upto 9.3/shoot under high day temperature of around 45°C during the summer. When rooted cuttings were exposed to a low temperature treatment (7°C, 2000 lux, daylength 16h, for 4 weeks) after a high temperature treatment (around 45°C, 5 h per day for 3-4 weeks), the number of axillary buds increased 5 to 6 fold, when grown at a night temperature of >15°C (Shin *et al.*, 1999b). Willits and Bailey (2000) observed that at reduced night temperature, time of flowering of heat tolerant

Table 12: Mean Temperatures and Light Integrals for each Glasshouse Compartment Calculated for the Period from Pinching to the Sixth Stage of Flowering (Flowering), from Pinching until Floral Initiation (Initiation) and from 14 Days after Pinching until Sixth Stage of Flowering (Development)*

Glasshouse Temp. (°C)	*Mean Light Integral (MJ m^{-2} d^{-1}; minimum total solar radiation)*			*Mean Temperature (°C)*		
	Flowering	*Initiation*	*Development*	*Flowering*	*Initiation*	*Development*
6	9.6	9.6	9.4	1.7	2.3	1.3
10	10.9	11.9	10.7	1.5	2.3	1.3
14	17.1	17.3	17.1	1.6	2.5	1.4
18	20.4	20.7	20.2	1.6	2.4	1.4
22	22.9	23.5	22.8	1.6	2.4	1.4
26	26.1	27.3	26.0	1.6	2.4	1.4

* Adams *et al.* (1998).

varieties was affected compared to heat sensitive varieties. But the inflorescence diameter increased in the latter case.

Carvalho *et al.* (2005) reported that the temperature (15°C to 24°C) applied to long-day period (phase I), start of short-day period to visible terminal flower bud (phase II), and end of phase II to harvest stage (phase III) resulted in increased number of flowers per plant, but decreased individual flower size due to higher temperatures. Flower characteristics were less sensitive to temperature applied during the long-day period. Number of flowers per plant was affected positively by temperature, mainly during phase III, whereas individual flower size increased with temperature during phase II, but decreased with temperature during phase III. Lower temperatures during phase III significantly enhanced flower colour intensity in 'Reagan Improved'.

Nagasuga *et al.* (2008) found that the development of flower bud after flower budding was delayed by high temperature at petal formation and flower budding stage, and the number of days to anthesis was increased. High temperature during the late stage of involucre formation and late stage of floret formation increased the weight of inflorescence and number of florets, and decreased the percentage of disk florets, resulting in preferable cut flower quality in cv. Iwa-no-hakusen. Mato *et al.* (2009) reported delay in anthesis was caused by low temperatures during flower differentiation or high temperatures during flower development.

On the basis of differential response to temperature, Cathey (1955a) classified chrysanthemum cultivars into three groups as given below:

Thermozero

Cultivars which flowered at any temperature between 10° and 27°C but, most consis-tently at 16°C night temperature, *e.g.*, Shasta.

Thermopositive

Cultivars in which continuous low temperatures between 10° and 13°C inhibited or delayed bud initiation. Minimum temperature required for bud initiation was 16°C and it was rapid at 27°C, but flowering was delayed, *e.g.*, Cameo.

Thermonegative

Cultivars in which bud initiation occurred at any temperature between 10° and 27°C but bud development was delayed at high temperature, *e.g.*, Defiance. Vince (1955b) tested several British cultivars for their response to temperature to classify them into above mentioned three classes.

7.2.3 Air Composition

Oxygen and carbon dioxide played important roles in the growth and flowering of plants grown extensively in greenhouses in the temperate regions.

7.2.3.1 Oxygen

During day time, the required amount of oxygen is available as a result of photosynthesis. During night proper ventilation is necessary for ensuring sufficient oxygen supply. Brenn and Leone (1972) found that O_2 at 40 ppm for 4 hours caused severe injury to one cultivar and moderate injury to cuttings of 7 out of 16 cultivars tested. Konjoian *et al.* (1983) grew stock plants at 2.0-5.0 per cent and 20.5- 21.5 per cent O_2 and found that at lower levels of oxygen twice as many cuttings were obtained in comparison to plants exposed to higher levels of oxygen.

7.2.3.2 Carbon Dioxide

CO_2 becomes a limiting factor in winter months in greenhouses because the ventila-tors are closed to keep the inside warm. By artificially enriching the greenhouse atmosphere with CO_2, production has been found to increase significantly. Sturm (1966) reported greater stem length by supplementary CO_2. Carpenter (1966) found early flowering by increasing CO_2 (700-900 ppm). An increase in the dry weight of shoot by 27-60 per cent was recorded as a result of CO_2 enrichment between 330 and 1000 ppm in case of rooted cuttings of cultivar Horim (Mortensen and Moe, 1983). Higher levels were of no use. Eng *et al.* (1985) also reported improved vegetative growth in the cultivar Dramatic due to CO_2 enrichment which also caused significant increase in yield of chrysanthemum. The CO_2-enriched leaves were more efficient at low irradiances (< 400 mmol $m^{-2}s^{-1}$) because of greater photosynthetic efficiency and at high irradiances because of greater CO_2 conductance (Peet *et al.*, 1991). Tanigawa *et al.* (1997) found increased stem length, number and weight of leaves, flower diameter and advanced flowering by 2-4 days with CO_2 enrichment (1000 ppm).

7.2.4 Interaction of Factors

Investigation on the role of environmental conditions on growth and flowering revealed that the response of a particular factor was often modified by one or more factors and these often interact in a complex manner (Schwabe, 1953).

7.2.4.1 Temperature and Light

Watson and Andrews (1953) showed that at 16°C the cultivar Gold Coast produced the maximum budding at high light intensity but failed to develop any bud at low light intensity.

Effect on Flowering

Temperature altered the response of critical photoperiod necessary for initiation and development of flower. Cathey (1957) observed that the photoperiod required for initiation of inflorescence was shortened by lowering and raising the temperature from 80° to 50°F (26.6° to 10°C) and from 50° to 80°F (10° to 26.6°C), respectively. In cultivar Shuckan, it was possible to substitute cold requirement (16 days at 20°C) with higher temperature (24 days at 27-32°C) when provided with continuous light of 100 w/m^2 (Blondon, 1982). At higher intensity of 270 w/m^2 flowering was recorded at 27°C instead of 22°C. In phytotron experiments under different light and temperature regimes varying from 14-20°C and 60 J $cm^{-2}day^{-1}$ to 455 J $cm^{-2}day^{-1}$, Jong (1982) concluded that cultivars responded differently to light and temperature combinations, indicating the need for developing cultivars adopted to a wide range of light and temperature conditions. Okada (1963) classified the Japanese chrysanthemums into 6 ecotypes on the basis of their response to photoperiod and temperature as given below:

Type	*Group*	*Photoperiodic Response*		*Temperature Response*
		Initiation	*Development*	
I.	Autumn flowering	Short day	Short day	Flower bud initiation at 15°C or at higher temperature. Development not inhibited at high temperature.
II.	Winter	Short day flowering	Short day	Flower bud initiation and development inhibited at high temperature.
III.	Summer blooming	Day neutral	Day neutral	Usually flower bud initiation takes place at 10°C.
IV.	August blooming	Day neutral	Day neutral	Minimum temperature for initiation of flower bud is 15°C. Development inhibited at low temperature.
V.	September blooming	Day neutral	Short day	Same response as in August blooming.
VI.	Okayam Heiwa type	Short day	Day neutral	Same as in autumn flowering.

Supplementary light altered the response time of different cultivars in presence of low night temperature (60°F). Plant height and number of flowers increased under low night temperature and supplementary lighting. Geiffith and Mc Intyre (1989) noted the optimum growing condition of chrysanthemum at 68-72°F (20-22°C) for 10 h (day) + 60°F (15.5°C) for 14 h (night) temperature. At a constant temperature of 20°C, the flower size increased linearly as PPF increased from 1.8 to 21.6 mol/day/m^2 (Karlsson *et al.*, 1989b). Nishio *et al.* (1989) observed that at 18°C temperature, optimum day lengths for flower bud differentiation and flower bud development were 10 h and 9-10 h, respectively. Longer photoperiod improved the plant shape through elongation of lateral flowering shoot. A photoperiod of 11 h with a dark period from 18.00 to 07.00 h was recommended until the appearance of flower colour to obtain best quality cut flowers. Zhou and Li (1993) stated that after cold induction, summer chrysanthemum cultivars required short-day cycles. Low light intensity dela-yed flowering and resulted in the uppermost axillary branches becoming vegetative.

The effects of photoperiod and temperature on flowering in seven *Dendranthema* species and two cultivars Spring Song and Shubo-no-chikara were investigated (Fukai *et al.*, 2000). All species flowered under daylengths of 10 or 12 hours while those under a 14-hour daylength produced crown buds. Spring Song and *D. indicum* flowered under 10-20°C, *D. japonicum* and *D. yoshinaganthum* required a minimum 15°C, while other species required 20°C. The chilling treatment of cuttings enhanced or delayed flowering depending on the species.

Negative DIF increased the number of days to anthesis under short days (Hansen *et al.*, 1996). Increasing levels of blue light significantly reduced the time to flowering (Khattak and Pearson, 1997).

7.2.4.2 Carbon Dioxide and Light

A positive correlation between CO_2 and light on growth rate in *C. morifolium* cv. Horim was also recorded by Mortensen and Moe (1983). Application of CO_2 at the light level of 268 $Em^{-2}s^{-1}$ increased relative growth rate by about 16 per cent, resulting in advancement of growth by about 5 days. Combination of supplementary lighting with high pressure sodium lamp (HPS) at 140 $\mu mol\ m^{-2}s^{-1}$ and CO_2 enrichment (1375 $\mu l\ l^{-1}$) was superior to either factor in improving vegetative growth in cultivar Dramatic (Eng *et al.*, 1985).

7.2.4.3 Carbon Dioxide and Temperature

Tanigawa *et al.* (1997) studied the effect of interaction of CO_2 enrichment (1000 ppm) and day temperature (15 to 25°C) on chrysanthemum plant and observed significant effects on the length of flower neck and number of tubular florets.

7.2.4.4 Carbon Dioxide and Humidity

High RH and CO_2 concentration in the atmosphere influenced the growth (increased RGR, increased DW of leaves, stems and roots, and increased leaf area on main and lateral stems) of chrysanthemum plants. Gislerod and Nelson (1989) concluded that increased RH might be the cause of wider opening of the stomata which in turn facilitated CO_2 absorption and its utilization.

7.3.0 Effect of Growth Regulators

Growth regulators and various other chemicals have been found to influence growth and flowering in chrysanthemum.

7.3.1 Growth Promotion

7.3.1.1 Gibberellins

GA_3 induced marked variations in respect of growth and development depending on the time of application and the stage of growth. When used in the 7th week of SD treatment, GA_3 caused elongation of stem and pedicel during the subsequent three weeks. Stem elongation was also reported by Sheu *et al.* (1998). Spraying with GA_3 after exposure to normal temperature for 4-5 weeks resulted in production of shoots as tall as those from plants chilled (kept at 2-3°C) for 5 weeks. Stems grew well at a night temperature of 14-16 °C (Fukuda *et al.*, 1987). Plant height was also found to be enhanced by GA_3 application (Talukdar and Paswan, 1994, 1996, 1998; Dutta and Seemanthini, 1998). Root number and activities showing growth promotion in the basal part of cuttings increased due to application of GA_3 (El-Antably *et al.*, 1991). Application of GA_3 also supplemented low temperature requirement and promoted flowering (Harada and Nitsch, 1959; Barbat and Oschesanu, 1964; Poole and Ying, 1966). Early flowering was induced by GA_3 application as reported by Koriesh *et al.* (1989b). Nagarjuna *et al.* (1988) obtained maximum flower diameter and early flowering by spraying 200 ppm GA_3. Treatment with GA_3 increased the FW and DW of inflorescence, flower size, flower yield and duration of flowering (Dehale *et al.*, 1993; Dutta *et al.*, 1993; Rajagopalan and Khader, 1994; Talukdar and Paswan, 1996; Dutta and Seemanthini, 1998). Dutta *et al.* (1993) found GA_3 (150 ppm) useful for enhancing the flower quality and shelf-life of cut chrysanthemum flower.

The experiment by Strefeler *et al.* (1996) and Nishio and Fukuda (1997) revealed that a combination of ethephon and GA_3 was found to delay flower bud initiation in chrysanthemum stock plants. In another experiment, El-Keltawi *et al.* (1996) found that GA_3 in combination with alar (daminozide) could overcome the depressive effect of salinity and enhance the growth and flowering of chrysanthemum. Zalewska (1998) reported that application of gibrescol (gibberellic acid) at 500 mg/l 3 times at weekly intervals lengthened the stem by 19.5-22.2 per cent.

Singh and Nigam (2015) reported that application of GA_3 (150 ppm) and CCC (3000 ppm) improved plant height, plant spread, diameter of main stem, number of branches/plant, number of leaves/plant and days to first flower bud appearance and flowering, flower size, flower weight/plant and number of flowers/plant.

Table 13: Effect of different Concentrations of GA_3 and Cycocel on Flowering Characteristics of Chrysanthemum cv. Paintball*

Tretaments	*Days to Flower Opening*	*No. of Flowers per Plant*	*Flower Size (cm)*	*Flower Persistence (Days)*	*Flower Fresh Weight (g)*	*Flower Dry Weight (g)*
Control	126.33 (±2.34)	19.33 (±0.67)	4.81 (±0.64)	35.33 (±1.45)	3.77 (±0.12)	0.67 (±0.05)
GA_3 @ 100 ppm	120.33 (±1.45)	24.67 (±4.67)	7.70 (±0.41)	38.67 (±2.03)	5.28 (±)0.13	0.90 (±0.03)
GA_3 @ 200 ppm	116.33 (±2.03)	30.67 (±0.67)	8.53 (±0.48)	42.00 (±1.16)	5.55 (±)0.20	0.95 (±0.03)
GA_3 @ 300 ppm	110.67 (±2.91)	34.67 (±3.72)	9.38 (±0.41)	45.67 (±1.20)	6.18 (±0.11)	1.00 (±0.00)
Cycocel @ 1000 ppm	127.67 (±1.77)	19.67 (±2.03)	6.98 (±0.21)	35.67 (±1.77)	5.43 (±0.20)	0.90 (±0.03)
Cycocel @ 1500 ppm	128.67 (±0.88)	21.33 (±1.77)	7.40 (±0.57)	38.67 (±2.03)	4.78 (±0.29)	0.85 (±0.03)
Cycocel @ 2000 ppm	129.33 (±1.45)	22.00 (±1.53)	7.86 (±0.34)	42.00 (±1.16)	4.81 (±0.19)	0.82 (±0.03)
$SED_{(0.05)}$	2.89	3.80	0.68	1.77	0.27	0.05
$LSD_{(0.05)}$	6.30**	8.28*	1.47**	3.86**	0.58**	0.10**

Data in parenthesis indicated standard errors within replicates. SED stands for standard error of difference among means. LSD stands for least significant difference among means. SED and LSD are calculated at 5 per cent level of probability. * and ** indicated significant statistical difference among treatments, respectively.

* Alhajhoj (2017).

Alhajhoj (2017) reported that application of GA_3 @ 300 ppm exhibited maximum plant height, plant spread, number of branches per plant, number of suckers, number of leaves per plant, leaf area per plant, plant fresh and dry weight, number of flowers per plant, flower size, flower persistence, flower fresh, dry weight and early flowering (Table 13); however, Gajbhiye *et al.* (2017) reported early first flower bud initiation, flower opening, 50 per cent flowering and first harvesting, maximum flowering span, flowers plant/plant, flower yield/plant, flower yield/plot, flower yield/hectare, weight of flower, flower diameter, disc diameter and shelf life the the same concentration of GA_3.

7.3.2 Ethylene Releasing Chemicals

Shanks (1969) observed that a combination of ethephon and a pruning agent increased branching in chrysanthemum. Spraying of ethrel thrice was found to increase the plant height, number of leaves, fresh weight of cut flower and delayed flowering in summer chrysanthemum (Nakayama and Yui, 1988; Rounkova, 1989; Stanley and Cockshull, 1989). Ethephon treatments (800 and 1000 ppm) were found to significantly increase axillary bud set, number of buds and ethylene evolution (Shin and Choi, 1996). Cockshull and Horridge (1978) reported that flower initiation was delayed and more leaves were formed below the terminal flower bud when plants were sprayed with either 100 or 1000 ppm CEPA (2-chloroethyl-phosphonic acid) at the onset of short days. The delay in flower initiation caused by long days was further increased by the application of CEPA.

7.3.3 Other Chemicals

Biswas *et al.* (1967) reported marked promotion in floral initiation in chrysanthemum by spraying ∞-sitosterol and lanosterol at 100 ppm, and glycocholic acid at 500 ppm.

Treatment with triacontanol (100 μg/l) in nutrient solution increased the dry weight of plant, number of inflorescence and number of superior quality flowers (Skogen *et al.*, 1982).

Tija *et al.* (1969) found morphactin useful for increasing the ultimate height of chrysanthemum. They (1973) found that morphactin treatment (20-100 ppm) resulted in delayed flowering, reduced leaf size, thickened stems and reduced cell division.

Carpenter and Carlson (1972) reported that spraying chrysanthemum at pinching time with BA (benzyladenine) at 40 ppm and PBA at 200 ppm significantly increased branching and ethephon delayed flowering. In hydroponic culture, plants were found to be more sensitive to BA treatment. BA increased stem thickness, leaf area and number of lateral branches but retarded stem elongation and root growth. It also reduced the chlorophyll content of plants (Kang *et al.*, 1995b). In 6 non-branching cultivars of chrysanthemum, branching in all cultivars was significantly increased by application of 300 ppm BA, in 4 cultivars it was also increased, but to a lesser extent, by application of 50 ppm GA_3 and in one cultivar branching was increased by application of 13.3 ppm IAA (Fuji and Sasaki, 2000).

Sharma *et al.* (1995) found that MH and NAA delayed bud break. They also noted that MH decreased plant height but enhanced number of flowers per plant with superior quality as against tall plants produced by NAA treatment. Flowering was delayed with application of MH (Nagarjuna *et al.*, 1988).

Application of phthalimides [1-(3-chlorophthalimidio)-cyclohexanecarboximide] at 250-1000 mg/l in the 4th week of SD caused elongated peduncles, higher peduncle dry weights and greater internodal elongation than on untreated plants (Mc Daniel, 1984).

Singh and Bala (2018) reported that foliar spray of Benzyl Adenine at 200 ppm at bud initiation stage and at fully developed flower buds enhanced keeping quality of chrysanthemum cv. Punjab Shyamli.

7.3.4 Growth Retardation

The growth retardants though primarily used to restrict stem elongation also cause other effects. In order to develop well shaped pot plants, growth retardants have proved very effective; SADH and phosfon are being used in commercial scale to develop more compact and well shaped pot plants of chrysanthemum.

7.3.4.1 SADH (Alar, B-nine)

Improvement in leaf colour (Gortz, 1965) and softening the intensity of pink-red flowers (Loeser and Essie, 1967) were observed in chrysanthemum using growth retardants. The latter workers also reported delay in flowering of the treated plants. Dicks (1972) found younger leaves absorbing larger amount of SADH than older ones and the chemical was more effective when applied to turgid plants with dry leaf surface. Buxton and Culbert (1967) demonstrated that spraying Yellow Delware chrysanthemum with SADH increased flower life upto 5 days. The longest flower life was recorded by the application of the chemical at 2500 ppm 3 weeks after the start of short days.

Treatment with B-9 (daminozide) at 5000 ppm gave the highest number of flowers per plant (Nagarjuna *et al.*, 1988). Foliar or soil application of alar significantly reduced inflorescence diameter, shortened the length of primary and secondary shoots, increased chlorophyll content of the leaves and reduced plant dry weight. Foliar application was more effective than soil drench, in shortening the stem, delaying flowering and increasing shelf life of pot plants (Shawareb and Qrunfleh, 1988). Zalewska (1989) also recorded reduction in plant height with foliar application of B-9, on pot mum. Foliar sprays of alar and cultar were found to reduce the lateral shoot length in chrysanthemum plants (Shawareb, 1989). Gregov (1992) observed that alar 85 and uniconazole effectively reduced the plant height, stem diameter, flower number/plant, flower head diameter and pedicel length and also delayed flowering by 2-3 days.

Sanderson *et al.* (1994) obtained greatest retardation of plant height when chrysanthemum plants were subjected to vacuum infusion treatments (VIT) of daminozide at 250 mg/l as compared to plants totally immersed in 1000 mg/l daminozide and control. On the other hand, Nam *et al.* (1995) recorded 30 per cent reduction in plant height by applying 4800 ppm and 233 ppm daminozide on 15th June and 14thJuly planting, respectively.

Application of B-9 or cycocel as soil drench, increased numbers of flowers and flower buds in cv. Cheasepeake (Pak *et al.*, 2005), however, B-9 (1,250 and 2,500 ppm) produced good spray form and long vase life in cv. Reagan Sunny (Kahar, 2008). Qureshi *et al.* (2018) revealed that foliar spray of cycocel and B-9 at 2 and 6 weeks after transplanting in cv. Flirt recorded significant increase in fresh and dry mass of whole plants, leaves and roots and increased inflorescence number, and early emergence of buds and inflorescences, whereas, no significant effect on number of laterals (Table 14).

Table 14: Effect of Graded Concentrations of Foliar Applied Cycocel and B-9 on the Number of Laterals, Number of Inflorescences and Inflorescences Diameter of Chrysanthemum Morifolium cv. Flirt*

TREATMENTS								
Control		*Cycocel*			*B-nine*			*L.S.D.*
D.W.	*0.1 per cent T.80*	*1000 mgL^{-1}*	*2500 mgL^{-1}*	*5000 mgL^{-1}*	*100 mgL^{-1}*	*250 mgL^{-1}*	*500 mgL^{-1}*	
Number of laterals plant^{-1}								
3.2	2.8	2.7	3.6	3.1	2.8	2.5	2.4	0.4
Number of inflorescence plant^{-1}								
6.8	5.5	6.5	8.5	9.4	6.1	7.3	6.1	1.0
Inflorescence diameter (cm)								
6.6	6.7	6.4	6.8	6.8	6.8	6.8	6.8	0.1

Each value represents the mean of five replicates.

* Qureshi *et al.* (2018).

7.3.4.2 Phosfon

It was particularly in wide use in Britain for potted plants of chrysanthemums. The granular form is mixed thoroughly in potting compost at rates ranging from 300-750 g/m3 depending on cultivar, compost material and season; the liquid form is used as soil drench. Spraying is not done as it is toxic to leaves. Mahamoud and Steponkas (1970) opined that the retardation of growth with phosfon was due to shortening of internodes as a result of reduced cell division while the size of cells remained the same. According to Sachs and Kofranek (1963) phosfon inhibited sub-apical cell expansion and division but stimulated transverse cell expansion and division which resulted in thicker stems. Poole (1965) and Newton (1965) reported that higher doses of phosfon delayed flowering and reduced flower size. It also reduced apical dominance and encouraged branching (Machin and Scopes, 1978). Phosfon improved the petal colour in some red and pink cultivars by affecting the protein metabolism in pigment.

The treated plants developed thicker cuticle on the leaves resulting in greater resistance to disease and drought.

7.3.4.3 Paclobutrazol (PP333, cultar)

This growth regulator is now being widely used for the production of dwarf plants suitable for pot culture of chrysanthemum. Paclobutrazol tablets, drenches, gels, capsules and sprays were effective to control height of plants. Sanderson *et al.* (1988a) recorded that drenches consistently reduced plant area and flower number per plant. PP333 and B-9 reduced the plant height of both potted and field grown chrysanthemum plants (Barrett and Bartuska, 1982; Qiu and Liu, 1989) but B-9 was significantly less effective than that of PP333. In order to obtain good pot mum, single spray of 100 ppm paclobutrazol one month after potting is most effective (Lozoya, 1994). Similarly, Yewale *et al.* (1998a) obtained more compact plants with paclobutrazol at 100 and 75 ppm. They (1998b) also recorded increased stem thickness and flower size and reduction in fresh and dry weight of plant with increased concentration of paclobutrazol. Gao *et al.* (1991), Pasian and Struve (1997) and Singh *et al.* (1999) also reported that paclobutrazol was very effective in retarding plant height in chrysanthemum plants. Million *et al.* (1999) suggested that the critical concentration of paclobutrazol affecting plant growth was 24 mg/l.

Velmurugan *et al.* (2010) studied the effects of artificial short and long days along with paclobutrazol on year round flower production in cv. Co.1. Combined application of paclobutrazol (50 ppm) with 14 h dark period during February, April and June planted crops, increased total N, P, K, Mg and carbohydrates, however, total Ca was maximum in 13 h dark period. Further, expanding the day length by 2 h coupled with paclobutrazol (50 ppm) resulted highest uptake of total N, P, K, Ca, Mg and carbohydrates in August, October and December planted crops. Nasri and Ghaderi (2015) reported that paclobutrazol (25 to 50 mg/L) reduced plant height, number of flowers and number of shoots in *Chrysanthemum frutescence.*

7.3.4.4 Uniconazole

This growth retardant is now being used in place of alar (daminozide) as dwarfing agent in pot chrysanthemums. Exogenous application of uniconazole reduced the plant height and flower diameter and delayed the opening of flowers in potted chrysanthemum cvs. Independence and Champ (Bassi *et al.*, 1987). Single spraying with uni-conazole at 20 mg/l, 2 weeks after pinching or 2 uniconazole application at 10 mg/l applied 2 and 4 weeks after pinching were effective in reducing plant height (Starman, 1990). Uniconazole was found to be as effective as alar 85 in suppressing plant height, stem diameter, flower and pedicel length and delaying flowering by 2-3 days (Gregov, 1992). Barrett *et al.* (1994) applied uniconazole with high volume spray and revealed that the efficacy of the chemical increased with increased stem coverage and amount of chemical reaching the medium. Uniconazole was found to retard stem elongation linearly upto 8 weeks after propagation (5 weeks after pinching). Uniconazole/dip + spray treatments at 2.5 + 10, 15 + 5 mg/l resulted in reduction in plant height upto 33 per cent than control while it increased the area and thickness of leaf (Schuch, 1994). According to Nam *et al.* (1995) spraying twice, once at the start of flower stalk elongation stage and then a week later with uniconazole at 20 ppm recorded reduction in plant height, number of flowers/plant and advanced days taken to flowering. Hosni (1996) obtained similar results with soil drenching of uniconazole at 10 or 30 mg/l. Yoo and Kang (1999) observed reduced plant height, leaf size, internode length and flower stalk length along with low C: N ratio in leaves and increased chlorophyll content with both foliar spray and soil drench of uniconazole at 0.25-1.25 mg/pot. Uniconazole treatment at 0.05 mg a.i./pot at 14 days after potting with pinching and short day treatment were recommended for pot plant production (Juug *et al.*, 2000).

Zheng *et al.* (2004) reported that spraying of uniconazole (25 ppm) one month after planting produced more ethylene on the third day. GA_1 and GA_3-like substance contents decreased markedly in plants sprayed with uniconazole or brushed. Thus, growth of chrysanthemum plants is controlled not only by GA activities but also by ethylene.

7.3.4.5 Other Growth Retardants

Many other chemicals have been tried with varying success in retarding the growth in chrysanthemum. However, Hosni (1996) suggested that a concentration of 1000 or 3000 mg/l was effective to obtain dwarf plants. Soil drenching with cycocel produced dwarf, sturdy and thick flowering stem. This treatment also delayed flowering and reduced the flower number (Koriesh *et al.*, 1989 a, b). Foliar application at higher concentra-tion caused phytotoxicity although growth retardation was more pronounced with increased concentration (Gregov, 1992). Short days were found to show greater dwarfing effect than CCC by Heike (1971) and Miranda (1973).

Ancymidol was found to reduce the plant height when grown in bark-free substrate (Barrett and Bartuska, 1982). Preplant dip of shoot, root or entire rooted cuttings in ancymidol (10-50 mg/l) slightly delayed flowering besides reducing plant height (Reiss and Lewis, 1984). Application of A-Rest (ancymidol) in a medium containing 15 per cent spaghnum peat moss caused retardation in height of chrysanthemum plants (Bhat and Tayama, 1989). Million *et al.* (1999) found that 10 mg/l ancymidol in subirrigation water resulted in 20 per cent reduction in plant size relative to untreated controls.

In trials with flurprimidol at 15 g/l, Hendricks (1993) obtained compact chrysanthemum plants with two sprayings @ 0.5-1.0 ml in dwarfing cultivars and 1-1.5 ml in vigorous cultivars. Application of 40 ppm and 11.9 ppm flurprimidol caused 30 per cent reduction in stem length of 15th June and 14th July planting, respectively (Nam *et al.*, 1995). However, flurprimidol treatment caused some leaf injuries, irregular internodal length and delayed flowering by > 1 week. Delayed flowering as well as reduction in plant height was also observed by Pobudkiewicz and Nowak (1997). It was recommended to apply flurprimidol before disbudding to regulate the length of flower stalk and decrease the Swan's neck effect (Anonymous, 1997).

Sanderson and West (1995) obtained significant reduction in plant height and delayed flowering by spraying cimectacarb at 150-600 mg/l as compared to daminozide (5000 mg/l). However, cimectacarb application changed the colour of red and purple flowered cultivars but did not show any phytotoxic effect.

Polyamine conjugates, ornithine decarboxylase and polyamine metabolism regulated flower initiation in normal and male-sterile plants (Aribaud and Martin-Tanguy, 1994). Shi *et al.* (1993) also noted that flower bud differentiation was regulated by the metabolism of nucleic acid and proteins.

Sugiura and Fujita (2003) found that sprays of ethephon (200 mg/l) increased the number of days to flowering, leaf number, stem length, weight of cut flowers and diameter of stems. Sakamoto and Takao (2007) expressed that spraying with ethephone delayed flowering and increased cut flower length and the number of leaves. The longer the period between pinching and ethephone treatment, the more prominent these ethephone effects became. Sumitomo *et al.* (2008) reported Ethephon briefly suppressed extension growth and flowering, and induced dormancy and rosette formation. Mato *et al.* (2009) found that spraying ethephon delayed anthesis mainly by inhibiting flower differentiation and flower development before flower budding.

Alhajhoj (2017) obtained delayed flowering time without affecting floral quality traits in chrysanthemum cv. Paintball by spray of cycocel. Singh and Bala (2017) found that spraying of daminozide (1000 ppm) on cv. Anmol, resulted into reduction in plant height in the order daminozide>ethephon>cycocel, reduced internodal length, highest plant spread, diameter of main stem and leaf area, earlier bud appearance, colour break stage, full bloom stage and highest number of flowers/ plant, however, extend flowering duration (Table 15).

7.3.5 Growth Regulating Substances and Greenhouse Covers

Responses to selected chemical growth retardants (daminozide, paclobutrazol, and prohexadione-Ca), and GA_1 and GA_3 under 5 photoselective polymethyl methacrylate greenhouse covers with various phytochrome photoequilibrium estimates were evaluated in growth chambers using 'Bright Golden Anne' chrysanthemum as the model plant to better understand the height control mechanism by far-red (FR)-light depleted environment (Tatenam *et al.*, 2000). Plant height linearly decreased as photoequilibrium increased from 0.72 to 0.83. The rate of height decrease of daminozide (3500 mg l^{-1}) treated plants was less than that of water (control or GA_3 (50 mg l^{-1})-treated plants. The rate of height reduction was not different between control and GA_3-treated plants among chambers with various phytochrome photoequilibrium. Both paclobutrazol (0.73 mg l^{-1}) and prohexadione-Ca (200 mg l^{-1}) reduced plant height regardless of photoequilibrium, but the height reduction by paclobutrazol was greater than that by prohexadione-Ca.

8.0 Propagation

Chrysanthemums can be propagated both by vegetative and sexual methods. The former is used to maintain the purity of cultivar while the latter method is adopted by breeders to develop hybrids. The vegetative methods include multiplication through sucker, cutting, air-layering and micropropagation. Among these, propagation through cutting is the most common and popular method.

8.1.0 Vegetative Propagation

8.1.1 Sucker

Propagation of chrysanthemum using suckers is still followed by some growers. Mukhopadhyay and Das (1976) noted that the tallest plants were produced from suckers in comparison with those propagated by tip cuttings. Suckers are separated

Table 15: Effect of Growth Retardants on Growth and Flowering of *Chysanthemum morifolium* cv. Anmol*

Treat-ment	*Plant Height at First Bud Appearance*	*Plant Spread*	*Inter-nodal Length*	*Diameter of Main Stem at Flowering*	*Leaf Area*	*Leaf Colour*	*Days to Bud Appear-ance*	*Days to Colour Break Stage*	*Days to Full Bloom Stage*	*Flower Diameter (cm)*	*Number of Flowers per Plant*	*Duration of Flowering (Days)*
T1	27.09a	13.07g	5.12a	4.01j	1.03h	YG GP145B	84.26c	106.03d	126.26e	3.76g	25.86e	20.56de
T2	25.46ab	14.22f	4.61b	4.22i	1.36g	YG GP146A	84.76c	106.20d	126.63d	4.94bcde	25.20ef	20.06efg
T3	25.18ab	14.67e	4.53b	4.44h	1.73f	YG GP146A	84.90abc	106.33d	126.83d	4.80cde	25.03ef	20.46def
T4	25.04ab	14.70e	4.5b	4.64g	1.86e	YG GP146A	85.13abc	106.40d	126.93d	4.73def	24.16g	23.66b
T5	24.77ab	15.29a	4.35bc	4.76g	2.55b	YG GP146A	85.50ab	106.86c	127.33c	4.62ef	23.80g	21.13d
T6	24.92ab	15.23a	4.21cde	4.95f	2.60ab	YG GP146A	85.63ab	107.10c	127.50c	4.35f	23.50g	21.6d
T7	25.34ab	14.90cde	3.86cde	5.11e	2.01d	G137B	82.86d	104.06e	124.50f	5.69a	27.60d	22.63c
T8	24.81ab	15.20ab	3.73e	5.56b	2.59ab	G137B	82.60d	103.80e	124.03g	5.21bc	28.40cd	23.09bc
T9	24.13b	15.36a	3.30f	5.90a	2.66a	G137B	80.56e	103.66e	123.56h	5.12bcd	31.06a	25.06a
T10	25.20ab	14.95bcd	4.28bcd	5.22de	2.33c	G138B	85.76ab	107.23bc	128.06b	5.38ab	28.63cd	19.70fgh
T11	25.14ab	15.10abc	4.07bcde	5.32cd	2.40c	G138B	86.00ab	107.53ab	128.56a	5.12bcd	29.50bc	19.46gh
T12	24.48b	15.25a	3.70e	5.45bc	2.52b	G138B	86.06a	107.73a	128.73a	5.06bcde	30ab	19.16h
F test	*	*	*	*	*		*	*	*	*	*	*

Mean values in each column with the same letter are not significantly different at P<0.05 according to DMRT. *significant at P<0.05.

T1: control	T4:CCC @1000ppm	T7: Daminozide@ 500 ppm	T10: Ethephon @ 250ppm
T2:CCC @500ppm	T5: 1250ppm	T8: Daminozide @ 750 ppm	T11: Ethephon: 500 ppm
T3:CCC@750 ppm	T6:CCC@ 1500ppm	T9: Daminozide @ 1000ppm	T12: Ethephon: 1000ppm

* Singh and Bala (2017).

from mother plant and directly planted in the field or in pots. The long roots of suckers are cut to about 1.25 cm and basal leaves removed before planting to avoid direct contact of leaves with soil. The early planted suckers, produce bigger plants. Jothi *et al.* (1987) obtained best rooting in suckers when planted during June-July. Terminal cuttings from vegetative shoots are also used as planting material. The number of suckers/plant was observed to decrease as the cutting date was delayed from October 20 to December 5. Plant height was more when grown from cuttings early in the season. Days to flowering and flower stalk length tended to decrease with the later cuttings. Kim *et al.* (1993) found October 20 to be the best date for removal of suckers. Rooted cuttings obtained from suckers exhibited more vigorous vegetative growth but delayed flowering. Early flowering was recorded in plants raised from these cuttings propagated and planted early in the season. Suckers were found to be better propagating materials than lateral shoots (SunZhao *et al.*, 1999).

8.1.2 Cutting

Propagation through terminal stem cutting is the common method followed by commercial growers. The modern method of chrysanthemum propagation consists of raising disease-free stock plants under controlled environment. Cuttings are taken and rooted under optimum environmental conditions.

8.1.2.1 Stock Plant Production

The use of sterilized soil (to avoid disease), vernalization to avoid rosetting (Schwabe, 1953; Furuta and Kiplinger, 1955), frequent pinching, adequate nutrition and application of growth retardants to impart sturdiness to cuttings are some of the necessary requirements for proper growth of stock plants. However, the temperature varied to some extent with the time of year. Light also influenced the growth of stock plants and the quality of cuttings. The quality of cuttings was also improved by increased fresh and dry weight and stem diameter. Similar observations were recorded by Machin and Scopes (1978) who obtained enhanced production of good cuttings through supplementary illumination of 0.68 MJ/m^2/day to the stock plants. Application of higher dose of N to the mother plant was found to increase the N content in cuttings, which in turn resulted in more number of roots/cuttings, and an increase in number, length and fresh weight of roots/cutting (Kadner and Zerche, 1997; Druge *et al.*, 1998; Zerche *et al.*, 1999).

Under hydroponic culture, PPH (plant plane hydroponic) plants were found to have high proline concentration (indicator of relative stress) which exhibited greater temporal variation and hence produced poor yield of cuttings and cut flowers. The stock plants of chrysanthemum grown under hydroponics were observed to have reduced growth under 7-days stress treatment with NaCl. Rooting of cuttings of stressed plants was also depressed (Prabucki *et al.*, 1999). It was found that such plants were subjected to more stress than non-hydroponically grown plants (Druge, 1997).

Tanigawa *et al.* (2010) reported the effects of soil or soilless culture of mother plants on the yield of cuttings and the quality of rooted cuttings in cv. 'Seiun'. The yield of cuttings in soilless culture was about two times that in soil culture. Stem

Propagation of Chrysanthemum

Seeds

Selection of terminal shoots for cutting

Planting of cuttings

Rooting of cuttings

Propagation by suckers

Micropropagation

Source: Kereša *et al.* (2012) *African J. Biotech.*, 11: 6027-6033.

lengths of rooted cuttings taken from mother plants grown in soilless culture were longer and the number of leaves greater compared with those in soil culture. The yield of cuttings grown in soilless culture was greatest for a planting density of 25.6 per m^2 and with an intermediate concentration (nitrogen concentration was 142 ppm). It was clear that the stem length of rooted cuttings became shorter when cuttings from the mother plant were taken later and the concentration of nutrient solution was lower. Mother plants in the greenhouse were shaded from ambient daylight (control) using cheesecloth between June and August.

Laishram *et al.* (2016) studied the rooting behavior of cuttings affected by integrated nutrient management during the production phase of the plants. Plants receiving treatment comprising of 22.5 g/m^2 each of NPK + Vermicompost (1 kg/m^2) + Biofertilizers (*Azotobacter* + PSB + VAM) produced the maximum number of cuttings/plant, length of the longest root, number of roots/plant, highest rooting percentage and maximum fresh weight of cuttings at harvest and after rooting, and took least number of days for rooting (Table 16).

Table 16: Rooting Characteristics of Chrysanthemum Cuttings as Influenced by Integrated Nutruient Management during Production Phase of the Plant (Data are the pooled mean of two years)*

Treatment	*Dys taken for Rooting*			*No. of Roots/Plant*			*Length of the Longest Root (cm)*		
	At Harvest			*After Rooting*			*Pur-nima*	*Ajay*	*Mean*
	Pur-nima	*Ajay*	*Mean*	*Pur-nima*	*Ajay*	*Mean*			
T_1: Control	22.83	34.70	28.76	11.83	15.90	13.86	3.60	4.03	3.81
T_2: NPK_{30} (ad-hoc recommendation)	21.70	31.63	26.66	25.66	25.96	25.81	7.20	8.13	7.66
T_3: NPK_{45}	20.86	31.96	26.41	26.53	22.36	24.45	5.63	6.93	6.28
T_4: $NPK_{22.5}$	22.93	33.96	28.45	24.30	21.36	22.83	5.83	5.13	6.48
T_5: NPK_{15}	22.43	35.96	29.20	20.76	20.36	20.56	5.70	5.23	5.46
T_6: NPK_{30} + VC (1 kg/m^2)	22.66	32.66	27.16	26.23	20.86	23.55	6.13	5.43	5.78
T_7: NPK_{30} + BF	20.80	32.86	26.83	26.03	21.16	24.50	6.73	5.66	6.20
T_8: NPK_{30} + VC (1 kg/m^2) + BF	20.61	32.06	26.33	27.06	22.20	25.23	7.10	5.90	6.50
T_9: $NPK_{22.5}$ + VC (1kg/m^2)	20.86	31.93	26.40	25.83	24.00	24.91	5.23	7.90	6.26
T_{10}: $NPK_{22.5}$ + Biofertilizers	20.06	32.86	25.96	23.93	28.13	26.03	6.36	8.09	7.63
T_{11}: $NPK_{22.5}$ + VC (1kg/m^2) + BF	19.73	30.66	25.20	29.40	28.50	28.95	8.26	8.90	8.58
T_{12}: $NPK_{22.5}$ + VC (1kg/m^2)	20.96	32.90	25.66	28.23	21.86	20.05	5.13	6.73	5.93
T_{13}: NPK_{15} + Biofertilizers	19.66	31.66	25.68	21.16	22.20	21.68	5.23	6.90	6.06

Treatment	Dys taken for Rooting			No. of Roots/Plant			Length of the Longest Root (cm)		
	At Harvest			After Rooting			Pur-nima	Ajay	Mean
	Pur-nima	Ajay	Mean	Pur-nima	Ajay	Mean			
T_{14}: NPK_{15} + VC (1kg/m²) + BF	19.80	31.56	25.68	16.20	28.10	22.15	4.96	8.06	6.78
T_{15}: VC (1kg/m²)	21.30	33.70	27.50	13.16	21.10	17.13	4.13	4.62	4.21
T_{16}: BF	20.83	32.76	27.50	13.60	17.90	15.75	4.13	4.98	4.53
T_{17}: VC + BF	19.87	33.10	26.83	15.90	19.93	17.91	4.53	4.99	4.65
Mean	20.97	32.82		21.69	22.46		5.63	6.38	
CD ((P=0.05)									
Treatments (T)	0.87	1.53	0.88						
Cultivars (C)	0.30	0.52	0,30						
T X C	1.23	1.16	1.01						

N-Nitrogen, P-Phosphorus, K-Potassium, VC-Vermicompost, BF-Biofertilizers (*Azotobacter chroococcum*+solublizing bacteria+VAM)

* Laishram *et al.* (2016).

8.1.2.2 Root Formation in Cutting

Quality of Cutting

The ideal cutting and optimum environment for rooting of cuttings have been thoroughly investigated in chrysanthemum, an ideal cutting being 4-6 cm long with basal stem diameter between 3.2 and 4.8 mm (Machin and Scopes, 1978; Butters, 1970).

Hoeven (1989) observed that larger and heavier cuttings initiated better rooting and promoted flowering than the smaller and the lighter ones. Hence, uniformity of cuttings (according to surface area) is very essential to obtain homogenous flowering. The quality of materials for cutting declines with the advancement in age of the stock plants. It was observed that removal of axillary buds along with some leaves produced better quality shoots for cutting than the untreated plants (Ruiter, 1993). Cutting quality increased when the ratio between the number of axillary shoots and the number of leaves decreased from 4: 16 to 1:16. The number of pith cells also increased when the ratio decreased thereby creating a greater sink for food materials which enhanced the growth of axillary buds (Ruiter and Tromp, 1996). *In vivo* studies showed that growth of axillary bud was better at lower positions in the plant and the time needed to produce a cutting tended to decrease with the age of axillary bud (Ruiter, 1996). Ruiter (1997) found that the axillary bud formation varied with the number of leaves available per axillary bud. He also found that high assimilate supply acceleratedaxillary bud development before cuttings were taken and this resulted in rooted cuttings with greater quality, size and weight.

Flowering influenced the formation of adventitious root in chrysanthemum cuttings as observed by DeVier and Genever (1997). Cuttings collected from stock plant bearing flower bud showed poor rooting. IBA could partially overcome the negative effect of flowering on root formation; cuttings taken from plants bearing fully opened flowers failed to root even after auxin treatment. However, it was observed that removal of buds from cuttings or continuous removal of flower buds during the growing period of stock plant increased the rooting response compared to the cuttings with flower buds intact. Furthermore, cuttings taken from the top 3 nodes of the stock plant containing flower buds rooted comparatively better than those from the lower stem portions with vegetative buds. The negative influence of flowering on root formation appeared to be due to the photoperiodic induction of the flowering stimulus rather than a direct competition for resources between the flowers and the developing roots. The optimum environment for rooting of cuttings can be considered in two parts, namely substrate environment and aerial environment.

Substrate Environment

The factors comprising environment of substrate are physical and chemical condi-tions of media, temperature and moisture and the presence or absence of pests and diseases. Several materials have been found suitable as substrate alone or in combinations. The basic requirements are good drainage, light weight and the absence of toxic substances. Rooting to the extent of 100 per cent was observed in the Ein Gedi System (EGS). Results obtained by Tomati *et al.* (1993) showed that composted materials favoured rooting, root initiation and root biomass development and these effects were particularly enhanced by earthworm casts. Verhagen (1993) on the other hand suggested the coarse graded peat substrate. According to Nishio *et al.* (1994), the best compost was a mixture of peat moss, perlite and vermiculite (1:1:1 or 3:1:1) with 3 g calcium hydroxide/l, although peat moss with 1-2 g calcium hydroxide/l was adequate. However, the addition of 60 mg N/l to the compost improved rooting, with nitrate giving better results than ammonium.

Geneve *et al.* (2004) found that that using the peat–perlite medium, water content had the greatest impact on rooting when the mist volume was low (600 $mLm^{-2} h^{-1}$). Wise *et al.* (2020) reported that Nutrifield's biostimulant complex Root Nectar® (1 mL/L) and biostimulant willow bark extract (1.06 µL/L) improved the formation of adventitious roots and root branching in semi-hard wood cuttings of chrysanthemum, can be utilized to speed up propagation of soft and semi-hard wood cuttings.

According to Buta *et al.* (2013) Klassman peat + perlite, followed by Klassman peat + NPK and Klassman peat + Osmocote was the rooting substrate. Blok and Verhagen (2009) in search of possible substitute of peat, there was an increase in the use of locally produced renewable carbon based substrates and in nutrient recirculation for soil grown crops. To address the peat harvesting for horticultural purposes in parts of Europe, Dutch growers, researchers, potting soil producers, an auction and traders initiated in 2003 a research program named the New Growing Media Project. The aim was to assess the technical and economical potential of peat alternatives. The most popular alternatives were coir products which increased from 17 to 40 per cent. Other alternatives used were various barks, rice hulls, perlite, wood fiber, composts, and rockwool granulate. Those alternatives were used in mixtures from 5 to 30 per cent.

Tymoszuk *et al.* (2009) compared the quality of chrysanthemum plants which were acclimatized after rooted *in vitro* and *in vivo* in selected substrates. Peat, perlite and a mixture of peat with perlite in the volume proportion of 2:1 (v:v) were used for direct rooting *in vivo* taking place simultaneously with acclimatization. Plants rooted *in vitro* had numerous short roots of high weight. The highest increase of the shoot length was observed in the plants previously rooted *in vitro* and in the plants rooted *in vivo* and directly acclimatized in perlite. The most of new leaves were formed by microcuttings rooted in vitro. The period of obtaining new *ex vitro* seedlings can be shortened and the costs of micro-propagation can be reduced (Table 17).

Table 17: Number of Roots on Chrysanthemum Microcuttings Depending on the Rooting Conditions and the Cultivars*

Rooting Conditions (B)		*Cultivar (A)*			*Means*
In vitro Rooting	*Substrate for Acclimatization*	*Bislet*	*Euro*	*Reddy*	
Yes	peat+perlite	18.2	21.04	20.84	20.03
No	peat+perlite	6.24	10.76	7.88	8.29
	peat	6.76	9.20	7.44	7.8
	perlite	8.24	12.52	20.20	13.65
Means	9.86	13.38	14.09	-	
$NIR_{0.05}$; $LSD_{0.05}$		A = 1.94, B2.47, B/A = 4.28, A/B = 3.89			

* Tymoszuk *et al.* (2009).

Rooting of chrysanthemum cuttings on ebb-and-flow hydroponic benches were investigated by Buwalda *et al.* (1995). Cuttings treated with 500 µg IBA/g were inserted into rockwool, jute plugs, glasswool and perforated polystyrene (PS) sheet. Cuttings rooted on PS sheets recorded the highest relative growth rate. Jute substrate addition with nutrient solution increased the relative growth rate. Grewal *et al.* (2005) reported significant higher rooting in tretated terminal cuttings of cv. Snowball with IBA (400 ppm), however, mean rooting was higher during first week of June (56.7 per cent) than in the first week of July (50.5 per cent).

Substrate temperature between 18° and 21°C has been reported to be useful for root formation in cuttings and below 16°C rooting declines rapidly (Machin and Scopes, 1978). Hentig and Fisher (1981), however, obtained optimum rooting when substrate temperature was 30°C. Roots tended to become thicker when root zone temperature decreased (Kang *et al.*, 1995a). The optimum pH range for rooting is between 7 and 8 (Parker and Kamp, 1958). Low pH possibly resulted in manganese toxicity. Paul and Smith (1966) recommended calcium saturation between 37.5 and 75 per cent while using peat as a medium. Raabe and Vlamis (1966) found that excess sodium or potassium caused complete or partial inhibition of rooting which could be prevented by adding calcium carbonate or calcium sulphate. Addition of trace elements and dolomite lime stone in peat for better rooting was also suggested by Gortz (1966). Paul and Thorhill (1969) reported inhibition of rooting due to magnesium toxicity when exchangeable Mg level was higher than 80 per cent.

In hydroponics, root hair development tend to increase as the concentration of the nutrient solution increases. Similarly, water holding capacity also influences rooting in cuttings. Sand having 75 per cent water holding capacity proved best for root formation while that with 35 per cent water holding capacity was unsuitable and resulted in death of most cuttings. Nutrient supply with irrigation was recommended by Hentig (1959) to promote rooting, but Riehl (1962) found that number of roots produced were inversely proportional to nutrients supplied by sub-irrigation. This is due to the fact that because of sub-irrigation, anaerobic condition and high salt concentration tend to develop in soil or other growing media used, thereby affecting the root growth. However, Zhang and Graves (1995) recorded 100 per cent rooting in cuttings under sub-irrigation with nutrient solution twice a day. Thus, it was concluded that sub-irrigation can replace mist during propagation of chrysanthemum. Nutrient solution promoted the growth of rooted cutting during hardening, and appropriate irrigation frequencies provided adequate water and nutrients to the cuttings. Leaf temperature and wetness of cutting was maintained properly by sub-irrigation method, which enhanced photosynthesis and the susceptibility to disease (Oh and Kim, 1998). Kim and Kim (1999) recorded better growth in fine perlite media supplemented with high irrigation frequencies. Nutrient mist too resulted in an increase in the number of roots and shoot growth of the cuttings (Wott, 1969; Wott and Tukey, 1969; Wille, 1974). An aero- hydroponic method showed the advantages of both types of culture. In this case, the cuttings were grown without mist. Root length and number of roots per cutting were greater in aero-hydroponics (Soffer and Burger, 1988).

Steam sterilization of the medium was practised as a routine preventive measure against diseases and pests (Machin and Scopes, 1978). Kher (1986) reported that heated and precooled sand reduced the loss of cuttings as compared to unheated sand. Treatment of the rooting medium (Gjaerum, 1963) or the basal ends of unrooted cuttings with fungicides like ferban (Tompkins and Middleton, 1950), zineb (Klaus, 1957), agrimycin (McFadden, 1958-59) and benomyl (Judd and Walton, 1973) was also suggested to reduce the loss due to diseases during rooting. Reed *et al.* (1986), however, noted that adventitious rooting in cuttings of cv. Golden Anne was inhibited by soaking or drenching the medium with fungicides. In general, botran, banrot and thiram caused significant reduction in rooting as compared to the controls in most replications; benlate, captan, daconil, lesan, and truban were intermediate, causing significant inhibition in some replicates but not in others; maneb, terrachlor and daconil + terrachlor usually were not significantly different from the controls.

Enhanced root and shoot growth of chrysanthemum cuttings inoculated with fungus *Trichoderma harzianum* was reported by Mac Kenzie *et al.* (1995). Peat bran amendment containing *T. harzianum* (isolate T-12) added to peat/perlite medium enhanced rooting of cuttings.

Aerial Environment

Temperature, light, humidity and air composition constitute the aerial environment. Borowski *et al.* (1987) observed that increasing the air temperature from 15 to 21°C accelerated rooting and increased the number of roots per cutting. In aeroponic culture, Molitor and Fischer (1993) observed optimum rooting at 27°C with continuous spraying and at 24°C with intermittent spraying. The combination of either a 5-30s spray every 4 min or 120s spray every 8 min also recorded good root growth. The optimum pH was 6.0.The fluctuation of root zone temperature between 43.3 and 8.3°C in the summer and 27.8 and 8.5°C in the winter produced consistently higher proline contents which in turn showed higher temporal variations and this affected the yield of cuttings (Druge, 1997).

Good light conditions have been found to be conducive for better root growth. Supplementary lighting increased the number of shoots per plant for cutting as well as the performance of the cuttings for rooting (Statens and Hagebruksskole, 1966; Waxman, 1967-68). Similarly, higher temperature (21°C) and high quantum flux density (200-420 $\mu Em^{-2}s^{-1}$) for 24 hours resulted in the greatest root and shoot weights (Stefanis and Langhans, 1982).

Cuttings under high temperature and high light intensity require high humidity. The effects of light intensity and mist interval on rooting and quality of chrysanthemum cuttings were studied by Song and Lee (1999). The best rooting (100 per cent) was achieved in unshaded cuttings misted at 1 and 1.5 h intervals. Growth retardation was observed under 50 per cent shading. All growth parameters were enhanced by decreasing the misting interval. Rooting percentage was found to decrease under shading conditions in a plastic greenhouse at 40 per cent shading. The unsha-ded treatment recorded higher percentage of rooting with more fresh weights of the rooted cuttings, volume of cuttings with roots, stem diameter, leaf number and plant dry weight. However, the rate of root formation, root number and length, root volume and fresh and dry weight of roots were enhanced at 40 per cent shade (Borys *et al.*, 1995). Chaitra *et al.* (2017) found that the cuttings under photoperiod recorded highest number of cuttings and rooted cuttings, rooting per cent, intermodal length of rooted cutting and root length. Further, concluded thatthe chrysanthemum varieties grown under photoperiod regulation helps in achieving year round production of quality planting material and subsequent commercial production of quality chrysanthemum flowers.

Growth Substances

Chrysanthemum cuttings root easily. Yet applications of some growth substances have been found to increase rooting. IBA treatment resulted in quicker rooting and faster roots production in chrysanthemum cv. Bright Golden Anne (Wodecki and Holcomb, 1989). Application of ethylene to chrysanthemum cuttings stimulated root development in cuttings held in high or low dissolved oxygen concentrations under aero-hydroponics culture (Soffer *et al.*, 1989). Spraying or immersing cuttings of several chrysanthemum cultivars in 50 ppm IBA stimulated rooting, especially during the dark winter months and resulted in a more compact root ball. Best results were obtained when the cuttings were stored for upto 7 days between IBA treatment and striking. When the resulting plants were used as mother plants, they gave a higher yield of cuttings than mother plants derived from untreated cuttings (Petters, 1992). Treatment with growth retardant SADH also promoted rooting in cuttings (Read and Hoysler, 1969; Shin and Lee, 1979). Cycocel, on the other hand, caused a marked depression in adventitious root production (Read and Hoysler, 1969).

Method of making Cutting

The method of making cuttings advocated now is to snap them from stock plant by hand (avoiding knife to prevent disease transmission). These should be inserted directly in the medium without removing basal leaves (Machin and Scopes, 1978). The practice of firming the medium immediately after insertion of cuttings is consi-dered unnecessary especially where benches are used. A spacing of 50 by 45 mm should be allowed in winter, 45 mm^2 in spring and autumn and from 35 **x** 30 mm to 35 mm^2 in summer was recommended (Machin and Scopes, 1978). Roots appeared within 5-10 days and cuttings were ready for planting within 3-4 weeks.

8.1.2.3 Storage of Cuttings

Pre-rooting storage of cuttings for 7 days at temperatures between 9 and 21°C under dark condition enhanced rooting. This saved space and time (Pol, 1988). Tip scorching occurred in cold-stored cuttings of glass-house chrysanthemum during winter months. Storage at 6-7°C was found to be better than 1-3°C in enhancing rooting (Hoeven, 1990). Rajapakse and Kelly (1995a) and Rajapakse *et al.* (1996) noted that low-temperature storage potential of chrysanthemum varied with cultivars. At 0-3°C, cuttings could be stored for 4-6 weeks without significant loss in quality and induced earlier flowering (Table 18). Chang and Huang (1995) reported that pre-rooting storage reduced the bench time and promoted efficient production of cuttings. Stored cuttings had better rooting efficiency and better growth (longer roots). In cultivars, where leaf necrosis occurred at low temperature, provision of light was found to be beneficial in delaying the development of leaf necrosis and maintaining quality. You *et al.* (1996) found that cuttings could be stored for three weeks at 5°C. The contents of carbohydrates in cuttings decreased with increasing temperature and storage period (Woo *et al.*, 1997). Cold storage of cuttings were found to reduce stem length and fresh weight of cut flowers, number and height of newly formed suckers while it induced early and uniform flowering and quality of cut flowers. Cold storage of cuttings could prevent or reduce rosette-like growth in the upper part of plants (Sun *et al.*, 1999).

Stored substrates had an adverse effect on the growth of chrysanthemum plants (Alt and Staat, 1997). In stored peat-clay mixture with high pH (7.0) and peat subs-trate 100 per cent survival of cuttings was recorded when stored for 30 days at 1 or 5°C and for 5 days at 25°C. But cuttings stored for 30 days at 1°C or stored for > 20 days at 5°C were infected by *Botrytis cinerea* and *Penicillium expansum* respectively.

Cuttings immersed in 10 ppm DCPTA (2-diethylaminoethyl-3,4-dichloro phenyl ether) for 10sec, stored overnight at 7°C, propagated under mist and grown to flowering under mist, produced taller plants with the highest number of flowers/ plant (Sanderson *et al.*, 1988b).

Table 18: Effect of Dark Storage Temperature on Poststorage Growth of ' Anna' and 'Debonair' Chrysanthemum*

Cultivar	*Temp (°C)*	*Flowers (No.)*	*Days to Flower*	*Flower Diam.*[z] *(cm)*	
Anna	Controly	23	48	6.9	
	0	25	43	4.5	
	3	26	41	4.6	Temperature
		NS	NS	NS	Control vs.
(0, 3)		NS	**	**	Debonair
	Control[y]	16	53	4.6	
	0	16	41	3.6	
	3	15	41	3.6	Temperature
		NS	NS	NS	Control vs.
(0, 3)		NS	**	**	

Z Average diameter of terminal flowers.

Y Control plants were not stored; numbers are the mean of 10 plants.

NS, *, ** Nonsignificant or significant at P= 0.05, or 0.01, respectively.

* Rajapakse *et al.* (1996).

8.1.3 Grafting

The flowering response of short day cultivars of *D. grandiflorum* (Paekulryeo, Seolpoong and Hwangulyreo) grafted on day neutral cv. Choonkwang was investigated by Lee *et al.* (1996). They observed that cultivar Choonkwang grafted on Hwangulryreo produced early flowering and more number of flowers. Graft union between day-neutral cv. Jeong Woon under inductive conditions and short-day cultivars Paekulryeo, Seopoong and Bonghwang under non-inductive conditions took 3-4 days. Flowering was induced in day-neutral cultivars under inductive conditions. Flowering was delayed in grafted stocks (Kim and Lee, 1998).

8.1.4 Micropropagation

8.1.4.1 Explant

Chrysanthemum can be multiplied at a fast rate by culturing shoot apex *in vitro* (Ben Jaacov and Langhans, 1972; Earle and Langhans, 1974 a) from somatic callus (Hill, 1968; Earle and Langhans, 1974 b), from peduncle and other floral parts (Roest and Bokelmann, 1975; Bush *et al.*, 1976) and from leaves (Sluarkiewicz *et al.*, 1982). Cossio and Bussi (1982) used 2-3 cm long terminal cuttings with at least one node as explant. These were further trimmed after sterilizing in 70 per cent ethanol followed by 4 per cent sodium hypochlorite, then placed on nutrient agar substrate in test tubes and kept at 23.2° C for 16 h/day under 4000 lux.

New cultivars of Higo-chrysanthemums were regenerated from florets of 3 lines cultured *in vitro*. Thus, floret culture was done to obtain new cultivars (Mizutani and

Tanaka, 1994). Chrysanthemum receptacles placed on MS medium supplemented with 0.5 ppm 2,4-D or NAA resulted in callus formation. Organogenesis was induced when calluses were transferred to medium supplemented with kinetin (Qian, 1996). The rate of differentiation was influenced by genotype. Shoot tip as explant wassuccessfully used for plant regeneration through somatic embryogeneis in chrysanthemum (Singh *et al.*, 2016; Ghosh *et al.*, 2018). Ray florets as explant was used for culture establishment, shoot initiation and proliferation (Misra and Datta, 2007; Padmadevi *et al.*, 2009; Batra *et al.*, 2011). It is generally used by the breeders in chimera management to establish the new cultivar in a better way.

8.1.4.2 Meristem Culture

Meristem culture was used for producing superior lines of chrysanthemum cultivars (Ohishi *et al.*, 1986). Clonal propagation through meristem culture *in vitro* resulted in production of disease free plantlets, which were resistant to most important viruses (TSWV) and viroids like chrysanthemum stunt viroid (ChSV). Both yield and quality of flowers were increased after *in vitro* culture (Angarita, 1999).

Furuya (1999) described a procedure for rapid multiplication of chrysanthemum cv. Shuho-no-Chikara by shoot tip culture. Shoot tips were cultured *invitro* in MS medium which contained various combinations of plant growth regulators (best results were obtained with 0.2 mg BA + 0.02 mg NAA/l). Multiple shoots in the same liquid medium increased about 7-fold in mass after 30 days of subculture. Multiple shoots transferred onto solid MS medium containing 0.2 mg BA/l also increased about 4-fold in mass after 30 days subculture. Shoots regenerated upon transferring the multiple shoots onto hormone-free solid MS medium containing 1.5 per cent sucrose. The regenerated shoots increased by aeration culture. About 240 plantlets could be produced in 130 days from a single shoot tip.

Shoot explants of chrysanthemum cultured in a 'Culture Pack' (a disposable box-shaped culture vessel covered in fluorocarbon polymer film) on agar medium or on rockwool resulted in larger plantlets which were more vigorous and having desirable morphological features (Tanaka *et al.*, 1992). Ihsanul *et al.* (1998) observed that the time required for bud sprouting ranged from 16 (BA at 1.2 mg/l) to 19 days (BA at 1.5 mg/l). Photoautotrophic cultures of chrysanthemum cv. Birbal Sahni were established under CO_2 enrichment (2 per cent v/v) from nodal explants, cultured on MS medium supplemented with 0.1 mg IAA + 0.2 mg BA/l. These cultures minimized the risk of contamination since the plant material did not require a carbon source in the culture medium. This also reduced production costs (Mitra *et al.*, 1998).

Deein *et al.* (2014) reported that 1.5 mg/L PBZ in combinations with 1.0 or 2.0 mg/L BA provided the best results with shoots in cv. Cayman. Pant *et al.* (2015) standardized *Psyllium* husk and market sugar as suitable alternatives to the conventionally used agar and sucrose, cutting down the production cost of tissue culture raised plantlets to over six times.

8.1.4.3 Regeneration

An overview of the regeneration of chrysanthemum by direct and indirect organo-genesis/embryogenesis from explants and embryo rescue has been

Promising Pot Mum Cultivars

Aideen Red Fire | Arlette Purple

Beverly Gold | Beverly Bronze

Cheryl Golden | Chelsey White

Hannah Orange | Wanda Lavender

Source: **Syngenta flowers, 2280 Hecker Highway, Gilroy, California 95020 USA, Syngentaflowers-us.co.**

enumerated by Rout and Das (1997). Datta *et al.* (2000) standardized the technique for large scale multiplication of cv. Kargil 99 through direct organogenesis from ray and disc florets. Disc florets were more organogenic than ray florets. The highest (75 per cent) organogenic response was obtained from disc florets in NAA (1 mg/L) and BA (5.0 mg/L).

Chitra *et al.* (2006) found that the high frequency of multiple shoot regeneration response for nodal segments (95 per cent) and shoot tips (80 per cent), when cultured on media containing MS + 20 mg/L BAP +1.0 mg/L IBA in cv. CO.1. Miler and Zalewska (2006) reported that for *in vitro* regeneration in cv. 'Lilac Wonder', green light showed the highest positive influence on stems length and fresh weight, length of internodes and propagation efficiency of plantlets. Using green light may be helpful in improving the efficiency of micropropagation of chrysanthemum.

Misra and Datta (2007) found that shoot bud differentiation could be achieved in ray florets in the presence of 0.5 mg/L NAA+ 2.0 mg/L BA in cv. Madam E Roger.

Thomas *et al.* (2009) reported stable micropropagation of 'Arka Ravi' for 2 years, with their resident endophytic bacteria in a covert form, was achieved on basal MS medium with >90 per cent shoot growth and rooting, a four-to-five-fold propagation rate at each 2-3 week sub-culture cycle, and with >90 per cent establishment of rooted plantlets *ex vitro*.

Padmadevi *et al.* (2009) found best response of ray florets with respect to culture establishment, shoot initiation and proliferation was obtained with BAP 2.0 mg/L which took minimum days for shoot initiation (10.50 days). Zalewska *et al.* (2010) reported unequal growth of plantlets during *in vitro* micropropagation which can be an effect of topophysis, and it is cultivar-specific in chrysanthemum.

Batra *et al.* (2011) screened different cultivars using the ray floret explants. The protocol A (MS Media (1 X), NAA (0.5 mg/L), BAP (1.25 mg/L), Sucrose (30.0 g/L) and Agar (8.0 g/L), followed by B (MS Media (1 X), NAA (0.5 mg/L), BAP (0.75 mg/L), Sucrose (30.0 g/L) and Agar (8.0 g/L) found to be the best protocols for plant regeneration (Table 19).

Nalini (2012) reported that shoot regeneration with 67.82 per cent success can be achieved through MS medium with Kinetin 3.0 mg/L + IAA 2.0 mg/L. Naing *et al.* (2013) stated first report describing culture conditions to induce secondary embryogenesis in cv. Borami. MS medium (1.0X) was more effective in promoting proliferation of somatic embryos than MS (0.5X). In addition, timentin was more efficient at inducing secondary embryogenesis than sucrose. The protocol facilitates large-scale clonal propagation and genetic transformation.

Telem *et al.* (2016) reported that maximum multiple shoots production occurred at 1.0 mg/L BAP and 0.1 mg/L NAA for leaf explants, 2.0 mg/L BAP and 0.2 mg/L NAA combination for shoot tip explants and BAP (4.0 mg/L) + NAA (0.1 mg/l) for ray floret explants. Therefore, shoot tip explants are the most suitable type of explants for plant regeneration of chrysanthemum through direct somatic embryogenesis. However, direct plant regeneration through ray floret explants will also be useful to recover the flower colour mutants.

Table 19: MS Composition of Four Media Used in Chrysanthemum Cultivars for Callus Induction, Shoot Differentiation and Root Induction*

MS Components	*Callus Initiation*	*Shoot Differen-tiation*	*Root Induction*	*Callus Initiation*	*Shoot Differen-tiation*	*Root Induction*
		(Medium A)			(Medium B)	
MS Media	1 X	1 X	1/2 X	1 X	1 X	1/2 X
NAA (mg/l)	0.5	0.5		0.5	0.5	0.0025
BAP (mg/l)	1.25	1.25		0.75	0.75	
Sucrose (g/l)	30.0	30.0	15.0	30.0	30.0	15.0
Agar (g/l)	8.0	8.0	6.0	8.0	8.0	6.0
		(Medium C)			(Medium D)	
MS Media	1 X	1 X	1/2 X	1 X	1 X	1/2 X
IAA (mg/l)	0.75	0.75		0.1	0.1	0.25
BAP (mg/l)	0.25	0.25	0.25	0.2	0.2	0.0025
NAA (mg/l)	0.2	0.2		0.2	0.2	
Kinetin (mg/l)	0.1	0.1		0.1	0.1	
Sucrose (g/l)	30.0	30.0	15.0	30.0	30.0	15.0
Agar (g/l)	8.0	8.0	6.0	8.0	8.0	6.0

* Batra *et al.* (2011).

Nugrahani *et al.* (2016) noted that the absence of growth regulators, microshoots and roots were able to grow *in vitro*, however, Trisnawaty *et al.* (2016) reported that the TDZ inhibited formation of shoots in callus culture. Mandang (2017) reported that the best shoot multiplication can be obtainsed through MS media in combination with IAA (0.5 and 1.0 ppm), BAP (0.5 ppm) and coconut water (30 per cent) in cv. Kulo. Kazeroonian *et al.* (2017) indicated significant superiority of shoot regeneration percentage from petiole explants (80.5 per cent) cv. Reagan Elite Salmon.

8.1.4.4 Rooting

Percentage rooting of chrysanthemum *in vitro* was high on media containing 0.01-1.00 mg PP333 (paclobutrazol)/l. Average number of roots and mean root length were also found to be increased (Tian *et al.*, 1993). Rooting of chrysanthemum on half-strength MS medium was promoted by 0.1 mg IBA, NAA or ABA. Rooting was precociously increased by addition of activated charcoal to the medium (Shen, 1990). Datta *et al.* (2000) found that shoots were best rooted on NAA (0.2 mg/L) containing half-strength MS medium in cv. Kargil 99. Misra and Datta (2007) reported that the differentiated shoots were best rooted in 0.5 mg/L NAA in cv. Madam E Roger.

Nalini (2012) reported that MS medium supplemented with IBA 1.0 mg l^{-1} was found to be the best with 82.12 per cent success for root production.Telem *et al.* (2016) found that IBA 1.0 mg/L for both leaf and shoot tip regenerated plants and 0.5 mg/L for ray floret regenerates proved to be the best for promoting root regeneration.

Mandang (2017) reported that good rooting can be achieved by the combination of IAA (0.5 and 1 ppm) with 30 per cent coconut water, and BAP (1 ppm) and 30 per cent coconut water.

8.1.4.5 Acclimatization

In vitro acclimatization of chrysanthemum plants were found to improve by application of paclobutrazol and by reducing the humidity. The addition of paclobutrazol to the growth chamber reduced plant height (Smith *et al.*, 1990; Ritchie *et al.*, 1991). Reduction in wilting was due to inhibition of gibberelin synthesis by the growth regulators. Roberts and Matthews (1995) recorded improved resistance of plantlets to desiccation (when transplanted) by treatment with paclobutrazol. *In vitro* culture in cellulose plugs (sorbarods) decreased wilting of the rooted plantlets when transplanted and exposed to an atmosphere of reduced RH. This was because of superior water uptake by the roots in sorbarods which was attributed to an increase in the number of roots, a reduction in damage they sustained during transplantation and improvement of contact roots and substrate (Roberts and Smith, 1990). *In vitro* propagated plants of cv. Westland and Escapade grew best in the greenhouse in fertile soil at 18-20°C (day) and 15-17°C (night). The optimum soil pH ranged between 6.0 and 6.5 and the soil EC between 1.5 and 1.6 mmhos (Tsvetkov and Krstanova, 1991). A medium composed of fertile and carbonized chaff (1:1 v/v) was found be a very effective medium for survival of transplanted *in vitro* plantlets. Survival of *ex vitro* plantlets was 89.5-95.5 per cent in the acclimatization room, throughout the year (Xia *et al.*, 1994). Liu and Zhang (1996) found that *in vitro* raised plantlets of most cultivars of mini-pot chrysanthemums, when transplanted 100 days before the Spring festival, were in full bloom on the festival days. The longer the subculture period, the lower was the percentage of flowering and smaller the flowers.

The genetic stability of micropropagated plants of the chrysanthemum (*Dendranthema*) cultivars Huang Shiu Fang, Netherlands Yellow and Blue Marble was studied by Chang and Huang (2000). The variation displayed by regenerated plants from shoot tip or callus depended on cultivar. Plants regenerated from shoot tip or callus of Netherlands Yellow and Blue Marble showed variation for plant height, the number of nodes and internode length. Also, all plants regenerated from callus had more disc floret and few ray florets. Tymoszuk *et al.* (2009) found thet perlite asbest substrate for direct acclimatization of chrysanthemum microcuttings.

Miler and Wozny (2016) reported that the survival rate in plantlets acclimatized after *in vitro* rooting was high (85–95 per cent) in varieties 'Bingo', 'Cool Time', 'Euro' and 'Vymini'. The light source did not influence the survival rates. However, under LED, the roots produced were longer in 'Vymini', chlorophyll content was higher in 'Cool Time' and 'Euro'.

8.2.0 Seed Propagation

Cultivars which bloom with conspicuous disc generally produce ample seeds. Poorly nourished plants are also known to seed better than well nourished ones. The seeds are collected after the blooms completely dry up. The small seeds require much care in handling before and after sowing. Fine compost is used to cover the seeds after sowing; watering is accomplished either by soaking from bottom hole of the seed pan or with the help of a fine rose so as not to allow the seeds to be exposed. Seeds germinate within few days of sowing. Treatment with thiourea promoted the germination of dormant seeds of *Chrysanthemum coronarium* (Nakamura *et al.*, 1955). The light, however, does not have any effect on germination. Pre-soaking of seeds in 1 per cent KOH for 30 min followed by washing also enhanced germination. This treatment reduced the level of inhibitory substances and increased the level of growth promoting substances (Chang and Park, 1997 a,b).

9.0 Open Cultivation

In recent years, the need for cut flowers of spray-type chrysanthemums of approximately 70/cm in stem length, referred to as "eco-mum", has been increasing to meet the bouquet processing demand from mass retailers. They stated that it is possible to harvest 240,000 "eco-mum" flowers per 10/a harvesting four times a year, which corresponds to twice the yield using conventional cultivation, through the use of efficient production conditions: suitable cultivar selection, planting style, and the effects of dwarfing-agent treatments on the cut flower weight (Imakiire *et al.*, 2017).

9.1.0 Soil and Climate

9.1.1 Soil

Chrysanthemum with a shallow but fibrous root system is sensitive to water-logging and prone to attack by diseases, such as root rot and wilt due to lack of aeration. Physical, chemical and biological status of soil is, therefore, an important factor which affects the growth of the plants. Clay and clay loam soils retain too much of moisture and thereby hinder proper aeration, resulting in rotting of roots. When dry, such soils become too compact and damage the tender roots. Sandy soils, on the other hand, drain too quickly and require frequent irrigation and also suffer from loss of nutrients due to leaching, though root growth is enhanced due to plenty of aeration. Sandy loam soils retain sufficient moisture and provide optimum aeration essential for proper root growth and therefore, are ideal for chrysanthemum growing.

Chrysanthemums thrive best in slightly acidic soils with pH ranging between 6.2 and 6.7. The nutrient status of soil is another factor for the satisfactory growth of chrysanthemums. Calcium, another essential macronutrient, is required to be at 100-150 ppm in soil. The critical levels of some micronutrients in soils suggested by Searle and Machin (1968) are: manganese 3 to 4 ppm, zinc 6 to 8 ppm, boron 20 ppm and copper 5 ppm.

9.1.1.1 Biological Characteristics

Soil is now regarded as a mass with a complex balance of minerals, humus, micro-organisms, water and air. Excessive use of chemicals in the forms of fertilizers, fungicides and insecticides without adding sufficient organic matter upsets this balance with an adverse effect on the crop. Besides, the organic matter also improves the physical condition of soil and acts as substratum on which useful microorganisms thrives. Chrysanthemum particularly thrives well in soils rich in organic matter.

9.1.2 Climate

Chrysanthemum is grown throughout the world. Light and temperature are the two important environmental factors influencing the growth and flowering, the former dominating in autumn-flowering and the latter in summer-flowering cultivars. As far as light is concerned, both photoperiod and the intensity have profound effects on growth and flowering of chrysanthemum. Chrysanthemums flower when the day length decreases and this leads to their classification as short-day plants. In general, they require high light intensity and plants grown under reduced light become taller and have thin stems and larger leaves.

Flower buds in chrysanthemums have been found to initiate and develop above a critical temperature, below which only vegetative growth occurs. Most of the cultivars need warm nights at the time of flower bud formation. At the start of short days, they require a minimum temperature of 60°F (15.5°C) at night. When the flower buds develop to a visible size the temperature is lowered one degree every two nights to 55°F (12.7°C) to improve the quality. This reduction of night temperature is very important, particularly in cultivars flowering in 12- to 15-weeks (Laurie *et al.*, 1968). Very high temperature, on the other hand, causes delay in the development.

The temperature requirement varies with the cultivars. The work of Post (1953c) sugges-ted that for the American cultivars, the critical night temperature for flowering was about 16°C. Working with some British cultivars, Vince (1955 b) also observed that plants required a minimum night temperature of 16°C for bud initiation, but for rapid and uniform development 10° to 12°C was desirable after the appearance of buds. The role of environmental factors on growth and flowering in chrysanthemum has been reviewed under 'growth and development'.

9.2.0 Selection of Cultivars

Varietal trials have been conducted at various places to select suitable cultivars for different purposes. Essig (1989) found the cvs. Liliane Hoek, Gerrie Hoek, Heidi, Ellen, Gompie and Beppie were suitable for growing in conventional culture. Wonkyo B 1-44, a newly bred chrysanthemum line derived from a cross between Target and Reagan and named as 'Noeulkukhwa' is an anemone type with the petals and flower centres yellow and bright red-brown, respectively and suitable for cut flower purpose (Shin Hakki *et al.*, 1998).

In Germany, Schmulling (1992) selected 'Ulysis' (a compact, small to medium sized plant with profuse yellow flowers) and 'Cream Star' yet another compact, small to medium sized plant with cream to white colour flowers, for landscaping. Barigdad and Patil (1997) recommended cultivar Indira for cut flower production as it produced more number of flowers and higher yield. Cultivars Bangalore, Karnool and Sarval too were found to have good market acceptability with good floral yield. Anguili (1999) evaluated the overall aesthetic values of some cultivars and described Korea a small flowered early flowering cultivar and Noir Violet a small to medium flowered late flowering cultivar to be the best.

9.3.0 Soil Preparation and Planting

Chrysanthemum requires well prepared soil for proper growth and development. Addition of organic matter in the form of peat or rotten manure improves the soil structure and helps in the development of the plant. The soil should be well drained. Solar sterilization of soil is effective to prevent soil borne diseases.

Stringheta *et al.* (1999) recommended that the growing of chrysanthemum in subs-trates containing urban solid waste compost (CLU) and carbonized rice husk (CAC) as conditioners. Plant height, leaf area and root fresh weight were influenced by the proportions of CLU and CAC, 66.66 per cent CAC + 33.33 per cent CLU gave the best results.

It is important to plant the cuttings immediately after rooting. Rooted cuttings after being graded according to size and those of same size are planted for uniform growth. Small plants surrounded by larger plants do not develop properly and remain weak and produce an indifferent crop. While planting, the roots are spread out in all directions and buried not deeper than 12 mm. Deep planting or planting in too wet soil deprive the young roots of the much needed oxygen hence, is carefully avoided. Florists' chrysanthemums are planted at different times to prolong the production and marketing, depending on the natural climate favourable for growth and flowering. In open cultivation, control of climate is not feasible. In climate controlled greenhouse, flower production in chrysanthemum the year round is not only standardized but computerized. However, investigations have been made on the influence of planting time on growth and flowering. Kiyatkin (1976) compared effects of 4 planting dates (30 March, 20 May, 2 June and 8 July) on growth and development and recorded reduced stem height, leaves/shoot ratio and flower size by late planting. May-planting resulted in well developed plants with good flowers. Barman *et al.* (1993) studied the effect of planting date 15th July to 30th September at 2-week intervals; chrysanthemum cv. Chandrama planted late showed early flower bud and longer flower life on the plant. Barman *et al.* (1998) also reported that planting on 15th or 30th July had the longest flowering season.

Many workers have reported that planting date along with effective concentration of growth regulators influence the flower production. Deotale *et al.* (1994) observed that the number of flowers/plant tended to increase with delayed planting dates in combination with increasing concentration of GA_3 upto 100 ppm. The maximum number of flowers (256/plant) was obtained from planting on 24th June and spraying 100 ppm GA_3 in small flowered pompons. Planting on 24 June along with a GA_3 spray produced the heaviest and largest flowers with longest vase life. Early planting date and pinching promoted flower number and plant height and increased the length of flower and stem and number of flowers. Pinching promoted flower production. Flowering was delayed by later planting of *D. boreale* and *D. zawadskii* (Shin *et al.*, 1995). According to reports obtained by Meher *et al.* (1999), cut flower production was the highest from May plantings and 50 ppm GA and 20 ppm IAA.

Planting date influenced the uptake of macro-and micro-nutrients at different stages of growth in chrysanthemum cv. Chandrama. The uptake of N, P, K, Mn, Fe and Cu concentrations increased with delayed planting whereas Zn showed the reverse trend (Barman and Pal, 1996).

The number of axillary bud on lateral shoots was the highest for rooted cuttings planted and pinched in December and this decreased as planting was delayed (branchlessness was promoted) (Shin *et al.*, 1996).

Uda *et al.* (1988) concluded that degree of juvenility controls the natural flowering time in cultivars that produce flower between July and September and manipulation of this time can regulate flowering time in chrysanthemum cultivars.

9.3.1 Substrate

Commercially desirable plants with highest number of flower buds were produced in a 1:1 mixture of spent compost and Speedle (commercially potting mixture) medium (Dallon, 1988). Putrasamedia and Sutapradja (1989) obtained best plant height and good flower diameter on substrates containing soil + bamboo compost in the ratio 1:1 and 1:2.

Best plant growth and flowering was obtained in a 1:1 mixture of composted ground bark and peat. Good results were also obtained in 1:1:1:1 and 1:1:2:2 mixture of bark, peat, soil and perlite (Strojny, 1989b). Holcomb and Walker (1995) found that coal gasification slag amended with peat or peat + bark was suitable for growing chrysanthemums. Incorporating a polyamide PA-6,6 in the growing medium showed better results with chrysanthemum seedlings as they were heavier and produced more leaves (Hetman *et al.*, 1996). A close linear negative correlation (R^2=0.77**; n =12) was found between the CAT (VDLUFA) soluble Fe and the shoot dry weight. Therefore, the Fe-contents might be a quality factor of peat to be used as a growing medium (Beckmann *et al.*, 2009).

Dutt and Sonawane (2006) suggested that the plants grown on compost amended substrates had adequate levels of all nutrient elements. It can be expected that adequate nutrient levels in compost containing substrates, due to the growth promoting abilities of compost influenced by presence of minor elements, organic matter, micro-organisms and growth regulating substances lead to optimum plant growth and flowering. Dutt *et al.* (2002) found that substrate cocopeat + soilrite, soilrite + compost and cocopeat + compost recommended as an alternative to soil for successful cultivation of chrysanthemum as it improved growth and flowering performance with more number of sprays/plant, earlier flowering and improved quality aspects like flower diameter, stalk length and girth, and number of flowers/ spray.

9.3.2 Planting Density

Planting density not only influenced the growth and development but also played an important role on production and quality of flowers. Results of the experiments revealed that optimum plant population per unit area varied according to type and cultivar, climate and management of the plants. The spacing suitable for plants intended for spray or standard cut flowers has been investigated. In England, distances of 12.5 x 12.5 cm in summer and 15.0 x 15.0 cm in winter gave good results (Machin and Scopes, 1978). Farina and Aatemiani (1982) reported that 2-stemmed plants at 10.1/m^2 gave greater stem and head weight and more fully opened double flowers than 3-stemmed plants. In an investigation with a small-flowered cultivar carried out by Yadav (1983), the maximum flower yield per hectare was recorded in plants spaced at 20 x 30 cm as compared to those at 30 x 30 cm or 40 x 30 cm.

A field experiment at Nagpur, India revealed that spacing at 45 x 45 cm gave the tallest plants and more primary and secondary branches while spacing at 60 x 45 cm took the minimum days to opening of flower after bud emergence (Belgaonkar *et al.*, 1996). They (1997) also suggested that spacing of 30 x 45 cm with 100 kg N and 200 kg P/ha produced flowers with the longest vase life and the highest weight of flowers/plant. Benedetto and Porto (1996) found that increase in plant density increased the flower quality and number of flower stems/unit area. Plant density had a pronounced effect on the growth of chrysanthemum plants as observed by Huld and Andersson (1997). Plants grown at low density were larger with respect to all measured para-meter. According to Barman and Pal (1999) conducting a field trial with cv. Chandrama, a large flowered cultivar, spacing of 30 x 30 cm increased plant height, stalk length and number of flowers per unit area.

Mali *et al.* (2017) found that the 45 cm x 60 cm spacing + N 200 kg/ha recorded the maximum plant spread, number of primary branches per plant, number of leaves per plant, leaf width, leaf length and duration of flowering while 30 cm x 30 cm spacing + N 200 kg/ha had the maximum plant height, flower yield per plot and flower yield per ha. Nehra and Munendra (2019) also suggested that the plant spacing of 30–45 cm produced taller plants but spreading of plants was more at 60–45 cm spacing.

Sarkar and Pal (2008) reported that plants with plant density at 12 plants/m^2 performed better in respect to all the growth and flowering parameter, however, leaves/plant was maximum in the treatment at 20 plants/m^2. Unpinched single stemmed plants recorded greater leaf area, showed early flowering and also produced larger and heavier flowers with longer and thicker flower stalks and maximum shelf-life in cv. Chandrama'.

9.4.0 Irrigation

Watering is an important aspect of growing chrysanthemum. The plants need adequate water during active vegetative growth when new leaves are formed. After the formation of flower buds no further leaves are formed and less amount of water is needed by the plants (Sheard, 1955).

Chrysanthemum roots are very sensitive to waterlogging but tolerate water stress appreciably. The best growth was observed in plants grown in soils that were kept constantly moist immediately after planting until the plants became established (Post, 1949).

Frequent watering is also needed during the period of maximum growth for maintaining the foliage and flower quality. Growth of chrysanthemums in benches with peat, sand and leaf mould increased as the pore saturation was increased to 90 per cent. Searle and Machin (1968) recommended 1.75-4.00 acre inch of irrigation in average weather condition of England. Rober and Hafeez (1982) grew chrysanthemums by subjecting them to moisture tensions at 0, 30, 60, 90 and 120 suction regulated by ceramic plugs or tensio-switches and observed that while flower and shoot weight increased with moisture content of the substrate, the flowers were of acceptable quality and stem length (80 cm) even at high moisture stress. Vase-life increased with decreasing substrate water content. The minimum water requirement to produce the largest number of marketable cut flowers in cv. Manatee Yellow Iceberg was found to be 35 cm with trickle irrigation. Linear responses for fresh weight, leaf area, leaf number and flower number between 13.6 and 40.7 cm of water supplied during production indicated that an additional 6 cm of water would improve marke-table stem quality (Harbaugh *et al.*, 1982). Plants of the same cultivar were grown from February to May in a fibreglass greenhouse with 0.16, 0.24, 0.33, 0.41 or 0.45 cm/day irrigation. Plant height, leaf area and marketable stem were the best at the 3 high irrigation rates (Stanly *et al.*, 1982). Different systems of irrigation are prevalent in various parts of the world. These include overhead mist spray lines and other types of sprinkler and self-travelling sprayers. The disadvantage of these methods is that both foliage and flowers get wet on each application. With a tall, closely planted crop, watering tends to become uneven and the plants remain wet for long periods after each irrigation with a serious danger of spreading fungal diseases.

With sub-irrigation methods, the disadvantage of wetting the foliage is avoided but other difficulties arise. In case of constant level method, anaerobic conditions and high salt concentrations tend to develop in soil or other growing medium used. With trickle irrigation and similar systems it is econo-mically possible to maintain adequate moisture with good aeration in soil under normal conditions of commercial production. Investigations carried out by Harbaugh *et al.* (1982) on cv. Manatee Yellow Iceberg revealed that a 90 per cent reduction in water requirement could be achieved with the use of trickle irrigation without loss of marketable yield as compared to the findings by Waters and Conover (1969) for overhead irrigation of cut-chrysanthemums. Trickle irrigation rates resul-ted in significant and positive linear responses on fresh weight, dry weight, leaf area, number of leaves and number of flowers at harvest. Eveleens and Blok (2014) evaluated three systems *viz.*, deep flow waterbed; aeroponic and peat in aluminium cassettes, for the year-round production of cut chrysanthemum in recirculating water without run-off. The deep flow waterbed and the aeroponic systems were found to produce constant good quality stems (1 g fresh weight per cm stem) and aeration of the deep flow waterbed system gave 10 per cent more fresh weight per stem.

9.5.0 Manuring and Fertilization

The growth and development of stock or production plants largely depend on proper feeding right from the beginning. Chrysanthemums are heavy feeders and have large requirements for both nitrogen and potassium. The emphasis at the early stage is on nitrogen. As the buds appear, the proportion of potassium should be increased. Plants need phosphorus throughout the growing period and it is best applied as basal dressing for gradual availability to the plant. The rate of nitrogen application should be reduced as buds appear. Hiroyasu and Marinaga

(1970) stated that omitting N from the nutrient solution supplied in sand culture to Oriflamme chrysanthemum during mid-August and September, when flower buds were developing, did not affect yield and quality.

Janipour *et al.* (2017) stated that to ensure proper growth of chrysanthemum branches in topiary form, furnishing of 200 mg N per kg of soil on vegetative growth stage and feeding reduction to 100 mg per kg of soil for the flowering process is appropriate and advisable. Sahana *et al.* (2019) found that 100 per cent RDF as water soluble fertilizer along with liquid biofertilizer through fertigation recorded maximum plant height, plant spread, least days taken for first flowering, longest duration of flowering, loose flower yield/hectare, cut flower yield/hectare, flower diameter, shelf life of loose flowers and cut flowers in cv. 'Marigold'.

Kumar *et al.* (2019) reported that 75 per cent RDF + Vermicompost @ 5t ha^{-1} along with foliar spray of 0.2 per cent humic acid recorded maximum plant height, number of leaves plant per plant, leaf area and number of branches/plant chrysanthemum cv. MDU-1. Nehra and Munendra (2019) found that a dose of Nitrogen at 160 kg/ha and potash 180 kg/ha produced better plant growth (plant height, number of primary and secondary branches) and dry matter accumulation in vegetative part which is the basis of better yield.

9.5.1 Role of Nutrients

The role of different elements, their deficiency symptoms and remedial measures have been described by Machin (1976). Nitrogen, phosphorus, potassium, calcium, magnesium and sulphur are required by plants in relatively large amounts. Micro- nutrients or trace elements like iron, manganese, boron, copper and zinc are, however, required in smaller amounts. These elements play a vital role in the plant growth and metabolism.

9.5.1.1 Nitrogen

Among the nutrient elements for commercial production of flowers, nitrogen has the most pronounced effect on growth of chrysanthemum. Lunt and Kofranek (1958) recommended maintenance of high levels of N during the first 7 weeks of the plant growth. They found that the plant suffered from moderate deficiency at that stage, subsequent N fertilization did not compensate flower quality. Maintenance of adequate level of N during the period of vegetative growth was suggested by Searle and Machin (1968).

Deficiency of N reduces plant growth and results in yellow-green foliage, reduced size of younger leaves and the plants remain almost erect. Flowering is delayed and the size of blooms and their number reduced (Machin, 1976; Roorda Van Eysinga and Smilde, 1980). Hosoya *et al.* (1978) also observed that effect of N application was most pronounced at the initial stage of growth. Deficiency of N at this stage delayed flowe-ring and resulted in lower fresh weight of all organs at the flowering stage. In extreme cases, leaf margins became yellow, reddish purple tints developed and lower leaves died prematurely. Watering with ammonium sulphate at the rate of 1.5 g/l was suggested to overcome N deficiency. Excess N resulted in very dark green leaves, premature wilting and delayed flowering. Leaching of N by adequate watering was effective but avoiding of waterlogging was recommended (Machin, 1976; Machin and Scopes, 1978).

Several investigations have been made to study the effect of N on growth and flowering of chrysanthemum. Chan (1960) stated that stem length, weight of tops and number of flower buds increased with higher rates of N. Improvement in plant growth and flower yield with adequate supply of N was recorded by Bik (1964) and Rober (1971). Waters (1965) and Carter and Dermott (1966), however, recorded decreased yield and postharvest keeping quality of flowers with high dose of N. Flower development was also delayed and the effect was enhanced with a lower K level and greater N ratio (Pawlowski, 1966).

The effect of N also varies with the source of nutrient. Ammonium sulphate or calcium nitrate appeared to accelerate flowering response compared to other N fertilizers while ammonium nitrate produced plants with stronger stems (Crater *et al.*, 1973; Tsujita *et al.*, 1974). Joiner and Knoop (1969-70) found the tallest plants with high nitrate N to ammonia N ratio at high N-K level. At low N-K level the reverse was true. The uptake of Ca was reduced by increasing NH_4 ions and that of P by increasing NO_3 ions. Hosoya (1979) reported that at the same concentration of N, plants absorbed more N from urea than from ammonium sulphate. Elliot and Nelson (1983) suggested that NH_4 added to NO_3 might stimulate growth by increasing transport of reduced N from roots to shoots, thus increasing the supply of reduced N available to support growth of shoot meristems. Kasten and Sommer (1990) recommended NH_4^+ N for obtaining quality cut flower. Schrader and Scharpf (1993) reported that supplying part of the N in the form of slow release fertilizer reduced the amount of residual mineral N in the soil after harvest, but soluble N is also needed at the beginning of the culture period.

Laishram *et al.* (2013) reported the effect of integrated nutrient management on soil physico-chemical properties and microbial population. Plots receiving 30 g/m^2 each of NPK + vermicompost (1 kg/m^2) + biofertilizers recorded highest available soil N, K and organic carbon content.

9.5.1.2 Phosphorus

This is one of the major elements limiting the growth and yield of flowers. Joiner (1967) evaluated the need of P during the periods of initiation and development of flower. An increase in P levels from lower to medium resulted in an increase in stem length and flower diameter. He also observed increased K absorption with more P.

A reduction in the size of new leaves is one of the first symptoms of P deficiency although leaf colour remains green. As the deficiency progresses, growth of main stem slows down and finally stops. New leaves become dark green and are reduced in size, lower leaves sometimes develop an orange-green colour, marginal browning develops rapidly and results in premature death of the affected leaves. Flowering is delayed with reduction in size and number of flowers. Feeding with 0.30 to 0.40 g monoammonium phosphate per litre water, 5 times, was recommended. Application of 30 to 60 g superphosphate per m^2 was also very effective to correct the deficiency. Excess P resulted in deficiencies of certain micronutrients like iron, copper, manganese and zinc (Machin and Scopes, 1978). Ushiro and Yamamoto

Nutritional Deficiency Symptoms

Boron deficiency

Calcium deficiency

Potassium deficiency

Iron deficiency

Training of Chrysanthemum

Sen Rin Tsukuri training

Cascade training

Cultivation of Chrysanthemum

Open cultivation of chrysanthemum

Planting of cuttings in polyhouse

In bud stage

Chrysanthemum in bloom

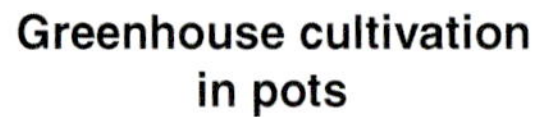

Greenhouse cultivation
in pots

Commercial cultivation of
chrysanthemum in greenhouse

(1996) also revealed that physiological disorders appeared as a result of excess phosphate due to an increase in plant P and low uptake of Ca. Soil amendment with $CaCO_3$ reduced the effects of high phosphate levels on plant inorganic composition and the crops grew well, presumably due to a reduction in water soluble phosphate. Laishram *et al.* (2013) reported that 22.5 g/m^2 each of NPK + vermicompost (1kg/m^2) + biofertilizers recorded maximum available soil P.

9.5.1.3 Potassium

Potassium is also an essential nutrient for proper growth and flowering of chrysan-themum. Kazimirova (1975) determined K requirement in chrysanthemum and observed that with vigorous growth, requirement of this nutrient increased and remained high till flowering.

Marginal browning and reduction of leaf size are the characteristic deficiency symptoms of this nutrient. Flowering is delayed and keeping quality gets impaired. Feeding with K_2SO_4 at the rate of 1.25 g/1 (5 applications) corrects the deficiency. Excess K causes dark green leaves, suppresses plant growth and results in Mg or Ca deficiency. Leaching of excess potassium by heavy watering and application of Mg (epsom salt) or calcium (gypsum and lime stone) were suggested as the remedial measures (Machin, 1976).

Hershey and Paul (1981 a,b) also described K deficiency symptoms in chrysanthemum cv. Bright Golden Anne as glazing and bronzing of leaf underside which resembled the injury attributed to the air pollutant, peroxyacetyl nitrate. However, the symptoms appeared first on just matured leaves during vege-tative growth and on the uppermost leaf during rapid flower development. They opined that leaf K levels of 0.5-0.7 per cent were associated with early sign of K deficiency, while a K level of 2.3 per cent in the recently matured leaves was associated with the maximum flower yield and 1.3 per cent leaf K with a minimum yield of 90 per cent.

9.5.1.4 Calcium

A constant supply of Ca is essential for normal growth in chrysanthemum and any breakdown in the continuity in supply shows immediate effect. Slight deficiency of this nutrient results in a complete cessation of top growth and death of roots, which become dark and adhere together (Machin and Scopes, 1978). At severely deficient level, yellowing of leaves and shortening of internodes become very marked; upper leaves form a rosette. Buds of growing points die. Lower leaves, which remain dull dark green, become stiff and brittle. Flowers become susceptible to heat injury. Spraying with a solution containing 3 g calcium nitrate dissolved in a litre of water is recommended. Excess Ca might cause boron, iron, manganese and copper deficien-cies and reduce availability of phosphorus (Machin, 1976).

9.5.1.5 Magnesium

Deficiency of Mg causes general loss of leaf colouration but veins and margins remain green, reddish purple colour on old leaves. Reduced flower size is another characteristic symptom. Foliar spray with epsom salt (10 g/l) applied fortnightly is effective to correct the deficiency. Xia *et al.* (1994) reported that the early flowering period is suitable for diagnosis of deficiency. At this stage, concentration of Mg in the leaf effectively indicates the availability of nutrient from the growing medium. They eventually concluded that inclusion of 14 mg/l in the nutrient solution was sufficient to prevent plants from developing Mg deficiency. Excess of the nutrient causes dark green leaves and reduces growth. Potassium and Ca deficiencies are also induced. Application of K and Ca and leaching through water have been suggested.

9.5.1.6 Boron

Boron deficiency results in brittle and curled leaves with pale green tips curling upwards. Petals become quilled or fail to unroll or reflex properly. Application of borax 0.02 g/l for 3-4 weeks alternately with irrigation corrects the deficiency. Excess B causes scorching of leaf tips and browning of margins.

9.5.1.7 Iron

Interveinal yellowing of leaves, reduction in size and delayed flowering with pale petal colour are the symptoms of iron deficiency. The deficient plants are also prone to fungal rot. Treatment with iron sequestrene every 7-10 days is effective. Excess iron results in deep red colour of leaf veins.

9.5.1.8 Manganese

General chlorosis and small necrotic spots near leaf edges and petiole are characteristic symptoms of Mn deficiency. Upper leaves are reduced in size. Foliar application of manganous sulphate at 0.6 g/l overcomes the deficiency. Excess Mn causes brown spots on leaves and application of 0.6 g/1 mono-ammonium phosphate followed by 1.5 g/1 calcium carbonate has been suggested (Machin, 1976).

9.5.1.9 Copper

Copper deficiency causes pale green leaves, excessive branching of laterals and delay in flowering. In extreme deficiency, interveinal chlorosis and failure of flower buds occur. Application of copper sulphate (0.35 g/l) as foliar spray has been recommended.

9.5.1.10 Sulphur

Sulphur deficiency symptoms include branchless roots, which age earlier, overall yellowing of new leaves and reddening on the leaf abaxial surface starting from older leaves and moving acropetally. Plants receiving no sulphur have smaller leaves, shorter stems, delayed inflorescence initiation and restricted inflorescence development. Without sulphur, plants do not produce flowers suitable for commercial use (Huang *et al.*, 1993).

9.5.1.11 Biofertilisers and Biostimulant

Biofertilisers and biostimulants are significantly improved the flower quality and yield of of chrysanthemum. Biofertilizers are important sources for plants' supplementing nutrition, which have significant effect on reduction of costs and adverse effects of mineral fertilizers on the environment (Nasri and Ghaderi, 2015), facilitated beneficial soil conditions reflected the significant increase in the flower yields (Laishram *et al.*, 2013). They stated that the application of 15 g/m^2 each of

NPK + vermicompost (1 kg/m^2) + biofertilizers recorded maximum *Azotobacter* population (10.50 x 106 cfu/g of soil) and phosphorus solubilizing bacterial count (12.83 x 10^6 cfu/g of soil).

Nasri and Ghaderi (2015) noted that the 1.5 L nitroxin + 25 mg/L paclobutrazol recorded the highest number of flowers, fresh and dry weight of shoots, fresh and dry weight of flower and number of shoots, highest amount of chlorophyll a and b, carotenoides, total soluble carbohydrates and protein content. However, 1.5L nitroxin/500 m^2 + 200 mg/L urea obtainedhighest number of shoots and flower per plant, highest chlorophyll a and b, carotenoids, total soluble carbohydrate and protein content. They concluded that nitroxin biofertilizer could reduce application of mineral nitrogen consumption in *Chrysanthemum frutescence.*

Vetrivel *et al.* (2017) found that Panchagavya 2 per cent + Seaweed extract 4 per cent recorded maximum plant height, number of leaves at bud appearance and peak flowering stage, internodal length, leaf area per plant at bud appearance and peak flowering stage, number of flowers per spray, stalk girth, flower diameter, 'A' grade stems and vase life. Thus, application of Panchagavya 2 per cent + Seaweed extract 4 per cent can be recommended for the commercial cultivation of chrysanthemum var. Amalfi.

Kanchan (2019) examined that the influence of microbial biofilm inoculants (Anabaena–Azotobacter, Anabaena–Trichoderma and Trichoderma–Azotobacter) on the availability of soil nutrients and structure of rhizosphere microbial communities in three varieties (White Star, Thai Chen Queen and Zembla). Dehydrogenase activity was highest when treated with the Anabaena–Trichoderma biofilm inoculants. The Anabaena–Azotobacter inoculant enhanced the availability of nitrogen, phosphorus and micronutrients in the soil, besides 40–50 per cent increase in soil organic carbon. The Anabaena–Azotobacter, and Anabaena–Trichoderma biofilm inoculants, proved superior.

9.5.2 Tissue Analysis and Nutrient Status

Symptoms of Zn toxicity were observed in plant tissueswhenindustrial waste was used as agricultural recycling var. Rage cultivated in a greenhouse.The high Zn mobility and concentration in the industrial waste suggests its utilization as a source of Zn (Costa *et al.*, 2004). The ratios NO_3^-/NH_4^+ in substrate affected the dry weight accumulation in the aerial portion of the plants. There were differences among the varieties regarding the tolerance to presence of NH_4^+ in the irrigation solution. When 50/50 ratio was used the concentrations of N, P, K, Ca, Mg and S were adequate for the development of the plants (Barbosa *et al.*, 2005).

9.5.3 Fertilization

Investigations on the requirement of different nutrients, carried out at various places, resulted in variable recommendations. Scharpf *et al.* (1985) suggested that chrysanthemum cultivars required 300-600 mg N/plant, depending on vigour, pot size or planting density for cut flowers, while pricked out plants required 100 mg N/plant and potted plants 200 mg N/plant. They suggested the optimum N:K ratio as 1: 1.2 or 1: 1.4. Lodhi and Tiwari (1993) obtained the maximum size and number of flowers with 30 g N along with 15 g P/m^2. However, the highest yield (in g) and dry weight of flowers were recorded with 30g N and 45 g P/m^2.

Before the soil is sterilized, certain fertilizers of low solubility should be incorporated thoroughly. The recommendation made by Kofranek (1980) was a moderate fertilizer application, *viz.*, 150 g of single superphosphate, 250 g of dolomitic limestone and 250 g of urea-formaldehyde per m^2.

Hwang *et al.* (1992) recommended an optimum rate of 20.9, 17.6 and 14.3 kg/10 acre N, P_2O_5 and K_2O application for the maximum cut flower production. An application at the rate of 10 g/m^2 was recommended for both nitrogen and phosphorus for optimum cut flower production (Jhon and Paul, 1999). Yassin and Pappiah (1990) recommended a combination of sheep manure and urea (75 kg/ha) for obtaining the highest yield.

Chaitra *et al.* (2018) reported that 150 per cent RDF with mulching resulted in increased plant height, leaf area, plant spread, number of laterals and suckers per plant, early bud initiation, 50 per cent flowering, increased flower diameter, number of sprays per plant, highest flower yield per plant and yield per ha, however, increased shelf life was recorded from 125 per cent RDF with mulching in cv. Marigold.

9.5.3.1 Micronutrient Application

Application of Fe-EDTA (75 ppm) significantly increased the number of leaves per branch, leaf area, stem length, stem diameter and stem dry weight. Zn-EDTA at 25 ppm also increased stem length, stem diameter and stem dry weight. Application of Fe and/or Zn produced flowers with the largest diameters. Duration of flowering was significantly influenced by application of Fe.

Vanlalruati (2019) noted that combined foliar spray of $FeSO_4$ + $ZnSO_4$ (0.5 per cent) recorded best results for number of branches per plant, minimum days for flower opening, petal length, petal width, flower bud diameter, flower longevity, single flower weight, plant dry weight, number of flowers per plant, flower yield per plant and flower yield/m^2 and maximum flower diameter and chlorophyll content was also recorded with individual spray of $FeSO_4$ (0.5 per cent) in cv. Mayur.

9.5.3.2 Coated Fertilizers

Waters (1965) used coated 14-14-14 (NPK) fertilizer for field-grown chrysanthemums. The granules were coated with a resinous polymeric membrane which regulated the rate of nutrient availability. An optimum yield was obtained when 1800-2100 lb/acre of heavily coated fertilizer was incorporated in soil before plan-ting. Simpson *et al.* (1975) reported that in a combination with liquid application of N and K, the coated fertilizers produced better crops. A single application of coated fertilizer was required; with other methods several applications were needed for similar effects. However, higher doses of coated fertilizers were found to reduce vase-life and increase postharvest deterioration of leaves and stem. A single pre-plant application of nutricote 100 (14:14: 14) or osmocote 100 (14: 14: 14) at 10, 12.5 or 15 lb/cubic yard to potted chrysanthemums cv. Yellow Mandalay yielded larger plants with more flowers as compared to the standard liquid fertilization which resulted in P deficiency. However, nutricote at 15 lb/cubic yard was found superior to osmocote (Boodley, 1981). Application of slow release fertilizer osmaform (agroblen 13-13-13) resulted in uniform production of flowers good quality (Sogni, 1989).

9.5.3.3 Liquid Feeding

Liquid feeding is preferred to dry feeding and the plants show apparent results within 48 hours of application of liquid fertilizer. The toxic effects of higher concentrations of fertilizer affecting the roots are minimized with application of liquid fertilizer. A concentration between 150 and 300 ppm N, P and K is safe for chrysanthemum. White (1965) indicated that chrysanthemum of superior quality could be grown in pots with the application of 150 ppm each of N and K alternately with 200 ppm of N, P and K. The timing of liquid feeding is also important. Seager and Champion (1984) found that commencing liquid feeding in the 1st week of SD increased plant height but decreased flower number compared to liquid feeding started 3 weeks later. Tissue analysis indicated excess uptake of N and P with early feeding.

9.5.3.4 Foliar Feeding

Machin (1975) enumerated the advantages of foliar feeding; other methods were not suitable due to either inadequate root growth or unfavourable soil condition. Foliar application is more efficient because higher percentage of nutrient is absorbed by plants as compared to other methods of feeding.

9.6.0 Mulching

Mulching is done to prevent fluctuations in the moisture content of soil. Peat or rotted manure was reported to be excellent material in greenhouses when applied to a depth of about 2-5 cm. Reflective film mulches (with reflective rates of 85.1-91.3 per cent) influenced the plant height, number of leaves, dry weight of chrysanthemum plants and quality production of cut flowers. Relative growth rate and leaf area ratio were also increased while net assimilation rate was decreased during vegetative phase but reversed in reproductive phase (Huh *et al.*, 1994).

Tayade *et al.* (2019) repored that the shade net used for controlling the effective parameter of crop growth. Film mulching is an important agricultural practice used to improve crop productivity. The yield of Chrysanthemum was increased in shade net than the open field. Film mulching significantly increased the soil temperature and water retention relative to the un-mulched treatments. Yield increases of 60-80 per cent were observed in shade net mulch treatment relative to the control. Shade net can be used to control environment. Chrysanthemum grow about 18-30°C and optimum light intensity was required to growth was 180000-300000 lux and relative humidity required 70-90 per cent.

9.7.0 Weed Control

Post-planting care concentrates on checking the weed growth and keeping proper moisture condition of the soil to promote sustained growth. Frequent shallow culti-vation should be done to conserve the moisture and control the weeds. The role of weedicides for controlling the weeds in chrysanthemum was studied by various workers. Under glass, the effective treatments were found to be nitrofen at 2 kg + either chlorthal-dimethyl at 7.5 kg or propyzamide at 1.5 kg or propachlor at 7.5 kg/ha, propachlor at 7.5 kg + either chlorpropham at 3 kg or propyzamide at 1.5 kg/ha and propyzamide at 1.5 kg + chlorpropham at 3 kg/ha. Treatments were made before or just after planting and emergence of weeds (Chapugier, 1979). Treatment with trifluralin, EPTC, nitrofen, DCPA (chlorthal-dimethyl) and diphenamid on the same day or the day after transplanting of rooted cuttings markedly reduced the weed population (Nishimoto *et al.*, 1979). Prevention of weed growth with the use of oxidia-zon, oxyfluorfen and oryzalin was also reported by Bing (1983). Bark mulching (10-40 mm) effectively controlled various weeds in chrysanthemum cultivation. Mulching was done immediately after planting (Grantzau, 1987).

9.8.0 Pinching

The operation of removal of terminal growing portion of stem pinching reduces plant height and promotes axillary branches (Machin and Scopes, 1978; Hendriks and Lemper, 1983; Reiss-Bubenheim and Lewis, 1986). Time and severity of pinching depend on the type of chrysanthemum and the desired objectives. The time of last pinching also influences blooming date besides affecting length of shoot. Pinching also helps in breaking the rosette. In an investigation with 18 single, 1 anemone, 2 pompons, 2 decoratives and 4 spider cultivars, optimum pinching date resulting in production of larger flowers at harvest was noted to be 5 July except for Yellow Tokyo where optimum pinching date was 20 June (Hong *et al.*, 1982). Hendriks and Lemper (1983) studied the effect of pinching 0, 7, 14 or 21 days after planting and observed that late pinching (at 14 and 21 days) produced better quality plants with shorter and more number of shoots in cultivars Golden Anne and Circus.

Gowda and Jayanthi (1988) followed a spacing of 20 x 30 cm and pinching twice, after 4 and 7 weeks of planting and obtained good flower yields. Removal of apical buds 15 days after growth recorded less accumulation of dry matter in stems, leaves and flowers and reduction in stem length as compared to plants grown without bud removal (Benedetto and Porto, 1996). Haynie and Klingaman (1999) did not find any significant effect to pinching twice. The plants receiving two successive pinchings delayed flowering for 6 days compared to single pinching.

9.9.0 Disbudding

Disbudding, the practice of removing undesirable immature flower buds to provide either a small number of large flowers or large number of small flowers is an important operation in the production of high quality plant and blooms. Buds are removed when they are amenable to be rolled out between the thumb and forefinger. Methods of disbudding vary according to the types of chrysanthemum grown. In standard types, the terminal bud is retained and all axillary buds removed as soon as they are large enough to handle. The intensity of disbudding of spray cultivars also varies with the type of spray produced. In this case, only the large apical bud is removed and the axillary buds allowed to develop.

Satisfactory results were, however, recorded using a substituted oxathiin (Cathey, 1976; Parups, 1976; Zacharioudakis and Larson, 1976). Fujime (1978) reported that tomino (90 per cent emulsion of ethyl toluene trimethyl benzene and C8-C12 fatty acid) killed almost all axillary buds, without injuring the terminal ones. He suggested application of spray when terminal flower buds were well developed and apical dominance was already suppressed.

Apu *et al.* (2019) found that disbudding significantly influenced stalk length and flower head size. The longest stalk and flower head size were achieved with allowing 4 bloom/plant.

10.0 Greenhouse Cultivation

Flowers are in demand throughout the year. However, it is not possible to produce quality cut-flowers throughout the year in the open. Hence, protected cultivation has been taken up extensively for year-round production of chrysanthemum. Temperature, humidity, light and carbon dioxide in the greenhouse are controlled to provide ideal conditions for growth and flowering.

Gaikwad and Patil (2001) reported that polyhouse planting of nine varieties recorded significantly better growth and yield (35-40 per cent more) than open planting. The variety Indira recorded medium height and spread with more number of branches, more number of flowers per spray and more number of sprays per plant and per plot. Location of a greenhouse affects the ambient light conditions. Growers have advantages from southerly locations with good sunshine.

10.1.0 Structure

Any structure to grow chrysanthemum is required to provide proper light during winter months and for this reason a minimum amount of framework is needed to be used. Aluminium is now replacing wood and steel for greenhouse construction. Greenhouse is usually covered with glass or polythene sheets. Improvements in orientation, design and roof geometry of greenhouse are made to meet all the requirements. East-west orientation of a single span greenhouse increases the transmission of light during winter. Machin and Scopes (1978) recommended 3.2 m multispan Venlo type of glasshouse which was found to be more advantageous than 20 m single Clearspan with a conventional roof and 20 m single Clearspan with Mansard type roof. The greenhouse is enclosed by constructing a wall 75-90 cm high, around the base and then glazing the rest of the framework with glass or plastic; fibreglass and rigid PVC are also suitable.

Traditional glasshouses have ventilators in order to regulate the inside temperature and humidity and to provide air movement around the plant. Modern trend is to use exhaust fans to draw air through moistened wood fibre pads on the opposite wall of the greenhouse. This system provides cool air in hot season where glasshouse get heated due to radiation of the sun.

Sethi and Sidhu (2005) observed that the total average solar radiation and light intensity entering the APS tunnels reduced by about 78 per cent and 68 per cent and 80 per cent and 67.7 per cent, respectively as compared to open field conditions. The average plant and soil temperature was also observed to be 5.5°C and about 2°C lower as compared to open field conditions. However, the average air temperature remained almost same as compared to open field conditions. The average plant height of chrysanthemum about 27 per cent more as compared to the plants in the open field conditions.

10.1.1 Benches

Benches inside the greenhouses are constructed for growing cut flowers as well as potted plants. For cut flower purpose, the benches are slightly elevated from the ground and are about 1.0-1.2 m in width. The benches for raising pot plants are about 75 cm higher and wider upto 2 m. For cut flower, the benches are built directly on ground with V-bottoms for good drainage. The raised benches for pot plants had flat bottom with drainage slots. Use of 2.5 cm welded wire mesh at the bottom for keeping the pots is ideal for drainage and air circulation. Walks between the benches occupy about 30 per cent of the greenhouse space. Usually a lengthwise wide central path is provided to transport wheel-barrows, sprayers, *etc.*; several crosswise narrow walks are also provided to get access to the benches.

Zhang *et al.* (2017) studied the effects of planting density on the growth of chrysanthemum in a greenhouse in cv. Sinma and Moonlight. A density of 12 cm x 12 cm recorded highest leaf number, leaf length, flower length, and leaf area, stalk height and diameter and flower number. Ratios of marketable production were 87.5 per cent (Sinma) and 88.0 per cent (Moonlight) under density of 12 cm x 12 cm.

Lee *et al.* (2002b) investigated the effects of planting date (season) and plant density (32, 48 or 64 plants m^2) on growth of cut chrysanthemum (Indicum group) by applying the expolinear growth equation. Final plant fresh and dry mass and number of flowers per plant decreased with plant density and this decrease was larger in summer than in winter. Stem length hardly responded to plant density and total dry mass production per m^2 linearly increased with plant density. Plant dry-matter content was not influenced by plant density. Plant dry-matter content, plant fresh and dry mass, number of flowers per plant, stem length and biomass production per m^2 were all higher in summer than in winter. Proportionality between incident light per plant and plant density, showing that plant density effects were primarily mediated through competition for light. Maximum relative growth rate (rm; assumed to be independent of plant density), maximum absolute growth rate (cm) and lost time (tb). rm was 2.4 times higher and cm was 4.1 times higher in summer than in winter and no effect of plant density on cm was observed, whereas tb decreased linearly with increased plant density.

10.2.0 Control of Environment

10.2.1 Temperature

10.2.1.1 Heating

Greenhouses located in temperate regions need some arrangement for heating, particularly during night to maintain proper temperature (at least 15.6°C). The heating consisted of laying pipes through which hot water or steam is passed. Steam is produced in boilers located at lower level passed through the pipes at about 5 lb/inch^2 in low pressure system and at 90 lb/inch^2 in high pressure system. The pipes are pitched slightly up so that condensed water return to the boiler by gravity. Hot water system also needs boiler usually kept at higher level than pipes and water is pumped back after it passes through the pipes.

Pipes are laid either below the benches along the side walls or overhead. The length of pipe line is determined by calculating the heat loss from the house. If a greenhouse of 7.5 x 30.0 m is to be heated with hot water upto 15.6°C it requires 540 linear metre of 5-cm pipe in hot water system and 480 linear metre of 3.1 m pipe in steam system (Kennard and Nelson, 1967). Thermostatically, operated automatic off-on heating systems are used by some growers. Winspear (1977) described a method of saving 15 per cent fuel in heating by use of thermal screens used for giving SD treatment to chrysanthemums.

Langton *et al.* (2004) used infra-red (IR) sensors to monitor plant canopy temperature (CT), and to control greenhouse heating and cooling on the basis of canopy temperature (CT). Potential benefits of controlling greenhouse operation on the basis of CT include avoidance of plant stress in summer, improved environmental optimization, more precise crop scheduling and, possibly, energy savings. Control based on CT is an example of the 'speaking plant' approach to crop management.

Lee and Cho (2011) suggested that higher night temperatures reduced cut flower length and number of leaves, but increased the number of young leaves attached to top part of the flower in cv. 'Baekma'. Higher ethephon concentrations and lower night temperatures increased cut flower length and the fresh weight of total, stems, and leaves due to the extension of vegetative growth period.

10.2.1.2 Cooling

For greenhouses located in warmer regions cooling is necessary specially during the day time. The mist and fan and pad cooling systems are usually recommended. For mist cooling, nozzles are spaced suitably throughout the greenhouse and water pressure (500 lb/inch^2) provided for misting to quickly evaporate without falling on the leaves. A reduction of 10° to 20°C is achieved by this method.

With fan and pad system, exhaust fans are placed on one of the greenhouse walls and porous wooden fibre pads installed on the opposite wall. Water is allowed to trickle over the pad with the help of a pump. As the exhaust fans operate air is sucked through the moist pads bringing down the temperature of the greenhouse by evapo-rative cooling. During night exhaust fans are used and the pads are kept dry. Large fans usually spaced at 3.75 m apart are used for this purpose; width of pad is kept about 60 cm if installed throughout the length of greenhouse facing the fans.

10.2.2 Light

Supplementary lighting is provided to chrysanthemums grown in greenhouse for normal growth of plants during periods of inadequate natural light and for delaying the blooming in year-round cut flower or pot-mum production.

Each bed is provided with overhead cables to facilitate illumination; at a time only one-third of the planted area is illuminated. According to an estimate, 730 metres of cable with 400 lamps of 100 watt each, requiring a load of 25 kw, is adequate to supplement light for an area of 0.4 hectare. Instead of incandescent lamps, fluorescent tubes have been found to be more effective (Cathey and Borthwick, 1964; Canham 1966 a, b). For large areas high pressure mercury fluorescent lamps (type MBFR/U) are also equally effective. But tubes and fluorescent lamps together are not suitable for cyclic lighting. Low pressure sodium lamps have an efficiency of 32 per cent compared to 7 per cent from incandescent lamps (Canham *et al.*, 1977). A minimum of 10-ft candle is supplied using 60-watt lamps, 4 feet apart and 2 feet above the plants.

Markvart *et al.* (2009) concluded that during the winter and periods of low light intensities, the daily carbon gain does not depend on the time of supplemental light application, but is linked to the total light integral. However, extended time with dark hours during day and night (24 h) might be a disadvantage because of longer periods with dark respiration and subsequent loss of carbon. Our results indicate that during times of low light conditions, it is not necessary to include factors such as the timing of supplemental lighting application to achieve higher net photosynthesis in climate control strategies. Zalewska and Wozny (2006) demonstrated that the blue light of the highest quantum irradiance (140 $\mu molm^{-2}s^{-1}$) both under long- and short-day conditions inhibited the plant growth, however, blue light under short-day conditions, slightly decreased the number of leaves in cv. Boulou White.

Carvalho (2002) found that the flower flower size and number of flowers per plant are important external quality aspects in cut chrysanthemum. The present work is conducted in a glasshouse and aims at investigating how these quality aspects can be predicted. Plants grown under no supplementary assimilation light, higher plant density, or with higher number of flowers per plant resulted in significantly lower individual flower dry mass and area. However, when no lateral flower buds were removed (control), higher assimilate supply resulted in more flowers rather than in larger flowers. Number of flowers per plant (including flower buds) showed a positive linear increase with total dry mass per plant. The combination of 32 plants/m^2 and no shade resulted in the highest number of flowers per plant (33 flowers) in contrast with 80 plants m^{-2} and 45 per cent light intensity (only 9 flowers).

Lee *et al.* (2002a) used a photosynthesis-driven crop growth model to determine acceptable plant densities for cut chrysanthemum throughout the year at different intensities of supplementary light. Dry matter partitioning between leaves, stems, and flowers was simulated as a function of crop developmental stage. Simulated light use efficiency (g additional dry mass MJ-1 additional supplementary light) was higher in winter (4.7) than in summer (3.5), whereas it hardly differed between the supplementary light intensities. This type of simulations can be used to support decisions on the acceptable level of plant density at different intensities of supplementary lighting or lighting strategies and on optimum supplementary light intensities.

Carvalho *et al.* (2003) quantified the effect of duration of long-day (LD) period and plant density on several chrysanthemum external quality aspects, a greenhouse experiment was conducted under summer conditions. Chrysanthemum 'Reagan Improved' was planted in May and grown under three durations (2, 9 and 16 days) of LD period combined with three plant densities (48, 64 and 80 plants m^{-2}). Plant height linearly increased with duration of LD period. Decreasing LD period from 16 to 2 days resulted in 25 per cent shorter plants, but a marketable height was always reached (>65 cm). Total number of flowers and flower buds per plant (NoF) increased with increased assimilate supply, measured as total plant fresh mass (TFM), whereas individual flower size (area and dry mass) was not affected. Plants received 16 days of LD period and grown at 48 plants m^{-2} were the heaviest and obtained the highest NoF. In contrast, 2 days of LD period and 80 plants m^{-2}, resulted in 57 per cent lighter plants with 67 per cent less flowers. A model to predict NoF was developed based on TFM (NoF= 0.353; TFM = 5.66; R2= 0.93). It was concluded that a similar TFM could be obtained using several combinations of plant density and duration of LD, without affecting either NoF or individual flower size. A photosynthesis-driven crop growth model was validated and used to quantify this trade-off, when aiming at a certain TFM. It was shown that such trade-off is dependent on the planting date throughout the year.

10.2.2.1 Use of Filters

Filters are found to regulate the quantity and quality of light penetration into the greenhouse, which also regulate the growth and flowering of greenhouse chrysanthemum. The spectral filter containing $CuSO_4$ is found to reduce photosynthetic photon flux density in red (R) and far red (FR) wave lengths and increase the R: FR and blue (B): R ratios of transmitted light relative to the water filter. Light through $CuSO_4$ filters reduces plant height, internodal length, leaf area and number of stomata per unit area, flower size and delays flowering compared to control plants, regardless of the season.

Altering light quality reduces water use and fertilizer demands. The spectral filters also affect the leaf pigments and morphology of leaves of chrysanthemum plants (Rajapakse and Kelly, 1993). Mc Mohan and Kelly (1995) recorded increased chlorophyll A and B and carotenoids content per unit area under $CuSO_4$ filters. Internode length and plant height were found to be reduced under $CuSO_4$ filters compared to plants grown under water filters, regardless of photoperiod. Plants grown under $CuSO_4$ filters that received artificial short days flowered earlier than those grown under long photoperiods. Floral induction was delayed under water filter. Under natural short days, a cool white fluorescent light as night break prevented or delayed flowering of plants grown under water and $CuSO_4$ filled filters. The results demonstrated that standard practices of photoperiodic control can be used to time chrysanthemum grown under $CuSO_4$ filters for pot-mum production (Mc Mohan and Kelly, 1999).

Filters with specific spectral characteristics can be used as alternative means of producing compact plants in the greenhouse. However, the delay in flowering and production of smaller flowers have limited their use for growth control of plants intended for flower production (Rajapakse and Kelly, 1995 b; Reddy *et al.*, 1997; Mc Mohan, 1999). Coloured plastic films have been found to alter the growth in chrysanthemum plants. In tunnels treated with blue pigment at 1 or 2 per cent growth was suppressed, flower numbers/spray were less and the flower bud initiated later, resul-ting in delayed harvest. However, stems were elongated in one cv. Suzy (Oyaert, 1996). Angus and Morrison (1998) observed that green light increased plant height whilst blue light did the opposite. Blue light also resulted in delayed flowering. The effect of blue cover on regulation of plant growth was similar to growth regulator such as gibberelic acid. The plastic covers (having spectral filters) with high UV absorption also reduced blackening of rose colour petals and diseases like *Botrytis cinerea*. The reflecting cladding materials can have a significant impact on the growth and development of chrysanthemum crop by filtering out wave-lengths of > 700 nm. Such materials also reduce internal temperature, conserve water, reduce the need for ventilation and allow CO_2 enrichment. Phthalocyanine derivatives have been incorporated into polymer films for use as spectral filters for the modification of plant growth. The unusual absorption characteristics of phthalocyanines, notably the narrow absorption band in the visible region of the electromagnetic spectrum, allow selective filtering of wave length necessary for excitation of either of the two isomeric phytochrome species. Chrysanthemum (short-day plant) grown under a far-red absorbing film showed a reduction in height and internodal length compared to those grown under a red-absorbing film or a control with no absorption in the visible part of the spectrum. These filters were effectively replaced with chemical plant growth regulators (Haeringen *et al.*, 1998). Verlodt *et al.* (1989) described a new photoselective film Astrolux (R) which helped in controlling the greenhouse climate better, specially reducing the extreme warming up during the hottest period of the day. This selective reflection of the interference film Astrolux (R) is based on the use of pearlscent pigments that selectively reflect a part of the green light and the near infra-red (NIR). It was designed specially for the regions with high irradiance. Chrysanthemum cuttings grown under this film has shorter internodes than those grown under the control film. Futhermore, the plant quality was not altered. The effect of using different coloured plastic films on stem elongation of chrysanthemum was evaluated by Hoffmann (1999). Coloured plastic film affect the red: far red ratio and the transmission of photosynthetically active radiation (PAR); so to achieve an effective reduction in stem elongation, plastic films which absorb far-red radiation but exhibit high PAR-transmission were recommended. A plastic film prototype (PMMA) with the above characteristics, strongly influenced the compactness of chrysanthemum. Thus, PMMA can be used as an alternative to plant growth regulators to reduce stem elongation. YXE plastic films were found to have growth retarding properties due to strong reflection of far red light, while SXE film reflected red light (thus improving stem growth). Drastic reduction in plant height was obtained under YXE 10 (Heemers and Oyaert, 1999). Khattak *et al.* (1999) also recorded significantly short plants under FR-absorbing spectral filters. When very high levels of nitrogen were added, the FR absorbing filter increased the leaf area.

The coloured filters were also found to alter plant habit drastically. The inhibition of stem elongation increased with increasing pigment concentration under the blue PE films. Fewer leaves were developed; the internodal length was significantly shorter irrespective of the use of coloured filters. The blue filters resulted in a lower number of axillary shoots, a smaller leaf area and a lower total dry weight. Another filter namely the vaporized film, which was characterized by the highest light transmission percentage, resulted in a relatively small growth reduction (Table 20). This film had no significant effect on branching rate, leaf area and dry weight and plant quality. If the vaporizing technique could be obtained, which would result in larger growth reductions, such a spectral filter would provide an environmentally friendly tool for growth regulation of ornamental crops (Oyaert *et al.*, 1999).

Table 20: Number of Leaves and Axillary Shoots, Average Internode Length of Chrysanthemums grown under Four Coloured Filters and Four Corresponding Control Filters*

Filter	*Number of Leaves*	*Internode Length (cm)*	*Number of Axillary Shoots*
Control 1	30.4 b	2.58 b	26.3 b
Blue 1 per cent	28.8 a	2.43 a	23.6 a
Control 2	29.9 a	2.45 b	24.4 b
Blue 2 per cent	26.7 a	2.36 a	18.2 a
Control 3	30.9 b	2.55 b	26.0 b
Blue 3 per cent	26.6 a	2.31 a	18.3 a
Control 4	31.6 a	2.41 b	28.3 a
Vaporised film	32.1 a	2.16 a	28.5 a

a Different letters are significant differences at 0.05 level (LSD-test).

* Oyaert *et al.* (1999).

Sethi and Sidhu (2005) observed that the total average solar radiation and light intensity entering the APS tunnels reduced by about 78 per cent and 68 per cent and 80 per cent and 67.7 per cent, respectively as compared to open field conditions. The average plant and soil temperature was also observed to be 5.5°C and about 2°C lower as compared to open field conditions. However, the average air temperature remained almost same as compared to open field conditions. The average plant height of chrysanthemum about 27 per cent more as compared to the plants in the open field conditions.

Khattak *et al.* (2004) Removal of far red radiation by spectral filters led to a reduction in final height at flowering by 19 per cent. This effect was most marked between internodes 6 and 11. However, no interaction occurred between far red filtering and plant density. Indeed, in chrysanthemum plant height did not increase with density when far red was present, as may be expected with a shade avoidance response.

10.2.2.2 Shading

When natural daylength is too long for bud initiation, artificial shade is provided to create SD conditions by covering the plants with some opaque material. Black polythene sheeting of 150 gauge is often used for this purpose. Shading is often done from 6 pm to 7 am by covering the sheet over inverted 'U' frames. For a year-round programme in an area of 0.4 hectare, approximately 1280 m long and 4.3 m wide sheet would be required to cover 1.5 m wide beds taking two-thirds of planted area under shade at a given period. Covering operation is being done by mechanized automatic time switch controlled system to save labour.

Shin *et al.* (1994) obtained more number of lateral branches, flower bud differentia-tion and flowering in spray-mum chrysanthemum by providing shading with day lengths of 11 or 12 h. Chang and Yih (1996) also studied the effects of shade on growth and cut flower quality and concluded that chrysanthemum plants grew faster under shade and recorded increased plant height, leaf area, flower size and chlorophyll content in leaf as compared to the open field culture. In order to avoid producing slender cut flowers, resulting in unfavourable packing ability and decreased quality, an optimum shading percentage of 50 per cent was suggested. Plants grown under low/high shading regime (low/high: 73 per cent shading when outdoor radiance was 200-500 W/m^2 and 87 per cent shading when outdoor radiance was > $500Wm^{-2}$) were taller with longer internodes (Huld and Andersson, 1997). Use of shading screens also reduced evapo-transpiration from plant canopies and enhanced the overall quality of chrysanthemum plants (Al-Arifi *et al.*, 1999).

Chrysanthemums cv. Baegkwang were planted on the 10 May and 4 June and subjected to different shading (0, 30 and 50 per cent) and mulching (using clear, reflective, black or green film) treatments. The quality of cut flowers and marketable yield of chrysanthemum planted on 10 May increased with 30 and 50 per cent shading and with the use of reflective film. The cut flower yield of chrysanthemum planted on 4 June was higher compared to those planted on 10 May. Flowering time was delayed with increasing shading intensity (Kim *et al.*, 2000).

Dhiman *et al.* (2017) found that high density polyethylene (HDPE) found best as black-out materials for off-season pot mum production of chrysanthemumfor Solan, Himachal Pradesh. Kobayashi (2005) stated that that photoselective 30 per cent red shadecloths was more effective in hastening flowering, highest proportion of the number of flowers to the combined number of flowers and buds showing color in cv. Arezzo.

10.2.3 CO_2 Enrichment

The need for carbon dioxide enrichment in greenhouses is well known. Carbon dioxide is produced by burning propane in burners preferably in the morning. Gardner (1966) suggested that the level should be raised from natural 0.03 to 0.1 per cent by adding CO_2 at a constant rate of 56 kg/ha/h at a temperature not less than 15.6°C.

10.2.4 Humidity

For chrysanthemum, humidity between 70 and 90 per cent is optimum. Relative humidity is raised by spraying water and be lowered by providing ventilation and heating. Schussler (1992) noted the effect of constant and fluctuating air humidity levels on shoot elongation and anthesis of chrysanthemum cultivars. Shoot elongation and plant height was found to be decreased at 60 per cent RH as compared to 90 per cent RH. Water consumption and days taken to anthesis were also enhanced at lower RH values. Reduction in total leaf area and leaf

DW as well as delay in flowering and flower development were observed under increased humidity from 0.1 to 1.1 K Pa. However, this range was not detrimental for commercial growing of chrysanthemum (Hand *et al.*, 1996). Best shoot length and total leaf area per plant was found to be increased at high level (95 per cent) of RH in plants grown hydroponically. Plants grown at such high RH, required less irrigation water due to reduced water uptake.

10.3.0 Selection of Cultivars

Right choice of cultivar is of paramount importance for success in chrysanthemum production. For greenhouse cultivation 7- to 9-week-response groups were suitable for growing in high light intensity areas or seasons, while 12- to 14-week types were chosen for those areas or seasons which received low light radiation. Cultivars in the 10-week-response group were suitable for most areas and some of them such as Snowdon, Hurricane, Fandango and Robeam belonged to the year-round groups. Shibata (1988) grouped the heat tolerant hybrids for year round production of chrysanthemums. These hybrids were able to flower early at longer day lengths than the autumn-flowering cultivars and showed a shorter reaction time to short day treatment. Market preference should also be taken into consideration. Spiders are preferred in the Netherlands and Germany, single sprays in Denmark, Sweden and Finland; in Britain, however, decorative types are favoured for cut-sprays. For standards, generally 10-week cultivars are used. Indianapolis White and its sports, Fred Shoesmith, Mefo, Snowdon and Escapade are some of the attractive and commercial cultivars grown throughout the world. Hanke (1996) observed the differences between the cultivars in their reaction time to short day treatment. The cultivars Yoder and Challet-Herault required 8-10 weeks respectively. Cultivars also differed in their response to different air temperature, modified by PPF (Schoellhorn *et al.*, 1996). Przymeska (1997) found that cv. Moonstone performed best under unheated plastic tunnel. Saud and Talukdar (1999) observed that spray cvs. Prof. Haris, Chaming and Rajkumari were suitable for cultivation in low cost plastic greenhouse.

Cermeno (1989) reported that all chrysanthemum cultivars tested (Butter Cap, Daymark, Pink Daymark, Yellow Daymark, Cream Daymark, Palaver and Wall Street) could be grown successfully in unheated greenhouses during the period from November to April in Gulf of Cadiz. Pinched plants produced larger size flowers but number of flowers per plant was more in unpinched plants. A closed nursery system comprising an ebb-and-flow irrigation system was found to be most economical for growing glasshouse chrysanthemum production. This system reduced the nutrient loss and provided hygienic condition to open system.

10.4.0 Substrate

In order to provide ideal conditions for the growth of chrysanthemum, various organic and inorganic materials have been used and found useful. In some countries several proprietory brands of composts such as JohnInns compost and GCRI compost are available. The basic requirements of these materials are to provide firm anchorage to roots, good drainage and uniform consistency. These composts also contain N, P or K fertilizers in slow or quick release forms and soil amendments like chalk or lime stone. Substrates like composted pine bark or 1: 1: 2: 2 bark: high peat (sphagnum): soil: perlite mixture were found to be equally suitable for year round production of chrysanthemum. Leinfelder and Rober (1987) planted chrysanthemums in closed system with fully automatic water and nutrient supply and obtained good growth. Substrates (TKS 1 and Einheitserde P) of 10 cm thickness, produced cut flowers of good length and quality. They (1989) also suggested the growing of mini-chrysanthemum (Improved Funshine, Funset, Funray, Cotton ball, Statesman and Butter Cup) in closed system on substrates like floraton 4 or peat + clay and volmatic drip irrigation system (14 nozzles/m^2), providing around 180 l/m^2 over the 10-week culture period and following a plant density of 100 plants/m^2 in beds. Klapwijk and Wubben (1991) used peat substrate for growing chrysanthemum in glasshouse. They also suggested hydroculture with constant flow system for growing of chrysanthemum. Good growth was observed on growing media comprising 60-100 per cent coir dust. K and Ca uptake were better in presence of high proportions of coir dust. Mixing of perlite: coir dust was found to be effective only when proportion of coir dust was more than perlite (Lee *et al.*, 1999a). Modified bed structures induced better growth, early flowering, quality and vase life of hydroponically grown chrysanthemum. When chrysanthemum was grown on B-2 (naturally aerated Biobed 2), it showed more rapid improvement of plant height, stem length and leaf area. Root zone temperature was higher in modified beds (B1 and B2). Thus, Biobed 2 (B-2) with perlite medium was considered the most effective structure for hydroponic cultivation of chrysanthemum (Lee *et al.*, 1999b).

Use of growing medium other than soil has become popular as an additional method of improving chrysanthemum plants. Soilless culture of glasshouse chrysanthemum was reported by Meer (1990). Malorgio *et al.* (1994) recommended the use of expanded clay, pumice and a pumice: peat mixture (1:1 v/v) together with irrigation at 1.87, 1.4 or 0. 93 l/m^2/day, 3-4 times/day or 1.87 or l/m^2/day, 10-12 times/day for growing quality flowers. Substrates comprising of either 30 per cent red peat, 30 per cent black peat, 40 per cent forest compost (composted bark) or 45 per cent red peat, 45 per cent black peat and 10 per cent sand gave higher yields of flowers than soil culture and were reused without any adverse effects or attack by *Verticillium dahliae* f. sp. *lycopersici* (Popescu *et al.*, 1996). Zhang *et al.* (1998) studied the effects of substrates comprising of perlite, peat, ceramsite (keramzite, clay heated at 200°C) and rice hull ash, either alone or in combination on growth, development and yield of chrysanthemum plants and obtained greater leaf area and higher yield. Zeolite based soilless media (NZ, EZ1 and EZ2) successfully reduced the nutrient leaching during crop production and application at low rates of soluble fertilizer was effective for potted chrysanthemum (Carlino *et al.*, 1998).

The effects of electrical conductivity (EC) of the nutrient solution and different growing media on the nutrition and growth of chrysanthemum cv. Japonica Yellow for cut flowers were studied by Zhao *et al.* (1999). For the medium composed of resin and sand (1:99), the most suitable EC was 2.6-3 mS/cm. The best EC value for the growing medium comprising of sawdust and cal slag was 2-2.5 mS/cm.

The soil of the bench requires plenty of organic matter in coarse form, *e.g.*, peat moss, straw, peanut hulls and inorganic coarse matter like gravel. The pH of about 6.5 is optimum. The beds should be perfectly levelled and kept 2.5 cm below the brim of the bench. Steam sterilization of soil is very important in greenhouse culture. Steam is injected into the moist soil to raise the soil temperature to 180°C for half an hour; the soil is covered with plastic sheet to prevent escape of steam. Vinyl covers are preferred over polythene.

Janakarin *et al.* (2006) found that growth parameters, yield and quality of flowers were better in the growing medium including soil+compost+cocopeat, with 40 kPa irrigation regime and fertigation with water soluble fertilizes under low cost polyhouse in cv. Yellow Regan. Fanourakis *et al.* (2011) reported very high plant density (344 plants/m^2) without affceting plant growth resulted in higher light interception and higher total shoot dry mass per m^2.

Blok and Vermeulen (2012) found that systems included a soil bed, a sand bed, a peat bed and a cassette bed. The soil bed was a 70 cm deep bed of the original soil on a water impermeable foil with a drainage system. The sand bed was a 15 cm layer of coarse sand with a 5-10 cm under layer of coarse clay pellets including a drainage system which also supplied irrigation water *i.e.* sub-irrigation. The peat bed was a 25 cm peat layer on a sub irrigation bench. The cassette bed was a 130 x 3x 15 cm (length x width x height) container filled with peat. The cassettes were hung on a sub irrigation bench. Chrysanthemum press pot plants were planted on soil and sand beds and bare chrysanthemum cuttings were planted in the peat based systems. Chrysanthemums were grown for the first of 6 crop cycles. Results showed a 5-15 per cent increase in dry matter production and 3-5 days shorter growing period in the peat beds and cassette beds. However, the economic performance is still marginally poor. Nevertheless, the systems tested are environmentally sound and comply with plant requirements for optimal growth. The sand bed and cassette bed may be further optimised by respectively EC control and top down irrigation.

Atikmen *et al.* (2014) indicated that 12.5 per cent fresh mushroom compost and 25 per cent exhausted mushroom compost were the best ratios for cultivation of chrysanthemum.

Prasanna *et al.* (2018) expressed that availability of nutrients in soil plays an important role in the productivity and quality of flowers in chrysanthemum. A set of novel biofilm inoculants-Anabaena-Azotobacter, Anabaena-*Pseudomonas fluorescens* (An-Psf) and Anabaena-Trichoderma (An-Tr) were applied as carrier based dry formulation or soil drench and their performance compared in two varieties of chrysanthemum (White Star and Zembla), in a climate-controlled greenhouse. Both the An-Psf and (An-Tr) inoculants enhanced glomalin related soil proteins in the rhizosphere of White Star, while in terms of polysaccharide content of soil, both these inoculants performed better in Zembla variety. Significant increases in the availability of selected macro and micronutrients in rhizosphere soil samples, in both chrysanthemum varieties were recorded, particularly when the inoculants were applied as soil drench. This investigation demonstrated the varietal effects on soil biological activities and significance of mode of application of microbial inoculants in influencing plant growth and rhizospheric metabolic activities.

10.5.0 Irrigation and Fertigation

Hydroponic culture is now being widely used for growing of chrysanthemums in greenhouse. Different hydroponic systems have been advocated by different workers. Buwalda *et al.* (1994) developed a familiar cultivation system applying ebb-and- flow irrigation in which only 15 cm^3 of substrate/plant was used in the form of a small, fibrous plug. High quality and productivity were achieved and the system showed no problems of root diseases. This system was recommended for year-round production. A simple ultrasonic aeroponic system was very effective in rooting experiments. Application of fog with this system increased rooting in cuttings. Aspirin and DCPTA (dichlorophenoxytriethylamine) treatment also showed promotive effects (Son *et al.*, 1994). Wilson and Finlay (1995) recommended the hydroponic systemfor all-year-round (AYR) chrysanthemum culture. The main advantage was slow release of fertilizer into the subsoil which improved the quality winter crop. Successful production of AYR chrysanthemums was achieved using a recirculating sand based system. The sand substrate successfully supported 7 successive crops without being sterilized, loss of yield or incidence of debilitating root disease. A high quality chrysanthemum was produced in the capillary mat system than in the deep flow technique (DFT) (Sakamoto *et al.*, 1998).

Pereira *et al.* (2005) determined the water consumption in two chrysanthemum cultivars (Kátia and Bi Time) under greenhouse condition. The average water consumption, for once and twice day irrigation, for cultivar Kátia was 118 and 108 mL/day and for the cultivar Bi Time was 98 and 113 mL/day, respectively. Fanourakis *et al.* (2011) found that the irrigation frequency should be three times per hour till week 4 after planting under soilless cultivation system (aeroponics).

Margaretha and Klaas (2006) studied the effect of different soil water levels and soil desiccation in a climate chamber and under changing light intensities in a greenhouse to assess the critical soil moisture status for optimal growth of Chrysanthemum plants. The efficacy of infrared (IR) thermometry as a useful tool for on-line monitoring and control. Water status of the plant can very well be monitored on-line in a greenhouse as changes in leaf temperature due to water shortage can be detected by IR thermometry, especially at summer radiation levels.We conclude that regression analysis of high frequency on-line infrared IR monitoring may be a useful tool for expost analyses of irrigation regimes at high light intensities.

10.5.1 Nutrient Film Technique (NFT)

Production of spray chrysanthemums on NFT system has been advocated by Herve (1990). Optimum growth (leaf area, leaf dry weight, stem length and flower weight) of chrysanthemum plants was noted in recirculating nutrient film system with root zone temperature regime of 20°C/16°C during day/night (Hicklenton, 1988, 1989).

Soilless culture (NFT) of chrysanthemum was reported to be effective for better growth, early flowering and quality cut flower production (Kang *et al.*, 1995c). Cut flowers obtained from plants grown in the half-strength solution were superior in quality (Kang *et al.*, 1995 d). In an experiment Ji *et al.* (1998 a), however, found that half-strength nutrient solution was good for early growth but standard nutrient solution was better at later stages. Irrigation twice a day resulted in good growth and longest vase life.

In hydroponic culture, strength of nutrient solution affected the different stages of growth and flowering of chrysanthemum, perlite being used as substrate. Application of standard nutrient solution resulted in elongation of stem and also increased photosynthesis and transpiration but did not show the same effect during the reproduction stage. Fresh weight and dry weight of each plant organ after flower bud formation were better with tap water treatment than with any nutrient solution. Plants grown in tap water also initiated early flowering (Ji *et al.*, 1998 b).

Hansen (1999) developed a modified nutrient film technique (NFT) for the production of cut chrysanthemums. The system included (i) rooting of the cuttings directly in a rigid growing channel without any growing media, (ii) small (3.2 x 3.2 cm internal cross section), cost effective channels, that allowed for high initial plant densities when channels are placed close together, (iii) channels fitted with the option of removal of plant support for the first 6-8 weeks that allowed for simple cleaning and reuse after harvest, and (iv) channels on transportable benches without a fixed line nutrient solution supply, thus allowing movement of the crop at any time, so that all labour inputs occurred in a central, customized location. These systems increased production of chrysanthemum stems per unit area of greenhouse upto 50 per cent. Molitor and Fischer (1999) recommended that the temperature of the nutrient solution in aeroponics should be between 20°-25°C.

10.5.2 Fertilization

10.5.2.1 Nitrogen

Grantzau and Scharpf (1986) observed that increasing doses of N increased fresh weight, reduced stem length and shortened the growth period. Increasing K appli-cation shortened the growth period but did not greatly affect other characters. It is recommended to start the plants in spring on substrate containing 200 mg N/l and watering is to be done 5 times weekly with 3 l/m^2 of a solution containing 0.8-1.2 g/l of a complete fertilizer with 15 per cent N. In the autumn and winter, the plants were watered twice a week with 2 g fertilizer/l. Kageyama *et al.* (1987) found that optimum range of N in the growing medium varied widely, the critical level being very low; for normal growth the plants were provided continually as much nitrogen as they could take up. Thus, they found that plants supplied with 12.5 - 50.0 g NO_3-N per m^3 soil (16 days after pinching) and given small amount of liquid NO_3-N daily for 15 days, grew well; higher rates of N inhibited growth.

Much reduced leaching resulted when ammonium fertilizer was used (Kasten *et al.*, 1990). Sawwan *et al.* (1999) found that 50 ppm N as $Ca(NO_3)_2$ gave the highest plant height, stem diameter, heaviest fresh and dry weight and the maximum number of inflorescences per plant. The time to full bloom was significantly delayed as N rate increased. Both urea and $NH4^+$ forms had an acidic effect and NO_3^- N an alkaline effect on the soil.

In a hydroponic culture experiment, King *et al.* (1995) studied the uptake of ammonium and nitrate nitrogen by chrysanthemum plant. Removal of nitrogen (mg/g tissue DW) was more during the first month of plant growth and decreased steadily upto 90 days. N-uptake decreased rapidly after the development of flower buds. Kageyama and Konishi (1996) found that weekly application of N at 100 ppm produced large leaves and flower stems compared to plants receiving N as per the 1.0-, 0.8- and 0.6 -fold N application curves. Based on N uptake and plant growth, a down-regulated N application curve (NAC) was developed by Yoon *et al.* (2000c) for producing economically marketable cut flowers. To produce a plant having two shoots of 70 g which is probably suitable for the market, 954-1142 mg N fertilizer was computed to be appropriate. At NAC 0.8 and 1.0 strengths, 90-100 per cent of the N, 90-98 per cent of the P and 85-97 per cent of the K applied were consumed, thus, confirming an efficient uptake of these nutrients. Zerche and Kuchenbuch (1995) noted sudden reduction of nitrogen in hydroponic solution that might be due to nitrification and/or denitrification.

10.5.2.2 Phosphorus

Hansen *et al.* (1998) reported that excessive P fertilization rates were detrimental to root activity as it led to increased root respiration. Reduced root respiration indicated changes in the source-sink relationship during the transition from vegetative to reproductive growth, making the roots less competitive than the developed flowers. Kato and Takei (1989) found that soils containing 200 - 300 mg P_2O_5/100 g produced the best plant growth and the highest yield of flowers and roots. A higher dose of P resulted in chlorosis during the vegetative growth. Excess P also caused root browning, rotting and reduced root activity. An upper limit of 300 mg/100 g was recommended for available P_2O_5 (30 mg/100 g water soluble P_2O_5). Kageyama and Konishi (1992) found that application of 100 mg phosphorous per plant was effective in improving growth and flowering of chrysanthemum.

The plant absorbed it within 7 weeks of application. Higher concentration showed toxic effects on plants. Soilless growing media have little capacity to retain P. Preplant amendments of triple superphosphate (TPS) and water soluble P applications are readily leachable. A soilless medium modified with aluminium sulphate was found to reduce P losses as evident from a study wherein P retention in the medium increased as the rate of Al addition increased (Williams and Nelson, 1996).

Light and humidity have been found to influence the uptake of phosphorus and rubi-dium by rooted cuttings of chrysanthemum. Uptake through both surface of leaves increased as light intensity increased. The greatest increase in uptake was recorded between the dark and 900 lux treatments. Air humidity was also found to affect the foliar absorption of the two ions. As the RH increased from 47-80 per cent, there was an increase in uptake. Air humidity influenced greatly the uptake of Ca-phosphate, probably due to its low solubility (Marczynski and Tukey, 1988).

10.5.2.3 Potassium

Kageyama *et al.* (1993) recommended that 25-30 kg K per 1000 m^2 of greenhouse space, is optimum for the production of 4000 chrysanthemum flowers.

10.5.2.4 Nitrogen and Phosphorus

Shima *et al.* (1996) recommended an optimum P:N application ratio of 12:100 along an N application curve, in an NFT system for high quality cut flower production of chrysanthemums.

10.5.2.5 Potassium and Phosphorus

Williams and Nelson (1997) suggested the use of zeolite precharged with K and PO_4 along with sphagnum peat and perlite for soilless culture of chrysanthemum. Precharged zeolite reduced the leaching loss of K and PO_4, thereby enhancing the plant growth.

10.5.2.6 Magnesium, Calcium and Sodium

Shima *et al.* (1995) suggested that the concentration of Mg should not exceed 100 mg/plant or 50 mg/flowering shoot. The use of 400 mg Ca per plant or 200 mg per cut flower shoot, was recommended by Kageyama *et al.* (1995).

Warmenhoven and Bass (1995) suggested that a high concentration of NaCl in a soilless ebb-flow system might lead to reduced growth in respect of plant height, leaf number, stem diameter and internodal length and increased leaf: stem ratio which can be controlled by irrigation (3 times/h).

10.5.2.7 Rate of Application

Strojny (1989 a) recommended 200 - 400 N, 50 - 100 P, 250- 500 K (mg/l) in pine bark and 100-200 N, 25- 50 P, and 125-250 K (mg/l) in other substrates, for obtaining optimum plant growth and flowering. The optimum concentrations of N, P_2O_5, K_2O, CaO, MgO and Na_2O in shoot for maximum biomass yield in chrysanthemum were found to be 2.45, 0.50, 4.40, 1.70, 0.45 and 0.20 per cent, respectively at 90 days after transplanting (Hwang and Yoon, 1994). The optimum soil DBS (degree of base saturation) values of chrysanthemum were 65-80 per cent at cation exchange capacity > 21 meq/100g, 75-85 per cent in the range 11-20 meq and 85-95 per cent at < 10 meq (Kato and Takei, 1994).

Slow release fertilizers were found to be better than liquid fertilizers as the slow fertilizers resulted in better and optimum uptake of nutrients. In liquid fertilizer treat-ment, there was greater root mass leading to quicker uptake of nutrients more than required for plant growth. Chrysanthemums were more efficient in N uptake as compared to liquid fertilizer treatments (Catanzaro *et al.*, 1998). Under aeroponics, Alt *et al.* (1989) recommended 25 mg K/l and 100 mg N/l for obtaining good growth and yield of chrysanthemum.

Under greenhouse conditions, fertilizer application was extended beyond the usual 2 months with a cautious application of K to avoid the high rate of K uptake. Macronutrient uptake/0.1 ha was calculated as 14.86 kg N, 4.28 kg P_2O_5, 24.55 kg K_2O, 11.29 kg CaO and 2.47 kg MgO in cv. Yellow Delta and 12.87 kg N, 3.13 kg P_2O_5, 21.28 kg K_2O, 12.96 kg CaO and 2.74 kg MgO in cv. Cerise Delta (Pilanali and Kaplan, 1999).

The effect of varying nutrient solution volumes (0.25, 0.50, 0.75, 1.00, 2.00 l plant -1) on mineral uptake in an NFT system on cut flower quality was investigated by Yoon *et al.* (2000a). The N, P and K uptake remained unaffected with the varying nutrient solution volumes. Nutrient solution volumes of < 0.75 l plant -1 resulted in considerable reduction in flower quality. Characteristics of mineral uptake in spray chrysanthemums grown under a hydroponic system in relation to growing seasons and developmental stages were also investigated by Yoon *et al.* (2000b). The amount of N, P, K, Ca and Mg taken up per plant differed markedly over the growing seasons and between the varieties. The rate of N uptake steadily rose after planting until flower buds become visible. The upward trend continued during bud development but decreased rapidly towards harvest. The The ratios of P/N and K/N decreased from 16.2 per cent and 134.2 per cent to 8.4 per cent and 69.5 per cent, respectively, from the vegetative growth to the flowering stages.

The nutrient solution with a 1.0: 0.3: 2.5 N:P:K ratio was found to be best for chrysanthemum cultivation in expanded clay of hydroponic technique. Plants grown under this treatment performed well for flower number per stem (6.42 flowers/ stem), stem length (46.51 cm), fresh weight of plants (60.3 g) and dry weight of roots, leaves, stems and flowers (Barbosa *et al.*, 2000).

10.6.0 Automation in Greenhouse

Automated control of environment in greenhouse, irrigation and fertilization devices for different crops have markedly improved the productivity and reduced the cost of cultivation. It is also now possible to evaluate the response of a crop upto production.

In automated control of greenhouse, electronic sensing and actuating equipment were used to create an optimum climatic condition inside the greenhouse as described by Juan and Carda (1988). High pressure sodium vapour lamps (150W) were found to be more economical than incandescent lamp for photoperiodic illumination of chrysanthemums in winter. These lamps not only reduced the cultivation period but also increased the number of flowers/stem (Reist, 1988). An improved light enhancing device, a reflecting venetian blind was found to increase (over 50 per cent light) the direct light on the floor of an east-west single span glasshouse in winter. Pot chrysanthemums grown under this reflector had greater dry weight than the control. Another light enhancing device, based on the refraction and total internal reflection of light by a prismatic device was also found to be useful (Bailey, 1989). A computer vision algorithm was developed for locating segments of stems of chrysanthemum micro-plants in a container and selecting suitable segments for a harvesting robot to grasp so as to carry out the harvesting operation quickly and easily (McFarlane, 1991).

DENAR, an energy saving specially designed glasshouse with several cultural systems including bare root irrigation with no substrate, drip irrigation crops on a film covered or coated concrete base or in film-covered basins containing different subs-trates was recommended for saving energy in glasshouse (Nunnink, 1991). Computer algorithms designs to spread the energy demand of greenhouse has been developed by McFarlane (1991) and Neilson (1995). To test the algorithms, chrysanthemum cultivars were grown in two greenhouses, employing a commercial climate computer to control room temperature, supplementary light and inlet water temperature to the heating system. One greenhouse was maintained at equal day and night temperature, as a reference. The other employed a low day temperature/high night temperature regime (negative DIF *i.e.*, low day and high night temperature). The negative DIF algorithms were able to level out the energy consumption peaks, transferring the use of energy to night time, while producing plants of unimpaired quality and unchanged production time.

An automated system was described measuring light compensation points (LPC) of plants or plant as an alternative to laborious determination procedures whose efficiency depended strongly on personal judgments of the research worker (Lieburg, 1997).

10.7.0 Models

Growth chambers capable of implementing various day and night temperature combinations and equipment to measure the internodal elongation in chrysanthemums and various environmental conditions on a less than 30 min scale have been developed. The system successfully measured the internodal growth and various environmental parameters with sufficient accuracy (Jacobson *et al.*, 1991). A model to predict the effect of different day and night temperatures on the final plant height of chrysanthemum was developed. Final plant height was taken as the product of the number of internodes and the average internode length. The final leaf number was determined by integrating the rate of leaf production, at a temperature over the duration of the vegetative phase. High night and low day temperature were considered as supraoptimal and suboptimal, respectively for internode extension (Pearson *et al.*, 1995). Jacobson *et al.* (1998) tested a system for measuring internode elongation and environmental conditions was developed and tested on 2 crops of chrysanthemum.

Larsen and Gertsson (1992, 1993) presented a model to calculate the effect of light integral and temperature on side shoot elongation in chrysanthemum cv. Garland. Daily increment in side shoot was calculated by multiplying the maximum daily increment with a light index (fl) and a temperature index (ft). The model was also used to find out the effect of day length on shoot elongation. A simple three-parameter-model was developed by incorporating the accumulated light integral and gave an excellent fit to data in all experiments. The fresh weight of flowers/ plant was allometrically related to 'productive' plant weight which was defined as total plant weight minus a fixed weight not contributing to flower production. The relationships showed that the proportional partitioning of fresh weight of flowers differed between the cultivars and varied between the experiments, being greatest under conditions of an increasing daily light integral (spring) and least under condition of a declining daily light integral (autumn). The development of a simple model to predict stem weights and gradients for a given spacing at particular times of year was also based on this model (Langton *et al.*, 1999).

A two dimensional mathematical model was developed to describe the time course of root growth and its spatial distribution for container-grown plants, using chrysanthemum cv. Bright Golden Anne. Potential root growth was considered as consisting of several concurrent processes, including branching, extension and death. Branching rate was assumed to be related sigmoidally to existing root weight density (RWD). A sensitivity analysis of modelled RWD to the estimated parameters indicated that the model was more sensitive to carbohydrate supply parameters than to root growth distribution parameters (Chen and Leith, 1993). A convective-diffused model of root growth and proliferation was developed and used to test some hypotheses on the response of roots to their environment with data on the growth of roots of potted chrysanthemum cuttings. Models were also developed to calculate the effects of environmental factors on growth and flowering in greenhouses (Acock and Pachepsky, 1996).

A combination model was developed to accurately estimate the evapotranspiration of a greenhouse grown potted chrysanthemum crop (Al-Shooshan *et al.*, 1991).

Jacobson and Willits (1998) developed a model to determine the relationships between the environmental variables and stem elongation in chrysanthemum for the use in the dynamic control of a greenhouse environment. The model used Richards' growth equation as its base.

A model to simulate flower development in chrysanthemum cv. Pert in response to daily light integral and ambient mean day temperature was described by Hiden and Larsen, (1994). In a validation trial the model predicted flower development well at constant temperature, but under estimated development of plants exposed to fluctuating temperature.

Control of elongation and time of flowering are major challenges in the production of high quality potted plants. The greenhouse CARE system was developed as a computer-decision-support tool to graphically track chrysanthemum crops. Benefits of this software included the delivery of models and expertise to end-users, new technologies for decision support, the ability to link biological models directly to the greenhouse environmental computer (Fisher and Heins, 1996). Mutiple Criteria Decision Making (MCDM) provided a set of mathematical techniques which helped to reach a decision in optimization of cropping systems where various factors were involved for year round chrysanthemum production (Bakker and Sanden, 1998).

Heuvelink *et al.* (2002) investigated the effects of plant density and light intensity on crop growth and yield of cut chrysanthemum experimentally and simulated with a generic explanatory crop growth model (HORTISIM). HORTISIM could accurately predict crop growth and yield at most light conditions, with measured climatic data, initial plant mass and time course of leaf area index being model inputs. However, in summer at 100 per cent light the model strongly overestimated dry mass production.

Carvalho and Heuvelink (2004) found that increased assimilate availability by higher daily incident photosynthetic active radiation (PAR, 4.2 to 18.3 mol m^{-2} d^{-1}), higher CO_2 concentration (345 to 623 μmol mol^{-1}) and lower plant density (80 to 32 plants m^{-2}) hardly influenced stem length (maximum increase of 10 per cent).

Nevertheless, the duration of the long-day period (0 to 21 days) and temperature (16 to 28°C) are required inputs when predicting stem length, since these strongly enhanced the number of internodes and the average internode length. A module is developed for predicting number of flowers excluding flower buds (NF) based on the total aerial fresh mass per plant (TFM) (NF = 0.210TFM – 1.25). However, when the incident PAR during the short-day period was under a threshold value (7 mol $m^{-2}d^{-1}$; *i.e.* autumn crops) individual flower size decreased linearly with decreasing PAR and higher temperatures resulted in smaller flowers.

Dai *et al.* (2008) developed prediction model for development and quality of standard cut chrysanthemum, can be used for decision making for the optimization of light and temperature management for standard cut chrysanthemum production in greenhouses.

Lessa *et al.* (2013) developed four models to describe the relationship between leaf growth and GDD accumulation in chrysanthemum 'Rage' grown in greenhouse without climate control. Mathematical models have been validated and can be used in chrysanthemum to relate leaf area and growing degree-days. The models showed RMSE of 88.0 cm^2 (summer), 29.44 cm^2 (fall), 55.61 cm^2 (winter) and 80.01 cm^2 (spring).

11.0 Pot Plant Production

In recent years, demand for pot plants for house decoration and for use in amenity horticulture has steadily increased. Chrysanthemum has been considered as number one among the major pot crops (Carter, 1980) and year-round pot-mums are grown all over the world. The pot-mum production has become a profitable form of commer-cial chrysanthemum growing. Economy of space, time and material has made this style of growing very promising; the excellent range of colours, forms, long-lasting quality of blooms and ease in handling have made them very popular.

An ideal pot-mum is one in which the diameter of the flowering head is equal to the height of the plant, as measured from the rim of the pot with plant height being almost three times the height of the half-pot used. Choice of suitable cultivar, ideal pot compost, high quality cuttings, proper potting technique, optimum environment, judicious feeding, watering and management are important for production of quality pot-mums.

11.1.0 Cultivars for Pot Culture

Important cultivars suitable for pot culture have been listed under the heading Species and Cultivars. The criteria for selection of cultivars for pot-mums are high bloom quality with optimum size, fullness, good shape, broad petals and vigorous growth with amenability to growth regulation by chemicals. Predictable response to temperature and light is also of utmost importance. Some of the commonly grown pot-mum cultivars belong to Princess Anne, Ridge, Ruby, Septa, Popsie and Red Star groups. In Princess Anne alone, there are about 20 shades. This group contributes to about 85 per cent of pot-mums grown in England (Machin and Scopes, 1978). Decoratives are also desirable. Waffenschmidt (1988) reported Fatima P, Steffi P, Birgit, P, Ulrike P, Anke P for pot culture. New cultivars include Clou P, Clou

Luck P, Sarina P and Yello Samba P which have been found to be suitable for pot-mum (Zalewska, 1994). Good plant growth with greater shoot and root dry weight and total biomass were observed in cvs. Grenadine, Nicole and Tolima grown in fibre containers impregnated with cupric hydroxide as compared to black plastic containers (Ruiter, 1996). The cultivars of Mini Nero, Prof. Haris, Charming and Rajkumari are also suitable for pot (Saud and Talukdar, 1999).

11.2.0 Growing Medium

The basic requirements of compost are to anchor the plants firmly and provide ideal conditions for root growth. Loam, peat, sand, saw dust and plastic foam are used for the purpose. A good growing medium should be loose and well drained. Peat and sand mixture are most commonly used and slow release or coated fertilizers have been found to be beneficial (Larson and McIntyre, 1964). Application of 10 g 14-14-14 slowly soluble coated fertilizer gave excellent results in 15-cm pots. The physical and chemical conditions of compost ideal for pot culture have been described by Bunt (1976 a, b). Plants grown in peatlite medium, consisting of vermiculite, sphagnum peat and sand (4.5: 4.5: 1, V: V: V) amended with superphosphate, $MgSO_4$ and dolomite lime at 2.9, 0.7 and 5.2 kg/m^3, respectively, produced the thickest and longest stems (Hicklenton, 1983).

Many growers used artificial mixes, uniform in physical and chemical properties, easy to handle and free from pathogens, insects and weed seed. Such mixes were found to be economical (Carter, 1980). Tayama and Carver (1992) suggested the use of compost having the recommended rates of resin coated sierra chrysanthemum mix 12-10-17 plus minors along with osmocote 19-6-12 (added to compost) or Peters 20-10-20 peatlite special for pot culture. Smith and Hall (1994) noted that pertite-based potting mixes were a good alternative to peat-based ones (Table 21). Works of Martinez and Gomez (1995) highlighted the chemical and biological quality of the compost and good flower yield was obtained in variants containing FYM in the ratio of 1:1. Altmann and Streitz (1995) recommended standard compost (ED 73) and bark for outdoor and lava for pot culture. Dubsky and Sramek (1999) studied the growth of chrysanthemum on peat-bark substrates amended with 33.3 per cent by volume of paper sludge and composted with decomposed bark.

Meinken (1988) described a new system 'Cultoplant' for pot culture of chrysanthemum. The cultoplant system consists of cylindrical pots 150 mm in diameter and 40-130 mm in height, supported in a polystyrene tray containing different substrates. Peat substrate was found to be the best as it allowed better root development. A pot density of 50/m^2 in the trays gave good results.

Mota *et al.* (2007) reported that EC of 2.13 dSm^{-1} during the vegetative stage and 2.57 dSm^{-1} during the bud stage provided the best visual aspect in pots plants cultivated in a greenhouse as well as the higher aerial part, higher leaf area, best formation and coloring.Terra *et al.* (2011) used rice husks as a sole or partial substrate promoted bud diameter (6 per cent) larger than the commercial substrate in cv. Funny.

Bala and Singh (2013) found that soil + sand + FYM + vermicompost (2: 1: 0.5: 0.5) improved vegetative and flowering for satisfactory for display of specimen as pot mum in cv. Yellow Charm. Kameswari *et al.* (2014) reported that cocopeat + sand + FYM+ vermicompost resulted maximum plant height, plant spread, number of branches per plant, duration of flowering, flower weight, spray length and number of flowers per plant in var. Punjab Anuradha.

Singh *et al.* (2016) reported that 50 per cent cocopeat and 25 per cent soil as a growing media amendment with composted sewage sludge (25 per cent) proved to be the best media mixture for ideal growth and performance of pot mums. Ari *et al.* (2018) reported that the growing medium of compost made from bamboo leaves could increase the number of leaves, the plant height, number of roots and the fresh weight of the plant in acclamtization of gamma ray induced mutants.

Thakur and Grewal (2019) suggested that light-weight growing media composition of cocopeat + vermiculite + FYM (1:1:1) was ideal for quality flower production of potted cv. Kikiobiory. Kala *et al.* (2020) found that potting media Soil: Sand: FYM: Vermicompost (2:1:0.5:0.5 v/v) was found best for pot mum chrysanthemum production. Thakur (2018) found thatlight-weight growing medium composition of cocopeat + vermiculite + FYM (1:1:1) is the best with better plant morphological development and sustained quality flower production in potted Chrysanthemum cv. Snowball. Choi *et al.* (2018a) investigates the potential of using biochar as a substrate component to replace pine bark (PB) to produce horticultural crops in containers. Chrysanthemum plants grown in PB mixtures with 80 per cent and 100 per cent biochar had significant higher shoot fresh weight and dry weight than plants grown in PB alone.

Choi *et al.* (2018b) investigated the potential of using biochar as a container substrate component to replace peat moss to produce horticultural crops. Biochar was incorporated into commercially available peat moss-based substrate (PM) at 0 per cent, 20 per cent, 40 per cent, 60 per cent, 80 per cent and 100 per cent (vol.) to grow chrysanthemum. Chrysanthemum grown in peat moss-based substrate (PM) mixes with 60 per cent, 80 per cent and 100 per cent biochar had significantly higher shoot fresh weight (FW) than in PM alone (0 per cent biochar). Asrar *et al.* (2014) found best vegetative growth and flower quality was recorded with application of the slow-release fertilizer Osmocote at 70 mg kg^{-1}.

Papafotiou and Vagena (2012) expressed that substitution of 50 per cent peat with cotton gin trash compost (CGC) in the commonly used medium for potted chrysanthemum commercial cultivation resulted in a reduction of at least 33 per cent of the dose of daminozide necessary to meet the needs of the market for compact plants, with multiple economic and environmental benefits. Terra *et al.* (2011) indicating that rice husks can be used as a sole or partial substrate for potted chrysanthemum cultivation in cv. Funny. Using rice husk as a substrate promoted bud diameters 6 per cent larger than the commercial substrate.

Sramek and Dubský (2009) found that composted bark and composts are the most readily available alternatives to peat for production of growing media in the Czech Republic. Compost-amended substrates had very high available potassium content. The content of available micronutrients was greatly influenced by bark or compost amendment but had no impact on their content in chrysanthemum and impatiens leaves. The highest plant dimensions, fresh weight, and dry weight were recorded in pure peat substrates but smaller size and more compact habit of chrysanthemum in bark-amended and compost-amended substrates meant better marketable quality. Individual peat types differed little in their effect on plant growth.

Wright *et al.* (2008) determined the comparativefertilizer requirements for chrysanthemum (*Chrysanthemum* x *grandiflora* 'Baton Rouge') grown in pine tree substrate (PTS) or a commercial peat-lite (PL) substrate. At any particular fertilizer level, substrate electrical conductivity and nutrient levelswere higher for PL compared to PTS accounting for the higherfertilizer requirements for PTS. Possible reasons for the lowersubstrate nutrients levels with PTS are increased nutrient leachingin PTS due to PTS being more porous and having a lower cationexchange capacity than PL, and increased microbial immobilizationof N in PTS compared to PL. This research demonstrates that PTS can be used to grow a traditional greenhouse crop if attentionis given to fertilizer requirements.

Table 21: Means of Subjective Plant Quality Criteria Scores for Pot Chrysanthemum Bright Golden Anne**

Substrate	*Fine perlite*	*Fine Perlite and Vermiculite*		*Coarse Perlite*		*Coarse Perlite and Vermiculite*		*Peat Analysis of Variance**	
Basal Reservoir	*No*	*Yes*	*No*	*Yes*	*Yes*	*Yes*	*No*	*p Value*	*SED*
Quality criteria:									
Evenness of growth-plan view	8.3	7.0	7.1	5.6	8.5	6.5	7.7	0.001	0.59
Evenness of growth-side view	8.5	8.3	8.1	7.2	7.9	7.6	8.5	0.005	0.36
Balance (ratio of height: breadth)	9.0	7.4	8.2	7.8	8.5	8.0	7.5	0.005	0.40
Flower abundance	8.3	7.5	8.0	6.6	8.9	7.9	7.9	0.005	0.51
Foliage colour and quality	8.4	5.4	8.1	6.0	7.0	7.6	9.5	0.001	0.53
Overall quality	8.5	6.8	7.8	6.2	8.1	7.3	8.0	0.001	0.45

* NS = not significant (p= > 0.05); SED = standard error of the difference between means.

**Smith and Hall (1994)

11.3.0 Potting

Proper grading of rooted cuttings according to length, thickness, leaf area, amount of root and fresh weight is essential before planting. Uniform cuttings should be planted for obtaining identical sets of pot plants and any variation in planting material will be reflected in growth and flowering and finally a set of uneven plants will be produced. Thicker cuttings were found to produce plants which flowered earlier and had more and larger flowers than those produced from thinner cuttings (Anon., 1971).

Correct method of potting is also important. Cuttings should be planted shallow. Plants should be spaced evenly around the edge of the pot at an angle of 45°. The size of pots varies according to the type and cultivar and the number of cuttings planted in each pot. Pot size ranges from 7.5 to 15 cm and one to seven cuttings may be planted in each pot. With 15 cm pot, Carter (1980) suggested planting of 4 or 5 cuttings during spring, summer and fall months and 5 cuttings in winter months. One cutting is suggested for a 7.5 to 10.0 cm pot. When cuttings are planted a thorough watering is necessary to settle the compost round the roots. Frequent misting in the first week after potting keeps them in turgid condition.

11.3.1 Spacing

Immediately after potting, pots are placed pot to pot in the nurse area and after a week they are moved into the intermediate area and spaced at a little distance. Later, pots are removed to their actual place of culture and spaced at proper distance. Pots of 1.4 cm size, after being placed pot to pot in the first week, are usually spaced at 20 x 23 cm for the next 2 weeks or 23 x 23 cm for the next 3 weeks and then shifted to the final spacing of 30.5 x 30.5 cm (Machin and Scopes, 1978). According to Carter (1980), a distance of 40 x 40 cm providing 0.2 m^2 space to each pot plant is ideal for the final spacing of 15 cm pots. Some of the ideal spacings used for different sizes of pots, as quoted by him were:

7.6 cm: Pot to pot: 12 x 12 cm, one plant per pot

10.0 cm: 18 x 18 cm, one plant per pot

15.0 cm: 30 x 30 cm, three to five plants per pot

18.0 cm: 36 x 36 cm, five plants per pot.

Mildenberger and Hendriks (1996) recommended a spacing of 26 plants/m^2 for 12 cm pots. With the spacing, there was 100 per cent coverage of the available space and only 20 per cent plants overlapped with each other. Plants produced linear growth at closer spacing (18 plants/m^2) and had fewer lateral shoots while those at wider spacing (36 plants/m^2), covered only 70 per cent of the available space. Sultanpuri *et al.* (2018) revealed that earlier visible bud formation stage was recorded in pots containing one plant per pot spaced at 9 inch apart and grown under controlled photoperiodic conditionswhereas visible bud formation stage took maximum time to reach when one plant per pot containing pots were spaced at 12 inch apart under natural photoperiodic conditions in selection 'UHFSChr-56'.

11.4.0 Environment and Regulation of Growth and Flowering

11.4.1 Environment

In order to obtain good vegetative growth and quality and quantity of flower, LD and SD treatments have been worked out for tall, medium and short cultivars for different chrysanthemum growing areas in various countries. Trials at Efford Experimental Horticultural Station showed that 2 weeks of SD treatment in rooting bench prior to potting reduced the cropping time by one week (Anon., 1970).

Photoperiod experiments revealed that only a few long days are necessary for producing a good pot plant. For 1.8 m wide benches a single line of 100 watt incandescent lamps spaced 1.8 m and hung 71 cm above the bench are sufficient to provide 50-100 lux light. Similarly, for shading the opaque sheet is supported at 71 cm above the bench by stretching 5 rows of wire over arches of galvanized pipes spaced 3m apart. The ideal temperatures are 18.3°C for the first two weeks after planting and 15.6°C afterwards, finishing at 13.3°C. Similarly, the required humidity is as high as 80 per cent in the beginning and low towards the end of the cropping cycle.

In Germany, Ludolph (1996) shortened the cultivation period of pot chrysanthemum to 8-10 weeks before the end of October and mid-March by intensive supplementary lighting. The short culture required a significant decrease in the juvenile phase and planting of good quality, even sized stocklings directly into the final pot to increase the root production and accelerate shoot growth and flowering. The light supply recommended was 40-120 klux/day and daylength (9-11 h) and temperature (18 or 21°C) to be maintained up to flowering.

The role of different environmental factors on growth and flowering of chrysanthe-mum has been discussed earlier.

Choudhari *et al.* (2018) conclude that, initiation of photoperiodic module P1 (12 h dark, 12 h light), after attaining standard growth size helped in increased vegetative growth, flower as well as flower quality of potted chrysanthemum cultivars. Abrol *et al.* (2018) reported controlled photoperiod found better for flowering over natural photoperiod. Short plants with improved quality were recorded under controlled photoperiodcompared to natural photoperiod.

11.4.2 Effect of Growth Regulators

The control of plant height is of utmost importance to the pot-mum growers. Pinching, disbudding and use of growth regulators are the common methods used to reduce the plant height. The guiding principle for the time of pinching is to leave 7-8 leaves on the plant. This stage comes in about 2 weeks in summer and 3 weeks in winter, after potting. Soft pinching is preferred to hard one. Disbudding is done to leave single bud per lateral shoot.

Although reduction in LD treatment, use of short duration cultivars, late pinching can reduce the height of plant, use of chemical growth retardants has been found to be very effective. They not only retard the rate of stem elongation but also impart stur-diness, increase colour depth of foliage and cause uniform growth without affecting the flower quality. As mentioned earlier SADH, A-Rest (ancymidol) and phosphon-D are the important growth retardants commercially used for the regula-tion of growth. Chemicals are used as foliar spray to the newly emerging shoots or visibly budding plants. The concentrations of the chemicals, time and method of application have been described earlier. Plants are sprayed with SADH (0.25 per cent solution) when the shoots attain 4 to 6 cm length after pinching. The pot-mums are treated about 2 weeks after pinching and a second application is made 1-2 weeks later. Application of Aqua-Gro wetting agent resulted in reduction of plant height of potted chrysanthemum. It was also effective in improving flowering and shelf life of flowers (Bhat *et al.*, 1990). Double foliar application of daminozide at 25 mg/pot or single application at 50 mg/pot after single pinching were found to be most effective treatments at reducing plant height in *C. zawadskii* s.sp. *naktongense* (Yoo *et al.*, 1999). Daminozide (succinic acid-2,2-dimethylhydrazide) at 4000 mgL^{-1} prolonged cycle and reduced leaf size in cv. Snowdon (Mainardi *et al.*, 2004).

Paclobutrazol is effective in reducing growth and made plants more suitable for growing in pots (Bañón *et al.*, 2012). Paclobutrazol @ 90 ppm showed maximum reduction in plant height and internodal length of pot mum, and Cycocel @1000 ppm reduced plant height, increase plant spread and number of flowers per plant, as well as prolonged senescence (Tripathy *et al.*, 2016). Best quality pot mum in cv. Ajaycan be achieved with spray of paclobutrazol (60 ppm) after one week of pinching. However, in 'UHFSChr-Collection 1' under natural photoperiod sprayed with daminozide 3000 ppm and paclobutrazol 60 ppm under controlled photoperiod produces highest quality pot mum (Abrol *et al.*, 2018).

Asrar *et al.* (2014) found that foliar spary of Alar at 1500 ppm recorded reduction in plant height, leaf area, node production or flowering and early flowering, increased number branches, number of leaves per plant, maximum duration of flowering, time to marketability, flower weight, flower diameter, and number of flowers per plant. It also recorded highest chlorophylls relative content, reducing and non-reducing sugars and N, P and K percentages in cvs. Kodiak 'yellow', Ivyridgf 'white', Auburn 'red' and Lansing 'pink'.

11.5.0 Watering

Hand watering for individual pots is still the most common practice adopted by small growers. The practice is, however, time consuming and the cost of labour becomes very high. Automatic watering has been found useful. For uniform plant growth it is necessary to ensure that each pot receives the same amount of water, which is achieved by spaghetti system; each pot receives water through a small bore tube connected to a hose running through the centre of the rows. Many growers use capillary system in which pots are kept on a 25 mm deep layer of sand or on mats, the moisture levels of the sand or mat is kept constant and near field capacity and water is taken by compost through the drainage holes. Spaghetti system is good for clay pots kept on benches with wire mesh tops. Capillary system is advisable for plastic pots (which had 3-4 drainage holes) with flat bottom when benches with solid tops are used. Volkering and Strauch (1987) found that ebb-flow system of watering was the best for pot plant production of chrysanthemum grown in EEP (Einheitserde, P. Standard clay-peat mix) substrate. Bursy *et al.* (1990) suggested the use of decarbonated water for watering pot chrysanthemum. Plant height increased by watering from below compared with watering from overhead. Farias and Saad (2011) reported that substrate water tension at -2, -6 and -10 kPa resulted in the highest number of pots with A1 quality and -30 kPa presented longer durability in cv. Puritan.

Ristvey *et al.* (2018) found that precision irrigation control of containerized plant material is essential to not only reduce the frequency of irrigation events of plants grown in porous soilless substrates, but also to reduce nutrient runoff to containment ponds and surface water bodies. Wireless sensor network using capacitance moisture sensors attached to monitoring and control nodes was used in production of containerized chrysanthemum on commercial scale.

Robson *et al.* (2019) reported that the 175 per cent crop evapotranspiration (Etc) depth in pots, which corresponded to an average daily depth of 10.69 mm, presented the best results (fresh mass of shoot system, dry mass of shoot system, fresh mass of the root system, dry mass of the root system, fresh mass of the flower, dry mass of the flower, plant height, stem diameter and flower diameter.

MacDonald *et al.* (2014b) stated that subirrigation is being adopted as an environmentally friendly strategy for managing the nutrition of potted greenhouse plants. They investigated two strategies for enhancing the remobilization of N during the development of the inflorescence in subirrigated potted chrysanthemum in an attempt to improve nitrogen (N) use efficiency. (1) The replacement of a portion of the nitrate in the nutrient solution with ammonium decreased the nitrate content, especially in the stem plus petioles, and increased the reduced N content early in the growth cycle, but did not improve N use efficiency. (2) The use of a lower N supply (200 vs. 400 mg total N per pot delivered over 3 and 5 wk, respectively), either as nitrate or ammonium nitrate, eliminated nitrate accumulation prior to inflorescence development, thereby improving N use efficiency. Inflorescence quality was unaffected; however, at the lower N level there was some evidence of chlorosis on the oldest leaves. Thus, there may be considerable potential to reduce the N supply in a commercial greenhouse setting as long as an adequate supply is provided early in the growing period.

11.6.0 Feeding

A proper fertilization programme is essential for production of high quality pot-mums. Plants need frequent and regular feeding immediately after planting. Feeding is done with each watering and several kinds of fertilizers such as soluble inorganic and slow release fertilizers are used for this purpose. Waters (1962) observed that growth responses of potted plants increased in a linear or curvilinear manner as the fertilizer rates were increased linearly at 2.66, 5.32, 7.98 and 10.64 g of 6-6-6 fertilizer per 15 cm pot every fortnight. The optimum treatment combination was 3.99 g of 6-6-6 or 6-0-6 fertilizer applied weekly. Application of 300 ppm N and 200 ppm K from mid-April to mid-October was found useful with each watering. For the remaining months of the year, feeding with equal parts of N and K, except in case of CO_2 enrichment where nitrogen predominated, was suggested. Care was taken not to overfeed with N in seasons with poor light and a feed containing 100 to 150 ppm N was satisfactory (Machin and Scopes, 1978). Carter (1980), on the other hand, recommended application of 200 ppm each of N and K with each watering after the initial fertilization. It was also necessary to maintain the correct balance between the essential elements and the right pH of the medium (5-5.5 in peat + sand and 6.0-6.5 in loam based compost). Mota *et al.* (2007) found thet EC level of 2.13 dSm^{-1} during the vegetative stage and 2.57 dSm^{-1} during the bud stage provided the best visual aspect in pot plantscultivated in a greenhouse. Beckmann *et al.* (2010) reported that increasing EC in substrate media inhibited plant development, but stimulated stem diameter. Salt leaching resulted in greater growth of the plant variables, except for stem diameter. An EC of 2.1 dSm^{-1} by periodical leaching of the substrate enables to obtain market quality standards in pot chrysanthemum cv. Miramar. Beckmann *et al.* (2013) further reported that increase in EC in nutrient solution inhibited the accumulation of micronutrients and of nitrogen, but stimulated the levels of P, K, Ca, Mg and S in the leaf dry mass of the cv. Cv. Miramar grown in pots. The leaching of salts increased the concentration Cu, Fe and Mn in the plant leaves. Among the macronutrients, except calcium and sulphur, no significant difference was observed for N, P, K and Mg relative to the leaching of salts.

Heidemann and Barbosa (2017) evaluated the production and quality of three varieties cultivated in pots with different NPK rates and found that stem length and EC were affected by NPK rates. Inverse relation between the increase in NPK rate and cultivation cycle, resulted precocity of flowering, therefore, NPK at 9 g/L found adequate for pot plant production in cvs. Sheena, Calabria and Indianapolis. Macdonald *et al.* (2014a) suggested that lower N supply (200 vs. 400 mg total N per pot delivered over 3 and 5 week, respectively), either as nitrate or ammonium nitrate, eliminated nitrate accumulation prior to inflorescence development, thereby improving N use efficiency in sub-irrigated potted greenhouse plants. Elimination of nitrate prior to inflorescence emergence can be an effective strategy for improving N use efficiency in subirrigated potted chrysanthemums without sacrificing quality.

Alexe *et al.* (2017) found that fertilization with ammoniacal nitrogen affects the maintenance of plants quality more strongly than nitric nitrogen. It was observed that the excess of nitrogen strongly affects the duration of the quality maintaining of the plants, comparative with the excess of potassium, optimum dose of nitrogen fertilization being much lower than the optimum dose of potassium fertilization.

12.0 Diseases, Pests and Disorder

Many diseases and pests infest chrysanthemum and severely damage the plants at various stages of growth. In the absence of adequate sanitary measures and lack of knowledge on effective control measures, these maladies often cause death of the plants. The measures include rouging, selection and maintenance of healthy stock plants, soil sterilization, spraying of suitable fungicides and insecticides and also use of biological control agents. The common diseases and pests of chrysanthemum, their description and control measures are as under.

12.1.0 Diseases

Several fungal, bacterial and viral diseases affecting the chrysanthemum were identified and their control measures recommended as follows.

Fungal Diseases

12.1.1 Root Rot

Root rot in chrysanthemum is caused by *Pythium* and *Phytophthora* spp. Lumsden and Haasis (1964) isolated 8 species of *Pythium* from chrysanthemum. Rotting of stem cuttings in rooting bed under warm moist conditions is most common; the fungus also enter established plants through pinch wounds. Sudden wilting of infected parts (root, stem and leaf) is the diagnostic feature of pythiuminfection. Control measures are (i) pre-planting sterilization of medium, (ii) good drainage, and (iii) quick removal of infected cuttings. Soil drench with thiram/captan mixture each at 2.5 g/m^2 prevents infection. Engelhard *et al.* (1972) found soil drench with 5-ethoxy-3-(trichloro-methyl)-l, 2, 4-thiadiazole effective.

Conditions that favour pythium also favour phytophthora infection with similar symptoms and the control measures are also the same.

12.1.2 Phoma Root Rot

Wilcox (1963) reported root rot in several cultivars caused by *Phoma* species. The disease is more likely to occur in unsterilized soils. The symptoms are first noticed on lower leaves which become chlorotic and show signs of wilting. The disease progresses upwards causing necrotic spots and limping of leaves. Peerally and Colhoun (1969) found that inadequate nitrogen and phosphorus favoured this disease. Pre-planting soil drench with nabam (140 g/100 l at 22 l/m^2) gave excellent control. This treatment given 2 weeks before planting prevented toxic effects. Of different *Phoma* spp., *Phoma chrysanthemi* was observed to cause the maximum damage (Metz, 1987). This species, however, attacks all above ground parts, thus causing collar rot. An early preventive fungicidal spray of sprortak alpha (carbendazim + borax) or sportax (prochlororaz) controlled the disease. Another species *Phoma chrysanthemicola* was found mainly in chrysanthemums grown in the same greenhouse all the year round. Differential responses of different cultivars towards susceptibility to *P. chrysanthemicola* were reported by Hauptmann and Schickedanz (1987). Biological control of *P. chrysanthemicola* by different *Trichoderma* spp. and *Gliocladium* spp. was reported by Granada *et al.* (1999).

12.1.3 Foot Rot

This disease, caused by *Rhizoctonia solani*, was reported to be particularly severe under warm moist conditions (Cox, 1970). Infection of cuttings in rooting beds leads to mushy brown rot of stem and leaves. In contrast to *Pythium* saturated soil conditions due to excessive water in soil reduces *Rhizoctonia* infection (Jackson and McFadden, 1961).

Chemical control is possible by soil drench with thiram (150 g/100 l) or dusting with quintozene (13 g/m^2). Spraying with or dipping in benomyl (240 g/100 l) has been reported to give protection to young plants (Machin and Scopes, 1978). Basal stem rot was inhibited by treating the cuttings with culture extracts of different antagonists like *Aspergillus*, *Gliocladium*, *Paecilomyces*, *Trichoderma* and *Bacillus* spp. (Tschen, 1991). *Rhizoctonia solani* also causes stem rot. Wu (1996) reported that biofungicide made from the combination of *Bacillus subtilis* and kaolin gave significant control of chrysanthemum stem rot caused by *Rhizoctonia solani*.

12.1.4 Stem Rot and Wilt

The symptoms of this disease caused by *Fusarium oxysporum* occur only after the appearance of flower buds or when they show colour, although, infection also takes place in the rooting bed. Stem near the soil level becomes dark brown and dry.

Diseases of Chrysanthemum

Stem rot

Septoria leaf spot

Pythium root rot

White rust

Bacterial leaf spot

Tomato spotted wilt

Diseases of Chrysanthemum

Chrysanthemum wilt

Powdery mildew

Chrysanthemum mosaic

Chrysanthemum stunt caused by chrysanthemum stunt viroid (CSVd)

Lower leaves turn yellow and the plants wilt during the day and recover at night for some days only to wilt permanently afterwards. Soil application of Dithane M-45 effectively controls the fungus at nursery stage. In later stage, the same treatment inhibited the growth of the disease. Kwon *et al.* (2013) reported first report of *Fusarium oxysporum* causing wilt in Chrysanthemum x morifolium mum in Korea. Jeyaram *et al.* (2019) reported that *Trichoderma, Pseudomonas* and *Bacillus* isolated from chrysanthemum rhizosphere were explored as an alternative to chemical fungicides. *Trichoderma viride, Pseudomonas fluorescens* and *Bacillus subtilis* effectively inhibited the growth of *Fusarium oxysporum* f.sp. *chrysanthemi* to an extent of 77.0, 59.6 and 48.3 per cent, respectively under *in vitro* conditions. Among the systemic fungicides tested, combination of carbendazim 12 per cent + mancozeb 64 per cent inhibited the fungal growth completely at all the four concentrations (500, 1000, 1500 and 2000 ppm) followed by azoxystrobin.

Chen *et al.* (2019) *Fusarium* wilt, a severe disease in chrysanthemum monoculture system. The effect on the chrysanthemum rhizosphere microbiome of the fumigant dazomet (DZ) and of a biofungicide based on *Bacillus subtilis* NCD-2 (BF). The results indicated that the BF treatment increased bacterial diversity by 4.2 per cent, while decreasing fungal diversity by 21.3 per cent. After two seasons of BF treatment, the abundance of microbes associated with disease suppression such as *Bacillus* spp. and *Trichoderma* spp. increased 15.1-fold and 4.25-fold more than that of the control, while the pathogenic *Fusarium oxysporum* was decreased by 79.20 per cent when compared to the control.

Besides, the DZ treatment reduced both bacterial and fungal diversity 7.97 per cent and 2.73 per cent respectively, when compared with the control. The DZ treatment controlled *Fusarium* wilt disease and decreased the abundance of *F. oxysporum* in the first year, but the abundance of the *F. oxysporum* was 43.8 per cent higher after two years in treated soil than in non-treated soil. Therefore, the application of BF has a great potential for the control of *Fusarium* wilt disease in chrysanthemum by changing soil microbiome structure and function.

12.1.5 Grey Mould

The disease incited by *Botrytis cinerea* may attack the plant at any stage of growth from rooting bed to flowering. Stem infection is characterized by girdling, resulting in death of upper part. Leaf infection often starts from margin and proceeds towards the centre and base showing occasionally semi-circular bands of infected tissue. Flower infection starts with brown water-soaked spots on lower petals. In pink flowers, the colour at the point of infection is intense (Jackson and McFadden, 1961). Control measures consist of good aeration by providing adequate planting distance in the field. Spraying captan (100 g/100 l), dichlofluanid (50 g/100 l), thiophanate methyl (50 g/100 l) or thiram (300 g/100 l) gave good results. The varieties differed in their susceptibility to *Botrytis cinerea* (Strider, 1994).

12.1.6 Ascochyta Blight

Ascochyta chrysanthemi or *Mycosphaerella ligulicola* causes 'black rot' disease; blackish lesions develop and grow bigger to make irregular black blotches on stem and the lower leaves. It also causes browning of petals and flower stalk which drop after becoming black. Spread of the infection to other petals causes distortion of flowers. Control measures necessitate hygienic practices like removing the debris quickly and rouging. Keeping the stock plants free from infection by resorting to regular fungicidal sprays is important. Differential response of varieties towards susceptibility to this pathogen has also been reported (Strider, 1994). *M. ligulicola* (*Didymella ligulicola*) is also reported to cause chrysanthemum ray blight (Fox, 1998).

12.1.7 Septoria Leaf Spot

In India, the most common and serious disease of chrysanthemum is caused by *Septoria obesa* and *Septoria chrysanthemella*. Brown spots, which appear on leaf, grow in size and number in humid weather. Leaves turn yellow and eventually die. Wetting of leaves facilitates the infection. Control measures consist of dry airy environment, picking and destroying leaves as they show signs of infection and spra-ying with antifungal chemicals. Foliar spray of bavistin or benlate (both at 0.1 per cent) effectively controls the disease. However, Madhu and Shyam (1992) recommended spraying of bavistin (0.05 per cent), 6 times at 15 day intervals for best control of the disease. On the other hand, multiple cropping of chrysanthemum with *Tagetes* and *Brassica campestris* sub sp. *chinensis* had lower incidences of *Septoria* as observed by Marwoto and Djatnika (1986). According to Honda *et al.* (1992), brown spot disease can be suppressed by controlling the light quality to manipulate the phototropic response of germ tubes of *Septoria obesa*. These differential light suppresses the invasion of the fungus through the stomata of leaves. Differential susceptibility of cultivars to *Septoria chrysanthemella* has been reported by Madhu *et al.* (1989) and Strider (1994).

Chandel and Chandel (2010) reported that Septoria leaf spot severity was high at 25°C and lowest at 35°C temperatures. Relative humidity more than 96 per cent influenced the spread of the disease to maximum by causing 57.20 per cent disease severity. Positive correlation of disease concerning rainfall and relative humidity. Contaf, SAAF, Roverol were found good in controlling the *Septoria* leaf spot of chrysanthemum in field, followed by Dithane M45, Dithane Z78 and benomyl.

12.1.8 Powdery Mildew

It is a very common disease in several countries and is caused by *Oidium chrysanthemi*. It is characterized by white powdery patches of spores on upper surface of the leaves. Control measures consist of a dry environment and chemical treatment with triforine (30 g/100 l), thiophanate methyl (50 g/100 l), pyrazophos (15 g/100 l) and carbendazim (25-30 g/100 l). Cho *et al.* (2017) reported first report of Powdery Mildew caused by *Golovinomyces artemisiae* on *Chrysanthemum zawadskii* var. Latilobum in Korea.

12.1.9 White Rust

This disease is also called the Japanese rust; it is a widely reported disease of chrysanthemum caused by *Puccinia horiana* and is characterized by whitish pustules on underside of leaves. This is one of the most dreaded disease of chrysanthemum. An efficient incubation and inoculation system for white rust was established using plantlets of chrysanthemum growing *in vitro*. Susceptibility to white rust in the plant-box evaluation showed a good correlation with the rating of sporulation on plants grown in a greenhouse (Takatsu *et al.*, 2000).

Sriram *et al.* (2015) reported that first report of *Puccinia horiana*causing white rust of chrysanthemum in India. The severity of occurrence coincides with the post-monsoon season followed by the winter spell (October to December) with disease incidence of 100 per cent in certain germplasm accessions while in others it ranged from 20-60 per cent. The first symptom was the appearance of yellow spots on the upper surface of the leaves (generally 2-5 mm diameter) with white to yellow pustules (telia) on the corresponding places on the lower side of the leaves. Dheepa *et al.* (2016) described the diseae as characterized with the presence of pale-green to yellow white rust spots, on the upper surface of the leaf. Centre of the spots turns brown and necrotic upon aging. On the corresponding lower surface, raised, buff or pinkish, waxy pustules (telia) were observed. Severely affected leaves wilt, droop and gradually dry up. Bonde *et al.* (2015) also demonstrated the systemic infection by *Puccinia horiana* in chrysanthemum.

Pedley (2009) developed conventional and real-time polymerase chain reaction (PCR) assays to detect *Puccina horiana* using primers designed to amplify portions of the internal transcribed spacer (ITS) regions of the nuclear ribosomal DNA (rDNA). The species-specific primers could detect the pathogen from 1 ng of DNA isolated from infected leaf tissue in conventional PCR assays and from 1 pg in real-time PCR assays. While both assays were capable of detecting P. horiana in symptomatic tissue, the greater sensitivity offered by the real-time PCR assay makes it more reliable for detecting the pathogen during the latent stage of infection.

Bonde *et al.* (1995) used myclobutalin (100 mg a.i./l) used as a dip treatment for chrysanthemum cuttings prior to planting for exclusion and eradication of the pathogen while Strider (1995) recommended fungicides myclobutanil, propiconazole, tridimethyl and triforine for control of this disease. Myclobutanil was also effective as a curative agent. Aragon-Caballero (1997) found that application of diniconazole gave the best control of white rust of chrysanthemum, but after treatments plants became stunted. In yet another experiment, azole fungicides desmel (250 g/l propiconazole) and alto 100SL (100 g/l cyproconazole) showed only protective effects. There was no prolonged residual action (Krebs, 1997) Fujii and Takamura (1998) recommended a new fungicide, difulmetorim for control of chrysanthemum white rust.

Epidermal polymer coatings like Nu Film also gave good control (Rodriguez-Navarro *et al.*, 1996). In the rainy season the white rust was more severe; so the best treat-ments to control white rust were potassium bicarbonate + Nu Film17 and copper sulphate. It was found that under different types of covers in greenhouse, the severity of this disease varied. Polythene resulted in 3-fold higher disease severity compared with blue-PVC and PVC (Freitas *et al.*, 1997).

Inorganic salts were also found to be effective in controlling the disease. Copper sulphate, potassium bicarbonate and sodium bicarbonate were found to control the disease effectively (Rodriguez-Navarro *et al.*, 1996).

Biological control of white rust by *Verticillum lecanii* has been found to be very effective. This can be used for integrated insect and white rust control programme on all-year round chrysanthemum (Whipps, 1993; Rodriguez-Navarro *et al.*, 1996). Stapel and Guerrand (2012) reported that acibenzolar-s-methyl (ASM) has an elicitor effect on chrysanthemum plants against white rust, while 'PrevB2' and 'Dipper' act mainly as a fungicide. It is a more effective and durable control measure against white rust in chrysanthemum. Dheepa *et al.* (2016) revealed that the foliar application of *Bacillus subtilis* (BS2) and *Bacillus amyloliquefaciens* (BSC7) at fortnightly intervals was most effective in reducing the incidence of white rust. It also increased plant height, marketable stems, and flower yield.

Screening of chrysanthemum germplams revealed the resiatant source againat white rust. Zeng *et al.* (2013) found accessions of *Chrysanthemum indicum*, *C. yoshinaganthum*, *C. makinoi* var. wakasaense, *C. nankingense*, *C. vestitum*, *C. lavandulifolium* and *C. crassum* were immune, and *C. japonense* and C.x *shimotomaii* had strong resistance among 11 Chrysanthemum species screened againt reaction to white rust infection.Thakur *et al.* (2019) reported that cultivars Flirt, CO.1, Fitonia, Rekha and Arka Ravikiran were moderately resistant and cvs. Arka Indira, Arka Kirti, Arka Pankaj, Arka Yellow Gold and Arka Red Gold were resistant to Chrysanthemum white rust in both *in vitro* as well as in pot culture.

Barhate *et al.* (2015) reported that curative spray of propiconazole was most effective followed by hexaconezole + mancozeb, tridemefon, captan + hexaconezol in controlling white rust. Dheepa *et al.* (2018) evaluated the bio-efficacy of fungicide modules, IDM module II, comprising of tebuconazole 25.9 per cent SC (Folicur), - BM, strain BS2, strain BSC11, azoxystrobin 23 per cent SC (Amistar) and Myclobutanil 400WP sprayed at weekly interval was most effective in reducing the incidence of white rust under protected cultivation and increased the number of marketable stems.

12.1.10 Verticillium Wilt

Verticillium wilt is a common malady in many countries including India. Symptoms of sudden wilt occur only after appearance of flower buds although infection, perhaps have taken place much earlier (Busch and Schoolboy, 1970 and Strider, 1994). The pathogens causing the disease are *Verticillium dahliae* and *V. albo-atrum*. At times it becomes very serious but introduction of routine soil sterilization culture, indexing of stock plants have helped the growers to minimize the incidence. Spraying or dipping in benomyl suspension (4 lb 50 per cent/100 gal) was recommended for immunization of rooted cuttings against *V. alboatrum* by McCain (1971). Drenching with carbendazim (25-30 g/100 l) was also recommended for preventing its spread besides removal of the infected plants.

12.1.11 Ray Speck

Dusky specks on ray florets are caused by either of the two fungi, *Stemphylium floridanum* and *Alternaria chrysanthemi* or due to their combined attack. Avoidance of overhead watering at flowering stage and application of zineb or maneb controls the disease (Tammen, 1965 and Sobers, 1965).

Stem Blight

Chen *et al.* (2018) reported first report of chrysanthemum stem blight and die back caused by tomato spotted wilt virus in cultivated chrysanthemum (*Chrysanthemum morifolium*) in China.

Foliar Blight

Fan *et al.* (2015) suggested that CmWRKY15, a group IIa WRKY family member, might facilitate *Alternaria tenuissima* infection by antagonistically regulating the expression of ABA-responsive genes and genes involved in ABA signaling, either directly or indirectly. Kavitha *et al.* (2013) reported that tebuconazole (0.1 per cent) recorded lowest percent disease index (6.06) and highest flower yield (73.49 q/ha) followed by propiconazole (0.1 per cent) and azoxystrobin (0.075 per cent) (15.48 and 16.24 PDI and 59.56 and 58.32 q/ha, respectively) in managing the foliar blight of Chrysanthemum.

Shamala and Janardhana (2015) found that the varieties Chandini and Karnel were more susceptible to blight showing highest disease incidence (100 per cent) and severity (100 per cent) recorded from Mandya and Mysore regions. The lowest disease incidence, severity and PDI of 18.40 per cent, 38.68 per cent and 24.89 per cent were recorded on Raja White variety in Bangalore region. *In vitro* evaluation of four fungicides on *Alternaria alternata* revealed that carbendazim + mancozeb was most effective with 95.65 per cent inhibition followed by carbendazim, mancozeb and metalaxyl + mancozeb with 68.26 per cent, 91.30 per cent and 91.30 per cent inhibition of the mycelium at 2.0 per cent concentration, respectively.

Bacterial Diseases

12.1.12 Bacterial Blight

Erwinia chrysanthemi causes blight disease in chrysanthemum and the symptoms start with wilting of one or more branches during a sunny day with recovery at night. With the spread of the disease, the stem tips turn brown and brittle and collapse. Stem becomes hollow with brownish streaks extending up to base. In rooting beds, the cuttings when infected either die or remain stunted. High temperature and humidity favour the growth of the organism. Control measures consist of soil sterilization, using disease-free cuttings and avoiding contamination during pinching. Use of streptomycin though useful is advised only in propagation benches (Jackson and McFadden, 1961).

12.1.13 Bacterial Leaf Spot

Circular to elliptical leaf spots appear and increase in size and coalesce to form large necrotic areas. The disease progresses from down upwards ultimately invading flower buds which turn dark and die prematurely. Jackson and McFadden (1961) recommended the use of disease-free cuttings and spraying with tribasic copper sulphate (4 lb/100 gal) at weekly intervals during the rainy season to control the disease.

Viral Diseases

The number of viral diseases affecting chrysanthemum are at least 20 according to Machin and Scopes (1978). Jackson and McFadden (1961) described some of them and the suggested methods of transmission and control:

Virus	*Symptoms*	*Transmission Methods*	*Control Methods*
Chrysanthemum stunt	Overall reduction in plant size, foliage pale in colour, flowers may open prematu- rely, red and bronze flowers often bleached. Leaf margins may fail to enlarge giving stiff and upright appearance.	During pinching.	Use of cuttings from virus-free indexed stock.
Tomato spotted wilt	Reduced growth, leaf may show ring and line pattern, plants may show one-sided development with necrotic streaks in stem and leaves.	Thrips, several weeds and other plants serve as sources.	Control of thrips and weeds.
Tomato aspermy	Flower breaking and distor- tion, flowers may be small with warty florets.	Aphids	Control of aphids and use of cuttings from virus-free indexed stocks.
Chrysanthemum flower distortion	Marked dwarfing and distortion of flowers, flowers may appear as un- opened buds with short incurved narrow ray florets.	Grafting, no insect vector known.	Use of cuttings from virus-free stock.
Chrysanthemum mosaic	Leaf symptoms mild or lacking, plant weak and small flowers may show breakdown of florets.	Aphids	Control of aphids.
Chrysanthemum rosette	Yellow mottling with mosaic pattern in some cultivars, enlarged veins wrinkle and rosette formation on terminal portion.	Grafting, no insect vector known.	Use of cuttings from virus-free indexed plants.

The abovementioned six viruses given have been the most widely reported ones (Brierley and Smith, 1951, 1955). Some other main viruses found on cut chrysanthemum flowers are potato X potex virus (PXV), potato Y potyvirus (PYV) and tomato aspermy cucumovirus (TAV). The best method to control these viruses is the meristem culture (Chen *et al.*, 1998). Zandarski (1998) enumerated the characteristics of the chrysanthemum stunt viroid and its control.

Heat therapy consisting of treating the plants at 37.2°C for 4 weeks has been found effective in eliminating aspermy virus (Fenton, 1969). Meristem culture after heat treatment was suggested by Asafani (1972) and Marani (1972) for production of virus-free planting material. Heat therapy, however, did not control all the viruses. While aspermy, English stunt and most other mosaic viruses are controlled; vein mottle, American stunt and some mosaic viruses can not be controlled by heat treatment.

Phytoplasma

Raj *et al.* (2007) reported that the association of 'Candidatus Phytoplasma asteris' with little leaf disease of *Chrysanthemum morifolium* was detected by nested PCR and identified by nucleotide sequence analysis of PCR products for the first time from India. Bayat *et al.* (2013) reported that the BLAST analysis of the chrysanthemum phytoplasma sequence exhibited 99 per cent identity to Candidatus *Phytoplasma* phoenicium (Ca. *P. phoenicium*) species of the 16SrIX group. The leafhopper (*Neoaliturus fenestratus*) may be a potential vector as it is an often encountered efficient transmitter vector of 16SrIX group phytoplasmas in Iran. Next to the susceptibility of chrysanthemum to members of aster yellows, stolbur, and Ca. *P. aurantifolia* phytoplasma groups. This is the first report of a 16SrIX group member infecting chrysanthemum in Iran. Ulrich *et al.* (2017) reported first Detection of *Phytophthora chrysanthemi* on *Chrysanthemum indicum* in Germany. Gamalero *et al.* (2010) reported that *Pseudomonas putida* S1Pf1Rif is able to alleviate Chrysanthemum Yellows Phytoplasma Infection, although it does not affect the presence of viable phytoplasma in young, developing leaves of the infected plants.

12.2.0 Pests

The common pests of chrysanthemum are aphids, hairy caterpillars, red spider mites, root cutting grub, thrips and nematodes. Leaf miner is another pest very extensively reported from many countries. A brief description of each of these pests and their control measures are as follows:

12.2.1 Aphids

Myzus persicae, Macrosiphoniella sanborni, Brachicaudus helichrysi, Aphis gossypii, Aphis fabae and Aulocorthum circumjlexum are the aphids attacking chrysanthemums.

Aphids, greenish to black dot-like insects, suck the sap from the growing stem tips and undersurface of leaves, causing loss of vigour. They are also carriers of virus diseases and they excrete honey dew which forms a substratum for fungal growth. The control measures include spraying with tobacco-soap decoction. Several insecticides, *viz.*, malathion, parathion, pirirmicarb and aldicarb have been used to control aphids. Wyatt (1966) and Gould (1966) demonstrated the resistance of *Myzus persicae* to organochlorine and organophosphorus insecticides. Overman and Poe (1971) found that 7 weekly sprays of D-1410 [S-methyl-(dimethyl carbamoyl)-N-(methyl carbamoyl) oxythioformimidate] reduced or eliminated aphid (*M. persicae*) population. Poe and Marousky (1972) reported success in controlling aphids on cut-chrysanthemums by postharvest absorption of azodrin and demeton. Scopes (1970) reported biological control and Scopes and Biggerstaff (1973) used a wasp parasite (*Aphidius matricariae*) successfully. Hall (1975), Sopp *et al.* (1990) and Vehrs and Parrella (1991) reported control of aphid by spraying spores of a fungus *Verticillium lecanii*. On the other hand, Flint *et al.* (1995) and Dreistadt and Flint (1996) found that adult lady beetle (*Hippodamia convergens*) provided augmentative control of relatively high aphid densities on small potted chrysanthemum plants. Varietal resistance to aphid has also been reported by Wyatt (1965) and Bethke *et al.* (1998). According to Bethke *et al.* (1998), fecundity, longevity and survivorship of aphids are significantly affected by the cultivar. Wang and Lin (1984) reported complete control of aphid by smoking with 98 per cent deltamethrine at 1 g/m^3 for one hour and fumigation with phosphine for another 5 hours in laboratory tests. Pasian *et al.* (1997) recommended the use of imidacloprid both granular and paint treatment for effective reduction in aphid population for at least 8 weeks. Neem seed extract was also found to reduce the aphid population (*Macrosiphoniella sarboni*) significantly (Opender, 1999).

12.2.2 Hairy Caterpillars

Hairy caterpillars (*Diacrisia obliqua*) attack the plants in rainy season and continue till winter. The pest is easily recognized by the presence of hair on the body. They multiply fast with gregarious habit during early stage. As they eat up the leaves from surface, papery skeletons are left which dry up subsequently. Manual collection and destruction in early stages can check heavy infestation. Spraying chlorpyriphos @ 0.25 per cent is recommended as control measure.

12.2.3 Red Spider Mites

Mites (*Tetranychus urticae*), red dot-like bodies, feed on the undersurface of the leaves causing white specks in the early stages. They occur in hot season and damage leaves and buds which give a pale appearance. Baranowski (1966) reported good mite control by soil application of systemic insecticide UC 21149. Conventional acaricides, *e.g.*, dicofol, tetradifon and aldicarb have been used with success in the past, but resistance is reported to have developed to these chemicals and many other organophosphorus compounds (Kollner 1973; Machin and Scopes, 1978). Promising insecticides are demeton S-methyl (22 g/100 l), diazinon (16 g/100 l) or quinomethio-nate (125 g/100 l) used as high volume sprays or vapourizing naled. Organotin compounds, *e.g.*, cyhexatin (25 g/100 l) are reported to give good control. Metasystox and kelthane are also effective. Acaricide resistance and application deposition rate are important factors for control of spider mites in chrysanthemum fields (Kunimoto *et al.*, 1998). Scopes and Biggerstaff (1973) and Stenseth (1988) found biological control very successful using *Phytoseiulus persimilis.*

12.2.4 Grubs

Grub (*Holtrichia* sp.) is a troublesome pest which remaims underground, particularly in shade under the trees. The underground portion of stem or roots is cut by the grub causing sudden wilting of healthy plants during dry hot months.

Soil application of aldrin, chloropyrifos or thimet has been found effective against the grub (Kher, 1986). Lever (1965) reported tropical cut-worm (*Spodoptera littoralis*) on chrysanthemum in the Canary Islands attacking the chrysanthemum cuttings. He could control it using 0.04 per cent dieldrin emulsion.

12.2.5 Thrips

Thrips tabaci, T. nigrophilosus, Frankliniella tritici and *Hercinothrips* sp. commonly attack the chrysanthemum. Of late, western flower thrips (*Frankliniella occidentalis*) is reported to cause menace in chrysanthemum, causing huge losses. Another species of thrips, *Thrips palmi* is also found to attack the flowers.

Thrips are slender white to black insects which feed on the growing point causing mottling and distortion of leaves and also leaf silvering due to separation

Pests of Chrysanthemum

Aphids

White fly

Thrips

Red spider mites

Leaf miner

Hairy caterpillars

of upper epidermal tissue from the rest of the leaf. The severity of leaf silvering was found to be influenced by daylength (Jager *et al.*, 1997).

Thrip population is less severe under high light intensity. Hence, the severity of the pest is found to be more in winter and under shaded condition (Kogel *et al.*, 1997). The insects also damage flowers of summer blooming cultivars. The affected flowers look as if dried due to scorching heat. Several insecticides *viz.*, malathion, lindane, metasystox have been found effective, specially if used alternately. Nicotine spray too has been found useful. Wang and Lin (1984) could control thrips by using deltamethrine and phosphine. Acephate, methamidophos, omethoate and dichlorol were also found very effective as reported by Heungeas *et al.* (1989). Seaton *et al.* (1997) recommended spraying of insecticides methamidophos, parathionmethyl and chlorpyriphos for effective control of western flower thrips. Other insecticides like lannate (methomyl), mesurol (methiocarb) and dicarzol (formetanate) also controlled thrips (Casteels *et al.*, 1996). Gill *et al.* (1998) observed good control of the thrips by spraying, a bacterial toxin formulation, zerotol [(active ingredients peroxyacetic (peracetic) acid and hydrogen dioxide (peroxide)] and sanmite [(active ingredient, pyriban (pyridaben)].

The occurrence and damage caused by thrips also vary with cultural practices. Low thrip population has been found with application of high amount of nitrogen and under high irrigation schedule (Schuch *et al.*, 1998). However, Chau and Heinz (2006) reported that by lowering N fertilization to 50 per cent of the recommended level (375 p.p.m. N) reduced 44 per cent *Frankliniella occidentalis* (western flower thrips)abundance and significant reduction in fertilizer input for the production system. Manipulation of fertilization can be a useful management tactic against *F. occidentalis*. Hoyle and Saynor (1993) and Parrella and Murphy (1996) recommended the use of trap crop for the control of western flower thrips. Pot chrysanthemum was found to be an effective trap crop for chrysanthemum crop. Coloured water traps especially blue and white traps also proved to be efficient in capturing thrips in chrysanthemum crops (Freitas and Bueno, 1998). Shibuya and Takeuti (1998) found that pink and blue trap boards were more effective than yellow ones for attracting thrips.

Under integrated pest management programme biological control of thrips was found to be very effective in controlling the insect population. Predatory mites *Amblyseius cucumeris* and *A. barker* were found to be very effective against thrips and hence recommended for their biological control by Hessein and Parrella (1990). The minute pirate bug *Orius insidiosus* was found to be a good predator of western flower thrips and effective in controlling the insect population (Fransen *et al.*, 1993). Stack *et al.* (1998) found that application of supplemental blue light in greenhouse enhanced the reproduction of the pirate bug thereby, significantly reducing insect population. Hence they recommended exposing infested chrysanthemum plants to a blue light supplemented long day. Parrella and Murphy (1996) and Gill *et al.* (1998) reported that 2 strains of *Beauveria bassiana* gave good control of Western flower thrips.

12.2.6 Leaf Miners

Although leaf miners (*Phytomyza syngenesiae* and *Liriomyza trifolii*) attract major attention in many countries where chrysanthemum is grown in greenhouses on large scale, Parrella and Jones (1984) described *Liriomyza* as a serious pest in California. The larvae of this insect make tunnels in leaf, and pupate at the end leaving charac-teristic trails behind. Various control measures are recommended. Alverson and Gorouch (1982) claimed 90 per cent control of *Liriomyza* by foliar treatment with methamidophos, cypermethrin and permethrin but found methamidophos phytotoxic. Ledien and Helyer (1983) found pyrazophos effective in reducing the damage by *Phytomyza* leaf miner by 27 per cent to that of control. Pyrazophos along with chlorpyrifos was also effective in controlling *Liriomyza trifolli* (Mora Medina and Mosquera, 1985). Nitenpyram granule was also recommended for the control of *Liriomyza trifolli*. The granules were added to the soil and this resulted in effective control of the leaf miner (Ozawa *et al.*, 1998). Knodel *et al.* (1985) showed that *L. trifolii* was attracted more to SD-grown chrysanthemums than LD-grown ones. Herbert *et al.* (1984) described three non-insecticidal methods of controlling *Liriomyza* leaf miner. The first method was trapping the adults on yellow sticky boards giving 50 per cent control, second by using fungicide pyrazophos at 0.5 ml/l concentration and third by interplanting *Vicia faba* (field bean) as trap crop. Among other non-insecticidal methods neem seed extract was found to be effective in reducing the insect population considerably as well as reducing the chances of reinfestation (Parkman and Peinkowski, 1990; Sanderson *et al.*, 1989). Integrated biological control of leaf miners (*Liriomyza trifolli*) by sprays of entomophogenic nematode *Steinernema carpocapsae* and eulophid parasitoid *Diglyphus begini* were effective (Harris *et al.*, 1990; Broadbent and Olthof, 1995; Ota *et al.*, 1996; Sher and Parrella, 1996). Varietal resistance to leaf miner has been reported by Baranowski and Dankowska (1985).

Yet another species of leaf miner, *Liriomyza huidobrensis* was found to cause severe damage to chrysanthemum plants (Bartlett, 1993; Cheek *et al.*, 1993). Control measures included spraying of insecticides abamectic and cypomazine (Staay, 1992, Cheek *et al.*, 1993).

12.2.7 White Fly

The important species causing severe damage is *Trialeurodes vaporariorum*. Control measures includes both chemical as well as biological control methods. Biological control includes parasitism by the aphelinid *Encarsia formosa* (Kollner *et al.*, 1991). Chemical control is achieved by spraying 2 per cent aqueous sunspray 6E plus which repelled adults and is toxic to the newly hatched and 3rd instar larvae (Larew and Locke, 1990). Pasian *et al.* (1997) recommended the use of insecticide imidaclo-prid for effective control of white fly.

12.2.8 Nematodes

The nematodes damaging chrysanthemum are *Pratylenchus penetrans*, *Aphelenchoides ritzemabosi*, *Belonolaimus longicaudatus* and *Meloidgyne incognita*.

Nematodes travel up the plant in film of water, enter the leaves through stomata causing triangular brown and black patches between the veins characteristic of eelworm infestation. Lower leaves are attacked first. In many countries, eelworm-free stock plants are used for propagation, use of sterilized compost and treatment with suitable insecticides keep the damage to negligible proportions. Use of parathion has been advocated by Hesting and Peachey (1963) and Razauska (1966). Overman and Raulston (1971) found mocap (O-ethyl S, S-dipropyl phosphorodithioate) useful. Bloom (1964) showed photonegative reaction of nematodes which moved up the stem in reduced light condition. Ivezic *et al.* (1982) found aldicarb (500 g/ha) applied before planting, effective against nematode (*P. penetrans*) in Yugoslavia. Nematodes were controlled by application of phospho-organic chemicals (Lamberti *et al.*, 1987). The best protective effect against chrysanthemum foliar nematode is obtained with granular nematicides at a dose of 1 g a.i./plant applied under the roots during planting of chrysanthemums (Wojtowcz and Labanowski, 1994). Ramakrishnan and Vadivelu (1995) found that quinalphos, neem cake, press mud, compost and intercropping with *Tagetes patula* reduced nematode populations.

12.3.0 Disorder

12.3.1 Crooked Neck

Bending of flower neck (crooked neck) in greenhouse chrysanthemum often occurs in the winter crop. There are 2 types of crooked neck in chrysanthemum: the bending of the flower neck which occurs in the uppermost (non-serrated) leaf node, and the adhering of the terminal flower neck to an axillary bud. In the former, the vascular bundles are twisted and disconnected in the stem, whereas in the latter, the cortical tissues of the terminal and axillary buds adhere to the remaining meristematic tissues scattered inside. The expanding axillary flower bud at the shoot apex, examined 9 days after the start of short day treatment under a 25/15°C (light/dark) regime, was found to develop 4-5 days later than that of the terminal flower bud. The adhering type of crooked neck was visible histologically 3 weeks after the start of short day treatment. A positive correlation was observed between the crooked angle of the flower neck and mean internode length of cut flowers (Tanigawa *et al.*, 1999).

13.0 Harvesting and Yield

13.1.0 Harvesting

The correct stage of harvesting depends on the cultivar, marketing and other facilities available to a grower. Single cultivars are harvested when the maximum number of flowers open before shedding of pollens from the outer row of the disc florets. Anemone cultivars are harvested before the central cushion in the topmost flower is fully developed whereas decorative types are harvested when the petals in the centre of the topmost flower is almost fully developed. In standards, harvesting is generally done when outer ray florets ceases to further develop (Machin and Scopes, 1978). Pot-mums are sent to the market with half to fully opened flowers. Singh *et al.* (2010) reported that the stems harvested when the ray florets had attained diameter of 1.0–1.5 cm, exhibited maximum floret longevity. Chang *et al.* (2018)

Harvesting of Chrysanthemum

Cut flowers of chrysanthemum

Loose chrysanthemum

Harvesting in open field

Harvesting stage in greenhouse

Packaging of Chrysanthemum

Polythene packing of blooms

Packing in card board box

Marketing of Chrysanthemum

Marketing of chrysanthemum in pots

International market of chrysanthemum

reported robotic harvesting of Chrysanthemum 'Hangzhou White'. The shape of Hangzhou White, the centroids of stamens were selected as feature points to match in the right and left images. This method recognizes 'Hangzhou White' with the accuracy of 85 per cent, which could meet the requirements of the localization accuracy of the harvesting robot.

Flowers are cut about 10 cm above the soil and one-third of stem from the basal end is stripped off leaves and immediately placed in water containing a biocide to prevent the growth of microorganisms. Silver nitrate at the concentration of 25 ppm has been found to be the most effective biocide for chrysanthemums (Kofranek, 1980). The presence of old stubs on flower stems affect the uptake of water and hence, should be removed. For the purpose of garland and other floral ornaments flowers are harvested without stems.

Standard chrysanthemums are harvested at unopened stage when only a few outer ray florets unfurl and the cut flowers store equally well in dry condition to be handled conveniently with the least damage and equal or superior post-storage quality. These cut-chrysanthemums open better under lighted (75 ft-candle) and warm (18.2-23.7°C) conditions. The bud opening solution is an absolute necessity (Staby *et al.*, 1978). An ideal bud opening solution is 200 ppm 8-HQC and 2 per cent sucrose. The buds stored dry at 0.6°C opened and lasted equally well compared to the buds which opened immediately after harvest (Marousky, 1971). Kofranek and Halevy (1972) recommended an opening solution with 2.5-3.0 per cent sucrose, 25-30 ppm silver nitrate and 75 ppm citric acid. Environmental conditions during the bud opening and size of bud at harvest usually determine the success of this technique. Singh *et al.* (2010) found that prolonged storage duration decreased floret longevity, hastened yellowing of leaves but the flower diameter was not significantly affected. BAP significantly delayed yellowing of leaves and the bioregulator at 50 mg/L was most effective.

13.2.0 Yield

The yield of flowers in chrysanthemum varies depending on the types and the culti-vars, the growing region and the environment, plant density, pinching, disbudding and other management practices. It also depends on the purpose of cultivation, *viz.*, production of loose flower, cut flower, pot mum, exhibition, *etc.* The average yield of loose flowers suitable for making garland and *veni* ranged from 7.5 to 15 t/ha while about 1,00,000 good quality sprays were obtained from one hectare (Kher, 1986). In a field trial, small-flowered types of chrysanthemum produced 28.6 to 41.4 lakh cut-flowers per hectare depending on the plant density (Yadav, 1983).

14.0 Postharvest Management

14.1.0 Grading

Cut flowers are graded into several grades depending on stem length and strength, colour and diameter of flowers, *etc.*

Export Standard for Chrysanthemum*

Parameter	*Standard*	*Spray*	*Dwarf*
Stem Length	88-100 cm	75-88 cm	25-38 cm
Weight	30g/stem of 90 cm	30g/stem of 85 cm	15g/stem of 30 cm
Number of flowers	Only 1 flower With 5 buds	10 flowers with 5-8 buds	10-12 flowers

*Anon. (2020e)

For the British markets stem length of blooms not less than 66 cm is necessary and those less than 51 cm are marked 'shorts'.

Bushy plants with good growth, 2.0 to 2.5 times as tall as their pots, having a minimum of 15 flowers and free from pests and diseases are preferred. A plant having 20-25 good quality flowers would be more desirable.

Chrysanthemums are traded in three quality classes, A1, A2 and B1, depending on the extent to which they meet quality and grading criteria. The minimum quality requirements for chrysanthemums are (Anon., 2020d):

- ✰ All flowers should be visually free of living manifestations of the Leaf Miner (Liriomyza and Phytomyza).
- ✰ All chrysanthemums should be 95 per cent free of the Leaf Miner.
- ✰ Spray chrysanthemums and Santini should have at least three bloomable buds.
- ✰ The maximum diameter of the fully open flower of Santinis is 40 mm.

  The minimum maturity for the disbud chrysanthemum is maturity stage 1 (half-opened bud). The minimum maturity for the spray chrysanthemum and Santini is maturity stage 1 (three half-opened buds).

  The following additional requirements apply:
- ✰ The lot should be free of growth defects, including irregularly formed disc florets or ray florets and forked branches.
- ✰ The lot should be free of heels (in particular the outdoor chrysanthemum).
- ✰ The branches in each lot should be visually uniform.
- ✰ All chrysanthemums longer than 60 cm should have the bottom 15 cm stripped of leaves (defoliated), with the exception that Santini chrysanthemums shorter than 60 cm should have 10 cm defoliated.
- ✰ An additional requirement for Santini states that the branches should have at least six flowers and/or buds. Fewer buds will result in a slight (in case of four to five buds) or substantial deviation (in case of three buds) from the requirements.

14.2.0 Packaging

Most often standard chrysanthemums are placed in sleeves and packed in display boxes measuring 91 x 43 x 15 cm. They are placed in the boxes according to the grades. If the blooms are large and loose, a pillow made from corrugated paper is placed under the necks of the bottom layer at each end of the box. This is not necessary in case of small and firm blooms. In all display packs it is recommended to hold the stems in place with a metal tie strip. For bulk packing of sprays, 10, 15 or 20 stems are placed in sleeves according to the grades. Six sleeves, three at each end, are generally packed in each box, measuring 80 x 50 x 23 cm. Pot-mums are packed by placing them in paper or polyethylene sleeves and then in cardboard boxes in groups of six.

Final retail sales mainly involve mixed bouquets and mono-bunches in either plastic containers or special containers from outlets such as bouquet producers, specialist florist's shops and supermarkets. Bouquets and flowers are sometimes pre-packed in plastic or paper sleeves, but they are mostly assembled and wrapped at the florist's shop.

Export and import trade is often in cardboard boxes. Flowers are repacked into plastic flower containers (buckets) at the auction in the Netherlands or at the importing wholesaler. There is a difference in requirements between spray chrysanthemums and disbud (single-stem) chrysanthemums, as well as between outdoor and indoor cultivated chrysanthemums.

Spray chrysanthemums from the glasshouse should be supplied in a box and:

- ✰ be supplied in bunches of five stems;
- ✰ be packed in a sleeve for each bunch;
- ✰ have 80 items per box (120 items per box for light chrysanthemums of 40 g or less).

Spray chrysanthemums from the glasshouse should be supplied at the auction in water in a plastic container and:

- ✰ have a maximum average weight of 40 grams;
- ✰ be supplied in bunches of five stems;
- ✰ be bundled in bundles of five bunches;
- ✰ be packed in a sleeve for each bundle.

Santini should be supplied in water in a container and:

- ✰ be supplied in bunches of 4 stems;
- ✰ be bundled in bundles of 5 bunches;
- ✰ be packed in a sleeve for each bundle;
- ✰ optionally be supplied in the small container (packing code 566) for Santini with a length of 50 cm or less.

Disbud chrysanthemums from the glasshouse and from outdoors should be supplied in a box and:

- ✰ be supplied in bunches of ten stems;
- ✰ be packed in a sleeve for each bunch.

Disbud chrysanthemums from the glasshouse and from outdoors should be supplied at the auction in a container and:

- be supplied in bunches of ten stems;
- be packed in a sleeve for each bunch.

Spray chrysanthemums from outdoors should be supplied in a box and:

- be supplied in bunches of five stems;
- have 80 items per box (100 or 120 items per box for light chrysanthemums);
- optionally be sleeved, although this is not a requirement.

Spray chrysanthemums from outdoors should be supplied at the auction in water in a container and:

- be supplied in bunches of five stems;
- optionally be sleeved, although this is not a requirement.

If chrysanthemums are traded in bunches (and not per item), the bunch should have at least five branches and all bunches in the lot should be uniform in volume (Anon., 2020d).

Kumari *et al.* (2017) showed that the LDPE 100 guage was most effective film in enhancing the vase-life and quality under both refrigerated wet storage and normal ambient conditions. The LDPE 100 gauge packed flower stems resulted in enhanced vase-life by maintaining flower and leaf membrane stability index coupled with lower physiological loss in weight. The percent decline in total soluble sugars and reducing sugar contents in leaf and ray floret tissues was less in LDPE packaged flowers.

Jagadale *et al.* (2019) revealed that pre-harvest sprays of BA 50 ppm at bud stage resulted in maximum shelf life in chrysanthemum varaities 'Solan Shringar' (3.73 days) and 'Surf' (4.27 days) in polyethylene packaging. Best freshness index maximum per cent moisture content, minimum per cent weight gain and per cent spoilage of flowers was also observed in the flowers packed in cardboard boxes lined with the polyethylene.

14.3.0 Storage

Chrysanthemum cut flowers can be wrapped in plastics and stored dry for 6 to 8 weeks at a temperature of -0.5°C. Temperatures for truck shipments across the country ranged between 2° and 4°C (Halevy *et al.*, 1978; Rij *et al.*, 1979). Proper precautions should be taken to control pests and diseases. A pre-storage dip in 600 ppm azordin for half an hour was found effective in controlling pests during long term storage (Laurie *et al.*, 1968; Staby *et al.*, 1978). Controlled atmospheric storage of cut flowers was not found better than low temperature storage. Similarly, low pressure and hypobaric storage also did not prove effective. In contrast, with modified storage, the flowers of chrysanthemum could be stored for 7 days without loosing any shelf-life as compared to those stored in water.

Ho and Chen (1995) investigated the quality changes in chrysanthemum flowers under various degrees of compression, deformation and transportation periods and observed that longer the chrysanthemum cut flowers spent in the postharvest process, the more broken leaves they had the longevity of flowers and leaves was also less. Even flowers if placed on lower layer in storage container, had more broken leaves and lasted for less time. Zheng and Guo (1998) studied the effect of different pretreatments and dry cold storage for 5-60 days on the quality of cut flowers of chrysanthemum. The best pretreatment for maintaining quality in storage was pulsing in a preservative solution containing BA+GA and removing all flower bran-ches from the solution 1 h before placing in storage. ABA and IAA contents of flowers increased with increasing period in storage. Storage of cut spray type chrysanthemums bymodified at mosphere packaging (MAP) using polyethelene (PE) and polypropylene (PP) film was studied. In the active MAP, the internal atmosphere was replaced with N_2 enriched 2 per cent O_2 gas provided with a gas separation membrane-N_2 generator; this established the desired atmosphere rapidly and successfully. Using the active MAP on cut chrysanthemums, harvested at a partly open stage, delayed further opening of the flower and bud during storage for 10 days at 10°C. The vase life of chrysanthemum stored in active MAP, was 13-15 days after exposure to the air at 20°C; those of flowers stored wet and in perforated bags were 6 and 7 days, respectively (Yamashita *et al.*, 1999). Storing the cut flowers wrapped in cellophane at 4 ± 1°C incoldstorage was found to be the best (Bhat *et al.*, 1999).

Pot-mums can easily be stored upto 2 weeks at a temperature of approximately 4°C; above 13°C the storage life decreases. Beneficial effects of light can also be achieved by raising the light intensity upto 400 ft-candle. Storage in dark deteriorates the quality of the bloom.

Srivastava *et al.* (2015) reported significant effect of wrapping material and storage condition were also observedon per cent weight loss, total water absorbed maximum flower diameter and vase life in cvs. Snowball Yellow and Snowball White. Wrapping of spikes in PP 200 gauge with refrigerated storage at 3-4°C for 3 days + 6 hours of simulated transit resulted in minimum per cent weight loss, maximum total water absorbed, flower diameter and maximum vase life in both the cultivars.

Jagadale *et al.* (2019) stated that maximum shelf life was observed in three day storage in Solan Shringar' and 'Surf'. More shelf life was observed in cold storage than the flowers stored at room temperature in both the cultivars. While comparing both cvs, *i.e.* 'Solan Shringar' and 'Surf' were found at par for shelf life and freshness index.

14.4.0 Vase-Life

The use of proper preservative solution throughout the period of postharvest handling is very important to prolong the life of cut flowers. Farnham (1973), suggested dipping of the stem for a very short period (5 seconds) in 1200-4800 ppm silver nitrate or soaking the stems in 1000 ppm silver nitrate for 10 minutes. Addition of 2 per cent sucrose to silver nitrate was found beneficial. It increased the vase-life from 12 days to 20 days. The type of preservatives used is very critical because 8-HQC even at level as low as 150-200 ppm can cause stem discolouration and 1-3 per cent sucrose may result in foliar chlorosis and/or necrosis (Staby *et al.*, 1978). Sodium dichloro-S-triazinetrione (SDT) in combination with sucrose was also found equally effective as silver nitrate and sugar. The flower stems should be placed into the preservative solution as soon as possible after harvesting or removing them from storage (Sacalis and Durkin, 1972).

There is significant association between postharvest life and ascorbic acid content in leaves and petals at fully opened flower stage in chrysanthemum (Hiremath *et al.*, 2018), spraying ascorbic acid to harvested cut flowers delayed early leaf yellowing and petal senescence in cv. Pusa centenary (Hiremath *et al.*, 2019).

Pretreatment with ABA or silver thiosulfate (STS) reduced transpiration and enhanced the longevity of flowers when shipped and stored under dark conditions (Rajapakse *et al.*, 1989). Hoogerwerf and Doorn (1992) found increased number of bacteria at the retail outlets often high enough to reduce flower longevity. Use of chlorine as an antibacterial chemical, resulted in reducing the bacterial population.

It was found that chrysanthemum flowers pulsed in 50 mg/l $AgNO_3$+ 200 mg/l HQS + 5 per cent sucrose for 20 h had the longest vase life with slow wilting of leaves and petals, greater increase in FW, without any decay of the stem base and slow wilting of leaves and petals (Ketsa, 1989). Song *et al.* (1995) and Lee *et al.* (1996) found that vase life was prolonged by precooling and shipping at low temperatures. Pretreatment with STS or chrysal RVB before shipping extended vase life of cut chrysanthemum flowers. A preservative solution containing sucrose (3 per cent) + 8-hydroxyquinolinesulfate (150 ppm) + $AgNO_3$ (50 ppm) also extended the vase life by about 10 days and improved the overall quality of flowers after shipping for 24, 28 and 72 h. Carbonate water (20 to 30 per cent) was also a good holding solution for promoting vase life. Song and Lee (1999) recorded extended vase life with pretreatment in 0.5 mM STS + 5 per cent sucrose for 5 h or hot water dipping of basal stem of cut flowers for 20s before shipping. Pretreatment with STS + 5 per cent sucrose + dipping in hot water and shipping at 3°C decreased CO_2 production and improved the overall quality of cut flowers. Vase life and fresh weight were found to decrease with increasing duration of shipping.

Holding chrysanthemum flowers in 100 ppm 8-HQC resulted in the longest vase life while holding in 200 ppm 8-HQS or 150 ppm ampicillin resulted in better water balance. The solutions could effectively control bacterial infection (Hussein, 1994). Vase life was found to be enhanced when cut flowers were kept in a preservative solution containing 200 ppm 8-HQC + 75 ppm citric acid + 5 per cent sucrose, kept under refrigeration (Arriaga and Guerrero, 1995). Bhat *et al.* (1999) found that cut flowers kept in a holding solution of 8-HQC (250 ppm) + sucrose (1.5 per cent) had the longest vase life, the greatest flower diameter and the lowest fresh weight loss in storage. The most effective pulsing solution was BA (0.025 mM) + STS (0.4 mM) + 8-HQC (250 ppm) + sucrose (5 per cent).

Gupta and Banerji (2008) recommended sucrose (2 per cent) + $AgN0_3$ (25 ppm) + citric acid (75 ppm) and sucrose (2 per cent) + 8HQC (500 ppm) as best holding solution containing for bud opening and increased vase life in cut Chrysanthemum, however, Suresh *et al.* (2010) found maximum vase life in cv. White Fizii and Peach Fizii with holding solution containing $AgNO_3$ (20 ppm) + sucrose (2 or 4 per cent). Extended vase life was observed with preservative solution containing 150 ppm citric acid + 500 ppm aluminium sulphate + 2 per cent sucrose (Jain, 2014), whereas, preservative solution containing 400 ppm 8HQ+400 ppm citric acid+ 3 per cent sucrose + one year old storage at low temperature resulted in improved vase life, maximum chlorophyll content with zero wilting and leaf yellowing and reduced leaf browning in cv. Thai Chen Queen (Jain *et al.*, 2016).

Mashhadian *et al.* (2012) found that applying acid (SA) and citric acid (CA) increased vase life, petal water content (per cent), initial fresh weight (per cent) and marketability, significantly. SA treatments increased leaf relative water content (RWC), petal water content (per cent) and initial fresh weight (per cent) by 49, 73 and 23 per cent, compared to the controls, respectively. The highest vase life (21.77 days) was observed for the treatments of SA (300 ppm). The significant increase (300 per cent) in vase life is considered to be due to plant regulating and anti-stress properties of SA and CA. Thereofre, SA and CA as natural, cheap, safe and biodegradable compounds are suitable alternatives for conventional chemical treatments in order to prolong vase life of cut flowers of chrysanthemum.

Rajagopalan and Khader (1993) obtained the longest vase life and the least flower damage by pulsing the cut flowers with 0.1 per cent aluminium sulfate treatment. Barman *et al.* (1996) and Banik *et al.* (1999) also recorded the longest vase life in chrysanthemum cultivar in 0.2 per cent aluminium sulfate + 3 per cent sucrose. Aluminium sulfate at 0.1 per cent + 3 per cent sucrose also gave good results.

Pretreatment (pulse solution) with BA resulted in longer vase life and better flower qualities than treatment with BA in the vase water (Guo *et al.*, 1997). Zhang and Guo (1998) also observed increased vase life and quality of cut flowers by BA treat-ment. BA delayed senescence by improving the membrane stability of the cut flower.

Chrysanthemum cut flower quality was substantially improved by changing the precooling method. A small forced air cooling facility was constructed, in which packed chrysanthemum flowers were cooled down to 5°C within 1 h (compared to > 10h by traditional room cooling). The treated chrysanthemum maintained good quality on the auction market ever after being fumigated (Huang, 1998).

Varietal sensitivity to ethylene showed leaf yellowing, resetted and elongating vegetative shoots. Pulsing with 0.2 mM anionic silver thiosulfate complex (STS) completely inhibited leaf yellowing in the ethylene-containing atmosphere (Doi *et al.*, 2003). Eexogenous ethylene promotes postharvest leaf yellowing in cut cv. Shuho-no-chikara (Doi *et al.*, 2004).

ABA and IAA could induce ethylene production, causing premature leaf wilting, yellowing, and shortened vase life. The effects of ABA are stronger than IAA. It showed that ABA and IAA regulated senescence cooperatively. ELISA also showed that iPA concentration decreased gradually, but the decrease was retarded by 6-benzyladenine (6-BA). Moreover, 6-BA promoted GA_{1+3} concentration, which increased consistently during vase life. 6-BA and GA_3 both prolonged vase life and delayed premature wilting. A total of five kinds of phytohormones affected the senescence process of cut chrysanthemum (Guo *et al.*, 2003). Amino-oxyacetic acid (5 and 10 mg/L) was most effective in prolonging vase-life, decreased floret and leaf senescence and improve the water uptake, membrane stability index, less reduction in total soluble solids and reducing sugar in flower stem of cv. Reagan White (Kumari *et al.*, 2016).

Sheu (1998) found that the quality of cut chrysanthemum flower was affected by the timing of rehydration and he recommended flowers for rehydration immediately (within 10 min) after cutting. Meeteren and Gelder (1999) suggested that an unidentified rehydration-inhibiting process starts after harvest, neither due to dehydration of xylem walls, or other stem tissues nor to the growth of microorganisms. They observed that fresh cut chrysanthemum flowers rehydrated fully after rapid water loss over 1 h at 2°C, 60 per cent RH and 14μ mol $m^{-2}s^{-1}$ light, the water loss being 5 per cent of their fully saturated fresh weight. For rehydration after desiccation, it proved important that flowers were cut at a minimum distance above soil level.

Tanigawa *et al.* (1995) proposed that the decline of (N + K): Ca ratio and the increase in starch and sugar contents in the leaves and stem as a result of CO2 enriched atmosphere effectively prolonged vase life of the cut flowers. Preservative solution containing sugar and germicides was found to prevent injuries caused by electron beam irradiation to cut flowers, thereby increasing the vase life of the cut flowers (Dohino and Hayashi, 1995; Hayashi and Todoriki, 1996). Meeteren *et al.* (2000) found detrimental effects on vase life of cut chrysanthemum flowers using deionised water instead of tap water. So they strongly questioned the appropriateness of deionised water as a control solution in vase life experiments.

Flower Drying

Ullas *et al.* (2018b) found that microwave oven drying is the best method in retaining highest total anthocyanins, total phenolic content and antioxidant activitiesin dried ray florets of cv. Red Gold.

15.0 Biotechnology

A large number of chysanthemum cultivars are micropropagated using various explants protoplast manipulation and transformation techniques are well advanced and molecular biology indicates promise by gene transfer for improving chrysanthemum. Embryo rescue and cryopreservation have potential for commercial exploitation (Rout and Das, 2000).

15.1.0 Tissue Culture

15.1.1 Embryo and Ovule Culture

The ovule culture technique provides an effective method in chrysanthemum. Watanabe (1977) compared shoot-bud regeneration from ovules with that from other culture systems. He also reported that ovary culture has been employed to overcome hybrid sterility and success was achieved in obtaining F1 hybrids and androgenic polyhaploids.

Anderson *et al.* (1990) reported a system for embryo rescue and seed set. The embryos exhibited higher percentages of seed set, germination and progeny reaching anthesis compared with normal development. The progeny developed through embryo rescue was significantly earlier in total generation time than the corresponding non-rescued seedlings. These two techniques have great potential for breeding chrysanthemum and other perennial crops.

Miler and Jedrzejczyk (2018) evaluated phenotypic as well as genetic variation of chrysanthemum plants regenerated from ovaries *in vitro*. The variation included variegated, marble-like, and lighter-green leaves, and changes in the morphology of inflorescences and ligulate florets, as well as changes in the shape of corymb. Variants with variegated and marble-like leaves were unstable. They concluded that variable regenerants did not originate from the gynogenic pathway, but they regenerated from somatic tissue and underwent somaclonal variation.

15.1.2 Protoplast Culture

Chrysanthemum protoplasts were isolated from a white spider cultivar and other two breeding lines by Schum and Preil (1981) and they succeeded in developing green calli from these protoplasts. Sauvadet *et al.* (1990) produced callus colonies from protoplasts. They observed colony formation in 16 and division in 18 out of 29 of the clones studied. Callus was obtained from colonies of five clones but buds were produced only in one case. The regeneration potential was retained over a prolonged culture period and regenerated plants were successfully grown in the greenhouse.

15.1.3 Somatic Embryogenesis

Somatic embryogenesis offers an alternative and efficient protocol for plant regeneration. The technique of somatic embryogenesis has also contributed information for the genetic, morphological and physiological manipulation.

Callus cultures of chrysanthemum have been established in many laboratories. Appropriate auxin and cytokinin levels are required for callus production from each species or variety. Kaui *et al.* (1990) regenerated adventitious shoots from leaf and stem explants of 11 chrysanthemum cultivars. MS basal medium supplemented with 5 μM NAA was optimum for both types of explants. They also reported that stem explants were superior to leaf explants. The morphogenic development was noted 4-6 days after the initiation of cultures depending upon the explant type. Bhattacharya *et al.* (1990) reported that a combination of 0.1 mg/IAA and 0.2 mg/1 BA was most appropriate for callus formation from nodal segments, shoot apices and leaf and for the regeneration of shoots from callus. An average shoot height of 3-4 cm (after 4 weeks of culture) with 2-4 leaves per cm height of the stem and production of at least 3-4 shoots from each explant was achieved.

Direct somatic embryogenesis and plant regeneration from leaf explants of several chrysanthemum cultivars were reported by Yantcheva *et al.* (1998). MS medium with 1 mg 2,4-D and 2 mg BA/1 was used for direct induction of embryoids without additional callus phase. Somatic embryo maturation was promoted by casein hydrolysate (100 mg/1). Somatic embryogenesis in chrysanthemum cv.

Iridon, Fortune, Goldmine, Helka and Zest has been obtained by using leaf midrib explants cultured on modified MS basal medium supplemented with 1.0 mg/1 2,4-D and 0.2 mg/1BA (May and Trigiano, 1991). They also noted that the induction of somatic embryogenesis depended on the photoperiod and sucrose concentration.

Zalewska *et al.* (2007) studied the explant type used for micropropagation affects the genotype and phenotype of chrysanthemum. Plants grown from shoot tips and adventitious buds formed on leaves were rooted in vitro, acclimatized and cultivated in glasshouse up to full-flowering. The variation appears only if non-meristematical explants were used. The adventitious buds technique might be useful in chrysanthemum breeding as a source of a new variability.

Mani and Senthil (2011) found that callus induction from leaf explant in MS medium containing 1.5 mg/L 2,4-D was 100 per cent and from petal explant in MS medium containing 2.0 mg/L 2,4-D was found to be 100 per cent. The best friable calli were subjected to suspension culture in MS media supplemented with 1.0 mg/L BAP for somatic embryos. All calli in suspension gave rise to somatic embryos, which were regenerated in MS media supplemented with various concentration of BAP.

Hemlata and Mahdi (2016) used petals from cv. White as explant to study indirect organogenesis (callus induction and its growth). Highest callus induction was recorded in MS supplemented with NAA (0.5 mg/L) and kinetin (10.0 mg/L). In shoot organogenesis, highest multiple shoot number (4.79) on callus was recorded in MS fortified with Kinetin (10.0 mg/l) and NAA (0.5 mg/l).

15.1.4 Somaclonal Variation

Sutter and Langhans (1981) reported that the plants derived from long-term cultures showed delayed flowering and the addition of GA_3 or IAA to the culture medium had no effect in restoring normal flowering. Somaclonal variation was more common among *in vitro*-cultured adventitious shoots of many species including chrysanthemum (Skiryin, 1978). Malaure *et al.* (1991) reported that shoots derived from ray florets of 16 cultivars of chrysanthemum showed more variation than plants regenerated from vegetative parts. Morphological changes were also observed in plants derived from petal segments (Ohishi and Sakurai, 1988). Subsequently, Ahloowalia (1992) developed 20 new variants which differed in height, leaf, flower shape and petal size and curvature. However, *in vitro* selection and somaclonal variation are random processes and have yet to be used to achieve specific goals in chrysanthemum improvement.

15.2.0 Molecular Markers

Wolff and Peters-van (1993) studied the genetic variation in chrysanthemum using a technique to generate random amplified polymorphic DNAs (RAPDs). Variations between cultivars were identified by using only two different primers. A family of cultivars derived from one original cultivar by vegetative propagation, had identical fragment patterns. The presence of high level of polymorphism and clonal stability, RAPD fragments were useful for cultivar identification. In order to study genetic variability at the DNA level in chrysanthemum, Pst I and Hind III genomic libraries were constructed. Probes from both libraries were tested for the presence of RFLPs. Of the probes from Pst I library, 91 per cent appeared to hybridize to low-copy genes, while only 31 per cent of these from the Hind III library appeared to do so. Genetic analysis was simplified by using locus specific polymerase chain primers to obtain simple polymorphic patterns in a number of cases. The RFLP probes and primers developed were used in marker assisted selection (Wolff *et al.*, 1994). Scott *et al.* (1996) recommended the use of DNA amplification fingerprinting (DAF) to study genetic relationships between closely related chrysanthemum cultivars. The evolution of genetic variability of cultivars within and between series was done using arbitrary octamer primers. Phenetic patterns were established by unweighted pair group cluster analysis using arithmetic means (UPGMA) and principal coordinate analysis (PCO) and the average distance between series was compared between cultivars within a series. DNA from all cultivars belonging to a series was bulked together to generate profiles containing unique amplified products for each series. Thus, DAF technique can be utilized for distinguishing clonal materials along with patent production, phylogenetic study and for identifying useful markers in breeding applications. Trigiano *et al.* (1998) reported that the cultivars especially mutants and sports are difficult to differentiate genetically even by DAF. But it was easily identified by using arbitrary signatures from amplification profiles (ASAP). This involved amplification of genomic DNA with 3-stranded octamer arbitrary primers, all of which produced monomorphic profiles. Products from each of these DNA fingerprints were reamplified using DNA fingerprints and subsequently reamplified with the help of 4 mini hairpin decamer primers. The number of ASAP polymorphisms detected helped to estimate the mutation rate in the mutant characters. It was concluded that the ASAP technique allowed clear genetic identification of somatic mutants and radiation induced sports that were genetically highly homozygous. So, they facilitated marker assisted breeding and protection of plant breeders rights.

Identification and characterization of new cultivars is essential to meet DUS testing, address IPR issues and their utilization and conservation.Traditionally, morphological markers were used for germplasm characterization as being simple and irreplaceable. Martín *et al.* (2002) and Kumar *et al.* (2014) characterised commercial varieties through RAPD analysis showed differences between varieties. The cultivars sharing higher similarity levels were Flirt with Yellow Gold, Pink Cloud with Korean Small and Sadbhawna with Jubilee. RAPD proved to be useful for the characterization of the genotypes for their efficient utilization, management and IPR protection. Kumar *et al.* (2015b) characterised varieties using ISSR markers. Cultivars Maghi Orange and Maghi Yellow had around 90 per cent similarity and were close to cultivar Maghi White at over 80 per cent similarity level. ISSR markers were proved to be useful for the characterization of the genotypes for their efficient utilization, management and IPR protection.Comparative transcriptome analysis of wild and cultivated chrysanthemum provides valuable genomic resources and helps uncover common and divergent patterns of genome and gene evolution in *C. morifolium* and *C. boreale*. An analysis of synonymous substitution rates (Ks) of paralogous and orthologous genes suggested that the two Chrysanthemum species commonly experienced the Asteraceae paleopolyploidization and recent genome duplication or triplication before the divergence of these species. Intriguingly, *C.*

boreale probably underwent rapid diploidization, with a reduction in chromosome number, whereas *C. morifolium* maintained the original chromosome number (Won *et al.*, 2017). Asha *et al.* (2017) characterized 30 chrysanthemum genotypes using ISSR and SSR markers to distinguish and to assess the genetic relatedness among genotypes. Molecular analysis of 50 loci using 17 SSR primer pairs resulted in 80.4 per cent mean polymorphism. Using dendrogram, three distinct groups were found based on SSR and ISSR indicating significant differences among the genotypes. There was a large genetic diversity exists among the genotype which helps in the selection of suitable genotypes for breeding. Both ISSR and SSR were found to be effective for characterization studies in chrysanthemum.

Hui *et al.* (2014) revealed that the molecular marker clustering analysis of wild and cultivated chrysanthemum and relatives were not fully congruent with the morphological classification. RAPD marker showed that *Chrysanthemum vestitum* was the most important species in the origin process of cultivated chrysanthemum. For classification and breeding of chrysanthemum, ISSR was best, and it was combined with the high polymorphism of AFLP, SSR and the commonality of RAPD. With the combination of molecular marker technology and traditional morphological approaches, breeding success will be improved. It will clarify the cross-compatibility among chrysanthemum cultivars and expand the range of breeding resources for genetic improvement. The breeding period will be shortened with the directional transfer of favorable genes and the accumulation of target genes. This also provides an accurate reference library for the protection of chrysanthemum germplasm resources and the innovation of taxonomy.

In order to the examining the expression of miRNAs associated with chrysanthemum embryo development, Zhang *et al.* (2015) sequenced three small RNA libraries to identify miRNAs and their functions. Study generated three miRNA libraries and gained information on miRNAs and their targets in the chrysanthemum embryo. These results enrich the growing database of new miRNAs and lay the foundation for the further understanding of miRNA biological function in the regulation of chrysanthemum embryo abortion. Aida *et al.* (2016) obtained chrysanthemum periclinal chimeras through regeneration from leaf explants using the fluorescent protein transgene as a selection marker. Yang *et al.* (2019) reported that four marker panels could be used for many fields of marker application, including germplasm evaluation, genetic mapping, marker-assisted selection (including genomic selection), and plant variety protection.

15.3.0 Genetic Transformation

Genetic transformation in *Dendranthema grandiflora* was done by cocultivaing the leaf disc with *Agrobacterium tumefaciens* strain LB A4404 carrying a binary vector containing a *GUS*-intron reporter gene (the presence of the intron ensured GUS expression in the plant and not the bacterium). After 6 days, GUS activity in leaf discs was determined by the presence of blue spots. Genetic transformation was undertaken to produce transgenic plants (Jong *et al.*,1990), Lemieux *et al.* (1990) developed a protocol for the genetic transformation of cv. Indianapolis White Giant. Transformation was achieved using *A. tumefaciens*-mediated gene transfer to peduncle explants. Transformation was confirmed by regeneration of calluses on medium containing kanamycin, GUS assays and polymerase chain reaction analysis (PCR). The same protocol when worked upon cv. Moneymaker with a chrysanthemum chalcone synthase (naringenin-chalconesynthase) gene, gave transformants with 2 flower colour changes, violet and white, not seen in the previous transformation. *A. tumefaciens* strain A281 was used to transform *D. grandiflora* leaf explants with binary vector pCP01-GT carrying the *B. thuirigiensis* endotoxin gene *Bt II* under the control of the TR-2′ promoter which resulted in resistance to larvae of *Heliothis virescense* (Wordragen and Honee, 1990). In another experiment, shoots from leaf pieces inoculated with pGA643 (containing selectable marker gene *npt II*) through *A. tumefaciens* LBA4404, were selected on kanamycin, and Southern analyses of shoots. Their ability to form root on kanamycin confirmed the presence of the *npt II* gene in the plant genome (Ledger *et al.*, 1991). Transgenic plants were recovered on MS medium containing hygromycin after *Agrobacterium* mediated transformation using the disarmed hypervirulent strain EHA101, containing a binary vector carrying the *nptII, gus* and hygromycin phos-photransferase genes. Four transformation protocols were evaluated and the optimal one resulting in 40 per cent regenaration involved inoculation of nodes (dropping 20 min. in the bacterial suspension) followed by their immediate transfer to a regeneration medium containing cefotaxime (500 μg/ml). Neoformation of buds occurred on selective medium within two months on almost 40 per cent of the inoculated explants. DNA-analysis of plants showing GUS expression confirmed the T-DNA integration in the plant genome (Renou *et al.*, 1993). Somatic embryogenesis from leaf segments of *D. grandiflora* cvs. Yellow Spider, White Snowdon, Orange Westland and Mistletoe was used for regeneration of plants following *Agrobacterium*-mediated transformation. A disarmed *A. tumefaciens* strain carrying the *gus* reporter gene and *npt II* gene as a selectable marker, and the wild type B6S3 carrying the original octopine Ti plasmid were used. Non-chimaeric transformed plants were obtained. Regenerated plants showed stable octopine synthase or GUS expression during vegetative propagation (Pavingerova *et al.*, 1994). Urban *et al.* (1992) attempted to transform three cultivars of *D. grandi-flora*. They reported success with 'Iridon1' using *Agrobacterium* EHA105 (pB121) on leaf explants. de Jong *et al.* (1993) reported enhanced regeneration ability of the leaf explants after cocultivation with *A. tumefaciens*. Subsequently, Urban *et al.* (1994) reported an efficient, high-frequency transformation protocol for cv. Iridon and regeneration protocols for cultivars Hekla and Polaris.

An efficient, high-frequency regeneration and *Agrobacterium*-mediated transformation system allowed todevelop the genetic engineering of 3 chrysanthemum cultivars. Transformed plants of the 3 cultivars were obtained following co-culture of leaf explants with *A. tumefaciens* strain Eha 105 harbouring the plasmid pBI121 contaning *npt II* and *gus* genes. Stable transformation of the 3 cultivars was verified via gus assays and Southern analysis (Sherman *et al.*, 1998). A 3′-end truncated crystal protein gene, derived from *B. thuringiensis* (*Bt*) sub sp. *aizawai*.7.21, encoding the toxic fragment of the insecticidal protein cryIA (b), was constructed. The *Bt* gene was inserted into a transformation vector, which was also carrying the *npt II* and *gus* genes and introduced in the oncogenic *A. tumefaciens* strain A281, harbouring the Ti-plasmid pTiBO542. The recombinant agrobacterium strain was used to transform leaf explants of chrysanthemum. The resulting tumours were kanamycin-resistant, exhibited GUS activity and produced agropine and mannopine. These tumour lines were tested against *Heliothis virescens* (tobacco bud worm). Several of these lines proved to be strongly inhibitory to larval growth. The results indicated that Bt-based insect resistance might be used as a tool in reducing the amount of pesticides used in chrysanthemum culture (Wordragen *et al.*, 1993).

Agrobacterium tumefaciens and biolistic transformation procedures were developed for Polaris and Golden Polaris, two important commercial chrysanthemum cultivars (Yepes *et al.*, 1999). The disarmed *A. tumefaciens* strains LBA4404, C58sZ707 and EHA105 containing the binary plasmids pBIN19 or pGA482GG were used for transformation. Transgenic plants were recovered using leaf and stem explants and a stepwise kanamycin selection procedure to optimize recovery of transformed plants. PCR and Southern blot analyses confirmed the integration of the *N* genes into the chrysanthemum genome.

Height control is a major consideration during commercial production of chrysan-themum. Plants of 'Iridon' chrysanthemum were genetically engineered to ectopically express a tobacco (*Nicotiana tabacum* L.) phytochrome *B1* gene under the control of the CaMV 35S promoter. The transgenic plants were shorter in stature and had larger branch angles than wild type (WT) plants. Reduction in growth caused by the ectopic expression of the tobacco phytochrome *B1* gene was similar to that caused using a commercial growth retardant at the recommended rate (Zheng *et al.*, 2001).

Shinoyama *et al.* (2012) had given an overview of the advances of an *Agrobacterium*-mediated transformation system. Valizadeh *et al.* (2012) incubated internode explants of 1581 and 4043 genotypes with *A. tumefaciens* strain AGL-0 containing pBIN plasmid with the npt gene as a selectable marker for kanamycin resistance and gcs gene as an aphicidal gene with rbcS promoter. Incorporation and expression of the transgenes were confirmed by PCR and RT-PCR analysis. Genotype 4043 has been transformed for the first time. Jo *et al.* (2015) developed GM chrysanthemum by introducing sense and antisense RNA sequences of CSVd into a commercial chrysanthemum cultivar. As a result, nine lines out of 16 tested GM chrysanthemum lines showed very strong resistance against CSVd infection. This is the first study which develops GM chrysanthemums resistant to CSVd using sense and antisense RNAs of CSVd. Mitiouchkna *et al.* (2018b) developed effective transformation system of chrysanthemum through somatic embryogenesis. The cultivars were transformed using *A. tumefaciens* strain AGL0: pBin-mGFP5-ER carrying neomycin phosphotransferase gene (nptII) and gfp gene with a leader sequence for expression in endoplasmatic reticulum. This developed transformation system provides a base system for genetic improvement of chrysanthemum. Mitiouchkina *et al.* (2018a) compared two different strategies for developing resistance in transgenic chrysanthemum to Chrysanthemum virus B (CVB). Transgenic chrysanthemum plants generated with constitutive expression of the full CVB coat protein gene in sense, double-sense and antisense orientation and also tested another promising strategy to create virus resistant plants - gene silencing via RNA interference (RNAi) technology. Three RNAi lines showed dramatic decrease in the levels of virus infection relative to the levels found in non-transgenic infected plants. Sen *et al.* (2013) developed a simple and efficient protocol for the *Agrobacterium* mediated transformation of rice chitinase gene in chrysanthemum cv. Snow Ball for improved resistance against leaf spot (*Septoria obesa*). Pak *et al.* (2020) suggested that PTA (Pinellia ternata Agglutinin) gene can improve aphid resistance of chrysanthemum and its aphid resistance can be generated to T1 generation stably. Noda (2019) developed technology based on genetic engineering to confer the pure blue flower color trait to chrysanthemum. The blue coloration of the chrysanthemum was accomplished by constructing functional genes producing flavonoid 3′,5′-hydroxylase and anthocyanin 3′,5′-*O*-glucosyltransferase. Accumulation of 3′,5′-*O*-glucosylated delphinidin-based anthocyanins synthesized by the transgenes led to the development of blue color via intermolecular copigmentation with flavone glycosides present in chrysanthemum petal.

At present, several agronomic and physiological traits have been introduced to chrysanthemums using genetic transformation. To protect chrysanthemums from the damage, Shinoyama *et al.* (2003) introduced the *mcbt* gene, which was a modified *cry1Ab* sequence (encoding *Bacillus thuringiensis* insecticidal protein *Cry1Ab*) that maintained the amino acid sequences. Shinoyama and Mochizuki (2006) obtained 1,586 transgenic chrysanthemum lines in 5 chrysanthemum cultivars and successfully induced 10 lines of 20 randomly sampled transgenic lines to confer a high lepidopteran insect resistance. Ohmiya *et al.* (2006) demonstrated white petals could be converted into yellow petals by suppressing the expression of carotenoid cleavage dioxygenase (CmCCD4a) using RNAi technology. Although carotenoids are initially synthesized in both yellow and white flowers, carotenoids are degraded by the carotenoid cleavage dioxygenase (CCD) protein in white flowers. Ohmiya *et al.* (2009) obtained 61 transformed plants carrying a CmCCD4a RNAi construct, but the petals of the transformed plants were very pale yellow, then chose the most yellow transformed plant to perform the second transformation with another CmCCD4aRNAi construct bearing different DNA sequences. They finally obtained 50 double-transformed plants, and more than half showed yellow-colored petals.

To eliminate the risks of transgene flow, Shinoyama *et al.* (2012a) tried to create male-sterile transgenic chrysanthemums using the modified melon ethylene receptor gene, CmETR1/H69A, over expression of the CmETR1/H69A gene delayed the tapetum degradation of the anther sac, resulting in a reduction of mature pollen grains in some of the transgenic lines. In 15 of the 335 CmETR1/H69A-overexpression chrysanthemum lines, male and female fertility was signicantly lower than in non-GM chrysanthemum. Li *et al.* (2019) provided more accurate and reliable qRT-PCR normalization for future studies on the expression patterns of genes regulating plant architecture of chrysanthemums.

To produce transgenic chrysanthemums with both insect resistance and temperature-independent, complete male-sterility, Shinoyama *et al.* (2012b) transformed the mcbt gene and a 582-bp fragment of chrysanthemum meiosis-specific recombinase gene, CmDMC1, as an RNAi trigger segment. We used the novel bi-directional maspromoter to regulate the expression of the mcbt gene and the CmDMC1-RNAi segment located at both ends of the promoters. Transgenic lines showed high resistance for lepidopteran pest insects of chrysanthemums, such as the tobacco budworm (*H. armigera*), the cotton cutworm (*Spodoptera litura*) and the beet armyworm (*S. exigua*). Zhang *et al.* (2013) suggested that flowering time in chrysanthemum is principally governed by main effect QTL but that epistasis also contributes to the genetic architecture of the trait. Ten pairs of loci epistatically determined the flowering time, but their contribution to the overall phenotypic variance was less than for the additive QTL, and the major QTL identified are useful to streamline the improvement of chrysanthemum. Su *et al.* (2017) provided insights into the complex genetic architecture of water logging tolerance (WAT), and the identification of favourable alleles represents an important step towards the application of molecular marker-assisted selection (MAS) and QTL pyramiding in chrysanthemum waterlogging tolerance (WAT) breeding programmes. To understand the roles of miRNAs and their target genes in chrysanthemum polyploidy breeding, three sRNA libraries of normal and abnormal embryos after hybridization were performed by RNA-Seq. miR172c was only detected in the normal embryo at 18 days after pollination, which means that miR172c mainly mediates gene expression in postembryonic development and these genes may promote embryo maturation (Zhang *et al.*, 2017).

15.4.0 Chimera

Courtney-Gutterson *et al.* (1994) reported that sense suppression was used routinely to develop novel varieties of chrysanthemum. A white flowering variety was produced by introducing chimeric (naringenin) chalcone synthase (CHS) constructs into the chrysanthemum cv. Money Maker along with a T- DNA vector lacking a CHS construct. The transgenic white flower accumulated higher levels of CHS mRNA. The effect of sense constructs appeared to be a function of co-suppression of endogenous gene expression.

15.5.0 *In vitro* Conservation

An effective method of *in vitro* storage of chrysanthemum would allow the breeders and nurseries to maintain a greater number of cultivars. Cryopreservation using preculture for 2 days, slow cooling (0.2°C/min) to -40°C with 10 per cent dimethylsulfoxide and 3 per cent glucose prior to immersion into liquid N_2 and rapid thawing resulted in higher survival and shoot regeneration rates. Shoot tips of chrysanthemum showed high viability even after a storage of 8 months in liquid N_2 (Fukai and Oe, 1991 and Fukai *et al.*, 1991). Fukai *et al.* (1994) observed that cryopreservation of shoot tips of a perclinally chimeric cultivar showed pink colour flower instead of original apricot. It was found that many of the shoots regenerated from the thawed shoot tips were adventitious in origin and were derived from the epidermis of the shoot tip. Ahn (1995) observed that preservation of excised axillary buds in a solution of dimethylsulfoxide (10 per cent) and glucose (0.5M) at 1°C for 1h was the best cryoprotectant.

Slow prefreezing (cooling velocity of 0.3°C/min) also promoted survival of buds. Maximum survival of axillary bud was recorded at -196°C. Kulus (2015) advocated vitrification-based techniques (focused on rapid freezing of the biological material) in chrysanthemum cryopreservation. Weilan *et al.* (2015) observed

Exhibition of Chrysanthemum

Exhibition in open field

Indoor exhibition

Display of Chrysanthemum

Display by the International Chrysanthemum Society

Flowering chrysanthemum for display

that 4°C low temperature is advantageous to wild *Chrysanthemum shiwogiku* var. Kinokuniensein *in vitro* plantlets preservation. After preservation, the recovered plantlets grow well and showed no differences in morphology and isoenzyme zymogram of peroxidase, ISSR-PCR compared with the control. In addition, low temperature *in vitro* plantlets maintain genetic stability.

16.0 References

Aabrandt, Z. (1987) *Gaertner Tidende*, **103**: 1410-1411.

Abrol, A., Dhiman, S.R., Sharma, P. and Sharma, M. (2018) *J. Hill Agric.*, **9**(2): 165-170.

Acock, B. and Pachepsky, Ya. A. (1996) *Plant and Soil*, **18**: 231-240.

Adames, A.H., Latado, R.R., Camargo, N.M. and Tulmann Neto, A. (1999) *Scientia Agricola*, **56**: 939-945.

Adams, S.R., Pearson, S. and Hadley, P. (1998) *Sci. Hort.*, **77**: 59-72.

Ahloowalia, B.S. (1992) *Mutation Breeding Newsletter*, **39**: 6.

Ahmed, K.Z. and Sagi, F. (1993) *Acta Phytopathol. Entomol. Hung.*, **28**: 143-159.

Ahn YoungHee (1995) *J. Korean Soc. Hort. Sci.*, **36**: 540-547.

Aida, R., Sasak, i K. and Ohtsubo, N. (2016) *Plant Biotechnol.*, **33**(1): 45-49.

Akihisa, T., Yasukawa, K., Oinuma, H., Kasahara, Y., Yamanouchi, S., Takido, M., Kumaki, K. and Tamura, T. (1996) *Phytochemistry*, **43**: 1255-1260.

Al-Arifi, A., Short, T. and Ling, P. (1999) *ASAE/CSAE-SCGR Annual Int. Meeting*, Toronto, Ontario, Canada, 18-21 July, 1999, St. Joseph, USA; Amer. Soci. of Agricultural Engineers (ASAE), p. 17.

Alexe, C., Vintila, M. and Caplan, I. (2017) *J. Hortic., For. and Biotech.*, **21**(1): 1-5.

Alhajhoj, M. R. (2017) *Pak. J. Life. Soc. Sci.*, **15**(2): 114-119.

Al-Shooshan, A., Short, T.H., McMohan, R. and Fynn, R.P. (1991) *Pro. Amer. Soc. Agricultural Engineers*, **91-4043**: 21.

Alt, A., Krasting, W. and Krupp, J. (1989) *Gartnerborse und Gartenwelt*, **89**: 958-961.

Alt, D. and Staat, G. (1997) *Gartenbauwissenschaft*, **62**: 267-273.

Altmann, A. and Streitz, D. (1995) *Gartenbaumagazin*, **3**: 12, 14.

Alverson, D. R. and Gorouch, C. S. (1982) *J. Econ. Ent.*, **75**: 888-91.

Anand, P., Vanlalruati, Kumar, G. and Kumar, S. (2020) *Indian J. Agric. Sci.*, **90**(3): 565–568.

Anderson, N. O., Leidi, B. E., Ascher, P. D., Widmer, R. E and Desborugh, S.L. (1988) *Self Plant Production*, **1**: 173-181.

Anderson, N., Ascher, P. D., Widmer, R. E and Ludy, J. L. (1990) J. L. (1990) *J. Amer, Soc. Hort. Sci.*, **115**: 329-336.

Andersson, N.E. (1990) *Sci. Hort.*, **44**: 163-169.

Andrews, P. S. and Watson D. P. (1952) *Proc. Amer. Soc. Hort. Sci.*, **59**: 516.

Angarita, A. (1999) *Acta Hortic.*, **482**: 197-201.

Anguili, E. (1999) *Colture Protette*, **28**: 59-63.

Angus, R and Morrison, R. (1998) *Landwards*, **53**: 19-22.

Anonymous (1970) *Rep. Efford Expt. Hort. Sta.* for 1969

Anonymous (1971) *Rep. Efford Expt. Hort. Sta.* for 1970

Anonymous (1997) *Lien Horticole*, **34**: 6-8.

Anonymous (2020a) http: //www.nbpgr.ernet.in: 8080/registration/ InventoryofGermplasm.aspx

Anonymous (2020b) http: //www.plantauthority.gov.in/pdf/Chrysanthemum.pdf

Anonymous (2020c) https: //www.statista.com/statistics/582064/value-of-the-import-and-export-of-chrysanthemums-in-the-netherlands/#statisticContainer

Anonymous (2020d) https: //www.cbi.eu/market-information/cut-flowers-foliage/chrysanthemums/europe#product-description

Anonymous (2020e) http: //ecoursesonline.iasri.res.in/mod/page/view.php?id=79647

Anonymous. (2016-17) Ministry of Agriculture and Farmers Welfare, Govt. of India. (ON1549) and Past Issues.

Apu, M.R.B., Ambia, K., Naznin A., Khan, F.N., Saha, K.C. and Halder, S.C. (2019) *Bull. Inst. Trop. Agr., Kyushu Univ.*, doi: 10.11189/bita.42.7.

Aragon-Caballero, L. (1997) *Fitopatologia*, **32**: 173-176.

Ari W., Bambang, S. and Rina, S. (2018) *J. Adv. Agric. Technol.*, **5**(3). doi: 10.18178/joaat.5.3.257-260.

Aribaud, M. and Martin-Tanguy, J. (1994) *Plant Growth Regulation*, **15**: 23-31.

Armbruster, J. (1967) *Dtsche Gartenb.*, **67**: 385-386.

Arora, J.S., Singh, P.J. and Sidhu, G.S. (2002) *J. Res.*, **39**(4): 640-640.

Arriaga, N. R. M. and Guerrero, J. E. (1995) *Revista Chapingo Serie Horticultura*, **1**: 103-107.

Asafani, M. (1972) *Berichte des Ohara Instituts fur Landwirtschaftliche Biologie*, Okayama Universitat, **1**: 169-79.

Asha, K.M., Sane A. and Kumar R. (2016) *Indian J. Agric. Sci.*, **86**(1): 103-112.

Asha, K.M., Sane, A., Kumar, R. and Biradar, M.S. (2017) *Indian J. Agric. Sci.*, **87**(1): 115-121.

Asrar, A.W., Elhindi, K. and Abdel, S. E. (2014) *J. Food Agric. Environ.*, **12**(2): 963-971.

Atikmen, N. C., Kutuk, C. and Karahan, G. (2014) *Bull. Univ. Agric. Sci. Vet. Med. Cluj-Napoca, Hortic.*, **71**(2): 233-239.

Bailey, B. J. (1989) *CNRE Bull.-European Cooprative Networks on Rural Energy*, Italy **21**: 41-45.

Bakker, E. J. and Sanden, P. A. C. M. vander (1998) *Acta Hortic.*, **456**: 509-514.

Bala, M. (2019) *Indian Hortic.*, **64**(2).

Bala, M. and Singh, K. (2013) *J. Ornam. Hort.*, **16**(1-2): 35-39.

Banerji, B. K. and Datta, S. K. (1991) *J. Nuclear Agric. and Bio.*, **19**: 252-256.

Banerji, B.K. and Datta, S.K. (2002) *Indian J. Agric. Sci.*, **72**(1): 6-10.

Banik, A.K., Dutta, P. and Barman, D. (1999) *Environment and Ecology*, **17**: 248-249.

Barakat, M.N., Fattah, R.S. A., Badr, M. and El-Torky M.G. (2010) *Afr. J. Agric. Res.*, **5**(8): 748-757.

Baranowski, R. M. (1966) *Proc. Fla. Sta. Hort. Soc.*, **79**: 478-481.

Baranowski, T. and Dankowska, E. (1985) *Roczniki Nauk Rolniczych*, **15**: 129- 137.

Barbat, I. and Oschesanu, C. (1964) *Naturwissenchaften*, **51**: 316-317.

Barbosa, J.G., Kampe, A.N., Martinez, H.E.P., Koller, O.C. and Bohnen, H. (2000) *J. Plant Nutrition*, **23**: 1327-1336.

Barbosa, J.G., Muniz, M.A., Martinez, H.E.P., Leite, R.A., Cardoso, A.A. and Barbosa, M.S. (2005) *Acta Sci-Agron.*, **27**(3): 387-394.

Barhate, B.G., Musmade, N.A. and Bahirat, J.B. (2015) *Intl. J. Pl. Prot.*, **8**(1): 134-137.

Barigdad, H. and Patil, A. A. (1997) *Karnataka J. Agric. Sci.*, **10**: 98-101.

Barman, D. and Pal, P. (1996) *J. Hill Research*, **9**: 40-42.

Barman, D. and Pal, P. (1999) *Hort. J.*, **12**: 51-59.

Barman, D., Banik, A.K. and Dutta, P. (1996) *Orissa J. Hort.*, **20**: 39-41.

Barman, D., Ghosh, S. and Pal, P. (1993) *Hort. J.*, **6**: 121-124.

Barman, D., Pal, P. and Upadhayaya, R.C. (1998) *Int. J. Trop. Agri.*, **15**: 65-73.

Barrett, J. E. and Bartuska, C. A. (1982) *HortScience*, **17**: 737-738.

Barrett, J. E., Bartuska, C. A. and Nell, T. A. (1994) *HortScience*, **29**: 893-895.

Bartlet, P.W. (1993) In: *Liriomyza*. Conference on Leaf-Mining Flies in cultivated plants. Paris, France, Centre de Cooperation Internationale en Recherche Agronomique Pour le Development (CIRAD), pp. 23-30.

Baskaran, V., Jayanthi, R., Janakiram, T. and Abirami, K. (2009) *J. Hortic. Sci.*, **4**(2): 174-176.

Bassi, P.K., Glazier, D.E. and Abernathy, S. M. (1987) *Proc. Plant Growth Regulator Soc. America*, pp.330-339.

Batra, A. and Dwivedi, A.K. (2015) *Agrica.*, **4**(2): 93-101.

Batra, A., Verma, A.K. and Banerji, B.K. (2011) *Indian J. Agric. Sci.*, **81**(11): 1078-1081.

Bayat, H., Verhoeven, J.T.J., Botermans, M., Peters, D. and Hassani-mehraban, A. (2013) *Plant Dis.*, **97**(8): 1110. doi: 10.1094/PDIS-01-13-0050-PDN.

Beckmann, C., Márkilla Zunete., Pivetta, Kathia, F. L., Cavalcante., Ítalo, H.L., Cavalcante, L.F., Bellingieri., Paulo, A., Campos. and Milton, C.C. (2010) *Rev. Bras. Ciênc. Solo.*, **34**(3): 747-756.

Beckmann, C., Márkilla, Z., Pivetta, Kathia, F.L., Cavalcante., Ítalo, H.L., Cavalcante, L.F., Silva, J. and José, V.D. (2013) *Rev Cienc Agron.*, **44**(4): 782-789.

Beckmann-Cavalcante M.Z., Lopes-Pivetta K.F., Meinken, E. and Roeber, R. (2009) *Acta Hortic.*, doi: 10.17660/Acta Hortic.2009.819.17.

Belgaonkar, D. V., Bist, M. A. and Wakde, M. B. (1996) *J. Soils and Crops*, **6**: 154-158.

Belgaonkar, D. V., Bist, M. A and Wakde, M. B. (1997) *J. Soils and Crops*, **7**: 90-92.

Benedetto, A. Di and Porto, P. (1996) *Revista de la Facultad de Agronomia* (Universidal buenos Aires), **15**: 131-135.

Benetka, V. and Kodytek, K. (1988) *Acta Pruhoniciana*, **55**: 3-13.

Ben-Jaacov, J. and Langhans, R. W. (1972) *HortScience*, **7**: 289-290.

Berninger, E. and Philouze, J. (1988) *Acta Hortic.*, **229**: 31-37.

Bertram, L. and Karlsen, P. (1994) *Sci. Hort.*, **58**: 139-150.

Bethke, J. A., Redak, R. A. and Schuch, U. K. (1998) *Entomologia Experimentalis et Applicata*, **88**: 41-47.

Bhat, A., Tripathi, S. N. and Sehegal, O. P. (1999) *Adv. Hort. and Fores.*, **6**: 125-131.

Bhat, N. R and Tayama, H. K. (1989) *Foliage Digest*, **12**: 1-2.

Bhat, N. R., Tayama, H. K., Carver, S. A., Prince, T. L. and Prince, T. A. (1990) *Ohio Florists' Association Bull.*, **723**: 3-6.

Bhattacharya, P., Dey, S., Das, N. and Bhattacharya, B. C. (1990) *Plant Cell Rep.*, **9**: 439-442

Bik, R. A. (1964) *Vakblad voor de Bloemisterij*, **19:** 1201.

Bing, A. (1983) *Proc. 37 Ann. Meet. Northeastern Weed Sci. Soc.*, pp. 362-366.

Biswas, P. K., Paul, K. B. and Henderson, J. H. M. (1967) *Nature*, **213**: 917-918.

Blok, C. and Verhagen, J.B.G.M. (2009) *Acta Hortic.*, **819**: 47-58.

Blok, C. and Vermeulen, T. (2012) *Acta Hortic.*, **927**: 865-878.

Blondon, F. (1982) *Acta Hortic.*, **125**: 119-126.

Bloom, J. R. (1964) *Phytopathol.*, **54**: 118-9.

Bonamino, V. P. and Larson, R. A. (1980) *J. Amer. Soc. Hort. Sci.*, **105:** 9-11.

Bonde, M. R., Peterson, G. L., Rizvi, S. A. and Smilanick, J. L. (1995) *Plant Dis.*, **79**: 500-505.

Bonde, M.R., Murphy, C.A., Bauchan, G.R., Luster, D.G., Palmer, C.L., Nester, S.E., Revell, J.M. and Berne, D. K. (2015) *Phytopathol*, **105**(1): 91-98.

Boodley (1981) *New York State Flower Industries Bull.*, **128**: 4-6.

Borowski, E., Hagen, P. and Moe, R. (1987) *Acta Agrobotanica*, **40**: 27-39.

Borthwick, H. A. (1947) *Year Book USDA*, p. 273.

Borthwick, H. A. and Cathey, H. M. (1962) *Bot. Gaz.*, **123**: 155-162.

Borys, M. W., Garcia-Alcala, M. and Leszczynso-Borys, H. (1995) *Serie Horticultura*, **1**: 109-112.

Bowen, H. J. M., Cawse, P. A. and Dick, M. J. (1962) *Road. Bot.*, **1**: 197.

Brenn, E. and Leone, J. A. (1972) *Plant Dis. Reptr.*, **56**: 85-87.

Brierley, P. and Smith, F. F. (1951) *Plant Dis. Reptr.*, **35**: 524-6.

Brierley, P. and Smith, F. F. (1955) *Phytopathology*, **45**: 2-7.

Broadbent, A. B. and Olthof, T. H. A. (1995) *Environmental Entomol.*, **24**: 431-435.

Broertjes, C. (1966) *Euphytica*, **15**: 156-162.

Broertjes, C. and Jong, J. De (1983) *Vakblad voor de Bloemisterij*, **38**: 47.

Broertjes, C. and Jong, J. De (1984) *Euphytica*, **33**: 433-434.

Broertjes, C., Koene, P. and Pronk, T. (1983) *Euphytica*, **32:** 97-101.

Broertjes, C., Roest, S. and Bokelmann, G. S. (1976) *Euphytica*, **25:** 11-19.

Brummitt, D. (1997) *Garden* (London), **122**: 662-663.

Bunt, A. C. (1976a) *Modern Potting Composts*, George Allen and Unwin, London.

Bunt, A. C. (1976b) In: *Chrysanthemums-The Inside Story*, National Chrysanthemum Society, London.

Bursy, M., Haas, H.P. and Rober, R. (1990) *Deutscher Gartenbau*, 44: 3047-3051.

Busch, L. V. and Schoolboy, H. D. (1970) *Canadian J. Bot.*, **48**: 1939-1941.

Bush, S. R., Earle, E. D. and Langhans, R. W. (1976) *Amer. J. Bot.*, **63**: 723-737.

Buta, M., Buta, E., Hort, D. and Cantor, M. (2013) *J. Hortic. For. Biotech.*, **17**(1): 99-103.

Butler, W. L., Norris, K. H., Siegalmann, H. W. and Hendricks, S. B. (1959) *Proc. Nat. Acad. Sci.*, **45:** 1703.

Butters, R. E. (1970) *Comm. Gr.*, **3879**: 28-36.

Buwalda, F., Baas, R and Weel, P.A. van (1994) *Acta Hortic.*, **361**: 123-132.

Buwalda, F., Frenck, R., Lobker, B., Van Den Berg-De Vos, B. and Kim, K. S. (1995) *Acta Hortic.*, **401**: 193-200.

Buxton, J. W. and Culbert, J. K. (1967) *Proc. Amer. Soc. Hort. Sci.*, **91**: 641-645.

Cairol, D. (1982) *Acta Hortic.*, **125**: 159-165.

Canham, A. E. (1966a) *Artificial Light in Horticulture*, Centrex Pub. Co., Eindhoven.

Canham, A. E. (1966b) *Expt. Hort.*, **16**: 53.

Canham, A. E. (1970) *Gdnrs' Chron.*, **168**: 18.

Canham, A. E., Cockshull, K. E. and Hand, D. W. (1977) *Acta Hortic.*, **68**: 63.

Carlino, J. L., Williams, K.A. and Allen, E. R. (1998) *Hort Technology*, **8**: 373-378.

Carpenter W. J. (1966) *III. St. Florists' Ass. Bull.*, **269**: 11.

Carpenter, W. J. and Carlson, W. H. (1972) *J. Amer. Soc. Hort. Sci.*, **97**: 349-351.

Carter, A. R. and Dermott, W. (1966) *Expt. Hort.*, **16**: 12-18.

Carter, G. D. (1980) In: *Introduction to Floriculture*, (ed. R. A. Larson), Academic Press Inc.

Carvalho, M.P.S., Heuvelink, E. and Kooten, O.V. (2002) *Acta Hortic.*, **593**: 33-38

Carvalho, S.M.P. and Heuvelink, E. (2004) *Acta Hortic.*, **654**: 287-294.

Carvalho, S.M.P., Abi-Tarabay, and Heuvelink, E. (2005) *J. Hortl. Sci. and Biotech.*, **80**(2): 209-216.

Carvalho, S.M.P., Dhonukshe, V.B. and Heuvelink, E. (2003) *Acta Hortic.*, **624**: 335-342.

Casteels, H., Goossens, F., Moermans, R. and Clerco, R. De (1996) *Parasitica*, **52**: 133-138.

Catanzaro, C.J., Williams, K.A. and Sauve, R. J. (1998) *J. Plant Nutrition*, **21**: 1025-1036.

Cathey, H. M. (1954) *Bull. N. Y. St. Flower Gr.*, **104**: 1.

Cathey, H. M. (1955a) *Bull. N. Y. St. Flower Gr.*, **119**: 1.

Cathey, H. M. (1957) *Proc. Amer. Soc. Hort. Sci.*, **69**: 485-91.

Cathey, H. M. (1974) *J. Amer. Soc. Hort. Sci.*, **99**: 17.

Cathey, H. M. (1976) *J. Amer. Soc. Hort. Sci.*, **101**: 599.

Cathey, H. M. and Borthwick, H. A. (1964) *Bot. Gaz.*, **125**: 232-236.

Cathey, H. M. and Borthwick, H. A. (1967) *Plant Physiol.*, **43**: 9.

Cathey, H. M., Yeoman, A. H. and Smith, F. F. (1966) *HortScience*, **1**: 61-2.

Cermeno, P. (1989) *Acta Horticulturae*, **24**: 121-129.

Chaitra, G.S., Seetharamu, G.K., Kumar, R. and Tejaswini, H.R. (2017) *The Bioscan.*, **12**(4): 1861-1863.

Chaitra, G.S., Seetharamu, G.K., Kumar, R., Munikrishnappa, P.M.and Shivanna, M. (2018) *Indian J. Agric. Sci.*, **88**(6): 871-876.

Chakrabarty, D., Mandal, A. K. A. and Datta, S.K. (1999) *J. Hort. Sci. and Biotechnology*, **74**: 293-296.

Chan, A. P. (1950) *Proc. Amer. Soc. Hort: Sci.*, **55**: 461.

Chan, A. P. (1960) *Progr. Rep., Hort Div.*, Central Exp. Farm, Ottawa, pp. 171-78.

Chan, A. P. (1966) *Proc. Amer. Soc. Hort. Sci.*, **88**: 613-20.

Chandel, S. and Chandel, V. (2010) *Indian J. Agric. Sci.*, 80(1): 54-58.

Chang, C. S. and Huang, S. C. (1995) *Bull. Taichung Dist. Agric. Improv. Sta.*, **49**: 9-18.

Chang, C. S. and Yih, M. S. (1996) *Bull. Taichung Dist. Agric. Improv. Sta.*, **52**: 1-11.

Chang, C., Qinghua, Y., Yi, X., Jun, F. and Guanjun, B. (2018) *Int. J. Agr. Biol. Eng.*, **11**(1): 88-95.

Chang, M. H. and Park, K. W. (1997a) *J. Korean Soc. Hort. Sci.*, **38**: 201-204.

Chang, M. H. and Park, K. W. (1997b) *J. Korean Soc. Hort. Sci.*, **38**: 489-493.

Chang, S.F. and Huang, M.C. (2000) *J. Chinese Soc. Hort. Sci.*, **46**: 21-34.

Chapugier, Y. (1979) *Compte Rendes de la 10e conference du columa*, pp. 604-17.

Chau, A. and Kevin, M. (2006) *Entomol. Exp. Appl.*, **120**(3): 201-209.

Cheek, S., MacDonald, O. C. and Bartlett, P. W. (1993) In: *Liriomyza. Conf. on Leaf-Mining Flies in Cultivated Plants*, Paris, France; Centre de Cooperation Internationale en Recherche Agronomique pour le Developpement (CIRAD), pp.79-86.

Chen Z.J., Kong, S.S., Song, F.F., Li, L.P. and Jiang, H.D. (2012) *Fitoterapia.*, **83**: 1616-1622.

Chen, D. X. and Lieth, J. H. (1993) *J. Amer. Soc. Hort. Sci.*, **118**: 181-187.

Chen, D.L., Luo, C., Liu, H., Li, X. M., Cheng, X., Li, M.Y. Wang, L.L. and Huang. C.L. (2018)*Plant Dis.*, **102**(5): 1047.

Chen, H., Zhao, J., Jiang, J, Sumei, C., Guan, Z., Chen, F, Weimin, F. and Shuang, Z. (2019) *Agric.*, **9**(12): e255. doi: 10.3390/agriculture9120255

Chen, J. Y., Yang, H., Fan, G. C. and Chen, J. Y. (1998) *J. Fujian Acad. Agricultural Sci.*, **13**: 12-14.

Chitra, R., Arulmozhiyan, R., Jawaharlal, M., Indhumathi, K. and Vadivel, E. (2006) *J. Ornam. Hortic.*, **9**(1): 58-60.

Chittenden, F. J. (1956) *Dictionary of Gardening*, The Royal Horticultural Society, Oxford Univesity Press.

Cho, S.E., Zhao T.T., Choi I.Y., Oh, H.T. and Shin, H.D. (2017) *Plant Dis.*, **101**(5): 840.

Choi, H.S., Peng, D., Zhao, Y., Gu, M. and Yu, F. (2018b) *Korean J. Hortic. Sci.*, **36**(4): 501-512.

Choi, H.S., Zhao, Y., Dou, H., Cai, X., Gu, M. and Yu, F. (2018a) *Hortic. Environ. Biotech.*, **59**(3): 345-354.

Choi, K., Hyung Kim, K. and Cho, W.K. (2015) *Sci. Hort.*, **195**: 17-24.

Choudhari, R., Kulkarni, B.S. and Shirgur, M. (2018) *Ann. Hortic.*, **11**(1): 70-74.

Cockshull, K. E. (1972) In: *Crop Processes in Controlled Environments*, Acad. Press, London.

Cockshull, K. E. and Horridge, J. S. (1978) *J. Hort. Sci*, **53**: 85-90.

Cockshull, K. E. and Hughes, A. P. (1968) *Nature*, **217:** 979.

Cockshull, K. E. and Hughes, A. P. (1971) *Ann. Bot.*, **35**: 915-926.

Cockshull, K. E. and Kofranek, A. M. (1993) *Sci. Hort.*, **56**: 217-234.

Cockshull, K. E. Langton, F. A. and Cave, C. R. J (1995) *Acta Hortic.*, **378**: 15-25.

Cossio, F. and Bussi, L. (1982) *Colture Protette*, **11**: 45-9.

Costa, A.C.S., Lopes, L.F., Doliveira, P.S., Silva, M.A.G., Gil, L.G. and Rocha, R.A.A. (2004) *Acta Sci-Agron.*, **26**(4): 407-411.

Courtney-Gutterson, N., Napoli, C., Lemieux, C., Morgan, A., Firoozabady, E. and Robinson, K.E.P. (1994) *Biotechnology*, **12**: 268-271.

Cowse, P. A. (1966) *Comm. Gr.*, **3660**: 381.

Cox, R. S. (1970) *Plant Dis. Reptr.*, **54**: 679.

Crater, G. D, Kiplinger, D. C., Tayama, H and Staby, G (1973) *Ohio Florist's Assoc. Bull.*, **529**: 10.

Cuijpers, L. H. M and Vogelezang, J. V. M. (1992) *Acta Hortic.*, **327**: 25-32.

Cumming, R. W. (1964) *The Chrysanthemum Book*, D. van Nostrand Co. Inc. Princeton, New Jersey.

Dai, J., Luo, W. and Yang, Z. (2008) *Acta Hortic.*, **801**: 1423-1430.

Dallon, J. Jr (1988) *Combined Proc., Int. Plant Propagators' Soc.*, **37:** 323-329.

Datta, S. K. (1985) *J. Nuclear Agric. Biol.*, **14**: 131-133.

Datta, S. K. (1986) *Sci. and Cult.*, **52**: 205-206.

Datta, S. K. and Gupta, M. N. (1981) *Bot. Progress*, **4**: 88-92.

Datta, S. K. and Gupta, M. N. (1983) *J. Nuclear Agric. Biol.*, **12**: 22-3.

Datta, S.K. and Gupta, V.N. (2012) *Sci. Cult.*, **78**(1-2): 71-77.

Datta, S.K. and Janakiram, T. (2015) *Indian J. Agric. Sci.*, **85**(10): 1379-95.

Datta, S.K., Mandal, A.K. and Saxena, M. (2000) *Indian J. Agric. Sci.*, **71**(10): 655-657.

de Jong, J., Rademaker, W. and van Wordragen, M.F. (1993) *Plant Cell Tissue Organ Cult.*, **32**: 263-270.

Dee-in, W., Thepsithar, C., Thongpukdee, A., Chanjirakul, K. and Obsuwan, K. (2014) *Acta Hortic.*, **1025**: 248-252.

Dehale, M. H., Deshmukh, P. P. and Moharkar, V. K. (1993) *J. Soils and Crops*, **3**: 135-137.

Deotale, A. B., Belorkar, P. V., Patil, S. R., Zade, V. N. and Keche, M. B. (1994) *J. Soils and Crops*, **4:** 148-151.

DeVier, C. L. and Genever, R. L. (1997) *Sci. Hort.*, **70:** 309-318.

Dheepa, R., Nakkeeran, S. and Renukadevi, P. (2018) *J. Mycol. Plant Pathol.*, **48**(2): 133-154.

Dheepa, R.S., Vinodkumar., Renukadevi, P. and Nakkeeran, S. (2016) *Biol. Control.*, **103**: 172-186.

Dhiman, S.R., Gupta, Y.C., Thakur, P., Kashyap, B. and Sharma, M. (2017) *J. Hill Agric.*, **8**(3): 283-287.

Dicks, J. W. (1972) *Amer. Appl. Biol.*, **71**: 313-326.

Dohino, T. and Hayashi, T. (1995) *Res. Bull.Plant Protection Service Japan*, No.31, pp. 95-100.

Doi, M., Aoe, K., Watabe, S., Inamoto, K. and Imanishi, Ho. (2004) *J. Jpn. Soc. Hortic. Sci.*, **73**(3): 229-234.

Doi, M., Nakagawa, Y., Watabe, S., Aoe, K., Inamoto, K. and Imanishi, H. (2003) *J. Jpn. Soc. Hortic. Sci.*, **72**(6): 533-535.

Donald, C. M. (1968) *Euphytica*, **17**: 385.

Dooreubos, J. and Kofranek, A. M. (1953) *Proc. Amer. Soc. Hort. Sci.*, **61**: 555.

Dowrick, G. L. and El-Bayoumi, A. S. (1966) *Euphytica*, **15**: 32-38.

Dreistadt, S. H. and Flint, M. L. (1996) *Environmental Entomol.*, **25:** 688-697.

Drewlow, L. W. and Widmer, R. E. (1971) *J. Amer. Soc. Hort. Sci.*, **96**: 633-636.

Drewlow, L. W., Ascher, P. D. and Widmer, R. E. (1973) *Theo. Appl. Genet.*, **43**: 1.

Druge, U. (1997) *Scientia Horti.*, **68**: 171-180.

Druge, U.; Zerche, S. and Kadner, R. (1998) *Gartenbaumagazin*, **7**: 24-25.

Dubsky, M. and Sramek, F. (1999) *Zahradnietvi*, **26**: 103-106.

Dutt, M. and Sonawane, P.C. (2006) *Indian J. Hortic.*, **63**(1): 66-69.

Dutt, M., Patil, M.T. and Sonawane, P.C. (2002) *Indian J. Hortic.*, **59**(2): 191-195.

Dutta, J. P. and Seemanthini, R. (1995) *J. Agri. Sci. Soc. North East India*, **8**: 29-32.

Dutta, J. P. and Seemanthini, R. (1998) *Orissa J. Hort.*, **26**: 70-75.

Dutta, J. P., Seemanthini, R. and Khader, M. A. (1993) *South Indian Hort.*, **41**: 293-299.

Dutta, J. P., Seemanthini, R. and Khader, M.A. (1996) *Orissa J. Hort.*, **24**: 1-4.

Dutta, J. P., Seemanthini, R. and Khader, M.A. (1998) *Karnataka J. Agri. Sci.*, **11:** 251-253.

Dwivedi, A.K., Banerji, B.K., Chakrabarty, D., Mandal, A.K. and Datta, S.K. (2000) *Indian J. Agric. Sci.*, **70**(12): 853-855.

Earle, E. D. and Langhans, R. W. (1974a) *J. Amer. Soc. Hort. Sci.*, **99**: 128-132.

Earle, E. D. and Langhans, R. W. (1974b) *J. Amer. Soc. Hort. Sci.*, **99**: 352-358.

El-Antably, H. M., Habib, S. A. and Rabie, K. A. E. (1991) *Annals Agric. Sci.* (Cairo), **36:** 69-83.

El-Keltawi, N. E., Mousa, G. T. and Makary, B. S. (1996) *Acta Horticulturae*, **426**: 657-669.

Elliot, G. G. and Nelson, P. V. (1983) *Physiol. Plant.*, **57**: 250-259.

Endo, M. (1990) *J. Japanese Soc. Hort. Sci.*, **59**: 613-620.

Endo, M., Saski, T. and Inada, I. (1991) *J. Faculty of Agriculture*, Iwate University, **20**: 91-103.

Eng, R. Y. N., Tsujita, M. J. and Grodzinski, B. (1985) *J. Hort. Sci.*, **60**: 389-395.

Engelhard, A. W. and Woltz, S, S. (1972) *Proc. Fla. Sta. Hort. Soc.*, **84**: 351-354.

Engelhard, A. W., Miller, H. N. and Neve, R. T. (1971) *Plant Dis. Reptr.*, **55**: 851-855.

Essig, W. (1989) *Zierpflanzenbau*, **29**: 568-569.

Eveleens, B. and Blok, C. (2014) *Acta Hortic.*, **1034**: 185-191.

Fan, Q., Aiping, S., Jingjing, X., Sumei, C., Jiafu, J., Yinjie, W., Xiran, Li. and Fadi, Chen. (2015) *PLoS One.*, doi: 10.1371/journal.pone.0143349.

Fanourakis, D., Heuvelink, E. and Verkerke, W. (2011) *Acta Hortic.*, **893**: 915-923.

Farias, M.F.D. and Saad, J.C.C. (2011) *Acta. Sci-Agron.*, **33**(1): 75-79.

Farina, E. and Aatemiani, T. (1982) *Annali dell' Istituto Sperimentale per la Floricolt*, **13:** 149-516.

Farnham, D. H. (1973) *Calif. Agric. Ext. Flower and Nursery Rep.*, July, pp. 1-5.

Fenton, T. (1969) *l0th Progress Rep. Exp. Husb. Fms. Exp. Hort. Sta.*, pp. 98-99.

Fiebich, D. and Henning, F. (1992) *Gartenbauwissenschaft*, **57**: 212-218.

Firoozabady, E., Lemieux, C.S., May, Y.S., Moll, B., Nicholas, J.A. and Robinson, K.E.P. (1991) *In vitro*, **27**: 96A.

Fisher, P. R. and Heins, R. D. (1996) *Acta Hortic.*, **417**: 41-45.

Flint, M. L., Dreistadt, S. H., Rentner, J. and Parrella, M. P. (1995) *California Agric.*, **49:** 5-8.

Fox, R. T. V. (1998) *Mycologist*, **12**: 135-136.

Fransen, J. J., Boogaard, M. and Tolsma, J. (1993) *Bull. OILB/SROP*, **16**: 73-77.

Freitas, J. C. De., Agostini, L. A., Matsuoko, S., Caldari Junior, P., Garcia, A. A. F. and Oliveira, S. H. F. De (1997) *Summa Phytopathologica*, **23**: 268-271.

Freitas, S. De and Bueno, Y. M. (1998) *Revista de Agricultura* (Piracicaba), **73**: 229-236.

Fryxwell, P. A. (1957) *Bot. Rev.*, **23**: 135.

Fujii, K. and Takamura, S. (1998) *Agrochemicals Japan*, **72**: 14-15.

Fujii, Y. and Sasaki, S. (2000) *Bull. Nara Agric. Exp. Sta.*, **31**: 42-43.

Fujime, Y. (1978) *J. Jap. Soc. Hort. Sci.*, **47**: 382-390.

Fukai, S. and Oe, M. (1991) *J. Japanese Soc. Hort. Sci.*, **59**: 383-387.

Fukai, S., Goi, M and Tanaka, M. (1994) *Sci. Hort.*, **57**: 347-351.

Fukai, S., Goi, M. and Tanaka, M. (1991) *Euphytica*, **54**: 201-204.

Fukai, S., Nagira, T. and Goi, M. (2000) *Acta Hortic.*, **508**: 337-340.

Fukai, S., Zhang, W. and Goi, M. (2000) *Acta Hortic.*, **515**: 167-172.

Fukuda, M., Nishio, J.and Arai, K. (1987) *Res. Bull. Aichi- ken Agric. Res. Centre*, Japan No.19, pp. 230-235.

Furuta, T. (1954) *Proc. Amer. Soc. Hort. Sci.*, **63**: 457.

Furuta, T. and Kiplinger, D. C. (1955) *Proc. Amer. Soc. Hort. Sci.*, **66**: 383.

Furuya, H. (1999) *Bull. Hiroshima Prefectural Agric. Res. Centre*, **67**: 31-39.

Furuya, M. (1998) *Bull. Hiroshima Prefectural Agric. Res. Centre*, **66**: 33-40.

Gaikwad, A.M. and Dumbre, P.S.S. (2001) *J. Ornam. Hortic.*, **4**(2): 95-97.

Gajbhiye, R.P., Kudmate, S.S., Munde, G.R. and Panchbhai, D.M. (2017) *Bioinfolet.*, **14**(1): 21-23.

Gamalero, E., Romina, D.A., Chiara, M., Simone, C., Barbara, P., Giovanni, D.A., Jin, D., Domenico, B., Cristina, M. and Graziella, Berta. (2010) *Phytopathol.*, **100**(8): 805-813.

Gao, Y., Mao, L.S., Zhao, H.Y. and Liu, X. J. (1991) *Plant Physiology Communications*, **27**: 192-194.

Gao, Y.K., Zhao, B., Ding, G.X. and Zhang, Q.X. (2001) *J. Beijing Forestry University*, **23**: 32-33.

Gardner, R. (1966) *Proc. 17th Int. Hort. Congr.*, p. 347.

Geest, V., Ieperen, W., Van., Post, A.G. and Schoutsen, C.G.L.M. (2012) *Acta Hortic.*, **953**: 345-349.

Geiffith, M., Mc Intyre, H.C.H. (1989) *Agroborealis*, **21**: 12-17.

Geneve, R.L., Kester, S.T. and Buxton, J.W. (2004) *HortScience*, **39**(3): 584-587.

Ghosh, S., Naika, M.B.N., Nishani, S., Shiragur, M. and Bhat, A. (2018) *Int. J. Curr. Microbiol. App. Sci.*, **7**(8): 3965-3971.

Gill, S. A., Reeser, R. and Raupp, M. J. (1998) *Grower Talks*, **62:** 46-48.

Gislerod, H. R. and Nelson, P. V. (1989) *Sci. Horti*, **38:** 305-313.

Gjmaerum, H. B. (1963) *Gartneryrket*, **53**: 1142-1147.

Goo, D.H., Yae, B.W., Han, B.H., Lee, S.Y., Song, H.S., Shin, H.K. and Ko, L.Y. (1999) *J. Korean Soc. Hort. Sci.*, **40**: 595-597.

Gortz, H. (1965) *Zierpflanzenba*, **5**: 117-118.

Gortz, H. (1966) *DCK Inf.*, **18**: 6-7.

Gould, H. J. (1966) *Plant Pathol.*, **15**: 109-112.

Gowda, J. V. N. and Jayanthi, R. (1988) *Current Res.*, University of Agricultural Sciences, Bangalore, **17**: 21-22.

Granada, E. G. De, Amezquita, M. O. De, Carvajal, I. M., Arbelaez, G., Salamanca, M. and Lee, R. (1999) *Acta Hortic.*, **482**: 163-168.

Grantzau, E. (1987) *Zeirpflanzenbau*, **27:** 805-806.

Grantzau, E. and Scharpf, H. C. (1986) *Zierpflanzenbau*, **26:** 934-936.

Gregov, Z. (1992) *Poljoprivredna Znanstvena Smotra*, **57:** 397-413.

Grewal, H.S., Kumar, R. and Chauhan, R. (2005) *J. Ornam. Hortic.*, **8**(3): 230-232.

Gu, Q., Chen, Y., Cui, H., Huang, D., Zhou, J., Wu, T. and Xu, J. (2013) *Planta Med.*, **71**(9): 871–876.

Guo W. M., Zhang Z. H. and Fang W. M. (1997) *Acta Hortic., Sinica*, **24:** 364-368.

Guo, W., Zheng, L., Zhang, Z. and Zeng, W. (2003) *Acta Hortic.*, **624**: 349-355.

Gupta, M. N. (1971) *Proc. Inter. Symp. on Use of isotopes and Radiation in Agriculture and Animal Husbandry Research*, JARJ., NewDelhi, pp. 164-174.

Gupta, V.N. and Banerji, B.K. (2008) *Bharatiya Vaigyanik Evam Audyogik Anusandhan Patrika.*, 68-73.

Haeringen, C. J.van., West, J. S. Davis, F. J. Gilbert, A., Hadley, P., Pearson, S., Wheldon, A. E. and Henbest, R. G. C. (1998) *Photochemistry and Photobiology*, **67**: 407-413.

Halevy, A. H., Bryne, T. G., Kofranek, A. M., Farnham, D. S, Thompson, J. F. and Hardenburg, R (1978) *J.Amer. Soc. Hort. Sci.*, **103**: 151-155.

Hall, R. A. (1975) *Proc. 8th British Insecticides and Fungicides Conf.*, p. 93.

Han, A.R., Bomi, N., Bo-Ram, K., Ki-Chang, L., Beom-Seok, S., Sang, H. K., Jin-Baek, K. and Chang, H. J. (2019) *Molecules.*, **24**: 329. doi: 10.3390/molecules24020329.

Hand, D. W., Langton, F. A., Hannah, M. A. and Cockshull, K. E. (1996) *J. Hort. Sci.*, **71:** 227-234.

Hanke, H. (1996) *TASPO Gartenbaumagazin*, **5:** 8.

Hansen, C.W., Lynch, J. and Ottosen, C. O. (1998) *J. Amer. Soc. Hort. Sci.*, **123:** 215-222.

Hansen, H.T., Hendriks, L., Ueber, E. and Anderson, A.S. (1996) *Gartenbauwissenschaft*, **61:** 188-196.

Hansen, R. (1999) *Acta Horticulturae*, **481**: 297-304.

Haouas, D., Ben Halima-Kamel, M., and Ben Hamouda, M.H. (2008) *Tunis. J. Plant Prot.*, **3**(2): 87-93.

Harada, H. and Nitsch, J. P. (1959) *HortScience*, **129**: 777-778.

Harbaugh, B. K., Staneley, C. D. and Prince, J. F. (1982) *HortScience*, **17**: 598-599.

Harris, M. A., Begley, J. W. and Warkentin, D.L. (1990) *J. Economic Entomol.*, **83:** 2380-2384.

Hattori, K. (1991) *Japanese J. Breed.*, **41:** 1-9.

Hauptmann, G. G. and Schickedanz, F. (1987) *Gb+ Gw*, **87:** 1.

Hayashi, T. and Todoriki, S. (1996) *HortScience*, **31**: 117-119.

Haynie, C. P., and Klingman, G. (1999) *Res. Series-Arkansas Agri. Exp. Sta.*, **466**: 62-68.

Heemers, L. and Oyaert, E. (1999) *Verbondsnieuws*, **43:** 32-33.

Heidemann, J. C. and José, G.B. (2017) *Ornam. Hortic.*, **23**(4): 426-431.

Heike, K. (1971) *Acta Pruhoniciana*, **24**: 1034.

Heinsvig, K., Thorup-Kristensen, K., Rosenqvist, E. and Aaslyng, J. Mazanti. (2007) *J. Hortic. Sci. Biotech.*, **82**(6): 867-874.

Hemlata, B., Patil, A. A. and Nalawadi, U. G. (1992) *Progressive Hort.*, **24:** 55-59.

Hemlata, C and Mahdi, A (2016) *Int. J. Sci. Res. and Rev.*, **5**(1): 35-44.

Hemsly, (1889) *Gdnrs'. Chron.*, **6**: 521-523.

Hendricks, L and Lemper, J. (1983) *Gb+Gw*, **83**: 751-753.,

Hendricks, L. (1984) *Gb+Gw*, **84**: 298-301.

Hendricks, L. (1993) *Gartenbau Magazin*, **2**: 63-65.

Hendricks, L. (1988) *Gb+ Gw*, **88:** 1578-1579.

Henry (1902) *Gdnrs'. Chron.*, **21:** 301-302.

Hentig, W. U. (1959) *Gartenbauwiss.*, **24**: 334-362.

Hentig, W. U. and Fisher, M. (1981) *Gb+Gw*, **81:** 119-122.

Herbert, H. J., Smith, R. F. and McRae, K. B. (1984) *Canadian Entomologist*, **116**: 1259-1266.

Hershey and Paul (1981a) *J. Plant Nutrition*, **3:** 935-943.

Hershey, and Paul (1981b) *HortScience*, **16**: 220-222.

Herve, A. J. (1990) *Combined Proceedings, Int. Plant Propagators' Soc.*, **39:** 66-70.

Hessein, N. A. and Parrella, M. P. (1990) *California Agri.*, **44:** 19-21.

Hesting, J. J. and Peachey, J. E. (1963) *Plant Pathology*, **12**: 180-183.

Hetman, J., Martyn, W., Wolski, T. and Szot, P. (1996) *Zeszyty Problemowe Postepow Nauk Rolniczych*, **429:** 143-147.

Heungens, A., Vermaerke, D. and Buysse, G. (1989) *Verbondsnieuws voor de Belgische Sierteelt*, **33**: 180-182.

Heuvelink, E., Lee, J.H., Buiskool, R.P.M. and Ortega, L. (2002) *Acta Hortic.*, **580**: 197-202.

Hicklenton, P. R. (1983) *Scientia Hort.*, **21**: 189-197.

Hicklenton, P. R. (1984) *J. Amer. Soc. Hort. Sci.*, **109**: 468-472.

Hicklenton, P. R. (1985) *HortScience*, **20**: 374-376.

Hicklenton, P. R. (1988) *Proc.Int. Cong. on Soilles Culture, Int. Soc. for Soilless Culture.* Wageningen, Netherlands; Int. Soc. for Soilless Culture, pp. 195-211.

Hicklenton, P. R. (1989) *Canadian J. Plant Sci.*, **69:** 690.

Hiden, C. and Larsen, R. U. (1994) *Scientia Hort.*, **58:** 123-138.

Hill, G. P. (1968) *Physiol. Plantarum*, **21**: 386-389.

Hiremath, V.M., Jain, R. and Sinha, N. (2019) *Indian J. Agric. Sci.*, **89**(9): 1498–1502.

Hiremath, V.M., Jain, R., Jain, N., Arora, A., Swaroop, K., Singh, M.K., Kumar, P. and Kumar, G. (2018) *Indian J. Agric. Sci.*, **88**(9): 1429-1433.

Hiroyasu, T. S. Y. and Marinaga, M. (1970) *Tech. Bull. Fac. Hort. Chiba*, **18**: 43-46.

Hisamatsu, T., Sumitomo, K. and Shimizu, H. (2008) *J. Hortic. Sci. Biotech.*, **83**(6): 695-700.

Ho, J. H. and Chen, J. M. (1995) *Bull. Taichung Dist. Agri. Improv. Sta.*, **46**: 47-57.

Hoeven, A. P. van der (1989) *Vakblad voor de Bloemisterij*, **44:** 156-158.

Hoeven, A.B. vander (1990) *Vakblad voor de Bloemisterij*, **45:** 50-51.

Hoffmann, S. (1999) *Gartenbauwissenschaft*, **64:** 253-260.

Holcomb, E. J. and Walker, P. N. (1995) *Communications in Soil Science and Plant Analysis*, **26:** 2497-2510.

Honda, Y., Kashima, T. and Kumagai, T. (1992) *J. Phytopathology*, **136:** 270-278.

Hong, Y. P. Kim, J. Y and Kun, Y. J (1982) *Res. Rept. Rural Development*, **24:** 49-53.

Hoogerwerf, A. and Doorn, W. G. van (1992) *Postharvest Biol. and Technology*, **1**: 295-304.

Horridge, J. S. and Cockshull, K. E. (1989) *J. Hort. Sci.*, **64**: 183-188.

Hosni, A. M. (1996) *Annals Agric. Sci.*, **41:** 381-385.

Hosoya, T. (1979) *Bull. Saitama Hort. Expt. Sta.*, **8**: 51-56.

Hosoya, T., Murai, C and Hiruma, H. (1978) *Bull. Saitama Hort. Exp. Sta.*, **7:** 49-53.

Hoyle, E. J. and Saynor, M. (1993) *Bull. OILB/SROP*, **16:** 102-104.

Hu L. H. and Chen Z. L. (1997) *Phytochemistry*, **44**: 1287-1290.

Huang C. C. (1998) *Special Publication- Tasichung Dist. Agric.Improv. Sta.*, **40**: 203-213.

Huang, H. C., Marshall, H. H., Collicutt, L. M., McLaren, D.L. and Kokko, M. J. (1993) *Plant Pathol. Bull.*, **2**: 103-105.

Huh, K. Y., Kim, K. S., Yamaguchi, T. (1994) *RDA J. Agric. Sci.*, 36: 422-429.

Hui X.J., Zhang Q.X. and Gao Y.K. (2014) *Acta Hortic.*, **1035**: 233-238.

Huld, A. and Anderson, N.E. (1997) *Acta Horticulturae*, **435**: 209-217.

Hussein, H. A. A. (1994) *Acta Horticulturae*, **368**: 106-116.

Hutchinson, J. F., Miller, R., Kaui, V., Stevenson, T. and Richards, D. (1989) *J. Cell Biochem.*, p. 261.

Hwang K. S., Yoon, J. H. and Park, Y. D. (1992) *Res. Reprt. Rural Develop. Administration, Soil and Fertilizer*, **34:** 44-50.

Hwang, K. S. and Yoon, J. H. (1994) *RDA J. Agric. Sci., Soil and Fertilizer*, **36:** 268-272.

Ihsanul Haq, Khan, J., Alam, M. and Khattak, M.S. (1998) *Sarhad J. Agric.*, **14:** 211-213.

Imai, H., Ohnishi, M., Kinoshita, M. and Kojima, M. (1995) *Biosci. Biotech. and Biochemi.*, **59:** 1309-1313.

Imakiire, S., Hakuzan, R., Watanabe, G., Ueno K.I., Nagayoshi, S., and Hisamatsu, T. (2017) *Hortic. Res., Japan*, **16**(1): 51-59.

Ivezic, M., Samota, D. and Dudar, A. (1982) *Savez. Drust. Zastitabilija Jugoslavije, Beorrad*, **3**: 337-339.

Jackson, C. R. and McFadden, L. A. (1961) *Bull. Univ. of Florida Agril. Exp. Sta.*, Gainesville, No. 637.

Jacobson, B. M. and Willits, D. H. (1998) *Transactions of the ASAE*, **41:** 825-832.

Jacobson, B. M., Willits, D. H. and Nelson, P. V. (1998) *Applied Engineering in Agric.*, **14:** 311-316.

Jacobson, B. M., Willits, D.H. and Nelson, P. V. (1991) *Paper- Amer. Soc. Agric. Engineers*, **91-4046**: 10.

Jagadale, S., Sharma, P. and Dhiman, S.R. (2019) *Int. J. Curr. Microbiol. App. Sci.*, **8**(6): 1472-1488.

Jager, C. M. De, Butot, R. P. T., Uiterduk, M. E. C. and Meijden, E. Van der (1997) *J. Economic Entomol.*, **90:** 188-194.

Jain R., Janakiram T. and Kumawat, G.L. (2016) *Indian J. Agric. Sci.*, **86**(11): 1489-1494.

Jain R., Janakiram T., Singh K.P. and Kumawat, G.L. (2014) *Indian J. Agric. Sci.*, **84**(10): 1250-1253.

Janakarin, T., Mahantesh, Murgod, I. and Prabhakar, B.S. (2006) *Acta Hortic.*, **710**: 321-328.

Janipour, B., Mirkohi, A.K., Jamali, F. and Khalighi, A. (2017) *Iranian J. Hortic. Sci.*, **48**(1): 237-247.

Jank, H. (1957) *Der Zuchter*, **27**: 223.

Jensen, H.E.K. (1994) *Gartenbauwissenschaft*, **59**: 280-284.

Jeyaram, M., Sankaralingam, A., Sangeetha, A. and Rajappan, K. (2019) *Ann. Plant Prot. Sci.*, **27**(3): 346-351.

Jhon, A. Q. and Paul, T. M. (1999) *Applied Biological Res.*, **1:** 35-38.

Ji E. Y., Oh W., Kim S. H. and Kim K. S. (1998b) *Korean J. Hort. Sci. and Technology* **16:** 247-250.

Ji E. Y., Oh W., Kim S. H. and Kim K. S. (1998a) *Korean J. Hort. Sci. and Technology*, **16:** 37-39.

Jiang, H., Xia, Q., Xu, W., and Zheng, M. (2004) *Pharmazie.*, **59**(7): 565–567.

Joiner, J. N. (1967) *Proc. Amer. Soc. Hort. Sci.*, **90**: 389.

Joiner, J. N. and Knoop, W. F. (1969-70) *Proc. Fla. Sta. Hort. Soc.*, **82:** 403-407.

Joiner, J. N. and Pickhardt, G. D. (1970) *Proc. 83rd Annual Meeting Florida Sta. Hort. Soc.*, **83:** 401.

Jong, J, De (1982) *Euphytica*, **31**: 285-292.

Jong, J. De (1984) *Euphytica*, **33**: 455-463.

Jong, J. De (1986) *Scientia Hort.*, **28**: 263-270.

Jong, J. De (1989) *Scientia Hort.*, **41:** 117-124.

Jong, J. De, Wordragen, M. F. van and Rademaker, W. (1990) *Eucarpia*, pp. 156-161.

Jordan, C., Reimann and Phillip, R. (1983) *Zeitschrift fur Pflanzenzuchtung*, **91**: 111-122.

Joshi, M., Verma, L.R. and Masu, M.M. (2009) *Asian J. Hortic.*, **4**(2): 292-294.

Jothi, L. J., Nanjan, K., Chezhiyan, N. and Khader, M. A. (1987) *South Indian Hort.*, **35**: 220-225.

Juan, V. C. and Carda, M. S. (1988) *Revista Espanola de Electronica*, **402**: 106-110.

Judd, R. W. Jr. and Walton, G. S. (1973) *Plant Dis. Reptr.*, **57**: 624.

Jung, E.K. (2009) *J. Bacteriol. Virol.*, **39**(2): 61-69.

Jung, S.S., Jeong, H.H. and Kim, K.S. (2000) *Korean J. Hort. Sci. Technol.*, **18**: 28-32.

Kadner, R. and Druge, E. (1999) TASPO *Gartenbaumagazin*, **8:** 24-25.

Kadner, R. and Zerche, S. (1997) *Gartenbauwissenschaft*, **62**: 184-189.

Kageyama, Y. and Konishi, K. (1992) *J. Japanese Soc. Hort. Sci.*, . **61:** 635-642.

Kageyama, Y. and Konishi, K. (1996) *J. Japanese Soc. Hort. Sci.*, **64:** 905-911.

Kageyama, Y., Hayashi, T and Konishi, K (1987) *J. Japanese Soc. Hort. Sci.*, **56:** 79-85.

Kageyama, Y., Nakagawa, Y. and Konishi, K. (1993) *J. Japanese Soc. Hort. Sci.*, **61**: 895-900.

Kageyama, Y., Shima, K. and Konishi, K. (1995) *J. Japanese Soc. Hort. Sci.*, **64**: 169-176.

Kahar, S.A. (2008) *Acta Hortic.*, **788**: 141-148.

Kala, D., Mahawer, L.N. and Bairwa, H.L. (2020) *Int. J. Chem. Stud.*, **8**(2): 1246-1251.

Kameswari, P.L., Girwani, A and Padmavaamma, A.S. (2014) *Prog. Hortic.*, **46**(2): 314-318.

Kanchan, A., Simranjit, K., Ranjan, K., Prasanna, R., Ramakrishnan, B., Singh, M.C., Hasan M. and Kavitha M., Srinivasulu B., Reddy M.L.N., Swarajyalakshmi K. (2013) *J. Ornam. Hortic.*, **16**(1-2): 1-4.

Kang, J. G., Chon, S. U. and Chung, S. J. (1995a) *J. Korean Soc. Hort. Sci.*, **36:** 548-554.

Kang, J. G., Seo, B. S. and Chung, S. J. (1995c) *J. Korean Soc. Hort. Sci.*, **36:** 83-89.

Kang, J. G., Son, J. K., Park, H. S. and Chung, S. J. (1995d) *J. Korean Soc. Hort. Sci.*, **36:** 747-754.

Kang, J. G., Whang, T. E. and Chung, S. J. (1995b) *J. Korean Soc. Hort. Sci.*, **36:** 401-409.

Karlsen, P. (1997) *Acta Horticulturae*, **435**: 33-45.

Karlsen, P. and Bertram, L. (1995) *Acta Hortic.*, **378**: 87-92.

Karlsson, M. G., Heins, R. D., Erwin, J. E. E and Hago, R. D. (1989a) *J. Amer. Soc. Hort. Sci.*, **114:** 234-240.

Karlsson, M.G., Heins, R. D., Erwin, J. E., Berghage, R. D, Carlson, W. H and Bierinbaum, J. A. (1989b) *Scientia Hort.*, **39:** 257-267.

Kasten, P. and Sommer, K. (1990) In: *Plant nutritionphysiology and applications, Proc. Eleventh Int. Plant Nutrition Colloquium*, Wageningen. Netherlands, 30 July

4- August 1989. (Eds. Beusichem M.L. van) Dredrecht, Netherlands; Kluwer Academic Publishers, pp. 530-537.

Kasten, P., Sommer, K. and Papenhagen, A. (1990) *Deutscher Gartenbau*, **44**: 1710-1713.

Kato, T and Takei, A. (1994) *Res. Bull. Aichi-ken Agric. Res. Center*, **26**: 247-255.

Kato, T. and Takei, A. (1989) *Bull. Aichi-ken Agric. Res. Center*, **21**: 230-238.

Kaui, V., Miller R. M., Hutchinson J. F. and Richards, D. (1990) *Plant Cell Tissue Organ Cult.*, **21**: 21-30.

Kavitha, M., Srinivasulu, B., Reddy, M.L.N. and Swarajyalakshmi, K. (2013) *J. Ornam. Hortic.*, **16**(1-2): 1-4.

Kavitha, P.S., Sudha, A. and Sriram, N. (2019) *Journal of Krishi Vigyan.*, **8**(1): 8-12.

Kazeroonian, R., Kalatejari, S., Mousavi, A. and Tohidfar, M. (2017) *Iranian J. Hortl. Sci.*, **48**(3): 527-534.

Kazimirova, R. N. (1975) *Bull. Gosjdau. Nikit. Botanich. Sada*, **3**: 74-77.

Kennard, S. Nelson (1967) In: *Flower and Plant Production in the Greenhouse*, The Interstate Printers and Publishers Inc., Danville, Illinois.

Ketsa, S. (1989) *Kasetsart J., Natural Sciences*, **23:** 1-7.

Khalid, N., Davey, M. R. and Power, J. B. (1989) *Scientia Hort.*, **38:** 287-294.

Khattak A.M., Pearson, S. and Johnson, C.B. (2004) *Sci. Hortic.*, **102**(3): 335-341.

Khattak, A. M., Pearson, S. and Johnson, C.B. (1999) *J. Hort. Sci. and Biotech.*, **74:** 206-212.

Khattak, A.M. and Pearson, S. (1997) *Acta Hortic.*, **435**: 113-121.

Kher, M. A. (1986) In: *Ornamental Horticulture In India*, ICAR, New Delhi.

Kher, M. A., Yokoi, M. and Kosugi, K. (1972) *Tech. Bull. Fac. Hort.*, Chiba Univ. **20**: 2599.

Kim, H. J., and Lee, Y. S. (2005) *Planta Med.*, **71**(9): 871–876.

Kim, H.T., Jung, W.Y. and Hee, L.J. (2020) *Plant Cell Tiss. Org.*, **140**(3): 677-683.

Kim, J.H., Lee, J.S:, Kim, T.J., Kim, H.H., Kim, S.D., Lee, H.D., Lee, C.H. and Yun, T. (2000) *J. Korean Soc. Hort. Sci.*, **41**: 631-635.

Kim, S. H. and Kim, K. S. (1999) *Korean J. Hort. Sci. and Technology*, **17:** 355-360.

Kim, Y. B., Moon, J. S., Song, C. H., Kim, K. H and Park, H. S. (1993) *RDA J. Agric. Sci., Hort.*, **35**: 427-435.

Kim, Y. H. and Lee, J. S. (1998) *J. Korean Soc. Hort. Sci.*, **39:** 784-788.

King, J. J., Peterson, L. A. and Stimart, D. P. (1995) *HortScience*, **30**: 499-503.

Kiyatkin, A. K. (1976) *Referativnyi Zhurnal*, **7**: 951.

Klapwijk, D. and Wubben, F. (1991) *Vakblad voor de Bloemisterij*, **46:** 55-57.

Klaus, H. (1957) *Gartenwelt*, **57**: 110-111.

Knodel, M., Lyons, R. E. and Poe, S. E. (1985) *HortScience*, **20**: 708-710.

Kobayashi, K. D. (2005) Hort Science, **40**(4): 1012D-1012.doi: 10.21273/HORTSCI.40.4.1012D.

Kofranek, A. M. (1980) In: *Introduction to Floriculture* (Ed. R.A. Larson), Academic Press Inc.

Kofranek, A. M. and Criley, R.A. (1967) *Grower*, **67**: 819-820.

Kofranek, A. M. and Halevy, A. H. (1972) *J. Amer. Soc. Hort. Sci.*, **97**: 578-584.

Kogel, W. J. De., Hoek, M. van der., Dik, M.T.A., Gebala, B., Dijken, F. R. van and Mollema, C. (1997) *Euphytica*, **97:** 283-288.

Kollner, V. (1973) *Nachrichtenblatt des Deutschen Pflanzenschutzdienstes*, **25**: 7-11.

Kollner, V., Wesche, A. and Cruger, G. (1991) *Gesunde Pflanzen*, **43:** 395-398.

Konjoian, P. S., Staby, G. L. and Tayama, H. K. (1983) *J. Amer. Soc. Hort. Sci.*, **108**: 582-585.

Koriesh, E. M., Abou-Dahab, A. M. and Ali, E. W. (1989a) *Assiut J. Agric. Sciences*, **20:** 27-41

Koriesh, E. M., Abou-Dahab, A. M. and Ali, E. W. (1989b) *Assiut J. Agric. Sciences*, **20:** 43-58.

Krebs, E. K. (1997) *Gartenbaumagazin*, **6**: 40-41.

Kulus, D. (2015) *Acta Hortic.*, **1087**: 225-232.

Kumar, A. and Poonam. (2008a) *Agric. Res. J.*, **45**(1-2): 105.

Kumar, A. and Poonam. (2008b) *Agric. Res. J.*, **45**(1-2): 106.

Kumar, A. and Poonam. (2008c) *Agric. Res. J.*, **45**(1-2): 107.

Kumar, A., Dubey P., Patanwar, M. and Sharma, R. (2015) *Trends Biosci.*, **8**(1): 175-177.

Kumar, G., Sindhu, S.S., Kumar, S. and Vanlalruati, V. (2017) *Indian J. Agric. Sci.*, 87(7): 958-963.

Kumar, G., Singh, K.P., Prasad, K.V., Rana, M.K., Namita and Panwar, S. (2014) *Indian J. Agric. Sci.*, **84**(11): 1323-1328.

Kumar, G., Singh, K.P., Prasad, K.V., Verma, N., Namita and Panwar, S. (2015b) *Indian J. Agric. Sci.*, **85**(4): 566-570.

Kumar, R., Rao, T.M. and Dhananjaya, M.V. (2019) *Indian J. Plant Genet. Resour.*, **32**(2): 267.

Kumar, R., Rao, T.M. and Dhananjaya, M.V. (2019) *Indian J. Plant Genet. Resour.*, **32**(2): 267-268.

Kumar, S., Sreedar, C., Sanjeev, K., Muralidharan, A. and Elakkuvan, S. (2019) *Plant Arch.*, **19**(2): 2743-2746.

Kumar, S., Sunil, M., Mukesh, K. and Omkar, S. (2015a) *Prog. Hortic.*, **47**(2): 271-274.

Kumari, M., Kumar, R. and Saha, T.N. (2017) *Indian J. Hortic.*, **74**(2): 258-263.

Kumari, M., Kumar, R., Saha, T.N. and Sindhu, S.S. (2016) *Indian J. Hortic.*, **73**(4): 582-587.

Kunimoto, Y., Nishino, S., Otuji, J. and Inda, K. (1998) *Japanese J. Applied Entomology and Zoology*, **42:** 135-140.

Kwon J-H., Choi, O. and Kim, J. (2013) *Plant Dis.*, **97**(8): 1118. doi: 10.1094/PDIS-01-13-0067-PDN.

Laishram, N., Dhiman S.R., Gupta, Y.C., Bhardwaj, S.K. and Singh, A. (2013) *Indian J. Agric. Sci.*, **83**(4): 447-455.

Laishram, N., Dhiman, S.R., Gupta, Y.C., Singh, A. and Bhardwaj, S.K. (2016) *Indian J. Agric. Sci.*, **86**(11): 1449–1454.

Lamberti, F., Tacconi, R., Marinari, A., D'Errico, F. P. and Basile, M. (1987) *Difesa della Plante,* **10:** 77-84.

Langton, F. A. (1976) In: *Chrysanthemums-The Inside Story*, National Chrysanthenum Society, London.

Langton, F. A. and Cockshull, K. F. (1976) *Acta Horticulturae*, **63**: 165.

Langton, F. A., Benjamin, L. R and Edmondson, R. N. (1999) *J. Hort. Sci. and Biotechnology*, **74**: 493-501.

Langton, F.A., Horridge, J.S., Holdsworth, M.D. and Hamer, P.J.C. (2004) *Acta Hortic.*, **633**: 145-152.

Langton, J. P. and Cockshull, K.E. (1997) *Scientia Hort.*, **69:** 229-237.

Larew, H. G. and Locke, J. C. (1990) *HortScience*, **25:** 1406-1407.

Larsen, R. and Gertsson, U. (1992) *Scientia Hort.*, **49**: 277-289.

Larsen, R. and Gertsson, U. (1993) *Acta Hortic.*, **328**: 171-178.

Larson, R. A. and McIntyre (1964) *N. C. Flower Grs. Bull.*, **8**: 4-10.

Laurie, A., Kiplinger, D. C. and Nelson, K. C. (1968) *Commercial Flower Forcing*, McGraw Hill, New York.

Ledger, S. E., Deroles, S. C. and Given, N. K. (1991) *Plant Cell Rep.*, **10:** 195-199.

Ledien, M. S. and Helyer, N. L. (1983) *Ann. Appl. Biology*, **102:** 275-279.

Lee B. S., Park S. G., Kang J. G. and Chung S. J. (1999a) *J. Korean Soc. Hort. Sci.*, **40:** 225-230.

Lee J.H., Heuvelink E. and Challa, H. (2002a) *Acta Hortic.*, **593:** 151-157.

Lee J.H., Heuvelink I.G. and Challa, H. (2002b) *J. Hortic. Sci. Biotech.*, **77**(2): 238-247.

Lee, B. S., Chung, S. J., Kang, J. G. and Seo, B. S. (1999b) *J. Korean Soc. Hort. Sci.*, **40:** 112-116.

Lee, Chang-Hee. and Cho, Myeong-Whan. (2011) *Korean J Hortic Sci.*, **29**(6): 539-548.

Lee, J. S., Kim, Y. H. and Ryu, B. Y. (1996) *J. Korean Soc. Hort. Sci.*, **37:** 802-804.

Lee, J. S., Song, C. Y., Wang, H. J., Kim, Y. A. (1996) *J. Korean Soc. Hort. Sci.*, **37:** 136-140.

Leinfelder, J. and Rober, R. (1987) *Deutscher Gartenbau*, **41:** 2890-2892.

Leinfelder, J. and Rober, R. (1989) *Deutscher Gartenbau*, **43:** 2412-2418.

Lemieux, C. S., Firoozabady, E. and Robinson, K. E. P. (1990) *Eucarpia*, pp. 150-155.

Lessa, M.A., Barbosa J.P.R.A.D., Paiva P.D.O., Paglis C.M., Barbosa J.V.C. and Nogueira D.A. (2013) *Acta Hortic.*, **970**: 319-327.

Lever, R. A. (1965) *World Crops*, **17**: 55-56.

Li, M., Zhang, X. and Chen, J. (1983) *Acta Hort. Sinicia*, **10:** 199-206.

Li Maoxue, Zhang Xiang, Chen Junyli (1983) *Acta Hortic., Sinica*, **10**: 199-206

Lieburg, M. J. van (1997) *Acta Horticulturae*, **418**: 229-232.

Liu, W. Y. and Zhang, M. P. (1996) *Acta Hortic. Sinica*, **23:** 269-273.

Lodhi, A. K. S. and Tiwari, G. N. (1993) *Fertiliser News*, **38:** 39-45.

Loeser, H. and Essie, W. (1967) *Dtsche Gartnerborse*, **67**: 31-34.

Lowe, J. M., Davey, M. R., Power, J. B. and Blundy, K. S. (1993) *Plant Cell Tissue Organ Cult.*, **33**: 171-180.

Lozoya, S. H. (1994) *Serie Horticultura*, **1:** 11-14.

Ludolph, D. (1996) *TASPO Gartenbaumagazin*, **5:** 13-14.

Lumsden, R. D. and Hassis, F. A. (1964) *Tech. Bull. N. C. Agric. Expt. Sta.*, **158**: 27.

Lunt, O. R. and Kofranek, A. M. (1958) *Proc. Amer. Soc. Hort. Sci.*, **72**: 487.

Mac Kenzie, A. J., Starman, T. W. and Windham, M. T. (1995) *HortScience*, **30:** 496-498.

MacDonald, W. N., Tsujita, M. J., Blom, T. J. and Shelp, B. J. (2014a) *Can. J. Plant Sci.*, 94: 643-657.

MacDonald, W. N., Tsujita, M. J., Blom, T. J. and Shelp, B. J. (2014b) *Can. J. Plant Sci.*, 94: 867-880.

Machin, B. J. (1973) *Ph. D. thesis*, University of Nottingham, UK.

Machin, B. J. (1975) *National Chrysanthemum Soc. Year Book*, 84.

Machin, B. J. (1976) *Chrysanthemum Nutrition, National Chrysanthemum Society*, London.

Machin, B. J. and Scopes, N. (1978) *Chrysanthemums-Year Round Growing*, Blandford Press, Dorset, UK.

Madhu, M. and Shyam, K. R. (1992) *Plant Disease Research*, **7**: 94-95.

Madhu, M., Shyam, K. R and Sandhu, G.S. (1989) *Plant Disease Research*, **4:** 90.

Mahamoud, T. A. and Steponkas, P. L. (1970) *J. Amer. Soc. Hort. Sci.*, **95**: 295-299.

Mainardi, Jucelma De Cássia C Tolotti, Bellé, Rogério Antônio and Mainardi, Leandro. (2004) *Ciência Rural.*, 1709-1714.

Makino, T. (1897) *Nihon Engei Zasshi*, No.23, pp. 1-3.

Malaure, R. S., Barclay, G., Power, J. B. and Davey, M. R. (1991) *J. Plant Physiol.*, **139**: 8-13.

Mali, G.L., Moond, S.K., Mishra, A., Sharma, M.K. and Singh, B. (2017) *Prog. Hortic.*, **49**(1): 75-78.

Malorgio, F., Lemmetti, S., Tognoni, F. and Campiotti, C. A. (1994) *Acta Horticulturae*, **361**: 495-500.

Mandal, A.K.A. and Datta, S.K. (2005) *Biol. Plantarum.*, **49**(1): 29-33.

Mandang, Jeany Polii. (2017) *Am. J. Agric. Biol. Sci.*, 123-129.

Mani, T. and Senthil, K. (2011) *Asian J. Pharm. Tech.*, **1**(1): 13-16.

Marani, F. (1972) *Fitopatholica*, **22**: 3-8.

Marczynki, S. and Tukey, H. B. Jr (1988) *Acta Horticulturae*, **226 (II):** 465-471.

Margaretha, B.Z. and Klaas, M. (2006) *HortScience*, **41**(1): 136-142.

Markvart, J., Rosenqvist, Eva., Sørensen, Helle., Ottosen, C.O. and Aaslyng, J.M. (2009) *HortScience.*, **44**(5): 1284–1290.

Marousky, F. J. (1971) *Marketing Res. Rep. USDA.*, **905**: 14.

Martín, C., Uberhuaga, E. and Pérez, C. (2002) *Euphytica.*, **127**(2): 247-253.

Martinez Silva, D. and Gomez Zambrano, J. (1995) *Acta Agronomica*, Universidad Nacional de Colombia, **45:** 79-84.

Marwoto, B. and Djatnika, I. (1986) *Bull. Penelitian Hortikultura*, **13**: 43-50.

Mashhadian, N.V., Tehranifar, A., Bayat, H., and Selahvarzi, Y. (2012) *J. Agr. Sci. Tech.*, **14**: 879-887.

Mato, M., Kudo, H., Yamagata, A., Sato, T. and Shibata, H. (2009) *Hortic. Res.: Japan*, **8**(2): 201-208.

McCain, A. H. (1971) *Plant Dis. Reptr.*, **55**: 347-348.

McDaniel, G. L. (1984) *HortScience*, **19**: 434-436.

McFadden, L. A (1958-59) *Proc. Fla. Sta. Hort. Soc.*, **71**: 419-425.

McFarlane, N.J.B. (1991) *Computers and Electronics in Agric.*, **6:** 95-106.

McMohan, M. (1999) *J. Amer. Soc. Hort. Sci.*, **12:** 483-487.

McMohan, M. and Kelly, J. W. (1995) *Scientia Hort.*, **64:** 203-209.

McMohan, M. and Kelly, J. W. (1999) *Scientia Hort.*, **79**: 207-215.

Meer, M. van der (1990) *Vakblad voor de Bloemisterij*, **45:** 82-83.

Meeteren, U. van and Gelder, H. van (1999) *Postharvest Biology and Technology*, **16**: 169-177.

Meeteren, U. van, Gelder, H. Van and Ieperen, W. van (2000) *Postharvest Biology and Technology*, **18**: 169-181.

Meher, S. P., Jiotode, D. J., Turkhede, A. B., Darange, S. O., Ghatol, P. U. and Dhawad, C. S. (1999) *Crop Res.* (Hisar), **18**: 486-489.

Meinken, E. (1988) *Gb+ Gw*, **88:** 154.

Metz, K. (1987) *Deutscher Gartenbau*, **41:** 113-114.

Meyer, J., Horn, W. (1995) *Acta Horticulturae*, **378**: 105-112.

Mildenberger, F. and Hendriks, L. (1996) *TASPO Gartenbau Magazin*, **5**: 54-55.

Miler, N. and Anita, W. (2016) Electronic Journal of Polish Agricultural Universities, 19(4). http: //www.ejpau.media.pl/volume19/issue4/art-02.html.

Miler, N. and Jedrzejczyk, I. (2018) *Turk. J. Bot.*, **42**(3): 289-297.

Miler, N. and Zalewska, M. (2006) *Acta Hortic.*, 725: 347-350.

Million, J. B., Barreti, J. E., Nell, T. A. and Clark, D.G. (1999) *HortScience*, **34**: 1103-1105.

Miranda, De. Larra, J. (1973) *Acta Horticulturae*, **43**: 375-80.

Misra, P, and Datta, S.K. (2007) *Indian J.Biotechnol.*, **6**(1): 121-124.

Mitiouchkina, T., Firsov, A., Titova, S., Shulga, O. and Dolgov, S. (2018b). *Acta Hortic.*, **1193**: 89-94.

Mitiouchkina, T.Y. and Folgov, S.V. (2000) *Acta Horticulturae*, **508**: 163-169.

Mitiouchkna, T., Smykova, N.V. and Dolgov, S. (2018a) *Acta Hortic.*, **1201**: 577-582.

Mitra, A., Bhattacharya, P. S., Dey, S., Sawarkar, S. K. and Bhattacharya, B. C. (1998) *Biotechnology Techniques*, **12**: 335-337.

Miyazawa, M., and Hisama, M. (2003)*Biosci. Biotechnol. Biochem.*, **67**(10): 2091–2099.

Mizutani, T. and Tanaka, T. (1994) *Proc. of faculty of Agriculture*, Kyushu Tokai University, **13:** 9-14.

Molitor, H. D. and Fischer, M. (1999) *Acta Horticulturae*, **481**: 179-186.

Molitor, H.D. and Fischer, M. (1993) In: *Proc. 8th International congress on soil less culture*, Wageningen, Netherlands; International Society for soil-less culture (ISOSC), pp. 223-239.

Mora Medina, H.R. and Mosquera Paris, F. (1985) *Revista Colombiana de Entomologia*, **11**: 36-38.

Mori, Y., Nakajima, T., Fujimoto, T., Tsunemi, T., Sumitomo, K., Hisamatsu, T., Goto, T. (2014) *Hortic. Res.: Japan*, **13**(4): 349-356.

Mori, Y., Sumitomo, K., Hisamatsu, T. and Goto, T. (2016) *The Hortic. J.*, **85**(3): 264–271.

Mortensen, L. M. and Moe, R. (1983) *Scientia Hort.*, **19**: 141-147.

Mota, Poliana, R.D.' Almeida, Bôas, Roberto, L. V., Sousa, Valdemício. F. de, Ribeiro and Valdenir, Q. (2007) *J. Brazilian Assoc. Agric. Engg.*, **27**(1): 164-171.

Moustafa, A. T. and Morgan, J. V. (1983) *Acta Horticulturae*, **133**: 13-24.

Mukhopadhyaya, A. and Das, P. (1976) *Indian Hort.*, **21**: 13, 15-16.

Nagarjuna, B., Reddy, V. P., Rao, M. R. and Reddy, E. N. (1988) *South Indian Hort.*, **36**: 136-140.

Nagasuga, K., Takayoshi, Y., Yamazaki, H., Inamoto, K. and Yamasaki, A. (2008) *Hort. Res., Japan*, **7**(1): 103-109.

Nagatomi, S., Degi, K., Yamaguich, M., Miyahira, E., Sakamoto, M. and Takaesu, K. (1993) *Technical News - Institute of Radiation Breeding*, **43**: 1-2.

Naing, A. H., Chung, M.Y. Lim, K.M. and Kim, C.K. (2013) *J. Hortic. Sci. and Biotechnol.*, **88**(6): 762-767.

Naka, T., Kenji, H., Naoki, T., Yuka, S., Yuri, T. and Tanjuro, G. (2018) *Hortic. Res.: Japan*, **17**(1): 105-114.

Nakai, T. (1928) *Bot. Magazine*, Tokyo, Vol. **XLII:** 492-495.

Nakajima, T. and Young, H. G. (1965) *The Art of Chrysanthemum*, John Weatherhill Inc., Tokyo.

Nakamura, S., Okasako, Y. and Yamada, E. (1955) *J. Hort. Ass. Japan*, **24**: 17-28.

Nakano, Y., Higuchi, Y., Sumitomo, K., Oda, A. and Hisamatsu, T. (2015) *J. Hortic. Sci. Biotech.*, **90**(2) 143–149.

Nakayama, M. and Yui, H. (1988) *J. Faculty of Agriculture, Shinshu University*, **25**: 1-13.

Nakayama, M., Koshioka, M., Shibata, M., Hiradate, S., Sugie, H. and Yamaguchi, M. (1997) *Biosci. Biotech. and Biochem.*, **61:** 1607-1608.

Nalini, R. (2012) *Int. J. Food, Agr. and Vet. Sci.*, **2**(2): 62-66.

Nam, S.Y., Kwon, Y. W. and Soh, C. H. (1995) *J. Korean Soc. Hort. Sci.*, **36**: 90-97.

Nasri, F. and Ghaderi, N. (2015) *J. Sci. and Tech. Greenhouse Cult.*, **5**(20): 111-124.

Nazeer, M. A. and Khoshoo, R. N. (1982) *Curr. Sci.*, **51**: 583-585.

Nedoluzhko, A. I. (2018) *Russ. Agric. Sci.*, **44**(5): 422-425.

Negi, R., Dhiman, S.R. and Dhiman, M.R. (2020) *Int. J. Curr. Microbiol. App. Sci.*, **9**(2): 2533-2536.

Nehra, M.K. and Sh. Munendra. (2019) *Ann. Hortic.*, **12**(1): 74-79.

Neilson, O. F. (1995) *Computers and Electronics in Agric.*, **13:** 315-335.

Newton, P. (1965) *Grdnrs' Chron.*, **157**: 103.

Nishimoto, R. K., Ranch, F. D., Hori, T. M. and Parvin, P. E. (1979) *Res. Rept. Hawaii Agril., Expt. Sta.*, **187**: 8.

Nishio, J. and Fukuda, M. (1987) *Res. Bull. Aichi-ken Agric. Res. Center*, Japan, **19**: 223-229.

Nishio, J. and Fukuda, M. (1997) *Res. Bull. Aichi-ken Agric. Res. Center*, Japan, **29**: 171-176.

Nishio, J. Yamaguchi, T. and Hara, M. (1994) *Res. Bull. Aichi-ken Agric. Res. Center*, Japan, **26**: 233-240.

Nishio, J. Yamaguchi, T. and Yonemura, K. (1989) *Bull. Aichi-ken Agric. Res. Center*, Japan, **1**: 211-216.

Niwa, T. (1936) *Chrysanthemum of Japan*, Sanscido Co. Ltd.

Noble, R. and Ellis, S. R. (1990) *J. Hort. Sci.*, **65:** 177-183.

Noda, N. (2019) *Acta Hortic.*, **1237**: 7-12.

Nugrahani, P., Moeljani, I.R. and Maskun, A. (2016) *J. Agric. and Biol. Sci.*, **11**(8): 347-350.

Nunnink, E. (1991) *Groenten+ Fruit, Glasgroenten*, **1:** 14-15.

Oda. A, Sumitomo, K., Tsunemi, T., Douzono, M., Motozu, T. and Hisamatsu, T. (2010) *Hort. Res., Japan*, **9**(1): 93-98.

Oh, W. and Kim K. S. (1998) *J. Korean Soc. Hort. Sci.*, **39:** 464-468.

Ohishi, K. (2000) *Res. Bull. Aichi-ken Agric. Res. Centre*, No. 32, pp. 161-167.

Ohishi, K. and Sakurai, Y. (1988) *Res. Bull. Aichiken Agric. Res. Cent.*, **20:** 278-284.

Ohishi, K., Yonemura, K. and Osuga, M. (1986) *Res. Bull. Aichi- Ken Agric. Res. Centre*, Japan. **18**: 168-172.

Ohmiya, A., Sumitomo, K. and Aida, R. (2009) *J.Jpn.Soc. Hortic. Sci.*, **78**: 450–455.

Ohmiya, A., Kishimoto, S., Aida, R., Yoshioka, S. and Sumitomo, K. (2006) *Plant Physiol.*, **142**: 1193–1201.

Okada, M. (1951) *J. Hort. Ass. Japan*, **20**: 23.

Okada, M. (1959) *J. Hort. Ass. Japan*, **28**: 209.

Okada, M. (1963) *Mem. Faculty Agric. Toyko Univ.*, **9**: 63-202.

Opender K. (1999) *J. Biosciences*, **24**: 85-90.

Ota, M., Saito, T. and Ozawa, A. (1996) *Proc. Kanto-Tosan Plant Protection Soc.*, **43**: 257-259.

Overman, A. J. and Poe, S. L. (1971) *Proc. Fla. Sta. Hort. Soc.*, **84**: 419-422.

Overman, A. J. and Raulston, J. C. (1971) *Proc. Fla. Sta. Hort. Soc.*, **84**: 413-841.

Oyaert, E. (1996) *Verbondsnieuws*, **40:** 42-43

Oyaert, E., Volckaert, E. and Debergh, P. C. (1999) *Scientia Hort.*, **79:** 195-205.

Ozawa, A., Ohta, M. and Kobayashi, H. (1998) *Proc. Kanto-Tosan Plant Protection Soc.*, **45**: 175-177.

Padmadevi, K., Jawaharlal, M., Kannan, M. and Ganga, M. (2009) *J. Ornam. Hort.*, **12**(4): 239-244.

Pak, Chun Ho, Kang Seung Won and Lee Chiwon W. (2005) *Hort Sci.*, **40**(4): 1046. doi: 10.21273/HORTSCI.40.4.1046A

Pak, Songhui, Han, Myonghun, Li, Hyonu, Pak, Hyongbom and Li, Jongnam. (2020) *Plant Biotechnol. Rep.*, doi: 10.1007/s11816-019-00587-4.

Pant, M., Lal, A. and Jain, R. (2015) *J. Appl. Pharm. Sci.*, **5**(7): 103-111.

Papafotiou, M. and Vagena, A. (2012) *Sci. Hortic.*, **143**: 102-108.

Parker, R. D. and Kamp, J. R. (1958) *III. St. Florists' Ass. Bull.*, **187:** 1-5.

Parkman, P. and Peinkowski, R. L. (1990) *J. Economic Entomology*, **83:** 1246-1249.

Parrella, M. P. and Jones, V. P. (1984) *Calif. Agric.*, **38:** 17-19.

Parrella, M. P. and Murphy, B. (1996) *Bull. OILB/SROP*, **19:** 115-118.

Parups, E. V. (1976) *Canadian J. Plant Sci.*, **56**: 531.

Parups, E. V. and Butler, G. (1982) *J. Amer. Soc. Hort. Sci.*, **107**: 600-604.

Pasian, C. C., Lindquist, R. K. and Struve, D. K. (1997) *Hort Technology*, **7**: 265-269.

Pasian, C.C. and Struve, D.K. (1997) *PGRSA Quarterly*, **25**: 177-186.

Paul, J. L. and Smith, L. V. (1966) *Proc. Amer. Soc. Hort. Sci.*, **89**: 626-630.

Paul, J. L. and Thorhill, W. H. (1969) *J. Amer. Soc. Hort. Sci.*, **94**: 280-282.

Pavingerova, D., Dostal, J., Biskova, R. and Benetika, V. (1994) *Plant Science* (*Limerick*), **97:** 95-101.

Pawlowski, H. E. (1966) *Z. Pflernahr Dung*, **115**: 100-113.

Pearson, S., Hadley, P. and Wheldon, A. E. (1995) *Acta Horticulturae*, **378**: 71-80.

Pedley Kerry F. (2009) *Plant Dis.*, **93**(12): 1252-1258.

Peerally, M. A. and Colhoun, J. (1969) *Trans. British Mycol. Soc.*, **52**: 115.

Peet, M. M., Willits, D. H., Tripp, K.E., Kroen, W. K., Pharr, D. M., Depa, M. A. and Nelson, P. V. (1991) *Proc. Indo-US workshop held on 8-12 January 1991 at New Delhi, India* (Eds. Abrot, Y. P., Govindjee., Wattal, P. N., Ort, D. R., Gnanam, A., Teramura, A.H.) New Delhi, India, Oxford and IBH Publishing Co. Pvt. Ltd., pp. 193-212.

Pereira, Joelma R. D., Carvalho, Jacinto de A., Miguel, Douglas, S. and Santana, Márcio J. de. (2005) *J. Braz. Assoc. Agric. Engg.*, **25**(3): 651-659.

Petters, S. (1992) *Gartenbau Magazin*, **1:** 68-70.

Pilanali, N. and Kaplan, M. (1999) *Turkish J. Agric. and Forestry*, **23:** 721-727.

Ploeg, V. D., Carvalho, S.M.P. and Heuvelink, E.P. (2009) *J. Am. Soc. Hortic. Sci.*, **134**(1): 31-40.

Pobudkiewicz, A. and Nowak, J. (1997) *J. Fruit and Orn. Plant Res.*, **5:** 43-52.

Poe, S. L. and Marousky, F. L. (1972) *Proc. Fla. Sta. Hort. Soc.*, **84**: 432-435.

Poesch, G. H. (1935) *Proc. Amer. Amer.Soc. Hort. Sci.*, **33:** 637.

Pol, P.A. van de (1988) *Acta Horticulturae*, **226**: 519-524.

Poole, R. T. (1965) *Diss. Abstr.*, **25**: 6915-6916.

Poole, R. T. and Ying, H. K. (1966) *Proc. Fla. Sta. Hort. Soc.*, **78**: 28-33.

Popescu, N., Roman, L., Chelu, M. and Lacatus, V. (1996) *Anal. Institutul de Cercetari Pentru Legumicultura si Floricultura*, Vidra, **14**: 451-455.

Popham, R. A. (1963) *Brookhaven Symp. Biol.* No.16.

Post, K. (1942) *Cornell Univ. Agr.Exp. Sta. Bull.*, **787**: 1.

Post, K. (1947) *Bull. N. Y. Sta. Flower Grs.*, **27**: 4.

Post, K. (1948) *Proc. Amer. Soc. Hort. Sci.*, **51**: 590.

Post, K. (1949) *Florist Crop Production and Marketing*, Orange Judd Pub. Co. Inc., New York.

Post, K. (1953b) *Bull. N. Y. Sta. Flower Grs.*, **99**: 3-4.

Post, K. (1953c) *Proc. 13th Int. Hort. Cong.*, p. 937.

Post, K. and Kamemoto, H. (1950) *Proc. Amer. Soc. Hort. Sci.*, **55**: 477.

Post, K. and Lacey, D. B. (1951b) *Bull. N. Y. Sta. Flower Grs.*, **76**: 4.

Post. K. (1953a) *Proc. 13th Int. Hort. Cong.*, p. 935.

Prabucki, A. Serek, M. and Andersen, A. S. (1999) *J. Hort. Sci. and Biotech.*, **74**: 132-134.

Prasanna, R., Saxena, G., Singh, B., Ranjan, K., Buddhadeo, R., Velmourougane, K, Ramakrishnan, B., Nain, Lata., Singh, M.C., Hasan, M. and Shivay, Y.S. (2018) *Open Agric.*, **3**(1): 478-489.

Przymeska, J. (1997) *Zeszyty Problemowe Postepow Nauk Robvia*, **449:** 161-172.

Punetha. P., Rao. V.K. and Sharma. S.K. (2011) *Indian J. Agric. Sci.*, **81**(9): 830-833.

Putrasamedia, S. and Sutapradja, H. (1989) *Bull. Penelitian Hortikultura*, **17:** 96-98.

Qian X. W. (1996) *J. Sanghai Agric. College*, **14:** 201-205.

Qiu, W. D. and Liu, K. B. (1989) *Plant Physiol. Communication*, **6**: 31-33

Qureshi, I.A., Gulzar, S., Dar, A.R., Rehman, R.U. and Tahir, I. (2018) *Indian J. Agric. Res.*, **52**(3): 319-322.

Raabe, R. D. and Vlamis, J. (1966) *Phytopathology*, **56**: 713-717.

Rai, H., Raju, D.V.S., Prasad, K.V., Singh, M., Kumar, G., Pandey, R.N. and Lekshmy, S. (2017) *Indian J. Agric. Sci.*, **87**(7): 870-877.

Raj, S.K., Khan, M.S. and Kumar, S. (2007) *Australas. Plant Dis. Notes.*, **2**(1): 21-22.

Rajagopalan, A. and Khader, J. B. M. M. A. (1994) *Indian J. Agric. Sci*, **64**: 240-243.

Rajagopalan, A. and Khader, M. A. (1993) *South Indian Hort.*, **41**: 61-63.

Rajapakse, N. C. and Kelly, J. W. (1993) *HortScience*, **28:** 999-1001.

Rajapakse, N. C. and Kelly, J. W. (1995a) *Acta Horticulturae.*, **405**: 427-434.

Rajapakse, N. C. and Kelly, J. W. (1995b) *J. Amer. Soc. Hort. Sci.*, **120:** 78-83.

Rajapakse, N. C., Miller, W. B. and Kelly, J. W. (1996) *J. Amer. Soc. Hort. Sci.*, **121:** 740-745.

Rajapakse, N. C., Reed, D. W. and Kelly, J. W. (1989) *HortScience*, **24:** 998-1000.

Ramakrishnan, S. and Vadivelu, S. (1995) *South Indian Hort.*, **43:** 174-175.

Ramatuelle, (1792) *J. d. Histoire Naturelle*, Tomll: 240.

Razauska, E. (1966) *Zasc. Rast. Vred. Bolez.*, **11**: 20.

Read, P. E. and Hoysler, V. C. (1969) *J. Amer. Soc. Hort. Sci.*, **94**: 314-631.

Reddy, V. K., Rajapakse, N. C. and Young, R. E. (1997) *Acta Horticulturae*, **440**: 292-297.

Reed, D. W., Sillivan, P. and Kazemzedeh, M. (1986) *HortScience*, **21**: 935.

Reiss-Bubenheim, D. and Lewis, A. J. (1984) *HortScience*, **19**: 434-436.

Reiss-Bubenheim, D. and Lewis, A. J. (1986) *Scientia Hort.*, **28**: 159-164.

Reist, A. (1988) *Revue Suisse de Viticulture et d'Arboriculture et d'Horticulture*, **20:** 299-300.

Renou, J. P, Brochard, P. and Jalbuzot, R. (1993) *Plant Science* (*Limerick*), **89:** 185-197.

Rieh1, G. (1962) *Zierpflanzenbau*, **2**: 105-106.

Rij, E., Thompson, J. F. and Farnham, D. S. (1979) *U. S. Dep. Agric. ATT-W-5.*

Ristvey, A.G., Belayneh, B.E., Leacox, J.D. and Ferd, J.I. (2018) *Acta Hortic.*, **1197**: 211-218

Ritchie, G. A., Short, K.C. and Davey, M.R. (1991) *J. Experimental Bot.*, **42:** 1557-1563.

Rober, R. (1971) *Gartenbauwiss.*, **36**: 189-200.

Rober, R. and Hafeez, M. (1982) *Acta Horiculturae.*, **125**: 69-78.

Roberts, A.V. and Matthews, D. (1995) *Plant Cell, Tissue and Organ Cult.*, **40:** 191-193.

Roberts, A.V. and Smith, E. F. (1990) *Plant Cell, Tissue and Organ Cult.*, **21:** 129-132.

Robson, P. P. (2019) *J. Exp. Agric. Int.*, **30**(2): 1-8.

Rodriguez-Navarro, J. A., Zavaleta-Mejia, E. and Alatorre-Rosas, R. (1996) *Fitopatologia*, **31:** 122-132.

Roeber, R., Kolb, D. and Ohmayer, G. (1995) *Acta Horticulturae*, **378**: 139-148.

Roelops, T. (1990) *Vakblad voor de Bloemisterij*, **45:** 48-50.

Roest, S. and Bokelman, G. S. (1975) *Scientia Hort.*, **3**: 317-330.

Roorda van Eysinga, J.P.N.L. and Smilde, K. W. (1980) *Nutritional Disorders in Chrysanthemum*, Centre for Agricultural Publishing and documentation, Wageningen.

Rounkova, L. V. (1989) *Acta Horticulturae*, **251**: 281-288.

Rout, G. R. and Das, P. (1997) *Scientia Hort.*, **69**: 239-257.

Rout, G. R. and Das, P. (2000) In: *Biotechnology of Horticultural Crops*, Vol. 3 (Eds. Parthasarathy, V. A., Bose, T. K. and Das, P.) Naya Prokash, Calcutta, pp. 86-107.

Roxas, N. S., Tashiro, Y., Miyazaki, S., Takeshita, A. and Oshima, T. (1993) *J. Japanese Soc. Hort. Sci.*, **61:** 919-924.

Ruiter, H. A. De (1993) *Scientia Hort.*, **56:** 43-50.

Ruiter, H. A. De (1996) *Ann. Bot.*, **77:** 99-104.

Ruiter, H. A. De (1997) *J. Hort. Sci.*, **72:** 77-82.

Ruiter, H. A. De and Tromp, J. (1996) *J. Hort. Sci.*, **71:** 607-612.

Sabine, J. (1822) *Trans. Hort. Soc.* London. Vol. **IV:** 326-534.

Sacalis, J. N. and Durkin, D. (1972) *J. Amer. Soc. Hort. Sci.*, **97**: 481-484.

Sachs, R. M. and Kofranek, A. M. (1963) *Amer. J. Bot.*, **50**: 772-779.

Sahana, A.S., Biradar, M.S., Hiremath, S.M. and Somanagouda, G. (2019) *J. Farm Sci.*, **32**(4): 497-500.

Sahlstrom, H. (1966) *Lantbr Hogsk., Medd.*, Uppsala, Ser. A. **49**: 53.

Sahu, M. and Sharma, G. (2014) *J. Ornam. Hortic.*, **17**(1-2): 32-37.

Saito, N., Toki, K., Honda, T. and Kawase, K. (1988) *Phytochemistry*, **27:** 2963-2966.

Sakamoto, H. and Tsuchiya, T. (2007) Hortic. Res.: Japan, 6(3): 411-416.

Sakamoto, Y., Nakashima, T. and Okano, K. (1998) *Environment Control in Biol.*, **36:** 77-84.

Sanderson, K. C., and West, M. S. (1995) *Plant Growth Regulator Soc. America Quarterly*, **23:** 141-148.

Sanderson, K. C., Martin, W. C. Jr and McGuire, J.A. (1988b) *HortScience*, **23:** 1008-1009.

Sanderson, K. C., Oetting, R. D. and Smith, D. A. (1989) *HortScience*, **24:** 856.

Sanderson, K. C., Smith, D. A. and McGuire, J. A. (1994) *HortScience*, **29:** 330.

Sanderson, K. C., Yokoyama, H. and Hearn, W. (1988a) *Plant Growth Regulator Bull.*, **16:** 8-9.

Sarkar, M.M. and Pal, P. (2008) *Nat. Prod. Radiance.*, **7**(2): 146-149.

Saud, B. K. and Talukdar, M. C. (1999) *J. Interacademicia*, **3:** 25-28.

Sauvadet, M. A., Brochard, P. and Boccon- Gibod, J. (1990) *Plant Cell Rep.*, **8**: 629-695.

Sawwan, J., Abu-Zahra, T., Khader, S. and Basher, E. A. (1999) *Dirasat, Agric. Sciences*, **26:** 364-370.

Scharpf, H, C., Hendriks, L. and Grantzau, E. (1985) *Dtsche Gartenb.*, **39:** 1923-1925.

Schmulling, M. (1992) *Plant Varieties J.*, **5:** 15-16.

Schoellhorn, R. K., Barrett, J. E. and Nell, T. A. (1996) *HortScience*, **31:** 74-78.

Schrader, R. and Scharpf, C. (1993) *Gartenbau Magazin*, **2:** 12-14.

Schuch, K. (1994) *J. Plant Growth Regulation*, **13:** 115-121.

Schuch, U. K., Redak, R. A. and Bethke, J. A. (1998) *J. Amer. Soc. Hort. Sci.*, **123:** 727-733.

Schum, A. and Preil, W. (1981) *Gartenbauwillenochaft.*, **46**: 91-93.

Schussler, W. K. (1992) *Acta Horticulturae*, **327**: 105-110.

Schwabe, W. W. (1950) *J. Exp. Bot.*, **1**: 320

Schwabe, W. W. (1953) *Proc. 13th Int. Hort. Cong.*, p. 952.

Schwabe, W. W. (1959) *J. Linn. Soc. Bot.*, **56**: 254.

Schwinn, K. E., Markham, K. R and Given, N. K. (1994) *Phytochemistry*, **35:** 145-150.

Scopes, N. A. E. and Biggerstaff, S. M. (1973) *Proc. 7th British Insecticide and Fungicide Conf.*, p. 227.

Scopes, N. E. A. (1970) *Ann. Appl. Biol.*, **66:** 323.

Scott, M. C., Caetano- Anolles, G. and Trigiano, R.N. (1996) *J. Amer. Soc. Hort. Sci.*, **121:** 1043- 1048.

Seager, J. C. R. and Champion, J. C. (1984) *Proc. 7th Int. Peat Congr.* Dublin, pp. 60-70.

Searle, S. A. and Machin, B. J. (1968) *Chrysanthemums-the Year Round*, BlandfordPress, London.

Seaton, K. A., Cook, D.F. and Hardie, D.C. (1997) *Austrailian J. Agric. Res.*, **48**: 781-787.

Sen, S., Kumar, S., Ghani, M. and Thakur, M. (2013) *Plant Pathol. J.*, **12**(1): 1-10.

Sethi, V.P. and Sidhu, G.S. (2005) *Agric. Res. J.*, **42**(3): 320-326.

Shabot, R. E. and Koths, J. S. (1987) *Connecticut Greenhouse Newsletter*, **138**: 10-16.

Shabot, R.E. and Koths, J. S. (1988) *Connecticut Greenhouse Newsletter*, **142**: 1-8.

Shamala, G. and Janardhana, G.R. (2015) *Int. J. Life Sci.*, **9**(3): 38-45.

Shanks, J. B. (1969) *HortScience*, **4**: 56-58.

Sharma, H. G., Verma, L. S., Jain, V. and Tiwari, B. L. (1995) *Orissa J. Hort.*, **23:** 61-64.

Shawareb, M. N. (1989) *Dirasat*, **16:** 178-179.

Shawareb, M. N. and Qrufleh, M. M. (1988) *Dirasat*, **15:** 54-68.

Sheard, G. F. (1955) *Agriculture*, London, **62**: 41.

Shen, H. J. (1990) *Acta Horticulturae Sinicia*, **17:** 65-70.

Sher, R. B. and Parrella, M. P. (1996) *Bull. OILB/SROP*, **19:** 147-150.

Sherman, J. M., Moyer, J. W. and Daub, M. E. (1998) *J. Amer. Soc. Hort. Sci.*, **123:** 189-194.

Sheu C. S. (1998) *Special Publication – Taichung Dist. Agric. Improv. Sta.*, **40**: 215-222.

Sheu C. S., Sheu C.Y. and Wu S. (1998) *Special Publication- Taichung Dist. Agric. Improv. Sta.*, **40**: 159-170.

Shi, W. L., Xia, Y. P. and Yao, Y. Q. (1993) *Acta Horticulturae Scinicia*, **20:** 99-100.

Shibata, M. (1988) *JARQ (Japan Agric. Res. Quarterly)*, **21:** 269-273.

Shibata, M. (1997) *Bull. Nat.Res. Inst. Vegetables, Ornamental Plants and Tea, Series A: Vegetables and Ornamental Plants*, **12**: 1-71.

Shibata, M., Amano, M. Kawata, J. and Uda, M. (1988) *Bull. Nat. Res. Inst. Vegetables, Ornamental Plants and Tea, A,* **A:** 245-255.

Shibata, M., Kawata, J., Amano, M., Kameno, T., Wagishi, H., Toyoda, T., Yamaguchi, T., Okimura, M. and Uda, M. (1988) *Bull. Nat. Res. Inst. Vegetables, Ornamental Plants and Tea,* **A:** 257-277.

Shibata, M., Kawata, J., Toyoda, T., Amano, M., Yamaguchi, T., Okimura, M and Uda, M. (1986) *Bull. Vegetable and Ornamental Crops Res. Sta., A,* **14**: 69-79.

Shibuya, S. I. and Tekeuti, M. (1998) *Annual Rept. Soc. Plant Protection of North Japan,* **49**: 164-166.

Shima, K., Kageyama, Y. and Konishi, K. (1995) *J. Japanese Soc. Hort. Sci.,* **64:** 177-184.

Shima, K., Kageyama, Y. and Konishi, K. (1996) *J. Japanese Soc. Hort. Sci.,* **65:** 355-361.

Shin Hakki and Choi, S. T. (1996) *J. Korean Soc. Hort. Sci.,* **37:** 299-303.

Shin Hakki, Choi, J. K. and Sang, T. (1996) *J. Korean Soc. Hort. Sci.,* **37:** 442-446.

Shin Hakki, Choi, S. Y., Kim, T., Kim, H. D., Jouong, H. Y., Kim, Y. Y., Ko, J. Y., Cho, M. W. and Park, H. S. (1998) *RDA J. Hort. Sci. (II),* **40:** 116.119.

Shin Hakki, Kim, J. Y. and Jin, S. S. (1995) *RDA J. Agric. Sci., Hort.,* **37:** 389-395.

Shin Hakki., Cho, S. R., Ko J. Y. and Park, H. J. (1999a) *J. Korean Soc. Hort. Sci.,* **40:** 231-234.

Shin Hakki., Kim J. Y., Choi, S. T. and Woo, S. J. (1999b) *J. Korean Soc. Hortic. Sci.,* **40:** 368-370.

Shin, H. J. and Lee, J. M. (1979) *J. Korean Soc. Hort. Sci.,* **20:** 111-116.

Shin, H. K., Kang, S. H., Jung, J. W., Yu, C. J. and Kwun, K.C. (1994) *RDA J. Agric. Sci. J. Agric. Sci., Hort.,* **36:** 417-421.

Shinoyama H., Aida R., Ichikawa H., Nomura Y. and Mochizuki A. (2012) *Plant Biotechnol.,* **29**(4): 323-337.

Shinoyama, H. and Mochizuki, A. (2006) *Acta Hortic.,* **714**: 177–184.

Shinoyama, H., Ichikawa, H., Saitoh-Nakashima, M., Saito, M., Aida, R., Ezura H., Yamaguchi, H., Mochizuki, A., Nakase, K., Nishibata, Y., Nomura, Y. and Kamada, H. (2012b) *Acta Hortic.,* **937:** 337-346.

Shinoyama, H., Mochizuki, A., Komano, M., Nomura, Y. and Nagai, T. (2003) *Breed Sci.,* **53**: 359–367.

Shinoyama, H., Sano, T., Saito, M., Ezura, H., Aida, R., Nomura, Y. and Kamada, H. (2012a) *Mol. Breed.* **29**: 285–295.

Simpson, B., Einert, A. E. and Hillman, L. H. (1975) *Flor. Rev.,* **156**: 27-28, 68-69.

Singh, D. B., Mehra, S. and Bensam, N. C. (1999) *J. Orn. Hort.* (New Series), **2:** 92-96.

Singh, J., Nigam. and Rashmi. (2015) *HortFlora Res. Spect.,* **4**(3): 273-276.

Singh, N. and Bala, M. (2017) *Indian J. Agric. Sci.,* **87**(5): 644-647.

Singh, R., Raghunath, S., Nirmal, M., Hussain, W.S., Sarkar, H.K. and Singh, N. B. (2016) *Int. J. Agric. Environ. Biotechnol.,* **9**(3): 345-352.

Singh, S., Dubey, R.K. and Kukal, S.S. (2016) *Commun. Soil Sci. Plan.,* **47**(16): 1856-1865.

Singh, T. and Bala, M. (2018) *Agric. Res. J.,* **55**(2): 386-388.

Singh. K., Ranjit, S., Ramesh, K. and Neena, C. (2010) *J. Ornam. Hortic.,* **13**(3): 233-236.

Skiryin, R. M. (1978) *Euphytica*, **9**: 241-266.

Skogen, D., Erikson, A. B. and Nilson, S. (1982) *Scientia Hort.*, **18**: 87-92.

Sluarkiewicz, J. A. Zenkteler, M. and Padlewska, B. (1982) *Acta Socictatis Botanicorum Poloniae*, **51:** 173-78.

Smith, C. A. and Hall, D. A. (1994) *Acta Horticulturae*, **361**: 159-166.

Smith, E. F., Roberts, A. V. and Mottley, J. (1990) *Plant Cell, Tissue and Organ Cult.*, **21**: 133-140.

Sobers, E. K. (1965) *Proc. Fla. Sta. Hort. Soc.*, **78**: 419-422.

Soffer, H. and Burger, D.W. (1988) *Acta Horticulturae*, **230**: 261-269.

Soffer, H., Mayak, S., Burger, D.W. and Reid, M.S. (1989) *Plant Physiology*, **89:** 165-168.

Sogni, S. (1989) *Colture Protette*, **18:** 27-30.

Son, K. C., Byoun, H. J. and Chae, S. C. (1994) *J. Korean Soc. Hort. Sci.*, **35:** 301-308.

Song, C. Y. and Lee, J. S. (1999) *J. Korean Soc. Hort. Sci.*, **40:** 491-594.

Song, C. Y., Lee, J. S., Ko J. Y., Shin, H. and Kwon, Y. S. (1995) *RDA J. Agric. Sci., Hort.*, **37:** 396-400.

Song, X., Kang, G., Guangxun, F., Xiaogang, Z., Zhilan, L. and Silan, D. (2018) *Hort Sci.*, **53**(9): 1258-1265.

Sopp, P. I., Gillespie, A. T. and Palmer, A. (1990) *Crop Protection*, **9:** 177-184.

Sramek, F. and Dubský, M. (2009) *Acta Hortic.*, **819**: 387-394.

Srilatha, V., Kumar, K.S. and Kiran, Y.D. (2015) *Agric. Sci. Dig.*, **35**(2): 155-157.

Sriram, S, Chandran, N.K., Kumar, Rajiv. and Krishna, R.M. (2015) **32**(8). *New Dis. Rep.*, **32**(8). http: //dx.doi.org/10.5197/j.2044-0588.2015.032.008.

Srivastava, R., Sharma, G., Chand, S. (2015) *J. Hortic.*, **2:** 123. doi: 10.4172/2376-0354.1000123.

Staay, M. van der (1992) *Mededelingen van de Faculteit Landbouwwetenschappen, Rijksuniversiteit*, Gent, **57:** 473-478.

Staby, G. L., Robertson, J. L., Kiplinger, D. C. and Conover, C. A. (1978) *Chain of Life, Ohio Florists Assoc.*, Ohio State Univ., Columbus.

Stack, P. A., Drummond, F. A. and Stack, L. B. (1998) *HortScience*, **33:** 710-715.

Stanley, C. J. and Cockshull, K. E. (1989) *J. Hort. Sci.*, **64:** 341-350.

Stanly, C. D., Harbaugh, B. K. and Price, J. F. (1982) *HortScience*, **17**: 812-813.

Stapel, O. and Guerrand, J. (2012) *Acta Hortic.*, **937:** 355-359.

Starman, T. W. (1990) *HortScience*, **25:** 935-937.

Statens and Hagebruksskole, R. A. (1966) *Gartnerysket*, **56**: 820-821, 825.

Stefanis, J. P. and Langhans, R. W. (1982) *J. Amer. Soc. Hort. Sci.*, **107**: 461-464.

Stefanis, J. P. and Langhans, R. W. (1983) *HortScience*, **18**: 202-204.

Stenseth, C. (1988) *Gartneryrket*, **78:** 383-385.

Stewart, R. N. and Derman, H. (1970) *Amer. J. Bot.*, **57**: 1061.

Stich, K., Halbwrith, H., Wurst, F. and Forkmann, G. (1997) *Zeitschrift fur Naturforschung, Sect. C, Biosciences*, **52:** 153-158.

Stich, K., Wurst, F. and Forkmann, G. (1994) *Acta Horticulturae*, **381**: 152-153.

Strefeler, M. S., Anderson, N. O. and Ascher, P. D. (1996) *Hort. Technology*, **6**: 251-253.

Strider, D. L. (1994) *Revista Chapingo, Serie Horticultura*, **1**: 131-136.

Strider, D. L. (1995) *Revista Chapingo, Serie Horticultura*, **1:** 109-112.

Stringheta, A. C. O., Cardosa, A. A., Lopes, L. C. and Gentes, L. E. F. (1999) *Revista Seres*, **264**: 175-188.

Strojny, Z. (1989a) *Seria B., Rosling Ozdobne*, **14**: 117-133.

Strojny, Z. (1989b) *Prace Instytutu Sadownictwa i kwiaciarstwa w skierniewicach*, Seria B, Rosliny Ozdobne, **11**: 79-97.

Stuart, N. W. and Cathey, H. M. (1959) *Plant Physiol.*, **34**: 16.

Sturm, M. (1966) *Dtsche Gartnerborse*, **66**: 669-671.

Su, J., Yang, X., Zhang, F., Wu, S., Xiong, S., Shi, L., Guan, Z., Fan, W. and Chen F. (2017) *Planta.*, doi: 10.1007/s00425-017-2833-2.

Su, J.K., Lee, K.T., Choi, H.E., Ha, T.J., Nam, J.H., Su, Y.H., .Kim, K.S. (2015) *Food Nutr. Res.*, **61**(1): 1–9.

Sugiura, H. and Fujita, M. (2003) *J. Trop. Agric.*, **47**(2): 109-116.

Sultanpuri, A, Dhiman, S.R., Gupta, Y.C., Kakar, R., Thakur, P., Kashyap, B. and Sharma, M. (2018) *J. Hill Agric.*, **9**(2): 149-152.

Sumitomo, K., Kunitake, T., Douzono, M., Onozaki, T., Shibata, M. and Hisamatsu, T. (2008) *J. Hortic. Sci. Biotech.*, **83**(6): 809-815.

Sun, W., Yang, X., Su, J., Guan, Z. and Zhang, F. (2019) *Sci. Hortic.*, **256.** https: // doi.org/10.1016/j.scienta.2019.108598

SunZhao, F., Li M., Chen, L., Li S. R. and Li C. S. (1999) *Acta Horticultural Sinica*, **26:** 391-396.

Suresh, G.J., Jayaprasad, K.V. and Chikkasubanna, V. (2010) *Asian J. Hortic.*, **5**(1): 96-98.

Sutter, E. and Langhans, R. W. (1981) *Ann. Bot.*, **48**: 559-568.

Takatsu, Y., Nishizawa, Y., Hibi, T. and Akutsu, K. (1999) *Scientia Hort.*, **82:** 113-123

Talukdar, M. C. and Paswan, L. (1994) *Hort. J.*, **7:** 141-144.

Talukdar, M. C. and Paswan, L. (1996) *Hort. J.*, **9:** 155-158.

Talukdar, M. C. and Paswan, L. (1998) *J. Orn. Hort.* (New Series), **1:** 11-116.

Tammen, J. (1965) *Sci. for the Farmer*, **12**: 13.

Tanaka, M., Hirano, T., Goi, M., Higashiura, T., Sasahara, H. and Murasaki, K. (1992) *Acta Horticulturae*, **300**: 77-84.

Tanigawa, T., Kobayashi, Y. and Matsui, H. (1997) *Environment Control in Biol.*, **35:** 107-115.

Tanigawa, T., Kobayashi, Y., Matsui, H. and Matsuda, Y. (1999) *J. Japanese Soc. Hort. Sci.*, **68:** 655-660.

Tanigawa, T., Kobayashi, Y., Matusui, H. and Sakai, Y. (1995) *J. Japanese Soc. Hort. Sci.*, **64:** 417-424.

Tanigawa, T., Kunitake, T., Nakamura, C., Yamada, A., Suyama, T. and Saeki, K. (2010) *Hortic. Res., Japan*, **9**(1): 31-38.

Tatenam, A., Rajapakse, N.C., Fernandez, R.T. and Rieck, J.R. (2000) *J. Amer. Soc. Hort. Sci.*, **125**: 673-678.

Tayade, R.G., Tipre, P.G., Dandge, A.R., Ingle, S.D. and Bhagat, A.K. (2019) *Int. J. Agric. Sci.*, **11**(2): 7732-7736.

Tayama, H. K. and Carver, S. A. (1992) *Hort. Technology*, **2:** 476-479.

Tayama, H. K. and Miller, R. O. (1963) *Ohio. Fl. Grs. Ass. Bull.*, **410**: 6.

Telem, R.S., Sadhukhan, R., Mandal, N., Wani, S.H., Sarkar, H.K. and Naorem, B.S. (2016) *Int. J. Agric. Environ. Biotechnol.*, 345-352.

Terra, S.B., Ferreira, A.A.F., Peil, R.M.N., Stumpf, E.R.T, Beckmann-cavalcante, M.Z. and Cavalcante, Í.H.L. (2011) *Acta Sci.-Agron.*, **33**(3: 465-471.

Teynor, T. M., Ascher, P. D., Widmer, R. E. and Luby, J. J. (1989a) *Euphytica*, **42:** 297-305.

Teynor, T. M., Ascher, P. D., Widmer, R. E. and Luby, J. J. (1989b) *Euphytica*, **42:** 199-207.

Thakur, N., Nair, S.A., Sriram, S. and Kumar, R. (2019) *Indian Phytopathol.*, **72**(3): 513–518.

Thakur, T. (2018) *Curr. Sci.*, **115**(6): 1198-1203.

Thakur, T. and Harpinder, S.G. (2019) *J. Plant Nutr.*, **42**(15): 1861-1867.

Thao, T.N., Nhat, N.D., Tinh, H.H., Thinh, N.H. and Tien, D.T. (2018) *The 1st Rajamangala Surin International conference "Academic Network Bridge through Research"*. AO1040.

Thomas, P., Panicker, B., Janakiram, T. and Sathyanarayana, B.N. (2009)*J. Hortic. Sci. and Biotechnol.*, 653-659.

Tian, X. F., Liu, Z. Q. and Zhang, J. F. (1993) *Acta Horticulturae Sinicia*, **20**: 101-102.

Tjia, B. O. S., Kiplinger, D. C. and Kozel, P. C. (1973) *J. Amer. Soc. Hort. Sci.*, **98**: 186-193.

Tjia, B. O. S., Kozel, P. C. and Kiplinger, D. C. (1969) *Flor. Rev.*, **144**: 21-23.

Tomati, U., Galli, E., Buffone, R., Cegarra Rosique, J. and Roig, A. (1993) *Acta Horticulturae*, **342**: 175-181.

Tompkins, C. M. and Middleton, J. T. (1950) *Bull. Torrey Bot. Club*, **77:** 287-293.

Trigiano, R. N., Scott, M. N. and Caetano – Anolles, G. (1998) *J. Amer. Soc. Hort. Sci.*, **123**: 642-646.

Tripathy, A., Mandal, T., Chakraborti, K. and Swathi, K. (2016) *Bioinfolet.*, **13**(3): 501-503.

Trisnawaty, A.R., Muhammad, R., Rinaldi, S., Arjunayanti, A. and Muhammad, D.R., Raham, S.K. and Feranita, H. (2016) *Int. J. Agric. Syst.*, **4**(1): 52-61.

Tschen, J. S. M. (1991) *Plant Protection Bull.* (Taipei), **33:** 56-62.

Tsevegsuren, N., Christie W.W. and Losel, D. (1998) *Lipids*, **33:** 723-727.

Tsujita, M. J., Kiplinger, D. C. and Tayama, H. K. (1974) *HortScience*, **9**: 294.

Tsvetkov, I. and Krstanova, S. (1991) *Rastenievdni Nauki*, . **28:** 57-61.

Tutty, J. R., Hicklenton, P. R., Kristie, D. N and McRae, K.B. (1994) *J. Amer. Soc. Hort. Sci.*, **119:** 138-143.

Tymoszuk, A., Miler, N., Zalewska, M. and Borawska, M. (2009) *Electron. J. Pol. Agric. Univ.*, **12**(3): #03. http: //www.ejpau.media.pl/volume12/issue3/art-03.html

Uda, M., Amano, M., Shibata, M. and Kawata, J. (1988) *Bull. Nat. Res. Inst. of Vegetables, Ornamental Plants and Tea*, **A**: 239-244.

Ueno, K., Nagayoshi, S., Imakiire, S., Koriyama, K., Minami, T., Tanaka, A., Hase, Y. and Matsumoto, T. (2013) *Hortic. Res. (Japan)*, **12**(3): 245-254.

Ukiya M., Akihisa, T., Yasukawa, K., Kasahara Y., Kimura, Y., Koike, K., Nikaido, T. and Takido, M. (2001) *J. Agric. Food Chem.*, **49**: 3187-3197.

Ukiya, M., Akihisa, T., Tokuda, H., Suzuki, H., Mukainaka T., Ichiishi E., Yasukawa, K., Kasahara, Y. and Nishino, H. (2002) *Cancer Lett.*, **177**: 7–12.

Ullas, S.P., Namita, Singh, K.P., Kundu, A., Panwar, S., Krishnan, G.S. and Kumar, G. (2018a). *Indian J. Agric. Sci.*, **88**(3): 393-399.

Ullas, S.P., Namita, Singh, K.P., Panwar, S., Kundu, A., Krishnan, G.S. and Kumar, G. (2018b) *Indian J. Agric. Sci.*, **88**(2): 228–233.

Ulrich, R., Werres, S. and Götz, M. (2017) **35**(6): 6.

Urban, L. A., Sheman, J. M., Moyer, J. W. and Daub, M. E. (1994) *Plant Sci.*, **98**: 69-79.

Urban, L. A., Sherman, J. M., Moyer, J.W. and Daub, M. E. (1992) *Proc. In vitro culture and Horticulture Breeding, June 28-July 2, 1992*, Baltimore, MD, p. 49 (Abstract).

Ushiro, T. and Yamamoto, N. (1996) *Bull. Hiroshima Prefectural Agric. Res. Center*, **64**: 45-53.

Valizadeh, M., Kazemitabar, S.K. and Jongsma, M.A. (2012)*Int. J. Plant Breed.Genet.*, **6**(4): 168-181.

Van der Ploeg, A., Kularathne, R.J.K.N., Carvalho, S.M.P. and Heuvelink, Ep. (2007) *J. Amer. Soc. Hort.Sci.*, **132**(1): 52–59.

Van Wordragen, M. F., Ouwerkerk, P. B. F. and Dons, H. J. M. (1992) *Plant Cell Tissue Organ Cult.*, **30**:

Van Wordragen, M. G., de Jong, J. and Dons, H. J. (1989) *J. Cell Biochem.*, 271 (Abstract).

Vanlalruati, Anand, P., Kumar, G. and Tiwari, A.K. (2019) *Indian J. Agric. Sci.*, **89**(1): 41-45.

Vanlalruati, Sindhu, S.S., Anand, P. and Kumar, G. (2019) *Indian J. Agric. Sci.*, **89**(8): 1282-1286.

Vehrs, S. L. and Parrella, M. P. (1991) *California Agric.*, **45:** 28-29.

Velmurugan, S., Vadivel, E. and Shanmugasundaram, M. (2010) *J. Ornam. Hortic.*, **13**(3): 216-222.

Verhagen, J. B. G. M. (1993) *Acta Horticulturae*, **342**: 221-227.

Verlodt, I., Verschaeren, P. and Oyaert, E. (1998) In: *14th Int. Cong. on plastic in agriculture*, Tel Aviv, Israel, March 1997(eds. Ben-Yehoshua, S) Jerusalem, Israel; Laser Pages Publishing, pp. 297-308.

Verma, A. K., Singh, M., Shukla, .S and Banerji, B.K. (2016)*J. Ornam. Hortic.*, **19**(1-2): 7-12.

Verma, A.K. and Sahukla, S. (2018) *J. Ornam. Hort.*, **21**(1-2): 31-38.

Verma, A.K., Prasad, K.V., Singh, S.K. and Kumar, S. (2012) *Indian J. Hortic.*, **69**(4): 562-567.

Vetrivel, T. and Jawaharlal, M. (2014) *Trends in Biosci.*, **7**(14): 812-1815.

Vetrivel, T., Jawaharlal, M., Arulmozhiyan, R. and Kannan, M. (2017) *Ecol. Environ. Conserv.*, **23**(4): 2297-2301.

Vince, D. (1953) *Agriculture*, London, **60**: 68.

Vince, D. (1955a) *J. Hort. Sci.*, **30**: 34.

Vince, D. (1955b) *Grower*, London, **44**: 1174.

Volkering, M. and Strauch, K. H. (1987) *Gb+ Gw*, **87:** 1898-1901.

Waffenschmidt, H. (1988) *Gartnerborse und Gartenwelt*, **88:** 1076-1077.

Walz, F. and Horn, W. (1997) *Acta Horticulturae*, **418**: 53-57.

Wang, C. L. and Lin, R. J. (1984) *J. Agric. Res. China*, **33:** 88-93.

Wang, F., Miao, M., Xia, H., Yang, L. G., Wang, S. K., and Sun, G. J. (2017) *Food Nutr. Res.*, **61**(1): 1-9.

Wang, W. K., Zhou, C. L. and Dai, S. L. (1999) *J. Beijing Forestry Univ.*, **21:** 92-95.

Warmenhoven, M. G. and Baas, R. (1995) *Acta Horticulturae*, **401**: 393-400.

Warrington, I. J., Norton, R. A. (1991) *J. Amer. Soc. Hort. Sci.*, **116:** 544-551.

Watanabe, K. (1977) *J. Hered.*, **68**: 317-320.

Waters, W. E. (1962) *Proc. F1a Sta. Hort. Soc.*, **75**: 460-463.

Waters, W. E. (1965) *Proc. Fla. Sta. St. Hort. Soc.*, **78**: 383-386.

Waters, W. E. and Conover, C. A. (1969) *Univ. Fla. Agr. Expt. Bull.*, No. 730.

Watson, D. P. and Andrews, P. S. (1953) *Proc. Amer. Soc. Hort. Sci.*, **61**: 551.

Waxman, S. (1967-68) *Comb. Proc. Int. Plant. Prop. Soc.*, **17**: 316.

Weilan., Xu, S., and Zhu. X. (2015) *Bangladesh J. Bot.*, **44**(4): 675-678.

Weimin, F. and Fadi, C. (2018) *Planta.*, **247**(4): 899-924.

Whealy, C. A., Nell, T. A. and Barrett, J. E. (1987) *J. Amer. Soc. Hort. Sci.*, **112**: 464-468.

Whipps, J. M. (1993) *Annals Applied Biol.*, **122**: 173-187.

White, J. W. (1965) *Pa. Flower Grs. Bull.*, **170**: 3-4.

White, J. W. (1968) *Grower Talks*, **31**: 1-7.

Wilcox, H. J. (1963) *Proc. 2nd Brit, Insecticides and Fungicides Conf.*, Brighton., pp. 291-299.

Wilkins, H. F., Healy, W. E. and Grueber, K. L. (1990) *J. Amer. Soc. Hort. Sci.*, **115:** 732-736.

Wille, D. (1974) *Rijksuniversiteit Gent*, **39**: 1520-1524.

Williams, K. A. and Nelson, P. V. (1996) *HortScience*, **31:** 381-384.

Williams, K. A. and Nelson, P. V. (1997) *J. Amer. Soc. Hort. Sci.*, **122:** 703-708.

Willits, D. H. and Bailey, D. A. (1994) *Amer. Soc. Agric. Engineers*, : 12.

Willits, D. H. and Bailey, D. A. (2000) *Scientia Hort.*, **83:** 325-330.

Wilson, D. P. and Finlay, A.R. (1995) *Acta Horticulturae*, **401**: 185-192.

Winspear, K. W. (1977) *A. R. C. Res. Rev.*, **3**:

Wise, K, Harsharn, G. and Jamie, S. (2020) *Sci. Hortic.*, **263**. doi: 10.1016/j.scienta.2019.109108.

Wodecki, M. J. and Holcomb, J. (1989) *Bull. Pennsylvania Flower Growers*, **391**: 1-3

Wojtowicz, M. and Labanowski, G. S. (1994) *Zeszyty Naukowe Instytutu Sadownictwa Kwiaciarstwa w Skierniewicach*, **1:** 139-146.

Wolff, K. and Peters- Van Rijn, J. (1993) *Heredity*, **71:** 335-341.

Wolff, K. and Peters-Van Rijn, J. and Hofstra, H. (1994) *Theor. Appl. Genet.*, **88:** 472-478.

Won, SY., Kwon, SJ., Lee, T.H., Jung, J.A., Kim, J.S., Kang, S.H. and Sohn, S.H. (2017) *Plant Mol. Biol.*, **95**(4-5): 451-461.

Woo, J.H., Han, Y. Y., Sim, Y. G., Choi, K. B., Choi, B. and Kwack, B. H. (1997) *RDA J. Hort. Sci.*, **39**: 101-106.

Wordragen, M. F. van and Honee, G. (1990) *Eucarpia*, pp. 146-149.

Wordragen, M. F. van, Honee, G. and Dons, H. J. M. (1993) *Transgenic Res.*, 2: 170-180.

Wott, J. A. (1969) *Diss. Abstr.*, 29: 3989-3990.

Wott, J. A. and Tukey, H. B. (1969) *Bull. N. Y. Sta. Flower Grs.*, **284**: 1-4.

Wright, R.D., Jackson, B.E., Browder, J.F. and Latimer, J.G. (2008) *Horttechnology.*, **18**(1): 111-115.

Wu Liang-Yu, Hong-Zhou Gao, Xun-Lei Wang, Jian-Hui Ye, Jian-Liang Lu and Yue-Rong Liang. *J.* (2010) *Med. Plant Res.*, **4**(5): 421-426.

Wu, W. S. (1996) *Plant Pathology Bull.*, **5:** 85-90.

Wyatt, I. J. (1965) *Ann. Appl. Bioi.*, **56**: 439-459.

Wyatt, I. J. (1966) *Proc. 3rd Brit. Insecticides and Fungicides Conf.*, p. 52.

Xia, W. X., Guo, D. and Jin, M. (1994) *Acta Agriculturae Zhejiangensis*, **6:** 171-175.

Xiaowei, L., Yujie, Y., Sagheer, A., Ming, S., Cunquan, Y., Tangchun, Z., Yu, H., Tangren, C., Jia, W. and Qixiang, Z. (2019) *PLoS One.*, doi: 10.1371/journal.pone.0225241.

Xu, W.H., Gao, H.Q. and Chen, H.J. (2000) *J. Zhejiang Forestry College*, **17**: 37-41.

Yadav, L. P. (1983) *Ph. D. thesis*, Bidhan Chandra Krishi Viswavidyalaya, India.

Yamakawa, K. and Ichikawa, S. (1970) *Tech. News. Inst. Rad. Breeding*, **6**: 2.

Yamashita, I., Dan, K. and Ikeda, H. (1999) *J. Japanese Soc. Hort. Sci.*, **68:** 622-627.

Yang M. F., Liu, X. D. and Pan, X. F. (1997) *J. Northeast Forestry Univ.*, **25:** 87-91.

Yang, X., Yangyang Wu, Jiangshuo Su, ni Ao, Zhiyong Guan, Jiafu Jiang, Sumei Chen, Weimin Fang, Fadi Chen and Fei Zhang. (2019) Mol *Breed.*, **39**(3): 1-12.

Yantcheva, A., Vlahova, M. and Antanossov, A. (1998) *Plant Cell Rep.*, **18**: 148-152.

Yassin, G. Md and Pappiah, C. M. (1990) *South Indian Horti.*, **38**: 232-233.

Yepes, L. M., Mittak, V., Slightom, J. L., Pang, S. Z., Gonsalves, D. and Fischer, G. (1999) *Acta Horticulturae*, **482**: 209-218.

Yewale, A. K., Belorkar, P. V., Bhattacharya, J., Chanekar, M. A. and Chimurkar, B. S. (1998b) *J. Soils and Crops*, **8:** 73-75.

Yewale, A.K., Belorkar, P. V., Chanekar, M. A., Bhattacharya J. and Chimurkar, B. S. (1998a) *J. Soils and Crops*, **8:** 82-84.

Yoo, Y. K. and Kang, S. W. (1999) *J. Korean Soc. Hort. Sci.*, **40:** 515-519.

Yoo, Y. K., Kang, S. W. and Kim, H. K. (1999) *J. Korean Soc. Hort. Sci.*, **40:** 598-602.

Yoon, H.S., Goto, T. and Kageyama, Y. (2000a) *Environ. Control Biol.*, **38**: 121-127.

Yoon, H.S., Goto, T. and Kageyama, Y. (2000b) *J. Japanese Soc. Hort. Sci.*, **69**: 255-260.

Yoon, H.S., Goto, T. and Kageyama, Y. (2000c) *J. Japanese Soc. Hort. Sci.*, **69**: 416-422.

Yoosumran, V., Ruamrungsri, S., Duangkongsan, W. and Kanjana, S. (2018) *Int. J. Agril. Tech.*, **14**(1): 73-82.

You, J., Chen, J. D. and Su Z. (1996) *J. Hebei Agric. Univ.*, **19:** 51-55.

Zacharioudakis, J. N. and Larson, R. A. (1976) *HortScience*, **11**: 36.

Zagarski, J. S. Ascher, P. D. and Widmer, R. E. (1983) *Euphytica*, **32**: 1-7.

Zalewska, M, Justyna, L.R. and Natalia, M. (2007) *Sci. Hortic.*, **113**(1): 70-73.

Zalewska, M. (1989) *Prace Instytutu Sadownictwa i kwiaciarstwa w Skierniewicach, Seria B, Rosliny Ozdobne*, **11:** 73-78.

Zalewska, M. (1994) *Zeszyty Problemowe Postepow Nauk Rdnicz*, **414**: 353-362.

Zalewska, M. (1998) *Folia Universitatis Agriculturae Stetinensis Agricultura*, **70**: 149-155.

Zalewska, M. and Wozny, A. (2006) *Electron. J. Pol. Agric. Univ.*, **9**(4): #44. doi: http: //www.ejpau.media.pl/volume9/issue4/art-44.html.

Zalewska, M., Miler, N. and Wenda-piesik, A. (2010) *J. Hortic. Sci. Biotech.*, **85**(4): 362-366.

Zandarski, J. (1998) *Ochrona Roslin*, **42:** 18-19.

Zeng. J., Jing, S., Yang, X., Fadi, C., Jiafu, J., Weiming, F. and Sumei, C. (2013) *Hort*, ., **48**(10): 1231- 1234.

Zerche, S. and Kuchenbuch, R. (1995) *Zeitschrift fur Pflanzenernahrung und Bodenkunde*, **158**: 393-398.

Zerche, S., Kadner, R. and Druge, U. (1999) *Gartenbauwissenschaft*, **64**: 272-278.

Zhang, F., Jingya, Z, Sujuan, X, Weimin, F., Fadi, C, Nianjun, T.M.G. (2017) *Int. J. Genomics.*, doi: 10.1155/2017/6790478.

Zhang, F., Wen, D., Lulu, H., Aiping, S., Haibin, W., Weimin, F., Fadi, Chen., and Nianjun, T. (2015) *PLoS One.*, doi: https: //doi.org/10.1155/2017/6790478.

Zhang, F., Sumei, C., Jiafu, J., Zhiyong, G., Weimin, F. and Fadi, C. (2013) *PLoS One.*, doi: 10.1371/journal.pone.0083023.

Zhang, H. Y. and Graves, W. R. (1995) *Hort. Technology*, **5**: 265-268.

Zhang, M., He, H., Qing, W. and Silan, D. (2018) *Hort Sci.*, **53**(4): 421-426.

Zhang, N., He, Z., He, S. and Jing, P. (2019) *Food Res. Int.*, **116**: 810-818.

Zhang, W., Fukai, S. and Goi, M. (1998) *J. Japanese Soc. Hort. Sci.*, **67:** 347-351.

Zhang, Y. S., Yu, C. M., Wei, L. N., Lu, N. L., Ling, Z. M., Shou, W. L., Chen, L. P. and He, S. M. (1998) *Acta Agriculturae Zhejiangensis*, **10:** 158-160.

Zhang, Y., Lee, Y.H., Nogoy, K.M., Choi, C.W., Kim, D.H., Li, X.Z. and Choi, S.H. (2017) *Korean J. Environ. Agric.*, **44**(4): 604-612.

Zhang, Z. H. and Guo, W. M. (1998) *Adv.* **2:** 697-701.

Zhao, J.Z., Chen, J.M., Chen, S.B., Liu, Q.H. and Liu, Z. (1999) *Acta Horticulturae Sinica*, **26**: 327-330.

Zheng, C, Jin, G., Ohno, H., Hara, T. and Matsui, S. (2004) *J. Jpn. Soc. Hortic. Sci.*, **73**(6): 568-573.

Zheng, C.P., Q. Dong, H.J. Chen, Q.F. and Cong, K.D. (2015) *Carbohydr. Polym.*, **130**: 113–121.

Zheng, L. and Guo, W. M. (1998) *Adv. Hort.*, **2**: 705-710.

Zheng, Z.L., Yang, Z.B., Jang, J.C. and Metzger, J.D. (2001) *J. Amer. Hort. Soc. Sci.*, **126**: 19-26.

Zheng, Y., Wang, X.S. and Fang, J. (2006) *J. Asian Nat. Prod. Res.*, **8**(3), 217–222.

Zhou, J and Li, H. J. (1993) *Acta Horticulturae Sinicia*, **20:** 379-383.

Zimmer, I. (1994) *Gartenbauwissenschaft*, **59:** 183-85.

Zimmer, K. and Preil, W. (1987) *Deutscher Gartenbau*, **41:** 1938-1941.

5

JASMINE

Sujatha A. Nair, S. Ramadas, G. B. Peter and S. Muthuswami

1.0 Introduction

Jasmine (*Jasminum* spp.), is one of the important traditional flower crops cultivated commercially for its fragrant flowers. Jasmine essential oil is widely used in cosmetics, the pharmaceutical industry, perfumery and aromatherapy. Jasmines are mostly shrubs, but some species are climbers. The diversity in the growth habit of jasmine plants along with attractive foliage and fragrant flowers make them a perfect choice as ornamental plant for landscaping.

2.0 Importance and Uses

Jasmines have been cultivated for various purposes since very early times in many tropical and subtropical countries. In India, they are cultivated mainly in the states of Tamil Nadu and Karnataka and to some extent in the states of Andhra Pradesh, Maharashtra, Uttar Pradesh, West Bengal, Rajasthan, Gujarat and in few of the north-eastern states. The area under jasmine in India was estimated to be about 25500 hectares and production 199000 MT in 2016-17 (Anon., 2020a). India exports jasmine flowers to the neighbouring countries like Sri Lanka, Singapore, Malaysia, the Middle east countries and to the United States.

Jasmine oil is predominantly extracted from the flowers of the Spanish jasmine (*J. grandiflorum*). The estimated world production is 15,000 kg jasmine essential oil, and the major suppliers are India, Egypt, China and Morocco. The two main producers of *Jasminum grandiflorum* concrete are India and Egypt. Together they account for about 95 per cent of the market share. In India the estimated production of concrete in 2014 was about 5.5 to 6 tons, while Egypt produced as many as 4.5 tons. (Fakhry, 2014). Morocco and China are the other two major producers. The estimated production of *Jasmine sambac* concrete in 2014 was about 3 to 3.5 tons. The annual demand of *Jasmine sambac* concrete is around three tons. While India is the primary source, China is also a producer. Egypt, though not a producer of *Jasminum sambac*, currently, used be an important producer in the 1970s. Belgium, Netherlands, Italy, Algeria, Turkey, France and Tunisia are the other important jasmine oil producing countries of the world. The international market prices of jasmine concrete and absolute are around Rs.40,000/kg and Rs.80,000/kg, respectively.

Flowers and buds are used for making garlands, bouquets, *veni*, used for decorating hair of women and for religious offerings. Jasmine oil is regarded as unique as it blends well with other floral extracts and it is highly valued throughout the world for production of high-grade perfumes. The oil is also used in production of *attars*, soap, perfumed hair oils, handkerchief perfumes, mouthwashes, dentifrices, bath salts, incense sticks, tobacco and in many cosmetics. The odour of jasmine flowers cannot be imitated by any known synthetic aromatic chemical or natural isolates thus giving it a unique status in the perfume world.

Jasmine blossoms have been in use in India, China and countries of the mystic orient for ceremonial purposes and as religious offerings since time immemorial. The flowers of the Arabian jasmine (*J. sambac*) are reported to be used in China for flavouring tea and extensively used for hair ornamentation by the women. In

Malaysia, the flowers are said to be used for scenting coconut hair oil (Anon., 1972). They also contain yellow pigments and hence used as a substitute for saffron. The flowers and other parts of the plant also find use in medicines. *Jasminum grandiflorum* L. is reported to be spasmolytic, anti-inflammatory, anti-microbial, antioxidant, antiulcer, cytoprotective, chemoprotective, wound healing and anti-acne (Pradhan, 2015). *Jasminum lanceolarium* Roxb. stems and roots have been used for the treatment of rheumatism and fever while the leaves are used as an anti-inflammatory agent to relieve pain in traditional Chinese medicine (Yan *et al.*, 2015).

Aqueous leachates of *Jasminum officinale* demonstrated promising allelopathic potential by inhibiting seed germination and radicle elongation of ragweed, ryegrass, and Johnson grass according to Rodino *et al.* (2017).

3.0 Origin, History and Distribution

3.1.0 Origin and History

The genus *Jasminum*, is distributed in tropical to temperate regions of the Old World and comprises of 200 species (Mabberley, 2017). This genus is found commonly in deciduous and evergreen forests as climbing shrubs with flowers generally in white, pink or yellow colours and is sweet-scented. Among the large number of species existing only four species have attained importance in commercial cultivation. The Arabian or Tuscan jasmine (*J. sambac*) is considered as a native of the East Indies, contrary to the opinion as to its original home being the region west of India (Anon., 1959). The Royal jasmine or common white jasmine or Poet's jasmine (*J. officinale*) is considered to be of Persian origin. The Spanish jasmine or Catalonion jasmine (*J. grandiflorum*) is some-times referred to as synonymous with *J. officinale* or *J. officinale* var. *grandiflorum*. Cooke (1905) stated that *J. officinale* is a native of Kashmir, Afghanistan and Persian while Baker and Bakhijean van der Brink (1905) mentioned it as a native of continental Asia. According to another report, *J. officinale* is a native of northern India and Persia and *J. officinale* var. *grandiflorum* is a native of the north-west Himalayas (Anon., 1972). However, the species status given to the Royal jasmine, *J. officinale* is itself questionable since Veluswamy *et al.* (1973) in their mutation studies have observed three white flowering plants from the mutated seeds of *J. grandiflorum* pink which paves way to claim that *J. officinale* could be a natural bud sport of *J. grandiflorum* type which has attained the species status as *J. officinale* but there is very little difference between these two species except for flower bud colour. The third species *J. auriculatum* is distributed in the west peninsula of India (Cooke, 1905), more specifically in Circars, Deccan Carnatic areas and in the southern districts of Travancore up to dry slopes of the Western Ghats (Gamble, 1957). According to Haines (1961), *J. auriculatum* is a native of South India and the Central Provinces. *Jasminum multiflorum* is native to India and south eastern Asia widely grown across the tropics for its white, mildly scented flowers. It has been introduced to numerous countries, having escaped cultivation and become established in the USA (Florida), Jamaica, Dominican Republic, Hawaii, Puerto Rico, Seychelles, Australia (Queensland) and perhaps other countries (Anon., 2019).

The plant is believed to have been brought from the southern foot hills of the Himalayas, their native habitat, to North Africa and Spain by the conquering Moors.

Uses of Jasmine

Jasmine border

In hanging basket

Jasmine in pot

In window boxes

Jasmine as climber

Perfume from jasmine

Uses of Jasmine

Jasmine in interior decoration

Jasmine on pergola

The name jasmine is of Arabic origin and is believed to have been derived from Yasmin or Yasmyn. In his book, *Records of the plants of the Southern Regions*, Chi Han of Tsin dynasty mentioned that jasmine was available in China, towards the end of the third century. It is reported that the height of its popularity reached its peak two to five hundred years ago at Canton and metropolis of southern China. The plant was probably introduced into France from Spain where *J. officinale* var. *grandiflorum* has been in extensive cultivation in the Grasse region of southern France for over two hundred and fifty years producing the world-famous jasmine oil. In the early 1900's, Charles Garnier began experimenting with this plant in Egypt and from there it spread to Syria, Algeria, Sicily, Calabria and Morocco where large plantations gradually sprung up, and today these countries have emerged as competitors to France in jasmine oil industry in the world.

3.2.0 Distribution

Though jasmines are distributed in tropical and subtropical countries of the world, a large number of species are centred around the regions comprising India, China and Malaysia (Anon., 1959). Among these, about 40 species are reported to occur in India. The distribution pattern of species in India, Malaysia and China provides a strong base to claim that India may be one of the principal centres of origin of *Jasminum* species. The references to the three cultivated species, *viz.*, *J. auriculatum*, *J. grandiflorum* and *J. sambac* in ancient Tamil literature belonging to the period of Sangam era prior to 400 BC suggest that South India may possibly be the home of at least some of these species. Gamble (1957) recorded 20 species in the former Madras Presidency State and some of these species are found in Mumbai, Bihar, Orissa, Chota Nagpur, upper Gangetic plains and sub-Himalayan tracts. Ridley (1922) listed 14 species occurring in Malay peninsula while Miquel (1956) reported 17 species in Java, Sumatra, Borneo and Celebes. Subsequently, Baker and Bakhijean van der Brink (1905) reported the existence of 7 new species in Java. The cultivated species of *J. sambac* and *J. grandiflorum* also exist in these countries. Most of them occurred only in wild forms.

Jasminum spp. are native to some and were introduced to other countries according to Anon. (2020b). The details are listed below:

Jasminum spp. are native to the following countries:

Afghanistan, Aldabra, Andaman Islands, Angola, Assam, Bangladesh, Benin, Bismarck Archipelago, Borneo, Botswana, Burkina, Burundi, Cambodia, Cameroon, Cape Provinces, Caprivi Strip, Central African Repu, Chad, China North-Central, China South-Central, China Southeast, Comoros, Congo, Cook Is., Djibouti, East Himalaya, Eritrea, Ethiopia, Fiji, Gabon, Ghana, Guinea, Guinea-Bissau, Gulf of Guinea Is., Hainan, India, Iran, Ivory Coast, Jawa, Kenya, KwaZulu-Natal, Laos, Lesser Sunda Is., Liberia, Madagascar, Madagascar, Madeira, Malawi, Malaya, Mali, Maluku, Marianas, Mozambique, Myanmar, Namibia, Nansei-shoto, Nepal, New Caledonia, New Guinea, New South Wales, Nicobar Is., Niger, Nigeria, Niue, Norfolk Is., Northern Provinces, Northern Territory, Oman, Pakistan, Philippines, Queensland, Rwanda, Samoa, Saudi Arabia, Senegal, Sierra Leone, Society Is., Socotra, Solomon Is., Somalia, South Australia, Sri Lanka, Sudan, Sulawesi,

Sumatera, Swaziland, Tadzhikistan, Taiwan, Tanzania, Thailand, Tibet, Togo, Tonga, Trans Caucasus, Turkey, Uganda, Vanuatu, Victoria, Vietnam, West Himalaya, Western Australia, Yemen, Zambia, Zaïre, Zimbabwe

Jasminum spp. were introduced into the following countries:

Alabama, Algeria, Argentina Northeast, Bahamas, Belize, Bermuda, Bolivia, Brazil Northeast, Brazil Southeast, Cayman Is., Christmas I., Costa Rica, Cuba, Dominican Republic, El Salvador, Florida, France, Guatemala, Haiti, Hawaii, Honduras, Italy, Jamaica, Korea, Leeward Is., Maldives, Mauritius, Mexico Southeast, Netherlands Antilles, Nicaragua, Ogasawara-shoto, Panamá, Portugal, Puerto Rico, Romania, Réunion, Trinidad-Tobago, Turks-Caicos Islands, Venezuelan Antilles, Windward Islands, Yugoslavia

4.0 Species and Cultivars

4.1.0 Species

The genus *Jasminum* belongs to the family Oleaceae and order Oleales. Evergreen or deciduous, climbing, trailing or erect shrubby; leaves are opposite or alternate, simple, trifoliate or pinnate; leaflets entire; flowers white, yellow or rarely reddish, sometimes solitary, more often in cymose clusters of three to many, usually fragrant; corolla tubular with four to nine lobes; stamens two; ovary 2-loculed with 1-4 erect ovules; fruit is a berry and black in colour.

There are various reports about the exact number of species belonging to the genus *Jasminum.* Baker (1877), Cooke (1905) and Duthie (1911) reported that the genus consisted of about 90 species while Bailey (1949) was of the view that there were 100 or more species in jasmines. According to Dickey (1970), *Jasminum* is a genus of more than 200 species. Many of these species are, however, synonyms and about 90 species are considered to be true in existence (Muthukrishnan and Pappiah, 1980). Srivastava (1987) reported 10 genera, 87 species and 15 intra-specific taxa belonging to the family Oleaceae, in India including the Himalaya, the northeast, peninsular regions, and Andaman and Nicobar Islands. Among these, 27 are endemic taxa. A total of 17 taxa of *Jasminum* are listed under various threat categories (Srivastava and Kapoor 1987). However, according to Green (2003) in India, *Jasminum* is represented by 37 species and 15 intra-specifc taxa, of which 12 species are considered endemic to India (Ahmedullah and Nayar 1986; Srivastava and Kapoor 1987; Singh *et al.*, 2015). In Andaman and Nicobar Islands, so far 12 species are known to occur, namely, *J. acuminatssimum, J. andamanicum, J. angustfolium, J. arborescens, J. atenuatum, J. auriculatum, J. caudatum, J. elongatum, J. fexile, J. multiflorum, J. ritchiei* and *J. syringifolium* (Green 2003; Pandey and Diwakar 2008), of which only one species, *J. andamanicum* is endemic to the Andaman group of Islands and is assessed as endangered as per the IUCN red list criteria (Karthigeyan and Murugan, 2018).

The comprehensive list of *Jasminum* spp. documented is listed below (Anon., 2020c):

J. abyssinicum Hochst. ex DC.

J. acuminatum (Lam.) Pers.

J. adenophyllum Wall. ex C.B.Clarke

J. agastyamalayanum Sabeena Asmitha Mulani E.S.S.Kumar & Sibin

J. alongense Gagnep.

J. amabile H.Hara

J. ambiguum Blume

J. amoenum Blume

J. andamanicum N.P.Balakr. & N.G.N

J. angolense Baker

J. angulare Vahl

J. angustifolium (L.) Willd.

J. annamense Wernham

J. anodontum Gagnep.

J. aphanodon Baker

J. apoense Elmer

J. arborescens Roxb.

J. artense Montrouz.

J. attenuatum Roxb. ex G.Don

J. auriculatum Vahl

J. azoricum L.

J. bakeri Scott Elliot

J. batanensis Kiew

J. beesianum Forrest & Diels

J. betchei F.Muell.

J. bhumibolianum Chalermglin

J. brachyscyphum Baker

J. breviflorum Harv.

J. brevilobum DC.

J. brevipetiolatum Duthie

J. calcareum F.Muell.

J. calcicola Kerr

J. calophyllum Wall. ex G.Don

J.campyloneurum Gilg & G.Schellenb.

J. cardiomorphum P.S.Green

J. carinatum Blume

J. carissoides Kerr

J. caudatum Wall. ex Lindl.

J. celebicum Merr.

J. coarctatum Roxb.

J. coffeinum Hand. -Mazz.

J. cordatum Ridl.

J. cordifolium Wall. ex G.Don

J. craibianum Kerr

J. crassifolium Blume

J. cumingii Merr.

J. cuneatum Jongkind

J. curtisii King & Gamble

J. cuspidatum Rottler

J. dallachyi F.Muell.

J. dasyphyllum Gilg & G.Schellenb.

J. decipiens P.S.Green

J. decussatum Wall. ex G.Don

J. degeneri Kobuski

J. dichotomum Vahl

J. didymum G.Forst.

J. dinklagei Gilg & G.Schellenb.

J. dispermum Wall.

J. domatiigerum Lingelsh.

J. duclouxii (H.Lév.) Rehder

J. eberhardtii Gagnep.

J. elatum Pancher ex Guillaumin

J. elegans Knobl.

J. elongatum (P.J.Bergius) Willd.

J. extensum Wall. ex G.Don

J. flavovirens Gilg & G.Schellenb.

J. flexile Vahl

J. fluminense Vell.
J. foveatum R.H.Miao
J. fuchsiifolium Gagnep.
J. gilgianum K.Schum.
J. glaucum (L.f.) Aiton
J. grandiflorum L.
J. greveanum Danguy ex H.Perrier
J. griffithii C.B.Clarke
J. guangxiense B.M.Miao
J. harmandianum Gagnep.
J. hasseltianum Blume
J. hongshuihoense Z.P.Jien ex B.M.Miao
J. insigne Blume *J. insularum* Kerr
J. ixoroides Elmer
J. jenniae W.K.Harris & G.Holmes
J. kajewskii C.T.White
J. kaulbackii P.S.Green
J. kedahense (King & Gamble) Ridl.
J. kerstingii Gilg & G.Schellenb.
J. kitchingii Baker
J. kostermansii Kiew
J. kriegeri Guillaumin
J. kwangense Liben
J. lanceolaria Roxb.
J. lasiosepalum Gilg & G.Schellenb.
J. latipetalum C.B.Clarke
J. laurifolium Roxb. ex Hornem.
J. ledangense Kiew
J. listeri King ex Gage
J. longipetalum King & Gamble
J. longitubum L.C.Chia ex B.M.Miao
J. mackeeorum P.S.Green
J. macrocarpum Merr.
J. magnificum Lingelsh.
J. maingayi C.B.Clarke
J. malabaricum Wight
J. malayanum Kiew
J. marianum DC.
J. melastomifolium Ridl.
J. mesnyi Hance
J. meyeri-johannis Engl.
J. microcalyx Hance
J. molle R.Br.
J. mossamedense Hiern
J. mouilaense Breteler
J. multiflorum (Burm.f.) Andrews
J. multinervosum Kiew
J. multipartitum Hochst.
J. multipetalum Merr.
J. narcissiodorum Gilg & G.Schellenb.
J. nardydorum Breteler
J. neocaledonicum Schltr.
J. nepalense Spreng.
J. nervosum Lour.
J. newtonii Gilg & G.Schellenb.
J. niloticum Gilg
J. nintooides Rehder
J. nobile C.B.Clarke
J. noldeanum Knobl.
J. noumeense Schltr.
J. nudiflorum Lindl.
J. nummularifolium Baker
J. obtusifolium Baker
J. octocuspe Baker *J. officinale* L.
J. oliganthum Quisumb. & Merr.
J. oreophilum Kiew
J. papuasicum Lingelsh.
J. pauciflorum Benth.

J. paucinervium Ridl.

J. pedunculatum Gagnep.

J. pellucidum Airy Shaw

J. peninsulare Kiew

J. pentaneurum Hand. -Mazz.

J. pericallianthum Kobuski

J. perissanthum P.S.Green

J. pierreanum Gagnep.

J. polyanthum Franch.

J. populifolium Blume

J. prainii H.Lév. *J. preussii* Engl. & Knobl.

J. promunturianum Däniker

J. pseudopinnatum Merr. & Rolfe

J. pteropodum H.Perrier

J. puberulum Baker

J. punctulatum Chiov.

J. quinatum Schinz

J. rambayense Kuntze

J. ranongense Kiew

J. rehderianum Kobuski

J. ritchiei C.B.Clarke

J. rufohirtum Gagnep.

J. sambac (L.) Aiton

J. sarawacense King & Gamble

J. scandens (Retz.) Vahl

J. schimperi Vatke

J. sessile A.C.Sm.

J. siamense Craib

J. simplicifolium G. Forst.

J. sinense Hemsl.

J. smilacifolium Griff. ex C.B.Clarke

J. spectabile Ridl.

J. steenisii Kiew

J. stellipilum Kerr

J. stenolobum Rolfe

J. x stephanense É.Lemoine

J. streptopus E.Mey.

J. subglandulosum Kurz

J. syringifolium Wall. ex G.Don

J. tetraquetrum A.Gray

J. thomense Exell

J. tomentosum Knobl.

J. tortuosum Willd.

J. trichotomum B.Heyne ex Roth

J. tubiflorum Roxb.

J. urophyllum Hemsl.

J. verdickii De Wild.

J. vidalii P.S.Green

J. vietnamense B.H.Quang & Joongku Lee

J. waitzianum Blume

J. wengeri C.E.C.Fisch.

J. wrayi King & Gamble

J. yuanjiangense P.Y.Pai

4.2.0 Cultivars

The existing variability in three commercially important species were catalogued at Tamil Nadu Agricultural University, Coimbatore (Raman, 1955b; Alikhan *et al.*, 1970; Veluswamy *et al.*, 1973) and important cultivars under different species are listed below:

J. auriculatum

Commonly known as Juhi, Jui, Malle, Mokggu, Mullai and Ambur Malliage, the flowers of this species of jasmine are utilized for garland and veni making as

Species of Jasmine

Jasminum auriculatum

Jasminum beesianum

Jasminum calophyllum

Jasminum grandiflorum

Jasminum flexile

Jasminum humile

Species of Jasmine

Jasminum multiflorum

Jasminum nitidum

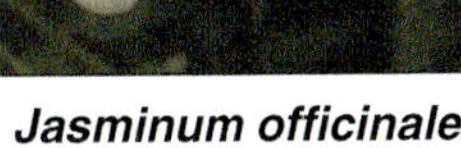

Jasminum officinale

Jasminum primulinum

Jasminum parkeri

Jasminum sambac

Cultivars of Jasmine

Argenteo-variegatum

Mysore Malligae

Parimullai

CO.1 Pitchi

Pink Pin

Double Mogra

well as in the perfumery industry. The plants produce numerous white scented blooms and are very good as loose flowers. It bears flowers from March- October. Peak flowering is in June and July. Five variants differing chiefly in flower bud characters were identified and named as Long Point, Long Round, Medium Point, Short Point and Short Round. A double flowered type of this species is commonly grown for garden decoration.

J. grandiflorum

Flowers are fragrant white, often tinged red purple outside, axillary or terminal and 5 petalled. It flowers from March to September and peak flowering is from July to October. A cultivar with leaves variegated yellow arose in nature and has been called Aureum (Dickey, 1970). Other cultivars are, Triploid bearing large size flowers with pinkish buds; highly sterile; chromosome number 39, Argenteo-variegatum having grey-green leaves with creamy white margins, Aureo-variegatum having leaves conspicuously marked yellow. Coimbatore, Lucknow, Local White, Thimmapuram are other popular cultivars.

Jasminum sambac

This is also called Arabian jasmine or Tuscan jasmine, Bela, Belphool, Malliage and Mogra. Peak flowering is from February to April in cultivars like Gundumalli, and gradually declines thereafter. Other cultivars peak in May or June. Flower buds are white to off-white with single or multi-whorled petals, used for garland making, landscaping and extraction of perfume Some popular cultivars are:Gundumalli (round flowers with good fragrance, high yielder, chromosome number 39); Ramabanam (long flower bud, high yielder, chromosome number 26); Madanban (long bold buds with short corolla tube, high yielder, chromosome number 26); Single Mogra (flowers with three or four whorls,chromosomenumber 26); Double Mogra (flowers with good fragrance resembling white rose, 8-10 whorls, chromosome number 52); Iruvatchi (flowers with shorter corolla tube with 3 to 4 whorls, chromosome number 26 and 39); Kasthurimalli (flowers with medium long tube, chromosome number 26); Oosimalli (long slender buds); Soojimalli (long buds, chromosome number 26); Grand Duke of Tuscany (flowers double like miniature gardenia).

Jasminum multiflorum

This species has been gaining popularity in commercial cultivation especially in areas where the night temperatures during winter months fall below 19°C, due to its prolonged flowering, availability of flowers during the winter months when flowering declines in the other commercial species. Also called Star/Pinwheel jasmine or Kakada, Kunda, Kundo, Kundphul, Kundamu, and Tundam. This species has white flowers borne in clusters and are not scented. This species flowers profusely almost throughout the year but for a decrease in yield during summer months. Kakada is a popular cultivar with many ecotypes exhibiting variability in corolla tube length, flower diameter, bud length and so on.

There are certain ecotypes are also been reported *viz.*, Mangalore malligae, Bhatkal malligae, Amboor Malligae, Elusuttina malligae, Ramanathapuram gundu malligae, which are specific to certain regions. Some of these are widely cultivated and exported.

Besides the above-mentioned cultivars public sector institutions like the Tamil Nadu Agricultural University, Coimbatore and ICAR-Indian Institute of Horticultural Research, Bengaluru have released some varieties listed in the Table 1.

Table 1: List of Jasmine Cultivars Released in India

Species	*Name of the Cultivar*	*Institution*	*Year of Release*
Jasminum grandiflorum	CO.1 Pitchi	TNAU, Coimbatore	1980
	CO.2 Pitchi	TNAU, Coimbatore	1991
	Arka Surabhi	ICAR-IIHR, Bengaluru	1993
Jasminum auriculatum	CO.1 Mullai	TNAU, Coimbatore	1980
	CO.2 Mullai	TNAU, Coimbatore	1988
	Parimullai	TNAU, Coimbatore	1972
Jasminum multiflorum	Arka Arpan	ICAR-IIHR, Bengaluru	1999
Jasminum sambac	Arka Aradhana	ICAR-IIHR, Bengaluru	2000
Jasminum nitidum	CO.1 Star Jasmine	TNAU, Coimbatore	2019

Jasminum grandiflorum

CO.1 Pitchi

It is a secondary clonal selection from germplasm collection. The average flower yield is 10 t per hectare in a year. The flower buds are pink tinged with long corolla tube. It is suitable for oil extraction with a concrete recovery of 0.29 per cent. The concrete yield is 29.42 kg per hectare (Anon., 2014).

CO.2 Pitchi

It is an induced mutant developed by treating the vegetative cuttings of CO.1 Pitchi with gamma rays @ 1.5 kR. This mutant is characterised by bold pink buds. The flower bud is 4.14 cm in length as against 4.00 cm in CO.1. The 100 buds' weight is 10 g in CO.2 as against 9.4 g in CO.1. It yields on an average 11.68 t of flower buds per hectare, which is 19.5 per cent higher than CO.1 Pitchi (Anon., 2014).

Arka Surabhi

It is a clonal selection from locally grown *Jasminum grandiflorum*. It yields 10 t of flowers/ha and 35 kg/ha of concrete and 17.5 kg of Jasmine absolute (Panda, 2005).

Jasminum auriculatum

CO.1 Mullai

It is a secondary clonal selection from a local type. The flower buds are white and bold with long corolla tube (1.50 cm) than Parimullai. The yield is 8.8 t of flower buds per hectare in a year. The jasmine concrete recovery is 0.34 per cent (Anon., 2014).

CO.2 Mullai

It is a clonal selection from progenies with desirable flower bud characters like long corolla tube and longer bud. The length of the corolla tube is 1.70 cm while it is 1.5cm in CO.1. It yields on an average of 11t kg of fresh flower buds. It exhibits complete field tolerance to the phyllody disease and gall mite infestation (Anon., 2014).

Parimullai

It is a clonal selection from a germplasm clone. The plants exhibit resistance to gall mite. The yield is 7.8 t of flower buds per hectare per year. The buds are white with moderate corolla tube length (1.25 cm). The concrete recovery is 0.29 per cent (Anon., 2014).

Jasminum multiflorum

Arka Arpan

It yields 6.19 t/ha. Buds are tinged red purple. Flowers are mildly fragrant. Besides the attractive foliage makes it a good landscape plant (Panda, 2005).

Jasminum sambac

Arka Aradhana

It is a clonal selection of *Jasminum sambac* cultivar Sooji malli, yields 8.0 t of flowers/ha and 14.95 kg/ha of concrete (Panda, 2005).

Jasminum nitidum

CO.1 Star Jasmine

It flowers throughout the year. It is promising for lean-season or off-season flowering. The flower buds bold, attractive, dark pink coloured with good keeping quality. The flower buds are easy to pluck and highly suitable for string-making due to bold buds with long corolla tube. Opened flowers are pure white and star shaped and emit mild fragrance. The attractive plant architecture makes the plant an ideal decorative ornamental also. Average flower bud yield is 7.5 ton/ha/year (Anon., 2020d). The flower buds remain unopened for 12 h under room temperature and for 60 h under cold storage at 7-8°C and the flowers possess mild fragrance (Ganga *et al.*, 2019).

4.2.1 Geographical Indication (GI) Status

Mysore malligae, Udupi malligae, Hadagali malligae from Karnataka and Madurai malli from Tamil Nadu have been registered under the Intellectual Property Rights (IPR) and accorded the Geographical Indication (GI) status which has provided exclusive rights to the local community to cultivate these.

Mysore Malligae, Udupi and Hadagalli malligae was accorded G.I. Registration in 2006 and Madurai malli in the year 2013.

Mysore Malligae

It is predominantly grown in and around Mysore of Karnataka (Longitude: 76° 23′ E to 76° 51′ E; Latitude: 11° 54′ N to 12° 24′ N), flowers are mainly used in garlands and other decorative purposes, buds have good keeping quality with shelf-life of 2-3 days and the unique fragrance is the main feature of this cultivar.

Udupi Malligae

It is predominantly grown in Shankarapura and surrounding villages in Udupi district of Karnataka (Longitude: 74° 45′ E to 75° 14′ E; Latitude: 12° 46′ N to 13° 13′ N), flowers are highly fragrant and used for making garlands, buds have good keeping quality with shelf-life of 3-4 days, fragrance is influenced by lateritic soils, heavy S-W monsoon rains, warm and humid conditions.

Hadagali Malligae

It is predominantly grown in Hadagali Taluk in Bellary District of Karnataka (Longitude: 75° 37′ E to 77° 11′ E; Latitude: 14° 33′ N to 15° 47′N), flowers are used mainly for garland making, strong fragrance of the flowers is its distinct trait, which is influenced by dry sandy soils and dry climate of the region.

Madurai malli

It originated in Thangachimadam on Rameswaram island and is predominantly cultivated in Madurai and Dindigul districts of Tamil Nadu (Latitude: 9°552 N Longitude: 78°072 E) in about 2,000 hectares. Flowering is continuous in summer, with daily average yields of 9 kg per acre. Mid-December to mid-February, when mist covers the region, the plants stop flowering. Madurai Malli has large consumer retail markets at Bangalore, Chennai, Hyderabad, New Delhi and Mumbai. On an average, two tonnes of flowers are exported every day to Singapore, Malaysia, Dubai, Canada and the U.S.

5.0 Genetics and Breeding

5.1.0 Genetics

Cytological investigations in jasmines were made by Krishnaswamy and Raman (1948), Raman (1955a), Alikhan *et al.* (1969) and Murthy and Khanna (1971). The genus with basic chromosome number of n =13 generally consists of diploid forms with 2n = 26. However, natural occurrence of forms with higher ploidy levels were also observed in *J. sambac* cv. Gundumalli (2n = 39), *J. flexile* (2n = 52), *J. primulinum* (2n = 39) and *J. Angustifolium* (2n = 52).

According to Sharma and Sharma (1958), cytologyical studies indicated a general resemblance in the chromosome morphology between species of *Jasminum*, indicating that the genus represents a natural assemblage. The different species and varieties have been found to differ from each other in minute details of the karyotype that suggests structural alteration of chromosomes in evolution. The minor part played by polyploidy and specially, allotetraploidy, before the species concerned lost its capacity for sexual reproduction, has also been pointed out. Variation in chromosome complements has been noted in the same somatic tissue containing normal nuclei and, as sexual propagation is mostly ineffective in this genus, the role of such somatic mutations in speciation of this genus through asexual means has been suggested. In *Jasminum* polyploidy has played a major role in speciation. The genus exhibits a wide spectrum of variation in ploidy level with diploids (2n = 26) predominating followed by triploids (2n = 39), tetraploids (2n = 52), pentaploids (2n = 65) and hexaploid (2n = 78). In the male-sterile taxa, speciation might have taken place by autopolyploidy rather than by allopolyploidy (George and Geethamma, 1992).

The studies carried out in *J. auriculatum* with seedling progenies raised from different clones have indicated the existence of a high level of phenotypic variation (64 to 92 per cent) for yield followed by length of primary laterals at flowering (43 to 82 per cent). The genotypic variation for yield was relatively low and ranged from 24 to 60 per cent, thereby suggesting the role of environmental influence on yield. In contrast to this, the variations observed for corolla tube length, bud length, bud diameter and weight of flower buds were predominantly genetic. Heritability was found to be maximum for corolla tube length followed by flower bud diameter (90 to 98 per cent) and length of flower bud (83 to 98 per cent). The genetic advance was lower than heritability for corolla tube length and diameter of flower bud. For the other characters, the genetic advance was higher. Heritability estimates varied with species and also among diploids and tetraploids within the same species. Among the different yield parameters, weight of flower buds exhibited high heritability in both diploids and tetraploids of *J. auriculatum* and in diploids of *J. grandiflorum* (Veluswamy, 1981).

According to Thangaraj (1977), the number of primary laterals and length of style were positively and significantly associated with flower yield in *J. auriculatum*. Length of flower bud, internodal length and number of days taken for flowering had significant association with length of corolla tube. Veluswamy (1981) also studied the relationship between the height of pruning and yield of flower buds and jasmine concrete recovery. The nature of association of these characters was found to be contrasting in the two species, *J. auriculatum* and *J. grandiflorum*. The genotypic and phenotypic correlations between height of pruned shoots and yield of flower buds and between jasmine concrete recovery and yield of flowers were negative in *J. grandiflorum* but positive in *J. auriculatum*. The negative direct effect of number of flower buds per shoot, number of pruned shoots and weight of flower bud on yield was recorded in *J. grandiflorum* but positive in *J. auriculatum*. The direct effects of length of shoot and bud diameter on yield of flowers were positive in *J. grandiflorum* whereas the effects were negative in *J. auriculatum* suggesting that different approaches are needed for crop improvement in these species.

5.1.1 Morphological Characterization

The morphological diversity that exists in *Jasminum* species have applications in crop improvement programme, besides having potential application in determining the Distinctness, Uniformity and Stability (DUS characteristics) of a variety which forms an important basis and a tool for protecting the breeders' and farmers' rights as custodians of the variety or landraces. In India, the DUS test guidelines have been developed for *Jasminum* spp. (Anon., 2015).

5.1.2 Molecular Characterisation

Molecular characterization is undertaken to identify and study genetic relationship among genotypes, to help the breeders in crop improvement programmes. In *Jasminum* spp., molecular characterisation studies have been conducted to assess the genetic diversity.

Mahmood *et al.* (2013) worked out the genetic relationship among the twenty-one accessions of *Jasminum* spp. using Randomly Amplified Polymorphic DNA (RAPD). Forty decamer primers from OP-S and OP-V were used and 195 alleles were detected. The PIC value calculated for these primers ranged from 0.438 to 0.698 which indicates a good level of genetic diversity among the accessions as also revealed by generated similarity matrix. The similarity values among the genotypes ranged from 0.263 to 0.776. The dendrogram divides the accessions into two distinct main clusters (A and B) at 0.40 similarity value. Cluster A has the most diverse jasmine accessions, while cluster B further divided the accessions into smaller groups.

DNA fingerprinting was performed for eight Jasmine species by the random amplified polymorphic DNA (RAPD)-PCR using 15 random primers. Primers OPZ8 produced a total 58 bands ranging from 7-11, OPZ9 produced a total 91 bands ranging from 10-13 and OPZ 10 produced a total 49 bands ranging from 5-8 for eight jasmine samples. Most of the bands were monomorphic with some polymorphic bands which can be used for marker development for these jasmine species (Sushant Shekhar and Prasad, 2014).

Ghehsareh *et al.* (2015) used 21 ISSR primers for molecular characterisation of 53 *Jasminum* accessions and these were divided into two major clusters. The first major cluster was divided into two sub-clusters; the sub-cluster A included *Jasminum grandiflorum* L., *J. officinale* L., and *J. azoricum* L. and the sub-cluster B consisted of three forms of *J. sambac* L. (single, semi-double, and double flowers). The second major cluster was divided into two sub-clusters; the first sub-cluster (C) included *J. humile* L., *J. primulinum* Hemsl., *J. nudiflorum* Lindl. and the second sub-cluster (D) consisted of *J. fruticans* L. Based on analysis of molecular variance, the amount of genetic variations among the eight populations was 83 per cent.

Nirmala *et al.* (2017) assessed *Jasminum* genotypes for thirty-nine qualitative and twenty quantitative morphological traits including vegetative and floral characters. In the dendrogram generated for qualitative characters and forty eighty genoptyes were grouped into three clusters. A wild species *J. rottlerianum*, distinct from all other genotypes constituted Cluster I. Another wild species *J. ritchiei* formed Cluster 2. Cluster 3 consisted two sub clusters. Sub cluster 3.1 consisted of variants of cultivated species *J. sambac*, two variants each of *J. multiflorum* and *J. grandiflorum* and one of the *J. auriculatum* along with seven species grown in home gardens (*J. angustifolium, J. rigidum, J. officinale, J. flexile, J. calophyllum, J. humile and J. primulinum* and four wild species. (*J. communis, J. cuspidatum, J.* sp. 2 and *J.* sp. 7). Sub cluster 3.2 consisted of another three wild species (*J.* sp. 1, *J.* sp. 3 and *J.* sp. 4). The variation in morphological characters is reflected in the clustering pattern as the cultivated and wild genotypes are found interspersed among one another. Forty-eight genotypes had a coefficient of variance of 42 per cent representing very high degree of variation among the genotypes.

Ghosh *et al.* (2020) explored the molecular diversity among 18 jasmine genotypes by employing two DNA based molecular marker techniques, *viz.*, random amplified polymorphic DNA (RAPD) and inter-simple sequence repeat (ISSR). Amplification of genomic DNA of 18 genotypes, using RAPD analysis using 20 RAPD resulted in a total of 248 products, out of which 234 (94.35 per cent) products were polymorphic. The polymorphism percentage ranged from 80–100 per cent with an average of 94.24 per cent polymorphism per primer. In ISSR analysis, among thirteen primers tested, a total of 595 products were produced out of which 562 (94.45 per cent) products were polymorphic. The polymorphism ranged from 82.60–100 per cent per cent, with an average of 94.45 per cent polymorphism per primer. The genotypes were categorized into several groups and subgroups which varied distinctly with *J. primulinum* forming a separate cluster.

5.2.0 Breeding

Introduction, clonal selection, half-sib evaluation, hybridization, mutagenesis and ploidy breeding has been attempted for identifying superior forms in respect of flower bud characters, yield, quality of fresh flower and essential oil content in the commercial jasmines.

5.2.1 Introduction and Clonal Selection

The clones of *J. grandiflorum* collected from different states of India showed considerable variations in respect to flower production as well as yield of jasmine concrete. The mean flower yield of the six clones varied from 4329 kg/ha to 10144 kg/ha while that of jasmine concrete showed a range of 13.85 kg/ha to 29.42 kg/ha (Veluswamy, 1980). This led to the release of cultivar Co. 1 Pitchi (*J. grandiflorum*) giving a yield of 10.15 t/ha of flower buds with a concrete recovery of 29.5 kg/ha.

The colour of flower bud is pink and attractive with pleasant fragrance suitable both for fresh flower trade and oil extraction. Similar work carried out in *J. auriculatum* led to the identification of a high yielding gall mite resistant clone Parimullai which was released by the Tamil Nadu Agricultural University in 1971 for commercial cultivation. The clone is characterized by long flowering season of nine months and a mean flower bud yield of 10 t/ha (Madhava Rao and Muthuswami, 1972). The selection, nevertheless, suffers from undesirable corolla tube length. Further, secondary selection has shown that the clone Long Round is more prolific in yield and possesses larger corolla tube as well as bud length and this was released as Co.1 Mullai which is capable of yielding 8800 kg flowers/ha with good consumer's preference. In an evaluation of 40 jasmine collections spanning ten species, *J. pubescens* and *J. auriculatum* were promising for flower yield while *J. grandiflorum* outranked others for jasmine concrete recovery (More, 1980). Veluswamy (1981) found clones of *J. grandiflorum* were superior to *J. auriculatum* in flower bud yield while in jasmine concrete recovery, *J. auriculatum* was better with 344 mg/100 g as against 288 mg/100 g in the former. Among diploid clones of *J.*

auriculatum, the clone Short Point was best in flower bud yield, while Thimmapuram and Lucknow of *J. grandiflorum* were adjudged as best jasmine concrete yielders.

5.2.2 Progeny Evaluation

In a study with seedlings obtained from open pollination in *J. auriculatum* four seedlings excelling the commonly grown cv. Parimullai were identified by Muthuswami *et al.* (1972). In a similar study, Thangaraj (1977) reported 175 to 284 per cent increased yield in elite seedlings over the standard Parimullai clone. Ten seedlings with superior morphological expressions for floral characters were identified. Veluswamy (1981) found that the progeny of Short Round in *J. auriculatum* was the best for flower yield as well as jasmine concrete recovery. The recovery was 413 mg/100 g of flower bud as compared to 363 mg/100 g from the normal diploid clones. He confirmed the view of Muthuswami *et al.* (1972) and Thangaraj (1977) that for synthesizing high yielding lines in *J. auriculatum*, selection from open pollinated seedlings offers a promising line of improvement, and the base material to start for seed collection is cv. Short Round since it has shown higher percentage of improved floral characters than other parents. This wide variation is due to high heterozygous nature of the parent with advanced level of heterostyly present in this particular clone and half-sibs of the clone are the best materials for crop improvement. However, in case of *J. grandiflorum* this has not shown any encouraging result for crop improvement and the clones collected from various places have shown wide variation suggesting extensive survey and assemblage of best types from geographically diverse areas and selection of improved types.

5.2.3 Hybridization

Inter-specific hybridization involving *J. auriculatum, J. grandiflorum, J. flexile, J. calophyllum* and *J. sambac* attempted in the past has shown that the success was very low ranging from 0.5 to 5 per cent. The seeds of direct as well as reciprocal crosses failed to germinate (Anon., 1974). However, a success of 3-16 per cent was recorded when an induced tetraploid of *J. auriculatum* was used as female and *J. grandiflorum, J. angustifolium, J. flexile* and *J. calophyllum* were used as donors. But the seeds of these crosses also failed to germinate. With the same female parent, crosses using *J. rigidum, J. arborescens* and *J. paniculatum* as male parents also ended in failure. Inter-specific hybridization among the wild species also met with the same fate. Veluswamy (1981) reported successful fruit set in the combinations involving *J. auriculatum* with *J. flexile* and *J. grandiflorum*; *J. grandiflorum* with *J. auriculatum, J. calophyllum* and *J. flexile* and *J. sambac* with *J. grandiflorum*. But in all these cases, the seeds failed to germinate showing a strong barrier for crossability and hybrid non viability existing among the different species and also the possibility of endosprem antagonism in operation. Existence of pollen sterility and seed sterility due to defective egg mechanism has also been indicated. The sterility factors operating in jasmines have been reported by Dutta (1952) and Murthy and Khanna (1971).

Intra-specific hybridization, on the other hand, showed partial success ranging from 0.5 to 35 per cent in *J. grandiflorum*. The crossed seeds registered a germination of 50-100 per cent. Cross incompatibility was also evident in some of the interclonal crosses. This indicates the genetic divergence of the clones in different degrees and the possibility of exploitation of such divergence for improving the economic attributes. In addition, reciprocal differences also seem to exist thereby indicating the operation of one-way incompatibility system. In *J. auriculatum*, success was obtained in Short Round x Short Point, Parimullai x Short Point and an induced tetraploid x diploid of *J. auriculatum*. Fruit set on crossing ranged from 0 to 27 per cent. However, none of the progenies proved of economic importance (Muthuswami and Abdul Khader, 1986).

Jasminum sambac is poor in seed setting due to abnormal development of embryo sacs and its restricted pollen viability which caused failures in fertilization (Deng *et al.*, 2016).

In *Jasminum sambac* (L.) pollen stored at -60°C over silica gel under vacuum showed better germination (40.60 per cent) after 48 weeks of storage Perveen and Sarwar (2011). Deng *et al.* (2017) reported crosses among single, double and multi-petal jasmine cultivars (*Jasminum sambac* Aiton) are unable to easily generate hybrids. Different reciprocal crosses exhibited varied pollen-stigma compatibilities according to the pollen germination rates. Although some pollen grains germinated normally on maternal stigmas, the pollen tubes were arrested in the pistils and were unable to reach the ovaries. Additionally, the embryo sacs remained unfertilized until degenerating and are affected by pre-fertilization reproductive barriers. Low pollen fertility and poor stigma receptivity are detrimental to pollen germination and pollen-pistil compatibility, indicating they are two factors affecting hybrid set. Ultrastructural observation of the pistil cells revealed that cell death occurred during flowering. Thus, the early and rapid senescence of pistils is likely responsible for the decreased pistil receptivity and inhibited pollen tube growth.

Deng *et al.* (2018) reported that *Jasminum sambac* (L.) rarely sets seed, and its poor pollen fertility significantly influences fecundity. The pollen development of jasmine was investigated at an ultrastructural level. Normally developed microspores are released at the late tetrad phase, whereas the abnormal ones formed large complexes of two to five tetrads with plasmodesmata. The sterility of some pollen grains might result from defective development of cell walls that failed to absorb their surrounding sporopollen in to form the exine. The secretory type of jasmine tapetum, the usually degenerated middle layer and thickened endothecium were confirmed during the process of pollen development. The epidermis, middle layer and endothecium exhibited no adverse effects on pollen development and dispersal. However, the asynchronous division and delayed degradation of tapetal cells resulted in abnormal development of some microspores and the formation of massulae and empty and flat pollen grains.

Pavithra *et al.* (2018) studied pollen viability, pollen germination, normal shaped pollen grains, length of pollen tube and viable pollen diameter of different species (Table 2). Pollen grains collected during evening (05:00 pm to 06:00 pm) at the time of anthesis showed best results. Pollen tubes revealed abnormalities such as bulging and coiling of tips in *J. sambac* and *J. grandiflorum* which might be one of the reasons for no fruit set when *J. sambac* was used as pollen parent.

Table 2: Pollen Viability, Germination and Associated Traits in *Jasminum* spp.*

Species	*Pollen Viability (Per cent)*	*Pollen Germination after 24 hrs 34.08 (Per cent)*	*Normal Pollen Grains (Per cent)*	*Length of Pollen Tube after 24 hrs (µm)*	*Viable Pollen Diameter (µm)*
J. sambac	34.27	34.08	70.63	412.17	42.75
J. auriculatum	71.32	71.09	96.93	848.15	38.25
J. grandiflorum	43.63	42.99	93.63	517.81	43.91
J. malabaricum	96.57	91.58	96.89	951.11	46.53
J. multiflorum	48.53	51.66	80.69	763.30	45.20
J. flexile	84.92	81.86	91.70	1170.68	44.77
CD (P=0.05)	2.94	2.67	4.35	35.96	1.91

* Pavithra *et al.* (2018).

Pavithra *et al.* (2019) studied the compatibility among six *Jasminum* species and among open, self and artificial cross pollination, the open pollination resulted in maximum degree of fruit and seed set proving the open pollination to be the best method of pollination. The best time for pollination was the time of anthesis. The species *Jasminum malabaricum* and *Jasminum auriculatum* were ideal as seed parents.

5.2.4 Selfing

Selfing did not result in any fruit set in *J. auriculatum*. The fruit set under natural conditions varied from 33.1 to 71.8 per cent. The seed germination ranged from 75 to 95 per cent. Veluswamy (1981) failed to get any fruit set on selfing in *J. grandiflorum* but open pollination yielded 18.0-52.1 per cent fruit set. The seed germination was satisfactory ranging from 68.8 to 74.1 per cent. Among the cultivated jasmines, only *J. sambac* cv. Soojimalli, *J. auriculatum, J. flexile* and *J. angustifolium* did set fruits.

5.2.5 Ploidy Breeding

In general, triploids have been found to be more vigorous and higher yielders than diploids while tetraploids are robust, hardy and produce bolder flowers.

Artificial induction of polyploids has been feasible in *J. auriculatum* but not in *J. grandiflorum* (Veluswamy, 1981). Seed or seedling treatment with 1 per cent colchicine recorded 11.4-23.1 per cent polyploidy in *J. auriculatum*.

Induction of polyploids was attempted in the diploid forms of *J. auriculatum* and *J. grandiflorum*. Seeds of Short Point clone of *J. auriculatum* were treated with 0.5 per cent colchicine for 12 and 24 hours. One tetraploid seedling was obtained from Parimullai seeds. In general, tetraploids were found to be poor yielders with low concrete recovery. Treatment of vegetative buds with 0.1-0.5 per cent of colchicine did not yield any useful polyploid. In *J. grandiflorum*, seed treatment with 0.5 per cent colchicine resulted in six tetraploids which were, however, not found to be economically useful. In *J. sambac*, tetraploids were more vigorous but produced a smaller number of bold flowers per plant (Alikhan *et al.*, 1969). The natural triploids were found to be vigorous with higher yield than diploids. The superior morphological expressions of tetraploids have the advantage as better forms in floriculture for utilization as cut flowers. However, the diploids were superior to induced tetraploids in most of the economical attributes, although some of the tetraploids proved superior in concrete recovery.

5.2.6 Mutation Breeding

Both chemical and physical mutagens have been tried in *J. auriculatum*. X-ray treatment of active vegetative buds produced no mutagenic effect. Seed treatment with gamma rays beyond 10 kR proved lethal. A mutant with longer corolla tube was obtained but it was susceptible to gall mite. Use of chemical mutagens showed that soaking in 0.1 per cent EMS for 72 hours proved lethal. However, with the use of this chemical, one mutant was obtained but it had to be rejected on account of its undesirable traits. In *J. grandiflorum*, gamma rays of dosage beyond 30 kR proved to be lethal. Although 10 mutants were isolated with lower dosages, none proved to be good for flower yield. Treatment of softwood cuttings with gamma rays (1 kR to 10 kR) also met with the same fate suggesting the limited role of mutagenesis in jasmine improvement. However, Nambisan *et al.* (1980) obtained two promising mutants in the M_2 generation with 25 kR gamma rays in *J. grandiflorum*. One was a dwarf mutant with short internodes and the other was found to be tolerant to leaf spot disease. But the yield of flowers and recovery of concrete were less than parent clone Thimmapuram.

Devaiah and Srivastava (1989 irradiated terminal cuttings of *J. grandiflorum* cultivars Pink Pin and Pink Thrum, *J. sambac* cv. Gundumalli, *J. calophyllum* and *J. flexile* with gamma-rays at 0.5 to 5.0 krad and found that the rooting percentage, number of roots/cutting and length and thickness of the roots were decreased with irradiation intensity. LD_{50} intensities resulting in 50 per cent rooting were 2.5, 0.5, 2.0, 1.0 and 2.5 krad for the respective genotypes.

Mekala *et al.* (2010) observed semi hard wood cuttings of *J. sambac* cv. Gundumalli treated with mutagens *viz.* gamma rays alone and in combination with ethyl methyl sulphonate (EMS) that the treated population had manifested reduced expression in most of the floral characters than the untreated population. The mean stalk length was minimum with 2.5kR+30mM EMS (0.89±0.021 cm). The floral characters like bud length and flower weight decreased with increase in the doses while the flower width did not follow a definite trend. Induced mutations produced higher amount of phenotypic and genotypic co-efficient of variation for number of flowers, flower weight indicating the scope for improvement of gundumalli for yield and novelty by selection.

Terminal cuttings of *Jasminum grandiflorum* cv. White Pitchi and semi hardwood cuttings of *J. nitidum* (Acc. Jn-1) and *J. multiflorum* cv. Arka Arpan were treated with ten different concentrations of EMS ranging from 5 mM to 50 mM. There was a gradual and significant reduction in survival of cuttings, shoot length, number of leaves, leaf length and leaf width with increase in dosage of EMS. The probit curve analysis revealed LD_{50} dosage of EMS to be 37.15 mM, 40.7 mM and 42.66 mM for White Pitchi, Arka Arpan and ACC Jn-1 respectively (Ghosh *et al.*, 2018a).

Ghosh *et al.* (2018b) studied the mutagenic sensitivity of jasmine species to gamma rays. Terminal and semi hardwood cuttings of three different jasmine species (*Jasminum grandiflorum*, cv. White, *Jasminum multiflorum* cv. Arka, Arpan and *Jaminum nitidum* Acc. Jn-1) were subjected to eight different dosages of gamma radiation ranging between 5 Gy and 40 Gy. The probit curve analysis based on the survival percentage and growth rate of treated cuttings revealed that the LD_{50} dosages of gamma rays for cv. White, Arka Arpan and ACC Jn-1 were 17.8 Gy, 28 Gy and 25.1 Gy, respectively.

6.0 Growth and Flowering

6.1.0 Flower Bud Differentiation and Floral Biology

Flowers of jasmine are hermaphrodite having 2 bilobed anthers. The gynoecium, has one ovule, simple style and bilobed stigma. Ovary is superior, bilocular having axial placentation. In jasmine, 4 patterns of anthesis have been reported *viz.*, 5 to 6 PM (early), 6 to 7 PM (medium), 7 to 8 Pm (late) and 8 to 10 PM (very late). Similarly, three patterns of dehiscence were recorded 3 to 11 PM, 4 to 5 PM and 5 to 6 PM. Period of blooming in jasmine vary from 3 to 4.5 months. Maximum number of flower buds opens from 6 to 7 PM. Floral dimorphism expressing long and short carpel have been reported in *Jasminum grandiflorum*. Distinct differences in flowering behaviour of these two types were observed. The long carpel (pin) type had prolific flowering, whereas short carpel (Thrum) type expressed shy flowering. In *Jasminum auriculatum* pin and thrum type flowers have been reported. Thrum type plants were reported to be superior to pin type in floral bud length, corolla tube length, and diameter of open flower.

Nedumaran and Shanmugavelu (1978) studied the ontogeny of *J. auriculatum* with reference to different pruning dates. The vegetative bud (apical meristem) is broadly convex, and differentiates into opposite leaves which protect the centre of the bud. The apical meristem, at the time of initiation of the inflorescence, undergoes a gradual transformation. The broad convex apical dome gradually becomes narrow and slightly constricts at its base forming a distinct articulation. From the articulated region, two lateral branches arise. At this stage, a trident-like apical bud is formed. The process is repeated to form a branched cyme with elongated peduncles. The apical meristem of each of the ultimate peduncle axes differentiates into a simple cyme with three distinct flowers subtended by their own bracts. It was also found that irrespective of dates of pruning from November to March, the flower bud initiation and differentiation followed a set pattern. The plants pruned from November to December took more than 45 but less than 60 days for flower bud initiation and differentiation. As pruning time was delayed, the time taken for flower bud initiation was shortened.

In *J. grandiflorum*, the vegetative phase, as reported by Subramanian and Shanmugavelu (1978), is characterized by a pointed, elevated and elongated meristematic apex in the centre which undergoes rapid cell division leading to the formation of a bulged apical meristem. The size of the apical meristem is gradually reduced, again bulges and is enclosed by two lateral leaves. The apical meristem soon becomes broad and flattened. This stage is recognized as the transition stage, from vegetative to reproductive phase. The meristematic tissue at the apex undergoes further cell division, it becomes slightly lobed with the base enlarged and apical meristem slightly pointed. As growth proceeds, the meristem becomes bilobed and the base of the primordia broad. The apical meristem gradually becomes trilobed and is enclosed by a well-developed leaf primordium on one side. At a later stage, lobes become more prominent with three to four lobe-like structures on either side of the apex where initials of sepal primordia are seen.

During the early reproductive stage, the sepals appear first as long elongated structures, thickened at the base and gradually tapering towards apex. The apex of the sepal is curved outwards with the apical meristem during the process of differentiation. The sepals elongate further and become more curved towards outside. The central meristem appears more pointed at this stage. The entire structure is enclosed by two leaf primordia and this stage is followed by the formation of petal primordia which arise from the central apical meristem enclosed by calyx lobes on either side. Petal primordia appear initially as separate lobed structures. As the reproductive stage advances, lobing of petal primordia become distinct. Five petal primordia conical in shape with blunt apex and a depressed notch at the base appear. The calyx primordia at this stage is considerably reduced in size, shortened, bulged and more thickened towards the base. Axillary buds start developing on either side of this reproductive primordia and these buds remain in dormant condition till the main floral primordium is completely differentiated.

The petal primordia become enlarged with the notch at the base being more prominent. As the development process continues, stamen primordia make their appearance at the base of the meristem while the tip of the meristem remains pointed and slightly elongated. Further differentiation occurs and stamen primordia become more prominent. Ovary primordium appears as a short-rounded structure around the staminal primordium. Carpels are not clearly visible at this stage.

Stamens and ovary develop further. The first, second and third pair of petals unite to form the corolla tube while the fourth pair of petals remain indistinguishable from the top petal. With further development, all the petal primordia unite to form the corolla tube which ultimately unites with the base and the calyx. Thus, flower bud initiation and differentiation occur in acropetal succession with sepals developing first followed by petals, stamens and ovary. Corolla tube development becomes complete only in advanced reproductive stage with differentiation of the ovary.

Isac *et al.* (2015) studied the morphological and anatomical structures responsible for floral scent production with scanning electron microscopy and light microscopy thin sections, respectively in *Jasminum sambac* cv. Gundumalli and reported that floral stomata were present on the abaxial surface and floral trichomes on abaxial tip surface.The unicellular trichomes on the abaxial surface of the petals are osmophores, rich in stored lipids, while the parenchyma surrounding the vascular bundles contains starch. Only freshly opened flowers produced odours, while the floral trichomes are in developing condition in its bud stage of the flowers. Gland anatomy and position of these scent glands on the flower petals plays a key role in release pattern of the fragrant volatile compounds.

Floral biology and flowering pattern of five lesser known *Jasminum* species namely, *J. calophyllum, J. flexile, J. multiflorum* (Pink), *J. nitidum* and *J. rigidum* were documented by Lakshmi and Ganga (2017). *J. flexile* and *J. multiflorum* (Pink) were profuse seed setters. This potential can be exploited in hybridization programmes, wherein such species can be ideal female parents.

Kalaiyarasi *et al.* (2019) reported that anthesis and anther dehiscence was recorded earlier in summer months, while, it was delayed during cooler months in *Jasminum* spp. However, the peak anthesis was observed from 6:00 to 7:00 p.m. and anther dehiscence from 4:00 to 7:00 p.m. The earliest anthesis was observed in *J. auriculatum* genotypes and anther dehiscence was observed in *J. auriculatum* and *J. multiflorum* genotypes. The stigma was receptive 12 h before anthesis and up to 24 h of anthesis. The maximum receptivity was observed from the time of anthesis to 12 h after anthesis.

Growth and flowering in jasmines are influenced by several factors like genotype, environment, pruning, nutrition and other management practices. Certain chemicals are also effective in regulating the growth and flowering. The effects of pruning, nutrition and other cultural practices are described separately and the role of other factors is discussed below.

6.2.0 Genetic or Inherent Factors

The genetic variation in jasmines is very large. There are both evergreen and deciduous species having climbing, trailing and erect shrubby habit. Flowering time, duration of flowering, yield, longevity of flowers, *etc.* vary according to species and cultivars. The peak flower production in *J. auriculatum* is generally confined during the month of June and July while in *J. grandiflorum* it occurs between July and October (Muthuswami and Adbul Khader, 1986).

Among the genotypes of *Jasminum sambac,* cluster bearing habit in terminal position was recorded in genotypes Iruvatchi and IIHR JS-5, whereas other genotypes recorded forked cymes in terminal and axillary positions. Three different shapes of flower bud were recorded among the genotypes, *viz.*, pointed and long, pointed and short, rounded and short. Flower bud colour and colour on flower bud opening (based on RHS colour chart) was white (155 B or 155 C) and pink tinge on flower bud was recorded in IIHR JS - 5. Star and round shaped flower were observed among different genotypes. Three different types of flowers were recorded, *viz.*, single, double and multi-whorled (3 to 6). Among the 22 genotypes, Arka Aradhana recorded longest bud and corolla. The genotype IIHR JS - 5 recorded the longest corolla tube and Iruvatchi recorded maximum number of forks per cyme. Maximum number of calyx lobe was recorded in Gundumalligae, whereas longest calyx lobe was recorded in Soojimalli. Bigger size of flower was recorded in Soojimalli and maximum number of petals was observed in IIHR JS – 1 (Kalaiyarasi *et al.*, 2018).

6.3.0 Environmental Factors

Muthuswami (1975) found that temperature played a major role in the production of flowers of *J. auriculatum*. Early pruned plants recorded high yield and longer duration of flowering as they received optimum heat units. These findings were also confirmed in *J. grandiflorum* in which maximum flowering occurred in plants pruned in later part of December. The requirement of heat units decreased as the pruning was delayed. Raman (1973) observed a direct correlation between yield and mean temperature and no relationship with rainfall or humidity in *J. sambac*. The relationship between temperature and flowering in jasmines and the variation that exists among the different species in their specific requirement of heat units have been brought out by the above investigators. Further studies to ascertain the possibilities of shifting the flowering pattern were undertaken by pruning on different dates from early November to late March at ten days intervals. Irrespective of the date of pruning, the months of June and July contributed the largest share to the total flower yield in *J. auriculatum*.

Similarly, in *J. grandiflorum*, the peak flower production was recorded between July and October under all pruning dates. Neither very early pruning in November nor late pruning in March did show any appreciable variation. In both the species, the monthly yield of flowers showed a negative association with Relative Temperature Disparity (RTD). Less difference between day and night temperature and relatively lower day temperature favoured a higher yield. Thus, diurnal temperature periodicity appears to have favoured more flower formation in particular. Muthuswami and Madhava Rao (1980) were of the opinion that the above two species could be regarded as responsive to thermoperiodism quantitatively and classified them as day-neutral plants requiring temperature alteration for flowering.

In *J. sambac* cvs. Gundumalli and Iruvatchi, there was a steep rise in yield from February to April (27 to 46 per cent) and gradual decline thereafter. In other cultivars the peak yield was in May or June. Although the commencement of flowering was at the same time, the quantum of yield and duration of flowering varied widely. This was attributed to photo-thermal influence. In the high yielding cultivars, the highest yield was noted also under a regime of long days and high day and night temperatures. These conditions proved favourable to produce more flowers per photo-thermal unit (PTU) than low yielders. Thus, the efficiency of different cultivars to produce flowers per PTU varied and depended on the interaction of both day length and temperature (Raman, 1973). Deng *et al.* (1986) reported that exposure to low winter temperatures caused ultra structural changes in *J. sambac*.

Growth and flowering in double petal (DP) and a multi petal (MP) type jasmine (*Jasminum sambac* Ait.) is largely affected by different levels of irradiance. In both types, net photosynthetic rate (Pn), stomatal conductance (gs) and transpiration rate increased with decreasing irradiance from 100 per cent to 20 per cent. Chlorophyll a, b and carotenoids contents in DP type shaded leaves increased whereas those in MP type decreased at 5 per cent irradiance (extreme shade). The maximum photochemical efficiency of photosystem II of DP plants showed subtle changes but that of MP plants declined by shading. Chloroplast ultra structure showed its best development in the leaves of DP and MP types mostly from 50 per cent to 20 per cent irradiance (considered as weak and moderate shade, respectively). At each shade treatment, Pn, gs and water use efficiency of DP-jasmine were always higher than those of MP-jasmine, thus the shade tolerance ability of the former was higher than that of the latter. Full sunlight and 5 per cent natural irradiance caused photoinhibition and light deficiency of jasmine plants respectively, and modulating chloroplast development by increasing the numbers of thylakoids and grana to contain more photosynthetic pigments is an important shade tolerance mechanism of DP type (Deng *et al.*, 2012).

6.4.0 Growth Regulators

In *J. auriculatum*, spraying with ethrel at 1000 ppm recorded maximum growth retardation, CCC at 1000 ppm induced early flowering while SADH produced greater number of laterals producing flower clusters (Muthuswami, 1975). Hore and Bose (1972), however, reported that application of SADH inhibited shoot growth in *J. sambac*. In *J. grandiflorum*, GA_3 at 25-75 ppm significantly increased the length of primary and secondary shoots whereas MH at 1000-3000 ppm, TIBA at 100-300 ppm, CCC at 500-1500 ppm, ethrel at 100-500 ppm reduced shoot length. The highest number of lateral shoots was observed on plants treated with CCC, followed by TIBA and MH. GA_3 treatment resulted in the least number of lateral shoots. MH increased the number of internodes and GA_3 increased their length. CCC significantly increased the number of leaves and flowers and induced early flowering whereas ethrel delayed flowering. CCC treated plants also flowered longest (Pappiah and Muthuswami, 1977).

In *J. sambac*, Raman (1973) noted that GA_3 promoted shoot length and delayed flowering while CCC and SADH retarded growth and increased flowering. Similarly, in *J. auriculatum* cv. Parimullai, GA_3 and ethrel delayed the onset of flowering and reduced essential oil content while CCC recorded the maximum flower yield through induction of more laterals and by extended flowering period (Pappiah and Muthu-swami, 1978). Pal *et al.* (1980a) reported that treatment with ethrel promoted the height of plants while CCC retarded growth in *J. auriculatum*. NAA application, in general, also resulted in reduced plant growth. Application of ethrel at 0.125 and 0.250 ml/l and NAA at 25 mg/l on *J. sambac* promoted more laterals resulting in increased flower number. The above concentrations of ethrel and NAA and CCC at 1.09 and 2.18 ml/l also increased flower yield.

In *J. sambac* cv. Malila removing 50 per cent of the leaves, covering the plants with transparent plastic film 1600 h to 0800 h or spraying with 100 ppm NAA resulted in 2.10, 29.1 or 4.0 flowers/plant compared with 18.6 flowers/plant in the control (Kitpow-song *et al.*, 1988).

Jasminum grandiflorum was sprayed with IAA, GA_3, NAA, MH or TIBA at 10, 100 or 1000 ppm; CCC, SADH at 100, 1000 or 5000 ppm, and ethrel at 500, 1000 or 2000 ppm. All concentrations of IAA and GA_3 increased shoot length, the most effective being IAA at 10 ppm which resulted in 12.6 per cent increase in shoot length compared with untreated control. All other chemicals, except NAA at 100 ppm retarded vegetative growth. The maximum number of flowers per plant per year was obtained with IAA at 10 ppm (5993.6), SADH at 5000 ppm, GA_3 at 10 ppm and CCC at 100 ppm increased the flower number by 31.8-49 per cent over the control plants producing on an average 4019.7 flowers per plant (Bhattacharjee, 1983a).

Induction of off-season flowering in *J. grandiflorum* cv. Thimmapuram was attempted (Vedamuthu, 1981). Spraying of growth regulators coupled with pruning round the year at monthly intervals revealed that pruning in January and spraying with SADH at 1000 ppm shifted the peak flowering to October. A shift in flower production to November was observed by pruning in February and spraying with CCC at 500 ppm. Three stages of peak production were seen in the off-season, with May pruning, however, the total production was very much reduced. Pruning from January to March in combination with SADH spray at 1000 ppm was found effective in extending the flower production in the off-season.

Krishnamoorthy (2014) reported that spraying of 20 per cent Nitrobenzene @ 2.5ml/l along with Tricontanol 0.2 per cent once in 15 days during November-April followed by pruning during July month recorded highest number of flower bearing shoots, number of leaves, flowering duration, number of flowers, flower yield per plant (748 g/year) and flower yield per hectare (4863 kg/year).

Sobhana (2014) observed an increase in yield in bush jasmine (*Jasminum sambac*) during April-May and October to December period indicating the favourable effect of the application of 20 ppm GA_3 followed by 1000 ppm cycocel in enhancing flower production during the cooler months of October, November and December coinciding with the off season. In *Jasminum sambac,* 24-Epibrassinolide (EBL) @ 2 muM improved morpho-physiological and bio-chemical traits like number of leaves, chlorophyll contents, leaf area, fresh and dry weight ratio, number of flowers per plant, photosynthetic rates, transpiration rate, stomatal conductance and sub-stomatal cavity. Plants treated with 3 mu M EBL showed better results for growth indices like plant height, number of branches per plant, fresh weight of flower, and dry weight of flower besides increase in essential oil contents (Younis *et al.*, 2014).

According to Kumaresan *et al.* (2017) soil drenching of paclobutrazol (300 ppm) significantly increased chlorophyll a, chlorophyll b, total chlorophyll, total phenol content, IAA oxidase activity, soluble protein, specific leaf weight, diameter of the flower buds, corolla tube length and weight of hundred flower buds in *Jasminum sambac*. In Spanish jasmine (*Jasminum grandiflorum* L.), application of 1500 ppm CCC enhanced the flower bud length, flower stalk length, total length of the flower and flower yield but did have any effect on shelf life of flowers (Sudhagar and Kamalakannan, 2017). According to Dhanasekaran (2018), application of 150 ppm GA_3 resulted in earliness in flowering, maximum duration of flowering enhanced growth and higher flower yield in *Jasminum sambac*.

Foliar spray of sea weed extract at 5 per cent recorded maximum plant height, number of primary and secondary branches, number of leaves/plants, leaf area and chlorophyll content in *Jasminum sambac* (Manimaran *et al.*, 2018).

7.0 Propagation

Jasmines are usually propagated by cutting and layering. Seed propagation, though uncommon, is necessary for crop improvement through planned hybridization. Multiplication through sucker, grafting, budding and tissue culture has also been found successful.

7.1.0 Seed Propagation

Inter-clonal variations in seed germination have been reported in *J. auriculatum* and *J. grandiflorum* and also among other species. In general, diploid species of *J. auriculatum* and *J. grandiflorum* have better germination potentiality (75-84 per cent) than tetraploids like *J. flexile* (55 per cent). Higher temperature and humidity hastened the germination and maximum germination occurred in saw dust and sphagnum moss. Seed dormancy has been noted in *J. auriculatum* and longevity of seeds does not exceed ten months (Veluswamy *et al.*, 1974, 1975, 1977).

Germination of seeds of *J. auriculatum* and establishment showed gradual decreases as the mutagenic treatment (2.5-20 kR doses of gamma rays or 10-30 mM ethyl methanesulfonate (EMS) or with combinations of both) was increased. The LD_{50} were 9.75 Kr gamma rays, 24 mM EMS, of a combination of 25 mM + 20 kR for germination percentage and 16 kR, 25 mM and 10 mM + 13.5 kR for establishment percentage (Chezhiyan *et al.*, 1988).

7.2.0 Vegetative Propagation

7.2.1 Cutting

This is the easiest method of propagation and has been attempted in certain species, particularly those grown on commercial scale. There are, however, several factors like type of cutting, rooting media, environmental conditions, *etc.*, which influence root formation in cuttings. The percentage of rooting and number of roots per cutting were noted highest in cutting of *J. auriculatum*, 15 cm long having two leaves and taken from the middle portion of the shoot; with softwood cuttings, 20 cm long ones were found better than others (Bose *et al.*, 1973). Improvement in root formation with the presence of leaves in cuttings was also confirmed by Mukhopadhyay and Bose (1979) who reported that rooting percentage as well as number of roots per cutting increased with the increase in leaf number. Leafless and budless cuttings, on the other hand, failed to produce roots even with the use of growth regulators. The success of rooting in semi-hardwood cuttings of *J. auriculatum* was 70 per cent, in terminal cuttings of *J. grandiflorum* 98 per cent and in semi-hardwood and terminal cuttings of *J. sambac* 92 and 94 per cent, respectively (Jayapal *et al.*, 1980). Paulas (1980) found single node cuttings with one leaf as the best material for quick, cheap and efficient multiplication of cv. Gundumalli of *J. sambac*. Lai (1995) found that *J. sambac* was easily propagated by twig cuttings. Zhang *et al.* (1997) obtained best rooting in cuttings from basal sections of *J. sambac*. Siddique (1973) reported that cuttings taken in the month of March rooted better than those taken in June, September or December whereas Bryzgalova (1974) recorded improvement in rooting in cuttings taken in July as compared to those prepared in January, April or October. In Tamil Nadu cuttings taken in May contributed higher success thereby showing the influence of the season on propagation (Anon., 1977).

Cuttings of *Jasminum auriculatum* and *J. grandiflorum* which failed to root or showed a low percentage of rooting under ordinary propagation facilities, showed high percentage of rooting under intermittent mist. Better survival and satisfactory growth after transplanting was also observed in cuttings rooted under mist (Bose *et al.*, 1973).

Treatment of cuttings with growth substances before planting has been found very useful for promoting root formation (Bose *et al.*, 1973; Bryzgalova, 1974; Singh and Motial, 1981). Sreelatha *et al.* (1991) reported high percentage of rooting in semi-hardwood cutting of *J. auriculatum* and *J. grandiflorum* under mist by treatment with IBA between 2000 and 4000 ppm. Veeraraghavathatham *et al.* (1985) reported better rooting in cutting of *J. auriculatum* by etiolation, girdling and treatment with IBA.

Netam *et al.* (2018) recommended treatment of the stem cuttings of Jasmine (*Jasminum sambac* (L.) Aiton) with 1500 ppm IBA for maximum survival percentage (88.33 per cent), lesser number of days to sprouting of cuttings, maximum number of leaves, length of shoot, number of roots and length of root per rooted cutting. Kumeresan *et al.* (2019) observed that terminal cuttings of *Jasminum multiflorum* (Pink Kakada) treated with IBA 500 ppm + NAA 250 ppm registered maximum survival percentage of rooted cuttings (85.36 per cent), along with increased length of shoots, number of roots, length of roots and number of leaves per rooted cutting.

In *J. grandiflorum*, Singh and Bhatnagar (1955) studied the effects of IAA and NAA on rooting of soft and hardwood cuttings. The highest rooting of 90 per cent was recorded in softwood cuttings treated with IAA 1000 ppm and in hardwood cuttings treated with NAA 500 ppm. Bose *et al.* (1973) also confirmed the effectiveness of IAA and NAA in rooting of cuttings of this species. IBA at 500 ppm stimulated rooting in 91 per cent cuttings.

In an investigation on the effects of IBA and IAA each at 1000-2500 ppm, SADH at 100-1000 ppm and Seradix on rooting of softwood cuttings of *J. auriculatum*, Pappiah and Muthuswami (1976) recorded the highest percentage (85) of rooting with IAA at 2000 ppm, followed by SADH at 500 ppm and IBA at 2000 ppm. Similarly, in *J. sambac* cv. Madhanban, IAA and NAA each at 4000 ppm resulted in 100 per cent rooting (Singh, 1979). According to Singh and Motial (1981), the best rooting (97.5-100 per cent) and survival of rooted cuttings (100 per cent) were obtained with the application of IBA at 4000 ppm.

Sand, vermiculite, moss and mixture of these are some of the commonly used media for root formation in cuttings of jasmine. The highest percentage of rooting and larger number of roots was recorded in cuttings of *J. auriculatum* planted in coarse sand followed by a mixture of sand and moss (1:1 or 1:3) and vermiculite. Root formation was adversely affected in garden soil (Mukhopadhyay and Bose, 1979). Maximum rooting (95 per cent) in cuttings planted in coarse sand was also recorded by Singh (1979). Hardwood cuttings of *J. sambac* and *J. auriculatum* rooted well in sand(Maurya *et al.*, 1996).

Matured leaves of *Jasminum sambac* harvested with petioles were soaked in water for 5 minutes and subsequently soaked in tender coconut water for 10 minutes. The leaves were placed in protrays in an amended media combination of soil+ coir dust + vermicompost (1:1:1) and rapid propagation was achieved with leaf cuttings (Baskaran *et al.*, 2016).

The number of days taken for rooting have been found to differ with different species, being 45 days in *J. grandiflorum*, 70 days in *J. auriculatum* and 68 days in case of *J. sambac* (Pappiah *et al.*, 1980).

Propagation of Jasmine

Seeds

Seedling of jasmine in flower

Propagation by cutting

Plant from cutting

Air layering

Propagation by grafting

Micropropagation

The number of days taken for rooting have been found to differ with different species, being 45 days in *J. grandiflorum*, 70 days in *J. auriculatum* and 68 days in case of *J. sambac* (Pappiah *et al.*, 1980).

7.2.1 Layering

Layering has been the mode of vegetative propagation in jasmines until recently. Simple layering during June-July to October-November gives good success in 90 to 120 days. Stem of *Jasminum nudiflorum* have been found to develop adventitious roots when the branch comes in contact with ground and these can easily be propagated from layers (Burns, 1953). This method of multiplication, however, involves more time and restricts the number of plants propagated from a bush. According to Cai *et al.* (2007) long term cutting/layering has found to cause varietal degeneration, lack resistance and decline in flower production.

7.2.3 Grafting and Budding

An interesting study on interspecific grafting was carried out by Nambisan and Krishnan (1980) who showed that approach grafting was successful with a take of 76 per cent in Gundumalli (*J. auriculatum*) on Mullai (*J. sambac*), 72 per cent in Mullai on Gundumalli and 90 per cent in Jathimalli (*J. grandiflorum*) on Mullai. The time taken for separation of grafts ranged from 16 weeks to 24 weeks. Other methods like T-budding and side grafting, however, failed. But an earlier experiment by Veluswamy *et al.* (1980) with *J. grandiflorum* as scion on eleven different species of *Jasminum* using shield and patch budding and side and cleft grafting revealed that *J. grandiflorum* is compatible with *J. sambac* cv. Madhaban, Mathuria and Single Mogra, provided patch budding is resorted to, and it takes a month for bud take. Lai (1995) reported that the survival rate of graft (*J. sambac*) was low but the plants grew vigorously.

7.2.4 Micropropagation

Tissue culture has emerged as a promising technique to obtain genetically pure elite population under *in vitro* conditions (Malik, 2007). Micro-propagation provides a fast and dependable method of production of a large number of uniform plantlets in a short time. Moreover, the plant multiplication can be continued throughout the year, irrespective of season and the stock of germplasm can be maintained for many years.

Apical and axillary buds, selected from plants at different times of the year, were cultured in MS medium with 4 different growth substance formulations. Axillary buds were dormant from June to about February-March and when cultured during this period always formed callus followed by roots, regardless of the medium composition. Outside the dormant period, axillary buds responded to treatments in a similar manner to the apical buds which were active through the year. On a medium containing 1 mg NAA and 10 mg/l BA, apical buds produced copious callus which surrounded the meristem and impeded differentiation. Roots were produced later, with 1 mg NAA (or 1.5 mg IAA) + 1 mg kinetin/l, a small callus formed after 20-30 days (or 15-20 days with IAA); meristem differentiation resulted in a green plantlet with numerous roots. With 1-5 mg IAA + 10 mg/l BA, callus and multiple shoots developed in 45-60 days followed much later by roots. Plantlets from the various treatments were transferred to MS medium containing 10 mg IBA/l and no cytokinin to complete their development before potting (Scaramuzzi and D'Elia, 1984).

Bhattacharya and Bhattacharya (1997) cultured nodal segments of *Jasminum officinale* L. on MS basal media with 17.76 µM 6-BA and 0.53 µM NAA supplemented with 3 per cent sugar for shoot proliferation. NAA added to the B5-medium in propagation of cape-jasmine had promoted rooting at concentrations from 0. 1mg/l to 0.8 mg/l, and NAA at 0.2 mg/l was the best concentration for the growth of shoot and root (Li Bin, 2000).

Rajasekaran *et al.* (2000) recorded that the leaf bit showed highest callusing ability in MS medium supplemented with 1mg IAA/l and 10mg BAP/l within 23 days and 3 mg NAA/l proved to be the best for rhizogenesis.

MS basal medium combined with benzyl adenine (2.0 mg/l) and NAA (0.1 mg/l) was considered as appropriate medium for optimum bud proliferation (Cai *et al.*, 2007). Optimum number of multiple shoots induction in axillary buds was recorded in MS basal medium combined with benzyl adenine (1.5 mg/l) and NAA (0.3 mg/l). They also reported that rooting can be induced in *Jasminum sambac* by inoculation of micro-shoots in half strength MS medium devoid of sugar and supplemented with NAA (1.2 mg/l).

Sun *et al.* (2009) reported the highest bud induction rate of 96.75 per cent was obtained in WPM medium supplemented with 2.0 mg/l 6-BA and 0.1 mg/l IBA. The optimal rooting culture mode was two-step rooting method, *i.e.*, the first step micro shoots were cultured on half strength WPM medium supplemented with NAA (0.1 mg/l) for 7 days. In the second step the micro shoots were the transferred to half strength WPM medium without adding any growth regulator to obtain profuse rooting. The rooting rate was up to 98.41 per cent.

Bhattacharya and Bhattacharya (2010) developed protocol for *in vitro* propagation of *Jasminum officinale* L. by shoot regeneration from existing as well as newly developed adventitious axillary buds via proper phyto-hormonal stimulation.

He *et al.* (2011) isolated *in vitro* grown micro-shoots from *Jasminum sambac* and were immersed in 450 mg/l NAA solution for 10 minutes. Later the NAA treated micro-shoots were sub-cultured in half strength WPM medium for rooting. According to him better rooting could be achieved in *Jasminum sambac* by treating with NAA and hence NAA was considered as the most effective auxin in inducing roots. Maximum number of shoots (11.1 per explant) was induced on medium containing 1 mg l^{-1} zeatin and 0.5 mg l^{-1} IBA.

U-Kong *et al.* (2012) reported that 50 per cent shoot buds proliferated from shoot tips when *J. sambac* strain 1 was cultured on modified MS medium supplemented with BA 4 mg/l and 60 per cent shoot buds were initiated on kinetin 1mg/l. Further, 75 per cent shoot buds were induced from shoot tip explants when cultured on BA 4 mg/l whereas 54.5 per cent shoot buds were initiated from shoot tips when cultured on modified MS medium supplemented with kinetin 1 mg/l from *J. sambac* strain 2.

Farzinebrahimi *et al.* (2014) used young stems (each contained one nodal part) of *J. sambac* for shoot development on MS medium supplemented with a combination of BAP (3.0 mg/l) and NAA (1mg/l) that showed 20 percent shoot regeneration. Young stems of *J sambac* inoculated in MS medium supplemented with 2,4-D (0.5 g/l) developed callus.

Lakshmanan and Gabriel (2015) used leaf and stem explants of *Jasminum angustifolium* for callus regeneration on MS medium supplemented with 3 per cent sucrose with different concentrations of 2, 4-D and BAP. Maximum callusing was obtained from leaf explants when cultured on MS medium supplemented with 2, 4-D and BAP at 2.0 mg/litre respectively. The callus obtained in 2, 4-D concentration was best and shows 73 per cent of callus growth than BAP.

MS medium supplemented with 4.0 mg/l BA+0.1mg/l 2,4-D and 6.0mg/l BA+ 0.1 mg/l 2,4-D recorded highest percentage of callus induction whereas highest number of buds (10.1) was recorded in the combination 4.0 mg/l BA + 2.0 mg/l Kin from the inter-nodal explants of *J. azoricum* (Salim, 2016).

Early bud break was observed (25-26 days) when MS medium was supplemented with BAP (2.0 mg/l), Kinetin (1.0 mg/l) and Ads (50 mg/l) in *Jasminum sambac* (Biswal *et al.*, 2016). In MS medium having BAP (2.0 mg/l.), Kn (1.0 mg/l.), Ads (50 mg/l) and 3 percent sucrose higher percentage of bud break and multiple shoots formation (83.7 and 90.5) within 4 weeks of subculture was observed.

According to Palai *et al.* (2017), *in vitro* propogation of *Jasminum sambac* (L) Aiton is done by using nodes, shoot apices, stems and leaves as explants collected from field grown mature plants. Ex plants are inoculated on Murashige and Skoog's (MS) medium for multiple shoot proliferation and callus induction. Callus induction from leaf and meristem explants has been achieved using different concentrations of auxins and cytokinins. Maximum percentage of cultures showing callus proliferation from leaf explants was observed in the medium containing MS basal salts supplemented with 3.0 mg/l BAP and 2.0 mg/l IAA in leaf explants. White and friable callus was obtained in the medium containing MS basal salts supplemented with 2.0 mg/l BAP, 2.0 mg/l and 2.0 mg/l IAA from meristem explants.

Rahman *et al.* (2018) developed a protocol for *in vitro* plant regeneration of *Jasminum grandiflorum* L. The highest elongated shoots (60 per cent) were achieved from axillary meristems using MS basal medium supplemented with 1 mg/l 6-benzyladenine (BAP) and 60 mg/l coconut water. After adding 1.0 mg/l BAP and 45 mg/l coconut water in the culture medium, the highest rate of shoot proliferation was exhibited after 4 weeks of culture. Rooting was found within 14-23 days after the cut end of shoots was soaked in 1.5 mg/l IBA solution for 3-7 minutes. The regenerated healthy rooted plantlets were transferred to small plastic pot containing garden soil and compost in a ratio of 2:1. Maximum in vitro rooted plants were survived in the shade house (75 per cent).

Thenmozhi (2019) reported that in *Jasminum grandiflorum* L., the best callus performance was observed in MS medium containing combination of 1.5 mg/l 2,4-D and 1.5 mg/l IAA.

8.0 Soil and Climate

Jasmines prefer well-drained rich sandy loam to clay soils. The ideal conditions for their successful cultivation are warm summer and mild winter with ample water supply and sunny days. *Jasminum* spp. are usually grown in the open for commercial flower production. Susini (1955) reported from Italy that in summer, some species require a cool greenhouse, though they can survive in the open in favourable years and in winter they should be placed in a greenhouse with minimum temperature of 14°C.Coolhouse species should be kept in a temperature of 45-55°F (7-13°C) andstorespecies at 55.6°F (13-18°C) from September to March. The coolhouse species benefit from shading outdoors in a sunny place from June-September. The hardy species should be treated as slightly tender in cold areas (Anon., 1969).

9.0 Cultivation

9.1.0 Planting

The soil should be thoroughly prepared and kept free from weeds before planting. Pits of 45 cm^3 are prepared depending on the type of soil and are exposed for a week. The pit is filled with top soil after mixing with about 10-15 kg of well rotten cattle manure or compost and then irrigated to settle down the soil properly. Chlorpyriphos/Thimet may be added to the soil in pit to prevent the attack of termites.

Plant density is one of the factors determining the ultimate flower production. It varies from place to place and depends upon the species and cultivars and also on the soil and environmental conditions. In France, Spanish jasmine is planted at closer spacing accommodating 100000 to 120000 plants per hectare (Guenther, 1952). In a 3-year trial, *Jasminum grandiflorum* planted at different spacings were assessed for flower yield. The closest spacing (2 x 1.5 m) resulted in the highest average yield per hectare (10.71 tonnes) but the lowest yield per plant (Sambandamurthi *et al.*, 1980). There was no marked variation among the treatments for concrete recovery. The effect of plant population ranging from 1736 to 4444/ha on flower yield in this species was also studied by Bhattacharjee (1985b) who recorded higher yield with the increase in plant density. Among the five spacings tested, a closer planting of 1.8 x 1.8 m was found most suitable for *J. auriculatum* while for *J. sambac* the spacing recommended was 1.2 x 1.2 m (Anon., 1969). Pal *et al.* (1985) reported that flower yield in *J. sambac* was maximum when planting was done with 40000 plants per hectare.

Plants of *J. sambac* cv. Mogra raised from cuttings were planted at 3333, 4444, 5000 or 6666 plants/ha. Plant height, number of shoots/plant and weight of flowers/plant were not affected by planting density. Flower yield increased with increasing planting density, from 321.93 kg/ha at the lowest density to 596.80 kg/ha at the highest density (Patel and Singh, 1995). Trupti *et al.* (2017) studied the effect of different intercropping systems on growth and yield characters in jasmine and obtained maximum yield was from sole crop of *Jasminum sambac* (124.79 g/plant) which was at par with Jasmine intercropped with African marigold (1: 1), Jasmine with French marigold (1: 1) and Jasmine with French marigold (1: 2).

Cultivation of Jasmine

Cultivation in open field

Pruning

Cultivation in greenhouse

9.2.0 Irrigation

Sufficient moisture in the soil is necessary for proper growth and flowering in jasmines. Normally, plants are irrigated by flooding once a week. A comparative study with sprinkler irrigation was made in *J. grandiflorum* and *J. auriculatum*. In *J. grandiflorum* 22596 litres of water as sprinkler and 32280 litres of water as flood irrigation per 648 sq/m plot were given. The plot receiving sprinkler irrigation recorded a calculated yield of 7920 kg flowers while flood irrigation recorded 9335 kg flowers per hectare.

Though there was a 30 per cent saving of irrigation water under sprinkler, there was 10 per cent reduction in flower yield. In case of *J. auriculatum*, sprinkler irrigation was given in an area of 312 sq m with 7100 litres of water and 10140 litres of water as flood irrigation. The field which received sprinkler irrigation produced 6788 kg flowers while the plots under flood irrigation recorded 9112 kg flowers per hectare. Though there was 30 per cent saving of water, sprinkler irrigation showed considerable reduction in yield of flowers (Pappiah *et al.*, 1980). In a 2-year trial the plants were irrigated when available soil moisture (ASM) reached to 60, 40 or 20 per cent of field capacity, with the control plants being irrigated weekly. Irrigation at 60 per cent ASM increased flower yield by 20.33 per cent over the control, whereas irrigation at 30 per cent ASM depressed the yield by 9 per cent (Thamburaj *et al.*, 1984).

Kumaresan and Jawaharlal (2017) recommended irrigation @ 20 litre per plant at 4-day intervals for vigorous growth, physiological and yield parameters in Jasmine plants. Raw water (RW) and distillery spent wash (SW) were mixed and potted plants of Jasmine were irrigated. It was found that the yields of plant were very good in 1:3 (RW: SW) irrigation, while very poor in 1:1 (RW: SW), moderate in 1:2 (RW: SW) and good in RW irrigation growth (Chandraju *et al.*, 2012).

9.3.0 Manuring and Fertilization

Although commercial cultivation of jasmines has a long history behind, systematic investigations to determine the nutrient needs of the crop have been attempted only since the last two decades. Nutritional studies revealed that application of different nutrients markedly improved the growth, flowering and oil content.

9.3.1 Frequency and Rate of Application

The results of a 4-year trial with *J. auriculatum* receiving N, P and K each at 0, 120 or 240 g/plant annually, in all possible combinations showed that flower yield was highest with N at 120 or 240 g/plant. The effects of P and K alone were not very pronounced, but the N at 120 g/plant there was a response to K at 120 g/plant. The level of nitrogen can be reduced to half the recommended dose of 120 g/plant if applied by means of foliar spray at monthly intervals (Muthuswami and Pappiah, 1976). In a trial with *J. grandiflorum*, Muthuswami and Pappiah (1977) reported that yield, flower bud diameter and weight of 100 flower buds were maximum with FYM and NPK applied together in 6 split doses. Natarajan and Madhava Rao

(1980a) studied the effect of frequency of fertilizer application on *J. grandiflorum* and observed that split application showed little effect on the vegetative growth or floral characteristics but the maximum flower yield (3.642 kg/plant) and essential oil content (9.69 g/plant) were recorded when 15 kg FYM + 60 kg N + 120 kg P_2O_5 + 120 g K_2O were applied per plant in 12 monthly applications followed by the same dose in bimonthly application. In another trial Natarajan and Madhava Rao (1980b) reported that application of N at 60 g and P_2O_5 at 120 g singly and together markedly enhanced the flower and essential oil yield. The best result, however, was obtained when these N and P rates were combined with 240 g K_2O and 30 kg FYM per plant.

A field experiment was conducted by Hugar and Nalawadi (1994) to study the response of *J. auriculatum* to pruning date and nutrition. Pruning in the last week of November along with applying a 60 g N + 120 g P + 120 g K/plant resulted in maximum number of sprouts and flowers/plant and flower yield.

In *J. grandiflorum*, N at 600 kg/feddan (1.038 acres) gave the highest flower yield followed by 300 kg N (Nofal and Marwan, 1982a). The flower concrete content was unaffected by the N level, but a slight increase in absolute content in the concrete was noted in plants receiving the highest N level (900 kg/feddan). Different rates of N (0, 50, 100 and 150 g/plant), P_2O_5 (0, 100 and 150 g/plant) and K_2O (0 and 100 g/plant) were applied in 24 treatment combinations, over a basal rate of FYM (10 kg/plant), to 3-year-old bushes of *J. grandiflorum*. Treatment with N (as urea) up to 100 g/plant significantly improved all vegetative characters, flower quality and yield, but no significant difference in flower production was noted between N at 100 and 150 g/plant; P_2O_5 and K_2O application increased flower yield and the number of branches/plant and P_2O_5 increased the weight of 100 flowers. Treatment with 100 g N, 150 g P_2O_5 and 100 g K_2O/plant annually resulted in the highest flower yield (Bhattacharjee, 1983b, 1985a). Application of NPK at 60: 120: 120 g per plant along with 10 kg of FYM applied in two splits was found to be optimum (Muthuswami and Abdul Khader, 1986).

The experiment by Natarajan *et al.* (1981) revealed that application of N and K each at 240 g per plant per year at bimonthly interval was best for flower yield in *J. sambac*. Kumar and Gill (1983) observed that plants of cv. JSL-1 receiving N at 30 g/plant in split doses, half the amount at pruning and remaining half at the end of the first flowering flush (late May), gave the highest yield of flower buds (635.8 g/plant); the yield in control plant was 353.7 g. Similarly, Pal *et al.* (1985) obtained the highest flower yield in *J. sambac* treated with N and P_2O_5 at 350 and 300 kg/ha, respectively. In an experiment with *J. grandiflorum* grown in soil with low available Fe (5.5 ppm), the highest yield of picked flowers (3132 g/plant) was obtained from plants receiving 20 kg FYM/plant at planting plus monthly foliar sprays with 1 per cent $FeSO_4$. The meancontrol flower yield was 1410 g/plant (Velu, 1988).

In *Jasminum auriculatum* application of *Azospirillum, Pseudomonas striata, Pseudomonas fluorescens* and *Trichoderma viridae* were used @ 8 kg/ha could substitute the recommended NPK fertilizers to the extent of 50 per cent without affecting various growth parameters, floral characteristics and flower yield (Jayamma *et al.*, 2008).

Application of 150 g/plant of nitrogen and 75 g/plant of phosphorus resulted in significant vegetative growth and flower yield of Jasmine (*Jasminum sambac* Ait.) cv. 'Double' (Shoram *et al.*, 2012). Sivakumar *et al.* (2014) recommended three sprays of a foliar formulation comprising of Ferrous sulphate (0.5 per cent), zinc sulphate (0.3 per cent), magnesium sulphate (0.3 per cent), salicylic acid (100 ppm), citric acid (0.1 per cent), borax (0.3 per cent), K_2SO_4 (0.5 per cent) and NAA (20 ppm) during the month of January, February and March in *Jasminum sambac* for increased yields and higher BC ratio. Varsha *et al.* (2015) recommended fertigation schedule of 60: 120: 120 g NPK per ha for *Jasminum multiflorum* in combination with $ZnSO_4$, $CuSO_4$, $MnSO_4$, $FeSO_4$ and Borax at 0.72 g each per plant. Minimum days taken for bud initiation, first flowering, 50 per cent flowering, maximum flower diameter, corolla tube length, flower bud length, shelf life, yield per plant (88.32 g), flower yield per hectare (6.00 t) and highest BCR (2.04) was recorded.

Application of 10 kg FYM + 120:240:240 g NPK per plant per year (RDF) in four equal split doses during February, May, September and December on September pruned plants augmented lean season flowering and quality of flowers during winter season (Nair and Venugopalan, 2016).

In *Jasminum sambac*, Chopde *et al.* (2017) reported significantly earliest sprouting and maximum sprouts plant^{-1}, length of primary shoot, secondary laterals primary shoot^{-1}, weight of 100 flower buds, flower buds plant^{-1}, flower yield plant^{-1}, diameter of flower bud and shelf life of flower were recorded when the jasmine plant treated with 120 kg/ha (100 per cent N through urea) + *Azotobacter* + PSB followed by 120 kg/ha (100 per cent N through vermicompost) + *Azotobacter* + PSB.

Diwivedi *et al.* (2018) recommended application of 25 per cent RDF + 15.5 kg FYM in *Jasminum grandiflorum* L. for maximum vegetative growth, yield of flowers and benefit cost ratio.According to Sowmiya and Karuppaiah (2019) in *Jasminum sambac* Ait. cv. Gundumalli application of 25 t FYM ha-1 + recommended dose of 60:120:120 kg NPK ha^{-1} along with three sprays of foliar concoction (MKP 1 per cent + 0.3 per cent Ferrous sulphate + 0.3 per cent Magnesium sulphate + 0.3 per cent Zinc sulphate + 0.3 per cent Borax + 0.5 per cent Seaweed extract + 100 ppm Salicylic acid) during the month of January, February and March was found to be the best in all growth, physiological, yield and quality characters. Paswan and Saravanan (2019) reported that plant growth and yield parameters of *Jasminum grandiflorum* was superior with application of 0.2 per cent Iron + 0.4 per cent Zinc along with the recommended dose of fertilizers. In *Jasminum sambac* cv. Mysuru Malligae application of 100 per cent RDF (60:120:120 NPK g/plant through fertigation) + Humic acid at 0.5 per cent, Chelated Zinc and Borax at 0.75g each per plant increased plant height, leaf area and chlorophyll content, bud diameter, bud length, corolla length and flower diameter (Keerthishankar *et al.*, 2019). Highest score for freshness, fragrance, minimum loses of physiological weight more shelf life of the flowers and flower yield was also recorded (Table 3).

9.3.2 Time and Method of Application

Application of fertilizers to the soil in split doses has been found very useful in improving the growth, flowering and oil content. The experiment with foliar

Table 3: Effect of different Levels of Fertigation and Foliar Spray of Micronutrients on Flower Quality Parameters*

Treatment	*Freshness Index (5-point scale)*	*Fragrance Index (5-point scale)*	*PLW (Per cent) (upto 24 hours)*	*Shelf Life (hrs)*	*Yield/Plant (g)*
Soil application 100 per cent RDF	3.90	3.98	18.20	26.13	518.67
100 per cent RDF through fertigation	4.27	4.23	13.25	29.33	615.73
75 per cent RDF through fertigation	4.13	4.10	16.50	28.38	556.09
50 per cent RDF through fertigation	4.33	4.25	21.50	26.35	506.33
Soil application 100 per cent RDF + 0.5 per cent Humic acid Chelated Zinc and Borax at 0.75g each/plant	4.42	4.33	15.20	27.37	566.50
100 per cent RDF through fertigation + 0.5 per cent Humic acid, Chelated Zinc and Borax at 0.75g each/plant	4.91	4.91	8.50	34.10	725.80
75 per cent RDF through fertigation + 0.5 per cent Humic acid, Chelated Zinc and Borax at 0.75g each/plant	4.83	4.82	8.86	33.51	640.50
50 per cent RDF through fertigation + 0.5 per cent Humic acid, Chelated Zinc and Borax at 0.75g each/plant	4.30	4.30	16.65	31.20	508.00
Farmer's practice (200g DAP, 200g MOP, 50g Urea/plant/ year)	3.85	3.87	31.50	24.30	458.33
C.D. (P=0.05)	0.28	0.26	0.74	0.21	8.94

Recommended dose of fertiliser (RDF) - 60:120:120 NPK g/plant/year.

* Keerthishankar *et al.* (2019).

and soil application of N revealed that soil application was better than foliar application in *J. grandiflorum*; foliar application, on the other hand, depressed the yield (Muthuswami and Pappiah, 1977). Experiment conducted on graded doses of NPK and split appli-cation of fertilizers showed that NPK applied in two split doses once in January and another in July every year gave a yield of 10.32 tonnes flowers per hectare (Muthuswami and Abdul Khader, 1986). In *J. sambac*, however, foliar application of nitrogen did not show any significant difference on flower yield compared with soil application. Treatment with N as foliar spray at 30 g per plant was reported to be more economical than the conventional method of applying nitrogen at 120 g per plant through the soil (Anon., 1975).

In *Jasminum sambac* cv. Ramanathapuram Gundumalli drip fertigation with 125 per cent recommended dose of fertilizer in combination with foliar spray of 0.4 per cent humic acid and 3 per cent panchagavya recorded the highest plant height, plant spread, dry matter production, minimum number of days taken for first flower bud emergence, highest number of cymes plant^{-1}, bud length, bud diameter and yield of flower buds. Highest BCR was obtained with the application of 100 per cent recommended dose of fertilizer in combination with humic acid (0.4 per cent) and panchagavya (3 per cent) spray (Sundar *et al.*, 2010).

Naik *et al.* (2019) recorded off season flower induction in *Jasminum sambac* L. through foliar application of 0.5 per cent $FeSO_4$ + 0.5 per cent $ZnSO_4$ + 20 ppm NAA to plants after pruning significantly increased plant height, plant spread early flower bud initiation, flower bud length and diameter, weight of hundred flower bud, shelf life of flower buds and flower yield.

9.4.0 Leaf Nutrient Analysis

The nutrient element (N, P, K, Ca, Mg and S) content of jasmine (*J. multiflorum*) leaves was investigated by Patil *et al.* (1998). Samples of 10, 20, 30, 40, 50 and 60 leaves were taken from nodes 5 to 8, from primary and secondary branches, and analyzed. N and K contents were lower for leaves on primary branches than on secondary branches, P, Mg and S were higher, and Ca was the same. There was no interaction between leaf position and sample size. It is suggested that sampling of 30-40 leaves on secondary branches would be sufficient for nutrient diagnosis.

In *Jasminum auriculatum* combination of FYM @ 25 t ha^{-1} plus 100 per cent of the recommended dose of inorganic fertilizers (120:240:240 kg NPK ha^{-1}) along with biofertilizers *viz.*, *Azospirillum* and Phophobacteria @ 2 kg ha^{-1} recorded the highest leaf nutrient content (Anburani, 2012).

9.5.0 Pruning

Pruning is an important practice for manipulation of growth and flowering as it influences growth, flower bud initiation, differentiation and ultimately the flower production. Normally, irrigation is withheld prior to pruning and plants are pruned by removing all the past season's shoots including dead and diseased branches. All the leaves in the bushes are stripped off. Following pruning all cut ends are smeared with Bordeaux paste to prevent infection. The height of pruning depends on the species in a particular environmental condition.

Muthuswami *et al.* (1973a) observed that pruning during the last week of January led to the maximum flower production of *J. auriculatum* and pruning beyond this time resulted in low yield. The higher yield associated with January pruning was attributed to greater production of vegetative shoots. The yield increase between early and late prunings ranged from 59.5 to 67.8 per cent. Pal *et al.* (1980b) studied the effect of time and height of pruning on growth and flower yield in the same species. For the cultivar Single, December and January pruning proved most favourable whereas the cultivar Double responded best when pruned in February and the optimum height of pruning was 30 cm. Pruning in the last week of November produced plants with a significantly lower number of leaves/ primary shoot and higher leaf area/plant (m^2), flower weight and flower number of *J. auriculatum* (Hugar and Nalawadi, 1993).

In *J. grandiflorum*, Muthuswami (1975) reported that the time of pruning influenced the quantum of flower yield in each month from February to October. Pruning as early as November did not help in shifting the peak production in June-July. Subramanian (1977) observed that early pruning in December enhanced the duration of flowering up to eight months while it was only five months in late pruned crop. The highest yield was obtained in plants pruned in December and the lowest yield from those pruned in March. The effect of pinching of new shoots, produced after pruning to stagger the flowering and thereby spread the period of productivity, was investigated by Muthuswami and Madhav Rao (1980b). They found that pinching delayed flowering by about 17 days in *J. grandiflorum* and 14 days in *J. auriculatum*. The overall growth of laterals produced after pinching was also more in both the cases compared with controls. A slight delay in flowering may be of some help to stagger the peak productivity. However, a second pruning of the spent shoots is September in the above two species was not helpful as the number of laterals produced were limited and they also took a longer time to flower.

In an experiment, *J. grandiflorum* was pruned on 5 December, 1 or 25 January. Average flower yield, from harvest between June and November, was highest after the earliest pruning and declined with successive pruning dates. A similar trend was noted with regard to yield of concrete and absolute and their quality (Nofal and Marwan, 1982b). In an investigation on the effect of different pruning height on growth and flowering of the same species, pruning at 90 cm from the base of the stem resulted in maximum number of shoots and the highest flower yield/plant (Bhattacharjee, 1985c).

Muthuswami *et al.* (1973b), working with *J. sambac* cv. Single Mogra, reported that flowering was at peak during March-May and declined thereafter with no flowering between October to January. By pruning different plants at six-month interval, staggered flower production could be achieved in this cultivar. Pal *et al.* (1980c) observed that plants of *J. sambac* cv. Khoya pruned at 40 cm height in January produced the highest yield.

Singh and More (1986), on the other hand, reported that pruning to 75 cm height in mid-December resulted in the highest flower and concrete yield. Effect of pruning of *J. sambac* of 75 cm above ground at monthly intervals on growth and flowering was also reported by Siddagangaiah and Rai (1990). The duration of flowering was significantly higher for bushes pruned in October and November. Flower yield was significantly higher for bushes pruned in December (380.10 g/bush), January (362.64 g/bush) and November (356.90 g/bush). Peak flower production occurred during March to June and was attributed to higher temperatures and longer hours of sunshine; it declined after June with falling temperatures, almost ceasing in the second week of July.

The usefulness of chemical defoliants for substituting manual pruning was investigated by Muthuswami *et al.* (1974). Chemical defoliants like paraquat dichloride, pentachlorophenol, potassium chloride and sodium chloride were employed on *J. auriculatum*. While paraquat dichloride caused quickest and heaviest defoliation; pentachlorophenol and sodium chloride application showed higher yield than the unpruned plants. Use of defoliants as an alternative to manual pruning was thus effective, however, it has some limitations.

Nair *et al.* (2009) sated that pruning time influenced lean season flower production in *Jasminum sambac* during winter months recording a decline in production and quality of flowers. Advancing the time of pruning to September yielded 72.9 g/plant of good quality flower buds flowers during winter months as compared to November pruned plants which produced only 25.51g/plant of marketable flower buds. The peak season yield was not affected by advancing the pruning time. September pruning enhances the yield under Bengaluru conditions and profitability due to higher prices realized in winter season. Chaitanya *et al.* (2016) recommended pruning *Jasminum sambac* at the height of 50cm from ground level in December under West Godavari conditions for highest flower yield is obtained with good quality produce.

In *Jasminum sambac* (L) maximum diameter of primary shoot, leaves primary shoot^{-1} and flower yield plant^{-1} were recorded when the jasmine plants pruned during 4th week of December under Nagpur conditions. With respect to severity of pruning, sprouts plant^{-1}, diameter of primary shoot, leaves primary shoot^{-1}, productive shoots plant^{-1}, weight of 100 flower buds, flower yield plant^{-1} and mean corolla tube length were maximum when the plants were pruned at 30 cm above ground level (Lokhande *et al.*, 2016). Pal (2017) recommended pruning of *Jasminum sambac* (L.) Ait. cv. Local in the last week of January at 45 cm height from the ground level for higher growth rate and yield under Western U.P. conditions.

9.6.0 Weed Control

Weeds cause enormous losses to jasmine crop and increase the cost of cultivation. Manual weed control is effective if done properly, but this is very expensive. Chemical weed control is comparatively economical, convenient and efficient in controlling weeds by one or two applications. Treatment with oryzalin controlled weeds for 10 weeks on silt loam soil. Good vegetative growth and flowering of jasmine was also recorded with the application of this chemical (Haramaki and Kuhns, 1983). Mulching also reduced the weed population.

Dilipkumar *et al.* (2019) recommended incorporation of imazethapyr at 10 g a.i. ha^{-1} with oil palm frond mulch powder (< 2mm) to effectively control goose grass, slender cyperus and coat buttons without injuring jasmine and other ornamental plants. The OPF mulch treated with imazethapyr at this rate acts synergistically and provides 80 per cent - 95 per cent inhibition of weed growth. However, increasing the rate of imazethapyr over 10 g a.i. ha^{-1} does not further improve weed inhibition, and results in antagonism on goose grass and slender cypress emergence.

10.0 Diseases, Pests and Disorders

10.1.0 Diseases

Fungal Diseases

Jasmines are affected by various diseases which affect growth and cause loss in flower yield. Some diseases also cause death of the plants. A short account of the important diseases and their control measures are summarized here.

10.1.1 Leaf Blight

Two fungi, *viz., Cercospora jasminicola* and *Alternaria jasmini* cause leaf blight disease. This disease occurs in severe form on *J. grandiflorum*. Reddish brown circular spots, 2 to 8 mm in diameter are produced on the upper surface of the leaves, spreading rapidly in the rainy season. Affected leaf margins show inward curling and become hard and brittle. In severe cases of infection, vegetative buds and young branches dry up. Since almost all the leaves are affected along with young branches, the plants look unhealthy with appearance of drying up from a distance. The disease appears in the month of May/June and the peak incidence occurs from August to November. Flower production is adversely affected and may cause 50 per cent loss in yield.

Screening studies conducted at Tamil Nadu Agricultural University, Coimbatore, India showed that *J. arborescens, J. paniculatum* and certain cultivars of *J. sambac* remained free from leaf blight and *J. grandiflorum* was highly sus-ceptible to this disease. *Jasminum angustifolium, J. auriculatum, J. communis, J. flexile, J. humile, J. pubescens, J. rigidum* and certain cultivars of *J. sambac* showed only stray incidence of blight. Fungicidal trials conducted by using the highly susceptible cultivar of *J. grandiflorum* revealed that 0.4 per cent Benlate, 0.2 per cent Dithane M-45, 0.1 per cent Bavistin and 1.0 per cent Bordeaux mixture were equally effective to control leaf blight. It was suggested to spray the fungicides at monthly interval commencing from May and continue up to pruning. The number of sprays may vary depending upon the weather conditions and severity of the disease (Ramkrishnan and Kandaswamy, 1980). Batra (1967) recommended spraying with Bordeaux or copper oxychloride to control the blight.Incidence of diseases in 5 species of *Jasminum* was observed by Thammaiah *et al.* (1997). *J. multiflorum* were highly susceptible to leaf spot (*Cereospora jasminicola*) diseases.

10.1.2 Rust

This disease is caused by *Uromyces hobsoni* and occurs in July-August during monsoon rains, in severe form on *J. auriculatum*. Other species *viz., J. sambac, J. flexile, J. communis* and *J. angustifolium* are also susceptible to rust. The leaves show the presence of orange coloured aecial cups on both sides, but predominantly on the lower surface. Numerous blisters are produced in advanced stages of infection causing yellowing and crinkling of the leaves. The stems and branches are also infected, causing splitting of barks and subsequent death of the branches. The disease can be controlled by dusting sulphur 20-25 kg/ha (Ramkrishnan and Kandaswamy,

1980). Spraying with Bordeaux or copper oxychloride was also recommended for the control of rust (Batra, 1967).

10.1.3 Wilt

This disease is caused by *Fusarium solani* or *Sclerotium rolfsii*, mostly in *J. sambac* (Thammaiah *et al.*, 1997). The early symptom is yellowing of lower leaves which gradually spreads upwards and finally resulting in death of the plant. In the field, the disease occurs in patches and roots show black colouration. In case of sclerotial wilt, in addition to the above symptoms, white mycelia are found generally girdling the roots and found adhering to the roots of the wilted plants. Drenching the soil around the plants with 1 per cent Bordeaux mixture controls this disease.

Ramamoorthy *et al.* (2000) recommended the application of *Trichoderma harzianum* @ 10 g/plant one day prior to planting for management of Sclerotium wilt (caused by *Sclerotium rolfsii* of jasmine (*Jasminum sambac*) to decrease the disease incidence to 21.88 and 25.00 per cent at 60 and 120 days after planting, respectively.

10.1.4 Root-rot

Pirone *et al.* (1960) observed root rot (*Clitocybe tabescens*) in jasmines. Root rot of jasmine caused by *Macrophomina phaseolina* was controlled by *Trichoderma viride* along with neem cake (Alice *et al.*, 1996).

Root and basal stem rots of jasmine were reported in Egypt by Hilal *et al.* (1987). The most commonly isolated fungi associated with infected plants and cuttings were *Botryodiplodia* spp., *Fusarium solani* and *Macrophomina phaseolina*. The most virulent pathogens on the cuttings were *F. moniliforme* (*Gibberella fujikuroi*), *F. oxysporum* and *Pythium* sp., followed by *Rhizoctonia solani*. These rots were most effectively controlled by dipping cuttings for 1 h in 2000 ppm suspensions of Benlate (benomyl), benomyl + sportak, rovral (iprodione), TMTD (thiram) + indole butyric acid (IBA) and topsin M (thiophanate-methyl) + IBA. Planting IBA-treated cuttings in nursery soil treated with benomyl, botran (dicloran), iprodione or thiopha-nate-methyl increased healthy survival over that obtained with the fungicides alone.

Jasmines are subject to several other diseases including crown-gall caused by the bacterium *Agrobacterium* (*Erwinia*) *tumefaciens* and a blossom blight caused by a species of *Phoma* (Pirone *et al.*, 1960).

Viral Diseases

10.1.5 Mosaic

The diseased plants show stunted and yellowish green appearance with small leaves. Yellowish green to chlorotic flecks of 1-2 mm in diameter appear irregularly on the leaf and these streaks form into a ring. The symptoms are more conspicuous on older leaves. The disease is graft-transmissible on *Jasminum simplicifolium* (McLean, 1960). *Jasminum sambac* and *J. pubescens* were found to be highly susceptible to the disease while *J. auriculatum* relatively tolerant. Control of insect vectors will prevent the transmission of the disease (Mariappan and Ramanujam, 1975).

Diseases of Jasmine

Leaf blight

Root rot

Wilting of jasmine

Jasmine rust

Mosaic virus in jasmine

Pests of Jasmine

Bud worm bore on bud

Caterpillar inside the bud

Infested flowers

Mealy bugs

Green plant hopper

Jasmine bug

Leaf roller

Leaf webworm

Red spider mite

White fly

Blossom midge

Phyllody

This disorder is suspected to be caused by mycoplasma. The affected plants produce malformed, greenish flower-like structure instead of fragrant white flowers on panicles which are highly congested and green in colour. The greenish corolla lobes become reduced and ovate in shape. Flower parts are transformed into leaf-like structures. The disease may be controlled by spraying with tetracycline hydrochloride (250 ppm). Cuttings from infected plants should never be used for planting (Ramkrishnan and Kandaswamy, 1980).

10.2.0 Pests

10.2.1 Bud Worm

Some important pests which cause considerable damage to jasmines have been reported by David and Kumaraswami (1975). *Hendecasis duplifascialis*, the bud worm which is a greenish larva with a black head bores into immature jasmine buds and feeds on floral structures and in severe cases webbing of buds is also noticed. Plots of 3-year-old jasmine (*J. sambac*) bushes infested with jasmine bud worm (*Hendecasis duplifascialis*) were treated with 7 insecticides. Deltamethrin at 25 g a.i./ha and cypermethrin (150 g a.i./ha) reduced infestation by 90.1 per cent and 90.2 per cent in the Ist trial and 87.3 per cent and 86.9 per cent in the 2nd trial, respectively (Dhandapani *et al.*, 1989). Amutha and David (1995) suggested that cypermethrin applied at 75 g a.i./ha was the most effective to control *Hendecasis duplifascialis*, followed by delta-methrin applied at 12.5 g a.i./ha, and diflubenzuron applied at 100 g a.i./ha. Endosulfan at 0.07 per cent alone or alternated with fenvalerate at 0.015 per cent was the most cost-effective treatment against *Hendecasis duplifascialis* (Chandramohan and Manoharan, 1990), which also had the lowest impact on its parasitoid Phanerotoma.

Kamala (2019) observed a significant reduction in incidence of jasmine budworm, *Hendecasis duplifascialis* in intercropped fields over monocrop of jasmine. Intercropping with marigold recorded significantly the lowest mean incidence of 15.21 per cent, followed by coriander (16.40 per cent) and gingelly (16.69 per cent) and jasmine as sole crop recorded highest budworm incidence (22.17 per cent). The population of natural enemies *viz.*, coccinellids, chrysopids, praying mantises and spiders were more in marigold intercropped jasmine. Moreover, the emergence of the larval parasitoids, *Bracon brevicornis* (3.10 no./five plants) and *Phanerotoma hendecasiella* (2.60 no./50 infested buds) were encountered in jasmine intercropped with marigold than sole jasmine crop. The pest defender ratio was highest in jasmine + marigold (1:3.50) and the order falls as jasmine + coriander (1:3.05) and jasmine + gingelly (1:2.34). The data indicated the supremacy of marigold in reducing the incidence of the jasmine budworm as well as attracting more natural enemy population.

10.2.2 Mites

The incidence of red spider mite is high during warm and dry weather, especially during summer. The mites are seen feeding on undersurface of the leaves which become yellow and drop off. Severe puckering and discolouration of leaves are caused by gall mite in *J. auriculatum*. The cultivar Parimullai of *J. auriculatum* released from Tamil Nadu Agricultural University, Coimbatore, is resistant to gall mite (Anon., 1973). The cyclamen mite, *Steneotarsonemus pallidus*, was found to severely damage *J. sambac*. In trials with 7 acaricides against *Aceria jasmini* on *J. auriculatum* treatment with 0.05 per cent ethion, 0.01 per cent fenpropathrin, 0.02 per cent dicofol, and a mixture of 0.05 per cent ethion + 0.03 per cent dimethoate were superior to the others (Mallikarjunappa *et al.*, 1991). In South India *Aceria jasminae* (*A. jasmini*) infestation of jasmine is present throughout the year. Untreated crops were severely affected during July to August (>90 per cent of branches affected) (Mallapur and Kubsad, 2000).

10.2.3 Mealy Bug

A field trial was conducted by Dhandapani *et al.* (1992) to evaluate the effect of 7 insecticides and coconut oil on the incidence of *Planococcus lilacinus* in *J. auriculatum*. Lowest populations of the *pseudococcid* were recorded with 0.036 per cent monocrotophos followed by coconut oil treatment.

10.2.4 Gallery Worm, Tingid Bug and Scales

The gallery worm, *Elasmopalpus jasminophagus* is also a serious pest which causes webbing of terminal leaves, shoots and flowers. The tingid bug, *Corythauma ayyari* is sometimes noted to be serious. The nymphs and adults suck the sap and cause yellowing of leaves which dry and drop off. Several species of scales infest jasmine. Plants should be sprayed with malathion to control these insects (Pirone *et al.*, 1960). Hung (1965) reported the occurrence of *Nauoinoe geomatralis* which can be controlled by spraying during blossoming with 1:4000-6000 solution of 50 per cent diptrex combined with hand picking of larvae and pupae was also recommended.

Topical application of *Beauveria bassiana* significantly reduced the survival rates of jasmine gallery worm, *Elasmopalpus jasminophagus* at both 2.5 x 10^7 and 2 x 10^9 conidia ml^{-1}. *Metarhizium anisopleae* at (2 x 10^9 conidia ml^{-1}) showed significant mortality percent in second (80.00), third (67.00) and fourth instar larvae (53.00). *Nomuraea rileyi* at conidial concentration 2 x 10^9 conidia ml^{-1} showed 73.3, 53.00 and 77.00 per cent mortality in second, third and fourth instars larvae respectively, and was on par with chemical check. In dry residue experiment, *M. anisopleae* showed higher mortality percent at conidial concentration of 2 x 10^9 ml^{-1} in all the stages of larvae and was on par with the mortality caused by chemical check, Indaxocarb and commercial Bt product (Pillai *et al.*, 2016).

10.2.5 Blossom Midge

Jasmine blossom midge (*Contarinia maculipennis* Felt Cecidomyiidae, Diptera) is a serious pest posing severe threat to flower yield. David *et al.* (1991) observed that 0.1 per cent monocrotophos was mosteffective against jasmine blossom midge, *Contarinia* sp. The lowest percentage of discoloured buds (2.62 per cent) was observed on bushes treated with monocrotophos (David *et al.*, 1992). Coccinellid predators like *Cheilomenes sexmaculata*, *C. transversalis*, *Scymnus* sp., *C. septumpunctata*, *Brumus suturalis*, *Illeis cincta* and *Chilocorus* sp. were present in the jasmine ecosystem with a mean population of 3.1 per plant. Neuropteran predators, *Chrysoperla zastrowi sillemi* and *Mallada boninensis* were found with a mean population of 2.2 per plant. A wide range of praying mantises and spiders with a mean population of 3.8 and 4.3 per cent respectively was present. A dipteran predator, *Systasis dasyneurae* played a vital role in natural suppression of the pest. Native weeds and flowering plants like *Parthenium hysterophorus*, *Rosa* sp., *Nerium oleander*, *Vernonia cinera*, *Datura metal*, *Lantana camara*, *Tagetus erecta*, *Ipomea carnea*, *Helianthus annus* and *Coriandrum sativum* provide nectar to the adult stages of the predators in jasmine ecosystem (Kamala, 2018).

Kamala and Kennedy (2018) evaluated the efficacy of microbials against jasmine blossom midge (*Contarinia maculipennis* Felt) under *in vitro* and field conditions. *Metarhizium anisopilae* @ $1x10^8$ spores ml^{-1} recorded maximum maggot mortality over untreated control (80.06 per cent) followed by *Beauveria bassiana* @ $1x10^8$ spores ml^{-1} spores ml^{-1} recording 67.57 per cent mortality. Further, field evaluation for confirmation revealed similar trend confirming the efficacy of *M. anisopilae* with minimum infestation of 8.32, 7.69 and 7.31 per cent followed by *B. bassiana* recording 8.86, 8.93 and 8.50 per cent and *L. lecani* recording 9.97, 10.26 and 11.77 per cent infestation after first, second and third round of application respectively.

Harini *et al.* (2019) evaluated the bio-efficacy of botanicals and insecticides against jasmine blossom midge, *Contarinia maculipennis* Felt infesting *Jasminum sambac*. Among the botanicals, pungam oil was the most effective treatment with least infestation followed by NSKE were able to reduce the infestation by 82.66 and 76.21 per cent, respectively. Flubendiamide 39.35 SC was the best treatment recording the greatest reduction of 85.47 per cent followed by chlorantraniliprole 18.5 SC, respectively. The results concluded that insecticides are more effective then botanic insecticides against blossom midge.

Suganthi *et al.* (2019) recommended integrated use of biocontrol agents and biopesticides for reliable and effective pest control in jasmine, reducing pest management cost and use of hazardous chemicals. Application of *Beauveria bassiana* (NBAIR formulation) at 5 g/litre of water three times along with 6 releases of *Trichogramma chilonis* @ 100,000/ha at 7 days interval from bud initiation stage during March was superior in checking the bud borer (50 to 86 per cent) and blossom midge damage (61 to79 per cent). Pirithiraj *et al.* (2020) recorded maximum incidence of blossom midge in *Jasminum sambac* L. during August (34.05 per cent Infested buds). Occurrence of midge was positively correlated with maximum temperature (r = 0.159) and minimum temperature (r = 0.505). The multiple linear regression on damage with weather parameter was worked out and it indicates that the cecidomyid pest was 85.4 per cent influenced by various weather parameters.

10.2.6 White fly

In China, Sengonca and Liu (1998) observed that realease of *Eretmocerus longipes* was successful biological control of whitefly (*Aleurotuberculatus takahashi*). Jasmines are also attacked by *Cacoeia pronubana* in Tripolitania. The insecticides recommended for control are phosphonic esters in combination with chlorinated product (Damiano,

1966). Some of the other pests which have been recorded on jasmine are the green plant hopper (*Flata oellata*), the jasmine bug (*Antestiopsis cruoiata*) and leaf rollers (*Glyphodes celsalis*).

10.2.7 Nematodes

Infestation of root knot nematode, *Meloidogyne incognita* in both *J. sambac* and *J. flexile* causing small swelling and enlarged rootlets leading to conspicuous pale yellowing of leaves and die back was recorded by Rajaram and Rajendran (1979). According to Nemec and Struble (1968), *J. nudiflorum* was very susceptible to *M. incognita*. *Psuedomonas fluorescens* @ 10 g/plant (6 x 10^8 colony forming units/g), gave maximum reduction of juveniles per 100g soil (69.7 per cent), number of adult females (70.1 per cent) and eggs per g root (87.2 per cent) of root-knot nematode, *Meloidogyne incognita*, gall index (2.6) and increased flower yield (36.6 per cent) in jasmine (*Jasminum sambac* L). Carbofuran, phorate, *Paecilomyces lilacinus* and neem cake were next in line in efficacy. Further, organic amendments and bio-agents had a positive influence on growth parameters *viz.*, plant height, shoot weight and root weight. The endophytic colonization potential of the bio-agents introduced into the soil was confirmed by re-isolating them from jasmine roots Seenivasan and Poornima (2010).

11.0 Harvesting and Yield

11.1.0 Harvesting

The stage of harvest depends on the use of flower to be harvested. For fresh flowers, fully developed unopened flower buds are picked early in the morning while for extraction of concrete only fully opened freshly picked flowers are required. Such flowers are picked preferably before 9-30 a.m. in Coimbatore, India. Delay in picking results in gradual reduction of concrete yield. Picking of the flowers after 11 a.m. will considerably reduce the yield and quality of the concrete (Muthuswami and Abdul Khader, 1986). The quantity and composition of essential oil from *J. multiflorum* flowers harvested at 08.00 and 18.00 h were compared by Ahmad *et al.* (1998). Harvest time did not affect the quantity of jasmine concrete extracted from the flowers but did affect its chemical composition. The concentrations of indole and *cis*-jasmone were higher in oil from flowers harvested in the morning than in that from those harvested in the evening. The oil from flowers harvested in the morning also contained higher concentrations of benzyl alcohol, linalool and benzyl acetate. Other aroma compounds such as eugenol, benzyl benzoate, farnesol, methyl palmitate and methyl salicylate were present in higher concentrations in oil from evening harvested flowers. Therefore, morning harvesting of jasmine flowers for oil extraction was recommended. Care should be taken to see that the flowers are not badly handled, wrinkled and damaged during harvest and transit. If this is not adhered to strictly, the quality of the fresh flower and concrete will be affected and the entire product may become unmarketable (Krishnaswamy, 1980).

11.2.0 Yield

Flower and concrete yield in jasmines vary considerably according to the species and cultivars, management practices, *etc.* The flower yield and concrete recovery in three important species of jasmine, as reported by Abdul Khader *et al.* (1984) are presented in Table 4.

Table 4: Flower Yield and Floral Concrete Recovery in different Species of Jasmine*

Species	*Flower Yield (kg/ha)*	*Concrete Recovery (Per cent)*
J. auriculatum (range for 5 clones)	4733 to 9152	0.28 to 0.36
J. sambac (range for 6 clones)	739 to 8129	0.14 to 0.19
J. grandiflorum (range for 6 clones)	4329 to 10144	0.25 to 0.32

* Abdul Khader *et al.* (1984).

The concrete recovery of *J. grandiflorum* ranged from 0.25 to 0.32 per cent. *J. auriculatum* recorded higher recovery of concrete between 0.28 and 0.36 per cent while *J. sambac*, which is mainly grown for flower, yielded lower percentage of concrete, the recovery ranging from 0.1 to 0.19 per cent. In a varietal evaluation, cultivar Co.1 Pitchi was found to be the highest yielder producing over 10 tonnes of flowers per hectare with an estimated concrete yield of 29 kg, the recovery being 0.29 per cent (Muthuswami and Abdul Khader, 1986). The chemical composition of four *Jasminum* spp. was analysed using gas chromatography- mass spectrometry and has been presented in Table 5.

Table 5: Floral Concrete Recovery and Major Chemical Components in different Species of Jasmine

Species	*Concrete Recovery (Per cent)*	*Major Chemical Components*	*Reference*
J. grandiflorum (range of 3 genotypes)	0.29 -0.34 per cent	Pentane, 3-ethyl-2, 2, 2-dimethyl-; 1-Pentanol; 4-methyl-2-propyl-; Triacontane; Nonacosane; Octacosane; Tetratriacontane and Tetracosane	Ranchana *et al.* (2017a)
J. auriculatum (range of 3 genotypes)	0.28- 0.36 per cent	17-Pentatriacontene; Octadecane, 3-ethyl-5-(2-ethylbutyl); 11-Pentanol, 4-methyl-2-propyl-	Ranchana *et al.* (2017b)
J. multiflorum (range of 2 genotypes)	0.31-0.33 per cent	Triacontane; Tetratriacontane; 1-Pentanol, 4-methyl-2-propyl and Nonacosane	Ranchana *et al.* (2017c)
J. nitidum (1 genotype)	0.32 per cent	1-Pentanol, 4-methyl-2-propyl; Pentane, 3-ethyl-2,2-dimethyl-; Pentane, 2,2,3,4-tetramethyl	Ranchana *et al.* (2017d)

12.0 Postharvest Management

12.1.0 Packaging

Though no standard grades are available for jasmines, flower buds should be graded according to their shape, size and freshness before packing in the boxes or baskets for marketing. Corrugated cardboard boxes are proper packaging material for distant markets while bamboo baskets may also be used for local markets.

Packaging should be functional, economical and attractive besides being acceptable in markets. Harvested flowers should also be given cold treatment before packing.

Thamaraiselvi *et al.* (2010) found treating the *J. sambac* flowers in boric acid 4 per cent, packing in polypropylene 60 µ and corrugated fibre board with 4 per cent vent packaging extended the shelf life of flowers to 192.32 hours. Visual quality and the physiological parameters were highly significant in this package when compared to other packages.

Jawaharlal *et al.* (2012) standardized packaging technology for export of jasmine flowers. Chemical treatment of flowers with 4 per cent boric acid, packing in aluminum-foil lined boxes and further packaging in thermocol boxes under gel-ice cold condition recorded a shelf life of 42.88h. This package also recorded maximum freshness index (70 to 90 per cent), minimum flower-opening index (10.5 to 50 per cent) and maximum colour retention index (77.77 to 88.88 per cent) of flowers. *Jasminum sambac* flowers stored in ambient condition (26-27°C and 50-70 per cent RH), after two days of packing in 38.1, 50.8 micron packaging material resulted in minimum ethylene evolution of 27.80 and 28.9 ppm with the chemical treatments of 2 and 4 per cent boric acid. In refrigerated storage (7-8°C and 80-85 per cent RH) condition, after eight days the minimum ethylene evolution of 22.21 and 26.43 ppm was recorded for flowers treated with 2 and 4 boric acid and packed in 38.1 micron, 50.8 micron LDPE material (Nagajjanavar *et al.*, 2012).

Jasminum sambac cv. Mysuru Malligae flowers treated with 4 per cent boric acid, packed in 60 micron polyethylene bags without ventilation and stored in thermocole box with aluminium foil lining in gel – ice pad cold condition extended the shelf life of flowers up to 48 hours with minimum physiological losses in weight of 0.034 per cent, high freshness score of 4.96, fragrance score of 4.65 and colour retention of 4.53 as compared to shelf life of 24 hours in control with physiological losses in weight of 11.94, high freshness score of 1.00, fragrance score of 1.67 and colour retention of 2.80 (Yathindra *et al.*, 2018).

12.3.0 Extraction of Essential Oil

Jasmine flowers owe their fragrance to a volatile oil present in the epidermal cells of the inner and the outer surface of both petals and sepals. In India the earliest work in jasmine oil production dates back to 1940. Experiments on a laboratory scale were carried out by the Tata Oil Mills Limited, Bombay and the Harcourt Butler Technological Institute, Kanpur. It revealed that Indian jasmines gave good quality oil comparable with that of French jasmines. Among the scented species *J. grandiflorum, J. auriculatum, J. sambac, J. angustifolium, J. officinale, J. humile* and *J. pubescens* were found useful. The Indian *J. grandiflorum* compared favourably with that of Spanish jasmine both in yield and quality of oil. By careful plant selection and quick processing of flowers, high quality material is supplied to the world market (Muthuswami and Abdul Khader, 1986).

12.2.0 Longevity of Flowers

Stage of harvest and pre- and postharvest handling markedly influence the quality of picked flowers. Treatment with certain chemicals, however, proved

Harvesting of Jasmine

Harvesting of flowers

Fresh harvested flowers

Value Addition of Jasmine

Loose flowers

Wreath from jasmine

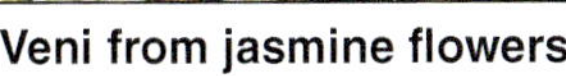

Veni from jasmine flowers

Oil from jasmine

Packaging of Jasmine

In card board boxes

Packing in polythene bag

Marketing of Jasmine

Jasmine loose flowers in local markets

Jasmine in international markets

to be useful in prolonging the life of the harvested flowers. Mukhopadhyay *et al.* (1980) reported that most of the flowers under control (soaked in distilled water) wilted within 48 hours while those treated with boric acid, aluminium sulphate and silver nitrate remained fresh up to 72 hours. The fragrance of the fresh flowers also remained unchanged for longer period due to the treatment.

Jasmine flower buds treated with anti-senescence chemical 20 ppm silver nanoparticle postharvest, packed in 200-gauge polypropylene bag without ventilation and stored in cold storage condition showed significant positive difference in freshness index, flower opening index, colour retention index and maintained the shelf life up to 11 days (Mohanasundari, 2018).

Fresh jasmine (*J. grandiflorum*) flowers were extracted with hexane, benzole or benzole + ethyl acetate (5 or 10 per cent), and the concretes obtained were stored under refrigeration until required. Jasmine absolute was extracted from the concretes using 95 per cent ethanol. The concretes and absolutes were also assessed by a trained sensory panel.

The concretes as a percentage of flower fresh weight were 0.253, 0.413, 0.467 and 0.536 per cent for the 4 solvents, respectively, and the absolutes as a percentage of the concretes were 50.57, 68.16, 68.19 and 69.69 per cent, respectively. The sensory panel judged the concrete and absolute extracted with hexane to have the best odour, being significantly better than those extracted with benzole, which in turn was better than those extracted with the benzole + ethyl acetate mixtures (Atawia *et al.*, 1988a). Atawia *et al.* (1988b) also reported that soaking of jasmine flowers in HCl solution increased the yield of concrete and absolute, but markedly lowered the odour quality. Freezing of jasmine flowers (*J. grandiflorum*) at –15°C for 24 h and thawing for 6 h before extraction increased the essential oil yield without any effect on quality (Atawia *et al.*, 1988c).

Verghese and Sunny (1992) reported seasonal variations in yields of concrete and absolute, and the composition of the absolute, for flowers of *J. grandiflorum* harvested at intervals between 30 June and 23 October. Gas chromatographic screening of the jasmine fragrances showed that the overall compositional pattern was similar to that of oils from other countries.

Younis *et al.* (2011) used supercritical fluid extraction (SFE) and found that flowers harvested at the open stage yielded more absolute oil than those harvested at the closed bud stage. Slight variations in physical properties like refractive index, congealing point, optical rotation, and specific gravity were recorded in the essential oil extracted from flowers harvested at different stages. Both harvesting stages yielded oil with differences in the percentage composition of each component, but flowers harvested at the open stage had higher yield and more components than those from the closed bud stage.

Hand held electronic nose technology (HEN) is a technique to identify the flower quality in non-destructive and quickest possible manner. HEN device generated Aroma Index (AI) score increased over the flower development stages and varied from 0.41 in immature bud (stage I) to 4.26 in matured bud (stage V). The comprehensive study on quantum of fragrance releasing pattern at different flower opening stages (physiologically matured bud to fully opened flower) over period of time interval showed that, minimum of 5.41 was recorded in an unopened closed bud stage which gradually increased up to 41.26 in the fully opened flowers (Isac *et al.*, 2016).

12.3.1 Composition of Oil

A total of 190 open pollinated progenies of *J. auriculatum* with different style lengths and flower bud forms were assayed for several biochemical traits (Indiresh *et al.*, 1989). The concrete recovery ranged from 0.25 to 0.29 per cent. The highest indole and methyl anthranilate content was found in the pin type with small buds. The thrum type with pointed buds had the highest benzyl acetate content. Gopalakrishnan and Narayanan (1988) found differences in the fatty acid composition of *J. grandiflorum* concrete, collected from 3 different regions in Tamil Nadu, in the proportion of saturated and unsaturated fatty acids. The major saturated fatty acid was 16:0 and major unsaturated fatty acid was 18:3. The quality of jasmine absolute, prepared from the concrete collected in these regions, varied between sources. High concentra-tions of linalool, benzyl acetate, (z)-3-hexenyl benzoate, *cis*-jasmone, indole and several sesquiterpenes were found in white jasmine (*J. sambac*) and red jasmine (*J. grandiflorum*), benzyl acetate, jasmine lactone and methyl jasmonate were more abundant in red than in white jasmine (Musalam *et al.*, 1988).

Guo *et al.* (1994) reported that immature flowers gave off slight fragrance, whereas mature flowers gave off a rich perfume with much greater amounts of alcohols (15.5 per cent) and esters (47.0 per cent). After withering of a flower, the ester content clearly decreased whereas the alcohol content slightly increased.

According to Rao and Rout (2003) analysis of the essential oil of the flowers of *Jasminum sambac* L. from three locations and the time of harvest showed considerable variation in the composition. The essential oil obtained from the leaves and flowers of *Jasminum pubescens* (Retz.) Willd. was analysed by GC-MS (Abeer *et al.*, 2009). Sixty-three and sixty-four components of the essential oils, for the leaves and flowers, respectively were identified. Main constituents were non-terpene derivatives, among which aldehydes (44.7 per cent) characterized the essential oil from the leaves. Besides aldehydes (14.3 per cent) other carbonylic compounds (acids, esters, and ketones, 38.1 per cent) were the main non-terpene compounds of the oil from the flowers.

The major constituents of jasmine essential oil are citronellol, phenyl ethyl alcohol, geranial, eugenol, farnesol, geranyl acetate, citrinyl acetate, 2-phenyl ethyl acetate, citral (mixture of cis and trans), and benzyldehyde (Younis *et al.*, 2011).

Prakash *et al.* (2012) prepared the concrete (0.35 per cent) of *Jasminum grandiflorum* L. flowers by extraction in n-pentane, and the absolute (0.27 per cent) by fractionation of the n-pentane extract (concrete) with cold methanol. Direct extraction of flowers with liquid CO_2 gave a relatively fat-free product in 0.26 per cent yield. The major compounds, such as benzyl acetate, (E, E)-α-farnesene and (Z)-3-hexenyl benzoate, along with compounds like indole, methyl anthranilate,

(Z)-jasmone, (Z)-methyl jasmonoate and (Z)-methyl epi-jasmonoate, are responsible for the high diffusivity of the jasmine fragrance.

Ye *et al.* (2015) standardised subcritical fluid extraction (SFE), for extracting fragrant volatiles from jasmine flowers which yielded the main aromatic compounds with significantly better quality. The extraction yield was 4.91 per cent and the key jasmine essence oil components, benzyl acetate and linalool, increased 7-fold and 2-fold respectively which lead to strong typical smell of the jasmine oil. It reduced the spicy components which lead to the essential oils smelling sweeter.

Sumangala *et al.* (2018) analysed volatile compounds of *J. malabaricum* flowers by solid phase micro extraction (SPME) and gas chromatography - mass spectrometry (GCFID and GC-MS). Of the 109 volatile compounds identified, terpenoids constituted major group accounted for 68.71 per cent followed by esters representing 17.18 per cent. Main constituents identified were 2- Methyl-2-bornene and Pentadecane (Hydrocarbons), β-Caryophyllene, α-Farnesene, trans-Ocimene and 3-methylene-p-menth-8-ene (Terpenoids), 2-Hexenyl propanoate and cis 3-Hexenyl tiglate (Esters), Linalool, Nerolidol and Cubenol (Oxygenated Terpenoids), 2-(5-Methyl-5-vinyltetrahydro-2-furanyl)-2- propanol and 1-Hexen-3-ol (Alcohols), cis-Jasmone and α-iso-Methyl ionone (Aldehydes and ketones), (2E)-5-Hydroxy-2-pentenenitrile and Indole (Nitrogenated Compounds).

The floral volatile emission from bud to senescence exhibited nocturnal maxima pattern for *Jasminum auriculatum* and *Jasminum grandiflorum*. *Jasminum multiflorum* and *Jasminum malabaricum* emitted its highest concentration at noon. The free endogenous concentrations of all VOCs were low when corresponding emitted concentrations were high. Enzymatic treatment of petal extract revealed that several aromatic volatiles including aromatic alcohols and monoterpenols are synthesized and stored in the flowers as water-soluble glycosides; which accumulate in higher amounts in flowers at late bud stage (Barman and Mitra, 2019).

13.0 References

Abdul khader, J.B.M., Chezhiyan, N. and Ponnuswami, V. (1984) *Proc. Seminar on Utilization of value added Agro-products*, Tamil Nadu Agricultural University, Coimbatore.

Abeer, T., Luigi, C.P., Guido, F., Alessandra, B. (2009) *Nat. Prod. Commun.*, **4**: 1729 – 1732.

Ahmad, S.H., Malek, A.A., Gan, H.C., Abdullah, T.L. and Rahman, A.A. (1998) *Acta Hortic.*, **454**: 355-363.

Ahmedullah, M. and Nayar, M.P. (1986) *Botanical Survey of India, Calcutta*, pp. 138–139.

Alice, D., Ebenezar, E.G. and Sivaprakasam, K. (1996) *J. Ecobiology*, **8**: 17-20.

Alikhan, M.W., Muthuswami, S. and Raman, K.R. (1970) *South Indian Hort.*, **18:** 25-32.

Alikhan, M.W., Raman, V.S. and Raman, K.R. (1969) *South Indian Hort.*, **17:** 79-85.

Amutha, S. and David, B.V. (1995) *Indian J. Environ. Toxicol.*, **5:** 69-70.

Anburani, A. (2012) *Adv. Plant Sci.*, **25**: 293-295.

Anonymous (1959) *The Wealth of India*, Council of Scientific and Industrial Research, New Delhi.

Anonymous (1969) *Floriculture Section Annual Report*, Tamil Nadu Agricultural University, Coimbatore.

Anonymous (1972) *Encylopaedia Britannica*, Encyclopaedia Britannica Ltd., Chicago.

Anonymous (1973) *First Annual Progress Report, Scheme for intensification of research on the development of rose and jasmine varieties with high essential oil content*, ICAR, Tamil Nadu Agricultural University, Coimbatore.

Anonymous (1975) *Fifth Annual Progress Report, Scheme for intensification of research on the development of rose and jasmine varieties with high essential oil content*, ICAR, Tamil Nadu Agricultural University, Coimbatore.

Anonymous (1977) *Fifth Annual Progress Report, Scheme for intensification of research on the development of rose and jasmine varieties with high essential oil content*, ICAR, Tamil Nadu Agricultural University, Coimbatore.

Anonymous (2014) *TNAU agritech portal, Varieties of flower crops.* http: //agritech.tnau.ac.in/horticulture/horti_TNAU_varieties_flowercrops.html

Anonymous (2015) *Jasmine.* http: //www.plantauthority.gov.in/pdf/Jasmine.pdf.

Anonymous (2019) *CABI Invasive species compendium. Datasheet Jasminum multiflorum (star jasmine).* https: //www.cabi.org/isc/datasheet/115016#tosummaryOfInvasiveness.

Anonymous (2020a) *Ministry of Agri. and Farmers Welfare, Govt. of India. (ON1549).* http: //www.indiaagristat.com/table/Agri.-data/2/jasmine/962546/964462/data.aspx.

Anonymous (2020b) *The International Plant Names Index and World Checklist of Selected Plant Families* http: //www.ipni.org and http: //apps.kew.org/wcsp/

Anonymous (2020c) *The International Plant Names Index and World Checklist of Selected Plant Families 2020.* http: //www.ipni.org and http: //apps.kew.org/wcsp/

Anonymous (2020d) *Varieties released.* http: //autoshelltnau.in/college-hcricbe/department-of-fl-technologies-developed/.

Atawia, B.A., Hallabo, S.A.S. and Morsi, M.K. (1988a) *Egyptian J. Food Sci.*, **16**: 213-224.

Atawia, B.A., Hallabo, S.A.S. and Morsi, M.K. (1988b) *Egyptian J. Food Sci.*, **16**: 225-235.

Atawia, B.A., Hallabo, S.A.S. and Morsi, M.K. (1988c) *Egyptian J. Food Sci.*, **16**: 237-247.

Bailey, L.H. (1949) *Manual of Cultivated Plants.* Macmillan Co., New York.

Baker, C.A. and Bakhijen van der Brink, R.C. (1905) *Flora of Java, Noordhoff Gereningin*, Netherlands, Vol. 2.

Baker, J.G. (1877) *Flora of Mauritius and Seychelles*, Reeves and Co., London.

Barman, M. and Mitra, A. (2019) *Physiol. Plant.*, **166**: 946–959.

Baskaran, V., Abirami, K. and Roy, S.D. (2016) *Int. J. Trop. Agric.*, **34**: 1093-1096.

Batra, H.V. (1967) *Indian Hort.*, **11**: 7, 34.

Bhattacharjee, S.K. (1983a) *Singapore J. Prim. Indus.*, **11**: 34-39.

Bhattacharjee, S.K. (1983b) *Pafai J.*, **5**: 21-27.

Bhattacharjee, S.K. (1985a) *Singapore J. Prim. Indus.*, **13**: 102-111.

Bhattacharjee, S.K. (1985b) *Indian Perfumer*, **29**: 177-180.

Bhattacharjee, S.K. (1985c) *Indian Perfumer*, **29**: 181-185.

Bhattacharya, S. and Bhattacharya, S. (1997) *Plant Cell, Tissue Organ Cult.*, **51**: 57-60

Bhattacharya, S. and Bhattacharya, S. (2010) *Protocols for In vitro propagation of Ornamental Plants Methods in Molecular Biology*, Springer Science and Business Media, New York: 117-126.

Biswal, M., Palai, S.K., Mishra, P., Chhuria, S. and Sahu, P. (2016) *Int. J. Curr. Res.*, **8:** 30164-30168.

Bose, T.K., Mondal, D.P. and Pramanik, D.K. (1973) *Prog. Hort.*, **5**: 43-50.

Boyle, W.W. and Harmoto, F.H. (1956) *Hawaii Farm Sci.*, **5**: 1, 7-8.

Bryzgalova, N.V. (1974) *Bjull. glav. bot. Sada*, No. 93, pp. 75-77.

Burns, W. (1953) *Trans. Bot. Soc. Edinburgh*, **36**: 84-98.

Cai, H., Chen, X-Q., Xiong, Z-M., Xie, L-I. and Zao, L-J. (2007) *Jiangsu J. Agric. Sci.*, **23:** 464-468

Chaitanya, R.P., Chandrasekhararao, C., Baburao, B. and Sankaradasu, C.H. (2016) *Adv. Life Sci.*, 5: 8817-8822.

Chandraju, S. Thejovathi, C. and Chidan Kumar, C. S. (2012) *Curr. Bot.*, **3**: 19-22.

Chandramohan, N. and Manoharan, T. (1990) *South Indian Hort.*, **38**: 293-294.

Chezhiyan, N., Ponnuswami, V., Khader, M.A. and Thamburaj, S. (1988) *South Indian Hort.*, **36**: 92-93.

Chopde, N., Patil, S., Kuchanwar, O. and Raut, V.U. (2017) *J. Pharmacogn. Phytochem*, **6:** 1201-1203.

Cooke, T. (1905) *The Flora of Presidency of Bombay*, Taylor and Francis, London.

Damiano, A. (1966) *Riv. Agric. Subtrop.*, **60**: 164-177.

David, B.V. and Kumaraswami, T. (1975) *Elements of Economic Entomology*, Popular Book Depot, Madras, pp. 159-160.

David, P.M.M., Hanifa, A.M. and Natarajan, S. (1992) *Madras Agri. J.*, **79**: 15-17.

David, P.M.M., Haniffa, M.D. and Natarajan, S. (1991) *Madras Agri. J.*, **78**: 82-84.

Deng, Y., Liang, L., Sun, X., Jia, X. and Su, J. (2018) *South African J. Bot.*, **114:** 69-77.

Deng, G., Li, X., Wang, Y., Wang, L., Tu, X., Zhou, S. and Seng, G. (1986) *Nat. Sci. J.*, Human Normal Univ., **9:** 73-81.

Deng, Y. Li, C., Shao, Q., Ye, X. and She. J. (2012) *Plant Physiol. Biochem.*, **55**: 93-102

Deng, Y., Jia, X., Liang, L., Gu, C. and Sun, X. (2016) *Sci. Hortic.*, **198**: 257-66.

Deng, Y., Sun, X., Gu, C., Jia, X., Lijian Liang, Jiale Su. (2017) *PLoS One***12**: e0176026. DOI: 10.1371/J. pone.0176026

Devaiah, K.A. and Srivastava, M.C. (1989) *Herba Hungarica*, **28**: 11-14.

Dhanasekaran, D. (2018) *J. Hort. Sci.*, **13**: 221-226.

Dilipkumar, M., Amirul, N.M., Bin, S.I. and ChuahTse, S. (2019) *Indian J. Hortic.*, **76**(1): 148- 154.

Dhandapani, N., Gopalan, M. and Sundarababu, P.C. (1989) *Madras Agri. J.*, **76**: 50-52.

Dhandapani, N., Gopalan, M. and Sundarababu, P.C. (1992) *Madras Agri. J.*, **79**: 54-55.

Dickey, R.D. (1970) In: *Flowering Vines of the World* (Ed. E.A. Meninger), Heartside Press Inc., New York.

Diwivedi, R., Saravanan, S., Shabi, M. and Kasera, S. (2018) *The Pharm. Innov. J.*, **7**: 683-686.

Duthie, J.F. (1911) *Flora of Upper Gangetic Plains and the Adjacent Siwalik and Sub-Himalayan tracts*, Superintendent, Govt. Printings, Calcutta, India.

Dutta, M. (1952) *Sci. and Cult.*, **17**: 528.

Fakhry, H. A. (2014) *IFEAT Int. Conference*, Rome, Italy

Farzinebrahimi, R., Rashid, K., Taha, R.M. and Ishak, K.A. (2014) *Int. Conference on Biological, Environment and Food Engineering*, Bali (Indonesia), pp. 4-5.

Gamble, J.S. (1957) *The Flora of Presidency of Madras*, Reprint Botanical Survey of India, Calcutta.

Ganga, M., Ranchana, P., Ganesh, S. and Kannan, M. (2019) *Acta Hortic.*, **1241**: 105-112.

George K. and Geethamma S. (1992) *Cytologia*, **57**: 27 -32.

Ghehsareh, G.M., Salehi, H., Khosh-Khui, M. (2015) *Mol. Biotechnol.*, **57**: 65–74.

Ghosh S., Ganga M., Soorianathasundaram K., Kumar A. (2020) *Indian J. Hortic.*, **77**: 149-157.

Ghosh, S., Ganga, M. and Joel, J.A. (2018a) *Electron. J. Plant Breed.*, **9**: 1002-1011.

Ghosh, S., Ganga, M. and Soorianathasundaram. K. (2018b) *Electron. J. Plant Breed.*, **9**: 956-65.

Gopalakrishnan, N. and Narayanan, C.S. (1988) *J. Oil Technologists' Assoc. India*, **20**: 45-46.

Green, P.S. (2003) *Kew Bulletin*, **58**(2): 257–295.

Guenther, E. (1952) *The essential oil*, Van Nostrand Company Inc., New York.

Guo, Y.J., Dai, L., Ren, Q. and Yang, L.P. (1994) *Chinese J. Chromatography*, **12**: 110-113.

Haines, H.H. (1961) *The Botany of Bihar and Orissa*, Superintendent, Govt. Printings, Calcutta, India, Vol. 2.

Haramaki, C. and Kuhns, L.J. (1983) *Proc. 37th Annual Meet. Northeastern Weed Sci. Soc.*, pp. 335-338.

Harini, K., Elanchezhyan, K., Murugesan, N. (2019) *J. Agric. Ecol.*, **8**: 19-23.

He L, He J, Xia B, Meng X. and Peng, F. (2011) *Propag. Ornam. Plants.*, **11**: 172-176.

Hilal, A.A., Shalaby, H.S., El-Ala, A.M.A., El-Deeb, A.A. and Mahmoud, B.K. (1987) *Proce. Ist Confe. Agri. Develop. Res.*, **3**: 192-203.

Hore, B.K. and Bose, T.K. (1972) *Indian J. Hort.*, **29**: 93-96.

Hugar, A.H. and Nalawadi, U.G. (1993) *Curr. Res. Univ. Agri. Sci.*, Bangalore, **22**: 65-66.

Hugar, A.H. and Nalawadi, U.G. (1994) *Karnataka J. Agric. Sci.*, **7**: 406-409.

Hung, R.H. (1965) *Acta Ento. Sinica*, **14**: 480-488.

Indiresh, K.M., Uthaiah, B.C. and Rao, J.V.R.B. (1989) *Indian Perfumer*, **33**: 44-47.

Isac, X. A., Rajadurai, K. R. and Jawaharlal, M. (2015) *Int. J. Trop. Agric.*, **33**: 1427-1431.

Isac, X. Alex; Rajadurai, K. R.; Jawaharlal, M.; Selvan, K. Arul Mozhi; Uma, D.; Roy, H. and Bhattacharyya, N. (2016) *J. Appl. Hortic.*, **18**: 19-24.

Jawaharlal, M., Thamaraiselvi, S. P., and Ganga, M. (2012) *J. Hortic. Sci.*, **7**: 180-189.

Jayamma N., Jagadeesh K.S., Patil V.S. (2008) *J. Ornam. Hortic.*, **11**: 275-280.

Jayapal, R., Sambandamurthi, S. and Vedamuthu, P.G.B. (1980) *Nat. Seminar Prodn. Tech. Comm. Flower Crops*, Tamil Nadu Agricultural University, Coimbatore, pp. 15-16.

Kalaiyarasi, A., Dhananjaya, M. V., Nair, S. A., Kumar, R., Yogeesha, H. S., Munikrishnappa, P. M., Devappa V. and Pavithra S. (2018) *Indian J. Agric. Sci.*, **88**: 1789–93.

Kalaiyarasi, A., Dhananjaya, M. V., Nair, S. A., Kumar, R., Yogeesha, H. S., Munikrishnappa, P. M., Devappa V. and Pavithra S. (2019) *Indian J. Agric. Sci.*, **89**: 983–8.

Kamala, I. M. (2018) *Indian J. Ecol.*, **45**: 367-372.

Kamala, I. M. and Kennedy, J. S. (2018) *J. Biol. Control*, **32**: 121-127.

Kamala, I.M. (2019) *Pest Manag. Hort. Ecosyst.*, **25**: 69-78.

Karthigeyan K., and Murugan P. (2018) *J. Threat. Taxa*, **10:** 12439-41.

Keerthishankar K, Kulkarni., Yathindra H.A., Taj, A., Gowda, M. A, Rashmi H.B. (2019) *Int. J. Chem. Stud.*, **7**: 1793-97.

Khoder, M., Villemur, P. and Jonard, R. (1979) *Comp. Rend. Hebdo. Seances Acad. Sci.*, **228**: 323-326.

Khoder, M., Villemur, P. and Jonard, R. (1981) *C.R. Acad. Sci.*, Paris, III, **293**: 343-346.

Kitpow-song, P., Geungchan, S., Sirimukdakul, S. and Taepongsorut, R. (1988) *Kaen Kaset Khon Kaen Agriculture J.*, **18**: 80-88.

Krishnamoorthy, V. (2014) *J. Krishi Vigyan*, **3:** 88-90.

Krishnaswamy, N. (1980) *Nat. Sem. Prodn. Tech. Comm. Flower Crops*, Tamil Nadu Agricultural University, Coimbatore, India, p. 30.

Krishnaswamy, N. and Raman, V.S. (1948) *J. Indian Bot. Soc.*, **27**: 77-83

Kumar, R. and Gill, A.P.S. (1983) *South Indian Hort.*, **31**: 20-21.

Kumaresan M, Kannan, M. Sankari, A. and Chandrasekhar, C.N. (2019) *Int. J. Chem. Stud.*, **7**: 935-939.

Kumaresan M, Rajadurai, K.R., Ganga, M. and Sivakumar, R. (2017) *Int. J. Chem. Stud.*, **5**: 2374-78.

Kumaresan S. and Jawaharlal M. (2017) *Int.J. Curr. Microbiol. App.Sci.* **6**: 5385-5394.

Lai, M.Z. (1995) *J. Fujian Agric. Univ.*, **24**: 291-294.

Lakshmanan, P. and Gabriel, J.J. (2015) *World J. Pharm. Sci.*, **3**: 1421-1425.

Lakshmi, J. and Ganga, M. (2017) *Int. J. Curr. Microbiol. Appl. Sci.*, **6**: 2811-2815.

LI Bin. (2000) *J. Gansu Agri.Univ.*, **2**: 235-238.

Lokhande, S., Chopde, N., Gaidhani, A. and Bhande, M. H. (2016) *J. Soils and Crops*, **26:** 127-130.

Mabberley, D.J. (2017) *Mabberley's Plant-Book. A Portable Dictonary of Plants, their Classification and Uses.* Fourth Editon. Cambridge University Press, Cambridge.

Madhava Rao, V.N. and Muthuswami, S. (1972) *Indian Hort.*, **20**: 1-2.

Mahmood, M.A., Hafiz, I.A., Abbasi, N.A. and Faheem, M. (2013) *Int. J. Agric. Biol.*, **15**: 505-510.

Malik CP. (2007) *Applications of biotechnology innovations in pharmaceutics and nutraceutics in multi therapeutic medicinal and special plants*, Aavishkar publishers, Jaipur. **2**: 243-265.

Mallapur, C.P. and Kubsad, V.S. (2000) *Insect Environment*, **6**: 18-19.

Mallikarjunappa, S., Mishra, R.K., Mithyantha, M.S. and Sivasankaran, K. (1991) *Curr. Res. Univ. Agri. Sci.*, Bangalore, **20**: 242-243.

Manimaran P., Rajasekar P., Lakshmi J. (2018) *Environ. Ecol.*, **36:** 262-264.

Mariappan, V. and Ramanujam, K. (1975) *South Indian Hort.*, **23**: 77-78.

Maurya, V.N., Singh, R.P., Mishra, R. and Kumar, R. (1996) *Recent Hort.*, **3**: 153-156.

McLean, D.M. (1960) *J. Rio. Grande Valley Hort. Soc.*, **14**: 187-188.

Mekala, P., Kumar, N., Kannan, M. and Vijayakumar. G. (2010) *J. Ornam. Hortic.*, **13**: 200-206.

Miquel, F.A.W. (1956) *Flora Van Nadearlanssch*, VaneDer Post, Amsterdam.

Mohanasundari P, Sivakumar T, Krishna Surendar K and Ganga M. (2018) *J. Agric. Ecol.*, **6**: 85- 93.

More, T.A. (1980) *Nat. Seminar Prodn. Tech. Comm. Flower Crops*, Tamil Nadu Agricultual University, Coimbatore, India, pp. 45-46.

Mukhopadhyay, T.P. and Bose, T.K. (1979) *Punjab Hort. J.*, **19**: 98-101.

Mukhopadhyay, T.P., Bose, T.K. and Maity, R.G. (1980) *Nat. Seminar Prodn. Tech. Comm. Flower Crops*, Tamil Nadu Agricultural University, Coimbatore, India, pp. 48-50.

Murthy, A.S. and Khanna, K.R. (1971) *Curr. Sci.*, **40**: 555-566.

Musalam, Y., Kobayashi, A. and Yamanishi, T. (1998) *Proc. 10th Intern. Cong. Essential oils, fragrances and flavors*, November 16-20, 1986 Washington, DC, USA, 18: 659-668.

Muthukrishnan, C.R. and Pappiah, C.M. (1980) *Nat. Seminar Prodn. Tech. Comm. Flower Crops*, Tamil Nadu Agricultural University, Coimbatore, India, pp. 1-3.

Muthuswami, S. (1975) *Ph. D. thesis*, Tamil Nadu Agricultural University, Coimbatore.

Muthuswami, S. and Abdul Khader, J.B.M. (1986) In: *Orna. Hort.*, India, ICAR, pp. 113-126.

Muthuswami, S. and Madhava Rao, V.N. (1980) *Nat. Seminar Prodn. Tech. Comm. Flower Crops*, Tamil Nadu Agricultural University, Coimbatore, India, pp. 25-26.

Muthuswami, S. and Pappiah, C.M. (1976) *South Indian Hort.*, **24**: 88-93.

Muthuswami, S. and Pappiah, C.M. (1977) *Indian J. Hort.*, **34**: 189-193.

Muthuswami, S., Alikhan, W.M. and Sayed, S. (1972) *Curr. Sci.*, **61**: 184.

Muthuswami, S., Alikhan, W.M., Sayed, S. and Pappiah, C.M. (1974) *South Indian Hort.*, **21**: 11-14.

Muthuswami, S., Pappiah, C.M. and Sayed, S. (1973a) *South Indian Hort.*, **21**: 70-72.

Muthuswami, S., Sayed, S., Pappiah, C.M. and Shanmugam, A. (1973b) *Madras Agri. J.*, **60**: 1695-1699.

Muthuswami, S., Thangaraj, T. and Pappiah, C.M. (1972) *South Indian Hort.*, **20**: 87-89.

Nagajjanavar, K., Thirupathi, V., Balakrishnan, M., Rani, C. I. and Kasturi, R. (2012) *Environ. Ecol.*, **30**: 806-809.

Naik, B. P., Bhatt, S. T., Bhatt, D.S., Patel, G.D., Gaikwad, S.S. and Gamit.D. (2019) *Int. J. Chem. Stud.*, **7**: 2022-26.

Nair, S.A. and Venugopalan. R. (2016) *Indian J. Agric. Sci.*, **86**: 321–5.

Nair, S.A., Sujatha, K. and Venugopalan, R. (2009) *Indian J. Agric. Sci.*, **79:** 857-60.

Nakasone, H.Y. and Bowers, F.A. (1956) *Hawaii Fm. Sci.*, **5**: 1-7.

Nambisan, K.M.P. and Krishnan, B.M. (1980) *Nat. Seminar Prodn. Tech. Comm. Flower Crops*, Tamil Nadu Agricultural University, Coimbatore, India, pp. 19-20.

Nambisan, K.M.P., Krishnan, B.M., Veeraraghavathatham, B. and Ramasamy, N. (1980) *Sci. and Cult.*, **46**: 427-428.

Natarajan, S. and Madhava Rao, V.N. (1980a) *Madras Agric. J.*, **67**: 207-210.

Natarajan, S. and Madhava Rao, V.N. (1980b) *Madras Agric. J.*, **67**: 252-254.

Natarajan, S., Nambisan, K.M.P., Krishnan, B.M. and Shanmugavelu, K.G. (1981) *Indian Perfumer*, **25**: 89-92.

Nedumaran, C. and Shanmugavelu, K.G. (1978) *Indian Perfumer*, **22**: 166-174.

Nemec, S. and Struble, F.B. (1968) *Phytopath.*, **58**: 1700-1703.

Netam, N., Shukla, N., Sharma, G. and Sahu, J.K. (2018) *J. Pharmacogn. Phytochem.* **SP1**: 614-617.

Nirmala, K.S., Champa, B. V., Hegde, M. (2017) *Mysore J. Agric. Sci.*, **54**: 822-31.

Nofal, M.A. and Marwan, A.E. (1982a) *Res. Bull. Fac. Agric., Ain Shames Univ.*, No. 1842, p. 13.

Nofal, M.A. and Marwan, A.E. (1982b) *Res. Bull. Fac. Agric., Ain Shames Univ.*, No. 1841, p. 17.

Pal, P., Maity, R.G. and Bose, T.K. (1980a) *Nat. Sem. Prodn. Tech. Comm. Flower Crops*, Tamil Nadu Agricultural University, Coimbatore, India, pp. 35-38.

Pal, P., Maity, R.G. and Bose, T.K. (1980b) *Nat. Sem. Prodn. Tech. Comm. Flower Crops*, Tamil Nadu Agricultural University, Coimbatore, India, pp. 27-29.

Pal, P., Maity, R.G. and Bose, T.K. (1980c) *Nat. Sem. Prodn. Tech. Comm. Flower Crops*, Tamil Nadu Agricultural University, Coimbatore, India, pp. 31-34.

Pal, P., Maity, R.G. and Bose, T.K. (1985) *South Indian Hort.*, **33**: 118-121.

Pal, S.L. (2017) *HortFlora Res. Spect.*, .**6**: 200-202.

Palai S.K., Madhuri G. and Biswal M. (2017) *Int. J. Farm Sci.*, **7**: 185-189.

Panda H. (2005) *Aromatic Plants Cultivation, Processing and Uses.* Asia Pacific Business Press Inc., New Delhi, pp. 129-137.

Pandey, R.P. and Diwakar, P.G. (2008) *J. Economic and Taxonomic Bot.*, **32**(2): 403–500.

Pappiah, C.M. and Muthuswami, S. (1976) *South Indian Hort.*, **24**: 30-32.

Pappiah, C.M. and Muthuswami, S. (1977) *South Indian Hort.*, **25**: 68-74.

Pappiah, C.M. and Muthuswami, S. (1978) *South Indian Hort.*, **26**: 66-71.

Pappiah, C.M., Sambandamurthi, S. and Jayapal, R. (1980) *Nat. Seminar Prodn. Tech. Comm. Flower Crops*, Tamil Nadu Agricultural University, Coimbatore, India, pp. 21-24.

Paswan, D.K and Saravanan, S.S. (2019) *The Pharm. Innov. J.*, **8**: 286-289.

Patel, N.L. and Singh, S.P. (1995) *Hort. J.*, **8**: 67-73.

Patil, P.L., Reddy, B.S., Patil, S.R. and Kulkarni, B.S. (1998) *Karnataka J. Agri. Sci.*, **11**: 813-814.

Paulas, D. (1980) *Nat. Seminar Prodn. Tech. Comm. Flower Crops*, Tamil Nadu Agricultural University, Coimbatore, India, pp. 17-18.

Pavithra S., Dhananjaya M. V., Nair, S.A., Kumar, R. Yogeesha H. S., Munikrishnappa P. M., Devappa, V., Halesh G. K., Kalaiyarasi A. (2018) *Indian J. Agric. Sci.*, **88**: 1864–8.

Pavithra S., Dhananjaya M. V., Nair, S.A., Kumar, R., Yogeesha H. S., Munikrishnappa P. M., Devappa V., Halesh G. K. and Kalaiyarasi A. (2019) *Indian J. Agric. Sci.*, **89**: 1543–6.

Perveen, A. and. Sarwar, G.R. (2011) *Pakistan J.Bot.*, **43**: 2109-2112

Pillai K. G., Nethravathi C. J., Visalakshy P. N. Ganga and Swathi C. (2016) *Pest Manag. Hort. Ecosyst.*, **22**: 67-70.

Pirithiraj, U., Soundararajan, R.P., Justin, G.L.C. and Lakshmanan. V. (2020) *J. Entomol. Zool. Stud.*, **8**: 680-684.

Pirone, P.P., Dodge, B.O. and Rickett, H.W. (1960) *Diseases and Pests of Ornamental Plants*, Constable and Company Limited, London.

Pradhan, S. (2015) *Int. J. Ayurveda Pharm. Chem.*, **3**: 79-92.

Prakash, O., Sahoo, D., Rout, P. K. (2012) Nat. *Prod. Commun.*, **7**: 89-92.

Rahman, M.S., Mouri, N.J., Nandi, N.C., Akter, S. Khan, M. S. (2018) *Bangladesh J. Sci. Ind. Res.*, **53**: 277-282.

Rajaram, B. and Rajendran, G. (1979) *South Indian Hort.*, **27**: 32.

Rajasekaran, Haripriya V, Armugam A. and Shakila. (2000) *Centennial conference on species and aromatic plants, challenges and opportunities, in the new century*, Calicut, Kerala, pp. 47-49.

Ramakrishnan, L. and Kandaswamy, T.K. (1980) *Nat. Seminar Prodn. Tech. Comm. Flower Crops*, Tamil Nadu Agricultural University, Coimbatore, India, pp. 107-108.

Ramamoorthy, V., Alice, D., Meena, B., Muthusamy, M. and Seetharaman, K. (2000) *Indian J. Plant Prot.*, **28**: 102-104.

Raman, K.R. (1973) *Ph. D. Thesis*, Tamil Nadu Agricultural University, Coimbatore.

Raman, V. S. (1955a). *Cytologia*.**20**: 133-147.

Raman, V.S. (1955b) *Bot. Mag. Tokyo*, **68**: 253-259.

Ranchana, P., Ganga, M., Jawaharlal, M. and Kannan, M. (2017 a) *Int.J.Curr.Microbiol. App.Sci.*, **6**: 1883-1891.

Ranchana, P., Ganga, M., Jawaharlal, M. and Kannan, M. (2017b) *Int.J.Curr.Microbiol. App.Sci.*, **6**: 1525-31.

Ranchana, P., M. Ganga, M. Jawaharlal and Kannan, M. (2017 c) *Int.J.Curr.Microbiol. App.Sci.*, **6**: 2229-2233.

Ranchana, P. Ganga, M., Jawaharlal, M. and Kannan, M. (2017 d) *Int.J.Curr.Microbiol. App.Sci.*, **6**: 5411-18.

Rao, Y. R. and Rout, P.K. (2003) *J. Essent. Oil Res.*, Taylor and Francis Online: 398-401.

Ridley, H.N. (1922) *The Flora of Malay Peninsula*, Vol. 2, Reeves and Co. Ltd., London.

Rodino, S., Butu, M. and Butu. A. (2017) *Bull. Univ. Agric. Sci. Vet. Med.*, **74**(1): 66-68.

Salim SAA. (2016) *J. Univ. Babylon Pure Appl. Sci.*, **24**: 795-802.

Sambandamurthi, S., Jayapal, R. and Vedamuthu, P.G.B. (1980) *Nat. Seminar Prodn. Tech. Comm. Flower Crops*, Tamil Nadu Agricultural University, Coimbatore, India, pp. 39-40.

Scaramuzzi, F. and D'Elia, C. (1984) *C.R. Acad. Sci.*, Paris, **298**: 107-112.

Seenivasan N. and Poornima K. (2010) *Pest Manag. Hort. Ecosyst.*, **16**: 34-40.

Sengonca, C. and Liu, B. (1998) *Zeitschrift fur Pflanzenkrankheiten und Pflanzenschutz*, **105**: 504-512.

Sharma, A.K. and Sharma, A. (1958) *Cytologia*, **23**: 172-185.

Shoram, Parekh, N. S., Upadhyay, N. V., Karapatiya, B. A. and Patel, H. C. (2012) *Asian J. Hortic.*, **7**: 52-54.

Siddagangaiah and Rai, B.G.M. (1990) *Mysore J. Agri. Sci.*, **22**: 492-497.

Siddique, M.A. (1973) *Indian J. Hort.*, **30**: 440-443.

Singh, P., Karthigeyan, K., Lakshminarasimhan, P. and Dash S.S. (2015) *Botanical Survey of India*, Kolkata.

Singh, R.P. and More, T.A. (1986) *Indian Perfumer*, **30**: 251-253.

Singh, S.N. and Bhatnagar, G.S. (1955) *Sci. and Cult.*, **21**: 210-221.

Singh, S.P. (1979) *South Indian Hort.*, **27**: 68-69.

Singh, S.P. and Motial, V.S. (1981) *Haryana J. Hort. Sci.*, **10**: 54-57.

Sivakumar, R., Chandrasekhar, C. N. and Srividhya, S. (2014) *Asian J. Hortic.*, **9**: 183-186.

Sobhana, A. (2014) *Asian J. Hortic.*, **9**: 160-163.

Sowmiya P. and Karuppaiah, P. (2019) *Plant Arch.*, **19**: 435-440.

Sreelatha, U., Gopikumar, K. and Aravindakshan, M. (1991) *Agri. Res. J. Kerala*, **29**: 67-70.

Srivastava, S.K. (1987) *Journal of Economic and Taxonomic Botany* **9**(1): 187–192.

Srivastava, S.K. and Kapoor, S.L. (1987) *Journal of Economic and Taxonomic Botany*, **9**: 173–177.

Sudhagar, R. and Kamalakannan, S. (2017) *J. Floric. and Landsc.*, **3**: 01-03.

Subramanian, R. (1977) *M.Sc. (Ag.) Thesis*, Tamil Nadu Agricultural University, Coimbatore.

Subramanian, R. and Shanmugavelu, K.G. (1978) *Indian Perfumer*, **23**: 38-41.

Suganthi, A., Saravanan, P. A., Sritharan, S. and Nelson, S. J. (2019) *J. Biol. Control*, **33**: 285-290.

Sumangala, H.P., Rao, V.K., Shivashankara, K.S. and Roy, T.K. (2018) *Int.J. Chem. Stud.*, **6**(2): 3008-3013

Sun YN, Tang FP, Fang WM, Tan GH. (2009) *Acta Agric. Zhejiang.*, **21**: 390-394.

Sundar, B.S.T., Kannan, M. and Jawaharlal. M. (2010) *J. Ornam. Hortic.*, **13**: 207-211.

Sushant Shekhar and Prasad, M. P. (2014) *World J. Pharm. Res.*, **3**: 829-836.

Susini, E. (1955) *Rev. Ortoflorofruttic. Ital.*, **39:** 460-463.

Thamaraiselvi S.P., Jawaharlal M., Ganga M. and Varadharaju N. (2010) *J. Ornam. Hortic.*, **13**: 171-181.

Thamburaj, S., Sambandamurthi, S., Chezhiyan, N., Abdul Khader, J.B.M. and Ponnuswami, V. (1984) *South Indian Hort.*, **32**: 363-364.

Thammaiah, N. Reddy, B.S., Gangadharappa, P.M., Kulkarni, B.S., Manjunath, M. and Kanamadi, V.C. (1997) *Av. Agric. Res.*, India, **8**: 19-20.

Thangaraj, T. (1977) *M.Sc. (Ag.) Thesis*, Tamil Nadu Agricultural University, Coimbatore.

Thenmozhi M. (2019) *Plant Cell Biotech. Mol. Biol.*, **20**: 106-111.

Trupti D.P., Patel G. D., Devi M. Sree, Dave P. P., Utsav, D. (2017) *J. Ornam. Hortic.*, **20:** 103-107.

U-Kong W, Wongsawad, P, Buddhark P. (2012) *Int. J. Appl. Agric. Sci.*, **7**: 17-26.

Varsha, A. P., Munikrishnappa, P. M. and Seetharamu, G. K. (2015) *Trends BioSci.***8:** 5888-91.

Vedamuthu, P.G.B. (1981) *M.Sc. (Ag.) Thesis*, Tamil Nadu Agricultural University, Coimbatore.

Veeraraghavathatham, D., Madhava Rao, V.N. and Shanmugavelu, K.G. (1985) *Indian J. Hort.*, **42**: 287-289.

Velu, G. (1988) *Res. and Development Reporter*, **5**: 103-104.

Veluswamy, P. (1980) *Nat. Seminar Prodn. Tech. Comm. Flower Crops*, Tamil Nadu Agricultural University, Coimbatore, pp. 5-6.

Veluswamy, P. (1981) *Ph.D. Thesis*, Tamil Nadu Agricultural University, Coimbatore.

Veluswamy, P., Muthuswami, S. and Thangaraj, T. (1973) *Indian J. Orn. Hort.*, **4**: 45-55.

Veluswamy, P., Muthuswami, S. and Thangaraj, T. (1974) *South Indian Hort.*, **22**: 127-128.

Veluswamy, P., Muthuswami, S. and Thangaraj, T. (1975) *South Indian Hort.*, **23**: 71-72.

Veluswamy, P., Muthuswami, S. and Thangaraj, T. (1977) *South Indian Hort.*, **25**: 20-25.

Veluswamy, P., Vijayakumar, M. and Muthuswami, S. (1980) *South Indian Hort.*, **28**: 156-157.

Verghese, J. and Sunny, T.P. (1992) *Flavour and Fragrance J.*, **7**: 323-327.

Vermeire, A. (1962) *Cult. Hand*, **28**: 65-66.

Yan, W., Zhang, J., Zhang, Y., Meng, D. and Yan, D. (2015) *J. Ethnopharmacol.*, **171:** 335-341.

Yathindra H.A., Keerthishankar K., Rajesh A.M., Harshavardhan M., Mutthuraju G.P. and Mangala K.P. (2018) *J. Pharmacogn. Phytochem.*, **7**: 257-259.

Ye, Q., Jin, X., Zhu, X., Lin, T., Hao, Z. and Yang, Q. (2015) *J. Oleo Sci.***64**: 645-52.

Younis, A., Khan, M.A., Akram, A. and Ashfaq, M. (2014) *Pak. J. Agric. Sci.*, **51:** 881-886.

Younis, A, Mehdi, A., Riaz, A. (2011) *Pak. J. Bot.*, **43**: 163-68.

Zhang, Y.S., Lai, M.Z., Chen, H.G. and Chen, J.M. (1997) *J. Fujian Agric. Univ.*, **26**: 44-47.

6

BIRD OF PARADISE

P. Bennurmath, Rajiv Kumar, A. Gurung,

B. K. Jana and D. Sanyal

1.0 Introduction

Strelitzia reginae, commonly known as bird of paradise or crane flower is an evergreen perennial herbaceous plant grown in the regions having moderate climate. In warmer areas, it grows as a specimen plant, while in frost prone areas, in a warm greenhouse. The brilliant colours and unusual appearance of the flowers made it exceptional cut flower. Therefore, the Bird of paradise cultivated in many parts of the world to produce cut flowers for both domestic and international markets.

It has much importance in domestic as well as international markets due to its attractive, remarkably shaped crested head of bird, a combination of orange and purple coloured ower clusters (Singh, 2006), blue green tropical foliage and distinctive flowers which appear periodically throughout the year (Gilman, 1999). It has been cultivated for the production of cut flowers, mainly by postharvest life, size, long stem and bright colors of their flowers (Pivetta *et al.*, 2007).

It is cultivated on a commercial scale in many parts of the world *viz.*, USA, Israel and South Africa. In North America, California is the largest producer followed by Hawaii (USDA, 2003). In temperate regions such as the Netherlands, Poland, China, Japan *etc.*, it is grown in controlled greenhouses with heating facilities. Costa Rica is an increasingly important supplier to the United States and Europe (ACOFLOR, 2003) and whereas Canary Islands for Europe (COLEACP, 2002).

In India, bird of paradise is grown in sub-temperate and sub-tropical regions such as Himachal Pradesh, Kalimpong and Darjeeling in West Bengal, Nilgiri hills and Western Ghats, Bengaluru and adjoining areas in Karnataka, Andhra Pradesh, Tamil Nadu, Kerala and Maharashtra (Sane *et al.*, 2020). As far as the production of agricultural activity is concerned, *Strelitzia* is one of increasing social and economic importance in the world (Da Silva *et al.*, 2012).

2.0 Importance and Uses

It occupies a place of pride in the garden and is an important choice for landscaping as an architectural plant and focal point (Van Jaarsveld, 2008). In the background of a herbaceous border, in front of a shrubbery or along the side of a tank or lilypool, a highly delightful effect is developed when in flower. It can also be grown in pots. The brilliant colour and unusual appearance of the flowers have made it exceptionally popular as cut flower. In beds of bird of paradise, amaryllis and other lilies flowering in semi shade can also be grown (Armbruster, 1973). It is the floral emblem of the city of Los Angeles.

The pigment Delphinidin-3-rutinoside was isolated from the petals and proanthoncyanidin polymers (flavonoids, antioxidants) from the leaves and the seeds are also used in the Cape to sour milk (Van Jaarsveld, 2008). The animal pigment, bilrubin found in *S. nicolai* arils, could be harnessed therapeutically to modulate certain pathological conditions (Dwarka *et al.*, 2017) and also is the primary aril pigment of *S. reginae* whereas, low concentrations found in its sepals (<44.0 ngg^{-1}) (Pirone *et al.*, 2010). The phytochemicals in owers of bird of paradise can be exploited for industrial uses. It has variable wax content (200–1400 µg/cm^2), carotenoids, anthocyanins *etc*. (Sane *et al.*, 2020).

3.0 Origin and Taxonomy

Strelitzia is indigenous to South Africa. The species, *S. reginae* is commonly known as bird of paradise, the name comes from remarkably shaped and coloured flower clusters, like the crested head of a bird. The genus is named in honour of Queen Charlotte Sophia, the wife of King George III, who was also the Duchess of Meck-linburgh-Strelitz, a patron of botany.

The genus *Strelitzia* belongs to the family Strelitziaceae. The roots are fleshy and finger-like. Plants are tufted, evergreen, acaulescent (stemless) perennial herb rhizomatous; leaves are distichous (opposite and arranged fan-like in one plane), large, long-petioled, margins sometimes red-edged; scape terminal or in the upper axils, short-exserted from the sheaths of the leaves; bracts large, spathe-like, boat-shaped, acuminate, solitary at the end of the scape or 2, slightly distant, perianth long-exserted; sepals free, long, carinate; petals dissimilar; stamens are 5; ovary is 3-celled, many-seeded.

The inflorescence stalk is tall with 4-6 flowers that emerge in succession in a boat-shaped spathe, producing a mucilaginous substance when in bloom. The flowers have orange sepals and blue/purple petals. The fruit is a hard woody capsule that splits from the apex in summer (August to February). The seeds are round, black to brown in colour with a yellow aril (a tuft of hairs) (Van Jaarsveld, 2008).

S. reginae, which is bird-pollinated, disperses its pollen in aggregates which form around multicellular threads. Bees are common visitors when the spathe is in flower (Van Jaarsveld, 2008). Examination of meiosis in *Strelitzia nicolai* revealed a somatic chromosome number of 22 and the formation of 11 bivalents. Previous records showed 2n = 14. The basic chromosome number was proposed as x = 11, with the basic number of x = 7, as in the more herbaceous species *S. reginae*, derived by chromosome reduction (Chen and Huang, 1990). Moore and Hyypio (1970) suggested the morphological characters for identifying *S. reginae, S. nicoli* and *S. candida*.

In the late 90's Mr John Winter, former Curator of Kirstenbosch, has made selections with desirable traits *viz.*, F_3 dwarf orange, F_2 dwarf broad-leaf Mandela's Gold, F_3 thin and long flower stock for cut flowers and the deep-red spathe (Van Jaarsveld, 2008).

4.0 Species and Cultivars

Strelitzia reginae is shown to be paraphyletic as *S. juncea* is nested within it. The highly localized and endangered *Strelitzia alba* is confirmed as a distinct species, as are *S. caudata* and *S. nicolai*, despite few morphological differences. Evolutionary trends are linked to changes in habitat and coevolution with pollinators. Climate change in southern Africa is thought to have restricted *Strelitzia nicolai* (or its ancestor) to the eastern coastal region, with subsequent allopatric speciation of *S. alba* and *S. caudata*, and relatively recent parapatric divergence of *S. juncea* from *S.*

Uses of Bird of Paradise (*Strelitzia reginae*)

Herbaceous border

Along side of a pool

Bird of paradise in pot

In flower arrangement

In flower vase

As specimen plant

Species of Bird of Paradise

Strelitzia alba

Strelitzia augusta

Strelitzia caudate

Strelitzia nicolai

Strelitzia juncea

Strelitzia reginae

reginae (Cron *et al.*, 2012). Two of the species, *Strelitzia nicolai* and *Strelitzia reginae*, are frequently grown as house plants.

Two easily recognized groups can be distinguished in the genus. A smaller stemless, tufted group (*S. reginae* and *S. juncea*) bearing elongated inflorescences protruding above the leaf canopy. A second group consisting of large shrubs or trees with elongated stems, huge banana-like leaves, and shortly stalked (and larger) inflorescences in the upper leaf axils (*S. alba, S. nicolai* and *S. caudata*). These three are very similar and closely related and difficult to distinguish when not in flower. Best known among them is *S. nicolai*, a coastal sand dune species. It has white sepals and blue petals. *S. alba* is the rarest. Both its petals and sepals are white (Xaba and Walt, 2011).

The genus *Strelitzia* comprises of about 5 species, some important ones are described below:

S. augusta

About 5 m high; leaves 60 to 90 cm long, oblong, acute; petioles 120 to 180 cm long; inflorescence composed of 2 spathes, shaped like bird's head, from the middle of which emerge white flowers; flowers on short purple pedicels; petals rounded at the base.

S. kewensis

The species is a garden hybrid developed from a cross between *S. augusta* x *S. reginae*. Up to 1.5 m high; leaves like that of in *S. augusta* but the blades 60 cm long and 35 cm wide; flowers vertical, pale watery yellow, more resembling those of *S. reginae*, but the small-hooded petal more like the other parent having lilac-pink patches at the base of the sepals.

S. nicolai

Clump forming perennial, vigorous; leaf-blade oblong, 1.5 cm long, rounded or heart-shaped at the base, leaf-stalks 2 cm long; spathes brownish red, 3-5, 40-45 cm long; flowers white, 20 cm long, corollas light purplish blue.

Interspecific and intervarietal hybridization in nature have caused a wide variation differing in flower colour as well as in inflorescence.

S. reginae

This species is the most widely cultivated, also known as queen of paradise, the flower-queen, bird of paradise and bananeirinha-of-garden (Karsten, 2009). It has been cultivated for cut flowers, mainly by postharvest durability, size, long stem and bright colors of their flowers (Pivetta *et al.*, 2007). Clump forming evergreen perennial shrub, up to 90 cm high; leaves bright green in two opposite rows (distichous); petioles green or with a slight greyish blue bloom; inflorescence stalk erect, firm and tall; flowering February to June; fruit oblong, triangular (triquetrous) woody capsule opens along the compartments (loculicidal dehiscence); seeds pea-sized, black and have a woolly orange aril (Xaba and Walt, 2011); the flowers orange and purple and very brilliant, emerging from the purplish spathes on a stem about

90 cm long; roots large, strong-growing. The plants are variable and among the varieties are – var. *humlis*: dwarf, dense clumps, leaves ovate-oblong. Suitable for containers; var. *glauca*: the leaves attractive, glaucous; var. *rutilans*: leaves midrib purple. *S. reginae* 'Mandela's Gold' is a form with yellow flowers. *S. reginae* sub sp. *Mzimvubuensis* is another subspecies of *S. reginae*. (Van Jaarsveld, 2008) which is favourite for garden and a choice cut-flower worldwide. The morphology of inflorescence, leaves, stem, root system, fruit and seed of this species have been described and illustrated by Szendel *et al.* (1976). The accession IC0624431 is having better growth, yield, physiological and biochemical traits (Sane *et al.*, 2020).

S. cauda

It is unbranched and multi-stemmed, 6 m tall, stem diameter 100 to 150 mm. The older stems are woody with leaf scars and often have suckers at the base. Leaves leathery, opposite, green to greyish; inflorescence with single purplish, boat-shaped spathe bearing several flowers in autumn to winter (Winter and Xaba, 2011). It is a good focal point for a mid-size to large garden.

S. juncea

This is a striking feature plant with upright cylindrical leaves without a leaf blade, growing from 1 to 2 m in height and producing large orange or yellow flowers borne on long, cylindrical scapes from May to October. Propagated by sub-division; inflorescence consists of a boat-shaped spathe bearing five flowers. The flower spathe is held below the level of the highest leaf tips compared to *S. reginae* where it is held above the leaves (Winter, 2004).

5.0 Growth and Flowering

5.1.0 Environmental Factors

Flower production is seasonal in most areas where *Strelitzia* is grown. In California, it blooms from September to May with an October-December peak (Besemer *et al.*, 1982), in South Africa during the fall-winter-spring peroid (Venter *et al.*, 1980), in Israel, a double peak in March-April and again in September (Halevy *et al.*, 1976), in the low elevations of Hawaii, from June-September (Kawabata *et al.*, 1984) and in the fall-winter months at higher elevations.

According to Kawabata *et al.* (1982) the flower yield in Hawaii showed an increase with the age of the plant and had annual waves; low yield in January to March, rising to a peak in July to September and dropping to low yield again in December.

In an investigation on 4-year-old plants, Hetman and Pogrozsewska (1983) observed the 1st, 2nd, 3rd, 4th and 5th year cropping of *S. reginae*. The plants started cropping in the 4th to 6th year, entering the reproductive phase when they had an average of 14.1 leaves. Flowering plants produced up to 4 leaves per shoot. In the 7th to 9th year of cultivation the plants had 2-4 ramifications. An inflorescence appeared in the leaf axil in about 53.7 weeks from the time of leaf appearance and flowered at 72.8 weeks. The flowering season lasted from June to May with the greatest concentration between December and March. According to Cai and Wang

(2002) the initial flowering of *S. reginae* started at the fourth year after planting and the full-bloom in summer and autumn; each inflorescence could produce 4-6 fruits and the mature period of fruit was 80-140 days.

5.1.1 Light

Plant height was reduced when plants were grown under 13 h photoperiods. The number of generative shoots was not affected by photoperiod treatments. The number of vegetative shoots was related inversely to the length of the photoperiod. The number of days to flower decreased as photoperiod increased (Healy *et al.*, 1993).

Fava *et al.* (2015) showed the effect of solar radiation in initial production of flower stalks of *S. reginae*. The environments with up to 30 per cent shade recorded highest number, height, diameter and fresh weight of stems. There is delay in the initial production of flower stalks due to shade. The radiation ensures the production process, but when it focuses directly on the plant, combined with other environmental factors like temperature and relative humidity, affects the pattern of trade of culture

5.1.2 Temperature

5.1.2.1 Leaf Emergence Interval

Haouala and Zouari (2001) recorded the development of inflorescence of *S. reginae* was directly related to temperature and the number of leaves. It is recommended that plants with 51-55 leaves should be favoured, by means of the practice of stripping off leaves.

Because each leaf normally bears one floral initial in its axil, the frequency of leaf emergence has a direct relationship to flower production. The results of the experiment in the phytotron in Israel showed that warm temperature caused rapid development of leaf (Halevy *et al.*, 1987). The interval between successive leaves was 5.9 weeks for plants grown at a 27/32°C (night/day) regime and 11.3 weeks for plants grown at 12/17°C. Between 22/27°C and 17/22°C, decreasing temperature lengthened intervals between successive leaf emergence.

5.1.2.2 Leaf Emergence to Flower Cut

It has been reported by Fisher (1976), and Criley and Kawabata (1984) that the time-frame for flower development spans more than a year from the first identifiable floral initial until the flower reaches anthesis. Measurable progress may be assumed to coincide with the emergence of the leaf which subtends the flower (Kawabata *et al.*, 1984). The least variation is shown in Hawaii with less than a year spent in leaf emergence to flower cut (LEFC), while in both South Africa and California, LEFC requires more than a year.

5.1.2.3 Flower Emergence to Flower Cut

Halevy *et al.* (1987) observed the fastest flower development in 8 weeks from the time the stalk was visible in both Hawaii and Australia in the summer months The slowest flower emergence to flower cut (FEFC) required about 28 weeks in South Africa through the fall-winter month and 9 weeks in spring. In Poland,

Pogroszewska (1996) studied the plant morphology, growth rate, flowering and flower quality of *S. reginae*. Plants were slow to mature and produce flowers, taking 7 years to give satisfactory results. They flowered from autumn to spring with peak flowering in December and January. Plants with elongated leaves, which were branched or very leafy tended to flower abundantly.

Kawabata *et al.* (1984) recorded that low temperature was mostly associated with flower production and an increase from 14 to 16°C retarded floral differentiation and subsequent elongation period to flower emergence. Halevy *et al.* (1987) in their international studies reported that low temperature inhibited growth and flower development, and air temperature below 13°C retarded the rate of leaf and flower production, while temperature above 17°C promoted the growth and development. They suggested that temperatures ranging between 17 and 27°C would provide optimum environmental conditions for rapid and uniform flowering.

5.1.3 Growth Regulators

When plants of bird of paradise were planted on a monthly basis, plant height showed a cyclic pattern with the tallest plants developing from spring and summer plantings and the shortest plants from winter planting. Spray treatments of ancymidol, cycocel or B-Nine were only moderately effective in controlling plant height while paclobutrazol was the most effective. The number of days to flower was increased by up to 5 days by paclobutrazol application (10 or 20 ppm) (Healy *et al.*, 1993).

Generally, date palm pollen extract used as an effective bio-stimulant. It is good nutritional, growth promoting and flowering enhancing because it contains growth substances as minerals, amino acids, auxin, tryptophan and higher concentration of indoles. In an investigation, the highest growth and flowering were recorded with 20 g/L extracts of date palm pollens and applying 10 g/L foliar spraying of extracts date palm pollens gave increased in pigment content and total carbohydrates percentage in leaves and inflorescences (Abou-Sreea and Yassen, 2016).

6.0 Propagation

It is commonly propagated by seed or clump division (Bautitz and Carvalho, 2007; Pivetta *et al.*, 2007). Usually, plant raised through seeds takes longer period to establish than the vegetatively propagated plants. Each inflorescence produces 1 to 6 fruits; each fruit contains an average of 30 seeds. The pollen is released before the stigma becomes receptive (protrandria), favoring cross pollination and a satisfactory production of seeds (Pivetta *et al.*, 2007).

6.1.0 Seed Propagation

Seed propagation would be helpful in hastening planting material multiplication as vegetative propagation is time consuming (Van der Walt, 2000). The flower pollinated by hands improved seed setting (Notten (2002) and germinated at 18-21°C. Cross-pollination between flowers of different plants is necessary to achieve >50 per cent pod set, correct pod growth and satisfactory seed production. There was no pod development on isolated flowers owing to the relative positions of anther

Propagation of Bird of Paradise

Fruits and seeds

Germination of seeds

Seedling in pot

Seedlings transferred in pot

Separation of suckers

Rooted sucker

Growth of plants

Cultivation of Bird of Paradise

Cultivation in open field

Growing bird of paradise in pots in greenhouse

Bird of paradisc flowering in greenhouse

and stigma and (probably) self-incompatibility factors. The yield of in fructescences with 5-6 pods was approximately 100 seeds, of total weight > 25 g (Malfa *et al.*, 1988).

El-Kholy and Hassan (1983) reported that seeds treated with tryptophan at 5 ppm showed the highest percentage of germination from 82.6 to 89.3 per cent compared with 15.3 to 20.6 per cent in the control. Seed germination in *S. juncea* was also increasingly stimulated by the O_2 concentration (0-100 per cent) in the incubation chamber.

Seed dormancy in *Strelitzia* spp. can be released by treatment of the seeds with high O_2 levels. Incubation of *S. juncea* [*S. reginae*] seeds in an O_2 atmosphere resulted in an increase in the glucose-6-phosphate dehydrogenase (GPDH) activity of crude embryo extracts on day 1; radicle protrusion started on day 5. Similarly, 6-phosphogluconate dehydrogenase (6 PGDH) activity increased over the first 4 d of incubation in O_2. The effects of O_2 supplied to dormant seeds, on the activity of the two key pentose phosphate (PP) pathway enzymes, support the hypothesis that oxygen shortage restricts PP pathway activity in dormant seeds (Meillon *et al.*, 1990a). They also found that ethanolic fermentation of the embryos of dormant normoxically incubated seeds resembled that of hyperoxically incubated seeds. Alcohol dehydrogenase (ADH) activity was similar over the first 4 incubation days, and therefore before the beginning of radicle emergence. Phosphofructokinase (PFK), inorganic pyrophosphate: D-fructose-6-phosphate: 1 phosphotransferase (PFP), pyruvate kinase (PK) and aldolase activities and glucose, fructose, glucose-6-phosphate and fructose-6-phosphate contents were compared for 4 d prior to radicle emergence in embryos of *S. juncea* (*S. reginae*) seeds incubated in air or O_2. PFK activity was similar for both treatments and tended to increase as the incubation period advanced (Meillon *et al.*, 1990b). Scarification of seeds with concentrated sulfuric acid for nine minutes favoured seedling emergence (Barbosa *et al.*, 2005). Seed germination in Bird of Paradise can be enhanced by exposing seeds to temperature regime of 20°C/30°C -16/8 hrs to 78 per cent and pre treatment of seeds with GA_3 @ 750 ppm for 24 hrs (Kumar, 2017).

6.2.0 Vegetative Propagation

Planting of rooted suckers is the common method of vegetative propagation and the plant flowers early.

6.2.1 Micropropagation

It is commonly propagated vegetatively by division or by seeds. Both these methods of propagation are slow (Karnataka, 2008). The development of a reliable *in vitro* method for propagating *Strelitzia* spp., through the culture of excised embryos and multiple shoot formation greatly contributing to the commercial production of *Strelitzia* spp. (North *et al.*, 2010).

In vitro tissue culture methods for the propagation of *S. reginae* from explants obtained from seeds (protocorms) and stem in different culture media and growth regulators were reported by Wang and He (2000). Regeneration was either direct from protocorm explants, or via shoot induction from callus derived from stem explants.

The use of tissue culture techniques for the propagation of *S. reginae* has failed due to oxidative browning of the wounded explant. The brown exudate that diffuses into the medium was found to be detrimental to further development of the explants which eventually became necrotic and died. Shreevatsa *et al.* (2004) reported that the browning of terminal buds of rhizomes explants was controlled effectively by dipping the explants in ascorbic acid at 150 mg/L and by the addition of 200 mg/L ascorbic acid to the medium. Culture establishment and multiple shoot induction were achieved on MS medium containing 0.5 mg/L IBA and 7 mg/L BAP, in addition to 200 mg/L ascorbic acid and sucrose 30g/L.

Yi-ru (2009) provided basic information about the establishment of rapid proliferation system of *Strelitzia reginae*. The most suitable method of sterilization was to soak explants into 0.1 per cent $HgCl_2$ solution for 10 minutes after pre-treatment with 70 per cent alcohol for 30 seconds and the medium MS + 6-BA 1.5 mg/L + ZT 1.0 mg/L + NAA 1.0 mg/L was effective to induce shooting and suitable for the first generation cultivation and for cluster shoot induction and secondary proliferation MS + 6-BA 2.0 mg/L + ZT 1.0 mg/L + NAA 0.5 mg/L is effective and the most optimum rooting medium was 1/2 MS+ IBA 0.6 mg/L, and the rooting rate was 90 per cent and had highest survival rate of 92 per cent. Interactive effects of higher NAA and BAP concentrations (0.1 mg/L NAA; 3 mg/L BAP and 0.5 mg/L NAA; 5 mg/L BAP) without wounding is the most effective in reducing explants discoloration at the media contact point. Interactions between antioxidants and wounding treatments resulted in the absolute absence of callus induction involving ascorbic acid (North *et al.*, 2012). In *in vitro* conditions, media composed of ½ MS+ Sucrose @ 1 per cent + Gelrite 3 g/L + Inositol 100 mg/L + Thiamine-HCl @ 1 mg/L + Biotin 2 mg/L + GA_3 @ 1 mg/L has increased the germination rate in the *Strelitzia reginae* zygotic embryos (Kumar, 2017).

7.0 Soil and Climate

In mild climate, *Strelitzia* grows and flowers well in the open, both in full sun or in slightly shady situation. They prefer well-drained loamy soil rich in organic matter. In cool climate, however, they can be grown in warm greenhouse either in borders or in pots having loam based composts. Gilman (1999) reported that the plant grows in partly shade/partly sun requires occasionally wet, clay, sand, acidic, loam, slightly alkaline soil. It is drought tolerant and has poor soil salt tolerance. Temperatures of 10-13°C at night and 20-22°C during the day are suggested (Auman, 1980).

8.0 Cultivation

The offsets usually take 3-5 years to come into flowering. However, with proper planning and correct use of fertilizers the plant can be made to flower earlier and in a comparatively shorter time, almost half. The desired effect can be achieved starting with fresh preparation of the soil and ending up with the supply of proper nutrition till the time of flowering. This crop can be grown as monocrop and as intercrop in low rainfall areas (Sane *et al.*, 2020).

8.1.0 Planting

Pots should be properly cleaned before use. Potting requires well-drained fertile medium and should be eventually adapted to a position in full sun. The recommended potting medium for *Strelitzia* is 2 parts loam, 2 parts sand, 3 parts 12 cm bark, 3 parts compost and 1/2 part bone meal, and 1 part Bounce back ® or 1 part 3.1.5 fertiliser and alternative application in May and November with a mixture of (a) 1 part superphosphates, 1 part 3.1.5, 2 parts Bounce back ® and 2 parts bone meal or (b) compost (Van Jaarsveld, 2008). Offsets are planted in the centre of the pot and watered to settle the medium. Roots of *Strelitzia* plants are wide spreading and, therefore, care should be taken that they do not cramp while planting.

The autumn is suitable for potting/reporting, transplanting and dividing plants. This allows the plant to focus its energy on recovering rather than flowering. Mature plants will normally flower due to stress when divided or transplanted, an in-built survival mechanism. Flower stalks can be cut to allow quicker recovery. When dividing mature plants it is important to have a minimum 60-cm clump diameter, so that recovery is faster, more or less two to three years maximum, to full bloom again (Van Jaarsveld, 2008).

For planting in ground, pits of 90 x 90 cm size are prepared and filled with soil and organic matter. Plants take long time to establish properly and therefore, should not be disturbed very frequently.

The *Strelitzia* is a plant that produces flowers throughout the year, especially during summer. They have good production in warm climates and optimum temperatures of close to 25°C, while the minimum temperature for better production is 10°C and relative humidity of 70 per cent (Lamas, 2002). Li *et al.* (2012) reported that under protected cultivation of *Strelitzia reginae*, seedlings took five or six years to blossom and reach full-bloom from October through April and also suggested that from Feburary through April, is the best period of manual pollination; April through September, 60 per cent to 80 per cent sunlight shade would generate a good habitat.

8.2.0 Irrigation

The first watering should be done immediately after planting and afterwards, depending on plant's demand and surrounding conditions. During summer the plants require abundant supplies of water. The beds should thoroughly be soaked at least once a week.

8.3.0 Manuring and Fertilization

If the compost, at the time of planting, has adequate amount of organic matter, plants grow satisfactorily. Plants of bird of paradise were grown (for cut flowers) in pots in 4 growing media, containing 3:1, 1:1 or 1: 3 sand: clay or clay on its own. Plants were fertilized with N, P and K (g/pot) at 0.49: 0.19: 0.33 (low), 0.99: 0.38: 0.66 (intermediate) or 1.48: 0.57: 0.99 (high). Within fertilizer treatments, the intermediate level produced plants with the highest numbers of flowers and leaves/plant, the longest leaves and flowers and the heaviest flowers (FW basis). Plants treated with the high level of fertilization contained the highest concentrations of N, P and K in leaves. It was concluded that the best growth of *S. reginae* was obtained in growing media containing and: clay at rates of 1:3 or 1:1 and fertilized with intermediate levels of NPK (El-Shoura and Hosni, 1996). In a two years greenhouse experiment, bird of paradise plants (*S. reginae*) were supplied with S at 0, 50 or 100 g/m^2 and NPK (18:18:5 per cent) at 0.0, 0.5 or 1.0 per cent with trickle irrigation. It was observed by Ali (1998) that S and NPK application significantly affected leaf length, number of leaves/plants, inflorescence length, number of inflorescences/plant, weight of the inflorescence, length of corolla, bract length, and weight and mineral content of bracts. The interaction effects were significant for these characters, with the exception of inflorescence length and length and weight of bracts. The highest values for the studied traits were recorded for the highest rates of S and NPK application in both years. In stunted plants contorted petioles and crumpled and flagelliform blades observed. Qui and Qui (1999) recommended application of additional molybdenum, sulphur-free urea and dihydrogen potassium phosphate are recommended to limit the occurrence of flagelliform leaves. Bhargava and Raghupati (1993) analyzed leaf nutrient status in bird of paradise showing variation in plant growth.

Siraj-Ali (1998) suggested that 3 per cent singral (NPK, 16:9:2) and 3 per cent nitrophoska (NPK, 18:18:5) fertilizers in a sprinkler irrigation system improved the growth and flowering under greenhouse conditions.

Jainag *et al.* (2011) recommended 80, 100 and 120 per cent dose of fertilizers (16:11:6 g NPK/plant/month) provided through fertigation in splits at monthly interval along with 8 and 12 litres of water per day per plant for 12 months. Higher fertigation level (120 per cent recommended dose of fertilizer + 12 litres of water/day/plant) also recommended for improved vegetative growth and flower production. Combined application of nitrogen @ 30 g and phosphorus @ 15 g/plant/year recorded significantly higher vegetative as well as flowering attributes and also improved quality and yield of flowers of bird of paradise (Patel *et al.*, 2017). A fertilizer dose of 62.5: 25: 62.5 g NPK/plant/year along with foliar application of $ZnSO_4$ (0.5 per cent) and boron (0.25 per cent) recorded maximum yield per plant, yield per hectare and benefit cost ratio (Table 1 and 2) (Navyashree *et al.*, 2017).

Table 1: Effect of NPK and Micronutrients Combinations on Cut Flower Yield per Plant and per Hectare per Six Months of Bird of Paradise*

Treatment	*Cut Flower Yield per Six Months per*	
	Plant	*Hectare (Lakhs)*
37.5: 15: 37.5 g NPK/plant/year	6.24	0.27
50:20:50 g NPK/plant/year	6.98	0.31
62.5: 25: 62.5 g NPK/plant/year	7.53	0.33
37.5: 15: 37.5 g NPK/plant/year + ZnSO (0.5 per cent)	7.24	0.32
50:20:50 g NPK/plant/year + ZnSO, (0.5 per cent)	7.72	0.34
62.5: 25: 62.5 g NPK/plant/year + ZnSO, (0.5 per cent)	8.17	0.36
37.5: 15: 37.5 g NPK/plant/year + Boron (0.25 per cent)	7.15	0.31
50:20:50 g NPK/plant/year + Boron (0.25 per cent)	7.60	0.33

Treatment	Cut Flower Yield per Six Months per	
	Plant	Hectare (Lakhs)
62.5: 25: 62.5 g NPK/plant/year+Boron (0.25 per cent)	7.83	0.34
37.5: 15: 37.5 g NPK/plant/year+ZnSO, (0.5 per cent) +Boron (0.25 per cent)	7.38	0.32
50:20:50 g NPK/plant/year+ZnSO, (0.5 per cent) +Boron (0.25 per cent)	8.35	0.37
62.5: 25: 62.5 g NPK/plant/year +ZnSO, (0.5 per cent) +Boron (0.25 per cent)	9.02	0.40
SEm±	0.15	0.01
CD @ 5 per cent	0.46	0.03

Control: 50:20:50 g NPK/plant/year.

* Navyashree *et al.* (2017)

Table 2: NPK, Zn and B in Leaf Content of Bird of Paradise as Influenced by NPK with Combination of Micronutrients (after 180 days of treatment imposition)*

Treatment	N (Per cent)	P (Per cent)	K (Per cent)	Zn (Per cent)	B (Per cent)
37.5: 15: 37.5 g NPK/plant/year	1.86	0.18	2.58	23.62	17.29
50:20:50 g NPK/plant/year	2.34	0.27	2.70	24.66	18.42
62.5: 25: 62.5 g NPK/plant/year	3.87	0.43	2.97	25.68	19.67
37.5: 15: 37.5 g NPK/plant/year + ZnSO (0.5 per cent)	2.04	0.22	2.57	37.44	17.83
50:20:50 g NPK/plant/year + ZnSO, (0.5 per cent)	2.38	0.32	2.75	38.68	18.58
62.5: 25: 62.5 g NPK/plant/year + ZnSO, (0.5 per cent)	2.93	0.45	3.02	40.56	19.81
37.5: 15: 37.5 g NPK/plant/year + Boron (0.25 per cent)	2.01	0.19	2.55	23.73	21.37
50:20:50 g NPK/plant/year + Boron (0.25 per cent)	2.36	0.28	2.71	24.86	22.30
62.5: 25: 62.5 g NPK/plant/year+Boron (0.25 per cent)	2.89	0.44	2.98	25.89	23.35
37.5: 15: 37.5 g NPK/plant/year+ZnSO, (0.5 per cent) +Boron (0.25 per cent)	2.06	0.24	2.62	37.51	22.13
50:20:50 g NPK/plant/year+ZnSO, (0.5 per cent) +Boron (0.25 per cent)	2.40	0.34	2.79	41.45	24.74
62.5: 25: 62.5 g NPK/plant/year +ZnSO, (0.5 per cent) +Boron (0.25 per cent)	3.19	0.48	3.11	43.37	26.27
SEm±	0.18	0.02	0.07	0.44	0.49
CD @ 5 per cent	0.55	0.06	0.22	1.30	1.46

Control- 50:20:50 g NPK/plant/year.

* Navyashree *et al.* (2017).

9.0 Diseases and Pests

Bird of paradise is quite hardy and less prone to diseases and pests, but damage of flower bud and flower due to fungal infection may be severe and it has low input requirement and highly economical with less pest and disease load (Sane *et al.*, 2020).

9.1.0 Diseases

Fungal Diseases

Studies on the diseases of flower buds and flowers of *S. reginae* showed that about 18 per cent of the plants grown for cut flowers were affected by diseases. Bud damage was attributed mainly to *Fusarium* and flower damage to *Botrytis cinerea, F. culmorum* and *F. avenaceum* (Orlikowaski, 1976). Orlikowaski (1977) also isolated 25 fungi from seeds and seedlings of this species out of which 51.8 per cent was *Fusarium* species. Root rot, caused by *Fusarium monoliforme* is a seed-borne disease. Pre-soaking of seeds in water at room temperature followed by hot water (135°F) treatment for 30 minutes controlled the disease (Pirone *et al.*, 1960).

Infection of *Strelitzia* with *Armillaria mellea* was reported by Davino (1984) from Sicily. Soil temperature, pH and microbial flora have been found to be important in the development of the disease. The use of disease-free soil was recommended as the best method of preventing further attacks. Control could also be achieved with the stress effect, *i.e.*, fumigating with methyl bromide, heat sterilization or solarization.

Incidence of *Phytophthora parasitica* has been reported. The disease can attack any part of a plant under a wide range of temperatures. Micropropagation of healthy plant material, meticulous hygiene, disinfection of tools and premises, and burning of infected plants are among the recommended protective measures (Orlikowski, 1999). First report of incidence of leaf blight and root and foot rot of a *Strelitzia* caused by *Phytophthora nicotianae* has been reportein Italy (Luongo *et al.*, 2010).

Symptoms of *Botrytis cinerea* infection were observed in *S. reginae* for the first time in Argentina (Wright *et al.*, 1996). Fungal infestation on the pollen wall and cytoplasm is common in *Strelitzia* after anther dehiscence (Skvarla *et al.*, 1997).

A fungal infection causing spots on the leaves and flower peduncles has been observed in Calabria and Sicily over the last few years. The causal agent was identified as a *Microsphaeropsis* sp. on the basis of its morphological and biometric characteristics. This is the first report of this genus attacking *S. reginae* (Agosteo and Pennisi, 1996).

A new basal stem rot was detected in *S. reginae* plants in a commercial nursery in Carrubba, Italy during 2015. The diseased plants displayed a dry rot of the basal stem which resulted in the detachment of the roots from the stem. General wilting and rot of internal foliage were also observed on the affected plants.

The species *Pleiocarpon strelitziae* was more aggressive than *Ilyonectria strelitziae*, resulting in the death of all inoculated plants (Aiello *et al.*, 2017).

Diseases and Pests of Bird of Paradise

Root rot

Bacterial leaf spots

Mealy bug attack

Scale insect infestation

Bacterial Diseases

Rodrigues *et al.* (2011) suggested that the seedlings of *Strelitzia* could be used as test plants for presumptive diagnosis for banana Moko disease.

Necrotic and soft rot symptoms on flower tissues of bird of paradise causes economic damage (Abd-El and Nofal, 2001). Four bacterial isolates were identified as *Pseudomonas syringae* pv. *delphinii* and the other four were identified as *Erwinia carotovora* subsp. *carotovora*. Pseudomonas was more pathogenic than erwiniae. Pseudomonas isolates were sensitive to both chloramphenicol and tetracycline, while erwiniae isolates were sensitive to chloramphenicol.

A bacterial leaf spot disease caused by *Pseudomonas syringae* pv. *syringae* was observed on *Strelitzia augusta* plants growing near citrus groves in Catania province, southern Italy, during the warm and humid spring of 1999 (Bella *et al.*, 2000). Sugiyama *et al.* (2017) reported incidence of bacterial leaf streak caused by *Paraburkholderia andropogonis* on *S. reginae* in Hawaii and can be managed with early recognition and rouging of diseased leaves.

9.2.0 Pests

Mealy bugs (*Planococcus citri and Pseudococcus adonidum*) and scales (*Aspidiotus camolliae*) are important pests of *Strelitzia*. Denmark and Osborne (1985) also noticed *Chaetanaphothrips signipennis*, a pest causing leaf roll and other foliar malformation in *S. reginae*. The use of vydate (Oxamyl) has been suggested to control the thrips in the greenhouse.

Two diaspidid species, *Chrysomphalus aonidum* and *Pinnaspis aspidistrae* have been reported to attack *Strelitzia* sp. Actellic 500 EC [pirimiphos-methyl], combined with the wetting agent citowett AL, are recommended for chemical control (Labanowski, 1999). The tineid *Opogona sacchari* and the noctuid *Sesamia nonagrioides* were observed causing damage to *S. reginae* in Puglia, Italy, in 1990-91. Both species lived endophytically; larvae of *O. sacchari* fed on the collar and roots of the plant, while those of *S. nonagrioides* fed on the flowers and leaves (Porcelli and Parenzan, 1993).

Association of various plant parasitic nematodes with this crop was reported by Petit and Crozzoli (1995). Gilman (1999) reported that the scale is the main pest problem in *Sterlitzia reginae*. The larvae of Bag worm defoliating while residing inside the bags (Pal and Sarkar, 2009).

10.0 Harvesting

The marketing of *Strelitzia* is increasing due to its beauty, exotic and high postharvest longevity (Queiroz *et al.*, 2012). Flowers should be cut in the morning when still turgid (Jaroenkit and Paull, 2003). However, Bayogan *et al.* (2008) suggested that the inflorescence should be harvested when the first floret was just about to emerge from the boat. The stalks harvested with closed buds had extended longevity, while the flower stalks harvested with the first opened button had the highest flower opening rate (Dias *et al.*, 2013).

Postharvest Management of Bird of Paradise

Packaging of flowers

Bird of paradise flowers in market

Jaroenkit *et al.* (2008) reported that harvesting before the inflorescence opened reduced the nectar and mucilage production, while, washing of the flower tongue with water containing detergent significantly reduced postharvest mould growth on the inflorescence.

The cardboard box packaging showed the lowest percentage of damage in the stems, the sleeve net protection most affected the visual appearance of the flowers (Dias *et al.*, 2013).

11.0 Postharvest Management

The major postharvest problems are saprophytic mold and is controlled by dipping inflorescences in benomyl or thiobendazole (TBZ) at 200 mg/L and senescence symptoms include floret desiccation and bract darkening (Jaroenkit and Paull, 2003).

11.1.0 Pulsing

Finger *et al.* (1999) observed that pulsing with 40 per cent sucrose for 24 h extended flower vase life and increased the number of opened florets. Addition of silver thiosulfate (STS) to the sucrose pulsing solution did not improve flower longevity over sucrose pulsing alone.

Table 3: Effect of different Pulsing Treatments and Pulsing Duration on Growth and Vase Life in Bird of Paradise*

Treatment	*Days taken to Open First Floret after Pulsing (Days)*	*Number of Florets Open/ Stem in Vase*	*Vase Life (Days)*
Control (distilled water)	4.15	1.10	7.35
20 per cent sucrose	3.32	1.23	8.17
20 per cent sucrose + 250 ppm 8HQC	2.56	1.63	9.43
20 per cent sucrose + 250 µM TDZ	2.16	1.77	10.17
20 per cent sucrose + 500 µM TDZ	2.23	1.67	9.80
20 per cent sucrose + 250 ppm 8HQC + 150 ppm citric acid	0.99	1.77	10.80
20 per cent sucrose + 250 µM TDZ +150 ppm citric acid	0.80	1.90	12.30
20 per cent sucrose + 500 µM TDZ +150 ppm citric acid	1.13	1.57	10.93
CD0.05	0.34	0.17	0.61

HQC: 8-hydroxyquinoline citrate; TDZ: Thidiazuron.

*Koley *et al.* (2016).

According to Jaroenkit and Paull (2003) major chemicals used in pulsing and holding solution for Bird of paradise at 8°C, sucrose (10 per cent w/v), citric acid (150 mg/L) and 8-hydroxyquinoline citrate (250 mg/L). Pulsing with 5 per cent sucrose overnight then shipped and upon unpacking held in 10 per cent sucrose, 200 mg/L 8-HQC and 10 mg/L silver nitrate improved postharvest life (Bayogan *et al.*, 2008). Queiroz *et al.* (2012) found that 40 per cent sucrose and 150 ppm of citric acid and cold storage (10°C) was best for the longevity of stem and petal color. However, pulsing treatment in cut stems for 48 hours with 20 per cent sucrose + 250 μM thidiazuron + 150 ppm citric acid recommended for increasing the vase life, postharvest quality and biochemical parameters (Table 3) (Koley *et al.*, 2016). Pereira *et al.* (2018) reported that pulsing with 2–6 mM SA resulted in significant reduction of transpiration, fresh mass maintenance and increased POD activity. In an investigation by Jeevitha *et al.* (2013) observed that the flowers harvested at commercial stage and pulsed with 10 per cent sucrose + 200 ppm 8HQS + 150 ppm citric acid for 48 hours had highest cumulative uptake of water, cumulative ratio of uptake of water and loss of water, less cumulative physiological loss in weight and highest vase life. Pulsing cut stems with 500 μM thidiazuron (TDZ) for 24 hours at 20°C extended floret longevity by 2.3-3.2 days (20-30 per cent) and pre-treatment with inhibitors of ethylene action (500 nl/L 1-methylcyclopropene or 0.2 mM silver thiosulfate) for 6 hours at 20°C reduced the ethylene-related loss in floret life (Macnish *et al.*, 2009).

11.2.0 Vase Life

Bayogan *et al.* (2008) reported the postharvest life of BOP inflorescence ranged from 6 to 16 days. El-Mokdem *et al.* (1994) reported that cut flower of *S. reginae* kept in 20 per cent sucrose solution produced the highest percentage of opened florets/ inflorescence, floret vase life and inflorescence duration were significantly increased, compared with controls, by either 20 per cent sucrose or 25-50 ppm GA_3. Paull *et al.* (2001) found no effect of BA on the vase life of bird of paradise. Periodically recutting the stem end increased the number of fully opened florets from 1.6 (controls) to 2.7 and extended the vase life from 7.6 days to 11.3 days (Campanha *et al.*, 1997).

Ali and Hassan (2014) recommended 8-HQS (250 ppm) or GA_3 (200 ppm) to maximize longevity as well as postharvest quality. Higher quality and the longest vase life of *Strelitzia reginae* cut flower spikes can be achieved using STS (0.463 mM) for 30 minutes then 10 per cent sucrose + 200 ppm (8-HQS) for 12 hours (Gendy and Mahmoud, 2012). Inflorescences of bird of paradise flower with stem diameter >12 mm, recut every 48 h, showed better vase life than the inflorescences with thinner diameter (Karsten *et al.*, 2012).

The best extension of post-storage longevity occurred after storing flowers up to 14 days at 10°C. Flower stored at 10°C for 28 days showed chilling injury symptoms in the bracts and sepals, and developed *Penicillium* sp. in the petals (Finger *et al.*, 2002). However, Da Silva Vieira *et al.* (2015) observed lower content of polyamines (putrescine and spermidine) a high content of polyamines (spermine) and an increase of peroxidase activity in sepals stored in the cold when the flowers were placed in recipients containing tap water and stored in a cold room, at 7.5°C and 90 per cent RH, for a period of 12 days.

12.0 References

Abd-El, H. and Nofal, M.A. (2001) *Arab Universities J. Agric. Sci.*, **9**: 397-410.

Abou-Sreea and Yassen (2016) *Int. J. PharmTech Res.*, **9**(12): 16-23.

ACOFLOR (2003) San José, Costa Rica, Flower Exports Report.

Agosteo, G.E. and Pennisi, A.M. (1996) *Informatore Fitopatologico*, **46:** 20-22.

Aiello, D., Polizzi, G., Crous, P. W. and Lombard, L. (2017) *IMA fungus*, **8**(1): 65-76.

Ali, E. and Hassan, F. (2014) *Sci. Agric.*, **2**(2): 77-82.

Ali, M.S.S. (1998) *Alexandria J. Agric. Res.*, **43:** 81-92.

Armbruster, J. (1973) *Gartenwelt*, **73:** 168-169.

Auman, C.W. (1980) In: *Introduction of Floriculture* (ed. R.A. Larson), Academic Press, New York.

Barbosa, J.G., Alvarenga, E.M., Dias, D.C.F.D.S. and Vieira, A.N. (2005) *Revista Brasileira de Sementes.*, **27**(1): 71-77.

Bautitz, F. and Carvalho, R. (2007) *Revista Acadêmica*, **5**(1): 47-55.

Bayogan, E.R.V., Jaroenkit, T. and Paull, R.E. (2008) *Postharvest Biol. Technol.*, **48**(2): 259-263.

Bella, P., Catara, V., Granata, G. and Cirvilleri, G. (2000) *J. Plant Pathol.*, **82**(2).

Besemer, S.T., Mc Nairn, R.B. and Halevy, A.H. (1982) *Flor. Rev.*, **170:** 16-18.

Bhargava, B.S. and Raghupathi, H.B. (1993) In: *Methods of Analysis of Soils, Plants, Waters and Fertilisers.* (Ed. H.L.S. Tandon). Fertilisers Development and Consultation Organisation, New Delhi, India, pp. 49-82.

Cai, B.P. and Wang, Z.Z. (2002) *J. Subtropical Plant Sci.*, **3**.

Campanha, M.M., Finger, F.L., Cecon, P.R. and Barbosa, J.G. (1997) *Revista Brasileira de Horticultura Ornamental*, **3:** 27-31.

Chen, Z.Y. and Huang, X.X. (1990) *Acta Botanica Austro Sinica.*, **6**: 47-49.

COLEACP(2002)www.coleacp.org/en/documentation/rapports/flowersplants/en_14_position.doc

Criley, R.A. and Kawabata, O. (1984) *J. Amer. Soc. Hort. Sci.*, **109**: 702-706.

Cron, G.V., Pirone, C., Bartlett, M., Kress, W.J. and Specht, C. (2012) *Syst. Bot.*, **37**(3): 606-619.

Da Silva Vieira, M.R., Lima, G.P.P., Vianello, F., de Souza, A.V., do Nascimento Simoes, A. and de Moura, F.B. (2015) *Australian J. Crop Sci.*, **9**(4): 313.

Da Silva Vieirasup, M.R., Limasup, G.P.P., de Medeiros, D.C. and de Oliveira, E.C.A. (2012) *J. Hortic. For.*, **4**(11): 178-180.

Davino, M. (1984) *Colture Protette*, **13**: 63-65.

Denmark, H.A. and Osborne, L.S. (1985) *Entomology Circulation Division of Plant Industry, Florida Dept. of Agri. and Consum. Services*, **274:** 2.

Dias, G.M., Queiroz, S.J., Sanches, J. and Tombolato, A.F.C. (2013) *Ornam. Hortic.*, **19**(2): 93-98.

Dwarka, D., Thaver, V., Naidu, M. and Baijnath, H. (2017) *Afr. J. Tradit. Complement Altern. Med.*, **14**(2): 253-262.

El-Mokdem, H., Khattab, M., Hassan, M.R. and Tarabein, A.M. (1994) *Alexandria J. Agri. Res.*, **39:** 231-241.

El-Shoura, H.A.S. and Hosni, A.M. (1996) *Ann. Agric. Sci. Cairo.*, **41:** 973-991.

Fava, C.L.F., Giachini, R.M., da Silva, J.L., Júnior, M.Z.A. and da Silva, T.J.A. (2015) *Ornam. Hortic.*, **21**(1): 39-46.

Finger, F.L., Campanha, M.M., Barbosa, J.G. and Fontes, P.C.R. (1999) *Revista Brasileira de Fisiologia Vegetal*, **11**: 119-122.

Finger, F.L., de Moraes, P.J., Barbosa, J.G. and Grossi, J.A.S. (2002) In *XXVI International Horticultural Congress: Issues and Advances in Postharvest Horticulture*, **628**: 863-867.

Fisher, J. B. (1976) *Canadian J. Bot.*, **54:** 578-592.

Gendy and Mahmoud (2012) *Aust. J. Basic Appl. Sci.*, **6**(5): 260–267.

Gilman, E. F. (1999) *University of Florida [Electronic resource].–Florida.*

Halevy, A.H., Kofranek, A.M. and Kubota, J. (1976) *HortScience*, **11**: 538.

Halevy, A.H., Koranek, A.M., Besemer, S. T. Venter van de, H.A., McKay, M. E. and Kawabata, O. (1987) *Acta Hortic.*, **205**: 89-96.

Haouala, F. and Zouari, M. (2001) *PHM Revue Horticole*, **426**: 30-33.

Healy, W., Klick, S., Roh, M.S. and Lawson, R.H. (1993) *Acta Hortic.*, **337**: 25-29.

Hetman, J. and Pogroszewska, E. (1983) *PraceInstytutu Sadownictwa w Skierniewicach, B.*, **8**: 7-23.

Jainag, K., Jayaprasad, K.V., Manohar, R.K., Shivanand, H. and Prakash, K. (2011) *Asian J. Hort.*, **6**(1): 118-121.

Jaroenkit, T. and Paull, R.E. (2003) *HortTechnol*, **13**(2): 259-266.

Jaroenkit, T., Chen, N.J. and Paull, R.E. (2008) *Postharvest Biol. Technol.*, **49**(3): 431-435.

Jeevitha, S., Sreeramu, B.S. and Raj, A.V.V. (2013) *Asian J. Hort.*, **8**: 150–153.

Karnataka, J. (2008) *J. Agric. Sci.*, **21**(2): 324-325.

Karsten J. (2009) *Dissertação apresentada a Universidade Federal de Viçosa (Mestrado em Fisiologia Vegetal), Viçosa, Minas Gerais.*

Karsten, J., Chaves, D.V., Costa, L.C., Ribeiro, W.S., Pereira, D.M. and Finger, F.L. (2012) In: *X International Symposium on Postharvest Quality of Ornamental Plants*, **1060**: 147-152.

Kawabata O., Criley, R.A. and Oshiro, S.R. (1982) *Proc. 21st Int. Hort. Congr.*, Hamburg, Abst. No. 1754.

Kawabata, O., Criley, R.A. and Oshiro, S. R. (1984) *J. Amer. Soc. Hort. Sci.*, **109**: 706-712.

Koley, T., Gupta, Y.C and Sharma, P. (2016) *Green Farming.*, 7(2).

Kumar, S.S. (2017) *Doctoral Dissertation*, College of Horticulture, Dr. YSR Horticultural University.

Labanowski, G. (1999) *Ochrona Roslin*, **43**: 14-15.

Lamas, A.M (2002) *Floricultura Tropical.*, p. 87.

Li, Y., Li, S.F., Wang, Q., Yu, G. and Li, S.J. (2012) *Forest Sci. and Tech.* (3): 28.

Luongo, L., Galli, M., Riccioni, L. and Belisario, A. (2010) *Plant Dis.*, **94**(1): 134-134.

Macnish, A.J., Reid, M.S., Marrero, A. and Jiang, C.Z. (2009) In: *VI International Postharvest Symposium*, **877**: 1763-1769.

Malfa, G. La., Romano, D. and La Malfa, G. (1988) *Rivista di Agronomia*, **22**: 214-220.

Meillon S. de, Venter H.A, van de and Small, J.G.C (1990b) *J. Exp. Bot.*, **41**: 885-892.

Meillon, S. de, Small, J.G.C and Venter H.A. van de (1990a) *J. Exp. Bot.*, **41**: 709-714

Moore, H. E. Jr. and Hyypio, P.A. (1970) *Baileya*, **17**: 65-74.

Navyashree, M., Munikrishnappa, P.M., Seetharamu, G.K., Krishna, H.C., Kumar, S. A. and Dayamani, K.J. (2017) *Environ. Ecol.*, **35**(3B): 2194-2198.

North, J.J., Ndakidemi, P.A. and Laubscher, C.P. (2010) *African J. Biotech.*, **9**(45): 7583-7588.

North, J.J., Ndakidemi, P.A. and Laubscher, C.P. (2012) *Sci. Res. Essays.*, **7**(10): 1118-1133.

Notten, A. (2002) *Strelitzia reginae* Aiton 'Mandela's Gold' 9from the SANBI's PlantzAfrika web-site http: //www.plantzafrica.com/plantqrs/strelitregmandelagold.htm).

Orlikowski, L. (1976) *PraceInstytutuSadownictwa w Skierniewicach*, B., **2**: 163-167.

Orlikowski, L. (1977) *PraceInstytutuSadownictwa w Skierniewicach*, B., **2**: 169-174.

Orlikowski, L. (1999) *Ochrona Roslin*, **43**: 18.

Pal, S. and Sarkar, I. (2009) *The J. Plant Prot. Sci.*, **1**(1): 98-101.

Porcelli, F. and Parenzan, P. (1993) *Entomologica*, **27**: 181-210.

Patel, D.K., Chawla, S.L., Patil, S. and Sathyanarayana, E. (2017) *Int. J. Curr. Microbiol. App. Sci.*, **6**(9): 1566-1570.

Paull, R.E., Chantrachit, T. and Theeranuch-Chantrachit (2001) *Postharvest Biol. Technol.*, **21**: 303-310.

Pereira, A.M., de Paula Gomes, M., Freire, A.I., Da Costa, L.C., Dos Santos, R.M.C. and Finger, F.L. (2018) doi: 10.5039/agraria.v13i3a5558

Petit, P. and Crozzoli, R. (1995) *Fitopato Logia Venegolana*, **8**: 41-44.

Pirone, C., Johnson, J.V., Quirke, J.M.E., Priestap, H.A. and Lee, D. (2010) *HortScience*, **45**(9): 1411-1415.

Pirone, P.P., Dodge, B.O. and Ricket, H.W. (1960) *Diseases and Pests of Ornamental Plants*. The Ronald Press Co., New York.

Pivetta, K.F.L, Barbosa, J.G. and Araújo, E.F. (2007) *Propagação de Plantas Ornamentais.*, pp. 43-70.

Pogroszewaska, E. (1996) *Ann. Univ. Mariae. Curie. Sklodowska. Sectio. EEE, Horti.***4**: 71-80.

Queiroz, S.J., Dias, G.M. and Tombolato, A.F.C. (2012) In: *X International Symposium on Postharvest Quality of Ornamental Plants.,* **1060**: 207-212.

Qui, YaFen and Qui. Y.F. (1999) *Zhejiang. NongyeKexue.,* **4**: 194-195.

Rodrigues, L.M., Destéfano, S.A., Diniz, M.C.T., Comparoni, R. and Rodrigues Neto, J. (2011) *Trop.Plant Pathol.,* **36**(6): 409-413.

Sane, A., Sujatha, S., Shilpa, K.N., Laxman, R.H. and Shivashankara, K.S. (2020) *Ind. Crops Prod.,* **151**: 112477.

Shreevatsa, K.S., Jayaprasad, K.V., Narayanaswamy, P. and Satyanarayana, B.N. (2004) *J. Ornam. Hort.,* **7**(3 and 4): 341-344.

Singh, A.K. (2006) *Flower crops: cultivation and management*. New India Publishing.

Siraj-Ali, M.S. (1998) *Indian J. Hort.,* **55**(3): 257-262.

Skvarla, J.J., Rowley, J.R. and Chissoe, W.F. (1997) *Taiwania,* **42** : 17-27.

Sugiyama, L.S., Bushe, B.C., Heller, W.P. and Keith, L.M. (2017) *Plant Dis.,* **101**(6): 1030-1030.

Szendel, A.J., Hetman, J. and Pogroszewska, E. (1976) *PraceInstututuSadownictwa w Skierniewcach,* B., **2**: 153-162.

USDA, 2003. www.usda.mannlib.cornell.edu/reports/nassr/other/ zfc-bb.84.

Van der Walt, L. (2000) South African National Biodiversity Institute (SANBI).

van Jaarsveld, E. (2008) http://pza.sanbi.org/strelitzia-reginae

Venter, H.A. van de, Small, J.G.C. and Halevy, A.H. (1980) *J. South Africa Bot.,* **46**: 305-311.

Wang, J.F. and He, X.L. (2000) *Acta Hortic. Sinica,* **27**: 300-301.

Winter, J. (2004) South African National Biodiversity Institute. pza.sanbi.org

Winter, J.P. and Xaba (2011) http://www.plantzafrica.com/plantqrs/strelitzjun.htm.

Wright, E.R., Rivera, M.C. and Palmucci, H.E. (1996) *Revista Brasileira de Hortic. Ornamental,* **2**: 59-63.

Xaba, P. and Van der Walt, L. (2011) http://pza.sanbi.org/strelitzia-reginae

Yi-ru , Yuan *et al.* (2009) *J. Anhui Agricultural Sci.,* **10**: 105-110.

Index

G

H

T

U

V